U.S. SMALL COMBATANTS

U.S. SMALL COMBATANTS

INCLUDING PT-BOATS, SUBCHASERS, AND THE BROWN-WATER NAVY

AN ILLUSTRATED DESIGN HISTORY

By Norman Friedman

Ship Plans by A.D. Baker III with Alan Raven and Al Ross

NAVAL INSTITUTE PRESS
Annapolis, Maryland

Copyright © 1987
by the U.S. Naval Institute
Annapolis, Maryland

All rights reserved. No part of this book may be reproduced
without written permission from the publisher.

Library of Congress Cataloging-in-Publication Data
Friedman, Norman, 1946–
    U.S. small combatants, including PT-boats, subchasers, and the
brown-water navy: an illustrated design history/by Norman
Friedman; ship plans by A.D. Baker III with Alan Raven and Al Ross.
      p.   cm.
  Bibliography: p.
  Includes index.
  ISBN 0-87021-713-5
    1. Torpedo-boats—United States. 2. Warships—United States—
History—20th century. 3. Warships—Design and construction.
4. Fast attack craft. 5. Gunboats. 6. Submarine chasers.
I. Title.
V833.F75 1987
359.3′258′0973—dc19   87-15893
                                                                                            CIP

This edition is authorized for sale only in the
United States, its territories and possessions,
and Canada.

Printed in the United States of America.

Contents

Acknowledgments   vii

1. Introduction   1
2. Subchasers and Eagle Boats   19
3. Second Generation Subchasers   47
4. Small Fast Craft and PT Origins   97
5. The Early PT Boats   115
6. The Wartime PT   153
7. The Postwar PTs   177
8. Postwar ASW Craft   197
9. Small Combatants for Counterinsurgency   221
10. Motor Gunboats and PTFs   243
11. Vietnam: Beginnings   279
12. Vietnam: Market Time and Game Warden   297
13. Vietnam: The Riverine Force   325
14. Vietnam: SEALs and STABs   365
15. The PHM   379
16. Post-Vietnam Small Combatants   391

Appendices

A. Gunboats   413
B. Minor Acquired Patrol Craft: SP, YP, PY, PYc   427
C. Crash Boats and Pickets   431
D. U.S. Small Craft Weapons: Guns and Rockets   439
E. Small Combatants for Export   455
F. Fates and Other Notes   461

Notes   507

Notes on Sources   517

Index   523

Acknowledgments

This book could not have been written without the assistance of many friends. I am particularly grateful to George Kerr, Kit Ryan, and Phil Sims of NavSea Preliminary Design; Ed Sloper and Sandra Buckley of Sea 941; Donald L. Blount and Elder L. Lash and the staff of the Naval Sea Combat Systems Engineering Station, Norfolk; Dean Allard and his staff at the Naval Historical Center, particularly Cal Cavalcante, Martha Crowley, Edward J. Marolda, Wes Pryce, and John Walker; John Reilly and Robert Cressman of Ships' Histories; Carl Onesty; Bob Rogers, formerly of Sea 941; Don Tempesco; Richard R. Hartley; Ken Spaulding; Dana Wegner and Colan Ratliffe of the navy model collection; John Forster of Lantana Boat and, formerly, Sewart; Philippe Roger of the French embassy, who provided material from the French Navy Historical Section; Chris Evers; Robert Scheina; and Bruce Vandermark. Boeing, Halter Marine, Lantana Boat, MonARK, Swiftships, and United Technologies supplied data on some of their craft. Special thanks go to Al Ross, for his generous assistance with PT data and for the PT drawings and photographs. I am also grateful to Alyce N. Guthrie of the PT Boat Museum.

Arthur D. Baker III supplied many of the photographs and executed most of the drawings in this volume. He also doubled as technical editor and conscience, catching many errors and making invaluable suggestions. Any remaining mistakes are, of course, my own responsibility. Alan Raven and Al Ross provided many of the other illustrations.

Other good friends deserve thanks for their generous assistance: Charles Haberlein, Thomas Hone, David Lyon, Norman Polmar, Stephen S. Roberts, and Christopher C. Wright. They made available significant material from their own collections and passed on invaluable comments and corrections.

I would particularly like to thank my editor, Constance Buchanan, for the swift yet careful completion of a complex manuscript.

Words are inevitably inadequate to describe the support, advice, and encouragement offered by my wife, Rhea, during the compilation of this book.

U.S. SMALL COMBATANTS

1

Introduction

This book describes a wide variety of ships and craft, united more by size than by function. Most have or had designators in the patrol (P) category, for example, ASW craft such as the PC, antiship attack craft such as the PT, and gunboats (PG), many of which were actually small cruisers. Others, such as those that fought the riverine war in Vietnam, never had P designators yet are clearly small combatants.[1] The scope is not broad enough to include amphibious craft, apart from those in Vietnam, where the boundary between combatant and amphibious vessels was blurred. The reader will look in vain for the many World War II fire support craft (LCS, LCIG, LCIM, LSSL, LSMR), which were, in effect, small gunboats. Similarly, mine craft have been reserved for a future volume, even though their hulls resembled those of many small combatant ASW vessels (PC, PCE, PCS) described in chapter 3 of this book. One clearly noncombatant type, the crash boat, is treated in some detail in chapter 4 and appendix C because of its close connection to the motor torpedo boat.

Thus, in addition to gunboats and some large ASW craft, the subject of this volume is small coastal craft, capable perhaps of crossing an ocean but not of fighting very far from shore. If, like the PT, they were designed for very high speed, that speed was generally obtainable only in the relatively calm waters of the coast. They were limited to protecting U.S. coasts or operating in overseas coastal or inland waters, often after being transported by larger ships. And so small combatants lie outside the mainstream of U.S. naval strategy, outside the balanced fleet of capital ships, escorts, and submarines designed to project U.S. power abroad. Because of this, they tend to fill extemporized roles in wartime and to be discarded in peacetime. This pattern makes the subject of small combatants an emotional one, as many of their advocates emphatically disagree with the fleet-oriented logic that results in the rejection of such useful craft during peacetime.[2]

The development of small combatants in the United States was driven by two quite separate factors: technology, particularly the technology of high speed, and national strategy. High speed, whether tactically useful or not, often appears so attractive that the possibility of attaining it, even at great expense, seems to justify the construction of new craft. Sometimes, as in the case of the PT, strategic conditions provided the necessary justification. Sometimes proposals for new craft turned out to be nothing more than "solutions seeking problems," and they died.

For high speed, a conventional ship requires both length and power. The power needed to achieve a given speed is a function of the speed-length ratio: the speed divided by the square root of the waterline length. A ship uses its power to overcome resistance created by its wetted (underwater) surface area and the waves formed as the hull pushes water aside. Thus, wave-making resistance arises from the mechanism that sustains the ship in the water—its buoyancy, that is, the water displaced by its underwater form.

Any ship with a speed-length ratio much over 1 is considered fast, so that 10 knots might be taken as the speed naturally associated with a slow 100-foot craft. Ships with comparable speed-length ratios should encounter similar amounts of resistance per ton.[3] For example, at 22 knots the 173-foot World War II PC was comparable to a 32-knot, 341-foot destroyer. The PC was relatively flimsy because its scantlings were approximately proportionate in weight to those of the much larger destroyer. The higher the speed-length ratio, the more power required

The navy scrapped its large PT force after World War II, but it built four prototypes to maintain expertise in PT design and operation. Two of them, PT 809 (*foreground*) and PT 811, are shown on early test runs. Both were designed to carry torpedoes (note the spaces marked on the deck of PT 809), but neither ever did. The 81mm mortar on deck (*to port*), opposite the twin 20mm cannon, failed its operational tests in 1952 but was resurrected for Vietnam. PT 809 herself still exists; she once served as the presidential chase boat *Guardian*.

required per ton, and thus the greater the weight of machinery a ship must accommodate. That is why World War II destroyers, PT boats, and even subchasers were filled almost completely with machinery, whereas battleships, which could attain similar speeds with much less power per ton, had much more weight and volume available for weapons and protection.

In this sense, virtually all the craft treated in this volume were fast, and many of them achieved their speed by resort to unconventional design and construction. The first important twentieth-century development in this sense was the internal-combustion engine; it was much lighter (per unit power) than a small steam engine because it did not require a separate boiler and power unit.[4] Moreover, fewer men were needed to operate an internal-combustion engine, a reduction in complement that saved further weight in structure (crew quarters and messes) and stores. However, internal-combustion engines tended to consume more fuel than steam engines. As a result, small combatants powered by internal combustion could reach high speeds, but they could not carry sufficient fuel for any great range.

Before and during World War II, motor torpedo boats and crash boats required particularly compact power plants, marine versions of aircraft engines that burned high-octane gasoline, which in turn could create an explosive vapor. Perhaps the most famous such engine was the Packard 4M 2500. It powered World War II PT boats and was modeled after the Packard 2500, itself a marine version of the World War I Liberty aircraft engine. Work on the 4M 2500 began in April 1939, when Packard was one of four prospective PT boat engine builders.[5] Eventually a total of 12,040 Packard engines was manufactured in wartime; the four postwar experimental PTs were all powered by an enlarged version. However, because the Packard was a very specialized engine, it soon went out of production. It could not, therefore, be considered for the next generation of high-speed boats, designed from about 1959 on.

The risk of using gasoline was aggravated in wooden boats such as the World War I subchasers, several of which succumbed to fuel fires. Diesel fuel mitigated the problem, but it was difficult to build a lightweight high-speed diesel engine. Although the Bureau of Engineering (BuEng) tried to develop a diesel to power PT boats, the United States and her allies had to rely on Packard gasoline engines in World War II. Even the postwar American PTs were so powered. Only the Germans had high-powered high-speed diesels for their motor torpedo boats. After the war the British developed the high-speed Napier Deltic diesel, which the U.S. Navy used in PTFs during the Vietnam War.

There is a limit to the work performed by a single fast-moving piston. The total power of a conventional internal-combustion engine is proportionate to the number of cylinders. Torsional vibration limits the number of cylinders that can be arranged in line or a V. One classic solution to this problem, applicable only to diesels, was the opposed-piston engine, in which each cylinder held two pistons driving separate crankshafts, one at each end. Combustion occurred in the space between moving pistons. Britain's postwar Deltic diesel represented an extreme application of this principle. It consisted of three interconnected opposed-piston engines—in effect, three coupled Vs. Viewed from the end, the engine resembled a triangle (a Delta) with crankshafts at its vertices. A boat driven by two Deltics was, in effect, driven by six more conventionally configured engines.

The U.S. wartime "pancake," the General Motors (GM) 16-184-A, used in 110-foot subchasers, was another ingenious solution to the problem of lightweight diesel power. The engine originated at a conference between representatives of BuEng and the General Motors Research Laboratory in the summer of 1937. BuEng badly wanted a lightweight diesel suited to mass production. Existing submarine engines weighed about 27.5 pounds per shp, far too much for a small surface ship. The conference finally decided to use four layers, each of four cylinders arranged radially around a vertical crankshaft, with the gear case at the bottom driving a propeller shaft through right-angle bevel gears. So radical an engine presented considerable design problems, and although its first trial run in June 1939 ended successfully, it did not pass navy acceptance tests until October 1940. Even then, it still required minor redesign, but the first subchaser powered by it, SC 453, was accepted in August 1941 and the engine entered mass production in February 1942. Weighing slightly more than 4 pounds per horsepower (hp)—a fourth to a fifth as much as previous marine diesels—it took up about one-third the space of more conventional marine diesels.[6]

However, the pancake was difficult to produce; the history of the U.S. subchaser program is a story of attempts to find acceptable alternatives. Perhaps the most ingenious was the "quad," four GM 6-71 diesels linked to a common bull gear by individual friction clutches, so that each diesel could be disconnected. Quads were also used in numerous wartime and postwar landing craft and, into the late 1960s, in standard motor gunboats built for foreign navies.

In small combatants there was a dearth of engine room space, particularly height, so piston strokes had to be short and, by way of compensation, they had to move fast. This resulted in vibration, which limited the length of (that is, the number of cylinders in) the engine. The powerful diesels of modern warships and merchantmen are practicable only because they are relatively slow-acting, long-stroke units.

The risk of gasoline fire explains why characteristics (staff requirements) for many small combatants of the late 1950s and the Vietnam era included nongasoline engines, generally either a Deltic or a gas turbine. After World War II, the gas turbine was the great small combatant power plant. Like a steam turbine, it could easily be adapted to large power loads by increasing the number of turbine stages. Torsional vibration was no great problem because the central (power) shaft did not have to absorb the alternating power strokes of the pistons, which could not be fully balanced. The only drawback was high fuel consumption. Many small combatants run on gas turbines had to accommodate two power plants, one for burst power and a smaller and more economical one for low-speed loitering or cruising.

A gas turbine might be imagined as a steam turbine using a much lighter gas generator instead of a boiler, and therefore operating at much higher temperature. After receiving a report of January 1939 on British and Swiss work, BuEng asked the National Academy of Sciences to study marine gas turbines. As a result of this study the bureau suggested that the navy order a 1,600-shp turbine operating at 1,500°F (the highest steam temperature then in use was 1,350°F). The Bureau of Ships (BuShips), which by then had absorbed BuEng, ordered a 3,500-shp unit from Allis-Chalmers in December 1940. Since the turbine was not to be installed in a ship, its size could be unrestricted. It was delivered in October 1944, first fired on 7 December 1944, and dismantled in 1951 after 3,700 hours of operation. Another wartime project, a Lysholm free-piston compressor proposed by the Elliott Company in June 1942, performed so well that in July 1945 BuShips contracted for two 3,000-shp units for a destroyer escort. They were never actually installed. In 1945 U.S. investigators in Europe learned that the Germans had also designed gas turbines, including a 10,000-shp unit for a fast torpedo boat.

After the war, the British were the first to apply gas turbines to warships, their Proteus (3,600-shp continuous power in its developed form) being installed in several American small combatants. BuShips retreated at first to small gas turbines comparable in power to standard boat diesel engines. Thus the fifth waterborne gas turbine was installed in an American LCVP, and the first U.S. watercraft to be powered solely by a gas turbine was a 24-foot personnel boat powered by a 160-hp Boeing gas turbine developed under BuShips contract. The vessel began trials on 30 May 1950. The first major BuShips gas turbine contract, let in September 1950, was for Solar generators to be used in minesweeping boats. The first U.S.-made small boat powered by high-powered gas turbines was the 52-foot LCSR, which used a Solar 10MV engine. Most of the gas turbines mentioned in this book were derived from aircraft jet or turboprop engines; the Solar is a major exception.

The AVCO Lycoming TF12 and TF14 were developed from the T-53, an army/air force helicopter engine. The TF14 was an upgraded version with four rather than two power stages and with variable-incidence inlet guide vanes. The TF12 was rated at 920 shp, the TF14 at 1,050. The TF20 (1,500-shp maximum continuous power) was an enlarged TF14. It led to the TF25; the uprated TF35 (up to 2,500-shp maximum continuous power) was a further development, based on the T55 aircraft engine. By 1979 the Super TF25 was rated at 2,500-shp continuous power (3,000-shp peak power), the Super TF35 was rated at 3,500 (4,050 peak). A further uprated Super TF40 powered the prototype air-cushion landing craft.

The General Electric (GE) LM100 was a marine version of the T58 helicopter engine (1,100 shp). It powered the Bell SK-5 Hovercraft used in Vietnam. The larger LM1500, a turboshaft adaptation of the J79 jet engine (15,000 shp), powered the *Plainview* and the *Asheville*s. The LM2500 was derived from the TF39 fan-jet used to power the C-5 transport (20,000 to 25,000 shp). The LM2500 has been adopted as the standard U.S. Navy gas turbine. It replaced the LM1500 originally specified for the Boeing NATO hydrofoil, which became the PHM.

The Pratt and Whitney FT4A, based on the J75 jet engine, won a 1961 BuShips design competition for an engine, 30,000 shp at peak power, 20,000 at normal, suitable for hydrofoils. It powered the Canadian *Bras d'Or* (FHE 400) and numerous conventional surface ships. The FT4A did not turn out to be as economical as the LM2500 that superseded it.

The big second-generation FT9 was modeled after the JT9D fan-jet. It was intended for the "3K SES" surface effect frigate and died when that was canceled. The FT9 would have competed with the LM2500.

Several riverine craft were powered by the United Aircraft ST6, a Canadian marine derivative of the PT6 turboprop. The ST6 produced about 550 shp in the form proposed for riverine craft.

Much of the story of small combatants, then, is taken up with the series of attempts to achieve higher and higher power in limited sizes and weights, culminating in the widespread introduction of gas turbines. As small combatants had to be built in enormous numbers during wartime mobilization, the search for a matching number of suitable power plants

1. Hullborne Propulsor
2. SSPU No. 2
3. Foilborne Propulsor
4. SSPU No. 2 Combustion Air Intake
5. Diesel and Pump Machinery Room Cooling Air Exhaust
6. Diesel Engine Exhaust
7. Diesel Engine Combustion Air and Diesel and Pump Machinery Room Cooling Air Intake
8. Foilborne Gas Turbine Exhaust Stack
9. SSPU No. 1 Combustion Air Exhaust
10. Auxiliary Machine Room No. 2 Air Exhaust
11. SSPU No. 1 Combustion Air Intake
12. Auxiliary Machinery Room No. 1 Air Exhaust
13. Gas Turbine Machinery Room Cooling Air Intake
14. Auxiliary Machinery Room No. 1 Air Intake
15. Gas Turbine Combustion Air Intake
16. Auxiliary Machinery Room No. 2 Air Supply
17. SSPU No. 1
18. Foilborne Gas Turbine Engine
19. Foilborne Propulsion Waterjet Inlet
20. Hullborne Diesel Engine
21. Hullborne Propulsor Inlet
22. Hullborne Waterjet Nozzle
23. Foilborne Waterjet Nozzle
24. SSPU No. 2 Combustion Air Exhaust

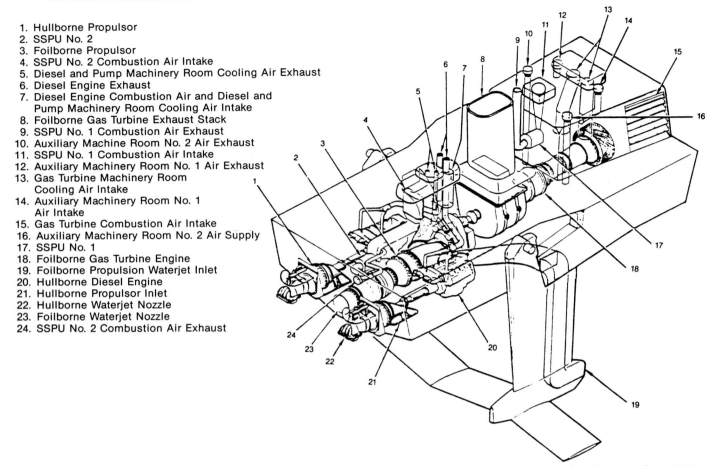

Hydrofoils also often require special propulsion arrangements. This is the main machinery of a *Pegasus*-class PHM, consisting of a gas turbine for foil-borne power, a diesel (*20*) for hull-borne propulsion. The PHM is propelled by water jet in both modes; water enters through the after shaft strut. Note also the big air intake typical of gas turbine ships (*15*) and the filter around the air intake of the turbine itself (*18*). (USN)

is also an important theme. Most World War II subchasers, for example, could not make their required speed of 22 knots because there were not enough pancake engines available. Engine production limited PC construction programs and inspired efforts to convert existing minesweepers to emergency ASW craft (PCE and PCS).

Given a compact power source, the usual limitations imposed by hydrodynamic laws can be overcome by exotic hull design, which reduces or even eliminates the wave-making component of drag. In both hydroplanes and hydrofoils, forces generated by the speed of the boat lift it partly or completely out of the water, reducing resistance largely or completely to friction on the lifting surface. The other alternative, the air-cushion or surface-effect vehicle, or hovercraft, uses a separate source of power to generate the low-friction air cushion on which it rides.

These three alternatives are limited to fairly small craft by the "square-cube law": Each square foot of surface area (proportional to the square of the craft's dimensions) can support only so much weight, but the total weight is proportional to the *cube* of the dimensions for geometrically similar craft. The ratio of square to cube falls as the craft grows. Thus, to make a larger craft plane or ride an air cushion, each square foot of lifting area must be made more effective. In the case of a hydroplane, that generally requires higher speed as well as more and lighter power. In the case of a hydrofoil, it may mean devising a supercavitating foil. In the case of the air-cushion craft, it means maintaining a higher cushion pressure by limiting leakage. Hence the use, first, of rubber skirts, then of rigid sidewalls and rubber air-cushion seals fore and aft.[7]

The three alternatives actually appeared before World War I, but only hydroplanes entered service in any numbers. They had the great advantage of simplicity. The basic planing concept was understood at least as early as 1840, but it did not become practical until the advent of lightweight engines. Probably the first of the type was built by a French engineer, Campet, in 1905. The first stepped planing boat (120 hp, 31.3 knots) appeared in 1908. In 1909

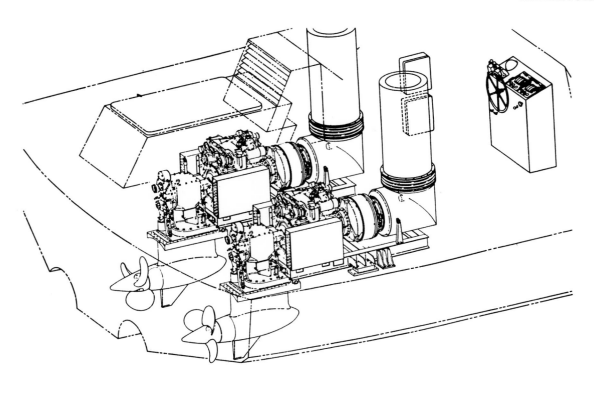

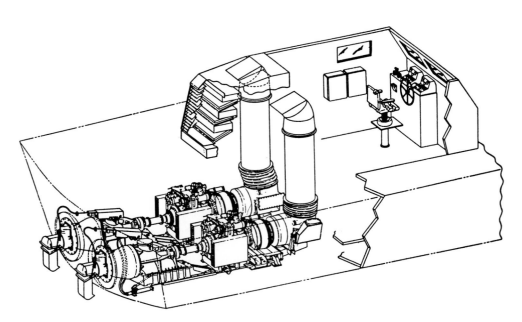

Warfare on rivers, which are shallow and often choked with weeds, poses special propulsion problems. These two solutions were developed as a result of experience in Vietnam. The water jet was designed for United Aircraft's ASPB Mark II prototype, the conventional package, with propellers in tunnels, for Sewart's ATC/CCB Mark II prototype. The engines are, respectively, the 1,050-shp (950-shp continuous output) Lycoming TF 1416-C1A and the 920-shp (800-shp continuous output) Lycoming TF 1260-C1A. Note that both craft use separate port and starboard propulsion packages, sharing only the system start and control batteries and the boat power supply. The use of a "Z"-drive reduced overall propulsion length in the ATC/CCB; it had the incidental advantage of lifting the engine well away from the bottom of the hull, thereby providing space for maintenance. To run the water jet in the ASPB prototype, the engine had to be as low as possible, and it was difficult to service. (From "Riverine Propulsion System Lessons Learned," presented to the American Society of Mechanical Engineers in March 1972 by T. B. Lauriat)

To achieve high speed, a small hull requires some unconventional means of support; planing is the most common. This is one of three half-scale craft built at Norfolk in 1938 to test proposed PT boat hull features. It has just planed off one wave and is airborne before crashing into another. The hard chine, sweeping up at the bow to create lift, is quite prominent.

the 49-foot, hard-chine, 5.75-ton *Ursula* (two 375-bhp engines) made 35.3 knots. A typical 55-foot British Thornycroft coastal motorboat (CMB) of World War I displaced 11 to 11.5 tons and made 40 knots with two 375-bhp engines.

At rest, a planing boat is supported entirely by its buoyancy (displacement), but if it can exceed a critical ("hump") speed, resistance falls off because most of the hull is lifted out of the water.[8] The hydrodynamic lift force generated by its motion suffices to balance all but 30 or 40 percent of its weight, and resistance is reduced largely to friction between the water and the planing surface.

The faster the boat, the greater the lift per unit area, and therefore the greater the proportion of total boat weight the planing area can carry. The planing effect is similar in principle to the lift generated by an airplane wing, and in this case the trim angle of the boat determines its angle of attack. For positive lift, that angle must be positive.

Planing lift is proportional to the available flat planing area, that is, to the square of the boat's dimensions and to the square of its speed. However, the required amount of lift is some fraction of the total weight of the boat, which is proportional to the cube of its dimensions. As the boat becomes larger, then, the ratio of lift to weight (square to cube) diminishes, and planing becomes more difficult. Given the relation between planing lift and boat speed, the critical or hump speed (for geometrically similar boats) rises in proportion to the *square root* of their length. That explains why large ships do not plane: if a 60-foot motorboat planes at 30 knots, a geometrically similar 240-foot ship would plane at 60.

Any sufficiently fast hull will plane, but the most efficient planing boat has a perfectly flat bottom. But a design that relied entirely on planing would have to accept extremely poor performance in waves: the boat would fly from one crest only to crash into the next. A displacement boat will cut through small waves, but it will also encounter much more resistance. In practice, small fast boats combine the two modes of operation, and terms like displacement boat or hard-chine planing craft merely reflect alternative emphases.

Beyond the hump, resistance increases much more slowly with speed. Planing made motor torpedo boats possible; it also made them bumpy. Small planing boats could outrun much larger conventional ships, but only in fairly calm water.

The earliest planing boats, built before and during World War I, had nearly flat bottoms for maximum planing surface. Typically, the forward bottom angled up slightly to provide initial lift for planing and met the nearly vertical side of the boat at an angle, the hard chine.

Many early planing boats such as the CMB had a step cut in their underwater hulls. They planed, at a positive trim or planing angle, largely on the area forward of the step, with part of the surface aft of the step immersed. Since the area immediately under the step breaks out of the water as the boat assumes a planing angle, the step helps the boat to break free of water suction, to get over the hump. Similar steps were and are standard in flying boat hulls, which must plane to reach takeoff speed. Many post–World War I racing boats had multiple steps; the Soviet *Sarancha*, an experimental hydrofoil missile boat of the mid-1970s, also had steps incorporated in its hull.

The hydrodynamics of the stepped planing boat are well suited to calm water. In rough, the step is only another feature vulnerable to wave action. As more powerful engines became available after World War I, stepped hulls were largely abandoned in favor of V-bottom hard-chine hulls. In such craft the line of the shallow V extends right aft from the bow, the chine running all the way along the hull. The same hydrodynamic forces lift the boat into a planing mode, but in the absence of a step the planing angle is less extreme and the boat can negotiate waves much more effectively. On the other hand, it does not enjoy the benefits of planing as much as the stepped boat, since more of its hull is immersed even when it planes. All of the U.S. Navy's World War II motor torpedo boats were hard chine displacement craft, compromises between the need for high speed (planing) and the desirable goal of decent rough-water performance.[9]

For their motor torpedo boats (*S-boote*, usually referred to as E-boats in Britain and in the United States), the Germans were unique in preferring a more conventional round-bottom displacement hull, which had better sea-keeping qualities. This hull performs as well as a planing boat up to a speed-length ratio of about 3.4, corresponding to a speed of 34 knots for a 100-foot boat, or about 27 knots for a 64-foot boat. The U.S. Navy's postwar PT program illustrates the comparative virtues of the hard-chine and round-bilge hulls (see chapter 8).

Another exotic displacement hull was the "sea sled" invented by W. Albert Hickman, a Nova Scotian, in 1911. It had the shape of a W with a tunnel running from the bow aft, deepest at the bow. Hickman characterized the sea sled as a V-bottomed boat cut in two lengthwise and then joined inside out. The bow wave ran into the tunnel and helped lift the boat, and an air-water cushion was formed beneath the tunnel to help absorb the normal shocks of planing and to reduce friction. In this sense the sea sled might be considered a hybrid planing/air-cushion craft. The original sea sled enjoyed some success in the United States before and during World War I, but it was later rejected because it gave a rougher ride than a hard chine, V-form hull, and because it was structurally weak. The Royal Navy bought three sea sleds during World War I, and a director of (British) naval construction, Sir William Berry, commented in 1927 that they showed poor sea keeping, and that their weight had to be aft because so little of their buoyancy was forward.

Modern analysis suggests that a sea sled has a particularly steep hump and that much of Hickman's success can be attributed to his efficient semisubmerged propellers.

The hydrofoil, which is supported by a wing immersed in water, predates World War I. It seems to have been conceived by several inventors during the nineteenth century.[10] An Italian, Enrico Forlanini, probably tested the first fully successful model in 1906. In 1911, H. C. Richardson and J. C. White built a dinghy with fully submerged, manually controlled foils at the Washington Navy Yard; the vessel lifted when towed at 6 knots.

That year, 1911, Alexander Graham Bell took a ride on Lake Maggiore and afterwards bought a license from Forlanini to build his type of hydrofoil boat. Bell saw the hydrofoil as a close relative of the airplane and developed what he called his ladder-foil craft, the "hydrodromes," testing them on Lake Bras d'Or, Nova Scotia.[11] He offered them to the U.S. and British navies, but with little success. His series culminated in the 5-ton HD-4, built in Nova Scotia in 1917–18. Initially powered by two 250-bhp Renault engines driving airscrews, it lifted at 18.3 knots and made 46.7. Refitted with two Liberty aircraft engines, it made 61.6 knots, at which speed it was supported by only 4 square feet of foil surface. In 1920, the HD-4 was tested by both the U.S. and Royal navies, carrying two 1,500-pound weights (corresponding to torpedoes) in 2.5- to 3-foot waves. It did not list even when there was only one off-center weight aboard, which corresponded to a fired torpedo. Although the U.S. Navy trials board, headed by Rear Admiral Strother Smith, concluded that at high speed in rough water the boat was superior to any type of high-speed motorboat or sea sled known, it was rejected, reportedly because of a tendency to porpoise in a rough sea. The Royal Navy bought two hydrodromes in the early 1920s, but both were destroyed during rough-water towing tests off Spithead.

Bell's colleague, Casey Baldwin, continued experiments between the wars, resulting in several small hydrofoil racing boats. During World War II, Baldwin built a towed hydrofoil to generate smoke screens (COMAX), an experience that led to postwar Canadian interest in large hydrofoils. In 1936 America's National Advisory Committee on Aeronautics (NACA, the predecessor of NASA) began work on hydrofoils, largely in connection with seaplanes, and built an experimental ladder-foil boat in about 1943. While developing PT boats in 1937–41, the U.S. Navy's Preliminary Design branch obtained hydrofoil data, but there is no evidence that it seriously considered anything other than a conventional hull form.

Other navies built hydrofoil combatants at this

time, although they never entered operational service. A British firm, J. S. White, built the 42-knot MTB 101 for the Royal Navy in 1939. During World War II the Germans built a variety of coastal hydrofoil patrol boats, minelayers, and even an 80-ton tank carrier, some of which fell into Allied hands in 1945.[12]

A hydrofoil needs relatively little speed to lift because water is about eight hundred times as dense as air. Since the boat proper rises above the waves, a hydrofoil can operate in fairly rough water. How rough depends on the hydrofoil's dimensions and on its control system. Here, too, there is a hump. Below a critical takeoff speed, the hydrofoil rides on its conventional hull. To an even greater extent than the hydroplane, it operates in two very different modes: hull-borne, where the drag of the foils makes it slower than a conventional boat of similar size, and foil-borne, where, flying, it requires much less power to achieve a given speed. Operation at intermediate speeds is difficult. Moreover, the lifting capacity of the foils determines the weight a hydrofoil can carry. Unlike the hull of a displacement boat, the foils cannot accommodate greatly increased hull weight simply by sinking deeper into the water to gain increased buoyancy. Lift can be increased either by increasing flying speed or by changing the lift characteristics of the foil, for example, using flaps or altering the angle of incidence.

Similarly, the hydrofoil may not be able to accommodate any great *reduction* in weight, such as that which comes of fuel consumption. The PC(H) *High Point* could not "fly" for more than six hours at a time, not because its fuel tanks were small but because it could fly only within a limited range of weight. This type of restriction affects all exotic hull forms; to a greater or lesser degree, all are designed for a narrow range of weight and speed. An overloaded planing boat will often be described as "too heavy to get onto the plane." An overloaded air-cushion boat may be too heavy for the air layer it generates. Conventional displacement hulls are much more forgiving, which is why they continue to be built.

Like the hydroplane, the hydrofoil obeys a square-cube law, although in this case the square applies to foil surface rather than to the bottom of the hull. Hydrofoils generally have planing hulls that help them break free of the surface so that they can fly.

These craft fall into two distinct categories, depending on whether their foils pierce the surface. Since lift is proportional to immersed plane area, the lift of a surface-piercing foil increases as it is pushed deeper into the water, that is, as more of its surface is immersed. A V-foil is inherently stable in roll, because the more deeply immersed side of the foil has greater lift and therefore rises back out of the water.

However, lift also increases as the foil passes through a wave, and the surface-piercing foil may be ill-adapted to rough weather. In a following sea, the rotary motion of the water may alternately increase and reduce foil lift as the speed of water over the planes and their effective angle of attack vary with wave motion. Hence the porpoising problem encountered by Bell's HD-4. A smooth ride, then, requires some means of varying the lift of the foils, such as flaps actuated by an autopilot.

Fully submerged foils were attractive because they operated in less disturbed water and produced much less drag. In about 1950 U.S. researchers also discovered that, as in the case of surface-piercing foils, the lift of foils near the surface varied with their depth due to free-surface effects nearby. Although this provided automatic stability in calm water, it did not provide trim or rough-water control.

For the kind of rough-water operation the U.S. Navy was interested in, then, submerged foils were attractive but required active, aircraft-type control; the lift of the foils varied, as in an airplane, by moving flaps or changing the angle of attack. The lift of submerged foils could be varied over a wide range, as much as 4 or 5:1 compared with 2:1 for surface-piercing foils. The larger the hydrofoil, the smaller the fraction of total cost represented by the control system, which is why the U.S. Navy was willing to forgo the relative simplicity of surface-piercing foils for its big craft (see chapters 8 and 10).

Christopher Hook, an Englishman, built the first modern submerged-foil boat, the *Hydrofin*, in 1947. He maintained altitude by installing surface "feelers," skis that measured height above the surface and were connected to the submerged foils to control their incidence. Hook later built a 39-knot, 14-ton hydrofoil landing craft, the *Halobates*, for the U.S. Navy. In France, Grunberg invented a boat with submerged foils aft and a stabilizer, often a hydroski, forward. In effect, the boat pivoted around the hydroski, automatically varying the angle of attack of the submerged foils. He patented this system in 1935 and it was used after the war, for example, in the Canadian *Bras d'Or*, whose stabilizer was a surface-piercing V-foil.

The U.S. Navy hydrofoil research program began in 1947 under the leadership of Admiral Armand Morgan, then assistant chief of BuShips. Morgan considered automatic sea-keeping control the key to the open-ocean hydrofoil, which was the only type the navy could effectively use; the hydrofoil's pilot could not be expected to maintain stability while steering. The necessary autopilot technology already existed in the aircraft industry.

A series of small craft tested concepts of control and configuration. Vannevar Bush, head of the war-

Big hydrofoils present special docking problems. The prototype patrol missile hydrofoil *Pegasus* is shown prior to launch, in November 1974, with bow and stern foils retracted.

time National Defense Research Committee, and his postwar Hydrofoil Corporation (Annapolis, Maryland) built the first submerged hydrofoil controlled by autopilot, the 20,000-pound *Lantern* (HC 4), in 1953. A pickup measured its height above water and exerted control in attitude and pitch. A wartime hydrofoil advocate, Bush had been trying to sell it to the navy for antisubmarine warfare (ASW). The MIT Flight Control Laboratory developed autopilots using a test bed built in 1952 and consisting of a frame, with plating, on which foils could be put in various places.

This work led to a successful rough-water submerged-foil boat, the *Sea Legs*, designed by Gibbs & Cox. It was essentially a cabin cruiser hull with a main submerged foil aft and a small stabilizing foil forward to control pitch. Although the *Sea Legs* was only 25 feet long (5 tons), she managed to operate in 4-foot seas with a 1.5- to 2-foot clearance between hull and waterline.[13]

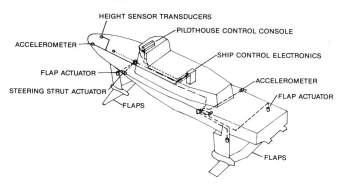

Automatic control systems made rough-water hydrofoils practical. This is the control system of a PHM in schematic form. The inputs are readings of both height above water and acceleration. Control was exerted through flaps on the foils. (Norman Friedman)

The *Sea Legs'* control system was applied to the first U.S. hydrofoil warship, PC(H) 1, the *High Point* (see chapter 8). Its autopilot, fed by two bow-mounted ultrasonic height sensors, used flaps to vary foil lift. The 120-ton PC(H) 1 was followed by a large hydrofoil, the 320-ton *Plainview*, AG(EH) 1, the lift of whose foils was controlled by varying their angle of incidence (attack). PC(H) 1 led directly to the two PG(H) prototypes and then to the modern PHM (*Pegasus*) of chapter 15. The AG(EH) was intended as a step towards large hydrofoil ocean escorts, or DE(H), which were, in the end, not built. They were superseded by projects for large surface effect craft (the abortive "3K SES" 80-knot frigate).[14]

BuShips and the Office of Naval Research (ONR) also financed 79-knot, 54-foot, 16,500-pound XCH-4, built in 1953 by the Dynamic Developments Corporation of Babylon, New York. In 1956 this company was absorbed by Grumman Aircraft, leading to that company's interest in hydrofoils such as in the gunboat *Flagstaff*. Grumman went on to build a 104-foot, 90-ton surface-piercing hydrofoil (the *H.S. Denison*) for the Maritime Administration under a February 1959 contract. The *Denison* was a hybrid, in that her foils had autopilot-controlled flaps; she was the first autopilot-controlled commercial hydrofoil. Grumman then designed the big submerged-foil *Plainview*, which was built by Lockheed, and a related smaller passenger hydrofoil, the *Dolphin*. The Grumman PGH *Flagstaff* was a militarized version of the *Dolphin*.[15]

Hydrofoils, with their engines well above the water, encounter special propulsion problems. Small hydrofoil boats could use more or less conventional propellers extending into the water, but big rough-water hydrofoils could not. The XCH-4, for example, used a pair of airplane engines driving air propellers. In the *High Point*, power had to be transmitted from engine to propeller by means of a Z-drive, a shaft kinked twice through bevel gears placed at right angles. Grumman demonstrated a 10,000-shp Z-drive in the *Denison*, and similar devices were used in the PC(H) and then in the AG(EH) vessels. After 1960 BuShips developed two parallel alternatives, a pump-jet and a water ramjet suitable for very high speeds. Boeing developed its own pump-jet, testing it aboard a small boat, the 50-knot *Little Squirt*, in 1961–62. Water was drawn up through the struts of the main (after) foils through a pump mounted in line with the gas-turbine engine. The later Boeing naval and civilian hydrofoils, including the *Tucumcari* and the *Pegasus*, used this type of propulsion. As in the case of planform, it is by no means clear whether the simplicity of the pump-jet outweighs the efficiency of the propeller. For example, the big navy hydrofoil designed (but not built) in 1972 had propellers.

The surface-effect craft rides an air cushion created either by its forward motion, as in the sea sled or a wing-in-ground-effect craft, or by a separate blowing engine. If the blower is powerful enough, the craft rides above the water even at zero speed.

The air-cushion concept began as a means of reducing the drag of more conventional craft by, in effect, lubricating their hulls with a layer of air blown through holes. Proposals can be traced back to the 1850s; some steam vessels actually had layers of air blown under and around them, apparently with little effect. In 1915 Lieutenant Dogobert von Mueller-Thomamuhl of the Austro-Hungarian navy proposed an air-cushion torpedo boat, using a 65-bhp fan plus three 120-bhp propulsion engines. A boat with four 120-bhp propulsion engines was actually completed in October 1915, but the fan proved inadequate to support its 6.3-ton weight. Even so, it is considered the ancestor of modern air-cushion craft.

All modern surface-effect craft are descended from the British Hovercraft of the 1950s, invented by Christopher Cockerell. Early ones were sometimes called ground-effect machines (GEM) or captured air bubble (CAB) craft, the bubble being the air cushion carried along with the vessel. Here the key design problem was to maintain the air cushion at minimum cost in power.

The earliest Hovercraft could achieve a clearance of one-twentieth or one-thirtieth of their size, so that they were unable to traverse rough water. However, in 1958 C. H. Latimer-Needham proposed a flexible skirt, which in effect provided additional height between the rigid portion of the craft and the surface by extending the craft downward. Early tests showed that it increased ground clearance by a factor of about ten. The skirt could also capture much of the cushion air and so reduce the need to blow more air into it. For example, the early skirtless (British) SR.N1 of 1959 could negotiate only 6- to 9-inch obstacles. Three years later, with 4-foot skirts fitted, it could cross 3-foot 6-inch obstacles and operate in 6- to 7-foot waves—at twice its original weight, without any increase in lifting power.

The U.S. Navy used Hovercraft in Vietnam (see chapter 13) and is currently commissioning 160-ton Hovercraft landing craft (LCAC), which will be included in a subsequent volume of this series. At this writing, the Soviets have a 350- to 400-ton amphibious Hovercraft, the "Pomornik" class, in series production.

The U.S. Navy also developed seagoing surface effect craft using rigid sidewalls (extending into the water) and flexible seals fore and aft to capture the air cushion and so reduce the load on lift fans. The current U.S. examples are the new SEAL support-boat project (SWCM) and the coastal mine hunter

Cardinal, construction of both of which was suspended at the time of writing (November 1986). Surface effect concepts figure prominently in some of the small-craft designs described in chapter 16.

Like a planing craft, the surface effect craft, made to ride the surface of the water, may ride poorly in rough water. The larger the craft and the deeper the cushion, the less this is a problem—but for most of the craft treated in this volume, it is.

The story of the motor torpedo boat exemplifies the potential conflict between an attractive technology and naval requirements. The first successful fast-planing motorboat (hydroplane) appeared in 1905. It seemed remarkable that a craft the size of a ship's boat could achieve destroyer speed, and several boat designers, including some Americans, suggested that it could carry torpedoes. The idea recalled the small steam torpedo boats of the previous century—inexpensive vessels capable of sinking the largest battleship. Their main drawback was a total lack of seakeeping ability. High (paper, or smooth-water) speed was not in itself sufficient to make them useful.

It took the particular tactical conditions of World War I in the North Sea for the Royal Navy and in the Adriatic for the Royal Italian Navy to make motor torpedo boats useful. The Royal Navy adopted motor torpedo boats, which it called coastal motor boats, because, with their shallow draft, they could pass over the minefields protecting German bases. In the calm Adriatic the Italians hoped that their boats, so small as to be almost invisible, could ambush Austrian capital ships within the protective range of shore defenses. Speed would make escape possible. In neither case did the unconventional weapon replace existing conventional ships; rather, it supplemented them in a tactical niche previously unavailable.

Similarly, when the General Board was asked in 1937 to consider motor torpedo boats for the U.S. Navy, it doubted that such coastal defense craft would be very useful, except in the Philippines, in the early stages of a war. To some advocates this might have seemed typical of the reactionary attitudes of a board of senior admirals obsessed with conventional large ships, but it is difficult to see how anyone in 1937 could have predicted the Solomons campaign of 1942–44, which provided one of the principal stages for American PT operations.

U.S. naval strategy explains the General Board's rejection of coastal craft. The central and consistent tenet of U.S. national security policy is to meet foreign threats as far away as possible, preferably at their point of origin abroad. In 1890, when America's modern naval strategy was first formalized, it could be assumed that the United States would enjoy substantial warning of any assault on the western hemisphere from abroad, and that a sufficiently powerful battle fleet could steam out to meet that threat in mid-ocean or even, preferably, in enemy waters. Later, what had been a tacit national strategic goal expanded to include the prevention of any single foreign power from uniting the productive forces of Europe into a potentially hostile and decisively powerful combination. This threat, alongside the fear that war would spill into the western hemisphere, compelled the United States to enter World War I. After 1940 the United States accepted entanglement in European affairs on the grounds that a European war would necessarily spill into the hemisphere. Fundamental strategy now required the presence of U.S. troops on the European continent. Similarly, the U.S. presence in the Pacific after 1945 was and is maintained to prevent any hostile power (the Soviets) from uniting the productive forces of Asia against the United States.[16]

Moreover, since World War II the threat has grown as the prospective warning time has shrunk. The same fundamental policy that, before 1940, justified isolationism, has since 1945 justified the forward basing of U.S. fleets, ground troops, and airmen.

Before 1941 America's defensive posture had to take account of the major overseas possession, the Philippines. The islands presented an almost impossible problem, being so close to the likely aggressor, Japan, and so far from the United States. The United States could never maintain a strong enough local force to protect a place so far from home. The oceangoing battle fleet, the chief instrument of the forward policy, had to be concentrated, and that concentration could be maintained only in home waters. Even Pearl Harbor was considered too far forward to support the fleet on a permanent basis.

The only viable solution to the problem was to design a battle fleet powerful enough to steam across the Pacific, defeat Japan, and thereby either recover the Philippines or relieve a besieged U.S. garrison.[17] This was the essence of War Plan Orange, which the U.S. Navy would in effect execute in World War II: the fleet would steam west, seizing operating bases en route, toward a decisive engagement, probably in Philippine waters.

Before 1941 the only legitimate continuous foreign naval involvement was the peacetime protection of U.S. nationals and trade in what would now be called the Third World: it justified the construction and operation of gunboats (which also helped maintain order in the Philippines).

While the Japanese threat to the Philippines disappeared in 1945, the need to maintain the U.S. presence in the Third World grew. A central logistical concern is still the difficulty of maintaining a per-

Underwater noise is a particular problem for a small, high-powered boat. In 1941, this air propeller was tried aboard a World War I–built subchaser, the SC 431. The propeller was rejected because of its considerable airborne noise, which entered the water. The boat and its engine are shown at Norfolk Navy Yard on 22 December 1941.

manent and sufficient American (or Western) power in forward areas.

More generally, economy of force dictates that a United States far distant from the potential sources of its problems rely on a concentrated and mobile battle fleet capable of carrying the fight to the enemy and destroying him. In 1914 that fleet consisted of battleships; since 1942 it has been built around carriers. In neither case did it seem wise for the United States to invest heavily in coastal forces in peacetime: unless U.S. strategy failed, coast defense would be of little importance. The great exception was ASW; hostile submarines could be expected to evade fleet operations in forward areas and attack U.S. ships close to home.[18] After World War I, the Orange strategy presented another exception, for a fleet steaming across the Pacific needed bases, which themselves had to be defended against Japanese submarines and surface raiders.

Before 1914 the U.S. Navy assumed that war, if it came, would most likely result from a German attempt to vitiate the Monroe Doctrine and seize colonies in the western hemisphere. Protracted German naval operations would have entailed the seizure and development of a forward base, an operation the U.S. battle fleet could have contested, probably from its own improvised forward base. (A major fixture of most pre-1914 U.S. naval war games was the race between the U.S. and enemy fleets to establish a forward base.) It seemed unlikely that the United States herself would be the enemy's immediate objective. The U.S. fleet would prevent the Germans from occupying territory on the American continent only because such occupation would in itself endanger the United States. The local defense of major U.S. cities was best left to local fortifications, although the navy did maintain a coastal submarine force. It also stationed submarines in the Philippines as a defensive screen against a Japanese landing there.[19] The closest thing the navy had to small combatants was the boats that battleships carried, which could help defend the forward base by laying defensive minefields. The battleships also carried the mines.

After 1914, however, it gradually became clear that German submarines could pass through virtually any blockade to operate off the U.S. coast. In line with its prewar strategic mind-set, the U.S. Navy reaction was to organize coastal patrol forces. After all, lacking any efficient means of directing themselves against targets scattered over the open ocean, U-boats were

U.S. subchasers are shown at Spalato, on the Adriatic, on 4 July 1919. They had helped patrol the Strait of Otranto.

most likely to operate relatively close to shore, in areas of shipping concentration. Large numbers of craft would be needed to patrol these waters, and the craft would have to be relatively small for quick production. Hence the 110-foot subchaser, a larger version of American-built British patrol motor launches.

It was not until mid-1917 that the U.S. Navy became aware of the success of the German submarine offensive, that is, of the failure of patrol tactics. The American contribution to the war against U-boats was in providing enough units to improve the level of patrol coverage (subchasers in European waters) and in making available much of the wherewithal, including destroyers, to build a successful convoy strategy. To be effective, ASW convoy craft would have to be truly seagoing. Hence the design of the eagle boat and of its planned 250-foot successor (see chapter 2).

After World War I, American attention turned to the Pacific; Japan was now the most probable future enemy. Small craft could play only a limited role in the Orange plan, particularly in its early open-ocean stages: they would ensure that the fleet could sortie from its bases in the face of Japanese submarine opposition. Special escorts such as the PC could convoy merchant ships and so release the more versatile destroyers for service with the battle fleet and the fast carriers. In the late stages of the war, when the fleet was operating close to Japanese bases, fast attack craft might become useful, but their value would depend in large part on the ease with which they could be carried overseas. It was not difficult to argue that submarines and aircraft (seaplanes and carrier types) would be more efficient means of torpedo attack against shipping.

When the Philippines became a semi-independent commonwealth in 1935, scheduled for independence in 1945, the U.S. Navy, enduring a period of fiscal stringency, showed no particular desire to buy specialized craft to defend it. The navy preferred to contribute submarines and patrol bombers, both universally more useful weapons. In practice, once the navy had bought PTs, it stationed them in the Philippines—but the decision to buy the boats seems not to have been predicated on considerations of Philippine defense.

World War II, like its predecessor, sprung many surprises on the U.S. Navy. Again abandoning her traditional policy of isolation, the United States fought a coalition war that entailed large-scale operations in European waters. Again mass-produced ASW craft were required to neutralize U-boats. This time, moreover, coastal patrol was not enough to counter the threat, and the two classes of ASW craft designed before the war, SC and PC, had to be supplemented by converted minesweepers (PCE and PCS) and by much larger ships, particularly destroyer escorts, the British corvette (PG) and frigate (PF), and large numbers of converted civilian craft. European warfare also included massive operations in the Mediterranean which had not figured in prewar U.S. strategic thinking; these involved many motor torpedo/motor gun boats.

The United States employed PTs throughout the

South Pacific and Mediterranean during World War II. Once they had been transported to forward areas, they proved effective. The question of their general relevance to U.S. naval requirements remained, however, and they were abandoned after the war. The United States did support PT construction by NATO coastal navies such as those of Denmark and Norway.

Coastal ASW craft were retained in the face of the continuing submarine threat, this time from the Soviets, but it appeared that existing types were too small and too slow to deal with the new generation of submarines. An attempt to develop a postwar PC led to the *Dealey*-class destroyer escort.[20] Funds were scarce, and throughout the postwar period ASW development concentrated on oceangoing ships; no new coastal ASW craft were built. Coastal ASW itself seemed important only in the context of a protracted World War II–style war, and that seemed more and more remote as the United States deployed more nuclear weapons.

Under President Eisenhower, the U.S. military did not envision a future large-scale conventional war involving coastal convoys. American naval forces did fight a coastal war in Korea, but that was dismissed as a special case, unlikely to recur. Instead, the two forward-deployed fleets (the sixth in the Mediterranean, the seventh in the Far East) were expected to support American allies by threatening nuclear air strikes.

In the event of a major protracted war, U.S. naval strategy increasingly emphasized forward ASW and convoy operations, to the virtual exclusion of coastal convoys. The new ASW strategy emphasized the gradual destruction of an attacker's submarine force, which did require some backup in the form of local defense forces to protect against submarines "leaking" through the forward ASW system early in the war. This concern echoed the prewar interest in local ASW craft to protect fleet sorties from enemy bases. Prototypes of replacement subchasers were included in early versions of several naval building programs of the 1950s and ultimately an inshore subchaser, the *High Point*—the first U.S. combatant hydrofoil—was built. However, this line of development died under the budgetary pressures exerted by the Vietnam War.

Throughout the 1950s, coastal and inshore warfare was important to many U.S. allies, particularly in the Far East, where Korea, Taiwan, and Vietnam directly confronted communist coastal fleets. The United States specifically developed a generation of coastal gunboats for allied navies and exported a variety of other coastal craft, some of them derived from World War II prototypes. After Fidel Castro's victory in Cuba, potential enemy surface warships infiltrated the western hemisphere for the first time since 1898. When Cuba obtained Soviet-built fast attack craft, the United States began to develop her own, the *Asheville* class in particular, to match them.

By the mid-1950s the navy's long-range planners could see the day, not far off, when the Soviets would possess a large enough nuclear arsenal to deter American attack, and thus when limited warfare on the Eurasian periphery would become much more probable. This guess was confirmed in Vietnam, where the navy once more had to fight a local protracted nonnuclear war. The confrontation was a coastal and riverine one, and to meet its special requirements the navy developed an entirely new generation of small combatant craft.

Like Korea, Vietnam was perceived as an isolated, special sort of war, and although a few of the small combatants used during it were retained afterward, they were gradually discarded. Now, as this is being written, small combatants are the province of the navy's special warfare units, the SEALs. Yet the logic of the long-range planners still seems valid: future war is much more likely to be a limited affair in the Third World than the all-out conflict for which the main U.S. fleet and units like the SEALs are designed. Does that mean that we will have to extemporize a new generation of small combatants in the next decade or so?

In keeping with their extemporized character, small combatants were procured in a variety of unconventional ways. Actual construction was always the responsibility of the materiel bureaus: construction and repair (C&R), BuEng, and ordnance (BuOrd). C&R and BuEng merged in 1940 and became BuShips.[21] In 1966 the bureaus, formerly independent and responsible only to the secretary of the navy, were reorganized as the Systems Commands under the chief of naval operations (CNO). BuShips became the Naval Ship Systems Command (NavShips). NavShips and the Naval Ordnance Systems Command (NavOrd) merged in 1974 to become the Naval Sea Systems Command (NavSea).

Above the bureaus sat the secretary of the navy and his civilian staff, including the assistant secretary, who was responsible for ship procurement. Franklin D. Roosevelt held this post through World War I. His experience with the subchaser program appears to have shaped his views on the World War II subchasers (see chapter 3). As president, he was deeply involved in the early stages of the PT program; again, his actions reportedly reflect his World War I navy experience.

For the period from 1908 through 1945, U.S. warships were designed to meet characteristics, or staff requirements, drawn up by the General Board, which consisted of senior admirals. The General Board began as a war-planning agency, and its deliberations often turned on issues of national strategy. Typically the board held hearings at which the views of the fleet could be determined. Given characteristics, the board reviewed preliminary designs presented by C&R (or, later, BuShips). The board was also responsible for major equipment such as guns and sonars.

In wartime the war planners were largely superseded by the operational command structure created in 1915, the Office of the Chief of Naval Operations (OpNav). Thus, although the General Board was involved quite early in the prewar design of the 110-foot subchaser, the earliest proposal for what became the 200-foot eagle boat originated in OpNav. The General Board reasserted its authority between wars, but by the late 1930s OpNav was increasingly important, partly because its chief, the CNO, acted as secretary of the navy when the civilian secretary was not available—as was often the case with Roosevelt's secretaries.

Just before and early in World War II, most decisions about small combatants were made either by special boards or, after Admiral Ernest J. King became CNO, by the vice chief of naval operations, Vice Admiral F. J. Horne. After the war the General Board, which survived until 1951, lost its characteristics-setting role to the Ships Characteristics Board (SCB), part of OpNav.

Unlike the independent General Board, the SCB included representatives of the materiel bureaus who voted on characteristics. OpNav representatives such as the chairman were responsible for shaping characteristics to meet strategic requirements. However, because the SCB was in part a creature of the bureaus, SCB hearings, unlike the earlier General Board hearings, tended not to be adversarial. Probably the most important SCB products described in this volume were the *Asheville*-class motor gunboat, the hydrofoil subchaser *High Point*, and the two hydrofoil gunboats.

Between 1955 and the late 1960s, OpNav views were also affected by the Long-Range Objectives group (LRO, Op-93), initially formed to assess the needs of the 1970s. LRO was responsible for the reorientation of U.S. naval strategy (with its increasing emphasis on limited, and then on proxy, warfare) in the mid- to late 1950s and was very active in early studies of fast motor and hydrofoil gunboats. Gradually the planning horizon of the LRO diminished, and by the time Admiral Zumwalt disestablished it, LRO had lost most of its power.

During the Vietnam War, craft were procured at first by the CIA and the Department of Defense (DoD), not the navy. Much of the early work in Vietnam was done by DoD's Advanced Research Projects Agency (ARPA). There was also the DoD/CIA Counterinsurgency Research and Development group. From 1961 to 1964 characteristics for craft such as the RPC were developed by a BuShips panel on unconventional watercraft. Characteristics for later riverine craft were developed partly by the Military Advisory and Assistance group (Vietnam) and partly by OpNav, drawing on advice from Amphibious Forces, Pacific, and from the Pacific Fleet. Some craft, most notably those used by the SEALs, were extemporized by operational commands.

After the war, relatively few combatant craft were developed. By this time the SCB was moribund; individual projects were usually developed within the offices of the deputy chief of naval operations (DCNO) (Op-03 was responsible for surface craft, and its subsidiary, Op-37, for small craft and mine craft). The CPIC was unusual: it was developed as a research and development project, carried out largely by the Naval Ship Research and Development Center rather than by NavSea's small boat design organization, and it was sponsored by a special office created by Admiral Elmo Zumwalt, the CNO (Op-OOZ).

Many of the designs originated with commercial yards specializing in small fast boats. C&R and its successor, BuShips, were best equipped to develop large combatants. Although they had considerable impact on the smaller combatants, it seems noteworthy that an experienced yacht designer, A. Loring Swasey, was brought in to develop the initial plans for the 110-foot subchaser during World War I. Although C&R conducted extensive work in PT boat design, the two major World War II production types, the Elco and the Higgins, were, respectively, developed from a British design (Scott-Paine) and a private design (Crouch). Andrew Jackson Higgins in particular had enormous influence on American small boat design during World War II; he will be discussed in greater detail in a later volume in this series on amphibious warfare and amphibious craft.

The reader should also be aware of the enormous effect on U.S. builders of the experience of the 1920s and early 1930s, when the Coast Guard battled the rumrunners. The official administrative history of the PT boat notes how commonly an engineer would preface some bit of technical advice with "when I was with the rummies...." Coast Guard efforts against the smugglers culminated in the 83-footer described in chapter 3.

On the other hand, the ASW types of World War II were all navy designs, as were the bulk of the Vietnam riverine craft.[22] The situation in Vietnam was complicated by a decision to move BuShips' small-craft unit from Washington to Norfolk in 1966. As a result, most of the designers in the unit left, and design continuity was lost at the time when the ASPB

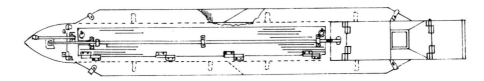

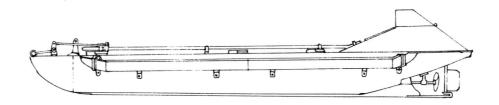

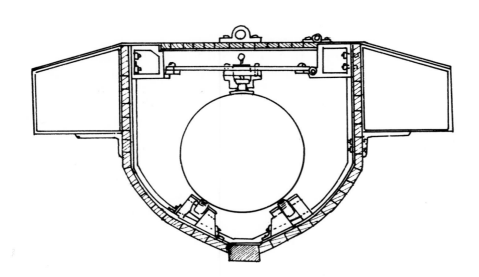

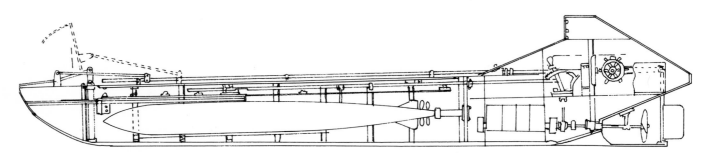

The Shearer One-Man Torpedo Boat exemplified unconventional U.S. small combatants. Begun at the marine railway depot of the Thames Towboat Company of New London, Connecticut, on 1 October 1917, it arrived at the naval base at New London for trials on 12 February 1918. Powered by a Scripps 35 HP 4-cycle engine, it could achieve 5 knots in semi-submerged condition, carrying its torpedo (in light condition, 9 knots in a calm and 7.5 in heavy weather). The inboard profile shows the arrangement for raising the bow to let the torpedo swim out, and the cross-section shows the two buoyancy tanks alongside the torpedo trough. For further details, see Chapter 4.

Table 1-1. U.S. Small Craft Designators, 1965–86

AB	assault boat (50-foot ASPB)	NS	nonstandard
AL	aluminum launch	PB	patrol boat
AR	aircraft rescue	PE	personnel boat
AT	armored troop carrier (mini-ATC)	PK	picket boat
BE	Bertram enforcer	PL	LCPL (landing craft, personnel)
BU	buoy boat	PP	plane/personnel boat
BW	Boston whaler (13- and 16-foot)		
BX	distribution box boat (for controlled minefields)	PR	patrol boat, riverine (PBR); also used for plane rearming boat
CC	commercial cruiser	RB	rescue (swamp) boat
CM	LCM (50-, 56-, or 74-foot)	RL	refuel launch
LH	line-handling boat	RP	river patrol boat (earlier designator for PBR)
MB	motor boat	TR	torpedo and weapons retriever
ML	motor launch	UB	utility boat (including STAB and LSSC in Vietnam)
MD	riverine minesweeping drone (MSD)	VP	LCVP
MR	riverine minesweeper (MSR)	WB	work boat
MW	motor whaleboat		

and LCM conversions were being developed. The Norfolk division of the Naval Ship Engineering Center was responsible for many of the post-Vietnam small craft described in chapter 16.

It is important to keep in mind the extent to which personalities shaped particular small combatant programs. President Roosevelt, for instance, was responsible for the World War II wooden SC; without him, there would probably have been only the steel PC. He was probably also responsible for the decision to select PT and PC designs by competition among navy and private designers, rather than relying entirely on C&R. And Admiral Elmo Zumwalt, as CNO, was personally responsible for the design and procurement of the *Pegasus*-class hydrofoil missile gunboat, the PHM; he was also the principal force behind the CPIC program.

Designations

The ships and craft described in this book fall into two categories: commissioned and noncommissioned. Commissioned ships and craft all had sequential hull numbers, and they are all listed in the tables in the appendix. In one important case, that of the subchasers, a variety of different designations (SC/PC/PCE/PCS) fell into the same numerical sequence. The other major series were eagle boats (PE), motor torpedo boats (PT), and fast patrol boats (PTF).

The noncommissioned boats fall into three categories. Until 1965, craft built in navy yards were listed in a purely numerical sequence. Special craft and craft built by contractors were designated in a C series (for example C-250). Unfortunately the C series is neither chronologically nor functionally arranged. It was probably originally assigned in blocks, but then the gaps left in and between blocks were filled in randomly. In 1965 this system was abandoned in favor of a functional and chronological one. Boat hull numbers consist of a length (in feet), a two-letter functional designation, then two digits for the fiscal year and a sequence number for that type within that year. Table 1-1 lists the principal functional designators.

Both the C series and the post-1965 lists are incomplete. Many locally or specially acquired boats (for example, those of the SEALs) were never registered or were registered several years after their purchase. In some cases units using such craft were unaware of the requirement to register and found out later that only through registration could they obtain funds for boat maintenance. It also seems likely that BuOrd and its successor, the Bureau of Naval Weapons (BuWeps), sometimes bought boats as equipment, so that they were never in the C-series. The bow numbers of the SWAB (chapter 10), for example, do not fall into any series the author could find. It also seems likely that many of the boats procured on an emergency basis for Vietnam in 1961–62 were never registered in any way, even though they were certainly produced in series.

The reader will note that many boats, particularly from the period before World War II, have their speeds given in statute, *not* nautical, miles per hour. This was a common practice; I have generally avoided translating into knots (1 knot is approximately 1.15 statute miles per hour) because that would conceal the use or nonuse of round figures.

2

Subchasers and Eagle Boats

World War I presented the U.S. Navy with a novel problem: the building up of a massive coastal patrol force. That meant acquiring numerous civilian yachts, but effective ASW patrol also required large numbers of specialized, specially built craft, the 110-foot subchasers (SC) and the 200-foot eagle boats (PE). They were the first modern U.S. small combatant warships, and as such were the direct ancestors of the large family of World War II ASW ships and craft. The submarine threat was also responsible for the earliest U.S. work on fast motor torpedo boats, which led to the World War II PTs and then to such fast craft as the current PHMs.

The subchasers and eagle boats were conceived as patrol rather than convoy craft, their role being to execute an ASW strategy already in use by the Royal Navy. In theory, if such craft were concentrated in U-boat operating areas, they would encounter and sink U-boats often enough to end the submarine problem. Before 1917, patrols were relatively ineffective, partly because individual boats had no effective means of detecting a submerged submarine except by observing its periscope. The U.S. Navy had adopted the patrol strategy—implicitly in 1916 when it formed a motorboat reserve and explicitly in early 1917 when the SCs were designed—in part because British reports overestimated the success of area patrol. Convoying, a very different strategy for which patrol craft were ill-suited, is generally credited with neutralizing the U-boats. Only in early 1917 did the ingredients of a successful patrol strategy become available with the development and production of hydroplanes and the mass production of depth charges.

Coastal areas, in which subchasers could operate, were the most likely locations of U-boat concentration, being the areas with the most targets. Even when convoying was introduced, its ASW component did not extend very much beyond these areas, because in midocean the principal threat to shipping was the surface raider, which could detect its targets at much greater range and close them more easily. This consideration still seemed valid before 1939, which is why both the U.S. and Royal navies concentrated on coastal and inshore ASW craft (the American SC and PC, the British corvette). After 1939, the Germans started using wolf packs supported by air reconnaissance and code breaking, which moved the ASW war to the mid-Atlantic and invalidated so many prewar measures—among them mass production of the successor to the World War I subchaser.

The reader may find it odd that submarines were almost always included with destroyers and subchasers as urgently needed *anti*submarine craft. Submarines could lie in wait athwart U-boat transit routes or in U-boat operating areas, ambushing surfaced U-boats. The U-boat, nearly immobile when submerged, could move considerable distances only on the surface. By forcing U-boats to stay submerged, Allied submarines enormously reduced their mobility, and hence their effectiveness. In the context of America's small craft program, the high priority accorded submarines was significant because they competed with subchasers and other small combatants for a limited supply of internal-combustion engines.

Low-frequency hydrophones, the chief ASW sensors, could detect the sound generated by a moving submarine. Effective listening required that the listening platform be silenced, generally by stopping engines and heaving to. Upon hearing a U-boat, the subchaser would dash toward its estimated position. Success depended on sea-keeping (hydrophone performance suffered if the boat rolled badly); the extent to which noise could be controlled, particularly that of engines, which had to be turned off and on suddenly and quickly; and dash speed. Hunting sub-

The ungainly eagle boats typified the U.S. mass-production response to the World War I submarine emergency. This is Eagle 32 after the war, probably in San Francisco.

chasers typically stopped about every ten minutes, which was possible only because they had internal-combustion engines. A steam plant, with its many auxiliaries, could not start up or shut down nearly so easily or frequently. For this reason, the subchasers were considered better listening platforms than destroyers.[1]

Even though submarines could avoid detection with hydrophones operating at low speed, about 3 knots, this immobilized and hence largely neutralized them in the face of patrol forces.

Late in 1915, the Royal Navy introduced a non-directional shipboard hydrophone that operated at 400 to 1,000 Hertz.[2] However, what was needed was a directional device, to lead an ASW ship or craft to the submarine. Its development was still incomplete when the United States entered World War I in April 1917. American research resulted in operational directional hydrophones by late 1917. The first was the SC-tube, a pair of rubber bulbs projecting below a ship's hull, one connected to each of the operator's ears. He could rotate the entire device, using the binaural effect to establish a bearing. Later the bulbs were replaced by a more sensitive and directional array, the MB-tube, with eight microphones and a 30-degree beam on each side. SC-tubes could detect sufficiently noisy vessels at ranges of 4 or 5 miles; effective ranges against submarines were much shorter.

The other major line of development was the fixed array. The U.S. K-tube was a triangle with microphones at its corners. It could be suspended at a depth of 40 feet from buoys, by means of which its bearing relative to the listening vessel could be observed. Its acoustic beam was rotated by shifting the relative delays between the microphones. The K-tube was credited with an acoustic range of 30 miles or more.

The MV-tube, developed from a French prototype, used a mechanical "compensator" to move its beam. It consisted of arrays placed on the port and starboard sides of a ship's hull. Unlike the earlier devices, it could operate from a moving ship; in tests the destroyer *Aylwin*, running up to 20 knots, tracked a 7-knot, 50-foot-deep submarine at ranges of 500 to 2,000 yards. MV was the most sophisticated system in use during World War I, and late in 1918 it was installed in the *Aylwin* and in one subchaser, the SC 252. It weighed about 2 tons; its two blisters were about 38.5 feet long (23.5 inches wide, 15.5 inches deep).

However sensitive they were, hydrophones were limited to determining the bearing of a submarine. U.S. subchasers therefore hunted in units of three to obtain cross-bearings. In European waters, they often hunted submarines initially located by radio direction-finder bearings (or decrypts) of U-boat messages.

None of these acoustic devices could detect a submarine lying on the bottom. U.S. subchasers were therefore equipped with an underwater electric potential detector, the "trailer," consisting of a bronze plate (later, simply the bronze rudder itself) trailed astern and a heavy chain at the end of a phosphor bronze wire. When the wire contacted the steel hull of a submarine, electrolytic inter-action with the bronze plate set off a buzzer in the wheelhouse. Trailers were reportedly used most effectively to establish the position of sunken U-boats on the sea floor.

The only weapon effective against a submerged submarine was the depth charge. Knowledge of a submarine's location was approximate at best, and the patrol boat had to fill that approximate volume with sufficient charges to achieve a reasonable kill probability. Depth-charge tracks were only a step in this direction, since they laid only a line of charges. Thus the development of depth-charge mortars, which could spread charges out on either side of that line, was an important one. The U.S. version was the "Y-gun," which used a single blank 3-inch charge to fire two charges at once, one to either side, to a range of about 50 yards. Their recoils were balanced.

Neither the Y-gun nor the depth-charge track could be aimed. The only way to "aim" an attack was to maneuver the entire ASW craft. Maneuverability thus became a major consideration in patrol boat design.

A small, inexpensive boat could carry enough depth charges to damage a submarine. Most U-boats were armed with deck guns, and they could, if on the surface, fight off patrol craft. To be effective, a patrol craft had to be armed with a gun similar in power to a U-boat's weapon. Sometimes the criterion for accepting a gun was its ability to penetrate a submarine's pressure hull. It was often argued that patrol craft were so numerous and dangerous that no U-boat would welcome the risk of engaging one on the surface.

Although several navies purchased motor launches before 1914, small combatants driven by internal-combustion engines reached maturity only during World War I. When war broke out motor yachts, most of them built in the United States, were fairly common in several European countries, and existing civilian craft were requisitioned for tasks such as harbor patrol. In January 1914, Britain formed a motorboat reserve consisting of pleasure boats displacing from 3 to 65 tons. A total of 198 such boats were called up after August 1914 and continued in service until motor launch deliveries began late in 1915. Motor craft were useful for harbor patrol but not seaworthy enough to operate farther out to sea.

The first official U.S. response to the submarine patrol problem was the purchase of two special boats to serve as models for interested private yachtsmen. This one shows A. Loring Swasey's 45-footer, similar to his earlier seagoing commercial craft such as the *Houp-La*. No photograph of the original boat appears to have survived.

Even so, the Admiralty found them extremely valuable and used American resources to produce a motor launch seaworthy enough and strong enough for naval use. The resulting program of American-built Elco motor launches led indirectly to the U.S. Navy's 110-foot subchaser program.

The first naval motor launch program in the United States was part of a larger preparedness campaign. Beginning late in 1914, many Americans argued that the United States should arm herself in case she was drawn into the European War. The old scenario of a German attack was revived in a popular movie showing European (obviously, but not explicitly, German) soldiers executing American Civil War veterans. The Wilson Administration, trying hard to stay out of the war, avoided such propaganda, but the assistant secretary of the navy, Franklin D. Roosevelt, ardently supported preparedness. An amateur yachtsman, he was well aware of the use of motor yachts by European navies. Many of his friends badly wanted to contribute to U.S. naval preparedness. In the late summer of 1915, while serving as acting secretary of the navy during a temporary absence of his superior, Josephus Daniels, Roosevelt announced a "department plan" to create a 50,000-man national naval reserve, made up mostly of civilian volunteers, to supplement existing state naval militias. The *New York Times* carried Roosevelt's account of his reserve plan, including a call for suitable motorboats, in its 3 September issue. Roosevelt claimed that "the Department has already endeavored to cooperate with the power squadrons." The "power squadrons" had been inspired by the appearance in 1915 of a particularly seaworthy 40-foot motor launch, the *Houp-La*, designed by A. Loring Swasey (see table 2-1). Stewart Davis, of Southampton, Long Island, had formed a volunteer patrol squadron with five of Swasey's 40-footers, each carrying a crew of four (a captain, two signalmen, and an engineer). Privately financed, the vessels would serve as scouts and patrol boats in wartime. In early 1916, additional volunteer boats were built by private yachtsmen, the Vanderbilt 72-foot yacht (later, Patrol Boat 8, SP 56) acting as mother ship. Swasey designed several more: a 45-footer for Nat Ayer (the *Lynx*, SP 2, built by Lawley), a 53-footer for J. L. Saltonstall (the *Scoter*, SP 20, built by George Lawley), and a 63-foot "scout cruiser" for H. Oelrichs (later, Patrol Boat 6, SP 54). Roosevelt reviewed the ten boats of the First Patrol Squadron at Boston on 28 April 1916. The volunteer boats participated in a training cruise in September 1916.[3]

Secretary Daniels could not disavow Roosevelt's blatant preparedness measure. In November, he asked the General Board for characteristics applicable to such craft and for designs that could be built for ten, twenty, or thirty thousand dollars. The board considered strength and a speed of over 16 knots most desirable. It included a torpedo tube and a 1-pound gun in characteristics for an experimental motorboat that could possibly enter fleet service, but doubted that private patrol boats could (or should) be armed with torpedoes, owing partly to difficulties of supply and maintainence. The boats, then, would be armed with 1- to 3-pound (37mm to 57mm) guns, for which decks would have to be specially strengthened, and they would remain on patrol with a crew of six for up to forty-eight hours. The smaller boats would have to be outfitted for hoisting.

Table 2-1. Patrol Boats, 1916

	Swasey	45-ft	66-ft
Length (ft-in)	40–0	45–5½	—
	—	44–9 (waterline)	66–0 (waterline)
Beam (ft-in)	8–9	8–8	13–3
Draft (ft-in)	2–6	1–7	4–6
Displacement (tons)	—	8.13	—
Engine	1 Sterling (6 cyl) 135 bhp	2 Van Blerck (6 cyl) 150 bhp	2 Van Blerck (12 cyl) 400 bhp
Speed (kts)	20–20.9	21.7	26
Gasoline (gals)	—	—	1,800
Endurance (nm/kts)	—	—	435/21.7
Guns	—	1 1-pdr	1 3-pdr

The board doubted the feasibility of a voluntary squadron of patrol boats, questioning whether owners would lightly sacrifice their needs for wartime efficiency. Even so, Daniels sought competitive designs for two boats suitable for both private use and patrol, one not less than 45 feet long, the other not less than 66 feet. They had much longer endurance than conventional motor cruisers. Swasey won the competition for the 45-foot boat, C 253, which was built by George Lawley and Son of Neponset, Massachusetts. After completion, it was sent to Newport Naval Training Station for examination by private yachtsmen. The larger boat, C 252, was built by the Luders Marine Construction of Stamford, Connecticut. Not completed until after war broke out in 1917, it was later reclassified as YP 66 and served in World War II. The 45-footer seems to have been less satisfactory. In 1919, Roosevelt discouraged a yachtsman who wanted to build one for future naval service: war experience had shown it to be too small.

The reserve was created in the summer of 1916. Daniels acquiesced to the idea, partly in hopes of enlisting specially skilled men, but Roosevelt's original emphasis survived and it was the yachtsmen who contributed their service and their boats. By September 1916, permanent boards for motorboat inspection had been ordered established in Boston, Newport, New York, Philadelphia, Norfolk, Charleston, Key West, Pensacola, New Orleans, San Francisco, Bremerton, and the Great Lakes Training Center.

Motorboats would be controlled by the naval districts. On 30 January 1917 OpNav directed the districts to form their civilian motorboats into scout patrols, six of which would form a scout division. One or more divisions plus a harbor entrance patrol would form a section patrol to cover a length of the U.S. coast. Boats were to be classified as follows:

- A (Slow). This would maintain station in a harbor in weather no worse than a moderate gale. Not less than 7 knots, it would be armed with one 1-pound and one machine gun.
- A (Fast). This would be capable of keeping the sea in a moderate gale. At least 40 feet long, it would have a speed of at least 16 knots, one 1-pounder and one machine gun, a crew of four, and provisions for four days.
- B (Slow). Capable of riding out a moderate gale at sea, this would be at least 60 feet long and have a speed of at least 10 knots. It would be armed with a dual-purpose (DP) gun, at least a 3-pounder, and at least two machine guns. It would carry a crew of eight, provisions for five days, and a radio and searchlight.
- B (Fast). This would resemble B (Slow) except in speed, which would be at least 16 knots.

The two prototypes corresponded, respectively, to A (Fast) and B (Fast).

Craft but not their crews, could be enrolled in peacetime. An owner wishing to participate in maneuvers had to have a reservist on board and in charge. In February enrollment slowed because of a shortage of guns, as the fleet badly needed its machine guns and there was not a large stock of 1-pounders. A single machine gun (forward) was adopted as an interim battery.

Meanwhile, in February 1916, the Greenport Basin Company offered the navy the standard 60-foot motor launch it was building for the czarist government (see below). It was rejected as not large or seaworthy enough, the navy preferring the 66-footer then under construction. At the New York Boat Show in February 1917, the company displayed the *Chingachgook*, a Russian type armed with a 3-pounder and a 0.30-caliber machine gun. She was later taken over by the navy as Patrol Boat 19 (SP 85).

By early 1917, OpNav estimated that the naval districts, with the exception of the tenth and eleventh (Great Lakes Training Center), needed a total of 2,200 patrol boats and 1,700 minesweepers. Of these, 1,430 patrol boats and 1,388 minesweepers would be needed to cover the Atlantic coast in case of unhampered enemy approach. As long as Germany kept busy at war in Europe, a minimum of 312 patrol boats and 119 sweepers, distributed as shown in table 2-2, would do, though the patrol would not be efficient. OpNav wanted a full patrol established for the First Naval District, headquartered at Portsmouth, New Hampshire, as soon as possible.

This model is Luders' winning 66-foot design of 1916. Note the pronounced hard chine, the 3-pound gun—and the absence of depth charges, which had not yet been invented.

As of 28 March 1917, only two seagoing and fifteen inshore or harbor patrol craft (and no minesweepers) had been enrolled. Once war began about 1,200 yacht owners offered their boats, but many proved unacceptable, including several of the prewar volunteer boats. Boats taken over for the duration were numbered in a section patrol (SP) series that included navy transports and special auxiliaries. The navy took over 288 yachts and patrol vessels and 454 motorboats. The Third Naval District, headquartered in New York, briefly considered taking over eight 78-foot boats under construction at Greenport for the Russian Red Cross, but Secretary Daniels released them, claiming that the needs of Allied governments were greater.

Before 1917, the most important small boat program in the United States was one undertaken for the Royal Navy. An Admiralty delegation visited the United States in the spring of 1915 and selected the Standard Motor Construction Company to build engines and the Electric Launch Company (Elco) of Bayonne, New Jersey, to build fifty antisubmarine motor launches (ML.1–50). Elco was famous for its mass production methods. Formed in 1892 to supply silent 36-foot electric launches to operate on the lagoons of the Chicago World's Fair, it introduced standardized construction to small boats with the 32-foot Elco Cruisette of 1914.

The British contract was signed on 9 April, with delivery of the first twenty-five, through Canadian Vickers, promised for 30 November. The Admiralty ordered five hundred more (ML.51–550) on 8 June 1915, shortly after completion of the prototype. All were to be ready by 15 November 1916. Three decades later Elco was still proud of its achievement: building 550 boats in 488 days.

Table 2-2. The OpNav Patrol Plan, March 1917

Naval District	Patrol Boats	Minesweepers
1 (Portsmouth)	60	14
2 (Newport)	48	14
3 (New York)	48	21
4 (Philadelphia)	36	21
5 (Norfolk)	48	28
6 (Charleston)	36	7
7 (Key West)	12	7
8 (Pensacola)	24	7

Forty (V.1–40) were transferred to France, and the Royal Navy ordered a final group of thirty (ML.551–80) in July 1917, delivery being completed by February 1918. In addition, twelve boats were built specially for France (V.62–73). Elco built another 110 boats for Italy (MAS 63–90, 103–14, 253–302, 377–96). Italian yards later built eighty-five additional boats of a modified "May 1917" design (MAS 303–17, 327–76, 377–96). Fifty more planned at the end of the war were not built. Two of the Italian Elcos were briefly taken over for training in April 1917.

Many wealthy yachtsmen built craft specifically for use as navy patrol boats in wartime. This is the *Mystery*, constructed by Luders Marine for Ralph Pulitzer of New York and offered to the government for one dollar a month; estimated value was thirty thousand. She was 71 feet long (overall), with a beam of 13 and a draft of 2.75 forward and 3.5 aft. Two 400-bhp, eight-cylinder Duesenberg patrol model engines drove her at 24 knots, and she was credited with a radius of action of 300 nautical miles (her designer claimed 750 to 1,000 at 20 knots). In navy service, she was armed with one 3-pounder forward, one 1-pounder aft, and two machine guns on the deckhouse aft, and she served in a section patrol as SP 428. She was returned to her owner in 1919. Note the similarity between her and the winning 66-foot design.

The Elco launches were designed by Irwin Chase and construction was supervised by Henry Sutphen. Both men remained with Elco between the wars and were involved in the Elco PT boat program of World War II.

The Admiralty required a speed of 19 knots fully loaded (with a 20,000-pound load) and an endurance (with 2,000 gallons, or 12,000 pounds, of fuel) of 800 nautical miles at full speed (1,000 at 15 knots or 2,100 at 11 knots). Boats were to be capable of maintaining station even in rough weather. Because they could not cross the Atlantic on their own power, they had to be small enough to be carried as deck cargo. A length of 75 feet was therefore chosen at first, although it was increased to 80 after completion of the prototype for ML.51 and later units. That allowed somewhat roomier internal arrangements without loss of speed, as the extra length balanced off the extra displacement.

The unprecedented requirement that 550 boats be completed in about as many days forced simplified design upon Elco. Even 19 knots was fast for such a boat, and it appeared at first that the requirement for seaworthiness would conflict with the requirement for speed. After model tests at the University of Michigan towing tank, Chase chose a hull with fine lines forward and with a flat after body. A more seaworthy alternative with greater deadrise aft (V-sections carried right aft) failed to achieve the equivalent of 19 knots in the tank.

As in all later specialized small craft, the choice of engines was crucial. Chase and the British commission were limited because they wanted reliability; they had to choose relatively massive medium-speed engines. They adopted twin 220-hp Standard air-starting and reversing gasoline engines.

Gasoline was a fire hazard. Seven of the British boats were lost by fire, not in action. In 1916 the fuel was changed to one part gasoline and two parts kerosene (paraffin).

Armament generally consisted of a new 3-inch/23 (13-pound) long-recoil Poole (Bethlehem Steel) gun. It proved too powerful for so small a boat, and by 1918 most motor launches carried Hotchkiss 3-poun-

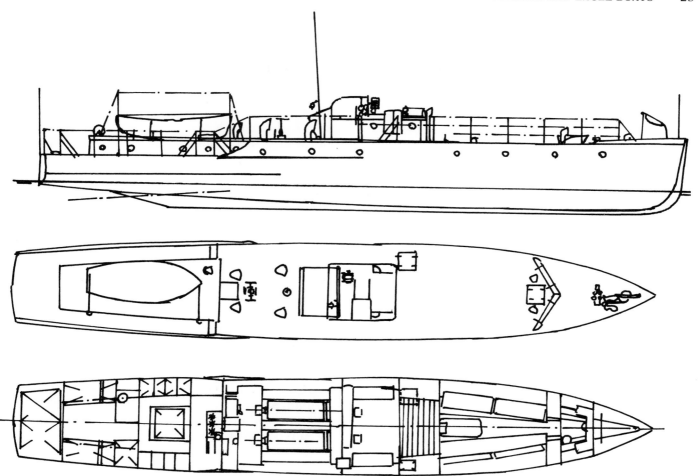

The Elco 80-foot motor launch built for the Royal Navy during World War I. Dimensions in feet-inches were 79-7 (overall), 78-9 (waterline) x 12-5 (extreme, over guards), 12-1 (on deck) x 3-1 (to bottom of deadwood). Displacement was 78,000 pounds. The boat is shown as delivered, unarmed. It was powered by two six-cylinder, 220-bhp Standard Motor gasoline engines and also had a 4.5-kw generator. A total of 2,100 gallons of gasoline sufficed for a radius of action of 750 nautical miles at 19 knots, or 1,000 at 15. On trial, boats achieved from 19.25 to nearly 21 knots. This drawing was adapted from one in *The Cinderellas of the Fleet*, an account of the subchasers published in 1920 by the Standard Motor Construction Company; the title could apply to most of the craft in this book. (Norman Friedman)

ders. The Poole gun was later adopted by the U.S. Navy for its 110-foot subchasers and as an antiaircraft (AA) weapon for destroyers and submarines.

Because they entered service before depth charges had been developed, the Elco motor launches carried several more primitive ASW devices: they towed explosive paravanes (Q-type) and carried small 14-pound bombs on the end of 4.5-foot handles. Ultimately they carried ten type D (420-pound) depth charges; there were no depth-charge throwers.

Motor launches were also used as minelayers, typically carrying four mines, and inshore minesweepers. In the Zeebrugge and Ostend raids of 1918, motor launches laid flares, made the smokescreens, and removed crews from the block ships.

Certainly, by 1916, the U.S. Navy was well aware of the versatility of the boats. The Royal Navy liked them but considered them too small to operate well in a seaway. They were better adapted to America's sheltered coastal waters than to the rougher waters around the British isles. In February 1917, when the U.S. Navy was considering buying some 80-footers itself, its attaché in London reported that the Admiralty would have preferred a boat 100 to 120 feet long.

The czarist government was the other major foreign customer for American-built motor combatants. During the Russo–Japanese War it ordered ten motor launches from Lewis Nixon. Assembled in Sevastopol in 1906 for shipment to the Far East, they were probably the first motor torpedo craft to serve in any navy. In 1906 they were transferred by rail to the Baltic Fleet for the defense of the Gulf of Finland. Originally designated patrol boats, they were redesignated torpedo boats on 10 October 1907. Although stricken in 1911, they were not discarded. The Russian navy placed them in storage between 1918 and 1923 and outfitted them as patrol vessels in the late 1930s. Six were broken up in 1940, one lasted until 1950. They displaced 35 tons, and their dimensions were 90 x 12 x 5 feet. Armament was one 47mm gun, one 18-inch TT, and two machine guns. Two single-shaft gasoline engines of 600 bhp drove them at 20 knots (18 knots in service).

In 1915 the czarist government ordered 18 fast (24-knot) SK-series subchasers from the Greenport Basin Company. Some later served as minelayers. Fourteen were assembled in Odessa in 1915, four in 1916, for the Black Sea Fleet. They appear to have required considerable modification, so that as of March 1916 only the first six were in commission. Even so, a further batch was ordered in late 1916. The series consisted of SK 311–18, 321–28, 331–38, and 341–47, thirty-one boats in all. SK 324 was converted to an MTB in the summer of 1917. According to the report of the naval inspector in New York, a typical unit was 60 x 10 x 2 feet 10 inches and weighed 14 tons; it was armed with one 3-pounder (47 mm) and one machine gun and had a crew of two officers and six enlisted men. The power plant consisted of three 8-cylinder 150-bhp Van Blerck gasoline engines with self-starters, dual ignition, and mufflers considered effective at 100 yards. Contract speed was 26 miles per hour (on trials two launches made 27.71 and 28.81 miles, respectively); radius of action was 300 nautical miles at full speed and 1,000 at economical speed, on 1,000 to 1,400 gallons of gasoline. Russian archives show a maximum speed of 22 to 24 knots and a range of up to 650 nautical miles (with 450 to 500 more common in service). The Russians claimed that they could operate in waters up to sea state 4. At least fourteen boats saw Soviet service after the Bolshevik Revolution, and four were taken over by the Whites. At least two boats, the *Hetman* (SP 1150) and *Russ* (SP 1151), as well as four privately owned Greenport boats of similar dimensions, the *Perfecta* (SP 86), *Quest* (SP 171), SP 689, and *Vitesse* (SP 1192), were acquired by the U.S. Navy.

The Russian boats were redesignated several times, making for some confusion. For example, SK 311 became SK 511 in September 1918, SK 11 in February 1919, SK 50 in 1921, and *Zhivuchii* in 1921. She was scrapped in 1925.

The czarist government bought a series of motor launches for the White Sea Flotilla, MI 5–8, MI 15–18, and PI 1–3. Soviet sources describe these boats, later taken into their service, as being 15.3 x 3.3 x 1.0 meters, with 120-bhp engines (18 knots), one 3-pounder and two machine guns, and a crew of eleven. It is not clear whether this is anything like a complete list of the series. MI 1–4 were built in England, but the missing numbers are not accounted for in any way. A photograph of an Elco boat apparently built for the czarists has survived, but without any accompanying description.

By early 1917, the U.S. Navy was uncomfortably aware that the country would soon be at war. On 31 January, the Germans announced the launching of their unrestricted submarine warfare. On 10 February, Secretary Daniels asked the General Board for recommendations on actions to take in the event of war with Germany. The board called for convoys and patrols of American waters, but President Wilson resisted any overt preparations. The CNO, Admiral Benson, wanted the mass production of antisubmarine patrol craft. Around 20 March, President Wilson decided that war was inevitable. It was not actually declared until 6 April 1917. A few days later, Secretary Daniels secretly assigned Rear Admiral William S. Sims as liaison with the Admiralty. Sims, who later became commander of U.S. naval forces in Europe, would have considerable influence on subchaser and eagle boat design.

Meanwhile work quietly began on the patrol boats required to enforce the General Board's plan. Should they be U.S. versions of the Elco 80-footer or something larger and more seaworthy? In February, Assistant Secretary of the Navy Roosevelt called a conference of naval experts who agreed that destroyers should be supplemented by hundreds of 110-foot wooden subchasers for open-sea patrol. Roosevelt, fearing an immediate assault by German submarines on unprotected American harbors, urged immediate construction of 50-footers, which he thought were the largest boats that could be built both quickly and in sufficient numbers.[4] Anything larger, he claimed, would suffer from bureaucratic delays. Roosevelt's biographers trace his perseverence with the 50-footer to his special fondness for small boats, his tenacity, and possibly also his friendship with Arthur Patch "Butch" Homer of Boston, the American representative of British Sterling Motors, whose engines would have powered it.

C&R and OpNav argued that the 50-footer was not worth building; it was little better than a converted motor yacht. Secretary Daniels agreed with his professional advisors, Chief Constructor David W. Taylor and then–Captain Hugh Rodman of the General Board, who favored the larger subchaser, but he was willing to allow his assistant a few 50-

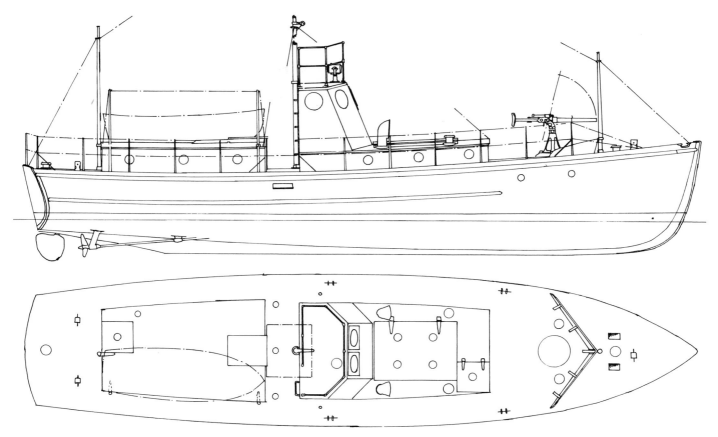

C&R's sketch design for a 50-foot subchaser, 1917. Note the absence of antisubmarine detectors and weapons, neither of which yet existed. The gun is a 3-pounder for dealing with surfaced submarines. The boat, like others of its time, would have been effective primarily as a means of keeping submarines submerged. The engine room, containing two eight-cylinder, 200-bhp engines, was just forward of the pilothouse; the crew was housed under the deckhouse aft. (Norman Friedman)

footers. On 24 March 1917 the General Board orally approved the 50-footer, and Roosevelt issued a contract, though without authority to do so. However, the next day, when it learned that 110-footers could be built rapidly and in sufficient numbers, the board reversed itself.

On 7 March 1917, Preliminary Design was ordered to work up both a 50-footer and a 110-footer. The former, based on the designs of a variety of motor launches, including the Greenport 50-foot torpedo boat, was armed with one 1-pounder and two machine guns. Power estimates showed 12 knots on 72 ehp. The goal was 18 knots and an endurance of 500 nautical miles at 12 knots. Work stopped on 24 March; Preliminary Design files show that Commander (later Rear Admiral) Furer, in charge of the subchasers, wanted the department's order to undertake the design suppressed.

In February 1917, C&R began design work on a subchaser larger than the Elco boat. It began as a 105-footer; early in March 1917, Preliminary Design developed both wood (58-ton) and steel (52.5-ton) versions. It also worked up an 80-footer of its own, similar in form to the proposed 105-footer.

The General Board wanted an inexpensive, robust wooden boat with a small displacement and capable of 17.5 to 18 knots. It was to have a cruising radius of 800 nautical miles, or 1,500 miles at 12 knots with emergency fuel, and stores for fifteen days, five of them emergency provisions. It would be armed with one 3-inch gun and one 6-pounder plus three 0.30-caliber machine guns and would have the most powerful possible radio. Other features included the ability to tow another ship of the same class, a raised lookout with access to a machine gun, and a protected steering station.

The C&R design organization was still entirely occupied with the large-ship program authorized by Congress the previous year, so A. Loring Swasey was brought in to develop the design; he was responsible for its general arrangement and principal features. Experience had already shown that subchasers would have to operate well out at sea, and that extreme speed such as that in the Elco boats was not as useful

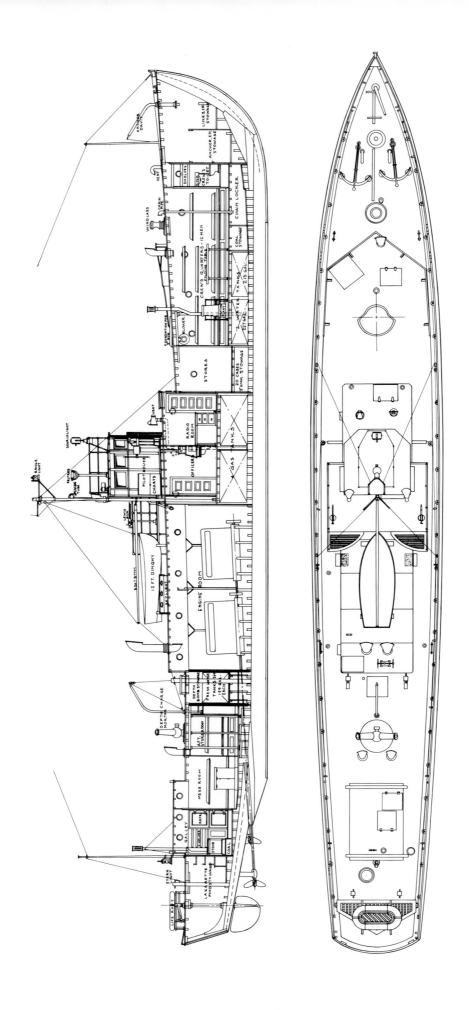

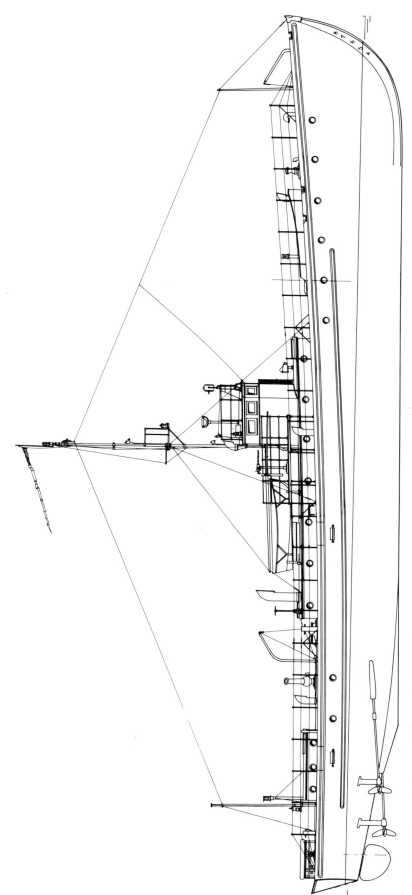

A 110-foot subchaser of the series SC 288–312, prior to installation of the 3-inch/23 gun. Only SC 310 had Lewis guns as original armament. (A. D. Baker III)

SC 235 shows a standard World War I subchaser battery: a gun (in this case, a 6-pounder) forward, a Y-gun abaft amidships, and a depth-charge track right aft.

as sustained but more moderate speed. Swasey therefore chose flare forward, for dryness, and a round-bilge form with roughly symmetrical ends to meet either a following or a head sea. On the basis of experience with rough-water craft such as those used by Gloucester fishermen, he cut away the forefoot as much as possible. That would make for better maneuvering in a following sea, as the bow would have a minimum hold on the water.

C&R's Preliminary Design division modified the hull form for optimum power, making considerable changes in the form coefficients originally selected. Two models were constructed and towed. Length had to be increased 5 feet (to 110 feet) to enable the boat to achieve the required speed on the expected power plant, two 300-bhp engines.

The stern was specially selected to make steering easier in a following sea. A flatter form like that of a torpedo boat would have been better for powering, but one with a marked rise of floor aft was chosen instead. Thus the hull form was described as similar to that of a whaleboat with a transom stern.

A model of an alternative hull, slightly filled out near the waterline, was towed and rejected.

Preliminary Design also evaluated the heavy keel and deadwood planned for the design. The keel added 10 percent to resistance at maximum speed, but it was retained because a boat of this general type rolled quickly, making for a poor and uncomfortable gun platform. The keel was also expected to help keep the boat on course when it was running before a following sea. Stability (metacentric height) was fixed by the requirements of gunnery and crew comfort, with a minimum of 18 inches established for safety.

Initial weight calculations showed a displacement of 59.5 tons, which was rejected as excessive. Scantlings were carefully pared down to get to 58 tons, with a metacentric height of 1.49 feet, a mean draft (body) of 3 feet $6\frac{5}{8}$ inches, and a trim of 5 inches by the stern.

In service, the SCs were considered very seaworthy—235 crossed the Atlantic under their own power in all but the worst weather. Their greatest disadvantage was small size; crews had to endure cramped quarters and lively motion in a seaway.

The General Board approved the C&R 110-foot design but asked that it be armed with two 6-pounders and three machine guns and that its fuel supply be increased to provide an endurance of 1,500 nautical miles at 12 knots. At this stage the boat was to have been powered by two 300-hp gasoline engines for a speed of 17.5 to 18 knots.

Although the subchaser would not be a satisfactory open-sea minesweeper, like the Elco it could be fitted for inshore sweeping.

With conventional (steel) shipyards already experiencing labor shortages, the board adopted wood construction so that the boats could be built easily and rapidly. A 110-foot boat required special steel stiffening in the form of metal bulkheads, but it was decided to make the after peak bulkhead wooden, on the theory that in a wooden boat, properly made, it would be more watertight than a light metal one. Even so, workmanship was sometimes faulty and many of the bulkheads leaked. As a result, the design of the later enlarged subchaser showed seven steel bulkheads instead of the six steel and one wooden bulkhead of the 110-footer.

Engines were the major bottleneck in this program, as in several World War II successors. Only three U.S. firms manufactured heavy-duty gasoline engines that could generate the required 300 hp at less than 500 rpm. Most high-powered engines ran much faster, hence were more subject to breakdown. A survey of the manufacturers showed that only the Standard Motor Company, which had produced the engines for the Elco boats could provide enough engines quickly enough. The choice would have to be Standard's 220-bhp model, which was already in production.

To achieve a sufficient power, three such engines had to be substituted for the two 300-bhp units initially proposed. That in turn required a change from twin to triple screws, a considerable complication at so late a stage in the design. It added a ton to the displacement and lowered the center of gravity of the machinery by almost a foot. Commander Furer, previously mentioned as being responsible for subchaser construction, commented after the war that production of 300-bhp engines would at best have been a matter of tens instead of hundreds of units in the time available; even at that, the production of engines proved later on to be the limiting factor in turning out completed boats.

The General Board showed no interest in a single-yard program like that which had produced the Elco boats. There was only one mass-production yard, Elco, and the board probably wanted it to concentrate on continuing production of 80-footers. Yards on the East Coast could lay down about one hundred 110-foot craft at once and deliver them in about six months. That was not good enough, which explains the board's interest in buying Elco boats.

On 6 March, Secretary Daniels approved all the General Board's recommendations except one: the boats were to mount only a single 6-pounder (there would be provision for another). He also ordered design work begun on a larger steel patrol boat for mass production. The mobilization of civilian motor boats commenced.

Assuming that numbers were most important, the General Board initially favored continuing production of the existing Elco 80-footer, arming it with a 6-pounder (57mm). Since that was not satisfactory, the larger C&R design was to be pressed as a replacement. Destroyers would be even better, but the board admitted that they could not be obtained fast enough in anything like sufficient numbers.

By 26 March, the General Board could happily report that the engine bottleneck had been broken. A new survey showed that at least five hundred large subchasers could be delivered by 1 January 1918. That killed off the proposed production of Elco boats as well as Roosevelt's project, since private motorboats would be just as good.

A conference of representatives of the principal boat builders in the East met in the chief constructor's office on 12 March. They rejected a C&R proposal to use diagonal metal lattice plate strapping between planking and the outside of frames. However, they endorsed the proposal to use steel bulkheads. The builders suggested other minor changes to enhance production. Plans were sent to bidders on 18 March 1917, and first contracts were signed on 3 April 1917. About this time the 3-inch gun was dropped temporarily from the design, and the 6-pounder took its place, although 3-inch gun foundations were retained fore and aft. Armament consisted of one 6-pounder (110 rounds), three 0.30-caliber guns (total 3,600 rounds), two "bombs" (150 pounds total), 250 pounds of small-arms ammunition, and 300 pounds of small arms.

The Brooklyn Navy Yard built a prototype, SC 6, which was commissioned in August 1917. The first privately built boat was completed on 17 September. Late in March 1917, the U.S. Navy contracted for 355 subchasers to be delivered by 1 January 1918; most were completed in time. Of this total, 50 were assigned to France, which ordered another 50 in January 1918. The U.S. Navy ordered another 42, for a total of 447. All but 7 (1 burned on the slip, another 7 were canceled) were complete by February 1919.

Although the SCs were designed for patrol along the U.S. coast, the U-boat situation in European waters was so serious that they were deployed abroad, the first sailing to Europe in August 1917 with a group of minesweepers converted from fishing vessels. With only about a third of the endurance required to cross the Atlantic, these early SCs had to be towed. Most of them were able to fuel at sea from their escorts. The typical route was from New London to Bermuda, on to the Azores, and thence to European waters.

By the end of the war, in November 1918, 135 SCs were operating in European waters. About fifty were based at Plymouth, thirty at Corfu (for the barrage across the Straits of Otranto, eighteen at Gibraltar, twelve at Brest, fourteen in the Azores, and ten at Murmansk. Others convoyed troops two-thirds of the way from Norfolk to Bermuda, through the U.S. coastal areas that the navy expected to be patrolled by U-boats. In the Mediterranean, subchasers typically hunted in groups of three, in line abeam, locat-

SC 159 shows the standard short 3-inch/23 gun of the subchaser. This photograph was probably taken after World War I, for no depth charges are visible. SC 159 was assigned to aviation patrol out of Charleston, South Carolina, from 1921 to 1926.

ing and tracking submarines by taking cross-bearings on sounds and triangulating. The usual attack employed patterns of eighteen depth charges.

The Otranto subchasers participated in one major surface action, the bombardment on 2 October 1918 of the Austrian naval base of Durazzo, on the Albanian coast.

The original designed battery of one 6-pound, nonrecoiling Davis gun and two Lewis or Colt machine guns was soon replaced by two 3-inch/23 guns and two machine guns. Later a Y-gun (depth-charge thrower) replaced the after 3-inch gun. Displacement increased considerably, so that by the end of the war 75 tons was a typical figure. At 66.5 tons, a boat could make 16.85 knots on three engines, 14.25 on two, and 9.4 knots on one. Cruising radius was 900 nautical miles at 10 knots on 2,400 gallons of gasoline.

The 110-footers were an undoubted production success and extremely valuable. Duties included training at Annapolis, where they were succeeded by specially built YPs (see appendix B), and auxiliary work such as surveying. But they were also much too specialized and flimsy to be retained in large numbers after the war. Thus only forty remained on hand by 1 July 1922. Another nineteen were transferred to the Coast Guard to fight the rumrunners (SC 431, 433, and 437 were returned in 1921), thirteen were transferred to the War Department (army), and three were loaned: SC 144 to the state of Florida, SC 188 to the marines at Quantico, and SC 428 to the city of Baltimore. Four (SC 274, 302, 311–12) were sold to Cuba in November 1918, and SC 306 was transferred to the Department of Justice in December 1930. A decade later only twenty-four, including the Baltimore boat, remained on the navy list. That had been reduced to fifteen by January 1938 (including the Baltimore boat and SC 192, on loan to the Sea Scouts of Baltimore), and to eleven (SC 64, 102, 229, 231, 330, 412, 428, 431–32, 437, and 440) by December 1941. SC 64 was converted to a water barge (YW 97) in November 1942. SC 229 and 231 were transferred to the Coast Guard in August 1942, which operated them as WPC 335 and 336. Only two boats, SC 412 and 437, remained on the navy list in April 1945.

Numerous SCs were bought and converted to civilian use, and several were offered to the navy in 1940–41. The Coast Guard operated SC 438 after World War I, sold it, and reacquired it as WIX 375. The Coast Guard also operated the former SC 238 and 258 (WPC 365 and 372).

At the outbreak of war, the typical navy SC armament was one 3-inch/23 and two 0.50-caliber machine guns; only SC 102 had a Y-gun. The approved battery, as of June 1942, included four machine guns and two depth-charge tracks (two charges each) or a Y-gun. By 1944, the standard battery was one 3-inch/23 gun, two 0.50-caliber and (in some) two 0.30-caliber machine guns, two depth-charge tracks (two charges each), a Y-gun (to be replaced by two K-guns), and two rocket launchers (Mark 20 Mousetraps). The WPCs had one 3-inch/23 gun, two lightweight (Mark 10) 20mm guns, two depth-charge tracks (four charges each), and two Mousetraps (Mark 20).

Twenty years after the SC program President Franklin D. Roosevelt, recalling the boat's weight growth and performance loss, would demand tight weight control in the new subchaser programs. The problems of growth in the early programs were reflected in an abortive attempt to develop an enlarged subchaser. On 3 December 1917 work began on a 123-foot or 130-foot subchaser, which was never built. Its features represented the perceived successes and failures of the 110-foot type. The principal changes included replacement of the after 6-pounder by a Y-gun with six depth charges (which the subchaser did not yet have), a heating system, and a crew of ten more men (thirty-one). The total weight increase was estimated at about 6 tons, 10 percent of the existing boat's designed trial displacement of 60 tons. However, actual weight growth would be much greater, since each extra ton of load would require more hull to carry it and more power to move it. The standard growth factor of 2.5 yielded an estimated trial displacement of 81.5 tons, or about 90 tons in emergency condition.

Matters were further complicated because, as in many other small combatant designs, power could not be increased arbitrarily. Either a significantly slower subchaser could be built using the three-engine plant of the 110-footer or the new boat could have four Standard Motor Company engines, each pair driving a shaft, as in a submarine. This latter possibility was abandoned because of its complexity, even though a similar arrangement was then being developed for the prototype fleet submarine, the *Schley* (which proved unsuccessful).

Internally, the great defect of the 110-footer was seen to be its semirectangular gasoline tank, placed forward under the stateroom. The tank was inaccessible, difficult to install properly, and imposed complicated bends on the pipe lines leading to the machinery compartment. In the new design, cylindrical tanks were substituted, and two of them were moved from beneath the stateroom to the wings of the engine room, under the deck. There, their weight, well outboard, would slow down rolling. It would be easier to gain access to and fill the tanks, and pipe connections to the engines would be simpler.

The new design showed less freeboard forward but more aft than the design for the 110-footer. It also had a fuller midships section (coefficient of 0.675 rather than 0.65) and a somewhat reduced ratio of beam to draft, in the interest of obtaining greater draft and beam for easier driving and increased stowage. The forefoot was cut away even more, the keel tapered forward to improve steering in a following sea. All of this cost. Because the forefoot was cut away, the sections just abaft the stem were sharply curved, and that in turn presented problems with planking and bending frames.

By 15 December, then, Preliminary Design had a sketch of a somewhat enlarged submarine chaser of 83 tons, 118 feet long on the waterline (123 feet 3 inches overall) and powered by three Standard Motor engines (15.75 knots). By now Assistant Secretary Roosevelt was asking about a still larger boat, and calculations for one with 125 feet of waterline length (130 feet overall), showing three 300-hp Standard engines for a speed of 16.5 knots, were prepared. A vessel this large might be powered by steam turbines.

The wooden subchaser was only a temporary expedient. In March 1917, when he approved the 110-footer, Secretary Daniels also called for the design of a larger steel subchaser. Draft characteristics drawn up about this time described it as a subchaser leader with better armament (but not equal to that of a submarine) and higher speed to deal with the 23- or 24-knot (later amended to 20- or 22-knot) submarines of the future. Sustained sea speed was, therefore, set at 25 knots (26 or 26.5 on trial in smooth water), later amended to 23 knots (24 on trial). Endurance would be 1,000 nautical miles at 12 knots (1,800 with emergency fuel). As in the wooden subchaser, stores for fifteen days would be carried. Note that at this time the U.S. Navy was working on its own high-surface-speed fleet submarine, which materialized after the war.

Desired armament was one 3-inch antiaircraft (AA) gun (150 rounds), one 6-pound rapid-fire (RF) gun (300 rounds), and three 0.30-caliber machine guns. The subchaser would also be equipped with an 18-inch searchlight.

The ship's stem would be cut away to facilitate the action of riding up over "mine ropes and other floating obstructions."

A second set of characteristics for a high-endurance steel subchaser was dated 27 April 1917. It would have a sufficiently powerful battery to engage a submarine (one 4-inch gun and one 3-inch AA gun, each with 100 rounds, and three 0.30-caliber machine guns), and its design would emphasize ruggedness, comfort, and good maneuverability. Sustained speed was cut to 20 knots (20.5 on trial), only slightly more than that of the earlier subchaser. However, endurance was increased to 1,500 nautical miles at 12 knots (2,500 with emergency fuel), with twenty-five days' worth of stores.

On 24 March 1917, W. Gardner submitted a sketch

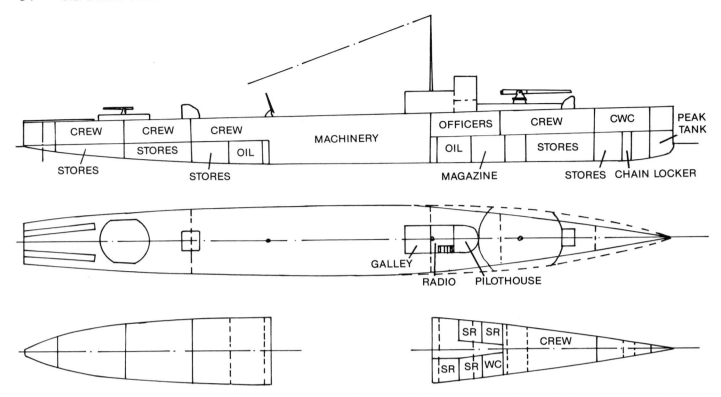

C&R's proposed "21-knot subchaser," the starting point for the eagle boat, 9 November 1917. It would have displaced 400 tons. (Norman Friedman)

design for a 156-foot "submarine destroyer" (see table 2-3), which someone in Preliminary Design marked as being "about what we want." It seems to have been used as a starting point for a series of studies of long-endurance, high-speed subchasers. Late in April, the chief of Preliminary Design, Naval Constructor Captain Robert Stocker, made a detailed critique of the Gardner design. He called for a robust hull, an endurance of 1,000 nautical miles at 12 knots (1,500 or 1,800 nautical miles on emergency fuel), a speed of at least 20 knots, and a battery of two 3-inch guns and one AA gun. Where the earlier weight estimates had been based on a normal displacement of 175 tons, he wanted a 145-tonner with estimates broadly derived from torpedo boat practice. The General Board soon killed this series of studies. By May, its "reliable" sources reported new 2,400-ton U-boats armed with three 5.9-inch guns.[5] They could be met only by something much closer to a destroyer.

At the time, Rear Admiral Albert W. Grant, commander of the Atlantic Submarine Force, was appointed senior member of a new board on submarine detection, which included Lieutenant Commander Clyde S. McDowell and Lieutenant Miles A. Libbey. The board submitted its first report, which was approved by Secretary Daniels, on 6 July 1917.

It called for the immediate construction of two hundred small ASW destroyers.[6] The board, concerned largely with developing listening devices, argued that only a destroyer had sufficient habitability, reliability, and endurance to hunt down and kill a submarine. As for U.S. submarines, the board saw them not as an important ASW device but rather as a hedge against failure in the U-boat war, and thus against German aggression in the western hemisphere. Better to solve the ASW problem at once.

The program was rejected that August because any shift away from the construction of standard destroyers was likely to slow down production. Any new intermediate patrol vessel would have to be built using resources not already committed to warships or oceangoing merchant ships, which were badly needed to make up for U-boat sinkings. There was some interest in mass-producing special escorts rather than merchant ships, on the ground that the contribution of the former to stopping merchant losses would more than make up for the losses themselves.

The special board concentrated on developing listening devices. Progress was extremely rapid; by August, the board announced that submarine location by underwater sound was feasible and it requested the services of a BuOrd officer for weapon develop-

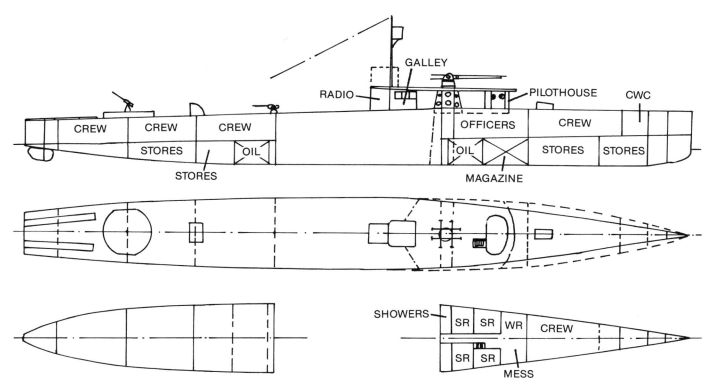

C&R sketch for a proposed 200-foot subchaser showing the raised gun platform characteristic of the eagle boat, 9 November 1917. Machinery is described as Still-Ackland or diesel. The forward gun was a 5-inch, the after one a 3-inch AA, and the amidships a 6-pounder (listed as a "torpedo defense battery"). The vessel would have displaced 400 tons. (Norman Friedman)

ment. The new C-tube demonstrated ranges greater than 4 miles against a 6-knot submarine in calm water.

The board envisioned a scouting line of subchasers that could maintain contact with a submarine. An accompanying pouncer destroyer, itself ill-suited to listening, would finish off the submarine by gunfire when it was forced to surface or by depth charge, using triangulation data from the subchasers.

Early in October 1917, trials at Newport with the SC 253 showed that it was strong enough to drop the Mark I-1 depth charge (50 pounds of TNT) at 25 feet and at 7 knots and above, and the Mark II (300 pounds of TNT) at 50 feet and at 15 knots and above. The Y-gun was also tested at this time.

Now the special board was convinced that the new listening devices could overcome the submarine problem—if they were deployed in sufficient numbers. The existing subchaser was deficient in only one thing: it could not, like the submerged submarine, operate in bad weather. In October, the board reported that if enough all-weather subchasers could be built, the U-boats would be defeated in 1918. The detection problem had been solved: now submarines could be discovered at ranges as great as ten miles (that figure turned out to be wildly optimistic). A sufficiently seaworthy listening vessel could, in theory, sweep a path twenty miles wide, and enough such vessels could form a chain across the Atlantic or, indeed, across any other body of water. A thousand vessels, each more seaworthy than a subchaser, could end the U-boat campaign. Although the board realized that the thousand-ship program was impractical, it did hope to have 250 special ASW ships by the spring or summer of 1918.

The special board hoped to build a 175-foot, 400- to 550-ton diesel or semidiesel vessel capable of 15 knots and armed with depth charges and a 4-inch gun. One hundred 20-knot boats, making an average speed of 10 or 12 knots (sprinting and drifting), could form a multiple listening barrier 200 miles long across the North Sea exit. The board estimated that only six of thirty-two submarines trying to penetrate it would survive.[7]

Further investigation showed that diesel power would suffice up to about 18 knots, but that turbines (which were easier to produce) would be needed for anything more. It appeared that steam auxiliary

Table 2-3. Subchaser Preliminary Designs, 1917–18

	Elco[4]	March 1917 Bureau	March 1917 Wood	March 1917 Wood
Length (ft-in, oa)	79-7	80	105	110
(wl)	78-9	—	—	105
Beam (ft-in)	12-2	12	13	15-4.75[5]
Draft (ft-in)	4-6	2.75	3.4	3.5
Depth (ft-in)	—	—	—	—
Displacement (tons)	34.8	32.00	—	57.11
Power[1]	2 x 220	2 x 220	2 x 300	2 x 300
Speed (kts)	18	—	—	16–17
Fuel (gals)	2,480	—	—	2,400
Endurance (nm/kts)[2]	—	—	—	1,000/12
Weights (tons)[3]				
Hull	11.0	11.70	27.30	27.30
Hull Fittings	3.5	3.50	4.90	4.90
Machinery	7.0	7.00	10.70	10.70
Battery	0.3	1.06	1.95	1.06
Ammunition	1.7	1.26	1.75	1.75
Outfittings	3.7	2.25	5.58	5.60
Stores	1.2	1.20	2.67	3.13
Fuel	2.8	2.80	—	2.67

1. Power is based on a propulsive coefficient of 0.50, as only ehp is given in the design book.
2. Endurance with emergency fuel (270 tons) was 7,500 nautical miles at 10 knots.
3. Weights in parentheses for the June 1918 boat are revised data of 25 July 1918, with two 5-inch guns, 4,000 nautical miles at 10 knots.
4. With a new armament: a 3-inch nonrecoiling (Davis) aircraft gun and three 0.30-caliber machine guns. No depth charges.
5. Extreme beam over guards, as built.
6. Weights were as estimated by Preliminary Design, based on two 3-inch AA guns (300 rounds each).
7. Data in parentheses refer to the final eagle boat design. Emergency oil capacity was about 148 tons.
8. Preliminary Design data, for a version armed with one 5-inch and one 3-inch gun, plus depth charges. The list of weights does not include 10 tons of reserve feedwater. A parallel diesel version displaced 500 tons, including 187 tons of machinery and only 16 tons of fuel (⅔ supply).

machinery could be stopped to silence the ship for listening, if reserve electrical power was available. Thus BuEng incorporated storage batteries in what became the eagle boat to ventilate the listening and radio rooms when engines and generators were stopped to listen. The draft of the new craft was set at roughly 7 feet to avoid torpedoes, which generally ran at a depth of 15 feet.

In Britain, Admiral William S. Sims, commanding U.S. Naval Forces, also concluded that a larger hunting vessel was needed. He wanted a 5-inch/51 gun to counter the large new U-boats, a speed of not less than 18 knots (with 20 preferred), and a range of about 4,000 nautical miles.

Both C&R and the General Board were extremely skeptical. Shipbuilding labor was scarce, the Shipping Board having soaked up most of what the destroyer and submarine programs left. Remaining pools of labor required training and direction by a core of experienced men. C&R noted, for example, that the new American Shipbuilding Corporation, on the Great Lakes, was taking two years to complete tugs of established design that would normally require only a year. The special board had not even sketched its ships. Moreover, machinery was in short supply. The special board hoped for diesel engines, but the United States had virtually no diesel industry, and the only marine diesels were those designed for and not yet installed in submarines. Nor did it appear likely that steam turbines would be available in the numbers required.

A C&R memo for the chief constructor argued that it would be much better to press for the development of listening devices to be used in destroyers and subchasers, both of which were beginning to appear in substantial numbers. The best way to multiply the number at sea would be to perfect their installation aboard seaplanes, since aircraft could be built quickly and in large numbers (a thousand seaplanes had recently been ordered).

The special board formalized its proposal late in October 1917. By this time intelligence reports of German submarines with 5.9-inch guns had prompted the General Board to call for 5-inch guns aboard destroyers expected to face such craft. Thus the earliest design studies for the special board ship showed such guns, and the General Board's preliminary characteristics of 1 December showed one 5-inch/51 gun, one 3-inch AA gun, one 6-pounder, and a speed of 20 or 21 knots.

OpNav, in the person of the acting CNO, Captain William V. Pratt, argued later in December that the

April 1917	December 1917	December 1917	June 1918 Patrol Boat	November 1917 Eagle Boat[7]
154-6	123-3	130	256-0	200-9
153-0	118-0	125	250-2.1	200-0
18-0	14-3(wl)	15.3(wl)	26-7.5	24-0
6-6	4-1	4.4	8-3.5	6-6
12-3	10-0	—	17-9	17-9 (18-6)
176	86	109	710 (emergency 941.2)	475 (480)
3,500	650	3 × 300 bhp	8,450	4,920
25	15.75	16.5	25	21
—	—	—	80	—
—	—	—	3,500/10	3,500/10
45	45.0	—	240 (250)	174.0 (201.34)[8]
8	7.0	—	40 (45)	28.9 (36.45)
70	14.0	—	198 (198)	93.0 (81.88)
16[6]	3.0	—	26.3 (27.2)	14.2 (15.66)
—	3.5	—	25.0 (27.0)	21.4 (23.96)
8	4.0	—	42.0 (42.0)	15.0 (17.06)
5	6.0	—	20.0 (20.0)	36.2 (26.91)
29	3.0	—	80 (120)	65.0 (60.00)

new ship should be conceived primarily as a hunter and destroyer of *submerged* submarines, designed around listening gear and armed mainly with depth charges. It would operate in company with destroyers, which could be expected to sink surfaced submarines but which were poor sound platforms. In the absence of destroyers, the new ships would hunt in groups, like subchasers, relying on numbers for strength against a craft that had to hunt singly. Pratt called for minimum draft for immunity against torpedoes, a sustained speed of 18 knots, a cruising radius of 3,500 nautical miles at 10 knots, and an armament of one 3-inch/50 and one 3-inch AA gun. Effective listening would require a machinery plant that could be stopped for listening and then restarted without difficulty; Pratt believed that a steam plant could be adapted for such operation. The complement would consist of only four officers and forty enlisted men.

The General Board disagreed with Pratt, maintaining that patrol craft should be able to operate without the support of destroyers or other powerful units. That was why it had specified a more powerful battery as well as a sustained speed, under favorable conditions, of 20 knots (not an emergency speed of 18 knots, as in the new design, which it saw as simply the ability to make 18 knots in short spurts). The General Board also wanted construction rugged enough to allow the new subchaser to operate well out to sea, as the U-boats were being driven from coastal waters by existing 110-foot subchasers.

Given the urgency of the situation, the General Board was willing to approve the existing C&R design with two changes: the ship would be armed with a 4-inch gun and a twin torpedo tube. The board later asked that the torpedo tube be replaced by a 4-inch gun and that the vessel be capable of "sustaining" 18 knots for at least four hours. The board placed on record its view that both the subchaser and the new patrol boat were emergency designs, neither of which would have been acceptable in a less urgent situation. This conclusion led to a further design study in 1918 (see below).

Pratt hoped that the new craft could be mass-produced on the Great Lakes and on inland waterways. But shipyard work forces were completely occupied, and new yards could not be made available. Henry Ford, whom President Wilson had asked to join the Shipping Board in June 1917, offered to solve the problem, suggesting to the secretary of the navy on 22 December 1917 that he manufacture new steel subchasers using the same mass-production methods he had so successfully applied to cars. He would, moreover, do so without using any existing shipyard labor. Ford later claimed that he was responsible for the concept of flat plates and the choice of steam turbines in what became the eagle boats; actually, his proposal for a new subchaser incorporated an unusual type of reciprocating steam engine. It is not clear to what extent he communicated with Chief Constructor Taylor on the structural issue prior to December 1917.

A contract for one hundred steel subchasers was placed with the Ford Company in January 1918, after a conference on 28 December in Taylor's office. Twelve

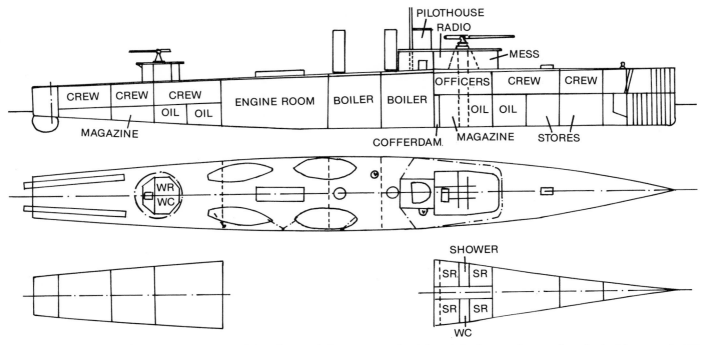

C&R sketch design for a steam-powered 200-foot subchaser armed with a 5-inch gun forward and an AA gun aft, 22 November 1917. It would have displaced 475 tons and made 21 knots on about 4,920 shp. An alternative diesel version could be distinguished by more widely spaced funnels. Note that the craft does not yet show the very straight lines associated with the eagle boat as built. (Norman Friedman)

more were later added for the Italian government; one in every five of the first units built would be turned over to the Italians. The first keel was laid on 7 May 1918, the first unit commissioned on 27 October. The contract called for the delivery of all units by 1 December 1918, but only three had been delivered by Armistice Day, and only seven by the end of 1918. The order was cut to sixty boats at the end of the war, and the Italian order was canceled.

The name of the eagle boat (PE) was taken from a *Washington Post* editorial calling for "an eagle to scour the seas and pounce upon and destroy every German submarine. On 5 November 1917, design work on what became the eagle began without any approved characteristics. Chief Constructor Taylor called for a speed of 21 knots and an endurance of 3,500 nautical miles at 10 knots, using oil (diesel) or gasoline engines. Armament was to consist of one 5-inch gun, one 3-inch AA gun, and one or more 6-pound guns, each with 250 rounds, plus twenty-four depth charges on two tracks at the stern. A flat stern would entirely cover the propellers and rudder, so that depth charges could be dropped straight over the stern. This would be a sacrifice in sea-keeping, since a ship with a broad, flat underbody aft would pound in rough waters.

The patrol vessel would have a flush deck for strength, good freeboard for sea-keeping, and a minimum of deck structure for rough weather. It would have a complement of seventy-five and a length of 150 to 200 feet.

Preliminary Design doubted that the crew and armament could be accommodated in much less than 200 feet, and it began with a displacement of 400 tons. An estimation of hull weight was derived from the weight of destroyers, with scantlings increased by 50 percent because yards mass-producing the new patrol vessel would hardly be up to maintaining destroyer standards of weight saving. Alternative hulls displacing 600 and even 800 tons were also considered, length being held to 200 feet for good maneuverability.

Initial sketches of a 200-foot, 400-ton patrol vessel were completed by 9 November. Chief Constructor Taylor now ordered two new sketch designs: one with the 5-inch gun mounted atop the deckhouse forward, where it would stay dry, and one with a 4-inch gun forward and aft. The vessel was to be powered by the only existing high-powered U.S. diesel plant, the 4,000-bhp plant designed for the new fleet submarine *Schley*. Its propellers would be widely separated, and Taylor ordered the transom widened and immersed more deeply, by about a foot. The transom would be straight, a change made partly in the interest of sim-

Eagle 58 is shown in April 1925. The flat-sided hull form is clearly visible, but it appears that the depth charges have been landed. Note the vertical arrow atop the bridge. It would have indicated the direction from which a submarine was heard, making it easier for accompanying hunting craft to triangulate.

plifying construction; Taylor explicitly ordered that the design involve the minimum number of different sizes and thicknesses of plate.

At this time the mass-produced, "flush-deck" destroyers were being designed for additional structural strength, both overall and locally. The patrol vessel design followed suit, with frames spaced only 18 (later 21) inches apart and stresses kept to 6 tons or less. Freeboard forward was only 12 feet, which still gave a greater freeboard-to-length ratio than the new 310-foot, 35-knot destroyers had (17 feet 6 inches forward). Since a full deck height of 8 feet had to be carried right aft, 8 feet was the minimum acceptable freeboard aft. A draft at normal condition of 6.5 feet was chosen to help the vessel hold its course in rough weather.

A metacentric height of about 18 inches was chosen as a compromise between what was required for a steady gun platform (minimum metacentric height) and what was required for the ship to survive battle damage. In fact, as the design progressed and machinery weight increased, this figure was reduced to 12 inches. Although the designers considered it sufficient, given the satisfactory operation of many British destroyers with metacentric heights of slightly more than 1 foot, they changed their minds after Eagle 25 was lost in June 1920, primarily because of open portholes too close to its waterline. All of the boats were subsequently ballasted, their loading strictly controlled.

The body plan was developed specifically to simplify framing and plating.[8] The designers tried to eliminate all double curvature in the shell plating, initially accepting several distinct knuckles in the waterline forward.[9] Plates would be curved only for the bilge strake, each section being run straight from the bilge to the keel, and from the bilge to the sheer (deck-edge) strake or flare plate. Model tests had already shown that, although an underbody such as this would encounter some additional resistance, it

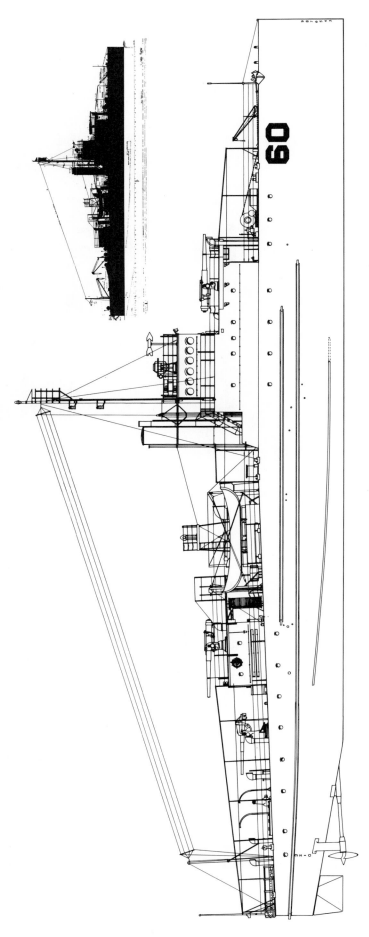

Eagle 60 as completed. Dotted lines at the forward end of the bilge keel show the larger keel fitted to eagles 1–3. Eagles 1–13 had their galley exhaust stack within the main funnel; eagles 14–60 had it placed as shown. The silhouette inset depicts Eagle 19 rigged as a minelayer in 1942. Note the mines on deck and the mine track on the stern. The silhouette also shows a minesweeping otter on the stern. (A. D. Baker III)

would not be excessive as long as the bilge was rounded.

Unlike the contemporary flush-deck destroyer, the patrol boat did not have enough freeboard for a straight sheer line from stem to stern. Instead, to provide enough headroom in the machinery spaces and enough freeboard at the forward quarter point, the deck was run flat back to about the bridge. Only then was it sloped down to the stern.

While it was being designed, the patrol boat had nearly vertical sides and was expected to be wet forward. However, any attempt to provide flare in the form of an outward slant forward, as in the subchaser, would complicate the design. Finally, one of the senior civilian constructors, Robert Stocker, decided that the effort would not be worthwhile, and Chief Constructor Taylor agreed.

When the design was being prepared for mass production, Ford asked that its hull be modified to provide as much parallel midbody as possible. No model with a lot of parallel midbody and such fine ends had been towed, but the model basin was able to show that a 15 percent parallel midbody was probably the maximum, increasing total resistance 5 to 10 percent. That was not enough to be worthwhile from a production point of view. However it was possible to arrange the waterline, forward for 60 feet and aft for 40 feet, in a straight line.

Arrangement and standards of habitability generally followed these for destroyers. One unusual feature was the location of most of the crew aft to protect them if the ship should run onto a mine.

Estimated displacement rose to 475 tons.

As in the case of the subchaser, the designers of the patrol boat encountered machinery trouble. Taylor had specified the *Schley* power plant, but the designers doubted whether it could drive the vessel at the desired 21-knot speed. As for the stated alternative, gasoline engines, the 300-bhp Standard was the largest reliable engine available in any numbers, and in the SCs it had been abandoned in favor of 220-bhp Standards, BuEng suggested a practical alternative, a single-screw geared turbine fed by two express boilers.

One peculiarity of the design was its electrical system, which had to function even while the main engines were shut down for listening. The solution was storage batteries (which weighed almost a ton) and an auxiliary lighting system, neither of which existed in earlier destroyers of similar size. The eagles had two 10-kw generators and a 24-inch searchlight, compared with a single 5-kw generator and an 18-inch searchlight in larger pre-1914 destroyers.

A sketch design of the diesel alternative prepared at this time showed two 4-inch guns; the steam alternative showed the 5-inch/51 gun. The diesel plant was expected to weigh 187 tons, compared with the steam turbine's 93 (plus 10 tons of reserve feedwater). However, with all available tanks filled (80 tons), diesel endurance would be 12,000 nautical miles at 10 knots. The diesel engine would require only 24 tons to make the required endurance of 3,500 miles; the steam engine would require 97.5 tons.

By early December there were still no fixed characteristics, although the 200-foot design was complete. OpNav representatives indicated that they would accept a reduced speed, 17 or 18 knots, which would considerably simplify production.

The story of the patrol boats recalled the evolution of the British "P-boats," designed in April 1915 to replace badly needed destroyers in subsidiary, particularly ASW, roles. Since Preliminary Design had in its ranks a British naval constructor, Stanley Goodall (later director of naval construction for the Royal Navy), it was natural to include him in a conference on the patrol vessel design on 26 December 1917. Goodall provided drawings of the British P-boat, which helped the meeting to conclude that the U.S. design was generally satisfactory. However, Naval Constructor McBride suggested that, as in the P-boat, torpedo tubes be added, in this case a twin 21-inch mount on the centerline, just forward of the after deckhouse. He also favored a Y-gun *abaft* the after deckhouse, on the centerline, and a twin-screw power plant, which would improve maneuverability and propulsive efficiency since water could reach the propellers more easily. However, Steam Engineering argued successfully for ease of production, and it continued to develop a 2,500-shp single-screw turbine.

McBride's armament suggestions were tentatively adopted and displacement rose to 525 tons, which included a considerable allowance for further weight growth.

The Ford program was celebrated at the time as a major innovation in shipbuilding.[10] Henry Ford himself had practically invented the technique of modern mass production; believing his method to be universally applicable, he could claim that eagle production would not have to draw on the limited pool of shipbuilding manpower in the country. In truth, although much of a ship's structure could indeed be built by mass-production methods, actual construction required numerous skilled shipfitters and electricians.

Ford had a full-scale wooden model of an eagle built in the main craneway of his Highland Park factory, and a new ship assembly hall was built at the Rouge River site. The model was used to refine the eagle design and develop precise automobile-style mass-production plans. Eagle boats were built in an enclosed shed, then moved onto a transfer platform and later a launching table.

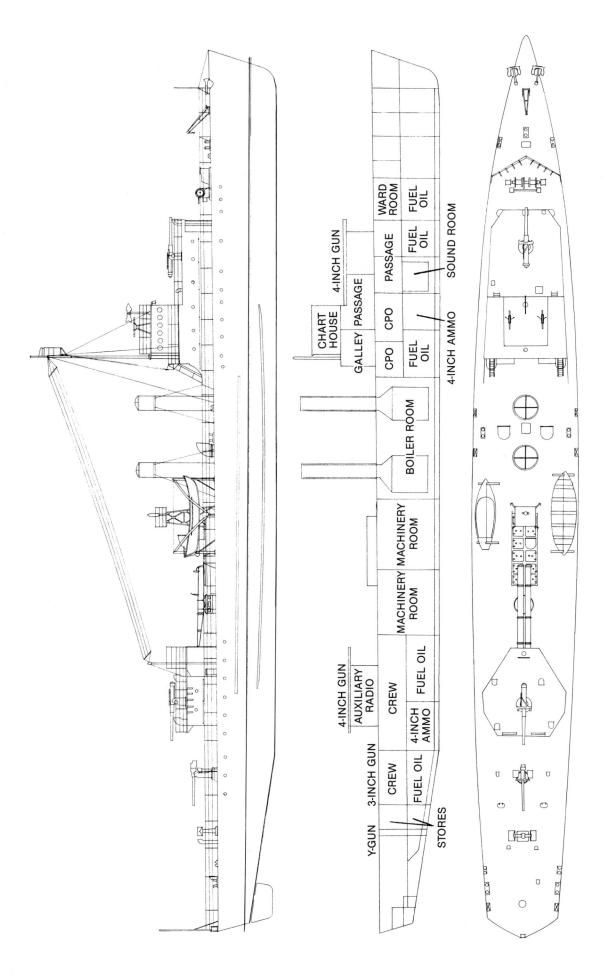

C&R designed this 250-foot subchaser in 1918 as an alternative to the eagle boat. It was never built. Note the two 4-inch single-purpose guns and the 3-inch/50 AA gun aft, with a Y-gun abaft it. Remarkably, the C&R contract drawing, which was never issued, shows no depth-charge racks. Nor does it show depth-charge stowage near the Y-gun. (Alan Raven)

Table 2-4. Patrol Boat Proposals, July 1918

	A General Board (200 ft)	B Sims (200 ft)	C OpNav (200 ft)	D C&R (250 ft)
Beam (ft)	24	24	—	—
Draft (ft-in)	6-6	—	Minimum	—
Displacement (tons)	475	—	—	650
Complement (off/enl)	—	4/50	4/40	—
Speed (kts)	20-21	Prefers 20	18	25
Radius (nm at 10 kts)	3,500	4,000	3,500	3,500
Hull	Steel	—	Steel	—
Engines	Steam or diesel	—	Steam; silent when stopped	Est 8,000 shp = 25 kts
Armament	1 5-in 1 3-in AA 1 6-pdr DCs	1 4-in 2 MG bomb thrower; 20 heavy, 40 light DCs	2 4-in 1 3-in AA 1 Y-gun DCs	2 4-in RF 1 3-in AA 1 21-in twin TT 1 Y-gun

The first boat was launched on 11 July 1918, after only two months on the stocks, but it was not commissioned until 30 October, and the last of the next dozen was not finished until mid-April 1919. These first boats were plagued by poor riveting and lesser quality-control problems. The Ford engineers apparently did not appreciate the extent to which ship construction differed from car or truck manufacture. Nor was Ford able to maintain the expected pace of production; as he discovered, the techniques of ship-fitting could not be standardized. Outfitting was further complicated because the boat was cramped internally. One lesson of the eagle story, then, is that small size in itself does not make for quick or easy construction. A larger eagle might well have been easier to build. Presumably, Ford's engineers, had they been more experienced in ship construction, would have made this point at the navy design stage. By late August 1918, Ford projected that only 26 eagles would be ready at the end of 1918, not the expected 112. The sixtieth eagle was delivered on 15 October 1919.

It is difficult to evaluate the program. At the time it was denounced as a failure, and there are some indications that Secretary Daniels kept it alive primarily to justify his faith in Henry Ford. Later Daniels argued that Ford's problems were no more than teething trouble, and that, had the war continued, output would have risen to match requirements. It is impossible to say what would have happened. Certainly the Ford Company had to press hard to meet the relaxed deadline of 15 November 1919 ultimately imposed by Daniels; however, it was able to launch the last boat only ten days after laying it down, and to complete it in less than four more months.

Some writers have compared the eagle unfavorably to the British P-boat, pointing out that they were the same size but that the latter was built faster by conventional shipyards. That may miss the point.

Mass production requires a preliminary planning period, which may seem wasteful in a short program. If that planning is carried out properly, the rate of construction can be quite high. And an overlooked advantage of the eagle program was that the boats, built in a yard and by a work force entirely divorced from the U.S. maritime industry, did not detract from other naval and merchant building programs.

When it was completed in October 1918, Eagle 1 displaced 488 tons in normal condition and drew 7 feet ¾ inches. Tactical diameter was 600 feet (three lengths) at 18 knots (18 degrees rudder), and on trials it averaged 18.32 knots at 480 tons. Because of a small metacentric height of 0.98 feet, the boat heeled an excessive 20 degrees when turning. C&R considered fitting a larger rudder, but that was no cure. The center of gravity was so high above the center of the underwater body, and the rudder's center of pressure so little below it, that achieving an appreciable righting effect would have required a rudder of impractical depth.

Operationally, the eagles appear to have been satisfactory, despite the early trouble with stability. By 1924 thirty had been laid up and twenty-two assigned to the naval reserves. Five more (Eagles 16, 20–22, and 30) were turned over to the Coast Guard's anti-rumrunner patrol in 1919. One was returned to the navy in 1923; the others were sold. Two eagles, 35 and 58, were employed in early U.S. sonar tests. Eight (19, 27, 32, 38, 48, 55–57) remained in U.S. service in 1941; of these, only Eagle 56 was lost. Eagle 19 was employed for sonar development.

At the beginning of the war, the eagles were armed with one 4-inch/50 single-purpose gun, one 0.50-caliber machine gun, and a depth-charge track (nine 300-pound charges). Eagle 19 was unique in also having a 3-inch/23 AA gun; Eagle 57 had a Y-gun. By late 1944, Eagles 19, 27, and 48 had been rearmed with a single 3-inch/50 dual-purpose gun, a Hedge-

Newly completed, the *Eagle 2* is shown off the Atlantic coast, 21 February 1919. The very sharp heel while turning, which is evident here, bedeviled the entire class. Also visible is the straight-line shape of the hull, with no curvature at all from bow back to bridge.

hog, two 0.50-caliber machine guns, and two four-charge depth-charge tracks. The other five ships were all scheduled to retain their 4-inch guns and be fitted with two Mark 20 Mousetraps; Eagles 32 and 38 had no 4-inch guns but did have Mousetraps. Ships retaining their 4-inch guns had not yet had Mousetraps fitted.

Neither Preliminary Design nor the General Board was really satisfied with the eagle boat, and early in 1918 work began on a faster patrol vessel. By late January 1918, tests on the original eagle model showed that, at 500 tons, 24 knots would require 4,000 ehp, roughly equivalent to 8,000 shp. However, as the original vessel had been designed for speeds of 16 to 20 knots, radical redesign would be necessary to reach 24. It would probably involve an increase of length to 225 feet, a displacement of 570 tons or less (ideally, 525), and an increased prismatic coefficient of 0.63 or 0.64 (the eagle's was 0.58 or 0.59). Increased power would probably require 40 to 60 tons of additional machinery. That in turn would require heavier scantlings and, probably, a more conventional hull form with curved surfaces for local strength.

Chief Constructor Taylor ordered a new design, a cross between the eagle and a destroyer, undertaken as a result of a meeting of the secretary of the navy's council on 6 June 1918. Work began on 13 June. The battery would match that of the eagle, except that the gun would be a 5-inch/51, and speed was increased to 25 knots. Scantlings would be equivalent to those of the eagle, and lines would be simplified for mass production. Captain Stocker required minimum length and the minimum number of plating sizes. He hoped to obtain the required speed on 5,000 shp. Two days later, the battery was changed to two 4-inch guns, with, if possible, a twin torpedo tube over the machinery space amidships. Taylor had initially hoped for a triple tube.

In light of data already obtained on the eagle boats, Preliminary Design tried a 240-foot, 750-ton ship with 6,500 shp. Powering estimates showed, however, that 10,000 shp would be required, prompting radical change. Two alternatives were tried: one with simplified eagle-like lines, the other a more conventional form with 4-inch guns on 2- or 3-foot platforms, using destroyer scantlings but the heavier shell plating of the eagle. Both alternative designs had a length of 250 feet, an shp of 8,000, and two destroyer boilers.

All of this still left open the delicate issue of production. The Herreschoff (yacht) Manufacturing Company expressed interest, and on 17 June its representative visited Preliminary Design to examine plans for the new patrol boat. C&R chose the conventional hull form specifically because Herreschoff seemed more willing to build it.

As in the cases of the other patrol craft, power was a problem. On 5 July BuEng proposed a triple-screw eagle plant, which would be heavy but also easy to produce. That brought down the center of gravity, and the forward 4-inch gun could be mounted atop the forward superstructure, as in an eagle. Details of this design are given in table 2-3. Armament was two 4-inch guns (200 rounds each), one 3-inch AA gun (300 rounds), and one twin 21-inch torpedo tube; the design book does not provide details of depth charges or their throwers. Normal displacement was 710 tons, emergency displacement, 941.2.

Table 2-4 compares patrol boats proposed by the General Board (A); Admiral Sims (B); OpNav (built by Ford) (C); and C&R (D). Sims reflected the General Board's views, and as a result A and B almost matched. C was 2 knots slower and had two 4-inch guns rather than the one 4- or 5-inch gun of A and B.

In mid-July, the General Board reviewed the C&R design. Given over 40 percent more displacement and a proportionately larger industrial effort, it would gain 7 knots over the eagle boat (5 more than the board had requested in December 1917). The board admitted that, for the present, designs already in production must be built; otherwise momentum would be lost. However, the emergence of the U-boat had demonstrated that patrol boats (and, for that matter, specially built minesweepers) would be needed in the foreseeable future, and the board wanted a satisfactory prototype. C&R had provided one, corresponding to requirements the board had just framed for the patrol and protection of the Atlantic and Gulf coasts and the Caribbean. These craft would have to be powerfully armed to cope with U-boats carrying 5.9-inch guns and capable of 20 to 25 knots on the surface. The patrol boats would have to follow the submarines out to sea, which meant long endurance. Sims had asked for 4,000 nautical miles at 10 knots.

Following Sims's wishes, the board rejected C&R's proposed armament. It wanted 5-inch guns to match the German 5.9s, and it noted that Sims preferred depth charges to torpedo tubes, considering the latter a waste of valuable top weight. As for speed, it found 25 knots attractive but knew that BuEng was reluctant to provide a new power plant, preferring a double eagle boat plant (twin screw) for about 22 knots. The board considered 22 acceptable. However, C&R itself rejected the BuEng argument on the grounds that the new vessels were crosses between 35-knot destroyers and 18-knot eagle boats, and that 25 knots was closer than 22 to an intermediate speed.

The final version of the design, dated 25 July, showed two 5-inch guns, no torpedo tubes, and a cruising radius increased from 3,500 to 4,000 nautical miles. BuEng wanted 10 inches more of hull depth for its engine. Secretary Daniels approved the General Board paper on 17 July 1918, but on the recommendation of OpNav he ordered that the new 250-foot boat not be built. Its virtues were obvious, but so was its major drawback—that it would compete with the destroyer program, already badly delayed because of a lack of building ways and the need to design and build new engines. The patrol boat program would probably also affect the eagles. The design was better, but not good enough to be worth the pain building it would entail. Daniels did agree to build a single unit for type development, should that be possible without too much delay. The unit was never built, and the U.S. Navy had to wait until 1942 to obtain its first destroyer escorts.

3

Second-Generation Subchasers

Although the World War I subchasers and eagle boats were clearly emergency expedients, they met requirements that could be expected to recur in a future war. According to the Orange plan developed by strategists between the wars, small ASW craft would be needed to protect bases and guarantee that the fleet would be able to sortie in the face of enemy submarines. As in World War I, the eagle boat would function as a kind of junior destroyer, releasing the more capable destroyers for fleet work. It would patrol offshore and escort coastal convoys. At first there was no need for destroyer substitutes; as a result of World War I mobilization, the United States had the largest destroyer fleet in the world. However, much of that fleet, aging and subject to the provisions of the London Naval Arms Limitation Treaty of 1930, was discarded in the late 1920s and the 1930s.

Base patrol craft (subchasers) appeared in the 1929 version of the Orange plan: 34 by M + 30, plus 59 in the M + 60 wave and 12 in the M + 90 wave, for a total of 102 boats. Three years later, with the destroyer force cut by 40, these patrol craft grew more important.

From the early 1930s on, then, replacement designs were considered. As commander in chief of the U.S. Fleet, Admiral A. J. Hepburn, painfully aware of the destroyer shortage, was responsible for the eagle boat replacement program that produced the 173-foot World War II PC. It was paralleled by a second-generation 110-foot subchaser. Problems of PC engine supply led to two classes of interim patrol craft, the 136-foot PCS and the 180-foot PCE, both converted minesweepers.

The nomenclature used may be confusing. In World War I, the subchasers were given PC-series *hull numbers* but they were *named* in an SC series. Thus PC 225 was, for example, named SC 225. In World War II, the PC/SC hull-numbering series continued, and all craft were initially listed as PCs. In October 1942, the series was split into the steel PC and the wooden SC, and for clarity this distinction is used below, even for the period before 1943. Later commissioned ASW craft, PCE and PCS, fell into the same numerical series.

Ironically, the first proposal for a replacement SC derived not from some strategic calculation but from a secondary SC role. Two subchasers, SC 223 and SC 383, were attached to the survey ship *Hannibal*, and in September 1932 her captain asked for replacements. OpNav was unenthusiastic, but the two materiel bureaus saw the request as the starting point for an SC replacement program. The existing subchasers were obsolete and practically worn out. Moreover, technical progress since 1918 had been such that neither the hull nor the machinery of the existing boats would be duplicated in a future emergency. C&R proposed a welded steel boat with diesel engines that would be suitable both for surveying and as a Naval Academy training boat (which had been requested in 1931). The new boat would last longer and would be easier to mass-produce.

Money was extremely tight in that Depression year. On 20 December 1932 the CNO, Admiral William V. Pratt, rejected any new subchaser program and suggested that the navy use existing or projected Coast Guard boats as wartime prototypes. After all, they had to operate under similar conditions.

BuEng, trying to develop a family of small lightweight diesels, was particularly interested in a new subchaser. The bureau prepared a table of possible SC replacements (table 3-1), recommending the Winton engine for survey boats. Unfortunately, the engine would be too slow for subchasers.

PC 449 was the first of the second generation of subchasers. It combined features proposed by Elco and Luders Marine but did not prove sufficiently better than the World War I 110-footers to be worth producing in quantity. The gun is a 3-inch/23.

Table 3-1. BuEng Subchaser Estimates, 1933

Length (ft)	Engines	Speed
75	2 Gasoline	13.5
78	2 Gasoline	21.7
100	2 Winton 4-cycle diesels (150 bhp)	10.0
125	2 Winton 4-cycle diesels (150 bhp)	10.0
165	2 Winton 4-cycle diesels (670 bhp)	16.5

President Franklin D. Roosevelt, whose connection with the World War I subchaser program was well known, encouraged interest in the new craft. By January 1934, Preliminary Design had prepared sketches of a 112-foot wooden subchaser and a 145-foot patrol boat.

The Depression motivated several private naval architects to seek navy work. Before the new sketch designs were presented to OpNav, a New York naval architect, Henry J. Gielow, repeatedly requested a look at them. In February, C&R provided sketches from which he could develop contract plans and calculations without putting the navy under obligation to follow through. Gielow hoped to obtain a production contract later. He offered schemes for diesel, turbine, and uniflow reciprocating-steam power. C&R, which was loath to support him but did consider his detailed design work useful, preferred the V-16 diesel alternative. At this time, the HOR (Hoover-Owens-Rentschler) Company was building a German MAN straight-8 1,400-bhp diesel under license. Subchaser design data were also furnished by C&R to the other two major U.S. marine diesel manufacturers, Fairbanks-Morse and Winton, a division of GM.

Gielow developed a 206-ton, 160-foot patrol boat. In September 1934, C&R sent lines for it and for a 105-ton, 112-foot subchaser to the model basin; models were tested for, respectively, 8 to 26 and 8 to 20 knots.

By November, C&R was concentrating on a 145-ton, 120-foot subchaser based on the wartime boat and powered by two 825-bhp diesels then being developed for submarines. The SC would be armed with one 3-inch gun, one Y-gun, and two 0.50-caliber machine guns. The hull would use several steel diagonals to maintain longitudinal strength. This design already showed the problems that would plague the later SC. Its machinery installation was relatively elaborate. It was already too large and, even with almost twice the power of the World War I SC, probably too slow (about 17.5 knots). It also required relatively sophisticated shipbuilding methods.

The new subchaser was developed, with the president's informal approval, as a mobilization expedient. C&R hoped to produce a simple, balanced design and to build a prototype in Fiscal Year (FY) 36 as a basis for further development. One would be built every year or so to keep the design up to date. On 24 November 1934, the bureau sent sketches of the 120-footer to the commandant of the Portsmouth Navy Yard, which would act as design agent. Speaking from wartime experience, Portsmouth suggested modified lines, with more sheer and more flare forward, slightly more waterline beam aft to reduce a tendency to squat, and flatter deadrise amidships for less roll.

In January 1935, C&R hired a civilian naval architect, W. Starling Burgess, to prepare new lines and to act as a consultant on hull construction. His plans and specifications were practically complete by the fall of 1937, but the design had little impact on work begun that year because the bureau doubted that his diesel engine would be suitable for a wooden patrol boat. When the design was well advanced, a description was prepared for boat builders, but apparently no one outside the navy showed much interest. Details are given in table 3-2, which compares the design of 1934 with two alternatives developed in 1937.

Lack of funds precluded any subchaser construction in FY 36.

The abortive subchaser program did have one significant result. BuEng suspected that, if it could be made compact enough, a steam engine would be the best power plant for a patrol boat. It would be more efficient than a diesel. In 1934, the bureau discussed this with the de Laval Steam Turbine Company and specifications were prepared, although no orders were placed. The ideas developed at this time were revived for PC 452 (see below).

The World War II program can be traced to two quite separate sources. President Roosevelt, who, as we have seen, had been intimately involved in the World War I subchaser program, revived the postwar SC program in the fall of 1937, when Japan invaded China and thus made war in the Pacific a serious possibility. The other source was a proposal made by Admiral A. J. Hepburn to the CNO, Admiral William D. Leahy, in September 1937. For some years, Hepburn had been uncomfortably aware of a shortage of destroyers, and he had little hope of any solution through new construction. In wartime, a seagoing boat, like the eagle boat, would release destroyers from patrol work at minimum cost in additional industrial and manning effort. Thus, it had to be a good sea boat capable of coastal escort. In peacetime, it would replace the eagle boat for reserve training. Both wartime and peacetime roles would demand simplicity of construction and operation, especially of machinery, and the wartime role would require minimum manning—no more than thirty-five or forty men.

Hepburn hoped that his patrol vessel could be built at 200 tons, a figure reflecting the likely limitations of wartime mass production. In effect, he sought something roughly equivalent to the wartime eagle boat—faster but with lighter guns and with much the same endurance on less than half the displacement. To Hepburn, ASW meant that the new patrol craft would have to be designed around new equipment, both passive and active. He believed that a specialized ship would always be a better sound platform than a destroyer. It would carry about forty (elsewhere he specified at least twenty-four) depth charges, launched over stern tracks as well as by projector. Hepburn hoped that, unlike the wartime Y-gun, the projector could be trained. As in World War I, successful ASW attack would depend in large part on maneuverability. The patrol vessel would also have to be able to deal with surfaced submarines and aircraft, which meant one dual-purpose (DP) gun. Since the existing 5-inch/38 was too large and the U.S. Navy had no 4-inch DP gun, the existing 3-inch/50 would have to do. In addition, a 1.1-inch gun and two 0.50-caliber machine guns would be helpful. Hepburn specified a speed of 24 knots with diesel power, but it was not clear at the time whether he meant sustained or trial speed. Though the patrol vessel would not often accompany the fleet, he sought an endurance of 2,500 to 3,000 nautical miles, probably at a fleet speed of 16.5 knots.

Hepburn visited Preliminary Design to discuss his modernized eagle boat, which he called a subchaser, and early in October, work on a sketch design began. The design was based on characteristics set out in a General Board memorandum of 5 October reflecting suggestions made in Hepburn's letter to the CNO.

By mid-November, Preliminary Design had two alternatives in hand. The first was a vessel with a displacement of 250 to 275 tons (full load), approximately 165 x 20 x 5.5 feet. Although the General Board asked for 3,000-hp diesels, Preliminary Design found that they would not provide sufficient speed in a hull large and heavy enough to be rugged. Instead, it suggested a lightweight, high-pressure steam plant. Convoy work would seem to require a larger vessel of about 600 tons, capable of maintaining speed in rough weather, but size had been held down in accordance with the General Board's emphasis on wartime mass production. The success of the wartime PCE seems to have confirmed Preliminary Design's views. This steam subchaser became scheme 1.

The second alternative was a vessel with a displacement of 200 tons and dimensions, in feet-inches, of 145–165 x 17–19 x 5–6. With this reduction in size and, therefore, capability, speed could be achieved with submarine-type diesels. It was not yet clear whether the sacrifices involved would be acceptable to the General Board. A few weeks later, Preliminary Design submitted two sketch designs of diesel-powered subchasers, schemes 2 and 3. Scheme 2, with two direct-drive submarine diesels (3,000 hp), met all the requirements except for speed. It would probably be much more expensive than scheme 1, because diesel engines were so costly. To achieve the required speed, an alternative scheme 3 used the same machinery but cut hull arrangements and scantlings to the level of the old and flimsy torpedo boats. Scheme 3 was presented not as a practical alternative but as an illustration of the dilemma that the General Board had presented.

Table 3-2. Subchaser Sketch Designs, 1934 and 1937

	1937		1934
	4	5	Wood Patrol Boat
LOA (ft)	110	110	120
LWL (ft)	106	106.3	—
BWL (ft)	13.6	13.7	17.75
Draft Trial (ft)	4.4	4.7	—
Draft Full (ft)	5.6	5.9	4.82
Depth, Side Amidships (ft)	8.7	8.7	—
Trial (tons)	82.1	90	135.0
Full Load (tons)	87.1	93.8	—
Machinery	Gasoline (2 Hall-Scott) ---Geared-drive--- ---2 shafts---		Supercharged diesel Direct-drive 2 shafts
SHP	1,100	1,100	1,650
V Trial (kts)	18	17.3	18.3
V Sustained Sea (kts)	16.8	17.3	—
Fuel Max (tons)	8.7	5.2	—
Endurance (nm at 12 kts)	1,000	1,000	2,395
Complement (off/enl)	2/20	2/20	—
1.1-in MG	1 single mount ---(1,008 rounds)------		1 1.1 2 0.50
DC racks	--1 (7 charges)--		2 (14 charges)
Provisions (days)	----------10-------		30
Weight (tons)			
Hull	38.8	41.0	59.6
Hull Fittings	7.7	7.7	13.0
Wet Machinery	12.0	20.0	29.4
Equipment and Outfit	6.0	6.0	6.4
Armament	2.0	2.0	2.9
Ammunition	2.7	2.7	3.8
Stores, Feedwater, Crew, Effects	6.1	6.1	7.2
Margin	0.5	0.5	3.7
Standard Displacement	75.8	86.0	126.0
⅔ Fuel Oil	5.8	3.5	9.0
Lube Oil	0.5	0.5	—
Design Displacement	82.1	90.0	135.0

The design was complicated by the uncertain stage of development of the sound equipment with which the patrol vessel would have to be equipped. The U.S. Navy had used low-frequency passive equipment since World War I; now it was beginning to install high-

SC 759 was converted into a minesweeper; note the davits aft and floats forward and aft. It carries neither depth charges nor Mousetraps.

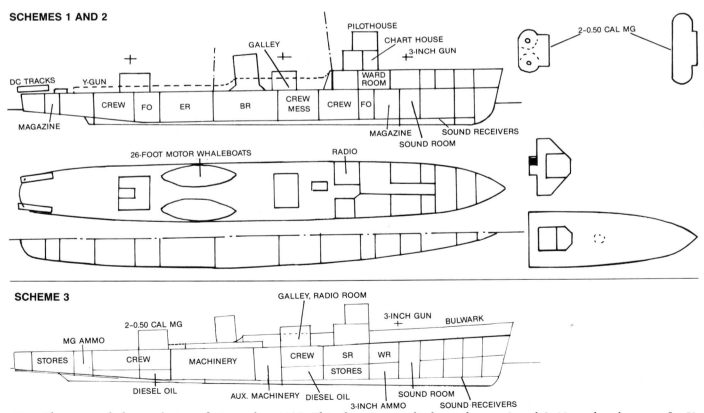

Top: The new subchaser design of November 1937. This drawing applied to schemes 1 and 2. Note the absence of a Y-gun. (Norman Friedman)
Bottom: Scheme 3 for a new subchaser (PC), 1937. The deep keel forward houses passive, low-frequency sound gear. (Norman Friedman).

SECOND-GENERATION SUBCHASERS

Table 3-3. Subchaser Sketch Designs, November–December 1937

	2 December			13 December	
	1	2	3	4	5
STD (tons)	213	213	166	139.6	154.6
LWL (ft)	---------------------------- 165 ----------------------------			-------------- 130 --------------	
BWL (ft)	20	20	18	18	17.5
Full Load Draft Forefoot/Body Amidships (ft)	8.5/5.5	8.2/5.2	7.5/4.5	4.65	5.25
Engines	Steam GT	Diesel direct	Diesel direct	Gas	Steam turbine
SHP	4,000	3,000	3,000	3,300	3,400
Full (kts)	22.4	19.7	22.4	21.4	20.4
Sustained (tons)	20.9	18.6	20.8	19.8	19.25
Fuel (tons)	45	28	20	27	24 (max)
Endurance (nm at 12 kts)	3,000	3,000	2,000	3,000	3,000
Complement (off/enl)	3/42	3/42	3/36	3/40	3/40
3-in/50	-- 1 --				
0.50	-- 4 --				
DC Proj	1	1	0	1	1
DC Tracks	-- 2 --				
DC	24	24	7	24	24
Endurance (days)	-- 20 --				
Hull Type	Fcstle	Fcstle	Flush	Fcstle	Fcstle
Weights (tons)					
Hull	80	80	55	46	46
Fittings	16	16	12	13	13
Engineering	65	65	65	3,045	
Battery	6	6	5	5.6	5.6
Ammunition	15	15	5	15	15
Equipment and Outfit	11	11	8	10	10
Complement	15	15	13	15	15
Margin	-- 5 --				
Standard Displacement	213	213	166	139.6	154.6
Reserve Feedwater	5	—	—	—	4
Lubricating and Fuel Oil	42	29	15	28	24
Additional Stores and Potable Water	6	6	5	6	6
Full Load	266	248	186	173.6	188.6
Trial Displacement	245	235	177	158	174
Trial Speed (kts)	22.4	19.7	22.4	21.4	20.4
Sustained Speed, 80 Percent Power (kts)	20.9	18.6	20.8	19.8	19.25

frequency active and passive systems, which would later be called sonars. Low-frequency equipment had to be large: the JL system planned for the subchasers required a passive array consisting of six magnetostrictive transducers (magnetophones) on each side. The arrays had to be mounted well forward, to minimize noise, and submerged as deeply as possible. Thus, the JL might easily be installed in a large ship at the turn of the bilge, but installation in a shallow-draft subchaser would be much more difficult. In any case, the JL system was as yet incomplete. The World War I MV array had required a 42-foot blister. By way of contrast, the QC/JK high-frequency/echo-ranging and listening device had a rotating transducer, 17 inches in diameter, that could easily be suspended beneath the keel. At this time the standard installation for destroyers, which would be repeated in patrol vessels, was one transducer on each side. BuEng, responsible for listening devices, suggested that the JL be installed in a streamlined fore and aft blister on the centerline, directly under the keel, with one of its arrays on each side. Any such underwater projection would have to be carefully faired to avoid eddies and other sources of noise.

Preliminary Design planned to carry a deep keel—extending about 3 feet below the hull proper, in way of the sound room—nearly to the forefoot, decreasing it in depth until it disappeared in a skeg ending aft. So unusual a hull form would be expensive in both power and maneuverability. Because it would add enormously to the side area of the hull, the deadwood aft would have to be cut away sharply to give the ship more maneuverability, which was considered essential. Sea-keeping required freeboard, particularly forward and amidships. Finally, ruggedness demanded that hull stresses be moderate, which precluded the use of light scantlings. Thus the hull would be relatively heavy, and the engine would have to be relatively light.

Preliminary Design freely admitted that none of

its three schemes was really small enough for mass production. Ideally, the new subchaser would be only slightly larger than the 1917 boat—110 to 120 feet long, and displacing only 90 to 120 tons. Such a boat would almost certainly not make her speed or operate effectively in rough weather. Similarly, the speed, increased over that of the eagle boat, seemed to rule out a simplified hull form of the type used in 1917–18. But the president wanted a less expensive subchaser, and so Preliminary Design tried some smaller designs. Scheme 1A of December 1937 was to have achieved 21 knots on a 2,800-shp geared steam turbine. She would carry the same military load as scheme 1, but her radius might be reduced to 2,000 nautical miles to save weight. Another alternative tried at this time was a boat with four 550-hp Hall-Scott gasoline engines, two per shaft. Neither was submitted to the General Board, but in mid-December C&R circulated the 130-foot scheme 4 (six Hall-Scott engines on three shafts, displacing about 100 tons less than scheme 1) and scheme 5, a similar small hull with scheme 1's high-pressure plant.

Reviewing the C&R proposals in December 1937, the president approved further development of a steel subchaser but added that this program hardly reflected his own desire for a much simpler craft, a true successor to the World War I SC. He believed that American boat yards could produce wooden boats in quantity and quickly, in five to eight weeks. He wanted wooden subchasers for employment in coastal and harbor waters and to protect fleet anchorages in event of war, and for the purpose of mobilization, he wanted the design developed to the point at which bids could be invited. Ironically, while it proved difficult to find boat builders for the World War II SC program, the country quickly mobilized to build steel ships, including the PC.

The wooden subchaser study was formally ordered by Admiral Leahy on 16 December. Preliminary Design began work using the estimated weights and lines and the inclining experiment report of the wartime subchasers. However, as in 1934, the design evolved in the direction of an enlarged hull, 120 feet long. Table 3-4 shows schemes 1, 2, and 3, submitted on 28 December, and how they would perform powered by gasoline, diesel, and steam engines. C&R considered diesels unsuitable because the vibration of their heavy pistons would put too much stress on a wooden hull. A lighter high-speed gasoline engine would be much better.

The General Board informally asked for alternative 110-foot wooden subchaser designs—schemes 4 and 5, submitted on 3 January 1938, of table 3-4. Again, the designs called for different power plants, the gasoline engine being considerably lighter. In both, a savings of 8 tons could be used to double armament and ammunition and thereby cut trial and sustained speeds, to 17.6 (scheme 4) and 16.5 (scheme 5) knots. Alternatively, the entire 8 tons could be used to double armament and ammunition and increase radius from 1,000 to 1,500 miles. Estimates were based on the performance of the wartime subchaser.

BuOrd investigated a variety of possible subchaser guns. It assumed that a typical submarine had a ⅜-inch conning tower, a ½-inch pressure hull, and ¾-inch ballast tanks. Possible weapons were the new 1.1-inch AA gun, the 3-inch/23 of the World War I subchasers, the standard 3-inch/50 dual-purpose gun, and the 6-pounder (57mm). The 1.1-inch gun could penetrate a submarine pressure hull at 1,500 yards, but a 6-pounder or better was needed at 3,000 yards, a 3-inch gun at any greater range (table 3-5).

The 1937 subchaser designs showed an armament of two 1.1-inch guns (one in the diesel-powered scheme 5), two depth-charge tracks, and fourteen depth charges. No single 1.1-inch gun existed. Preliminary Design developed a preferable alternative armed with the single 3-inch/23, the weapon that had armed the old subchasers. The General Board adopted this battery.

By January 1938, the gasoline-powered scheme 4 had been tentatively chosen, although it was still uncertain whether military load would be increased and speed accordingly reduced. Compared with the wartime subchaser, scheme 4 had modernized stern lines for reduced resistance and to suit two rather than three screws, a shorter engine compartment to open up space elsewhere in the boat, better accommodation, and easier access to fuel tanks. Displacement was considerably greater than that of the earlier subchaser, but speed was also greater because of the higher power.

In March, BuEng representatives inspected 72- and 80-foot Coast Guard cutters powered by marine Liberty engines, clutched two to a shaft to produce about 1,700 bhp. In a subchaser, a Liberty engine would produce 21.6 knots, but BuEng considered it unacceptable because it would require a degree of skill in maintenance that was not available in wartime. The bureau sought instead a single modern engine geared to each of two shafts, probably producing a total of 1,100 bhp for 17.6 knots. This limit affected the General Board's overall evaluation of the subchaser program.

The schedule of further design work depended on the priority assigned to subchaser projects. If, for example, the ships were to be built under the FY 40 program, then detailed designs would be needed early in the summer of 1939. But if, as some expected, they were to be built sooner, under a proposed small-craft clause in the FY 39 Naval Expansion Bill, C&R would

Table 3-4. Wooden Subchaser Designs, 1937

	1	2	3	4	5
Standard Displacement (tons)	117.8	135.8	128.8	—	—
LWL (ft)	120	120	120	110 (loa)	110 (loa)
Trial Displacement (tons)	134.6	146	146	82.1	90
Full Load Displacement (tons)	145.3	153.3	155.8	87.1	93.8
Power Plant	Gas	Supercharged diesel	HP steam turbine	Gas (1,100 hp)	Supercharged diesel
Trial (kts)	20.4	19.7	19.7	18	17.3
Sustained (kts)	19	18.4	18.4	16.8	16.3
Endurance (nm at 12 kts)	---------------------------------- 2,500 ----------------------------------				1,000
Complement (off/enl)	---------------------------------- 3/24 ----------------------------------				2/20
1.1-in	---------------------------------- 2 single (2,016) ----------------------------------				1 single (1,008)
0.50	---------------------------------- None ----------------------------------				
DC tracks	---------------------------------- 2 ----------------------------------				1
DC	---------------------------------- 14 ----------------------------------				7
Stores Endurance (days)	---------------------------------- 14 ----------------------------------				10

Table 3-5. BuOrd Penetration Data, January 1938

Range	1.1-in	3-in/23	3-in/50	6-pdr
1,500 yds	1.2 (0.8)	1.83 (1.23)	3.55 (2.0)	1.91 (1.26)
3,000 yds	0.7 (0.4)	1.35 (0.93)	2.07 (1.38)	1.19 (0.81)
4,000 yds	0.5 (0.4)	1.14 (0.81)	1.59 (1.07)	0.99 (0.67)

Note: Figures in parentheses are for a 40° angle.

have to complete design work during the summer of 1938. The latter seemed likely in the first months of 1938, and design work proceeded. C&R circulated its SC design early in April 1938, seeking comments by 25 April so that a contract design could be completed in time to advertise for bids for FY 39.

The small craft program, which would also have included torpedo boats, was eliminated from the Naval Expansion Bill and made the subject of a separate experimental program, authorized at the same time, in May 1938. This program differed from the one originally proposed in that virtually all of the craft were to be built in private yards from private designs chosen by competition.

The plan had originally been to build two wooden 110-footers along with one 59-foot and two 70-foot aluminum motor torpedo boats at Norfolk; two steel 165-footers at Philadelphia; and a single 59-foot aluminum motor torpedo boat in a private yard. At that time, early 1938, it still appeared that Norfolk would also be building two 70-foot motor torpedo boats for the Philippines.

After considering Preliminary Design's alternatives, the General Board prepared characteristics for the large and small subchaser. These were submitted on 15 January 1938 and approved by the secretary of the navy on 25 January. The boats would be used in wartime as submarine hunters near naval bases and areas of concentrated shipping, as ASW escorts, and to provide ASW protection for the fleet while it moved to and from its bases. In peacetime, the boats would be valuable for the development of ASW tactics and listening gear as well as for the training of sound operators.

The General Board recommended the construction of two twin-screw PC prototypes—one diesel- and one steam-powered—and of two SC prototypes. The 215-ton PC would be armed with two depth-charge racks on the stern, a Y-gun with a total of twenty-four depth charges, a 3-inch/50 gun (200 rounds), and four 0.50-caliber machine guns (20,000 rounds). The boat was to make 22.5 knots on trials (with two-thirds of its fuel and two-thirds of its reserve feedwater), and it was to have an endurance of 3,000 miles at 12 knots. The complement was to consist of three officers and forty-two enlisted men, provisioned for twenty days.

C&R circulated a sketch design for its 165-foot PC at the end of March 1938, but owing to the press of work in its design branch, it had to farm out more detailed work to the Philadelphia Navy Yard. The preliminary design showed a hull form based on that of the destroyer *Mahan* (with a modified transom), because 22.5 knots at a length of 165 feet corresponded to destroyer speed at destroyer length. A forecastle provided freeboard forward for dryness and achieved maximum internal volume on minimum hull weight. A destroyer-style bulwark extending from the after end of the forecastle to the stern would keep the deck as dry as possible in a seaway. With the vertical sail area of the forecastle sides, it might be difficult for the boat to turn in wind, so it would be supplied with twin rudders 80 percent larger than the single rudders of recent destroyers. C&R expected to improve maneuverability further by properly positioning the propellers.

High speed required a lightweight hull, but because the PC would be designed for mass production by inexperienced shipyard workers, weight could not be saved by welding. Preliminary Design hoped to improve matters by providing bulkheads with spe-

Table 3-6. Prototype Subchasers, 1940–41

	PC 449	PC 450	PC 451	PC 452	PC 453	PC 461	PC 497
Hull Material	Wood	Wood	Steel	Steel	Wood	Steel	Wood
Length (oa, ft-in)	110-0	110-0	169-7	173-8	110-6.5	173-8	110-10
(wl, ft-in)	106-10	106-1	163-0	170-0	107-6	170-0	107-6
Beam (ft-in)	18-4	15-2	21-6	23-0	16-9	23-0	17-11.5
Standard Displacement (tons)	96	86	270	284	86	293	98
Draft (trial, ft-in)	6-1.75	5-10.5	7-8	6-6.5	5-2	6-2.5	5-8
Trial Displacement (tons)	106.2	101.0	305.8	329.9	100	330.0	116
Engine	Cooper-Bessemer	GM	GM	De Laval Turbine/ Beasler Boiler	GM	—	GM
Power (bhp)	2 824	2 600	2 2,000	2 2,000	2 1,200	2 2,000	2 1,200
Total Power (shp)	1,648	1,200	3,760	4,000	2,400	3,600	2,400
Trial Speed (kts)	17	16.9	21.8	22 Est	22 Est	22 Est	22 Est
Endurance (nm/kts)	1,500/12	1,500/12	3,000/12	3,000/12	1,500/12	3,000/12	1,500/12
Complement (off/enl)	2/20	2/20	3/43	3/55	2/20	3/55	2/20
Sonar	----------QC/JK----------				----------JK/QC----------		
Echo Sounder	NM	NM	NM	NM	NJ3	NJ3	NJ3
Passive Sonar	JL	JL	JL	JL	(lightweight)	(heavy)	(heavy)
Weights (tons)							
Hull	43.34	38.36	—	—	—	108.96	41.73
Fittings	8.52	7.70	—	—	—	28.80	8.14
Machinery (wet)	26.63	22.18	—	—	—	103.40	19.57
Armament	2.35	2.35	—	—	—	11.15	3.49
Ammunition	6.07	6.07	—	—	—	8.70	9.64
Equipment and Outfit	3.95	4.37	—	—	—	—	4.47
Complement	2.36	2.36	—	—	—	—	2.36
Stores and Feedwater	4.53	4.39	—	—	—	—	4.45
Standard	97.75	87.78	—	—	—	—	93.85
Fuel Oil	11.67	10.65	—	—	—	—	19.91
Full Load	109.42	98.43	—	—	—	372.9	113.76
Stability (metacentric height)							
Light	3.65	1.75	2.02	3.44	—	—	—
Trial	3.43	1.66	2.31	3.21	—	—	—
Full Load	3.45	1.69	2.16	3.06	—	—	—

Note: PC 461 weights refer to late-production craft (PC 1546 series). Other data are design estimates.

cial diagonal stiffeners in addition to the usual vertical stiffeners—and also by keeping stresses to a minimum.

At this stage the PC was 165 feet long on the waterline (168 feet 3 inches overall), with a molded beam of 20 feet 4 inches and a draft of 5 feet 11.5 inches (mean) at 290 tons fully loaded (220 tons standard, 265 on trial). Model tests showed that 4,000-shp steam or diesel engines could drive the vessel at the required 22.5 knots, and that 53 tons of fuel oil would suffice for 3,000 nautical miles at 12 knots, allowing for a 25 percent increase in cruising power because of fouling. The design still showed low-frequency magnetophones as well as paired sonars. C&R noted that the steam plant would require five additional enlisted men, for a total complement of three officers and forty-seven enlisted (forty-two for the diesel).

Admiral Claude C. Bloch, Hepburn's successor as commander in chief of the U.S. Fleet, questioned the need for long endurance. Several foreign subchasers of about the same displacement (but apparently with shorter range) seemed to carry much more impressive armament. C&R replied that the boats would often cruise at speeds much higher than their endurance speed, thereby reducing their effective endurance. Admiral W. T. Tarrant, commander of the Scouting Force, agreed with Bloch. He also feared that submarines would easily outshoot the PC, which therefore could not function as an unsupported ASW hunter. Finally, several senior officers wanted ballistic protection (STS), at least for the pilothouse, chart house, radio room, and machinery. The added weight would have been prohibitive, for at this time the minimum thickness of STS was half an inch (20 pounds per square foot). The General Board's reply was, again, that weight was limited: the size of the PC had been held down deliberately to make mass production relatively inexpensive. The board preferred a single-purpose surface gun for weight saving; four 0.50-caliber machine guns would suffice for AA defense.

In August, OpNav's war plans division summarized fleet views and added some comments of its own. Analysis of the plots of several hundred ASW attacks made over the previous three years by experimental sonar-equipped destroyers showed that the probability of hitting a submarine would double if two Y-guns and one stern track replaced the pro-

PC 451, the winning, 165-foot, private subchaser design, was ultimately rejected as too tight to accept further improvements.

jected single Y-gun and two stern tracks. The weight of an extra Y-gun, 1,310 pounds, hardly seemed prohibitive, although it could be expensive in terms of scarce centerline space. The war plans division also feared that the diesel engine itself would create excessive noise, requiring special silencing measures.

This division was responsible for planning America's island-hopping attack across the Pacific envisioned in the Orange plan. The PCs were attractive to the staff because "shallow draft, high maneuverability, low freeboard, relatively clear deck space, armament, and ample speed . . . will [suit them] for use as troop ferries and in close support during landing operations and in the defense of atolls. If available in sufficient numbers these craft will probably prove more suitable for these purposes than would the proposed converted destroyer transports [APD], and will certainly be more generally useful." Thus, the PC would best be served by a 3-inch/50 gun with an AA mount (despite the weight penalty of 1,323 pounds) and a shoal-water depth finder, which would complement the sonar.

The wooden SC would make 17.6 knots in trial condition and have an endurance of 1,500 nautical miles at 12 knots. Her complement would be about half that of the PC, two officers and twenty enlisted men, provisioned for ten days. Although characteristics indicated commercial gasoline engines, the secretary of the navy hoped that at least one SC would be diesel-powered. C&R sketch plans of April 1938 showed paired sonar transducers (one of which was deleted a month later) and a low-frequency array recessed into each side of the keel forward. At this time, BuEng wanted to use two geared 560-bhp diesels. C&R feared that this would not suffice to maintain the required 17.6 knots. Ironically, that speed had been specified, not for any operational reason, but merely because C&R had guaranteed it in earlier sketch designs. Weight was a problem; C&R had already had to increase estimated standard displacement from 82 to 89 tons, owing to a 5-ton increase in machinery weight and a 2-ton increase in hull weight. The new SC was expected to displace about 100 tons fully loaded.

The fleet was not entirely pleased with the SC. Admiral Bloch considered the 3-inch/23 gun inadequate against either submarines or aircraft. It might be improved with better sights, but an effective 5,000-yard range against surface targets was needed. Admiral Tarrant observed that submarines would be armed with 5-inch guns and would have a surface speed of 21 knots; they would be able to outrun *and* outshoot the SC. Why fear it? He wanted both SCs to be diesel-powered, not only for safety but because the U.S. Navy urgently needed to develop satisfactory diesel engines. At this time considerable effort was already going into submarine diesels.

As estimated machinery weight grew, the design team at the Philadelphia Navy Yard found it impossible to remain within the required weight limits of the 165-footer. To achieve a trial displacement of 265 tons, they had to keep machinery weight (wet) down to 65 tons, yet in October 1938, they reported that the diesel plant would probably weigh 107.55 tons, for a trial displacement of 309.52 tons. This increased draft from 5.6 to 6.3 feet, reduced freeboard, and would probably make it impossible to meet the

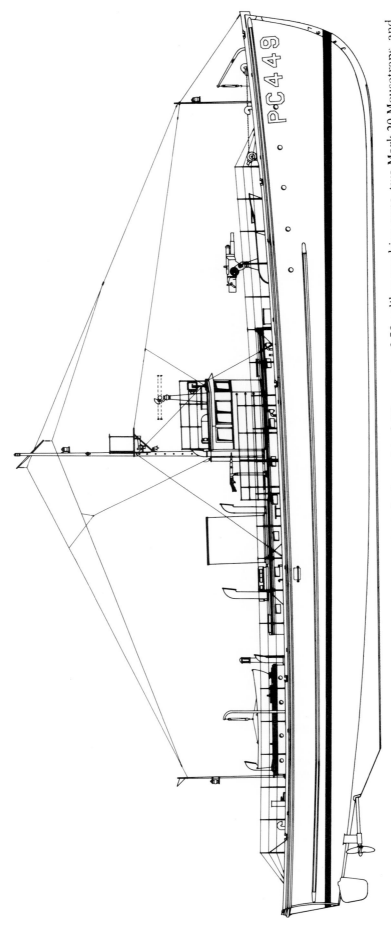

PC 449 as completed. At the end of the war, this boat and PC (SC) 450 had one 3-inch/23 gun, two 0.50-caliber machine guns, two Mark 20 Mousetraps, and two depth-charge racks. (A. D. Baker III)

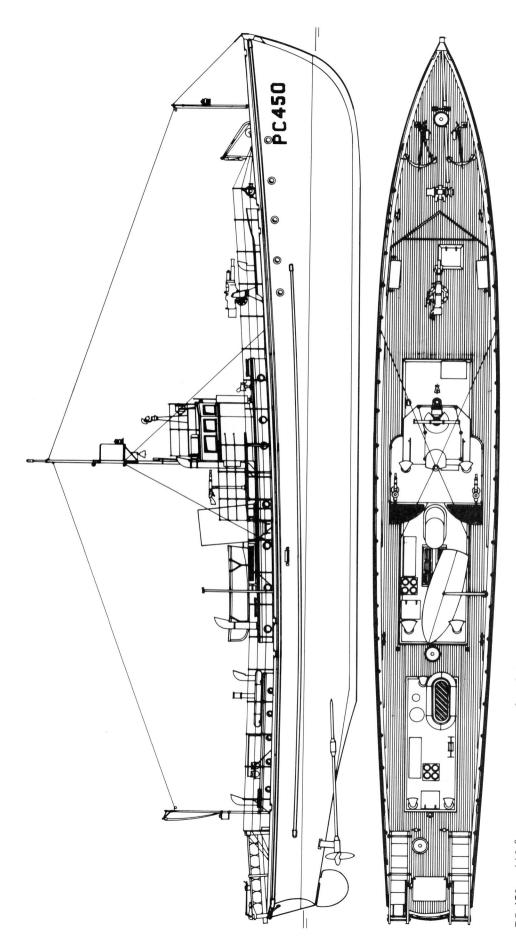

PC 450, a 110-foot prototype, as completed. (A. D. Baker III)

PC 450 was a modernized version of the World War I subchaser. It is shown running trials, before receiving armament.

required speed. The steam plant was lighter, but estimated trial displacement was still 297.52 tons. Moreover, the steam plant was not likely to realize the required endurance on the available weights. It was credited with 2,412 nautical miles at 12 knots, compared with 3,600 for the diesel.

To achieve even these weights required a light hull, and the design team feared that, manned by reservists, the PCs would suffer badly in service. With their shallow draft and considerable sail area, they might easily be driven into piers. And yet, given the constraints of required speed, which limited displacement, and overweight machinery, little could be done. By 1939, the PC 452 design had been changed to a flush-decker, which was stronger than a vessel with forecastle hull for a given hull depth and scantling weight and which provided additional headroom in the machinery spaces.

The competition, which ended in March 1939, attracted eight 110-foot and eleven 165-foot proposals. Two diesel-powered 110-footers and five 165-footers (two of which were considered underpowered) were selected for further work. The competition ended in March 1939. Luders Marine and Elco won the 110-foot competition, and C&R decided to combine their best features in a single design, which Luders built (SC 449) to Elco lines. It was powered by a pair of commercial (C.B. EN-8) diesels with a total of 1,200 bhp). The other boat, SC 450, was a government design, essentially a modified World War I 110-footer, with the GM 8-268A diesel, which powered many of the later 110-footers.

The 165-foot winner was the diesel-powered PC 451, designed by Sidney A. Vincent of Newport News Shipbuilding in collaboration with several other naval architects. It was one of the first U.S. surface warships to employ a new lightweight high-speed diesel developed specifically to meet a 1932 BuEng requirement for larger engines compatible with submarines and weighing less than 27.5 pounds per bhp. These characteristics would produce an engine suitable for railroad use. Thus, the diesel manufacturers found that BuEng seed money, which encouraged them to design lightweight units, brought them into an expanding market for diesel locomotives. For example, an eight-cylinder version of the navy Winton engine would eventually power the Burlington Railroad's Zephyr streamliner. As the railroads shifted from steam to diesel, they paid for further engine development and production tooling, and the navy benefited enormously.

Five companies entered the 1932 competition; Winton Diesel (GM) won with the 950-bhp 2-cycle model 201 V-12. Later the company built a series of more powerful V-16s (GM 16-2588). In 1934, Fairbanks-Morse and HOR, not in the original competition, built prototypes of their own that were later adopted. Fairbanks-Morse's 1,300-bhp, 2-cycle, opposed-piston design (38D8 1/8) was similar in principle to the German Junkers Jumo aircraft diesel engine, and HOR's R-99DA of 1936 was a license-built German MAN diesel.

The first GM lightweight diesel, the Winton 16-201A, was installed in the *Perch*-class submarines completed in 1937. Many of the PCs, including PC 451, used the GM 16-2588. These big standard engines were used widely during World War II—in submarines, destroyer escorts, large auxiliaries, and LSMs. Fleet minesweepers were powered by a variety of diesels of about 900 bhp. LSTs used similar engines. Smaller minesweepers (YMS) were powered by GM's 500-bhp straight-8 (8-268A) diesel.

C&R persevered with its parallel design, which became PC 452, in order to test the experimental steam plant conceived in 1934. When ordered from de Laval in 1938, the plant had been taken a considerable step beyond what BuEng was just then introducing into service. It demonstrated the principle that the higher the temperature and pressure, the more compact the plant.

PC 452 was powered by two 2,200-shp, 15,000-rpm turbines operating at 950 pounds per square inch (psi) and 800°F. For compactness, the high- and low-pressure units of each turbine were mounted side by side in casings cantilevered to the forward face of the gear casing, eliminating the usual separate turbine foundation, and the high-pressure exhaust casing and low-pressure inlet were a single unit. To fit the narrow space, two of the three low-pressure turbine wheels had to be designed so that steam would flow radially from their edges.

The boiler also had to be compact. Built by Combustion Engineering, it was a forced-circulation, air-encased unit of high pressure and temperature with an integral superheater. The boiler worked at 950 psi and 870°F.

The hull of PC 452 was delivered without machinery on 15 October 1941. It arrived at Philadelphia Navy Yard for completion and was formally commissioned on 26 July 1943. The turbines, tested by de Laval early in 1942, were delivered to the navy a year later for installation. The prototype boiler was delivered to the Naval Boiler and Turbine Laboratory in September 1942. PC 452 underwent dock and river trials in September 1943 but was not completed until 1 May 1944; it arrived at Miami for shakedown the next day.

Because the project was experimental and would not affect wartime machinery, it was terminated after initial tests and postponed until after the war. PC 452, decommissioned on 7 December 1944, was used to develop procedures for putting ships in mothballs. It was reclassified as IX 211 (the *Castine*) on 10 March 1945, declared surplus in February 1946, stricken on 23 June 1946, and sold on 27 January 1947.

After reviewing the winning designs in March 1939, President Roosevelt called for the immediate completion of plans so that bids could be made as soon as possible. He wanted a wooden deckhouse substituted for the steel one in the SC; that way, small yacht yards could proceed without waiting for fabricated steel, which, he feared, might be difficult to obtain in an emergency. He also wanted a second 3-inch/50 gun added to the PC. Finally, recalling his World War I experience, the president specifically ordered that particular care be taken during detailed design to avoid adding weight (that is, draft) and thereby losing speed.

At this time, the 22-knot requirement was explained as a 10-knot margin over a notional 12-knot speed in convoy. The endurance of 3,000 nautical miles suited a convoy run from San Diego to Balboa in the Canal Zone and more than sufficed for the run from the West Coast to Hawaii. Given the somewhat vague basis for the 22-knot requirement, it is difficult to understand why it was taken as seriously as it was.

The speed requirement automatically divided the subchasers into two categories. The 165-footer clearly met characteristics, and the General Board and OpNav supported it enthusiastically. None of the 110-footers came anywhere close to meeting the speed requirement, and by 1940 the General Board, among others, suggested doing away with them altogether. The 110-footer was saved only by the appearance of a new design and a new engine—the GM pancake.

Of the large steel craft, the 165-foot PC 451 seems to have been favored at first. C&R's PC 452, with its steam power plant, was substantially larger, yet probably wetter owing to its lower bow. Ultimately its size made it more attractive, allowing more margin for growth. The PC 452, rather than the tighter PC 451, could easily accommodate the extra crewmen needed to serve additional AA weapons, beginning with the president's 3-inch/50. But there was a drawback: the PC 452's experimental steam plant could not be mass-produced.

On 12 July 1940, BuShips recommended using the PC 452 hull with PC 451 diesel engines. In September, this judgment was supported by a special board inspecting the first two prototypes, SC 449 and PC 451. It reported that PC 451 had reached its limit in complement and would encounter serious problems if more AA weapons had to be mounted, as seemed probable.

The diesel-powered flush-decker series began with PC 461, although PC 471 was the first to enter service. This choice was not entirely popular within the fleet, because it was thought that, while the PC 451 was a good sea-keeper, the flush-decked PC 461 was not. It may also have been significant that PC 451 entered service more than a year earlier than the flush-deckers; it was commissioned on 12 August 1940, and the first flush-decker, PC 471, did not appear until 15 September 1941. PC 451 also benefited from a considerably lighter load, because of limits imposed by her internal volume and because major additions to AA and depth-charge batteries were not ordered until after she had been completed.

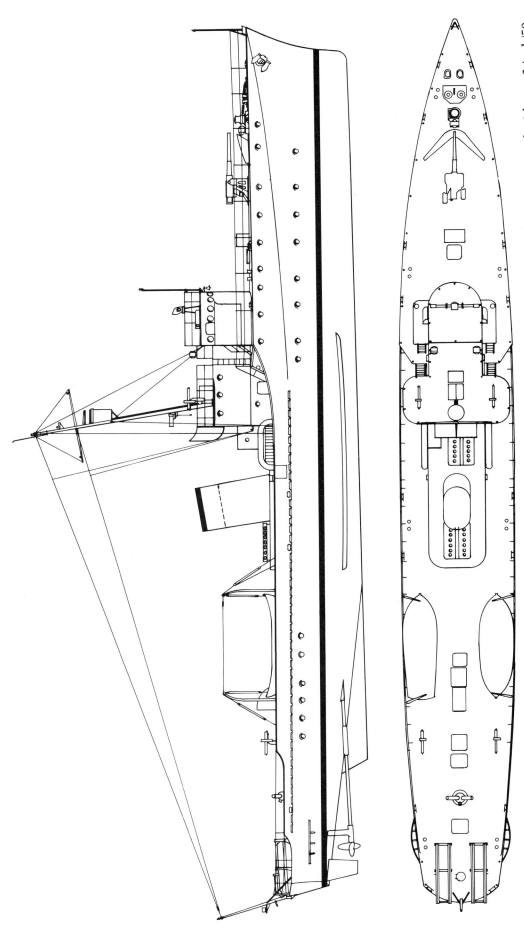

PC 451 as completed, August 1940. The planned shorter stack is indicated by a dotted line. At the end of the war, the boat was armed with one 3-inch/50, two twin 20mm and two single 20mm (Mark 10) guns, two Mousetraps (Mark 20), two K-guns, and two depth-charge racks. (A. D. Baker III)

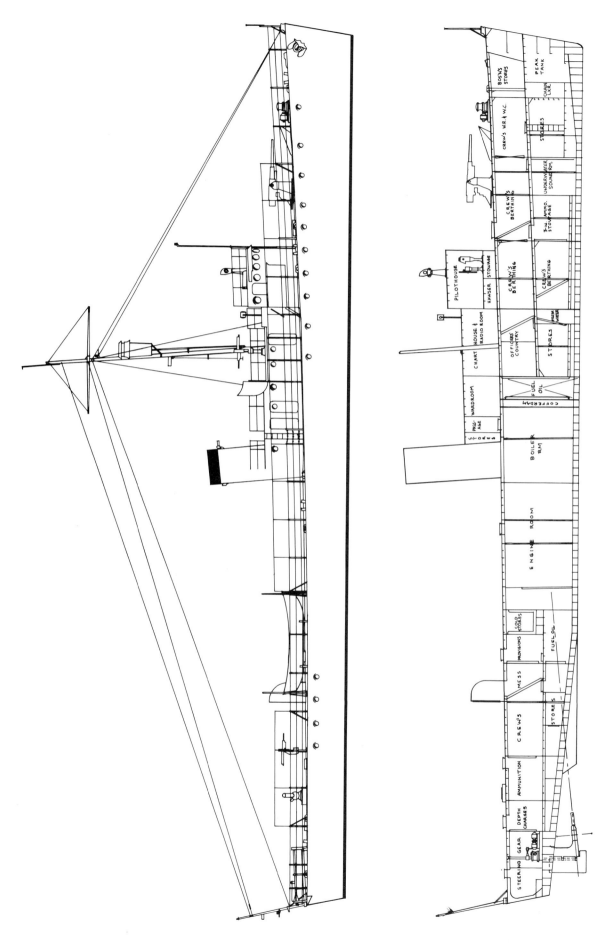

PC 452 as designed. (A. D. Baker III)

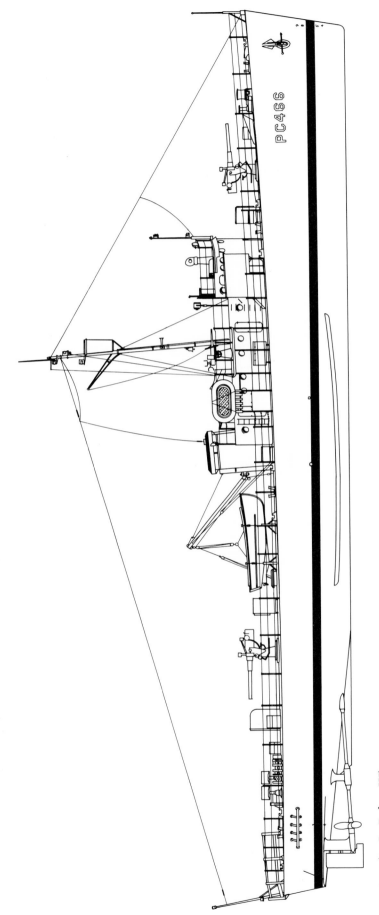

PC 466. (A. D. Baker III)

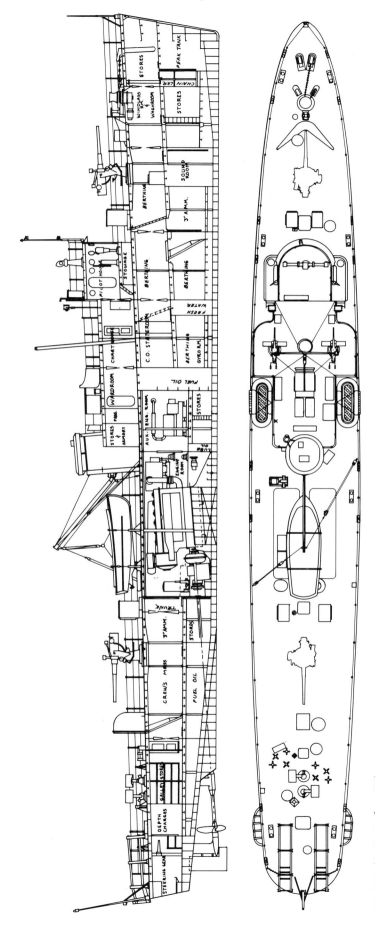

PC 466. (A. D. Baker III)

The General Board considered the 173-foot PC the ideal coastal escort. This is PC 546 off the East Coast in 1942, with the original main battery of two 3-inch/50s. The radar is an SF-1, an early American microwave surface-search set. It was to have complemented an SA-1 air-search set, which had not been installed when the picture was taken. Most later PCs had only a microwave surface-search set, installed at the masthead. The davits aft loaded K-guns; there were as yet no roller-loaders.

In March 1941, the captain of PC 451 reported that it was a better sea-keeper than the old flush-deck destroyers. Its flared bow and high forecastle were generally dry (although the deck aft was often covered by several inches of water in rough weather). But the PC 451, with its more rapid periods of pitch (2 to 4 seconds) and roll (4.6 seconds), was considerably harder on personnel. Moreover, since commissioning, it had lost speed by gaining displacement. An underwater exhaust (which the PC 461 class shared) increased resistance and limited available power. Thus, although the vessel made 21.99 knots at 306 tons on trial, it only made about 19 knots loaded, with its main engines operated as in PC 471. PC 471 made 20.4 knots on trials, its engines making less than their rated power. The boat was expected to match the PC 451 in speed as soon as its mufflers were properly adjusted to reduce back pressure.

Because of its performance and light scantlings, the director of the naval districts division of OpNav complained to the CNO that the new 173-footer was too light and too poorly constructed; that its seaworthiness was highly questionable; and that it would probably make no more than 19 knots in service. BuShips replied in November 1941 that the Board of Inspection and Survey had praised the workmanship that went into the boat and that, once engine problems had been corrected, it was expected to exceed 20 knots on trials. The bureau expected it to be a good sea boat considering the limitations of its size. As for light scantlings, Captain E. C. Cochrane, then head of design, noted that 22 knots in a 173-foot ship corresponded to 32 in a destroyer, thus requiring correspondingly lighter construction.

Finally, there was the proof of experience. In November 1941, PC 471 encountered bad weather, seas of 15 feet, en route from Quebec to Boston. It rolled 40 degrees and was shaken up but suffered no structural damage. The roll was always sharp, without any tendency to hang. The captain concluded that PC 471 was a good sea boat as long as it did not have to be driven into a sea at high speed. Similarly, the commander of the eastern sea frontier, Rear Admiral Adolphus Andrews, wrote in 1943 that the PC behaved well in a moderate sea. Although it could be expected to roll and pitch badly in a heavy sea and to ice over in rough northern waters, it was still seaworthy.

Wartime modifications consisted of the addition of equipment and electrical load but not generating capacity.[1] As a result, the two 60-kw generators tended to cut when there were extra loads, stopping auxilary electrical equipment. The PC did not have enough air capacity to restart the engines.

Authorizations for the early PCs covered SCs and PTs as well. The two experimental ships were included in the special FY 39 program. None was included in the FY 40 or FY 41. In February 1940, apparently

Off Norfolk on 31 August 1942, PC 472 displays newly installed K-guns and depth-charge loading davits. Note the line drawn around them. The radar at the masthead is a Canadian SW-1C. Note also the prominent diesel exhaust port at the waterline, just abaft the false funnel casing.

unaware of the PC program, the Bureau of Navigation (BuNav) asked for twenty-six new coastal vessels for the naval reserves: eight eagle boats, six World War I subchasers, six YPs, and five cabin motor launches. It wanted miniature destroyers with 5-inch guns. The General Board proposed that the new PCs be used instead. The main drawback would be their diesel machinery, which did not match the steam engines of the fleet. However, reservists could expect to spend two weeks each year cruising in the big steamships. As of April 1940, the CNO, Admiral Harold G. Stark, could promise two PCs from the existing FY 39 program, two more in FY 42, and two per year thereafter until all twenty-six had been built.

Three months later, with the fall of France, the appropriation floodgates opened.

In the account that follows, note that PCs 454–60 were converted yachts, as were PCs 509, 510, 523, and 826.[2]

The first major small craft program was settled in September 1940. In an act of 19 July 1940, Congress fixed the total amount to be spent. In August, the proposed program consisted of twenty PCs, twenty SCs, and seventy-five PT boats. However, the General Board disliked the slow, 16.5-knot SC typified by SC 450, which it considered little more than a warmed-over World War I 110-footer. It therefore proposed that the program be revised, to thirty PCs, twenty-four PT boats, and no more than twelve SCs. On 16 September 1940, the secretary of the navy approved a program of thirty-six PCs (461–96), twelve SCs (497–508), and twenty-four PT boats (21–44).

The General Board suggested that a new steel SC be designed, but a fast wooden one was built instead (see below). With the success of the new SC, the secretary of the navy changed the program from twelve to thirty SCs. In December 1940, he increased the number of new SCs to thirty (511–41, with 523 an ex-yacht).

Minesweepers were so important at this time that the program approved in December 1940 included eighteen minesweepers modeled after the 173-foot PC design (AM 82–99) but with reduced power, as well as thirty-two standard 220-foot sweepers (AM 100–131). The program also included continued production of the 136-foot wooden minesweeper (YMS), which had been conceived earlier in the year as a sweeper version of the SC but was now a very different design.

Standard PC and SC production continued: thirty SCs (511–41) under a December 1940 directive; thirty-six PCs under a January 1941 directive (542–77); fifty PCs (578–627); and forty-eight SCs (628–75) under a June 1941 directive from the secretary of the navy.[3]

These orders were awarded as part of the major ("two-ocean navy") expansion authorized by Congress after the fall of France. Other programs, such as those for submarines, minesweepers, and even small seaplane tenders, competed with the PC program for supplies of powerful lightweight diesels. BuShips began to consider alternative power plants. In the summer of 1941, it studied the possibility of substituting the 900-bhp diesels used in fleet minesweepers. An engine would be geared to each of the two shafts and located in each of two engine rooms. The total length of the engines and gears, 54 feet, was 9 feet greater than that occupied by the existing 2,000-bhp diesels, and the alternative machinery arrangement would weigh 29 tons, or 27 percent, more. Any such radical redesign would slow the production of PCs, so the idea was dropped early in August. Instead, BuShips found alternative sources for 2,000-bhp diesels.

The engine problem came to the forefront when Congress authorized up to four hundred miscellaneous light craft, at a total price of not more than $300 million, on 21 November 1941. At this time, the District Craft Development Board was responsible for deciding which types of mine and patrol craft to buy. It was told that, because of other demands on diesel production, no more 20-knot, 173-foot PCs could be bought until after January 1944. Redesigning the 173-footer for a very different engine would take too long; the board sought an alternative while BuShips investigated redesign in greater detail.[4]

Meanwhile Preliminary Design was developing an austere, 180-foot fleet minesweeper originally conceived for lend-lease. It did not compete with the PC for high-powered submarine diesels and so might be easier to produce. However, at about 15.7 knots, it was substantially slower than the PC, and its bluff hull limited the possibilities for improvement. With 3,600 bhp, the PC 461 could be expected to make 20.65 knots on trials (at 370 tons, carrying twenty-two 300-pound depth charges). With the same machinery, a 180-foot PC, which would carry eighty depth charges and eight rather than two throwers, would make 17.4 knots at 844 tons. This comparison was based on a condition of sufficient fuel for an endurance of 4,000 nautical miles at 12 knots, and a battery of two 3-inch/50 and two light machine guns (0.50-caliber for PC 461, 20mm for the 180-footer).

The District Craft Development Board, which included A. Loring Swasey, recommended a program that included the following:
- Fifty yachts or other privately owned small vessels for the use of local defense forces
- Thirty-six PTs
- Eighty SCs
- Thirty 180-foot minesweepers (they were simpler to build than the *Auk* class then in production)
- Twenty 180-foot PCs, derived from the minesweepers

The General Board was unhappy enough at the prospect of slow patrol craft to recommend abandoning them altogether; the navy could build 150 wooden SCs per year and invest the remaining resources in 200 destroyer escorts per year. Only Admiral Ernest J. King, then commander of the Atlantic Fleet, disagreed; he liked the 173-footer. The General Board reconsidered in January 1942, when it was told that, in fact, sufficient engines had been found. It therefore proposed 150 PCs, 100 SCs, and 100 destroyer escorts per year.

The new (late 1941) program included another 100 SCs (676–75) and 50 PCs (776–825), as well as the first 180-foot sweepers (AM 136–65), but no 180-foot PCs.

Once the United States entered World War II, BuShips developed a "maximum effort program" to make full use of available shipbuilding slips. This program merged with the one that had originally authorized 1,799 lend-lease craft for Britain.[5] Production of 180-foot minesweepers continued, and, after the District Craft Board's proposal for the 180-foot PCE was approved, initially for lend-lease, production of 150 PCEs (827–926) was started. The program also included 100 SCs (977–1076) and 189 PCs (1077–1240). Because of a shortage of steel, 56 PCEs (921–76) were canceled on 27 April 1942. PCE 921–60, reordered under the 1945 small combatant program, were later canceled again.

During the Battle of the Atlantic, more small ASW craft were approved. A FY 43 program of 800 patrol and mine craft included 110 SCs (1266–1375). In addition, 90 units of a subchaser version of the 136-foot YMS, PCS (sometimes explained as PC, sweeper) 1376–1465, were ordered. The Canadians supplied several ASW craft under reverse lend-lease, including eight Fairmile subchasers (SC 1466–73).

A continuing wooden shipbuilding program, the result of a directive of 16 April 1943, included seventy-two SCs (1474–1545). The World War II program was finished with a small combatant program of 1945 that included forty PCs (1546–85). The eighteen 173-foot minesweepers were converted to PCs (1586–1603).

By the late summer 1943, the Battle of the Atlantic was clearly being won, and in September President Roosevelt ordered a series of cancellations. The 110-foot subchasers planned for 1944 were canceled, excluding only those whose construction had proceeded too far. The president also cut the production of steel ships to increase landing craft construction by 35 percent. That eliminated 205 of the 1,005 planned destroyer escorts, 40 steel subchasers (PC 1106–18, 1150–66, 1570–75, 1582–85), and 37 SCs

(1494–95, 1500–1501, 1509, 1513–16, 1518–20, 1521–45).[6] Another 36 PCS (1393–95, 1398–1401, 1406–12, 1415–16, 1427–28, 1432–40, 1443–44, 1447–48, 1453–54, 1456, 1462–63) were reordered at the end of September as YMS 446–81; they would be needed to clear invasion beaches. Forty more PCEs (905–34, 951–60) were canceled on 27 September 1943. PCE 905–20 were reordered as minesweepers (AM 351–66). Twenty more PCEs (887–90, 935–50) and twenty more PCs (1092–1105, 1576–81) were stopped on 5 November 1943. Six more PCEs (861–66) were canceled to release engines for landing craft on 21 March 1944. These figures are summarized in table 3-7.

Fortunately, engine production kept pace much better than expected, and few PCs were underpowered. Work on alternative power plants continued through early 1942. For example, in January, BuShips considered three-, four-, and eight-engine (500 bhp) power plants, all of which would be considerably heavier than the original two-engine plant. PC 1083–85 were unique in receiving lower-powered Alco 540 diesels, which produced a total of 1,440 bhp.

Table 3-7 The World War II Patrol Craft Program

	SC	PC	PCS	PCE
Total	475	403	90	134
Canceled	37	60	31	66
Net Program	438	343	59	68
Conversions				
PGM	8	24	—	—
AGS	—	—	4	—
ACM	—	—	—	1
AG	—	—	—	1
YDG	—	—	—	3
PCE(R)	—	—	—	13
Sonar Vessels	—	—	28[1]	—
AMC	6	—	2	—
Weather	—	—	—	17
Control Craft	70	35	13	10
Dan Layers	—	—	4	—
Total Conversions	84	59	51	—
Lend-Lease	104	46	—	15
China	—	—	—	2
Greece	—	1	—	—
Mexico	3	—	—	—
Netherlands	—	1	—	—
Norway	3	1	—	—
Uruguay	—	1	—	—
Brazil	8	8	—	—
France	50	32	—	—
Russia	78	—	—	—
Britain	—	—	—	15
Cost Each ($)	500,000	1,600,000	723,000	1,960,000

Note: PGMs were motor gunboats; AGSs were surveying ships; ACM was an auxiliary (practice) minelayer; AG was a miscellaneous auxiliary used as a personnel transport to San Clemente Island; YDGs were degaussing vessels; PCE(R)s were convoy rescue escorts; the sonar vessels were sonar school ships; AMCs were coastal minehunters (minesweepers in the case of SCs); and control craft were converted to control boat waves in amphibious landings.

1. Fourteen were converted by their builders, fourteen after commissioning.

As in virtually all other classes, wartime modifications added weight, reduced performance, and further restricted space. Growth was particularly severe in the case of the PC. In August 1940 the General Board proposed a battery of two 3-inch/50 dual purpose guns, eight 0.50-caliber machine guns (or four twin 0.50-caliber or one single 1.1-inch), four single depth-charge throwers, and two stern tracks (with twenty-four 300-pound or twelve 600-pound charges). In November, General Board characteristics showed four 0.50-caliber machine guns (with two quadruple mounts if possible), two single depth-charge throwers, and two stern tracks (with twenty-two 300-pound, or twelve 600-pound and eight 300-pound, charges). Endurance at this stage was still 3,000 nautical miles at 12 knots. Ammunition supply was 200 rounds per 3-inch gun and 5,000 per 0.50-caliber gun. Eight more depth charges were added in October 1941.

During 1941, U.S. warship armament was revised, and, wherever possible, 0.50-caliber machine guns were to be replaced by 20mm Oerlikons when the latter became available. In the PC, the light battery changed to two 20mm guns. A third was added in June 1942. As weight compensation, the two seven-charge tracks aft were to have been replaced by four-charge tracks (this decision was reversed before it could be put into effect, and by August 1942, the "ultimate" planned battery showed only two 20mm guns). Many ships were completed early in 1942 with two 0.30-caliber machine guns and without 20mm guns.

In September 1942, BuOrd proposed the addition of two more K-guns for a total of four. BuShips found that existing stability would suffice, as long as the total number of charges did not exceed twenty-two. However, it might be difficult to strengthen the deck sufficiently. Moreover, as the bureau argued, the maximum desirable displacement had been reached with the fitting of one 20mm gun atop the pilothouse. The bureau particularly protested against the fitting of two extra 20mm guns, for a total of five, except in ships expected to operate within range of the enemy's land-based aircraft (it cited the case of PC 1119–24). BuOrd retorted that the addition of the guns had been balanced with the replacement of the usual 20-foot motorboat by a 16-foot wherry, and that substitution of the new lightweight Mark 10 mount for the earlier heavy Mark 4 would save 1,000 pounds per mount. Therefore it wanted five 20mm guns in each PC likely to be subject to heavy air attack.

In November, the vice chief of naval operations (VCNO), Rear Admiral F. J. Horne, established a standard battery of three 20mm (Mark 10) guns and four K-guns, each with four depth charges. Vessels assigned to an area of heavy enemy air activity (which was later translated to mean the Pacific) would be fitted with five 20mm (Mark 10) guns and only two

PC 600 at Mare Island on 22 January 1946. Note the folded-down Mousetraps right forward, the four 20mm cannon (in the sponsons atop the pilothouse and on deck abaft the false-funnel casing), the roller-loaders, and the tracks holding streamlined, "teardrop" depth charges aft. The object atop the false funnel—often labeled a vent trunk in plans—is a ventilator air intake.

K-guns. Separate Pacific and Atlantic armaments were eliminated late in July 1943, and most PCs served out the war with three 20mm guns and four K-guns.

At the end of the war, U.S. warships began to exchange their single 20mm guns for the new lightweight twins. Assigned an "ultimate" battery of four twin 20mm guns, few of the PCs were actually so fitted.

In October 1942, because of a temporary shortage of 3-inch/50 guns, OpNav ordered unpowered 40mm AA guns temporarily installed in place of the after 3-inch/50s. This battery, one 3-inch/50 and one 40mm, was later made permanent.

For ASW, the most important development was the forward-throwing weapon, originally the British Hedgehog. It was a heavy, roll-stabilized spigot mortar, which generally replaced one destroyer gun mount. In July 1942, Admiral Horne asked whether it could be mounted in a PC. BuOrd replied that that would require the elimination of the forward 3-inch/50 mount. The alternative was a rocket launcher developed in the United States, the Mousetrap, which was much lighter and which, therefore, would not require elimination of the gun. The Mousetrap was approved for virtually all U.S. ASW craft, even though it was not stabilized against rolling. The heavier Hedgehogs were reserved for destroyer escorts and ships of similar size. PCs and SCs were initially fitted with the Mark 20, which fired twelve rockets, and then with the double-layer Mark 22.

In addition, fuel was added to increase endurance to 4,000 nautical miles at 12 knots (a realistic wartime figure was about 3,800 miles at 15 knots). As a result, displacement increased considerably and speed was lost.

There was one major wartime conversion, the motor gunboat (PGM). The earliest PGMs were converted 110-foot subchasers (see below) intended primarily

PC 618 remained active after World War II as an experimental sonar ship. She is shown around April 1961 with a Hedgehog in place of her 3-inch/50 forward. By this time, she had been named the *Weatherford*.

to support barge-busting PT-boats in the southwest Pacific, but PCs were larger and hence could be more heavily armed. PC conversions were first planned in December 1943. As with the SC, the superstructure (except, in this case, the forward 3-inch/50 gun) was removed as well as all ASW gear. A small protected pilot/chart house was installed abaft its previous location. Six 20mm guns and one twin power-driven 40mm were installed on deck in splinter shields. SO-8 radar and TCS radio, which was in the PTs, were also fitted. Special effort was devoted to reducing silhouette and top weight in the interest of speed.

The prototype, PGM 9 (ex–PC 1548), was converted while under construction and completed on 29 June 1944. In August 1944, OpNav ordered twenty-three more PCs (805, 806, 1088–1091, 1148, 1189, 1255, 1550–59, and 1565–68) converted to PGM 10–32. At this time, the authorized battery was increased by one twin 0.50-caliber machine gun and a 60mm mortar for illumination fire. The mortar, of limited usefulness, was eliminated in November 1944. PGM 9 was unique in having a sextuple rocket launcher aft. PGM 10–32 were completed between October 1944 and March 1945.

The other wartime PC conversion was the PCC, for use in amphibious control. Similar conversions were extended to the SC, PCE, PCS, and YMS (see below for details). In each case, extra accommodation and communications equipment had to be added. The 20mm mount on the PC's flying bridge was eliminated as weight compensation and to provide space for audio and visual signaling equipment and personnel. Berths for four officers and six enlisted men were installed, and the on-deck depth-charge allowance was reduced to eighteen Mark 9 teardrop charges. Total stowage was fifty-two 300-pound charges. Then, in June 1945, the four remaining single 20mm guns were ordered replaced by twin mounts. As weight

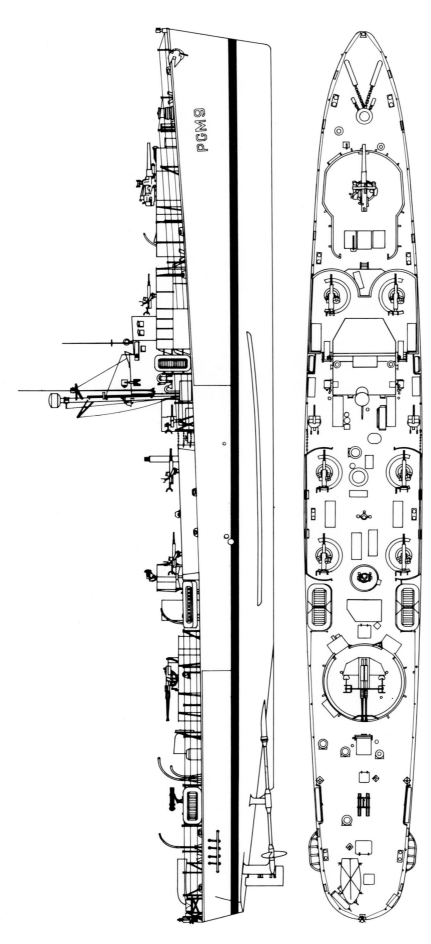

PGM 9 as completed, 1 July 1944. PGM 10–32 had portholes rather than square windows in the pilothouse and lacked any provision for the portable sextuple rocket launcher shown aft. The launcher could also be mounted in the forecastle, off center to port of the anchor capstan. Shields were later added to the 20mm guns (other ships were completed with them). The mast was slightly to port of the centerline. (A. D. Baker III)

compensation, four depth charges were removed from their stowage on deck and the existing eight-charge racks were reduced to four charges. Total stowage was reduced to forty-eight depth charges. In addition, PCCs being fitted with SU radar in place of the earlier SF or SL had four more charges removed as weight compensation. Thirty-five PCs were converted: 462, 463, 466, 469, 549, 555, 563, 578, 582, 589, 598, 802, 803, 804, 807, 1079, 1080, 1081, 1125, 1126, 1127, 1136, 1137, 1168, 1169, 1177, 1178, 1180, 1230, 1231, 1244, 1251, 1260, 1599, and 1601. Four more conversions were planned but not carried out.

Finally, in June 1943, the navy requested proposals for the installation of two 1,350-shp gas turbines with contrarotating variable-pitch propellers in a PC hull. Though an important indication of future developments, the idea was not carried through.

The PC is difficult to evaluate. It was well liked because it handled like a destroyer. However, mobilization requirements kept it small, limiting its flexibility and ability to meet postwar ASW requirements, particularly for larger hull sonars. Variable-depth sonars were conceived after the war largely as a means of overcoming this limitation. They were initially tested aboard PCs. Certainly early postwar plans showed continuing appreciation of PC qualities (see table 3-8). The earliest such plan, dated May 1945, showed 64 active, 4 reserve (reduced complement), and 44 inactive PCs in the Pacific Fleet, plus 22 active, 4 reserve, and 34 inactive ones in the Atlantic. In addition, 18 were assigned to each of four overseas sea frontiers—the Caribbean, Panamanian, Alaskan, and Hawaiian—and 12 were assigned to the Philippines. Thus, of 284 PCs on hand, only 38 were viewed as surplus. The PGMs were originally to have remained active as police forces, mainly in the Philippines sea frontier.

All versions of these plans showed that the 257 SCs would be entirely eliminated. They were very limited in endurance and sea-keeping, and only one actually sank a submarine during World War II.

Two further versions of the PC appeared after World War II. The United States financed a French design of a considerably modified 174-foot PC. This became PC 1610–18 of the U.S. FY 52 program. PC 1613, 1614, and 1617 went to Portugal; 1615 to Yugoslavia; 1616 to Ethiopia; and 1618 to West Germany as an ASW trials ship (it was later transferred to Tunisia). France built five of these modified PCs under her own 1955 naval program, six under her 1956 program. Portugal, under the U.S. FY 52 program, built PC 1635–37 to the same design in her own yards, plus two more for which she herself paid.

BuShips developed its own slightly modified PC for Turkey, with a big SQS-17 scanning sonar in a 100-inch dome and a trainable, stabilized Hedgehog (Mark 15) in place of the 3-inch gun forward. A single depth-charge track replaced the two stern tracks, and the depth charge battery was cut from fifty to thirty-six charges. An aluminum deckhouse was used to cut top weight, but displacement still increased by 15 percent, and speed could not be improved. The United States financed the Turkish ships as PC 1638–43 (FY 61), and Chile built a similar unit as PC 1646 (two more projected units were canceled).

Table 3-8. Postwar Patrol Craft Plans, 1945–46[1]

Class	On Hand	Active	Reserve	Lay Up	Total	Excess
PF	49	0	0	0	0	49
	48	0	0	0	0	48
	45	0	0	0	0	45
PG	18	1	2	0	3	15
	8	0	0	0	0	8
	3	0	0	0	0	3
PE	8	0	0	0	0	8
	5	0	0	0	0	5
	1	0	0	0	0	1
PCE	33	31	2	0	33	0
	33	31	2	0	33	0
	33	33	0	0	33	0
	21	—	—	—	—	—
PCER	13	0	7	5	12	1
	13	0	7	1	8	5
	12	0	5	0	0	7
	8	—	—	—	—	—
PC	284	152	16	78	246	38
	282	92	16	76	184	98
	253	40	1	72	113	140
	126	13	0	120	123	3
PGM	30	22	0	0	22	8
	29	20	0	0	20	9
	22	14	0	0	14	8
PT	409	48	4	0	52	357
	218	0	4	0	4	214
	4	0	4	0	4	0
	8	4	0	0	4	4
PY	16	0	0	0	0	16
	12	0	0	0	0	12
	3	0	0	0	0	3
PYc	14	0	0	0	0	14
	15	0	0	0	0	15
	5	0	0	0	0	5
SC	257	0	0	0	0	257
	218	0	0	0	0	218
	57	0	0	0	0	57
PCS	52	0	30	3	33	19
	52	8	22	3	33	19
	43	8	2	18	28	15
	16	—	—	—	—	—
Total	1,179	254	61	86	461	778
	933	—	—	—	—	651
	481	95	—	—	197	284

1. In this table, the first line of each class refers to postwar plan no. 1 of May 1945, the second to plan no. 1A of December 1945, and the third to plan no. 2 of 1946. The fourth line, where there is one, indicates numbers as of January 1950, at the bottom of the postwar slump. "On hand" numbers include ships under construction, and in some cases the decline reflects postwar cancellations. The PF entry does not include lend-lease vessels or the twenty-eight ships transferred to the Soviet Union; twenty-seven were returned late in 1949, and thirteen were reactivated for the Korean War.

Twenty-four PCs were converted into small gunboats (PGM). The small pipe on PGM 17, photographed 12 November 1944, has replaced the orginal false funnel of the orginal PC. The hooded weapons are six single 20mm guns. Positions between the bulwark and pilothouse were to have been occupied by single 0.50-caliber machine guns, the mounting rings for which are visible in PGM 17.

The other postwar PC/SC numbers all referred to ships funded by the United States for friendly navies under the FY 52 program: PCE 1604–9, slightly modified versions of the wartime PCE built for the Netherlands (FY 51); PCE 1619–21, a series based on an Italian (Ansaldo) export design and built in Italy, and PCE 1626, built for the Italian navy; and PCE 1622–25, built for the Danish navy. The U.S. Navy acquired the rights to the Italian design and much later ordered further units for Iran (PF 103–6, two under FY 61 and one each under FY 66 and FY 67) and Thailand (PF 107–8, in FY 69 and FY 72). Ansaldo built two very similar ships, the *Pattimura* class, for Indonesia.

As in wartime, the PC/PCE series was interleaved with an SC series: SC 1627–31 were built in and for the Netherlands to a Dutch design under the FY 52 program, and under the FY 53 program, SC 1632–34 were built in the United States for Thailand to the World War II SC design.

PC 1644–45, completing the series, were the large Danish-built frigates *Peder Skram* and *Herluf Trolle* provided for in the FY 62 program.

The 16- to 17-knot SC did not generate much enthusiasm, and by mid-1940 the program was clearly in trouble. The General Board considered 16 knots too slow for use against modern submarines or for effective escort. Three of the existing supercharged 900-bhp Hall-Scott gasoline engines could, in theory, drive the existing wooden hull at 23.5 knots, but it would not be able to stand their vibration. The board therefore suggested that a better high-power, high-speed diesel be developed, and a new steel subchaser designed around it. That would have killed President Roosevelt's subchaser concept.

Swasey saw a solution: a wooden hull driven by the new pancake diesel. He later said that President Roosevelt personally brought him out of retirement to undertake a fifty-boat program of 23-knot (22.5 knots' sustained speed) subchasers. At this time GM was developing three pancake engines. One was canceled to provide funds for a third FY 39 subchaser, which became PC 453. The new SC also had new hull lines with a wide transom stern. The secretary of the navy formally approved the proposal for the new prototype subchaser on 11 May 1940. The builder, the Fisher Boat Company in Detroit, was chosen for its proximity to the GM laboratory developing the pancake diesel.

The boat was completed only on 12 August 1941, so that in the fall of 1940 SC production plans still referred to the earlier 17-knot wooden SC. With 3 inches less beam, PC 453 was slightly smaller than the production boats, and fully loaded it displaced 134 rather than 148 tons. PC 499, the first boat with production pancake engines, completed its successful trials in March 1942.

The program was soon expanded, but pancake production was considerably slower than expected. In November 1941, it appeared that most SCs would be powered instead by General Motors straight-8 diesels (8-268A, 500 instead of 1,200 bhp). Their trial speed was about 15.6 rather than 21 knots. By April 1942 matters had considerably improved, and the pancake was reinstated wherever possible. Out of 435 SC 497–class 110-footers, 243 had pancake diesels.

In 1943, the commander of the eastern sea frontier evaluated the SC favorably. It was a good sea boat, quick to handle, and could operate effectively at sea for up to four days. On the other hand, it was wet, gave a rough ride, and had to seek shelter in heavy weather. Quick motion made it an unsteady gun platform. In any kind of sea, reloading its K-guns with the boom and topping lift required, was impractical.

The combination of the new pancake diesel and a variable-pitch propeller seemed satisfactory, driving SC 662 and 672 at 20.5 knots. By way of contrast, the straight-8 diesel vibrated excessively. Although its output was much less, it was heavier and so increased draft about 18 inches aft, lowering maximum speed to about 14.8 knots. Even that was available only briefly, as the injector spray tips tended to foul.

Armament was initially set at one 3-inch/23 gun, two twin 0.50-caliber machine guns, and two depth-charge tracks (fourteen 300-pound charges). Because BuOrd had no twin 0.50-caliber mount in production, single guns were substituted as an interim battery. Late in October 1940, OpNav asked that two lightweight depth-charge throwers be added. This change was made in December 1940, and in October 1941, the SC depth-charge battery was increased to allow eight more 300-pound charges for the throwers. Thus the initial wartime depth-charge battery was two K-guns and two depth-charge tracks, seven 300-pound charges each.

By June 1942, the SC program was plagued by a shortage of 3-inch/23 guns. BuShips initially denied that the wooden SC hull could take the weight of the 3-inch/50 but found otherwise after recalculating. In August, the gun was formally substituted for the smaller weapon, except in SC 505–6, 532–35, and 1013–22. The secondary battery was set at two 0.50-caliber machine guns, but in October 1942, they were replaced by two twin 0.50s in SC 675 and below; for higher-numbered ships the secondary battery was two single 20mm guns.

Neither type of 3-inch gun was really satisfactory. In October 1942, the VCNO ordered the army-type 40mm gun tested aboard SC 508. If the gun was suc-

The wartime PC design was repeated after the war in slightly modified form for the Turkish and Chilean navies. This is PC 1638, newly completed in April 1964. The wartime 3-inch/50 forward has been replaced by a big, stabilized, trainable Mark 15 Hedgehog, coupled with a large medium-frequency sonar, the SQS-17. PC 1638 became Turkey's *Sultan Hisar*.

PC 453 combined new hull lines and the General Motors "pancake" diesel. It was the prototype World War II subchaser, although it had a rather different superstructure.

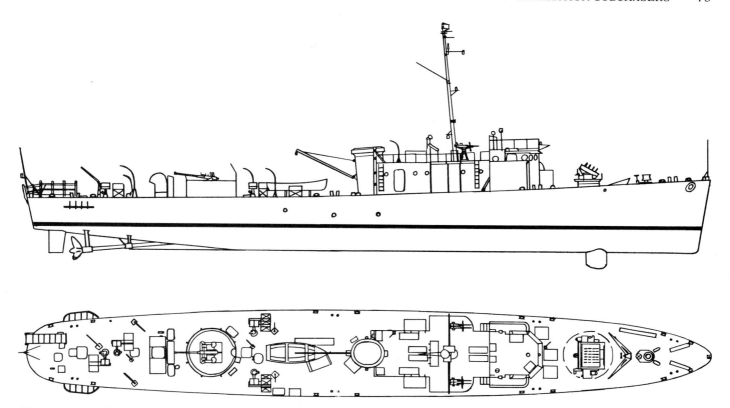

The PC 1638 class was built after the war for Turkey and Chile. Note the large SQS-17 sonar dome and the trainable, stabilized Mark 15 Hedgehog, which replaces the 3-inch/50 of wartime PCs.

cessful, as it turned out to be, it rather than the 3-inch/50 would replace the 3-inch/23. In November, faced with a temporary shortage of 3-inch/50s, the VCNO ordered 40mm guns installed on 3-inch/50 foundations in new SCs. Battle experience showed that the 40mm owing to its high rate of fire and tracers, was much more effective. In May 1943, for example, SC 521 reported the success it had putting up a barrage to drive off Japanese dive-bombers trying to attack an LST off Guadalcanal. The four bombers circled, out of 20mm range, but separated and fled when 40mm shells began to burst among them. The 3-inch gun might have reached farther, but it did not have the requisite rate of fire. The commander complained only that the gun had no armor-piercing round for use against surfaced submarines.

By March 1943, the standard "ultimate" SC armament was one 40mm gun, three 20mm (Mark 10) mounts, six single-depth-charge release chocks (tracks type C), two K-guns, and two Mark 20 Mousetraps. Before Mousetraps became available, SCs would be armed with two 20mm guns and ten C tracks.

SCs in Alaskan waters had special modifications, such as a spray shield for their 40mm gun.

In November 1943, in response to requests by PT commanders in the southwest Pacific, eight pancake-powered SCs (757, 1035, 1053, 1056, 1071–72, and 1366) were ordered converted to motor gunboats (PGM 1–8). They were completed in the late fall of 1943, with their superstructures cut down and with heavy gun batteries, open bridges, and PT-type radios and radars installed. The planned battery was originally two single 40mm and four twin 0.50-caliber machine-gun mounts, all protected by 10-pound STS armor, one 60mm army-type mortar, and two FM smoke generators. Forward area commanders were to have had the option of substituting a single 3-inch/23 for the forward 40mm gun, but as of January 1944, all boats had the latter installed. At that time, their planned ultimate battery also included a sextuple 2.36-inch (bazooka) rocket launcher, a weapon also associated with PTs at this time. The PGMs were well liked, but they were much too slow to keep up with PTs; the PT squadrons had to rely instead on PT gunboats.

The SCC (amphibious control) conversion entailed replacement of the single 40mm gun forward by a 20mm cannon, the 40mm magazine space being used for radios, leaving four 20mm guns. Two 0.50-caliber machine guns were added. One extra bunk was added to officers' quarters, and all double bunks in enlisted quarters were converted to triple bunks. Two Mark

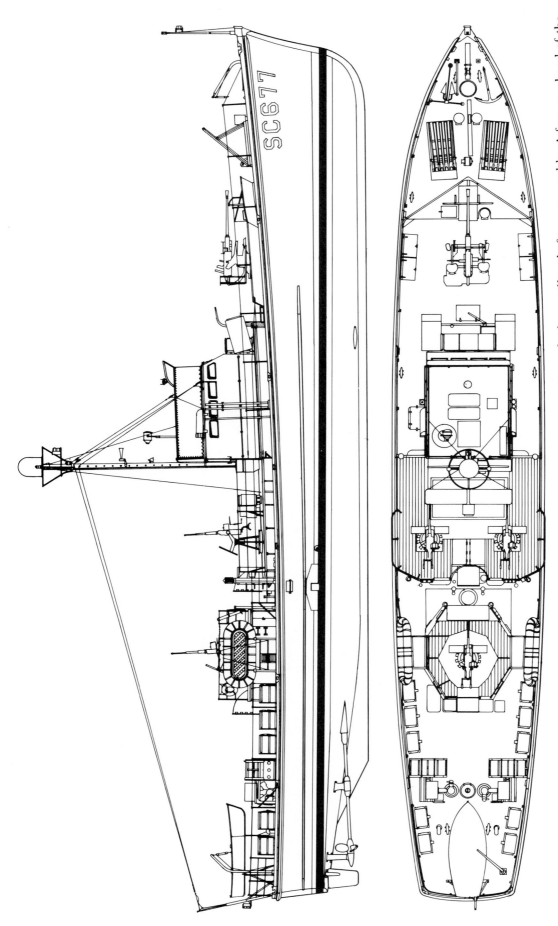

SC 677 late in World War II. A few of the SCs had tripods supporting their masts. Late versions had propeller shafts supported by A-frames ahead of the propellers. Early SCs had two depth-charge racks aft and lacked centerline 20mm guns. By the end of the war, all SCs had the armament shown here, except for SC 670, 1025, and 1042, which retained their 3-inch/23s, and SC 723, 739, 745, 749, and 1065, which had 3-inch/50s. Ships armed with 3-inch guns had no centerline 20mm gun aft. Several SCCs had 20mm guns forward as well (1272 and 1298) or no gun forward (1306, 1311–12, 1314, 1323, 1338, 1341, 1343, 1350, and 1474). SCC 724 had a 3-inch gun forward. (A. D. Baker III)

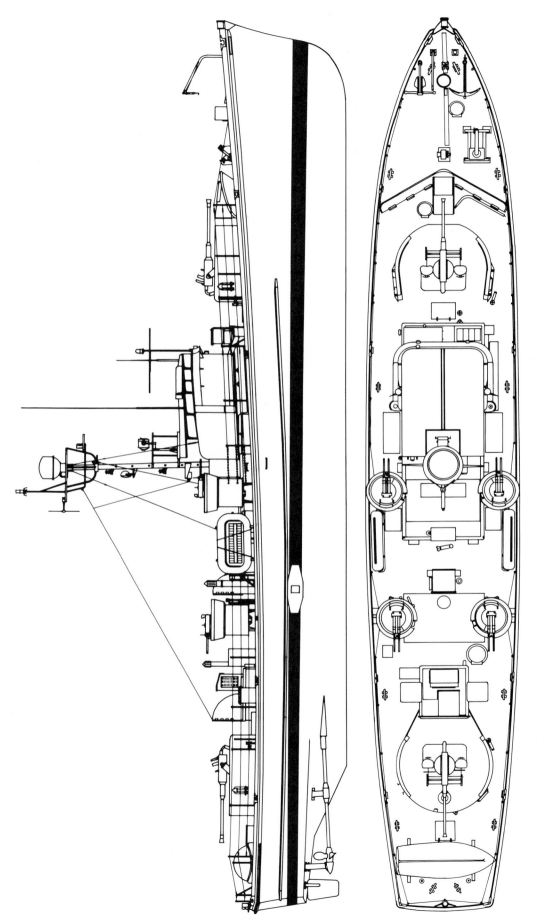

PGM 33, 1953. (A. D. Baker III)

The 110-foot SC was the other mass production subchaser designed before the war. SC 712, photographed in May 1943 in early configuration, has a Canadian SW-1C radar, a short 3-inch/23 gun forward, and two short depth-charge tracks along with guns aft. The machine guns are 20mm Oerlikons.

The first SCs were armed with water-cooled 0.50-caliber machine guns. SC 661, newly completed, is shown on 6 October 1942. K-guns have not yet been installed aft, although the base for one is visible alongside the sailor. The mast and boom would have been used to load them. Note the small, open bridge.

SC 717, photographed off the Atlantic Coast on 2 July 1944, was typical of late-war subchasers. Depth-charge rails have been removed in favor of remote-control side launchers. The two K-guns have been retained, and the 40mm gun is an unpowered army type. Mousetraps, forward, are folded down.

Photographed on the Delaware River on 9 August 1943, SC 1078 displays its side-dropping gear in detail.

SC 1049 on the slip, 1 May 1943. Note the waterline exhaust and twin rudders, installed for maneuverability.

PGM 7, 7 January 1944. This boat typified the earlier SC gunboat conversions, with a 3-inch/23 forward and a 40mm gun aft. Postwar equivalents built for the Philippines had 40mm guns fore and aft.

smoke-generator racks (without the pots) were installed at the fantail. The conversion was complex, and not all SCCs had their 40mm guns removed. In all, seventy SCs were selected for conversion to control vessels, although not all were ultimately redesignated as SCCs (SC 504, 514, 521, 630–32, 636, 667–68, 686, 712, 724, 727, 729, 760, 999, 1004, 1012, 1018, 1020, 1049, 1052, 1066, 1272–73, 1278, 1281, 1298, 1306, 1309, 1311–12, 1314–16, 1319–20, 1323, 1325–28, 1338, 1341, 1349–50, 1360, 1364–65, 1369, 1374–75, 1474).

SCs were also used for shallow-water minesweeping, beginning with six boats in the Mediterranean in 1943: SC 498, 535, 655, 770, 978, and 979, 770 being the prototype. They were fitted with captured German sweep gear, intended for R-boats, and with U.S. Oropesa size-5 towed-wire sweeps. Six more (SC 633, 645, 671, 759, 1034, and 1036) were converted for the Pacific Fleet in the summer of 1945, another six (SC 658, 716, 773, 1363, 1368, and 1372) in September 1945.

From late October 1944 to early 1945, BuShips considered converting some SCs as ferries for big Pacific anchorages, carrying up to 230 passengers. In August 1945, OpNav decided instead to use modified LSILs and smaller landing craft, despite BuShip's objections that they had flimsy steel hulls, as opposed to sturdy wooden ones, and were powered by eight rather than two diesels. The OpNav decision was due in part to the absence of SCs from the postwar plan—probably the clearest verdict on their value. Seventy-two were transferred to the Coast Guard as air-sea rescue boats in 1945.

The U.S. Navy also operated eight Canadian-built Fairmile B subchasers, SC 1466–73. Production to British plans began in Canada in 1940, and the builder, the Hall Scott Motor Car Company, invited U.S. interest. However, the boat was generally dismissed as a British adaptation of the U.S. World War I subchaser; in fact, it was a wholly new design, in part a reaction to what was perceived as the poor sea-going behavior of the 80-foot Elcos. The Fairmiles were powered by Hall-Scott gasoline engines and delivered with one 40mm gun forward. In October 1942, their authorized battery was set at one 3-inch/23 gun (in place of the 40mm originally provided), two 20mm guns, two K-guns (six charges), twelve depth-charge chutes, and two Mark 20 Mousetraps. By late 1944, all remaining U.S. units except SC 1472 had exchanged their 3-inch/23s for army-type 40mm guns. At the end of the war, SC 1474, the amphibious control ship, had three 20mm guns, four rather than ten or twelve depth-charge chutes, two K-guns, and two Mark 20 Mousetraps. Three Fairmiles, SC 1466, 1469, and 1471, were transferred to Mexico under lend-lease.

The larger 180-foot PCE can be traced to a British proposal of November 1940 to mass-produce steel-hulled escorts in the United States. Visiting Fairmile, Britain's subchaser production organization, Captain E. C. Cochrane of BuShips was shown the preliminary design for a 205-foot, 500- to 700-ton escort. He declared it quite unsuitable for U.S. use. The following month, under OpNav pressure to build minesweepers more quickly, Cochrane's design branch began work on an austere sweeper that became the PCE. At the time, all BuShips could offer for quick sweeper production was the modified PC, which cost about $1.5 million, half as much as a full 220-foot fleet sweeper. However, the PC was clearly deficient in sea-keeping, and BuShips continued to consider alternatives such as a modified fleet tug.

Early in 1941, a new design was started, this one to displace 600 to 800 tons on a length of 170 to 190 feet and to be powered by the two 900-bhp geared diesel engines of the modified PCs. In addition, there would be a separate magnetic-sweep generator engine. The designers soon chose a length of 180 feet and a battery of two 3-inch/50 guns, two Oerlikons, and ten 300-pound depth charges. Endurance was 4,000 nautical miles at 15 knots, which was slightly worse than that of the existing fleet sweeper (4,500 miles at 16.5 knots) but much greater than that envisioned for the PC. Plans were submitted to the secretary of the navy in May 1941.

BuShips described its sea-keeping as intermediate between the standard 220-footer and the converted PC. A forecastle would increase freeboard forward.

By this time, the Lend-Lease Act had been passed. Any warship procured through the U.S. Navy and designed for Britain had to fit American specifications in order to be available for U.S. service. The General Board and the secretary of the navy approved characteristics for the 180-footer in July 1941 and suggested that one or more of the existing minesweeper contracts be modified to permit early construction of the new design. Although it was not yet clear whether the 180-footer would be built for the U.S. Navy, the design was offered to the British. They asked for diesel-electric machinery to accommodate the frequent starting and stopping action of sweeping and a protected British-style open bridge like the one in the new destroyer escorts, with the helmsman below rather than on the level of the conning officers. As built, the ships received the new bridge but retained their geared diesels.[7]

In August 1941, the Admiralty rejected the 180-footer because of an unfortunate experience with the *Bangor*, a British-designed sweeper of similar size. Captain Cochrane objected; he considered the 180-footer superior to the U.S. 220-footer, which had been conceived as a hybrid sweeper-minelayer (although

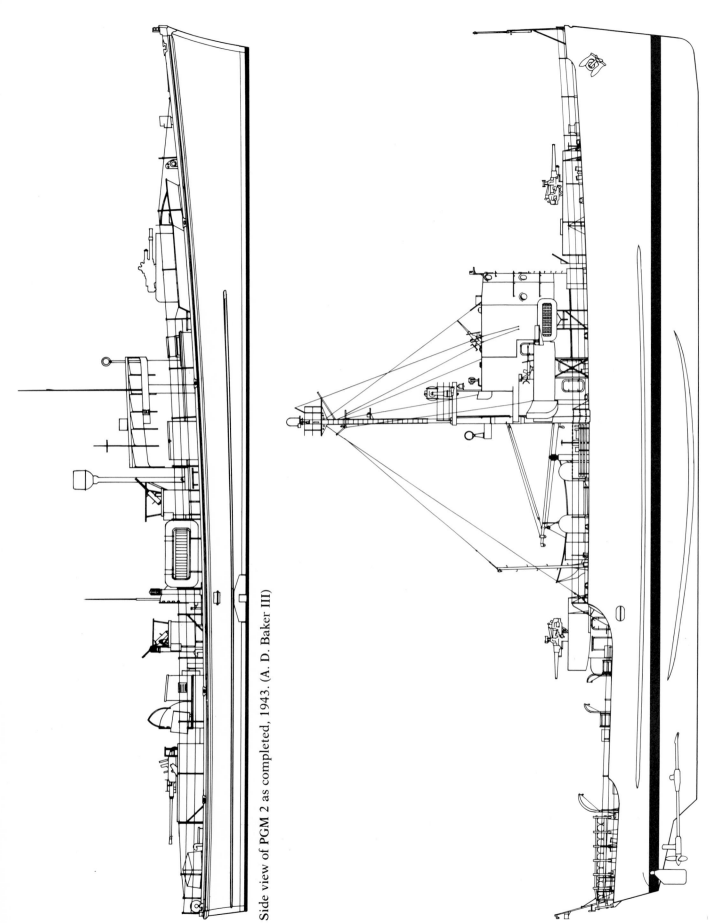

Side view of PGM 2 as completed, 1943. (A. D. Baker III)

The PCE 827 class as conceived, April 1943, with two 3-inch/50s, three 20mm guns, a Hedgehog, four K-guns, and two depth-charge racks. There is no stack. (A. D. Baker III)

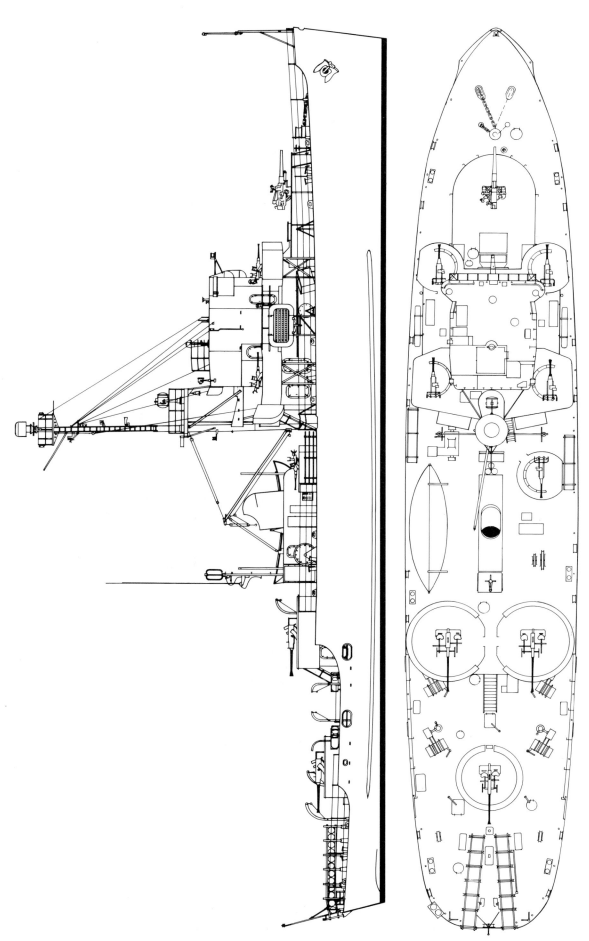

The **PCE 827** class as completed for the Royal Navy late in 1944 with three single 40mm guns aft; in U.S. ships, these guns were later replaced by twin 40mm guns with Mark 51 directors in adjacent tubs. The American ships also had five single 20mm guns. This particular unit has the enclosed sonar hut not across the bridge face but, instead, at the aft end of the bridge. (A. D. Baker III)

Inclined at the yard of Harbor Boat Building, PGM 4 (ex-SC 1053) displays the details of conversion. Note the PT-style 0.50-caliber gun tubs and folding radar mast, and the austere, open, armored cockpit. PGM 4 was armed with 40mm guns fore and aft.

none was ever fitted with mine rails). The 180-footer was specifically designed to survive the shock of ground mine explosions, and with only two diesels in the diesel-electric version, it would be easier to build and maintain.

The design survived as a U.S. minesweeper that could be used as an alternative PC, carrying a much heavier depth-charge load; it was much more seaworthy than the faster 173-footer. In January 1942, it was offered to the British as a PC. They required ten depth-charge patterns, with sixty or more charges, two stern tracks, and two throwers. Fifteen ships of the PC version were transferred to Britain. The Royal Navy never accepted the 180-foot minesweepers. For those, the only lend-lease customers were China and the Soviet Union, which received four and thirty-four, respectively.

The first 180-foot PCs, built to British staff requirements, were officially designated on 16 January 1942 as PC 827–86. At that time, their estimated speed was 16.6 knots on trials, using a maximum output of 2,400 bhp; fouling would reduce that to about 16 knots. With a normal output of 1,800 bhp, they would make about 15.9 knots on trials, fully loaded (830 tons). Estimated endurance was 4,000 nautical miles at 15 knots, full power.

In 1942, the projected battery was two 3-inch/50 DP guns, four 20mm guns, two tracks and track extensions, and four K-guns (80- to 300-pound charges); provision would be made for future installation of a Hedgehog with magazine and ready-service stowage for 144 charges (six patterns). By April 1943, the light AA battery had been modified to three 20mm guns, two in the bridge wings and one atop the flying bridge. The centerline 20mm mount would have interfered with 3-inch fire control and been masked in all directions except 30° on either bow. It was therefore relocated aft, just forward of the depth-charge racks. As a result, only one gun could bear on either surface or air targets from 10° to 30° on either bow; there was a maximum of two guns on all bearings except dead astern.

By mid-1943, the planned PCE battery showed the after 3-inch/50 replaced by two single 40mm mounts, and there were four 20mm guns, two depth-charge tracks (eleven charges), four K-guns, and a Hedgehog. Pacific units would have six 20mm guns and two K-guns. PCEs were generally completed with three single 40mm and five 20mm guns. The next step was to replace the single 40mm with twin mounts by early 1944, and by the end of the war, the projected ultimate battery was one 3-inch/50 and three twin 40mm and four single 20mm mounts. The amphibious control ship (PCEC) had three twin 40mm and four twin 20mm guns.

The British PCEs had both 3-inch/50s and lacked 40mm guns.

In June 1943, the VCNO ordered twenty-five PCEs (842–66) completed as convoy rescue ships. Originally designated APR, they were soon redesignated PCE(R) and were described as escorts with secondary roles of interisland personnel transport and rescue. They could be distinguished by their extended forecastles. PCER 842–47 were reordered as standard PCEs, and PCER 861–66 were canceled in March 1944, leaving thirteen PCERs to be completed. The longer forecastle added top weight, so the PCERs had reduced light AA batteries—two single 40mm and six single 20mm guns. In 1945, conversion to six twin 20mm mounts was being considered. After the war, several PCERs were used for sonar research, their longer forecastles providing valuable internal space.

Ten PCEs were ordered converted into PCECs; nine were actually converted (PCEC 878, 872–73, 877, 882, 886, 891, 896, and 898). Unlike the smaller amphibious control craft, they sacrificed few weapons for greater accommodation and more communications equipment (the latter in a new compartment at the 02 level abreast the mast).

In October 1944, OpNav ordered eleven PCEs (882, 884–86, 897–900, 902–4) converted to weather ships at Puget Sound for duty off Hawaii. They were completed in August 1945, and six more (PCE 880–81 and 893–96) followed in October. An aerologic office was constructed on the navigating bridge abaft the chart room, and the 20mm gun was removed from the starboard side amidships and replaced by a balloon inflation room, 10 feet square and 12 feet high.

Finally, there were three noncombatant conversions, PCE 876, 879, and 883, which became deperming tenders YDG 8–10. PCE 878 became the *Buttress* (ACM 4), a drill minelayer and mine recovery ship. A heavy boom was installed aft, sufficient to lift a 36-foot mine-recovery motor launch. She was also provided with portable tracks sufficient to lay twenty mines and with new mine detection and location (electromagnetic mine-hunting) gear. PCE 901 was converted while under construction to the *Parris Island* (AG 72), a logistic support ship for San Clemente Island. Armament was deleted, the superstructure was pushed aft, cargo booms were rigged forward and aft, and refrigerator machinery was installed to accommodate 27 tons of refrigerated cargo (the boat carried 58 tons of dry cargo). She could seat 200 passengers.

The PCEs seem to have been very satisfactory; the first postwar plan retained all of them in active service.

The PCS was the other, less successful expedient escort. Ironically, its design began, in May 1940, as a minesweeper version of a prototype subchaser, SC 449. The final design was considerably beamier, somewhat longer, and displaced about twice as much.

The PCE was the most successful of the extemporized subchasers. This is the newly completed PCE 893 on 21 July 1944. Note the crease built into its false funnel for stowage of the boat boom and the porthole in the side of its sonar hut, forward of the British-style open bridge. Pipe rails around the single 40mm guns aft indicate that they were not director-controlled; they did not have cutout cams to keep them from firing into the ship.

SECOND-GENERATION SUBCHASERS

PCE(C) 882 displays the final wartime battery of twin director-controlled 40mm guns. Note the absence of pipe rails. The amphibious control role is evident in the HF whips prominent on the bridge and the false funnel. Less visible are numerous VHF dipoles at the masthead.

PCE 899 was a weather ship with a balloon inflation room on the upper deck, abaft the bridge. The ship is shown newly completed on 27 March 1945. Note that, in contrast to other PCEs, it has an air-search radar (SA) at the masthead to complement the usual surface-search set.

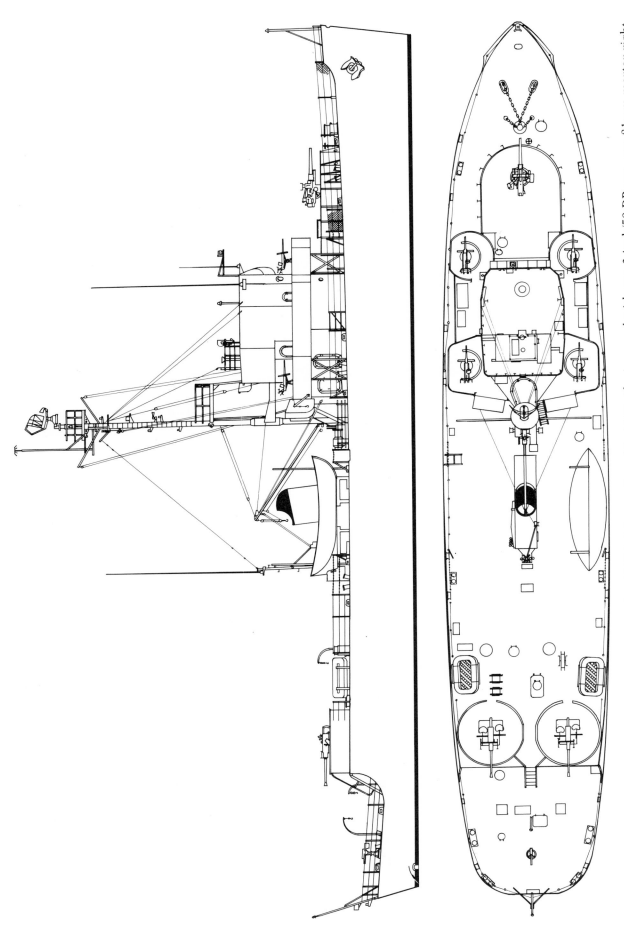

PCER 853, ex-USS *Amherst*, as the Vietnamese *Van Kiep II* (HQ 14), transferred in 1970. She is armed with one 3-inch/50 DP gun, one 81mm mortar right aft, two single 40mm Mark 3s, and four single 20mm guns. Her sonar and all her ASW weapons were removed during modernization at Guam. The radar is an SPS-5. She has berths for forty-five passengers at the after end of the extended forecastle, portside. The *Brattleboro* (PCER 852) was similarly converted, as *Ngoc Hoi* (HQ 12). The *Van Kiep II* escaped from Vietnam in May 1975 and was sold to the Philippines on 5 April 1976, after which she became the *Datu Marikudo* (PS 23). (A. D. Baker III)

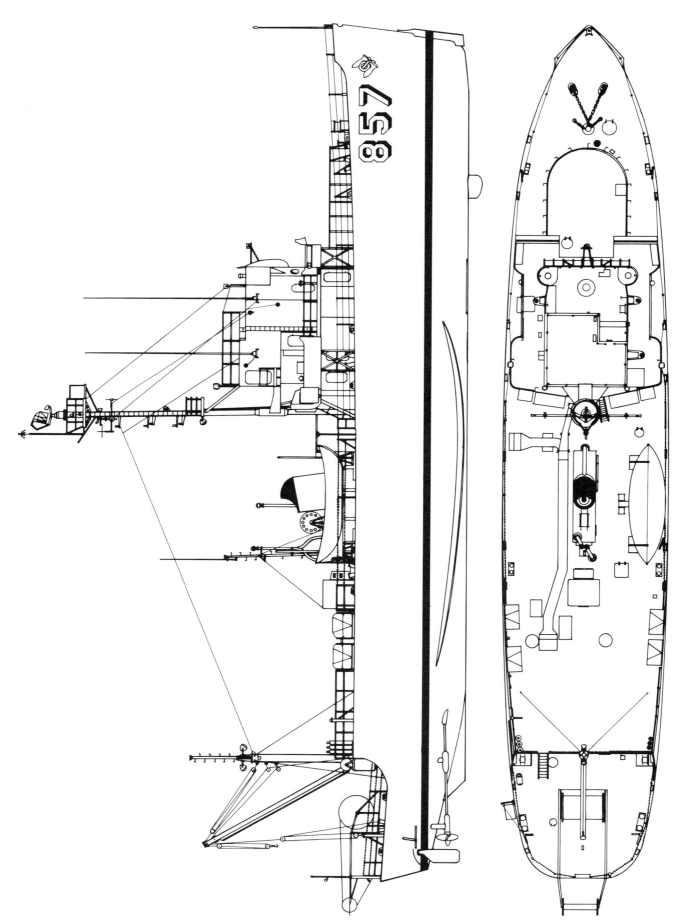

The *Marysville* (E-PCER 857) in 1963. Her conversion to "sound laboratory ship" was completed in September 1948. The other five E-PCERs were differently configured, according to mission. The *Fairview* (E-PCER 850) had an even longer forecastle, a tripod mast aft, and a coffin-shaped deckhouse abaft the stack. None of the ships was armed. (A. D. Baker III)

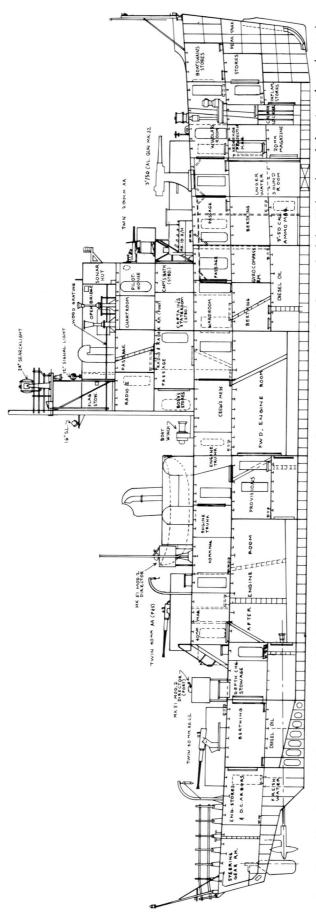

Inboard profile of PCE(C) 886 in January 1952. PCEC conversions had a second radio space added at the after end of the 02 level, and the signalman's station was raised one deck. Armament at this time was one 3-inch/50 DP, three twin 40mm, and four twin 20mm guns, one Mark 10 Hedgehog, and two depth-charge racks. K-guns were removed in 1948. (A. D. Baker III)

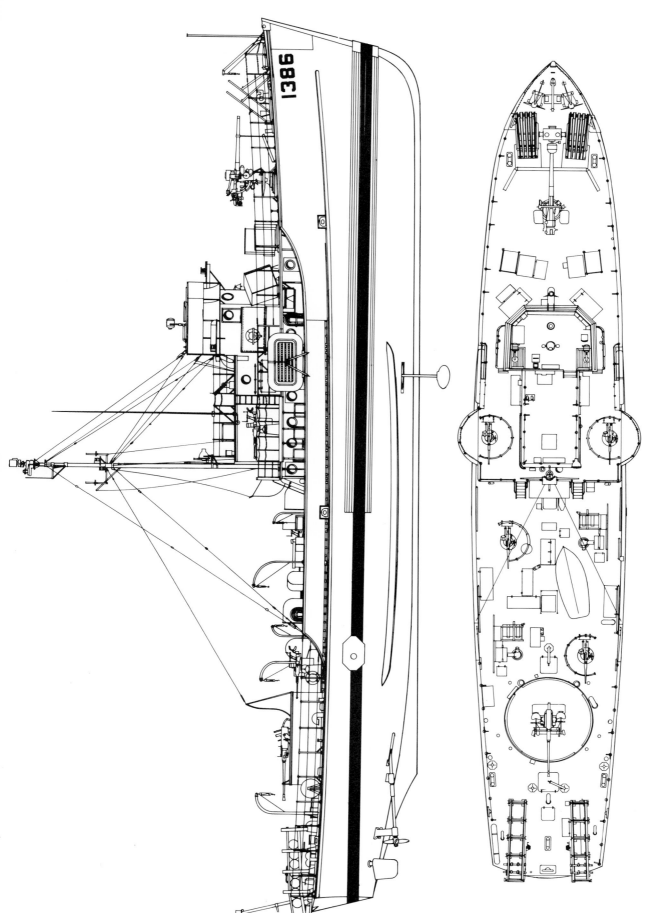

PCS 1386 as completed, 4 November 1944. By August 1945, nineteen PCSs had had their 3-inch/50 guns replaced by Hedgehogs and their Mousetraps deleted: 1383–87, 1392, 1397, 1399, 1400, 1407, 1423–26, 1430–31, 1441–42, 1444–49. Thirteen had become PCSCs: 1379, 1389–91, 1402–3, 1421, 1429, 1448, 1452, 1455, 1460–61. They had four 20mm guns and two K-guns as here, as did PCS 1414 and 1422. The remaining units had four K-guns and only the two 20mm guns on the forecastle. PCS 1431 ended the war with neither a 3-inch gun nor a Hedgehog. A few dispensed with the folding, semicircular platforms abreast the forecastle 20mm guns (shown here in working position). (A. D. Baker III)

Above and right: The other major extemporized subchaser, in addition to the PCE, was the PCS, converted from a 136-foot wooden minesweeper hull. PCS 1461 is shown soon after completion, on 7 February 1944. The Mousetraps, right forward, are shrouded. There are pipe racks for the single 40mm gun and two single 20s, and two depth-charge throwers just forward of the after end of the bulwark.

On 20 June 1942, the VCNO ordered BuShips to build 100 subchasers based on the YMS hull and machinery. Removal of the minesweeping gear and the 540-kw magnetic-sweep generator made space for enough fuel to increase endurance 50 percent, to about 3,000 nautical miles at 12 knots. Armament was initially set at two 3-inch/50 guns (250 rounds each), two 20mm cannon (4,000 rounds each), two stern tracks (eight 300-pound charges each), four K-guns (four 300-pound charges each), and a Hedgehog (ninety-six charges, four patterns). Including magazine stowage, the PCS would carry sixty-two depth charges. However, as of mid-1943, the planned PCS battery consisted of one 3-inch/50 gun, one single 40mm and two 20mm guns, four K-guns, and two depth-charge tracks (seven charges each). The Hedgehog was much too heavy; ships would ultimately be armed with two Mark 22 Mousetraps. Pacific units would have four 20mm guns and two K-guns. The Atlantic-Pacific distinction was soon abandoned. In 1945, the projected ultimate PCS armament was one 3-inch/50 gun and one single 40mm and two 20mm guns. The PCSC had its 40mm guns removed and received four 20mm mounts as well as shorter depth-charge tracks (with a capacity of three rather than six 600-pound charges, or four rather than eight Mark 9 300-pound teardrop-shaped depth charges).

The major PCS conversion program was a series of sonar school ships, their 3-inch/50s replaced by single Hedgehogs (PCS 1377–78, 1383–87, 1392, 1397,

1399–1401, 1417, 1420, 1423–24, 1426, 1441–42, 1444–46, 1448–49, and probably four others).

In December 1944, PCS 1464–65 and four small coastal sweepers (AMC) were ordered converted to special mine hunters. PCS 1465–66 became AMC 203–4 (AMC 204 became AMCU 14 in 1952).

Thirteen PCSs became amphibious control craft (PCSC): PCS 1379, 1389–91, 1402–3, 1418, 1421, 1429, 1452, 1455, and 1460–61. A bunkhouse replaced the single 40mm gun, and the 40mm magazine was rearranged to house a radio transmitter. The depth-charge allowance was reduced to leave sixteen Mark 9 teardrop charges on deck for a total of fifty-two. Two smoke-generator racks without pots were installed on the fantail.

Four others, PCS 1388, 1398, 1404, and 1457, became special hydrographic survey ships (AGSC). PCS 1405, 1413, 1425, and 1450 were converted to dan buoy layers at Pearl Harbor in October 1945; they supported mine-clearance operations by marking swept paths.

Two other wartime subchaser programs deserve brief mention here, the Coast Guard 83-footer and the PTC version of the 63-foot crash boat (see appendix C). The 83-footer culminated a series of fast prewar Coast Guard boats, many of them designed specifically to combat rumrunners. Reportedly, the 83-footer was designed to counter the opium trade on the West Coast—it had to be large enough to reach ships that dropped floating packages into waters well off shore. But because of its size, the 83-footer also served as an offshore subchaser and, off Normandy, as a rescue boat.

Wheeler Shipyard of Brooklyn, New York, built a total of 230 of these boats for the Coast Guard (CG 83300–529). The first was completed in 1941. Nine-

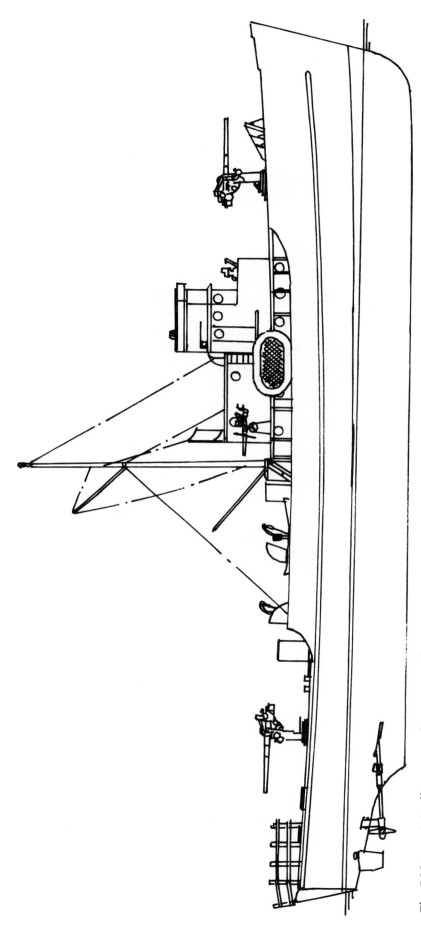

The PCS as originally proposed in 1942 with 3-inch/50 guns fore and aft and only two 20mm guns on heavy Mark 4 mounts. Note the handwheels that elevated and depressed the guns. The only visible ASW weapon is the depth-charge track aft, but the davits amidships loaded four K-guns. Lifelines have been omitted for clarity. (Norman Friedman)

Some PCSs were modified as sound-school craft, with single Hedgehogs forward so that students could carry out dummy attacks. This is PCS 1445, whose two K-guns and two after 20mm cannons have been eliminated.

teen of the gasoline-powered version were transferred to Latin American navies: eight to Cuba, two to Colombia, six to Peru, and three to Mexico. Another twelve, these diesel-powered, were built specifically for transfer: four to Cuba, three to the Dominican Republic, one to Haiti, and four to Venezuela. After the war, ten modified 83-footers were supplied to Burma in February 1951 and four to Turkey in June 1953 (see chapter 9). They were all converted to diesel power, the Burmese boats and some of the Turkish ones being distinguished by their prominent funnels. The Burmese boats were gunboats, armed with 40mm mounts fore and aft. The Turkish boats were subchasers, armed as in World War II. Six apparently unmodified units were supplied to France as coast surveillance ships for Indochina.[8]

As subchasers, the 83-footers were armed with a single 20mm gun, two Mark 20 Mousetraps, and four depth-charge racks (two 300-pound charges each).

The 63-footer was designed specifically as a crash boat for lend-lease, but it was adapted as a subchaser for the Soviets and redesignated RPC. Eighty RPCs were built. RPC 50 was redesignated a small boat; RPC 51–80 were redesignated PTC 37–66. All but PTC 50–53 were delivered to the Soviet Union. The standard battery was two twin 0.50-caliber machine guns in Mark 17 scarf ring mounts, one Mark 10 20mm gun, and eight type C depth-charge chutes (one 300-pound charge each).

After the war, 63-footers were supplied as patrol boats to several countries: two to Cambodia (C26656 and C36275) in December 1961; two to Guatemala in 1964–65; three to Jamaica (C77466, C77467, and C26584) in 1964; and three to Taiwan (PTC 35, 36, and 37) in June 1969. Ten were transferred to Japan in 1958–59 as air-sea rescue boats.

4

Small Fast Craft and PT Origins

The story of American motor torpedo (PT) boats illustrates well the conflict between the requirements of U.S. strategy on the one hand and the perceived potential of new technology—in this case, the internal-combustion engine. The issue is a recurring one in the history of weapon systems development: To what extent should technology push that development? To what extent should requirements, presumably related to strategy, pull development and the choice of weapon systems? Since funding is always limited, the development of any new system has to come at the expense of others that may be better suited to the strategic needs of the navy. The problem arose as early as 1909, at the dawn of the motorboat era, appeared again in 1937 when development of the PTs actually began, and returned in 1945, when they were discarded.

American intelligence about foreign naval technical developments before 1941, except in 1917–18, when cooperation with the British was intense, seems to have been poor at best. Prior to World War I, the navy was not always informed even of the vessels constructed in American yards for foreign powers. For example, information on the impressive performance of a cruiser built in 1899 for Russia by the leading U.S. warship yard, Cramps, came not through domestic channels but from the American naval attaché in Paris. Intelligence on small craft was harder to come by. It was therefore relatively easy for adherents of motor torpedo boats to support their claims by citing supposed foreign developments or interest.

There is another important theme in the story of the PT boat. It is true that the U.S. Navy avoided development of the craft between World War I and 1937 while several foreign navies maintained active interest. However, that did not mean that the United States gave up the development of fast boats; when the PT's time came, the U.S. Navy had much of the technological base needed for its design. Besides commercial yachts, the navy invested in a series of "crash" (air-sea rescue) boats that were unquestionably relevant to U.S. requirements. These boats had to be fast because every minute lost after an aircraft crash could mean a drowned aviator. Since several of the naval air stations faced open water, crash boats also had to be good sea boats. Both features corresponded to those demanded of the PTs developed after 1937. For this reason, although the crash boat was hardly a combatant craft, its development is described in some detail in this chapter. (The large, 63-foot crash boat of World War II was actually used in several wartime and postwar combatant roles.)

The British motor torpedo boats of World War II, closely related to the U.S. PTs, were initially developed because crash boats built for the Royal Air Force showed that fast craft could be made seaworthy enough to be useful. Hubert Scott-Paine, who built Britain's first motor torpedo boats, was initially financed by the Royal Air Force. Scott-Paine is central to this story: the Elco PTs of the World War II were all derived from his 70-foot British torpedo boat of 1939.

By 1908, many yachtsmen had bought motorboats. Compared with existing steam launches, they required fewer and less skilled men in the engine room and so could be either smaller or faster. Motorboat builders, particularly Lewis Nixon, pressed the navy to use such craft, primarily as ships' boats. It was not yet generally understood that war operations would require the assistance of hundreds or even thousands of small patrol craft, even though one navy that had fought a modern war, the czar's, had invested heavily in motorboats, particularly in the United States. Reportedly, these included the motor torpedo boats mentioned in chapter 2, although

Fast motor torpedo craft were possible because it took so little equipment to launch a torpedo. In Shearer's "one-man torpedo boat," which was tried in both world wars, torpedoes swam out of troughs. This is the World War II version (with the viewer looking forward), opened from above as if for reloading.

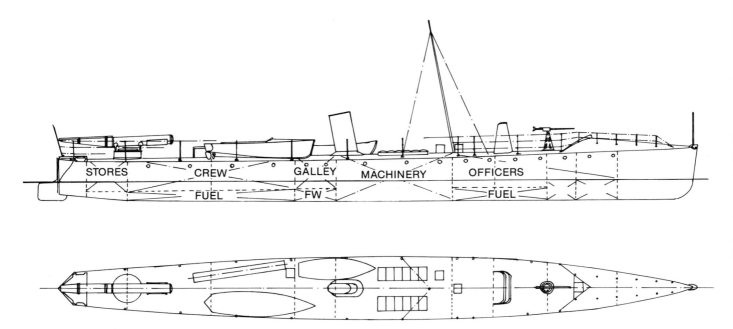

The Satterlee torpedo boat as sketched by Preliminary Design, March 1909. It resembles contemporary steam torpedo boats and includes provision for a reload torpedo. This sketch almost certainly marked the first official U.S. interest in a motor torpedo boat. (Norman Friedman)

no contemporary official navy references to such craft have been located. In 1908, the Office of Naval Intelligence (ONI) compiled a list of foreign naval motorboats, commenting that the Russians had the largest such fleet but not drawing the conclusion that it had been important in warfare.

ONI did report on foreign (private-venture) interest in motor torpedo boats. An 8-ton, gasoline-powered Yarrow-Napier motor torpedo boat could achieve 24 knots with a radius of action of 250 nautical miles, while a 12-ton steamboat was capable of only 18 knots with a radius of 108 miles. The weight was important, as such boats were intended to be redeployed by rail to reach threatened areas of a coast. Another report characterized the motor torpedo boat as 100 feet long, with a 600-hp Napier engine driving it at 21 knots. In January 1909, ONI reported erroneously that Britain was building one hundred 60- to 70-foot gasoline picket boats for coastal defense, each armed with a "fair-sized" gun and each powered by a 120-hp engine.

H. L. Satterlee, assistant secretary of the navy, argued that the new motorboats represented a major opportunity for harbor and river defense because they could be operated easily by semiskilled personnel. He recalled that militiamen, mobilized to operate steam-driven torpedo boats during the Spanish-American War, had often mishandled their machinery enough to reduce speed.

On 31 December 1908, Satterlee officially called upon the Board on Construction to develop a sketch design for a motor torpedo boat with a waterline length of 150 feet. He envisioned storing such boats, possibly disassembled, along the coast to meet the needs of some future mobilization. They might, therefore, best be built of noncorrosive material like bronze. The 300-hp engines then available might, he hoped, drive the boat at 25 miles per hour. Satterlee appended the ONI report to his proposal.

The Board on Construction was about to die. It had long been criticized for its lack of interest in strategic considerations, its concentration on purely technical issues. In 1908–9, the General Board, the war-planning agency concerned with U.S. strategy, was only beginning to dominate the development of the basic characteristics of U.S. warships. It was C&R that proceeded to develop a pair of sketch designs of Satterlee's boat, with accompanying estimates of costs of steel and bronze versions. It submitted the designs on 9 March 1909. In both cases, the power plant consisted of two 300-hp gasoline engines.

More powerful engines of 500 hp would increase speed to, respectively, 24 and 26 miles per hour. Displacement would increase by 7 tons, and cost would be disproportionately greater. In any case, C&R strongly recommended steel-hull construction.

The General Board was appalled. It did not matter how well or how poorly such a boat could be made to perform—the idea of mobilizing small craft to defend the U.S. coast against foreign surface warships directly contradicted the navy's fundamental strategy of sending the fleet out to meet an enemy in distant waters. The board killed Satterlee's idea and, by virtue of its increasing power over the basic requirements for U.S. warship design, precluded any further study of motor torpedo boats.

However, the idea that internal-combustion engines and torpedoes could be combined to form an inexpensive form of coastal defense was attractive. The torpedo was the first nonrecoil weapon capable of killing a large warship, the first weapon by means of which a small boat could become a giant killer. The fact that naval districts maintained substantial numbers of motor and steam launches in peacetime further encouraged the idea.

Table 4-1. C&R Sketch Designs for Satterlee's Motor Torpedo Boat

	1	2
Displacement (tons)	80	45
Dimensions (ft)	150 x 13.75 x 4	115 x 12 x 3.5
Guns	2 6-pdr	1 1-pdr
Torpedo Tubes	1 18-in	1 18-in
Gasoline (gals)	4,000	2,000
Speed (mph)	21	22
Cost ($)		
Steel hull	59,000	37,000
Bronze hull	75,000	44,000

In June 1910, the Navy Department announced its intention to supply the districts with dropping gear (frames) and torpedoes; after testing aboard suitable depot ships or ashore, the torpedoes could be used to arm small craft in wartime. BuOrd tested a method of firing torpedoes from boat falls (weighing about 200 pounds, excluding the davits) and designed a special dropping cage of 2,200 pounds for smaller craft at the Washington Navy Yard, its main development plant. Both methods used a torpedo director in the boat, and in both the torpedo was dropped and the starting lever automatically thrown to the rear by a single operator. The methods were successfully tested in 1915, but Admiral Strauss of BuOrd cautioned that trained personnel were needed to handle torpedoes. Without them, it would be dangerous and useless to issue the weapons to small craft. This caution seems to have killed the project, since none of the motorboats taken over during World War I was fitted to fire torpedoes.

Technology continued to develop. Between 1908 and 1914, several designers in the United States and Europe produced fast motorboats, including early versions of the hydrofoil and the air-cushion craft. The inverted-V "sea sled" invented by a Canadian, W. Albert Hickman, seemed particularly promising, performing better in rough water than its flat-bottom planning contemporaries. There was, moreover, a new naval mission for fast boats: air-sea rescue.

In 1913, C&R ordered a demonstration of a 20 x 5-foot, 65-bhp Hickman sea sled with a speed of 30 miles per hour. A test off Boston on 21 September was sufficiently successful for the navy to order two sea sleds (one designated C224) of 24 feet 6 inches x 6 feet, powered by paired Sturtevant 75-bhp engines. Built to a C&R specification in 1914, they achieved 36.14 and 36.45 miles per hour on trial, exceeding their guaranteed speed by 4, and turned out to be better sea boats than some, more conventional motorboats tested at the same time. Both were employed as rescue boats at the new naval air station at Pensacola. From 1915 on, the navy bought between forty and fifty more sea sled air-sea rescue boats (according to Hickman—the recorded numbers are C233, C273, C275–83, and C1131–40). There were two models, one 32 x 7 feet 6 inches, the other 32 x 8 feet, each powered by two 90-bhp Van Blerck engines. The 32- x 8-foot model, with a guaranteed speed of 37 miles per hour, made 41 to 42 on trials. At the time, these sea sleds were the only fast motorboats in U.S. Navy service; several proposals to buy more conventional boats failed.

With the outbreak of war in 1914, Hickman considered a more combatant role for his boat. He sketched a 54-foot 6-inch x 12-foot motor torpedo boat, powered by four 350-bhp Sturtevant engines, and proposed relevant tactics, which he later claimed foreshadowed those used by Britain's Thornycroft coastal motor boats (CMB). The material bureaus suggested that, as the United States was not yet at war, Hickman should approach the Admiralty with his ideas. He does not appear to have done so, for the CMB, so similar to his motor torpedo boat, has always been credited to three junior British officers.

The General Board was then interested in a motor torpedo boat because, although small, it was by no means limited to coastal defense. Small enough to be carried aboard major fleet units, it might participate in fleet actions, at least under suitable weather conditions. On 17 July 1915, the board proposed formal characteristics for a motor torpedo boat small enough to be hoisted aboard battleships, cruisers, and scout cruisers, each of which might carry two. The boat would scout against submarines, collect information on the whereabouts and movements of the enemy, and launch torpedo attacks in appropriate weather. Maximum smooth-water speed would not be less than 40 knots, and endurance would be 150 nautical miles at full speed. Armament would consist of one 18-inch torpedo and one 1-pound gun.

The board suggested that one boat be bought for experimental purposes, and Secretary of the Navy Josephus Daniels approved on 3 August. General Board characteristics reflected those of Hickman's new sea sled *Viper*, built for John Astor, which could

Table 4-2. Bids on Schedule 9009

Builder	Cost ($)	LMLD (ft, molded)	LOA (ft)	Beam (ft)	HP	Smooth-Water Speed (mph)
Murray & Tregurtha	59,746	46	48.5	11-3	1,300	44
Alt(a)	43,835	42	45	11-3	860	40
Alt(b)	29,980	42	45	11-3	760	38
Thomas B. Taylor	50,000	50	—	12-0	1,600	45 (40 ocean)
Greenport Basin	33,000	60	—	10-0	1,000	41
Alt (a)	19,900	50	—	10-0	600	41
Alt (b)	18,000	42	—	10-0	500	41
Gas Engine Company	39,500	48.5	—	10-0	—	41

make 40 knots in smooth water and 30 in rough. However, in October the navy called for competitive bids (schedule 9009).

The Murray & Tregurtha boats were all sea sleds. Hickman later stated that he had initially proposed a 54.5- x 12-foot boat with four Murray & Tregurtha model J's, a total of 1,600 hp, and a speed of 37 miles per hour, guaranteed at $55,000. The Gas Engine Company bid required the development of a new engine.

The navy chose Greenport's alternative (a), a $19,900 conventional V-bottom boat that became C250. The contract called for a guaranteed 41 miles per hour in smooth water and 34 in a moderate sea, to be sustained for three hours. Endurance would not be less than 150 miles at full power. The boat would be rejected if it failed to reach 38 miles in smooth water, and a penalty of $125 would be paid for every quarter knot below 41 miles. As in many later motor torpedo boats, engines proved to be a major obstacle. Greenport used two experimental 400-bhp Duesenberg airplane engines with problems that took almost a year to fix.

The boat, originally scheduled for completion in 1916, was not delivered until 29 June 1917. On trial, it behaved well but had to stop after 1 hour and 48 minutes when lines to the gasoline tanks, reacting to the working of the hull, broke. Nor did it make the designed speed. However, in view of the generally satisfactory performance and the urgent need for small craft, Assistant Secretary of the Navy Roosevelt accepted C250 on 22 June 1917. The 34-mile rough-water requirement was waived, the boat being assigned to the commander of the Submarine Force for further experiments. Further tests in Long Island Sound showed that the boat, pounding badly in rough water, had to reduce speed drastically.

A second motor torpedo boat (C378) was then ordered to the Hickman sea sled design, but it was canceled at the end of the war in 1918.

In 1915, when it framed schedule 9009, the U.S. Navy was probably unaware of British and Italian work on motor torpedo boats. The CMB was proposed only in the summer of 1915, the first boat launched on 6 April 1916. The Italians built their first two motor torpedo boats in 1915. However, by 1917–18 it was clear that two quite distinct types were emerging. One, typified by the British Thornycroft CMB, relied on very high speed for surprise in attack and evasion afterward. The other, typified by the Italian craft, relied on stealth to get into an attacking position. In effect, the British made maximum use, in terms of power per pound, of the compactness of the gasoline engine. The Italians were sometimes more concerned with the compactness of the *crew* of a gasoline-powered boat, which did not have to include engine-room personnel. In its ultimate form, the Italian idea led to a torpedo boat with a crew of only one.

The U.S. Navy's Greenport boat, although unrelated to British work, might be considered equivalent to the Royal Navy's CMB. The American equivalent of the stealthy torpedo boat was W. Shearer's one-man torpedo boat, tested in 1918. It was only 27 feet long, with a 10-foot beam, and under the best conditions could make only 9 knots. Its forward part was submerged, the torpedo being carried in a floodable well from which it could swim out. A special board found the boat's performance sufficiently impressive to recommend construction of numerous one-man torpedo boats in Great Britain. However, the General Board rejected a proposal that a flotilla of such boats be used against German bases, arguing that labor and material would be better employed turning out destroyers, submarine chasers, submarines, and aircraft.

Shearer's ideas were revived two decades later, when the PTs seemed to have abandoned their origins of small size and stealth. Some felt that his one- or two-man boat promised a return to small size and expendability. In the spring and summer of 1941, the British reported that the Italians and Germans were using one-man explosive motorboats pointed toward their targets and then abandoned by their operators at the last moment. HMS *York* was supposed to have been sunk this way at Suda Bay, Crete, on 26 March 1941. Manned torpedoes were also reported at this time; BuOrd made a proposal to develop a one-man torpedo, asking that BuShips be responsible for a one-man explosive motorboat. Preliminary Design

Greenport Basin built the first U.S. motor torpedo boat, the 50-footer shown here. The hard chine is visible at the waterline, the conventional torpedo tube abaft the open conn. The USS *Rivalen* (SP-63) is in the background.

made weight calculations for the boat in September 1941; its hull was based on that of one of the 25-foot experimental boats designed to aid in PT boat development.

BuShips reported its studies in March 1942. Its 6,500-pound 20-footer could carry 500 to 600 pounds of explosives at 27 knots. A 7,500-pound 25-footer could carry a heavier charge. Meanwhile, ONI reported on the chariot, a two-man torpedo that approached its target just awash, then laid a large explosive charge under it. The Italians used chariots in their successful raid on Alexandria on the night of 18–19 December 1941, bottoming the British battleships *Queen Elizabeth* and *Valiant*.

Although the first notes on the explosive motorboats suggested that Higgins be consulted, someone clearly remembered Shearer and his tiny torpedo boat. In January 1942, the General Board suggested that several miniature semisubmersible torpedo boats be built for immediate trial and development. They would be as small as possible, strong enough to ride over harbor-defense booms, and designed for manufacture at automobile plants, the parts shaped mainly by presses. Delivered to their operating areas by mother ship, they would partially submerge at reduced speed, leaving only the conning tower visible. They would plane at a maximum (escape) speed of 25 knots, and their range would be about 250 miles. Armament would consist of one, two, or three torpedoes, and the crew would be limited to two men.

A parallel set of characteristics for a miniature submarine specified a similar battery, a surfaced speed of 10 knots, and an endurance of 10 hours submerged or surfaced.

In February the CNO, Admiral Stark, recommended that four miniature semisubmersibles be built. He rejected the chariot or miniature submarine. Secretary Knox agreed, but BuShips decided to order only two at first, from Shearer; it would hold off on the other two. The contract was placed on 27 May 1942, and two 45-foot semisubmersible torpedo boats, C 25464 and 25465, were built at City Island, New York, under subcontract to Shearer. Delivered to New York Navy Yard on 1 July 1943, they were stored there until their sale in April 1945.

The small fast boat was also considered for specialized antisubmarine work. In October 1915, with the schedule 9009 vessels in mind, C&R proposed testing a fast boat against existing U.S. submarines.

The stealthier side of torpedo boat tactics was played by Shearer's one-man torpedo boat, presented here in its World War II version (1943). A contemporary PT commander felt that its designers had reverted to the origins of small torpedo craft, in which stealth was the most important virtue.

Although existing submarines could make only about 14 knots on the surface, the bureau was working on the design of a 25-knot fleet submarine. A new Hickman boat, the *Viper* (built for Vincent Astor by Murray & Tregurtha in South Boston), could make 40 knots in smooth water, 30 in rough. C&R suggested that she be tried in the antisubmarine role, and Astor lent her to the navy for that purpose.

In the test, ordered on 19 October, it was assumed that the fast (semidisplacement) boat would carry a single gun, and therefore that it would attack surfaced submarines. Depth charges did not yet exist, and it is not clear whether the U.S. Navy was aware of British work on explosive paravanes. For purposes of the test, it was assumed that the enemy did not control the surface of the ocean, so even small ASW craft could operate freely.

Three situations were envisioned:

(1) A submarine is forced to surface and recharge her batteries. She uses her periscope to check whether the area is clear, which alerts the fast motorboat. Remaining beyond visual range of the periscope until the submarine surfaces, the boat runs for the submarine at full speed, trying to get into effective gun range before the submarine can dive.

(2) A submarine is surfaced, charging batteries, when it is sighted. It attempts to crash-dive before the motorboat comes within gun range.

(3) A submerged submarine, using its periscope, prepares to attack a slow merchant ship with its gun. The motorboat tracks the submarine by observing the movement of the periscope, then attacks with its gun when the submarine surfaces to sink the merchant ship. The submarine tries to escape by submerging.

Trials were held in Long Island Sound on 20 October 1915 against the submarines G-4 and D-2. The results were encouraging:

(1) The boat sighted G-4 at 3-mile range and headed toward it at 40 miles per hour. The boat in turn was sighted at 2 miles' distance. The submarine was completely submerged when the boat was about 1,000 yards away.

(2) D-2 sighted the boat at 2.5 miles and headed toward it at 40 miles per hour. In 30 seconds D-2 began to flood her tanks and, inside 2 minutes, when the boat was still about 1 mile away, the submarine had disappeared.

(3) With the weather hazy, the boat was not observed by D-2 until it was a little over a mile away; it reached the submarine before the conning tower had submerged. This was held to show the need for great speed.

To the General Board, these experiments confirmed reports from the war zone on the value of fast motorboats, in combination with nets and trawlers, for offensive action against submarines. (Fast motorboats at this time probably meant Elco boats.) However, the board cautioned that technology should not be developed in a strategic vacuum. The United States was far from its potential enemy, Germany, and as a result, it seemed unwise to lavish resources on specially built ASW motorboats. Nonetheless, the fleet might well operate overseas, and if so, it would have to be protected against submarine attack when at anchor or in a blockading position. Fast motorboats, which, as the board saw it, could be carried overseas by fleet units, would be a valuable defense. The board proposed that C&R develop appropriate plans after trials had been conducted.

Finally, the Americans considered using small, fast boats to launch attack aircraft from the North Sea. The Royal Navy had already towed aircraft-bearing lighters from destroyers, launching fighters at speeds of about 35 knots. The original plan was to tow seaplane bombers astern of destroyers. Sixty special lighters, designed to flood so that seaplanes could float free, were ordered. After prototype lighters were observed by German Zeppelins, the project was canceled. Existing lighters were converted to cargo barges for use in congested European ports.

Captain H. C. Mustin suggested a variation on this theme, a fast self-propelled lighter from which a land bomber could take off. He had been impressed by the smooth ride of a 50-foot sea sled and was assigned to develop a larger version from which a 10,000-pound bomber, which took off at 55 miles per hour, could fly. Hickman's airplane carrier was 55 feet long, with a 12.5-foot beam and sponsons extending to a width of 18 feet. It was driven by four Murray & Tregurtha model J aluminum-base engines producing a total of 1,800 bhp. Two were built, but the contract stopped with the armistice.

The U.S. Navy did not order any more domestically built, fast motor craft until 1928. It did buy two British-built boats. In May 1919, hoping to find a postwar market, Thornycroft approached U.S. Navy headquarters in England with a proposal to sell three CMBs representing all three existing variants: the 45-foot, 55-foot, and 70-foot. CMBs had just succeeded spectacularly against the Russian fleet in Kronshtadt Harbor, and their wartime record had been publicized since the end of the war. At the same time, Thornycroft offered several destroyer leaders, the construction of which had just been canceled by the Admiralty. C&R and BuOrd wanted to buy the CMBs, but BuEng disagreed. It recalled the General Board's reasons for rejecting the Shearer one-man torpedo boat and argued that the same logic applied to the CMB.

The other two materiel bureaus persisted, carrying the argument to the secretary of the navy, Josephus Daniels, in the fall of 1920. CMBs would be carried aboard U.S. warships to overseas points of operation, as in the raid on Kronshtadt. The General Board suggested that purchase of experimental boats would be justified in view of the wartime record of the CMB and the probability of its future value. But adverse comments on the big 70-footer made by the Admiralty and Thornycroft itself prompted the bureaus to reduce their proposal to two boats, the 45- (C1608) and 55-foot (C1609). They ran acceptance trials in England, reaching the United States in 1922. Trials before a special board that April showed maximum speeds of, respectively, 32.25 and 35.13 knots, with a 200-nautical-mile endurance at such speeds. They were considered good sea boats in moderately rough coastal waters.

Trials were marred only by the failure of attempts to fire torpedoes with their gyros angled. The Thornycroft fired its torpedo over the stern, tail first, swerving out of the way to let the weapon pass. The method worked because the torpedo took some time to accelerate and thus overtake the boat. Torpedoes fired this way took 500 to 800 yards to settle at their preset depth, so a CMB could not hope to attack from very short range. Gyro angle on the torpedo would allow the boat to fire at a much lower, hence less

Greenport's 50-footer can be contrasted with Thornycroft's 45-foot CMB, shown here as a postwar crash boat in March 1934, with a light derrick forward. As a torpedo boat, the CMB launched its weapons tail first down stern ramps.

detectable, speed. Both boats were therefore assigned to the naval torpedo station at Alexandria, Virginia, for further tests. The station found them reliable; in January 1923, it reported that the 45-foot CMB had run for 135 hours with nothing worse than spark plug trouble. The 55-foot CMB did, however, require a full engine overhaul.

Meanwhile, the special trials board reported that the U.S. Navy ought to maintain a small number of CMBs for peacetime development. Unlike the flimsy Thornycrofts, they should be durable craft, built of steel. The board considered the CMB a special type that could attack under favorable conditions. Aircraft, the alternative torpedo carriers, were more easily detected and might even be used as decoys against enemy defenses during a CMB attack. The CMB could be expected to stop and creep and feel her way in fog better than a seaplane.

The board suggested that, in wartime, future CMBs be carried aboard auxiliaries rather than warships. It rejected the 45-foot CMB as too small. Finally, because the CMB had aircraft engines, and for that matter, was constructed by aircraft methods, it would require the level of maintenance expertise typical of that used for air squadrons. It seemed most natural, in any case, for future CMBs to work with naval aircraft.

OpNav's war plans division found the CMB particularly attractive. In January 1924, it advised the CNO that this craft would be valuable for defending outlying U.S. Pacific possessions and the advanced bases the fleet would need as it executed the Orange plan. Considering base defense crucial, the division urgently recommended that the two Thornycroft boats then in storage at Norfolk be reconditioned and shipped to the Asiatic Fleet, first for operational tests and then for Philippine defense. They were, instead, shipped to the U.S. Fleet at San Diego to have their strategic and tactical value tested.

The U.S. Fleet, however, did not conduct further tests. The two Thornycrofts were briefly tested by the naval air station at San Diego, and about December 1926, they were assigned to Destroyer Squadrons, U.S. Fleet, for torpedo recovery. They were laid up in June 1927.

For the next decade, the U.S. Navy developed fast

boats only for air-sea rescue. This did not spell the end of fast boat construction in the United States, though. People still wanted pleasure and racing craft, and rumrunners bought wartime CMBs and built new fast craft. To catch them, the Coast Guard bought destroyers and fast patrol boats. Although most Coast Guard hulls were not well adapted to high speeds, the program to defeat the rumrunners gave rise to one of the important small boat engines of this period, the Hall-Scott Defender.

The crash boat program involved both engines and new hard-chine hulls. Although most of the 32-foot sea sleds bought for rescue work during World War I appear to have been discarded soon after its end, as late as 1928, Quantico Marine Air still had one such sled, and Pensacola had six of the 21-foot version. For virtually all rough-water rescue, the coastal naval air stations had to rely on slow whaleboats. The Bureau of Aeronautics (BuAer) began to seek fast rough-water crash boats around January 1926. That September it asked the major naval air stations to frame requirements.

Speed, 25 to 30 knots in 2- to 3-foot seas, was essential. An injured pilot could easily drown before a slow boat reached him. He could easily die if he did not reach a hospital in time. A crash boat would also have to carry divers and a light, hand-operated salvage crane, powerful enough to lift wreckage to get at a pilot. The new boat would have a crew of two as well as two medical attendants. Its open cockpit would have to accommodate at least two stretchers.

Successful boat design would be a combination of a seaworthy hull and a compact, powerful engine. BuAer wanted to use one of its aircraft engines—the existing Liberty, the Wright T-2 (which had been adapted for marine use), the Curtiss V-1400, or the Packard 1500—to simplify maintenance. The Wright and Curtiss were the only of these engines not to enter U.S. naval service. The Liberty was developed into the Vimalert, which figured in several designs for American small-attack boats. The Packard, enlarged, became the 4M 2500 (V-12), which powered U.S. PT boats in World War II.

In January 1927, OpNav estimated that the six major naval air stations (Hampton Roads, Anacostia, Coco Solo, Pearl Harbor, San Diego, and Pensacola) and the aircraft of the battle and scouting forces would require a total of sixteen crash boats. Four were provided under the FY 28 program, and another two were planned for FY 29.

In March, C&R produced three sketch designs, for 40-, 45-, and 55-foot crash boats. Based on the wartime Thornycroft CMBs, the boats would carry salvage equipment that would probably weigh the same as the torpedoes for which the CMBs had been designed. Experience with the Thornycrofts showed that they could operate in rough water without serious damage, and C&R proposed that they form a point of departure for future development. The converse was also true: the crash boat was much more heavily loaded than civilian speed boats, and its load was not much different (in magnitude) from the *military* load of a future motor torpedo boat. Although U.S. PT-boats did not develop directly from BuAer crash boats, the hydrodynamic experience gained in crash boat development was applied to the U.S. PT program. British crash boats, however, did lead directly to motor torpedo boats and, incidentally, to U.S. Elco boats.

The two Thornycrofts were ordered surveyed for conversion to crash boats, and the 45-foot boat was shipped to Pensacola in 1928, where it proved very satisfactory, serving until 1934. The 55-foot CMB, not converted, was condemned in 1934.

It appeared that the C&R proposal would take at least two years to develop; BuAer had needed crash boats as much as a year earlier. It wanted to buy existing commercial boats, saving money by using its own engines. Three Marshall hard-chine, V-bottom 34.5-foot motorboats were bought at once (C1656–58) to supplement standard navy-issue 35-foot motorboats, which were slow. The Marshall boats had a maximum draft of 3 feet (for shallow water) and could carry four men at 35 knots (40.3 miles per hour). Maximum capacity was thirty-five men. The boom on the bow, with an outreach of 4 feet and a maximum hoist height of 5 feet above water, could lift 3,000 pounds. Powered by single Liberty engines with a bhp of 368 (the Thornycrofts had 360), they could sustain 30 knots or more for the stipulated three hours. In general, they typified the requirements of the time, though, with a tendency to dig their noses into a sea, they were much poorer seaboats than CMBs.

On sea trials at Pensacola, the 45-foot CMB performed impressively because of its greater length and a planing surface that began close to the bow. Pensacola reported that the way it behaved at high speed in the bay's rough waters "was a revelation to all who witnessed it." The CMB also handled much better than the Marshall. However, BuAer, instead of buying more CMBs, sought better hard-chine commercial boats. It bought eight in 1929 and, in addition, ordered the conversion of two fast, hard-chine torpedo retrievers, C268 and C269, built by the Albany Boat Corporation in 1917. They had made 35 miles per hour when powered by Van Blerck's J-8 163 215-bhp engines, which were no longer in production; these engines would be replaced by the Liberties.

The big Marshalls were relatively expensive, and BuAer needed numerous boats to serve its shore sta-

The Luders 45-footer won the 1936 crash boat competition. Note its portable crane aft and the stretcher inside the after cabin. These boats were unarmed, being employed only in American coastal waters. The wartime 63-footers were armed because they often had to operate in contested waters.

tions as well as the fleet proper. In 1929, it began to buy 26-foot Dodge Water Car runabouts (C1667–74). They were too small, although they were considered sufficiently maneuverable and fast. BuAer split its requirements into small craft for inshore work and larger ones, like the Marshalls, for rougher water. In July 1930, the bureau estimated that it needed nineteen large and five small boats; the navy's inventory was four large and twelve small boats (all Water Cars). In the future, then, BuAer wanted to buy only large crash boats.

In 1930, it set requirements for the boat: a minimum high sustained speed of 35 knots in moderately rough water, a 3-hour cruising radius at maximum speed, good maneuverability at all speeds, and ample stability and seaworthiness so that it could put to sea in fairly heavy weather. It would have to accommodate four injured personnel, a doctor, a medical corpsman, a coxswain, and a rescue crew. The bureau also discontinued the term "crash boat" in favor of "aircraft rescue boat" and "aircraft salvage craft."

By this time, C&R was working on its own 35-foot rescue boat. Bids for two boats approximately based on this design were accepted in September 1930. Scruggs Boat Company built the 36-foot C1704, powered by a single Sterling engine. It attained 33 miles per hour. The C1705, built by Robinson Marine, made a similar speed on its Hall-Scott engine. Both boats could carry twelve men, more than enough to accommodate a crew of three, a doctor, a corpsman, and the five-member crew of a crashed patrol bomber. BuAer complained that they were the two slowest boats it had bought since 1917.

The navy's urgent requirements were not being met. Since 1919 it had obtained only five new large boats and one conversion, the CMB, which was ten years old. Making matters worse, the two Scruggs boats performed poorly. Based on the recommendations of the naval air stations and the fleet, BuAer decided early in 1933 that it needed something substantially larger, a 40- to 45-foot boat powered by two Liberties, capable of 45 miles per hour (sustained speed, 38), and with a cruising radius of 125 nautical miles. C&R objected that such a boat would be too large and expensive, but the new requirements were accepted.

C&R designed a 45-foot hard-chine crash boat, powered by two 650-bhp Packard Vimalert (3A 2500) conversions, as a basis for comparison with commercial designs submitted by bidders. In March 1932, Luders Marine had already proposed a larger rescue boat, citing foreign work on 55- to 65-foot boats capable of 50 to 60 miles per hour. It won a contract for two 45-footers (C1732 and C1733) in 1933, and BuAer bought a second 36-footer, the "Seagull," from Robinson.

Luders' first 45-footer arrived at Norfolk on 30 December 1935. It had two Packard engines, was

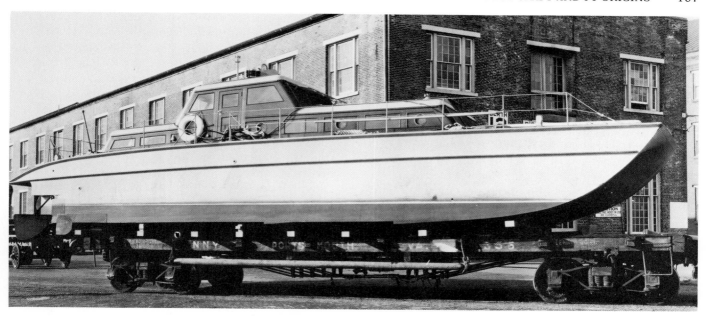

The sea sled was probably the most uniquely American contribution to the development of small fast craft. This large sea sled was entered into the crash boat competition in 1936. The broadside view shows the bow end of the tunnel running between the two hulls.

capable of 1,300 nautical miles at 40 knots, and displaced 12.5 tons. Dimensions were as follows: 44 feet LWL/45 feet LOA x 10 feet 4 inches x 2 feet 7 inches. This boat, the 36-foot "Seagull," and a later 29-footer for inland and river bases were the standard U.S. Navy crash boats at the outbreak of war in 1941. The Luders 45-footer, moreover, performed so well that its hull form was the basis for the C&R motor torpedo boat design work begun in 1937.

The Hickman sea sled was the only real competitor of these hard-chine designs. Hickman had turned over his rights to the design to a boat builder immediately after World War I, and several small sea sleds were built as pleasure craft in 1918–32. He regained control in 1933 and again approached the navy. The sea sled apparently had a good reputation; in October 1934, BuAer considered purchasing one for comparison with its new 45-footer. The sea sled, which displaced 16.4 tons, had four 1,080-hp Hall-Scott gasoline engines, and was capable of 39 knots, was delivered in January 1936. At this time, the chief constructor, Admiral Emory S. Land, was sufficiently impressed by the value of the sea sled to forbid a license for its construction in Japan.

The hard-chine boat won the trials, conducted off Hampton Roads, decisively. The Luders handled well at all speeds, giving a much smoother and softer ride than the Hickman Sea Sled. Its principal drawback was a tendency, in a quartering sea, to shoulder in and slew around at speeds exceeding 40 miles per hour. The Hickman yawed badly and was sluggish in answering the rudder at high speeds. At low speeds, the sea sled was practically unmaneuverable in any wind. In a chop, it pounded so sharply and severely that the trials board considered it unsuitable for transporting injured personnel. The trials board concluded that it was defective for several reasons: pounding, the excessive loads that its design exerted on its structure, limited vision, sluggishness, and poor handling qualities. C&R and its successor, BuShips, never again considered buying sea sleds, and they did not figure in the PT program.

Hickman did not give up. The C&R records suggest that he was personally acquainted with both the chief constructor and Admiral King, then chief of BuAer, and that the 1936 trials themselves had been made only at his insistence. He did not submit a design for the July 1938 torpedo boat competition, on the ground that the stated requirements excluded the inverted-V or sea sled type. The navy, forced to modify requirements for the projected short boat, offered to build a 25-foot sea sled at its own expense to test for maneuverability and sea-keeping. However, when the navy advertised for builders to construct Philippine torpedo boats in 1938, it did exclude inverted-V hulls.

Hickman returned in 1942, asking for another competitive test. The navy officially refused, but Captain Swasey and Commander R. B. Daggett rode a 36-foot sled. BuShips observed that the sea sled hull structure was fragile, and that floating objects would tend to be sucked into its tunnel and then into its propellers. In December, it rejected Hickman's offer to build a 40-foot sled with the lines of a 75-foot torpedo boat. But the sea sled did have strong adherents. Both the Mexican and the Soviet governments asked for sea sleds under lend-lease, ony to be told that procurement would be limited to standard U.S. Navy types.

The Army Transportation Corps bought its own boats, including small ASW craft. It apparently found sea sleds attractive. In March 1943, it bought a sea sled antisubmarine boat, 36 x 9 x 4.5 feet in depth, powered by two 150-bhp gasoline engines (army design 303). This was followed in June by design 386, an experimental ASW sea sled, 39 x 9.5 feet, powered by two 150-bhp engines. The army also bought thirty smaller sea sled rescue boats of the 27-foot type—in feet-inches, 31-10 x 7-6 x 3-9⅛. They were completed between December 1943 and September 1944 as design 387.

Finally, the Army Transportation Corps ordered a 70-foot sea sled. From January through March 1945, the sea sled and other navy, army, and Coast Guard types were tested under severe weather off Rockland, Maine. Unfortunately, the army operator changed the interlocked throttle and steering control, causing the boat to skid, and a man was lost overboard. En route the big sea sled grounded and, entering harbor, ran over a nan buoy that damaged it. Although the sea sled was repaired, it had to be withdrawn in sinking condition.

Hickman now complained to Senator Saltonstall of Massachusetts, relenting only when he was told in September 1945 that no more motor torpedo boats were to be built. Even that was not the end of the story. A further series of crash boat tests was held in 1952–53 in Chesapeake Bay. This time, a 55-foot sea sled, C7075, ordered in June 1952 and powered by two Packards, was severely damaged and could not complete the tests. It was declared surplus in 1956 and sold in March 1957.

The British crash boat story is relevant to this history because it led directly to one of the two standard U.S. World War II PT types, the Elco. In the British case, the Royal Air Force (RAF), with responsibility for shore-based maritime aircraft, was theprimary user. As in the United States, just after the war the fastest British boat capable of carrying a considerable load was the Thornycroft CMB, and the firm that made it continued to supply fast launches, of non-CMB design, from 1927 to 1935. Another fast-boat builder appeared, British Power Boats. It was founded in 1927 by Hubert Scott-Paine, former owner and managing director of Supermarine, who initially sought to mass-produce civilian motorboats. Scott-Paine was also well connected with Napier, the aero-engine manufacturer, through work at Supermarine, and he was a founding director of Imperial Airways, a market for motorboats (flying boat tenders).

In 1930, Scott-Paine offered the RAF a 37-foot, 23-knot, hard-chine launch—about the size of the U.S. boats but considerably slower. Like the U.S. Navy, the RAF came more and more to demand good seakeeping performance. By 1934, it was seeking eighteen considerably larger boats, and in June 1935, it issued a specification for a 35-knot, 60- to 70-foot boat with a range of 500 nautical miles. The winning design, from British Power Boats, had a range of 500 miles at 32 knots, 800 at cruising speed; it was powered by three 500-bhp Napier Lion aircraft engines. Compared with a CMB, this rescue launch carried a considerably larger crew and was designed to remain at sea for forty eight hours or more.

By this time, the Admiralty had already ordered a motor torpedo boat version of the same boat. It formed a coastal motor boat committee in October 1931, and in September 1933, the committee proposed that a new CMB be designed for possible wartime production. The project would also stimulate the design of suitable high-powered marine engines. The recommendation was rejected on the ground that a single boat would prove very little, while existing civilian efforts would suffice to maintain progress in specialized small craft. Even so, a year later Scott-Paine sold two 60-foot motor torpedo boats, essentially armed rescue launches, to the Admiralty. These broke with their CMB forebears in being too large for transport overseas in davits. Four more were ordered when Italy invaded Ethiopia a year later, and all six were commissioned in June 1936. By that time, they had been reclassified as motor torpedo boats rather than CMBs, reportedly to reflect their much superior open-ocean performance. In June 1937, the First MTB Flotilla, with six boats, steamed directly from Britain to Gibraltar and Malta, a widely publicized performance.

These boats showed that an MTB could be a good sea-keeper, but they did not yet realize the potential of the hard-chine hull. By August 1936, the Admiralty was seeking a larger boat capable of carrying two 21-inch or four 18-inch torpedo tubes. British Power Boats and Vosper competed, and the latter won. The losing design became the basis for the U.S. Elco PTs. Vosper's winner was later built in the United States for Britain under lend-lease.

The Royal Navy also developed a much larger and slower seagoing Camper & Nicholson MTB, which, had little if any impact on later U.S. development.

One other U.S. development of this period deserves mention as the last gasp of the World War I concept of the MTB. While the first of the new battleships, the *North Carolina*s, were being designed in 1936, there was some interest in providing them with one or two small boats that could be rigged alternatively as ferries carrying thirty officers or as torpedo boats carrying two torpedoes or nine depth charges. C&R produced sketch design, dated 30 December 1936, but the General Board never approved the concept. For Preliminary Design, the experience was valuable because it led to the collection of information on fast planing hulls. Aluminum construction was considered for both the framing and the hull skin itself. Also, the existence of the design probably convinced the General Board that a small motor torpedo boat was practicable, and so influenced its choice of characteristics for the PTs.

The 1936 boat would have had the following dimensions in feet-inches: 53-0 (lwl), 55-0 (loa) x 10-0 x 2-11 (20 tons). One Hall-Scott and two Allison engines, 2150 bhp, would have driven it at 34.8 knots. Armament would have comprised one 0.50-caliber AA machine gun and two 22-inch (aircraft-type) torpedoes or nine 300-pound depth charges. The complement would have been six enlisted men.

On 19 June 1936, while the 53-foot study was proceeding, the acting CNO, Captain William S. Pye, asked C&R to consider a CMB that might be designed by either the navy or civilian naval architects.[1] The origin of this proposal is unclear, although it may have been a consequence of contemporary British work. The U.S. CMB would be useful for the following:
- Local defense of U.S. coastal or island bases, in which the CMB would guard channels and patrol likely submarine operating areas. It would need at least eight 420-pound depth charges, a machine gun (Pye specified a single 0.50-caliber weapon with ten thousand rounds of ammunition), and a sound or radio submarine detector.
- Attacks against surface raiders and scouts. The boat would need high speed, at least 40 knots, not fewer than two aircraft-type torpedoes, which could be replaced by depth charges, and a smoke generator.
- Support of landing operations, for which it would need a smoke generator, machine guns, and high speed, if possible.

Pye doubted that one size of boat would suffice. He proposed two, a large one for continental coastal waters and a small one that the fleet could transport overseas.

He sought a radius of action, at full speed, of at least 50 nautical miles, and a radius of 350 miles at an economical speed of not less than 6 knots. His boat would carry stores and water for five men for four days. Its radio would permit communication with a base or with similar craft at ranges of up to 100 miles. Finally, it would be suitable for hoisting aboard larger ships by sling.

C&R made a preliminary study, details of which are not recorded, and later in June replied that Pye's tentative characteristics could be attained, but only using a light hull that might not be suited to open-sea operation, not "even in the trade winds met in most of our strategic areas." The speed requirement would have to be reduced to 37 knots.

To ensure against too flimsy a hull, C&R proposed that the full-power, sustained-speed requirement be stiffened. The bureau proposed that the new boat be required to make 80 miles at 90 percent power in seas of up to 4 feet. It would have to maintain its course in seas this size on any heading. Moreover, this performance would have to be attained using existing marine engines.

In any case, the economical cruising radius would almost certainly determine fuel stowage, so the proposed 75-minute full-power radius was clearly too short. The bureau proposed 80 miles (two hours). On the other hand, to reduce loading, it might be well to cut 0.50-caliber ammunition from ten to five thousand rounds (saving 2,000 pounds), that is, to ten minutes of continuous fire. At this time, the standard U.S. Navy shipboard requirement was one thousand rounds per 0.50-caliber machine gun.

Nothing more was heard of the American CMB until December 1936. That may have been owing in part to a presidential order calling for large funds to be reserved against the possibility of deep budget cuts. As a peripheral project, the CMB would certainly have been dropped. On 5 December, C&R suggested to OpNav that a new study of motor torpedo boats would be worthwhile. The bureau cited foreign developments showing that such boats could be extremely effective in local defense and noted that motor torpedo boats assigned to the defense of an advanced base might release scarce, larger seagoing craft for other duties—virtually the same argument that Admiral Hepburn would make for PCs nine months later. This C&R letter was probaby the result of the small MTB study mentioned above.

The bureau proposed an experimental program of two motor torpedo boats per year, one built to a private design and one to a bureau design. BuEng reported that it was already interested in applicable engines, partly for crash boats. It hoped to receive funds to develop a diesel for motor torpedo boats.

Pye, again acting CNO, submitted the issue to the General Board via the secretary of the navy on 5 January, since formal characteristics would be required before development could begin. Again, it is not clear just what startling foreign developments C&R had in mind; most of the European navies, notably the British, had virtually abandoned CMBs after World War I. The earliest German S-boats were known (C&R listed a 92-foot, 50-ton boat), as were Thornycroft's export boats and the new British Power Boats craft (12 tons, 60 feet long). The bureau emphasized the development of hard-chine displacement boats, as opposed to the stepped-hydroplane CMBs, and mentioned one important U.S. commercial development, the 33-foot, 38-knot boat *Tintavette* (1936). She was designed by J. Starling Burgess as a large-scale model of a 750-ton, aluminum-hull destroyer proposed in November 1936 by the Aluminum Company of America (Alcoa). The *Tintavette* reportedly had excellent sea-keeping qualities. C&R did not mention its ongoing work with crash boats, although they figured prominently in internal memoranda.

The General Board, in turn, viewed torpedo boats themselves as valuable primarily in the later stages of a war—when the U.S. Fleet had crossed enough of the Pacific for its advanced bases to be within range of Japanese forces, and when base defense would represent a considerable burden. The precise course of a future war could not be predicted; it might indeed be possible for such small craft to be used offensively. However, the board was not so sure that the local defense role, which, at least early in a war, would mean ASW above all, would require anything as complex as a torpedo boat. Small civilian craft might easily be converted, as in World War I.

In peacetime, then, the General Board would not sanction development beyond prototypes; there were too many other urgent requirements. The board foresaw two types: one small enough to be transported overseas by auxiliaries or by cargo ships, and a larger one with better endurance for offshore work. The larger type might well be derived from the Philippine patrol boat then under development (see below).

Although the board never said as much, the smaller one would clearly be related to the 1936 53-foot design. The boom capacity of fleet auxiliaries, or that likely to be available in other ships, set the maximum stripped weight at 20 tons. Length would be about 60 feet, and speed about 40 knots, which is what Pye's early correspondence indicated. The boat would carry one or two torpedoes, with depth charges and listening gear an alternative. Since maximum rough-weather performance was desired, the hull would be a hard-chine displacement type rather than a stepped hydroplane. The board gave no endurance requirement, but it specified stores for five days at sea for a crew of two officers and eight enlisted men. It supported the C&R experimental program and recalled the importance of private designers in World War I patrol boat programs. It would, therefore, be important for the navy to keep in touch with U.S. boat designers and builders so that they could be mobilized easily in an emergency.

Meanwhile, C&R began work on what was to be its first serious MTB project, a patrol boat for the Philippines. The course of the design illustrated some of the compromises that had to be made in all later designs. At this time the available engines were the 550-bhp Hall-Scott Defender, the 600-bhp Vimalert (soon boosted to 760 bhp), and the 800-bhp Packard converted aircraft engine. Hall-Scott also manufactured a smaller 275-bhp Invader.

The major threat to the Philippines was clearly invasion. General Douglas MacArthur, commander of Philippine forces, could not afford a conventional navy, but motor torpedo boats were an attractive substitute. He proposed an offshore patrol of about thirty MTBs. In peacetime, they would form a coast guard. In wartime, they would attack transports attempting to unload troops. They would be built over the ten-year period leading to the islands' scheduled independence.

MacArthur visited Washington in February and March 1937, at which time he asked the secretary of the navy for assistance. His naval assistant, Lieutenant S. L. Huff, had already submitted a sketch design for a fast motor speed boat; it was presented on 1 March at a meeting in the office of the chief constructor, with representatives of BuEng and BuOrd present. Not long afterward, the secretary approved MacArthur's request that C&R prepare a design for which it would be paid three thousand dollars by the Filipino government.

Huff envisaged a relatively long-range metal boat with a minimum radius of action of 500 nautical miles at full speed, about 45 knots, and 700 miles at a cruising speed of 10 to 15 knots. Complement would be the same—two officers and eight enlisted men—as that called for in the tentative General Board characteristics, and the boat would carry at least two torpedoes (18- or 21-inch), two 0.50-caliber machine guns, and two depth charges. It would be 60 to 70 feet long, with a beam of about 14 feet and a shallow draft, about 3 feet 6 inches. A power plant consisting of not more than three engines would be controlled from grouped one-man controls at the bridge, and the boat would have a minimum of four watertight compartments. MacArthur wanted the boats built either in the United States or at a Philippine navy yard, preferably Cavite.

At the conference, it was pointed out that 45 knots was impracticable, and the speed requirement was scaled down to 40. High speed had been considered essential for the boats to escape destroyers. The specified cruising radius would have entailed an inordinately large fuel load, so it was cut to 250 nautical

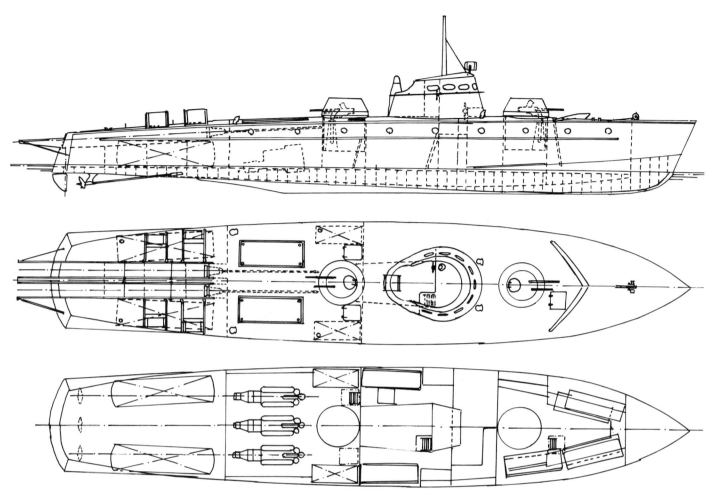

General MacArthur's Philippine patrol boat, designed in the Philippines by MacArthur's naval assistant, Lieutenant S. L. Huff, USN (Ret.), and presented to MacArthur in January 1937. Dimensions were 70-0 x 14-0 x 3-6 feet-inches. Note the enclosed machine-gun turrets. (Norman Friedman)

miles at top speed, 250 at cruising. The possibility of using an aluminum hull was mentioned, and it was concluded that this would be satisfactory, provided that the difference in cost between metal and wood was not too great. The shallow-draft requirement was relaxed to 4 feet. Another requirement, for protection in the form of bullet-proof steel and revolving shields for the machine guns, was given up, as it would have required too much extra weight.

The three-engine requirement was dropped because initial estimates called for 4,500 bhp, which was well outside the performance of any three available engines. Given the possibility of using more engines, however, BuEng reported that gasoline engines under development could provide power in excess of that required to achieve MacArthur's performance. Later boats could be powered either by the liquid-cooled gasoline aircraft engines then under development or by the diesels the bureau had mentioned three months earlier in connection with a U.S. motor torpedo boat.

In a small, high-powered boat, the power plant arrangement effectively determined length and beam, as the length for the general arrangements forward and for the torpedo stowage aft was, within narrow limits, the same for all alternatives. By placing the machinery units three abreast (two engines per shaft), a length of 70 feet would suffice on a minimum acceptable beam of 15.1 feet. Using 550-bhp engines, a total of 3,300 bhp, a 70-foot boat would make about 40 knots.

The lines of this 70-foot boat were derived directly from those of the Luders 45-foot crash boat, but with slightly less beam in proportion to length. Model tests at this time confirmed that such a hull was indeed as efficient (hydrodynamically) as the much less seaworthy CMB-type stepped planing hull. The stepped plane would indeed be superior at a high-enough speed, but that point would be reached only well beyond the range contemplated and only at the cost of severe pounding in any sea.

Even so, there was some fear of severe pounding in rough water, and an alternative boat, longer and

Bunker- or barge-busting was a problem in both World War II and Vietnam. The sextuple bazooka, a 2.36-inch rocket launcher, was one possible solution. Here it is mounted abaft the mast of PT 354 (MTBRon 25) at Coco Solo in the Canal Zone, 17 December 1943.

thus more seaworthy, was studied. Dimensions were 96 x 12.4 x 3.84 feet, instead of 70 x 15.5 x 3.26; displacement, 94,000 instead of 87,000 pounds; and metacentric height, 4.8 instead of 7.3 feet. The boat's proportions were similar to those of the successful Burgess yacht. However, Preliminary Design felt more comfortable with the Luders hull, which had been thoroughly tested. The Burgess boat was considered successful, but the bureau did not yet have its lines. The longer boat was roomier, with a better engine room layout. It could be expected to behave better in a seaway, and its lower center of gravity might reduce heel on turning. The shorter boat would be a more maneuverable, smaller target. Estimated costs were $112,000 for the shorter boat, $122,000 for the longer. At full load, it was estimated that the shorter boat would make 40 knots, the longer, 38.

Captain Chantry, head of Preliminary Design, rejected the longer boat on account of its reduced maximum speed and greatly increased length, although he was willing to try an 80-foot boat as a compromise.

Preliminary Design also tried an alternative engine arrangement in which the wing propellers were driven by four 800-hp Packards, two per shaft in the forward engine room, and the after engine room contained two 550-hp marine engines abreast, driving the center shaft. These two engines would suffice for cruising speeds up to about 16 knots. The total of 4,300 bhp would fit in a 79- x 12-foot hull and would make for a maximum speed of over 40 knots. This design was carried as far as that of the 70-foot boat.

Preliminary Design soon found that only limited general data on planing boat hulls were known, and it outlined a model basin test program. The model basin had actually started an even more elaborate series of experiments in 1925 but had to stop because of other business. With 80 percent of the necessary models completed, work could begin at once. Early experiments, run on 24 May, suggested that the 79-foot boat would require about 20 percent more power for 40 knots; it was therefore abandoned.

As for the power plant, BuEng decided that six 550-bhp Hall-Scotts would badly crowd the engine room. The alternative was four 800-bhp Packards or 760-hp Vimalerts on two shafts, with a small cruise plant for fuel economy on the center shaft. Captain Chantry suggested using two 275-hp Hall-Scott Invaders, geared to a centerline shaft and driving a variable-pitch propeller. With 760-bhp Vimalerts, that gave a total of 3,590 bhp. This plant was approved in mid-June 1937. The final version of the preliminary design showed dimensions, in feet-inches, of 70-0 (lwl), 72-0 (loa) x 15-1 x 3-1 for a displacement of 40 tons and a speed of 40 knots. Armament was two 0.50-caliber machine guns (four thousand rounds), two 21-inch torpedo tubes, and four 420-pound depth charges. Complement would be one officer and twelve enlisted men.

In July 1937, Lieutenant Colonel James B. Ord, assistant to General MacArthur, informed the secretary of the navy that the Philippines wished to buy two boats in the United States, with an initial investment of $35,000. That November, C&R circulated plans to interested builders. They showed an all-aluminum alloy hard-chine boat with a displacement of 106,400 pounds and a draft of 3 feet 7 inches fully loaded. Armament consisted of two 18-inch torpedoes fired stern first, as in World War I CMBs, two 0.50-caliber machine guns (with half the ammunition stowed in a magazine and half in ready service), and up to four depth charges. The cumbersome six-engine power plant, required if U.S. engines were to be used, would drive the boat at 40 knots without any reserve of power; the two 275-bhp Invaders would drive it at 15. Estimated range was 250 nautical miles at maximum speed and 300 at 12 knots.

C&R wanted the first one or two boats built at Norfolk Navy Yard, to gain experience; the yard estimated costs at $128,800 for the first and $106,500 for the second boat, without armament.

All of this data was sent to the Philippines, but MacArthur showed no great interest in it. He was asked to return the specifications the following May.

Meanwhile, he negotiated with Thornycroft, ordering one 65-foot and two 55-foot boats in April 1938 at a unit price of less than $70,000. The 65-foot *Luzon* (Q III) made 41.19 knots on trials and was shipped to the Philippines via Antwerp in April 1939. The 55-foot *Abra* (Q I), delivered in January 1939, made 46.533 knots on trials. The three Thornycrofts, under army control, participated in the defense of the Philippines alongside U.S.-built PTs. Q III, sunk at Cavite in May 1942, was raised by the Japanese and recommissioned as no. 114 on 12 April 1943. It was sunk in the Philippines in 1944 or 1945.

Another two boats, Q 2 and Q 3 (the *Abray* and the *Aguasan*), were ordered on 10 April 1939. They were requisitioned by the Admiralty after the outbreak of war that September, then released on 1 October. In January 1940, they were sold to Finland and replaced by two more boats. By that time, the Philippine government had already ordered three more, and a total of five were being built in the summer of 1940. The Admiralty took over, first, the two ex-Finnish boats, then the five Philippine boats (MTB 67, 68, and 213–17), allowing Thornycroft to lay down five replacements. The replacements themselves were finally taken over in May 1941 (MTB 327–31). Thornycroft also contracted for further construction in the Philippines, going so far as to send a technical expert in 1940. No boats appear to have been completed, however.

From the fall of 1937 on, C&R would concentrate on motor torpedo boats—PT boats—for the U.S. Navy.

5

The Early PT Boats

In January 1937, Admiral Pye and his associates thought of the motor torpedo boat as a potentially useful adjunct to more conventional fleet units. Scarce funds were earmarked for larger ships such as destroyers, which would take several years to build, on the theory that with a suitable prototype design MTBs could be turned out quickly in an emergency. The first U.S. boats were purchased under a special emergency program in 1938. As war approached in 1939, the navy began industrial mobilization. Motor torpedo boats became attractive because, under the appropriate conditions, they could deliver torpedo attacks much less expensively than destroyers. That was particularly important for the defense of such Pacific bases as Manila Bay and Guam, which would surely be attacked early in an Orange war.

After late 1939, then, the program shifted from the experimental development of a few prototypes to urgent production, initially, of a boat, the Elco, based on a developed British type. Production expanded rapidly through 1940, when twenty-four were ordered, and through 1941, when a further sixty were ordered before the United States entered the war.

This effort reflected urgent requirements. In January 1941, the Asiatic Fleet asked that a squadron be assigned to the Sixteenth Naval District, the Philippines, to defend Manila Bay. The naval districts division of OpNav argued that the boats would make an attack on Manila Bay hazardous.

Tactically, there was always the question of the extent to which motor torpedo boats could operate with the fleet itself. In 1940–41, there were several proposals for fleet MTB carriers. BuShips killed the idea by comparing the MTBs with carrier-based torpedo bombers. Twelve MTBs would carry forty-eight torpedoes, a load similar to that of an aircraft carrier, but they would require the hangar space of a 15,000-ton carrier like the *Ranger*, and without providing any of the carrier's other valuable services.

The boats would, therefore, have to operate from bases. From 1941 on, tenders (AGP) were commissioned, but still, an OpNav conference on PT boats in May 1941 concluded that elaborate shore bases would be necessary to maintain the torpedoes and the aircraft-like MTB propulsion plant. As this chapter shows, without considerable maintenance, the delicate boats deteriorated fairly quickly.

At this time only the 70-foot Elcos, with their 18-inch torpedoes, were available, and they were still carrying out valuable experiments off the U.S. Atlantic coast. OpNav could only hold out the hope of providing later Elcos, and on 28 September, six 77-foot Elcos of MTB Squadron (MTBRon) 3 arrived in Manila. To bring the squadron up to full strength, six more boats were at Pearl Harbor on 7 December aboard the tanker *Ramapo*. Meanwhile, the Pacific Fleet asked for a squadron to help defend its major base at Pearl Harbor. Six boats of MTBRon 1 were present for the 7 December attack. MTBRon 2 was fitting out in New York; it was dispatched to the Canal Zone for local defense. MTBRon 3 was the famous "expendable" squadron that brought General MacArthur and Philippine President Manuel Quezon out of Manila. MTBRon 1 was at Midway, having run the 1,385 nautical miles from Pearl Harbor under its own power, fueling from patrol vessels and a submarine tender at three islands en route. MTBRon 2 and a reconstituted MTBRon 3 moved to Guadalcanal during the summer of 1942, when, as predicted five years earlier, the two fleets found themselves in such close proximity that small combatants could operate offensively.

No one expected the austere PT to end this way, massively armed with guns and rockets. This is the Higgins PT 631, newly completed, on 1 March 1945. The gun is an army-type 37mm automatic. The rocket launchers fired 5-inch spin-stabilized rockets. Elevated guns further aft are four 0.50s and one 20mm Oerlikon; a 40mm Bofors can barely be seen right aft. Torpedoes were dropped from the racks along the deck edges. Remarkably, all of these weapons fit a hull designed in 1940 to accommodate only two twin 0.50s and four torpedo tubes.

The development of these early U.S. motor torpedo boats began in 1937. Preliminary Design already had a tentative design for a large boat in the work it was doing on the Philippine patrol boat (see chapter 4). Beginning in May 1937, it therefore concentrated on the small boat defined by the General Board.

At this time, the navy expected to order the first two experimental boats, as C&R had proposed, under the deficiency bill scheduled for congressional action late in 1937 for the FY 39 program. In fact, they were not built until later, and the program was expanded and stretched out. Even so, through most of 1937, Preliminary Design worked on an in-house torpedo boat design for comparison with private designs.

The designator by which all the U.S. boats would be known, PT, came the following year; on 9 February 1938, the acting CNO, Admiral J. O. Richardson, announced that they would be torpedo boats (T) within the patrol (P) category.

Captain A. J. Chantry of Preliminary Design sought to keep length as near 50 feet as possible, weight within the critical 20-ton lifting limit. The General Board had not produced formal characteristics, so Preliminary Design had to estimate a radius of action based on the requirement that the boat be able to sustain itself at sea for five days. As in the Philippine boat, Chantry suggested a cruise-plus-boost plant, two Packards and two Invaders. As it had earlier in the year, C&R proposed a parallel program of navy and private designs.

BuEng had no entirely suitable engine, as the story of the awkward machinery arrangement of the Philippine boat demonstrated. In July, it sought $500,000 for prototype FY 39 engines, among them a GM 1,200-bhp diesel then in the blueprint stage. The bureau argued that the diesel would be extremely valuable, a bargain at $250,000. The budget was extremely tight, and the CNO, Admiral William D. Leahy, deleted the engine money a week later. Secretary of the Navy Charles Edison restored it for fear that delays in engine development would defer construction of motor torpedo boats to FY 40. However, in September, as a result of preliminary budget hearings in Congress, both boats were canceled.

A sketch design dated 24 May showed dimensions in feet-inches as follows: 54-0 (lwl), 56-0 (loa) x 15-0 x 2-7 and a displacement of 33 tons; two diesels were expected to drive the boat at 33 knots. Armament was listed as two 0.50-caliber machine guns and two 18-inch torpedoes with six 420-pound depth charges as an alternative. The crew consisted of two officers and eight enlisted men.

Preliminary design continued. In November, BuEng had to admit that, since its diesel would not be ready for about eighteen months, it would have to accept existing engines. It fixed on the new 1,000-bhp (1,250-bhp in March 1938), supercharged Vimalert gasoline engine. A small torpedo boat might be too crowded to accommodate three such engines, but the Vimalert design was attractive because it could be built if Congress authorized construction before the favored power plant, the diesel, was ready. The diesel was never used, and all American torpedo boats were powered by gasoline engines. Model tests showed a requirement for 2,270 bhp at 72,000 pounds for 41 knots. At this time, the 54-foot hull design was compared with the 70-foot Philippine boat, the Luders 45-foot rescue boat (and a version with buttocks straightened), a 64-foot CMB, a 40-foot boat designed in 1918 by Professor George Crouch, and several alternative 54-foot hulls displacing 55,000, 65,000, and 72,000 pounds. A comparison with stepped-hull forms showed that a step was not useful at speed-length ratios below 5.4, about 40 knots for a 54-foot boat (48 knots for an 81-foot boat).

Further model tests run in March 1938 were based on a 43,500-pound hull with dimensions as follows: 54 feet LWL x 13 feet BWL x 2 feet 9.5 inches. It would be powered by two 1,250-bhp Vimalert engines.

Word of the project soon spread among civilian small-boat designers, who began asking C&R for basic data beginning in mid-August. The bureau welcomed their proposals but cautioned that they should not expect payment. Procurement would be decided on the basis of competitive trials. By late 1937, W. Starling Burgess, John H. Wells, and Herreshoff Manufacturing had all received data packages, and the list of competitors grew in early 1938.

In September, C&R began to circulate its outline requirements. The motor torpedo boat could be up to 60 feet long and displace up to 20 tons. Armament would comprise two single 0.50-caliber machine guns in separate centerline cockpits plus two 18-inch torpedoes in stern tracks. The boat would also have to carry listening gear for her alternative ASW role. The bureau envisaged a hard-chine hull that would combine high smooth-water speed with seaworthiness, stability, and maneuverability in comparatively heavy weather. In a chop, with waves 5 feet high from hollow to crest, or in a ground swell with waves 150 feet long from crest to crest, the boat would have to be able to maintain a speed of at least 30 knots indefinitely without damaging its hull or unduly exhausting its crew. Metacentric height would have to be at least 5 feet to ensure ample rough-water stability. The bureau emphasized the need to avoid rolling or porpoising when planing, and it wanted a small turning circle and no outboard heel when turning at maximum speed with the rudder hard over.

The hull could be steel, aluminum, or wood. If

wood, it must have at least six transverse metal bulkheads to resist racking stresses. C&R wanted diesel engines to eliminate the fire hazard of gasoline. No definite cruising radius could be set, but the bureau wanted the boat to be able to cruise at maximum speed for at least 10 hours and to cruise at 12 knots for at least 60 hours. For a 40-knot boat, that meant 400 nautical miles at full speed and 720 at 12 knots, figures far beyond what was actually achieved.

In October 1937, C&R estimated that each boat would cost $50,000, a figure that would be used as a statutory limit when asking for the appropriation.

Meanwhile, the bureau continued to seek more information about hull forms. Preliminary Design doubted that conventional towing tank tests would provide sufficiently accurate information about the most important aspect of torpedo boat performance—the ability to maintain speed in rough water. Around December 1937, a new approach was conceived: large boats, each about half scale, would be used to test alternative torpedo boat hull forms, following tank tests of 54-inch ($1/12$-scale) models.

The approach was so successful that a year later it was described as a standard method to be used in the future. Three hard-chine boats (25-7$1/8$ x 6-4$1/4$ x [depth] 3-3$1/4$ feet-inches at the side) were built at Norfolk, followed by two more at Philadelphia early in 1939. Each was powered by two 140-bhp Gray Fireball engines.

The first of the three original boats was conventional, with a low chine forward, moderate transom immersion, moderate deadrise (14° at $5/8$ length, measured from the bow). The center of buoyancy was 58 percent of the length from the bow. The median buttock (halfway between keel and chine) had a reversal or hook aft to hold the bow down. The transverse sections from keel to chine were moderately concave.

Boat 2 was a rather extreme type, with the chine kept very low far forward and with less deadrise throughout its length (13° at $5/8$ length). It showed greater transom immersion, and the center of buoyancy was at the 60 percent point. Transverse sections were quite concave. The low deadrise, which increased dynamic lift, carried weight better. It also made the boat dryer because the forward sections picked up buoyancy faster. On the negative side, the deadrise made for harder riding, and the low chine forward reduced maneuverability. The hooked buttock was eliminated.

Boat 3 had a very high chine forward and a low chine aft for greater transom immersion. Deadrise was much greater than in the other two boats, 17.5° at $5/8$ length. The center of buoyancy was further aft, at the 63.5 percent point. Transverse sections were less concave than in either of the others. The buoyant bow made for dryness, and the high chine forward allowed the boat to bank more steeply, thus improving turns.

All three boats were much more rugged than typical racing boats, weighing about 6,000 as opposed to 3,000 pounds. They were, in effect, miniature PT boats with frames spaced closely, 6 inches apart, the main frames being sawed, the intermediate ones bent. The bottom was of double thickness ($5/8$ inch total), the sides of single ($7/16$ inch). Fuel jugs were calibrated to measure consumption.

The scale tests were somewhat difficult to interpret. Because the boats were lightly loaded compared with full-scale torpedo boats, an apparently low coefficient of resistance might not result in an efficient hull form. One model with just such performance, for example, had too large a planing surface (too much wetted area) and hence needed too much power to get over the planing hump.

To estimate a satisfactory hump speed, the designers looked at a boat whose performance they considered borderline, the sea sled, which had difficulty getting over its hump. It seemed reasonable that hump speed power should not exceed three-quarters of maximum available power for a hump speed 55 to 60 percent of maximum speed. Otherwise, the boat would be difficult to plane. The tests suggested that it would be best to have the hump at relatively high speed, providing there was ample margin between hump and top speed for latitude in full-power operation.

The full-scale tests confirmed that if the line of the chine near the bow were too low, it would affect turning. The run of the buttocks (their fore-and-aft angle) determined the boat's angle of attack and thus its planing angle. A hook down at the stern, as in the first boat, would keep the bow down when planing, which was desirable; the flatter the planing angle, the better. Three to 4° seemed reasonable. Similarly, boats seemed to operate best when trimmed initially by the stern by about 1 or 2°.

A comparison of high and low deadrise models and boats showed that greater deadrise made for an easier ride, closer to that of a round bottom, whereas lower deadrise (a flatter bottom) made weights easier to carry. As for the shape of the sections, between the keel and the chine, a straight or slightly concave section seemed best to prevent pounding. Greater transom immersion increased maximum speed—but also hump speed. The center of buoyancy was best located 60 to 65 percent of the length aft.

The boats were intended to maintain maximum speed in $1/8$ L waves at a speed-length ratio of about 6 (54 knots for an 81-foot boat, 44 knots for a 54-foot

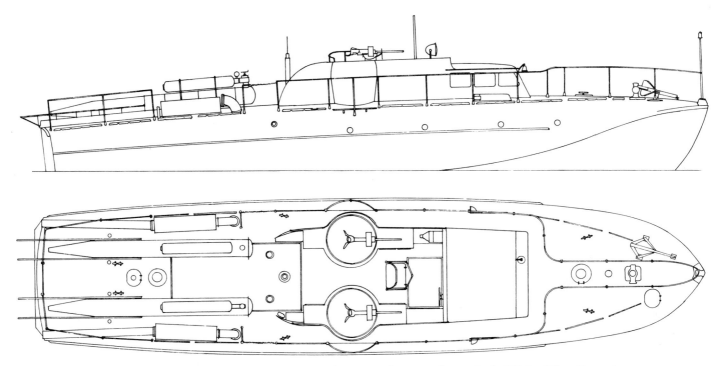

PT 1 as planned by C&R, April 1939. The guns are single 0.50s; the torpedoes, Mark VII-4s. (Alan Raven)

boat). At maximum speed, wetted (largely planing) area would determine resistance. Tests showed that this area was to be the minimum consistent with stability, a loading of about 200 pound/square feet being the maximum feasible figure. For a 100,000-pound boat, that amounted to a wetted area of 500 square feet, equivalent to a 10- x 50-foot rectangle.

Tests showed that a requirement for a tactical diameter of about six lengths or less at maximum speed could be met.

Given these data, Preliminary Design could develop its own high-speed hull forms. It could also critique the private designs it expected to receive, although, given the state-of-the-art design, it could not really be sure of performance until full-scale hulls were built. Ironically, the careful series of model and half-scale tests had less impact than might have been expected on the earliest U.S. MTB designs. However, the results were resurrected when new designs were developed in 1944–46 (see chapter 6).

By mid-1938, it was clear that the navy would hold a large competition for experimental small craft, including both the motor torpedo boats initially envisioned and the new subchasers. The terms of the competition were drafted late in June 1938, and invitations to builders and designers were published on 11 July. There were two phases: a preliminary design competition from which finalists would be chosen, and a final-design phase. All finalists would receive a prize of $1,500, and the designs actually chosen would earn $15,000 each. The finalists would turn over all rights to their designs to the navy. The preliminary design phase ended on 30 September 1938 (it had been scheduled for completion on 24 August), and winners were announced on 21 March 1939. Boats were built under the Second Deficiency Bill of 1938, which provided $15 million for experimental small craft.

Two designs were planned: a 54- and a 70-foot one. The conditions for the 54-foot design generally followed those Preliminary Design had circulated from August 1937 onward, except that cruising radius was set at 120 nautical miles at full speed and 120 at cruising speed (15 knots). For movement to forward areas, the boat would have to be strong enough to be hoisted by sling. The characteristics of the larger boat would approximate those of the Philippine patrol boat (see chapter 4), with a speed of 40 knots on trial and an endurance of 275 miles at full speed, 275 at cruising speed.

In all, twenty-four designs were submitted from twenty-one designers for the short torpedo boat; three finalists were chosen. Thirteen designers each submitted one design for the long torpedo boat; five became finalists, but one dropped out before submitting a final entry. Based on its model tests, C&R required four of the final competitors to provide more power. The 54-foot winner was Professor George

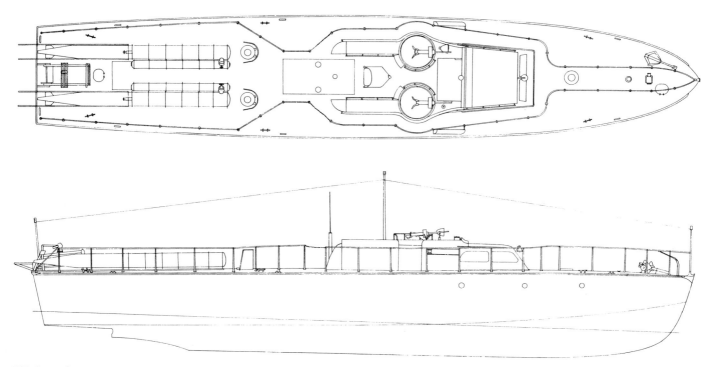

PT 5 as planned by C&R, April 1939. Note the depth-charge rack between the torpedoes. (Alan Raven)

Crouch of Henry B. Nevins, the 70-foot winner, the yacht design firm of Sparkman and Stephens.

Eight boats (PT 1–8) were ordered, four of them 59 feet and and four, 81. Of the 59-footers, two were designed (one with two supercharged Vimalerts, 2,400 bhp total, and one with three supercharged Vimalerts, 3,600 bhp) by Miami Shipbuilding, then Fogel Boat. They became PT 1 and PT 2. Two built from C&R designs by Fisher Boat Works in Detroit became PT 3 and PT 4 (one with two Packards, 2,700 bhp, and the other with three, 4,050 bhp). The supercharged Vimalerts were delivered late because the manufacturer, Stirling, was being reorganized, so that the hulls of the first two short PTs were completed about eighteen months before their engines arrived. They never entered service as PTs. Instead, they were converted to torpedo retrievers, each with a single tube mounted diagonally across its deck, and operated as uncommissioned boats (C6083 and C6084). The former PT 1 served through the war as a familiarization hull for motor machinists at the PT training base in Melville, Rhode Island; its Vimalerts were replaced by standard PT Packard 4M2500s. The former PT 2 became a service launch at the naval torpedo station in Newport.

Of the 81-footers, three were wooden, one aluminum. Two of the wooden boats, PT 5 and PT 6, were built by Higgins to the winning design and powered, respectively, by three supercharged Vimalerts (3,600 bhp) and three supercharged Packards (4,050 bhp).[1] They were meant to cruise on the centerline engine with the other two propellers trailing. The other two boats, the wooden PT 7 and the aluminum PT 8, were built at Philadelphia Navy Yard. C&R justified government construction because it wanted to keep the engine installation in the wooden boat secret and because the experimental aluminum structure required extended research and close supervision. PT 7 was described as powered by three special engines of 3,600 bhp, presumably the new BuEng diesels; it was actually completed with four 900-bhp supercharged Defenders. PT 8 was powered by two 2,000-bhp Allisons (actually paired V-engines connected on common-base plates to form X-engines) plus one Hall-Scott 60-bhp Defender. Both PT 7 and PT 8 were considered experimental machinery installations, and they suffered from numerous breakdowns; Packard engines had to replace PT 8's original engine installation. PT 8's aluminum hull, on the other hand, was quite successful. Even though it received little care in wartime, it was still in excellent condition at the end of the war (by which time it had been redesignated YP 110), so BuShips decided to use aluminum in the four postwar PTs.

The PT 5 hull was found to be too narrow, and a 1-foot sponson had to be built near the waterline to solve the problem. PT 6 was sold to Finland, then at war with the Soviet Union, and was ultimately taken

The first modern U.S. PTs were designed to fire their weapons over the stern, like the World War I British CMB. This is PT 3, 15 September 1940, showing two torpedo racks with their firing mechanism, a muffled engine exhaust, and smoke generators. The gun tubs carry single 0.50-caliber machine guns.

PT 8, cruising at about 40 knots near Philadelphia on 29 June 1941, was significant for its aluminum hull, the success of which inspired the use of aluminum in the four postwar PTs. Note the two massive conventional torpedo tubes aft.

THE EARLY PT BOATS

Table 5-1. The U.S. PT Prototypes

	PT 1/2	PT 3/4	54 ft	PT 5/6	PT 7	PT 8
LOA (ft-in)	58-11	58-5	57-7¾	81-3	80-7¾	80-7¾
LWL (ft-in)	55-6	55-6	54-0	77-5	75-0	75-0
Molded Beam (ft-in)	13-9	13-8	15-4¼	16-8	16-8	16-8
Beam (waterline, ft-in)	12-7¾	12-9	14-6¼	14-4¼	15-5½	15-5½
Beam (chine, ft-in)	12-3¼	12-7½	12-1½	14-3½	15-4½	15-4½
Hull Depth (at side, ft-in)	8-10	9-6	9-1	10-8½	10-7	19-7
Full Load Displacement (lbs)[1] (1)	67,250	66,200	57,567	106,800	114,500	108,850
(2)	73,950	73,000	—	—	—	—
Trial Displacement (lbs)	69,900	59,900	56,107	94,230	101,800	94,350
	67,350	66,350	—	—	—	—
Draft (fully loaded, ft-in)	3-2¼	3-1¼	2.89	5-5	3.28	3.19
	3-4½	3-1½	—	—	—	—
Draft (trial, ft-in)	3-0	2-10⅞	2.84	5-2	3.06	2.93
	3-2¼	3-1¼	—	—	—	—
Metacentric Height (lightship, ft)	4.53	3.81	3.77	4.75	—	6.30
	4.33	3.79	—	—	—	—
Metacentric Height (trial, ft)	3.47	2.93	3.51	4.37	4.67	4.74
	3.19	2.85	—	—	—	—
Metacentric Height (fully loaded, ft)	3.18	2.85	3.43	3.95	4.36	4.27
	3.06	2.80	—	—	—	—

1. Second figure is ballasted.

over by the Royal Navy as MGB 68. Higgins built a new PT 6 at his own expense (it was bought by the navy), incorporating his own ideas on strength, speed, and hull form, which he claimed were based on experience building boats for the Coast Guard during Prohibition. Like subsequent Higgins boats, it had concave buttocks, their radius of curvature uniform from stem to stern on either side.[2] This "tunnel" on each side of the keel was intended to trap the air and water and thereby make the propeller more efficient. It was also meant to reduce pounding and provide a cushioning effect in head seas. This second PT 6 also ended in British service, as MTB 268.[3]

C&R was still conducting its model experiments. It decided to build a ninth boat, in effect a full-scale version of model 3 (above) at Norfolk. This craft never received a PT number because it was built for eventual service as a crash boat. It was tested with sufficient ballast to simulate PT loads.

The entire program was approved by Secretary of the Navy Charles Edison on 29 March 1939. Details of the boats are given in table 5-1.

BuOrd armed the boats with single 0.50 caliber machine guns in single, open, manual scarf-ring mounts. All carried their torpedoes aft; in PT 1–6 the weapons were dropped stern first, as in the early British torpedo boats. PT 7 and 8 had conventional forward-firing tubes. Although stern-first launching had been standard in the early motor torpedo boats, there were major disadvantages to using the method. The boat had to drop the torpedo and then outrun it, which meant that it could not fire in the event of engine trouble. Nor did it seem particularly desirable to run before a field of torpedoes, risking damage if one ran erratically. Only one torpedo could be launched at a time, since two launched together might collide. Ejection would have to be by air impulse, entailing a heavy air cylinder, since powder impulse ejection would involve a blow directly to the torpedo warhead. There were several major advantages to using bow-launched torpedoes. Fired well forward, they would easily clear the bow wave and the sides of the boat, suffering no wake interference. Duds could easily be observed. The tubes could be trained out from the bow by a small angle, the torpedoes made to follow the base course of the boat for easy and accurate fire control. Because all of a boat's tubes could be used in a salvo, a target could be subjected to much more intense torpedo fire. Moreover, weight could be saved on each tube by using powder for the launch impulse.

Only later was it discovered that the flash of the powder revealed the position of a boat making a quiet night attack. Higgins Industries developed an alternative compressed-air launcher, accepting the extra weight involved.

The small boats, PT 1–4, were equipped with the 18-inch torpedo, the larger ones with the elderly Mark 8 21-inch weapon. Neither torpedo was altogether satisfactory. The British used a modern 42-knot, 18-inch aircraft torpedo. The U.S. Navy, however, had replaced its 18-inch Mark VII aircraft torpedo with the 22.5-inch Mark 13 some years before. The existing Mark VII was considered unreliable, and with a 419-pound warhead, it had a range of only about 3,000 yards at a speed of only 33 knots. Thus a boat might approach at 46 knots, but it would fire a much slower weapon, with attendant tactical complications. The newer Mark 13 matched the torpedo boat in speed (6,000 yards at 46 knots) and carried a much more powerful warhead of 813 pounds. Tactically, the match in speed between torpedo and boat would mean that the PT could be set on a collision course with its target, then fire.

At this time, PTs were to have used the 21-inch Mark 14 developed for submarines; it was rated at 9,000 yards at 31 knots, or 4,500 at 46 knots, carrying 487 pounds of explosives. In 1940, it replaced a version of the older Mark 8 and required changes in the torpedo tubes planned for the Elco boats (see below), which was why the issue of the Mark 13 did not arise. In fact, Mark 14s were so scarce that they were never even test-fired from PTs. Instead, PTs were issued with obsolete torpedoes, primarily versions of the Mark 8, which was typically rated at 10,000 yards at 27 knots and carried only a 300-pound warhead. The PTs had little use for its long range.

These considerations doomed the 59-footers even before they had been completed. In their designs, weight was so critical that none was available for conventional torpedo tubes. Yet, conventional torpedoes had to be launched upright, so that their gyros would not tumble; they could not be rolled over the side. That meant that the 59-footers would have to use stern tracks, firing their weapons propeller first over the transom—a procedure that BuOrd and the General Board totally rejected.

The experimental boats would not be completed until mid-1940. Meanwhile, Elco, the mass builder of World War I motor launches, imported a Scott-Paine 70-footer. The secretary of the navy, Charles Edison, had served as governor of New Jersey and thus was aware of the Elco yard at Bayonne and its World War I success. After the war, the yard had developed high-speed craft for the Coast Guard and its rumrunning adversaries powered largely by converted Liberty engines. The company avoided the navy design competition, considering it too risky, but by late 1938, Henry R. Sutphen, vice president of Electric Boat and head of its Elco division, was in contact with British Power Boats. He saw production of the advanced British boats as a way of entering the U.S. motor torpedo boat market. At this time, the British were beginning to place large production contracts in the United States, and it is possible that Elco saw the possibility of repeating its World War I triumph.

On 13 January 1939, well before the conclusion of the design competition, Edison proposed to the General Board that the U.S. Navy acquire a British boat to gain design experience. Sutphen and his designer, Irwin Chase, were scheduled to visit England in February to inspect the British boats. On the sixteenth, the board agreed with Edison's idea of obtaining a 70-foot British torpedo boat and suggested that the navy might also acquire the British equivalent of the U.S. short torpedo boat. On the nineteenth, Sutphen was asked to attend a navy conference discussing a possible purchase, and he accepted, enclosing a letter from the New York representative of British Power Boats. Sutphen and Chase left for England at Elco expense on 10 February and inspected existing Thornycroft, Vosper, and British Power Boats craft there. The Royal Navy had just chosen the Vosper as its future standard type, Power Boats' 70-foot private venture (PV 70) design having lost out. Existing Thornycroft boats were relatively small and could not compete effectively with either Vosper or Power Boats.

Reportedly, Scott-Paine convinced Elco on the strength of his superb boathandling and his boat's excellent behavior in the rough waters of the English Channel. The U.S. naval attaché, who witnessed these tests and who knew of official interest in them, described the performance of the Scott-Paine boat as revolutionary. Elco had, after all, been interested in his boat for some time. Sutphen bought the 70-footer on 17 March for delivery in June and offered it to the U.S. Navy. Some reports suggest that Edison had guaranteed Sutphen before he left that, if the boat proved satisfactory, the navy would buy it. On 3 April, President Roosevelt personally approved the purchase on condition that the Scott-Paine boat be as cheap as the American 70-footers (they were actually 81-footers). That was possible because the price of the American boats included substantial amounts for engine development.

Elco formally proposed the boat on 25 April, and on 1 June 1939, a contract was signed for delivery on 1 November. The 70-foot Scott-Paine boat became PT 9; it was bought under the 1938 experimental craft appropriation. PT 9 was delivered to Elco in New York on 4 September 1939 and retained for trials and then as a pattern for further construction. Commander Swasey, who had designed the World War I subchasers, was brought back by the navy from retirement to observe the trials of the new boat.

A division of the Board of Inspection and Survey tested PT 9 off New London on 9–10 October 1939. It was then powered by three marinized Merlin aeroengines, later replaced by Packards. C&R liked the airplane-type construction; the boat was described as obviously well and strongly built, able to stand the abuse characteristic of PT service. Moreover, its habitability compared "more than favorably" with that of the World War I subchaser. The boat was so well soundproofed that there was "no nervous strain from noise, and it [was] possible to converse in normal tones anywhere outside the engine room, even at high speed."

The boat behaved well and was dry in 3- to 4-foot waves, but a C&R representative described her as rough riding in comparison with some of the C&R

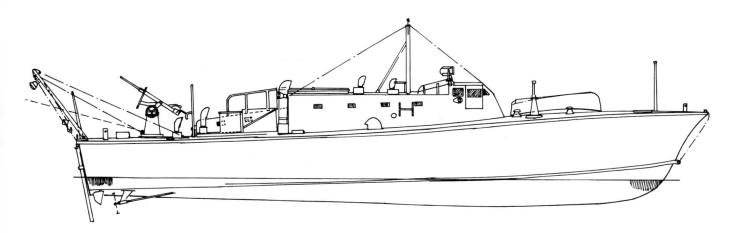

A Canadian-built Power Boats 70-footer converted to a crash boat for the Royal Navy, November 1943. The similar U.S. Elco 70-footers transferred to Britain were retained in Canada as air-sea rescue boats, though in somewhat different configuration, and four Canadian 70-footers entered U.S. service as PT 368–71. They were part of a series of sixteen (TM 22–37) ordered for the Royal Netherlands Navy, of which the first eight were built in Canada, the last eight at Fyff's Shipyard in the United States. This drawing applied to British MTB 333–34, 336–37, and 340–43. MTB 332–43 were transferred from the Royal Canadian Navy to the Royal Navy in 1941. Note the A-frame erected aft; the object running down the stern is a propeller guard. The ship is armed with a twin 0.30-caliber machine gun forward (only the pedestal is shown) and a 20mm gun aft. It was powered by two gasoline engines, not the three common in later U.S. torpedo boats. (Norman Friedman)

models. Captain Robert B. Carney, who was in charge of the board, preferred to say that "it is true we were slapped around considerably but it was not an inordinately rough ride considering the speed. I was not impressed by the maneuverability of the boat and from my limited experience would say that she turns slowly; she is, however, dry on turns and not tender."

There were some detailed criticisms. Navigational equipment was insufficient, and the boat missed its return landfall on 10 October. Scott-Paine followed the British practice of keeping a man in the engine room; the U.S. Navy, doubting that anyone could stand the noise level for long, preferred grouped, one-man remote controls. The boat had no torpedo director.

Captain Carney considered PT 9 an excellent yardstick, but not so superior to other U.S. designs as to justify scrapping the latter. It was strong, dry in a seaway, and habitable, but rode fairly hard as a result of a compromise between dryness and riding qualities and seemed rather slow. Carney judged it to be good "but no miracle." He even doubted that, with the heavier U.S. Packard engines, it would make the required 40 knots; the speed-rpm curve rose so steeply over 40 knots that it seemed unlikely that a considerable increase in power would add much speed.

However, the key consideration was that the Elco boat already existed, whereas its American counterparts were still more than six months away. Given the ongoing international crisis, the existing good, if not spectacular, boat was infinitely preferable to the possibly better future boat. Thus, even before the rough-water trials, it was tentatively decided late in September to buy enough Scott-Paines to gain tactical and operational experience. A final decision was delayed until after the 9–10 October trials, but on 3 October the assistant secretary of the navy suggested that more boats be bought using funds remaining from the 1938 experimental craft program. It appeared that eighteen to twenty vessels could be bought at $271,000 each, a figure later considerably reduced.

Elco presented a formal proposal for further construction on 26 October, and the coordinator of shipbuilding agreed to release $5 million for construction. It received a contract for twenty-three boats, successors of PT 9, on 13 December 1939. This award for a foreign design was sharply criticized, but Secretary Edison could reasonably reply that, time being short, it was much better to have boats in service than to wait for better designs. Elco could mass-produce Scott-Paine boats because it had not only plans but also a completed boat to use as a pattern.

Because the original concept had been to use PTs for both torpedo attack and local ASW, the General Board recommended that half the Elco boats be completed as fast ASW craft, which were designated PTC. Thus, the order split into PT 9–20 (of which PT 20 was built to a lengthened 77-foot design) and PTC 1–12. Like PT 9, the 70-foot Elcos were armed with four 18-inch torpedo tubes and two bomber-type, twin-machine-gun turrets, enclosed in plastic to protect the weapons and operators from spray. However,

PTC 1, the ASW version of the Elco PT, in Florida in 1941. The long tray along its side led to a pair of Y-guns, each flanked by short hoists to lift the charges. The tarpaulins abaft the Y-guns covered conventional depth-charge tracks. (USN)

This photograph reveals the unique PTC depth-charge system in greater detail. Y-gun arbors were stowed alongside the superstructure, and charges moving aft along the rails were hoisted up the inclined racks. (USN)

where PT 9 had twin 0.30-caliber guns (unlike other U.S. PTs), PT 10 and later boats had 0.50s.[4] The Dewandre turret itself was not well liked. Its power controls were not smooth enough, and the guns were limited to a 27-round magazine, enough for only one burst. The commanding officer of PT 10 called for a better turret with belt-fed machine guns, which had 150 rounds of ready-service ammunition.

BuOrd had had no choice in the matter, as the Elco prototype had been bought as a unit. But the bureau had no intention of continuing to buy an inferior weapon. Although without a power turret of its own, it argued that nothing so elaborate was really needed for such a light weapon. The experimental PTs were armed with simple scarf rings, which were 54 inches in diameter as opposed to the Dewandre-Elco turret's 60. Surely a simple, fixed-cylindrical shield running to the height of the trunnion would protect a gun from wind and spray. By April 1941, Colt was completing a suitable twin scarf ring 55 inches in diameter, which could easily replace the Dewandre. BuShips agreed, and ordered all boats after PT 21 delivered without guns, pending delivery of scarf rings. The PT squadrons complained that they might have to fight with no guns at all, so the power turrets remained at least through PT 44. In the Philippines, PT crews removed the plexiglass shields and cut the power lines on their Dewandres.

Boats built during the war were equipped with Bell Aircraft scarf-ring mounts (Mark 17) that had shock absorbers.

The PTC was a subchaser version of the PT 10 type, equipped with a hull sonar, two Y-guns aft (with eight additional arbors, for a total of twelve), and twin depth-charge release tracks aft. Two single 0.30-caliber Lewis guns were mounted on pedestals just forward of the superstructure.

Sonar was required, but its installation in so specialized a hull was difficult. In January 1940, BuShips used a NACA high-speed towing tank, normally used to test seaplane hulls, to compare alternative sonar domes. A cylinder 17 inches in diameter and 30 inches high made the Elco hull form unstable, causing it to yaw and roll violently and increasing resistance 65 percent at a model speed corresponding to a full-scale speed of 45 knots. A streamlined shape at the end of a vertical shaft proved more satisfactory.

As for armament, the General Board wanted a gun that could actually penetrate a submarine's pressure hull. BuOrd suggested the army 37mm automatic gun, which weighed about as much as the two twin 0.30-caliber machine guns of Scott-Paine's prototype boat. Compared with the existing navy 1.1-inch machine gun (which did not exist in a single mount), it fired a heavier shell (1.34 as opposed to 0.92 pounds) at a greater range (9,300 instead of 7,300 yards) and at a higher rate of fire (120, not 150, rounds per minute). Unfortunately, the 37mm was larger than each of the twin 0.30s, and it was not clear whether the 37mm could be protected against spray.

In theory, the PTC would function as part of a larger ASW defense system, adapted from the British loop/HDA (harbor-defense asdic) system. The magnetic loops would detect a submarine attempting to penetrate a base, and the HDA would track it, feeding data to a central control station. The loops and sonar gave approximate location; the subchaser would close at maximum speed, slowing to search and attack. The PTC seemed attractive for its high speed and shallow draft, which would allow it to take shortcuts denied to larger craft.

Eight of the twelve boats, PTC 5–12, were transferred to the Royal Navy upon completion. The first four units were commissioned in January 1941, equipped with experimental sonars, and tested at Key West. They turned out to be extremely noisy, and their sonars were effective only when they were stopped, as with World War I subchasers. However, when stopped, they suffered from a short, sharp roll and could not receive echoes effectively. Base defense was the province of slower subchasers, which suffered from less violent motions in a moderate seaway.

Note, however, that immediately after the Pearl Harbor attack, the PTs of MTBRon 1 were in great demand for ASW patrol because they were the only available craft fast enough to drop depth charges in the harbor's shallow channel without damaging themselves. Depth charges were part of the specified armament of the PTs of this period, even though the boats were not equipped with any form of sound gear. In this role, they exchanged their two after torpedo tubes for a set of eight single-depth-charge tracks, a modification made later to many other PTs.

The PTC series continued with PTC 13–24, planned in September 1940 as part of a second series of Elco boats (the others were PT 21–32). These boats were each to have carried two 21-inch torpedo tubes, sonar, one four-charge depth-charge track, and one depth-charge thrower. They reverted to PT status (PT 33–44) only in March 1941, when they were required to replace early boats lend-leased to Britain. By that time, PTs were urgently needed, and the appropriate PTC sound gear was not expected until October. Even then, the navy expected to continue building PTCs, both for itself and for the Royal Navy. In March, the same war plans division paper arguing that the Elcos be redesignated also argued that future PTs be built in lots of fifty (twenty-five for the Royal Navy), and that the Royal Navy be asked whether it wanted PTs or PTCs. Of the third series of twenty-four Elco boats, half were initially to have been PTCs (PT 45–56, PTC

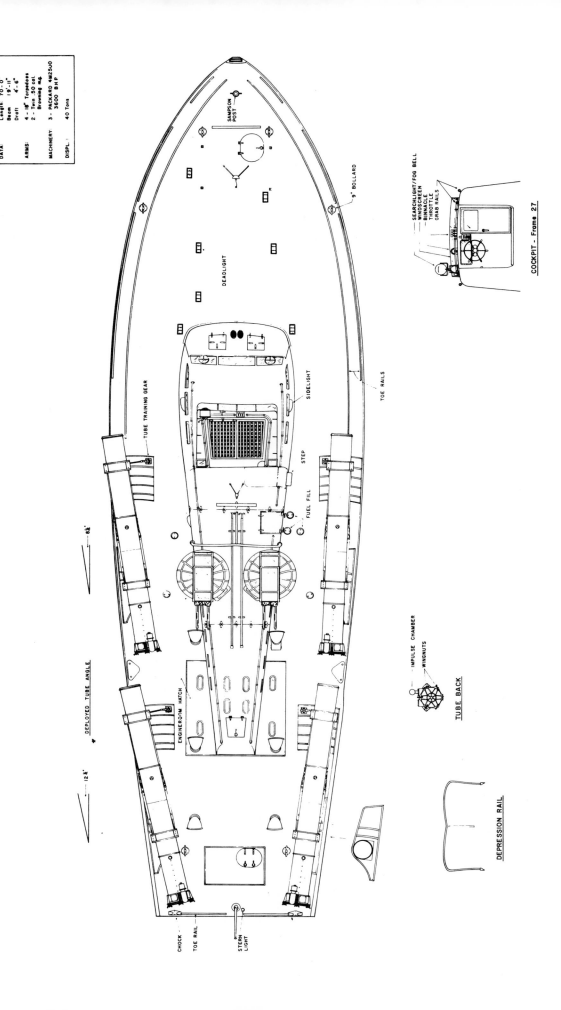

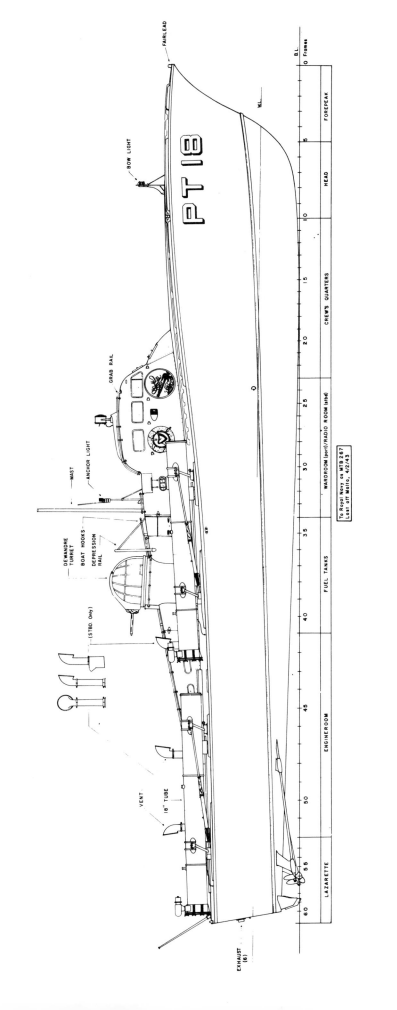

PT 18, a 70-foot Elco. (Al Ross)

PT-34
APR 42

SERIES: PT20-44
BUILDER: ELCO
Bayonne, NJ
DATA
Length 77'-0"
Beam 19'-11"
Draft 4'-6"
ARMS:
4 - 21" Torpedoes, MK.8
2 - Twin .50cal mg
2 - 30cal LEWIS mg
MACHINERY 3 - PACKARD 4M2500
3600/4050 BHP
DISPL 40 Tons

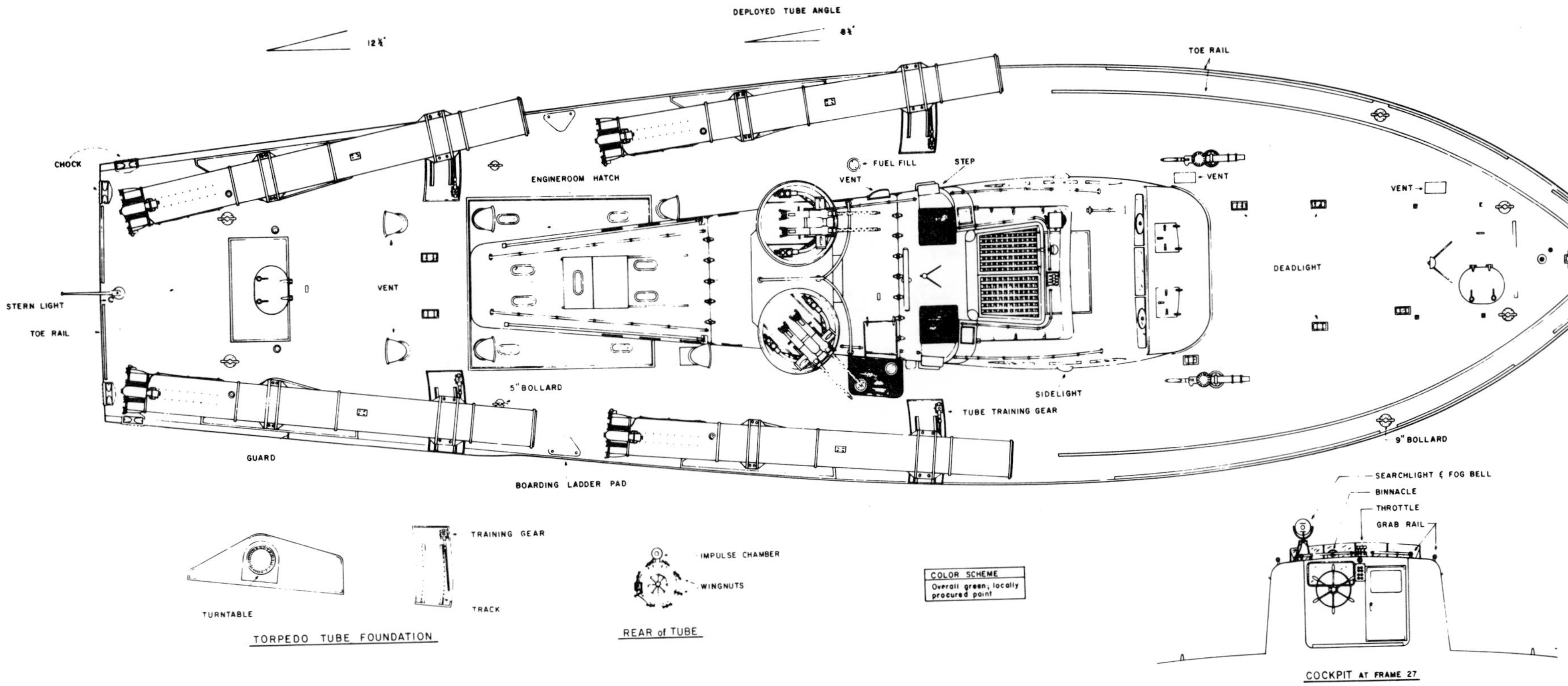

VENT
9" BOLLARD
SEARCHLIGHT & FOG BELL
BINNACLE
THROTTLE
GRAB RAIL
COCKPIT AT FRAME 27

DEADLIGHT
TOE RAIL
VENT
STEP
FUEL FILL
VENT
SIDELIGHT
TUBE TRAINING GEAR

COLOR SCHEME
Overall green, locally pressed paint

DEPLOYED TUBE ANGLE
8°
IMPULSE CHAMBER
WINGNUTS
REAR of TUBE

ENGINEROOM HATCH
5" BOLLARD
BOARDING LADDER PAD
TRAINING GEAR
TRACK

VENT
12½°
GUARD
TURNTABLE
TORPEDO TUBE FOUNDATION

CHOCK
STERN LIGHT
TOE RAIL

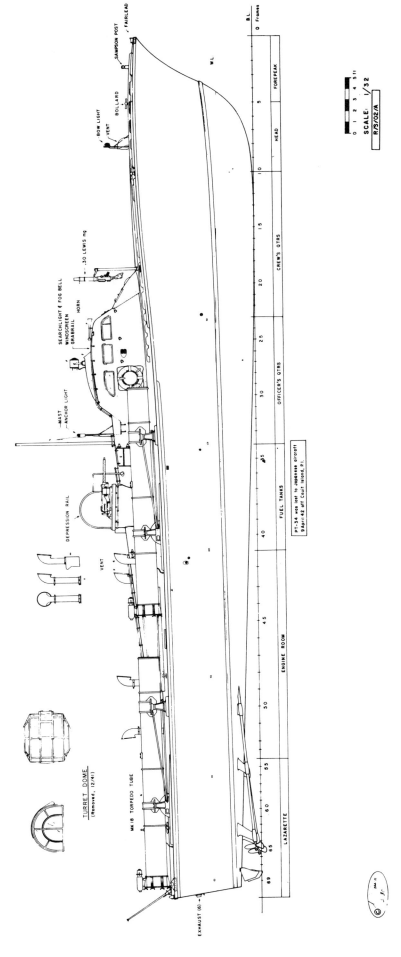

PT 34, a first-series 77-foot Elco. (Al Ross)

PT-65
MAR 43

SERIES: PT 45-48, 59-68
BUILDER: ELCO
Bayonne, NJ
DATA:
Length: 77'-0"
Beam: 19'-11"
Draft: 4'-6"
Displ: 46 Tons
ARMAMENT:
2 - Twin .50 cal.
1 - 20mm Mk 4
2 - 21" MK 18 MOD 1 Torpedo Tubes
8 - Depth Charges
MACHINERY: 3 - PACKARD 4M2500 3600/4050 BHP

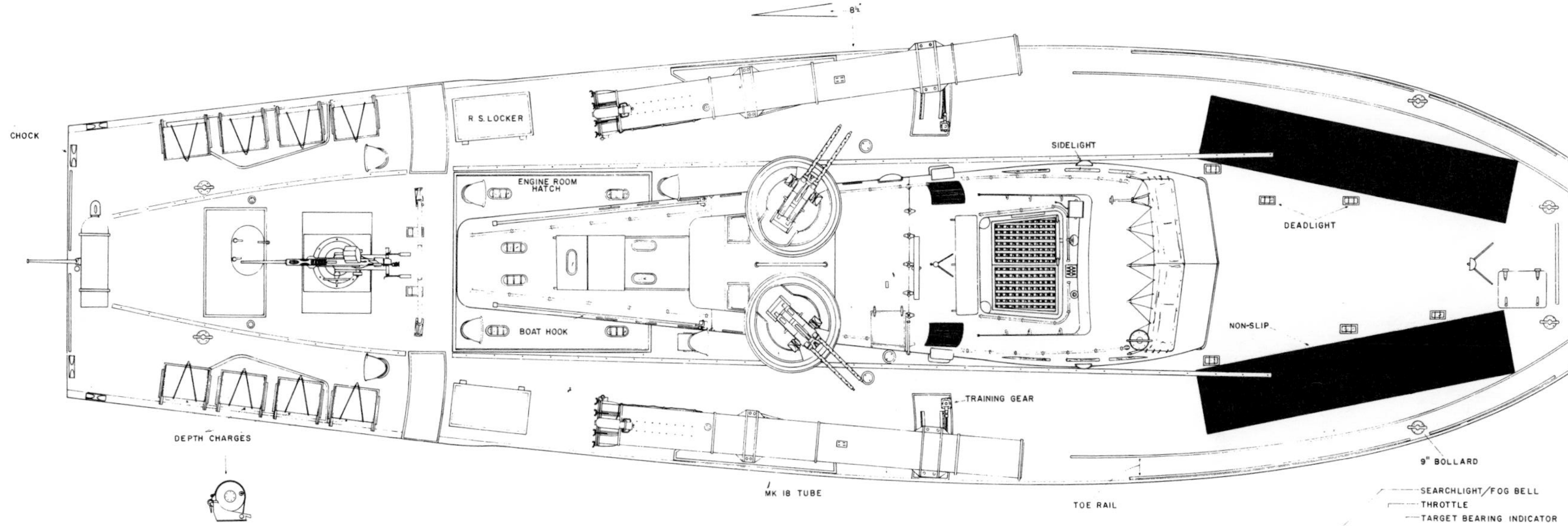

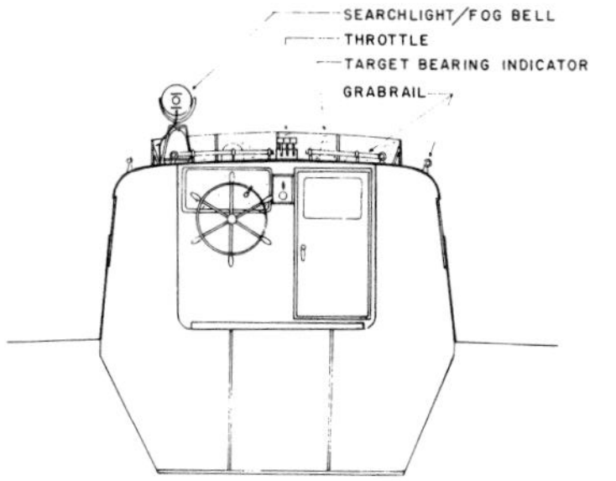

COLOR SCHEME
5-O (Ocean Gray) overall,
"COPPEROID" anti-fouling

Labels: 9" SEARCHLIGHT/FOG BELL, THROTTLE, TARGET BEARING INDICATOR, GRABRAIL, BOLLARD, DEADLIGHT, NON-SLIP, SIDELIGHT, TRAINING GEAR, TOE RAIL, MK 18 TUBE, R.S. LOCKER, ENGINE ROOM HATCH, BOAT HOOK, DEPTH CHARGES, CHOCK

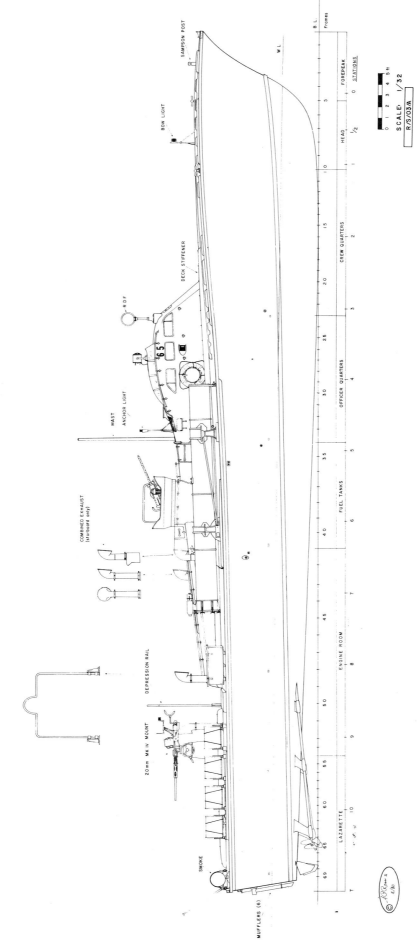

PT 65, a second-series 77-foot Elco. (Al Ross)

PT-59
OCT 43

SERIES	PT 45-48, 59-68
BUILDER	ELCO Bayonne, NJ
DATA	Length 7'-0" Beam 19'-11" Draft 4'-6" Displ. 40 tons
ARMAMENT	2 – 40mm MK3 14(7) – 50cal. MG
MACHINERY	3 – PACKARD 4M-2500 3600/4050 BHP

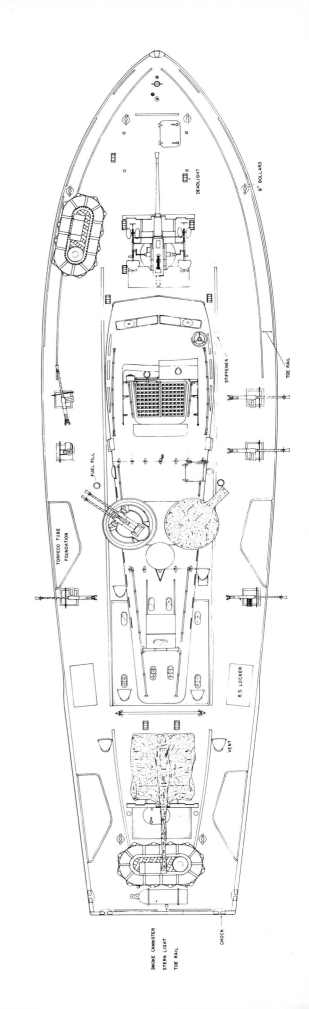

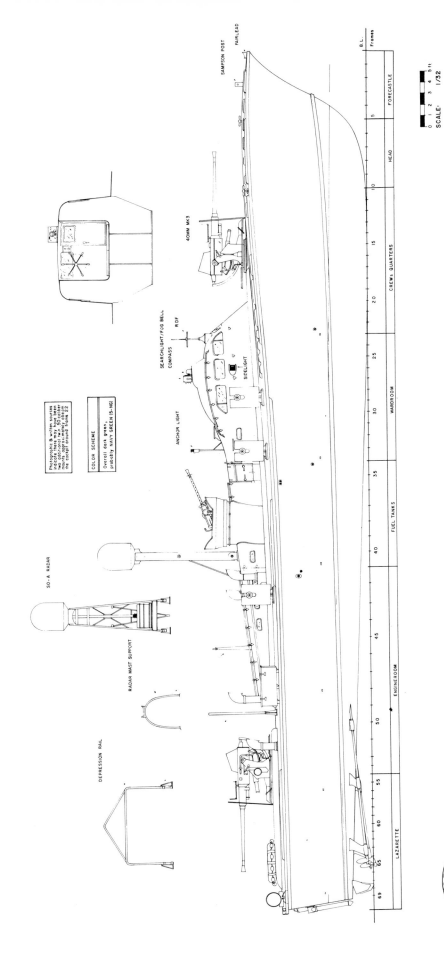

PT 59, a second-series 77-foot Elco converted to a gunboat, 1943. (Al Ross)

25–36), but in June 1941, the PTCs were redesignated PTs (57–68). The PTC designation was later applied to 63-foot crash boats converted to subchasers for transfer to the Soviet Union (PTC 37–66). Moreover, the Coast Guard's 83-foot patrol boats assigned to local ASW defense in the Philippines in 1945 were organized as PTC Flotilla 1.

The Elco 70-foot PTs were unique among the long-hulled boats in being armed with 18- rather than 21-inch weapons. In July 1940, the General Board decided not to rearm them. It argued against incurring cost to change the contract. A total of 462 suitable 18-inch torpedoes were available—they could not be converted to aircraft torpedoes and were not required for other vessels—while 21-inch weapons were scarce. The new construction program was so large, and the consequent requirement for 21-inch torpedoes for destroyers and submarines so great, that shortages would continue despite the existing supply of 6,000 and an annual production rate of 1,350. Nonetheless, the merits of the 21-inch torpedo were so obvious that it would be standard in all future PTs, thanks to the expansion of torpedo manufacturing facilities then under way. Like PT 9, future PTs would have four tubes, but they would be 21-inch tubes (BuOrd had argued that four 18-inch torpedoes were superior to two 21-inch, but that four 21-inch would form an ideal battery).

PT 20, the last of the original Elco series, was ordered lengthened by seven feet to accommodate the two extra tubes. It thus became the prototype production boat and, in fact, the first Elco boat for which the navy could effectively set requirements. This caused some difficulty. First, PT 9 and her half sisters did not have a smoke generator; PT 20 and later boats clearly needed one to cover their escape after firing. Second, there was some confusion about endurance. The General Board's characteristics required 500 nautical miles at 20 knots, but the detailed specifications—which governed acceptance—showed 650 miles at 22 knots.[5] BuOrd required that the torpedo tubes be lengthened 6 inches to take a heavier Mark 15 torpedo; this increased net weight to 860 pounds. The bureau also wanted more 0.50-caliber ammunition. Elco protested these weight increases on the ground that they might keep the boat from making its guaranteed speed. As a result, the guaranteed speed was reduced from 40 to 39 knots in January 1941.

As noted above, twenty-four Elco four-tube 77-footers were ordered in September 1940. While they were being built, the United States began to supply military equipment to Britain under lend-lease. The Royal Navy badly needed motor torpedo boats, and in March 1941, nearly all of the earlier boats were scheduled for transfer: PT 3–5, 7, 10–15, 17–19, and PTC 5–12. Most were transferred in April 1941, but PT 16–19 were delayed until July so that they could be used for training. PT 1 and 2 were still waiting for engines, and PT 6, 8, and 16 were retained, pending comparative trials against representative Elco 77-footers. Only in April 1941 did the U.S. government formally decide to retain the experimental aluminum PT 8. The trials would be used to determine the design of the next batch of twenty-four boats, which were more Elco 77-footers.

If the Elcos had any major flaw, it was their lack of strength. As early as November 1940, the naval attaché in London reported persistent structural problems in British Power Boats' 70-foot MASBs (similar to the U.S. 70-foot Elco); the fastenings connecting the deck to the gunwale bar or clamp often failed. The Admiralty planned to order the company to use a hard-wood clamp and thorough fastenings on future boats, and on finished boats to add a steel shear plate and a deck stringer with the brackets between frames.

The problem was highlighted during a heavy-weather run by MTBRon 2 from Key West to New York on 22–31 March 1941, just before its shipment overseas. The weather became foul after the squadron, which consisted of the 70-foot Elcos but without PT 16, 18, and 19, left Charleston; heavy head seas brought with them 8- to 10-foot waves. While running even at moderate speed, the boats pounded heavily and seas continuously broke high over their bows. Operating personnel reported extreme discomfort and fatigue.

Off Savannah, several outer planks cracked in PT 10, 12, and 15 in the vicinity of the forward engine-room bulkhead on both sides, and there was noticeable working between deck and sides in several boats. The squadron had to put into Savannah and run up the Intracoastal Waterway to Charleston for temporary repairs. In better weather, between Norfolk and New York, the boats cruised in the open sea at 33 knots, demonstrating an endurance of 576 nautical miles at 28 knots (555 miles at 33 knots).

All the boats suffered from some sort of structural failure. The forward chine guards were ripped away by the force of the water under them. In several cases, the bottom framing under the bows broke. In three boats, side planking near the forward engine-room bulkhead cracked, demonstrating the need for more longitudinal strength. As a result, the seam between the deck and the side opened, admitting water when the boat sagged. That the side planking broke outward in one instance and inward in another showed, not planking failure, but a lack of rigidity in this part of the boat. These hulls had no longitudinal shelves to serve as strength members. The side planking and chine worked relative to the clamp, the deck, and a thin piece installed in place of the shelf.

BuShips expected additional framing, which had

THE EARLY PT BOATS 135

The shape of the future: the Elco PT 19 makes a high-speed run, probably in late 1940. The enclosed 0.50-caliber machine gun turrets proved unsuccessful in service.

PT 30, an Elco 77-footer, shortly after the Battle of Midway. Tripods forward of the bridge were built to take Lewis guns. There is an additional AA weapon mounted between them; it may have been the 20mm cannon authorized for many PTs at this time. The shrouded object right aft is a smoke-screen generator. Note, too, the depth charge between the torpedo tubes on each side. By this time, the plastic 0.50-caliber turret tops had been removed, leaving the mechanism vulnerable to corrosion; hence the shrouds. The wooden strip extending up the side of the boat beyond the front of the superstructure is a deck stiffener.

been successful in the case of PT 13, to solve the problem of the broken framing in the impact area forward. The separation between deck and side, about a quarter of the way aft, was more serious. It could be traced initially to the shearing of the vertical fasteners connecting the deck to the frame. Once they had given way, the deck could move relative to the ship's side, opening cracks in the outer hull planking. The airplane-type construction of the boat saved weight by depending on the skin for much of its strength. Cracks opening in the outer planking followed the seams in the inner planking, and in some cases, the cracks almost reached the chine. Ultimately, a boat would break in two. As in the British case, the proposed solution was a stronger gunwale (shelf): 1¾- x 8 inches of oak rather than 1½ x 3¾ inches of spruce. Admiral Cochrane, chief of BuShips, suggested that the problem was primarily one of shear strength, and molded plywood angles were installed to absorb longitudinal shear stress.

Further experience showed that, although bottom frames forward would crack in waves of a particular height and length, a boat could run through somewhat more severe seas without difficulty. Later, after the forward bottom framing had been reinforced and additional frames inserted, the chine became the weak point, and it too had to be reinforced. Other weaknesses were gradually discovered and fixed. Still, there were problems: weight was critical, and it was impossible for designers to analyze the hull structure mathematically. They began, then, with a reasonably robust structure, reinforcing it only when required so as not to introduce unnecessary weight.

Meanwhile, MTBRon 1 operated the 81-foot Higgins PT 6, which was clearly the best of the American-designed boats. The squadron's report of April 1941 was enthusiastic: the Higgins showed such good sea-keeping that further purchase of Scott-Paine boats was unnecessary. Early in 1941, BuShips lent Packard engines to both Huckins and Higgins, which wanted to build competitive boats at their own expense. The bureau agreed to buy the boats if they proved successful. The Huckins boat became PT 69, the Higgins 76-foot "dream boat," PT 70. Inspecting the experimental 76-footer at Higgins in May 1941, the commanders of MTBRons 1 and 2 were impressed by its extremely rugged construction. The boat, they felt, would be capable of effective operation in all sea and weather conditions without risking damage to the hull. They recommended the design as a prototype for future MTB construction. The Higgins 76-footer was designed to carry four torpedoes, ten depth charges (in two depth-charge tracks), and three twin 0.50-caliber machine guns—a heavier battery than the Elco's.

One more batch of Elco 77-footers was to be built. Plans initially called for twelve PTs (45–56) and twelve PTCs (25–36). However, only PT 45–48 were ordered on 10 April, the other twenty boats being ordered under lend-lease as BPT 1–20. All were renumbered in the PT series in December 1941, when the United States formally amalgamated lend-lease and other U.S. programs. PT 49–58 did enter British service, the United States retaining PT 59–68. These ex-British boats were the first in U.S. service with self-sealing tanks.

This final batch of 77-footers differed from earlier boats in several ways. The deck was stiffened, and the rear of the cockpit was reshaped for better all-round visibility. The enclosed power turrets were finally discarded.

An OpNav PT conference held in May 1941 and including representatives of BuShips, BuOrd, and the MTBRons, the Interior Control Board, and OpNav discussed the characteristics of future PTs. It was generally agreed that no PT in service combined the desired military characteristics with acceptable hull strength and sea-keeping. Since all craft prior to the 77-foot PT 20 had been found defective, it was by no means certain that the 77-foot Elco would be a sufficient improvement. Progress to date did show the possibility of designing a satisfactory boat.

Desirable characteristics included a maximum length of 75 to 80 feet; four 21-inch tubes firing forward; two 20mm guns in single, power-operated, stabilized mounts (with 0.50-caliber guns, an interim battery, in the 20mm positions); self-sealing gas tanks; ammunition for a single operation aboard (one thousand rounds per barrel); a speed of 40 knots at full war-service load minus a third of the fuel; a cruising radius of 500 nautical miles at 20 knots; one smoke generator; provisions sufficient for forty-eight hours; emergency rations for five days; and a suitable radio installation. These requirements were much like those the General Board had suggested a year earlier, except for the self-sealing tanks. The conference decided to save weight by reducing provisions and eliminating cold storage.

Finally, the conference recommended a series of comparative tests, the "plywood derbies," using the 81-foot Higgins PT 6, the new Higgins 76-footer, a Higgins 70-footer being built for the British, the new Huckins 72-footer (which became PT 69), the aluminum PT 8, and several Elco 77-footers (PT 20, 26, 30, 31, and 33). PT 20 had been specially strengthened and had special propellers. The conference strongly recommended that no more Elco 77-footers be ordered until the tests had shown that they were indeed satisfactory.

The Board of Inspection and Survey rated the experimental PTs on hull strength, habitability, internal accessibility, internal arrangement from the point of view of control, communication facilities, tactical diameter, sea-keeping, and speed. Because rough-water performance was so important, the derby was run in July 1941 at maximum speed over a 190-mile course—from New London, around Block Island and Fire Island lightship, then around the Montauk Point whistling buoy and back to New London. Since the Elco boats were the only ones with full armament, the others had to be ballasted with matching weights, which were not, however, evenly distributed over the hull. In this first series of trials, the Higgins 76-footer developed cracks across the decks and the shearing in nearby frames. The trials board attributed this to the concentrated deck weights. Moreover, the Elcos also betrayed structural weaknesses; deck stiffeners (strongbacks) were later installed in the 77-footers to prevent their decks from cracking.

Both on the measured mile and over the course, the Elco boat PT 20 won. It made 39.72 knots over the 190-nautical-mile course, 45.3 knots lightly loaded, and 44.1 knots fully loaded. The other Elco boat to finish came in second, the Huckins PT 69 third, and the Higgins PT 6 fourth; PT 70 had been forced to withdraw. On the measured mile, PT 70 made 41.2 knots light and 40.9 knots heavy, compared with 43.8 knots light and 41.5 knots heavy for the Huckins PT 69; PT 6 made only 34.3 knots light and 31.4 heavy. Even in a specially light condition, PT 8, the slowest, made only 33.9 knots and was criticized for being unmaneuverable. The Elco proved the least maneuverable; even the big PT 8 turned more tightly, although she was extremely slow (see table 5–2). Finally, in the relatively moderate seas encountered in July, PT 8 proved the smoothest riding of the boats (vindicating the scale-model experiments of 1938–39), the Elco the roughest. The Huckins and the 81-foot Higgins were intermediate, in that order.

Table 5-2. Maneuverability of Plywood Derby Boats

	TACTICAL DIAMETER (yds)	
	Turn to Port	Turn to Starboard
PT 69 (Huckins)	336	274
PT 6 (Higgins)	368	256
PT 8 (Bureau)	443	340
PT 20 (Elco)	432	382

Note: With her rudder full over, the Huckins boat slowed materially, to the point of ceasing to plane, and heeled outward. The others slowed but kept planing and heeled inward. The Huckins boat behaved like the others at rudder angles of less than 20°.

The aluminum PT 8 was clearly the strongest boat, but it was also unacceptably uncomfortable—hot in summer and, the trials board supposed, extremely cold in winter. That left the wooden boats, of which the Higgins PT 6 seemed the strongest, followed by the Higgins British boat, the Huckins, the Elco, and, worst of all, the dream boat. The board rated the latter best in habitability—at least before it developed structural problems and had to be shored up internally—followed by the Higgins PT 6, then the Huckins. The Elco was rated less habitable even than the aluminum PT 8; the small British Higgins rated at the bottom in this category.

In these boats, accessibility meant protected passage through the vessel. The dream boat started out best but ended worst when modified; the board justified the modifications on the ground that they added considerable strength. The Higgins PT 6 scored highest in accessibility, followed by the roomy PT 8 and the Huckins. The Elco was worse than the British Higgins.

For ease of control during attack, the dream boat was best, then the Huckins, the big Higgins, PT 8, the Elco, and finally the small British Higgins. For internal communication, the boats rated as follows, from best to worst: the Elco, the British Higgins, the dream boat, PT 8, and PT 6.

Relative costs are listed in table 5-3. The boats were attractive largely on account of their reasonable price, so the cheaper Higgins dream boat was preferable to the Elco.

Table 5-3. Relative Costs of Plywood Derby Boats[1]

	Hull	Machinery	Total
PT 6 (Higgins)	61,000	70,500	131,600
PT 70 (Higgins)	120,000	70,500	190,500
PT 69 (Huckins)	118,000	94,000	212,000
PT 20 (Elco)	157,600	70,500	228,100
PT 8 (Bureau)	268,200	413,000	681,400

1. Figures given in contemporary dollars.

A second series, tested over the same course reduced by 5 nautical miles, was held in August. By this time, PT 6 had been transferred to the Royal Navy under lend-lease, limiting the field to PT 8, 21, 69, and 70, with a second Elco, PT 29, assigned to pace PT 8 and take accelerometer (ride quality) readings at comparable speeds. This time the weather was much worse, with heavy 6- to 8-foot cross-swells off Montauk and waves as high as 15 feet between Block Island and Montauk Point. PT 8 gave the least satisfactory ride, her worst acceleration being twice that of the next worst boat off Block Island lightship. PT 69 suffered structural damage. Elco again won, but the effect of rough seas showed in a speed of only 27.5 knots, with Higgins close behind at 27.2. PT 8 made 25.1 knots, the Higgins British boat, 24.8. Even the winners had structural problems: PT 21 developed minor skin cracks, and some of the planking and the deck fastenings in PT 70 pulled loose.

Observers were impressed by the fact that the destroyer *Wilkes* completed the course only twenty-five minutes ahead of the winning PT. The board concluded that the endurance of the crew, rather

Opposite and above: PT 117 was typical of early wartime Elco 80-footers. Unlike the 77-footers, they had their machine guns mounted en echelon so that both could bear on either beam. In the stern view, note the six mufflers (not present in the 70- and 77-foot boats) and the powder-firing chambers atop the torpedo tubes. Tubes were trained by hand, turning on rings placed in the after pad under each tube and along training arcs in the forward pad. They could be fired electrically from the bridge or by hitting the striker knob, at the forward end of each firing chamber, with a mallet. Also visible in the stern view is the armor plate abaft the cockpit, which was often removed when boats reached forward areas. The bipod mast, soon to be surmounted by a radar antenna, could be folded down. Many boats carried four depth charges in place of each after torpedo tube.

than the structural strength of the hull, would determine a boat's ability to operate in rough water.

The Huckins design was considered good enough to justify immediate production. Although the board did not recommend as much, the boat had to be lengthened because when tested it carried only two torpedo tubes. Clearly, the board felt, Higgins had sacrificed too much strength to gain heavy armament in his dream boat; the earlier 81-foot PT 6 was better. PT 6 was, after all, large enough to accommodate the four-torpedo battery. The board therefore recommended that Higgins shorten it to fit the assigned battery. BuShips also had to modify its lines for higher speed. Finally, the Elco 77-footer would be acceptable if it was strengthened and its lines were modified to reduce pounding. The Elco had its chine raised forward to take care of the pounding, and length was increased by three feet.

By October, money was available for thirty-two more PT boats (71–102). The decisions of the conferences at the May gathering and the results of the trials were incorporated into a new series of specifications, presented at a BuShips conference with the principal builders, Elco, Higgins, and Huckins, on 6 October 1941. The new requirements generally paralleled those recommended at the May conference, except that an upper length limit was set at 82 feet, based on available transportation. Specifications

A typical Higgins boat, PT 200, newly completed in January 1943. The mast has been folded down, and the fixed torpedo tubes show Higgins compressed-air bottles for impulse launching. The bridge, closer to the bow than in an Elco, is flanked symmetrically by two twin 0.50s, which block the view. The small, streamlined object atop the forward superstructure is a siren.

called for silent low-speed operation, but there would be no special cruising engine. The complement would consist of one officer and eight enlisted men, with space for another officer. BuShips wanted more structural strength than existing boats had and simplicity for mass production. Armament would be four 21-inch torpedo tubes that could be replaced by depth charges, two twin 0.50-caliber machine guns with one thousand rounds per barrel, and a smoke generator. Since the PTs, owing in part to poor visibility, never used their internal steering stations, the new boats would have an external steering position.

Contracts were awarded on 19 November 1941 to Higgins, for PT 71–94, and to Huckins, for PT 95–102. Elco did not receive its next contract, for PT 103–38, until 17 January 1942, but that was still part of a prewar program authorized in November 1941. The Higgins 78-footer (PT 71) and the Elco 80-footer (PT 103) were the standard wartime types.

The Huckins was much less successful. Only eighteen were built, PT 95–102 and 255–64, and they served only in Hawaii and at the PT training base at Melville, Rhode Island. Despite the Huckins' success in plywood-derby tests, it proved too weak for hard service. With its wide transom, it tended to yaw excessively in a following sea. Following another PT, it would often fall off the other boat's bow wave rather than cut through it, causing the helmsman to lose control. Given the existence of two satisfactory designs, there was little point in redesigning the Huckins.

Wartime experience generally favored the Elco over the Higgins. Despite its higher freeboard, the Higgins was substantially wetter, throwing spray over the conning station. The Elco had a much wider foredeck, and its sections were much flatter below the forward chine, which tended to give the water more horizontal than vertical velocity. This lower silhouette was considered tactically valuable. Moreover, because its machine-gun turrets were arranged on either side of the conn (rather than en echelon, as in the Elco), the Higgins boat suffered from poorer visibility. The Higgins, however, made a tighter turn because it had larger rudders, greater draft forward, and less drag aft.

It does appear that the conventional structure of the Higgins stood up better in service than the Elco's airplane structure (almost monocoque, or longitudinally framed). The Higgins paid with greater structural weight and lower maximum speed.

The MTB training center at Melville (MTBSTC) conducted the first comparative trials of the standard war types. In September 1942, it reported difference of opinion over whether the Elco 80-footer,

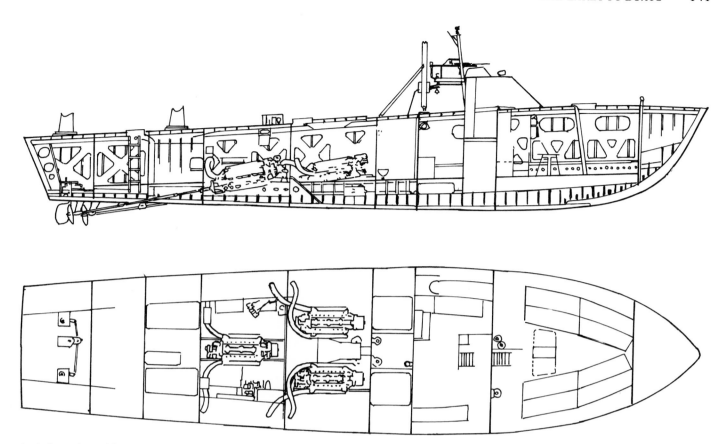

An inboard profile and partial internal view show the basic arrangement of a wartime Higgins boat. This drawing applied to PT 265–313. Note the divided gasoline stowage and the heavy framing in the inboard view, both normal frames and longitudinals extending through much of the hull's depth. In an Elco, the engines were further aft; gasoline stowage was concentrated forward of the engines just abaft the bridge, which was further aft than the bridge in a Higgins. (Norman Friedman)

the new standard, was really better than a reinforced 77-footer. The consensus ranked the 80-footer ahead of the 77-footer, which was followed by the Huckins and then the 78-foot Higgins. The 77-footer was lighter, smaller, and 3 to 5 knots faster; it accelerated faster, had a longer cruising radius using the same three-engine power plant, and enjoyed a shallower draft and smaller silhouette—all classic torpedo-boat virtues. The 80-footer was drier, heavier, and stronger (and therefore could sustain its speed in a seaway); it offered a softer ride; and it had better firepower ahead, a smaller turning radius, superior arrangement below decks, and an armored steering station. These virtues made the new PT 103 class an acceptable successor to the earlier Elcos.

Tests at Melville showed that the Huckins was 1 to 3 knots slower, the Higgins (displacing 126,000 pounds as opposed to the Elco's 106,000), eight to ten knots slower.[6] The Elco was the most economical of the three, its range exceeding that of the Huckins by 75 nautical miles and that of the Higgins by 150.

Both the Higgins and the Huckins showed materially larger silhouettes, and neither could bring guns to bear dead ahead (the Higgins not within 30 degrees of the bow, the Huckins not within 20 degrees). The Elco also had a much better internal arrangement; watertight doors in the bulkheads provided access throughout. In the other boats, which were closer to standard warships, the crew had to come on deck to proceed from one compartment to another. The tighter turning circle of the Higgins boat was discounted because it was so much slower.

MTBSTC wanted the Higgins abandoned altogether, an upsetting prospect with 131 Higgins and 158 Elco boats either built or on order. Given the pressure for sheer numbers of PTs, and the existence of two mass-production plants already tooled up, Elco and Higgins, there was little chance that either design would be abandoned or, for that matter, that a radically different PT would be ordered—unless one design or the other was disastrously bad.

The VCNO had never promulgated an official speed

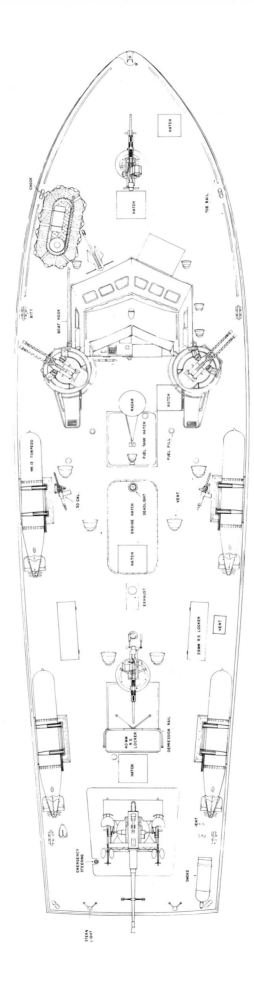

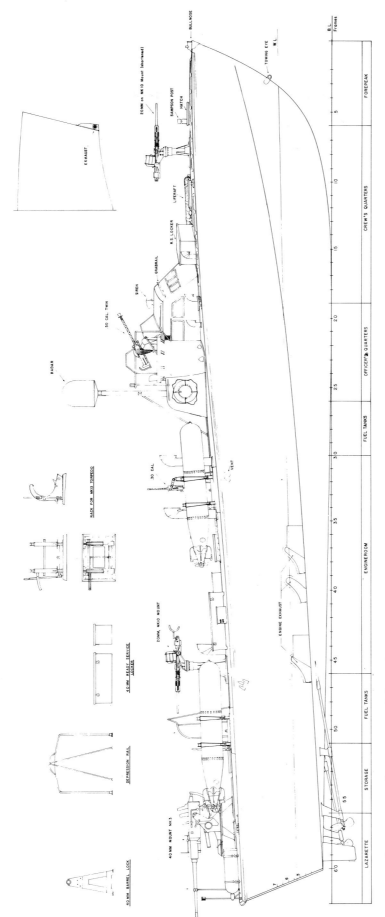

PT 209, a first-series 78-foot Higgins. (Al Ross)

PT-109
AUG 43

SERIES:	PT 103-196
BUILDER:	ELCO Bayonne, NJ
DATA:	Length 80'-0" Beam 20'-8" Draft 5'-3"
ARMS:	4 - 21" MK8 Torpedoes 2 - twin 5 cal MGs 1 - 20MM MK4 2 - MK6 Depth charges #1 - 37MM M3
MACHINERY:	3 - PACKARD 4M2500 petrol engines 4500/4050 B.H.P
DISPL:	38-40 Tons

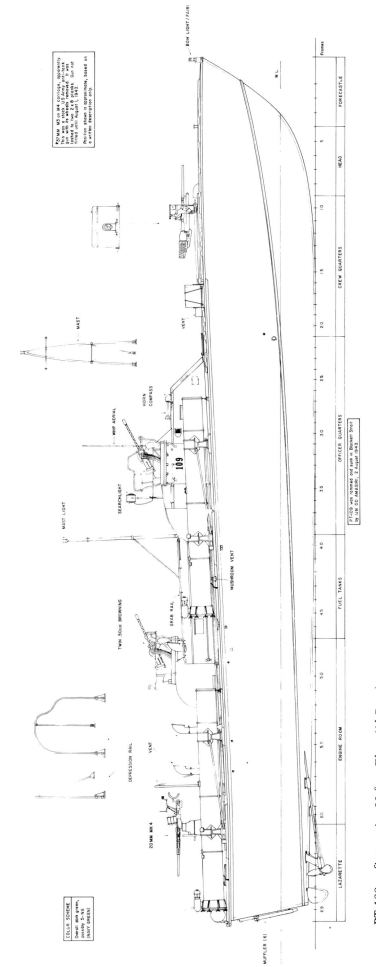

PT 109, a first-series 80-foot Elco. (Al Ross)

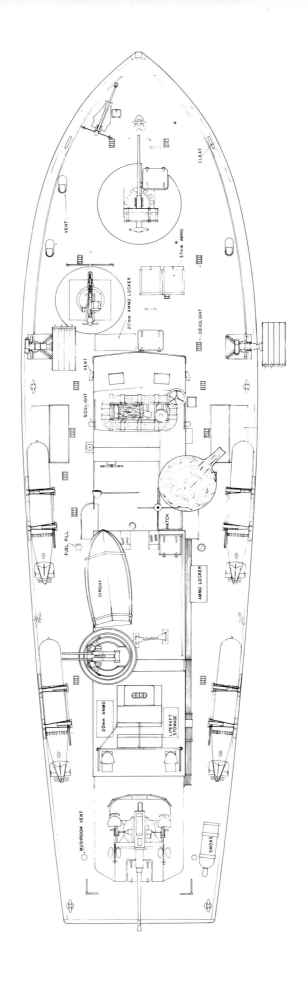

PT-596 — AUG 45

SERIES: PT 565-622
BUILDER: ELCO, Bayonne, NJ
DATA: Length 80'-0"; Beam 20'-8"; Draft 5'-3"; Displ. 50 tons
ARMAMENT: 4 – 21" MK13 torpedoes; 1 – 40mm MK3; 1 – 37 mm; 1 – 20 mm; 2 – twin 50 cal; 2 – 5" rocket launchers

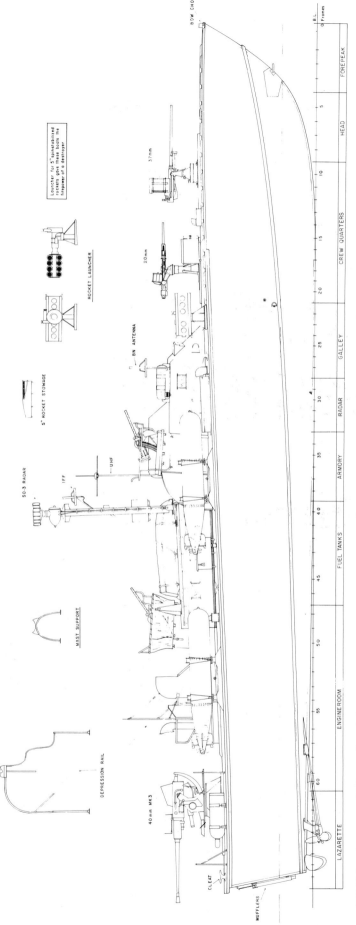

PT 596, a late-war 80-foot Elco. (Al Ross)

Here is PT 95, newly completed, with two torpedo tubes and dropping cradles for eight depth charges. It carries the new standard gun battery of two twin 0.50s and one 20mm cannon aft.

An unidentified Higgins PT at Salerno, 29 September 1943, has a small rocket launcher forward, to port, as does the PT in the background. Note the Higgins compressed-air torpedo tubes, the two added 0.30-caliber machine guns abaft the twin 0.50s on either side, and two 20s aft, the radio direction-finding loop atop the bridge, and the dazzle camouflage.

standard for the PT, but it was understood that a boat incapable of about 40 knots was not acceptable. Higgins had to suspend production—substantially revising its design to cut weight—until the Board of Inspection and Survey witnessed speeds above 39 knots in salt water. This suspension probably explains the disparity between wartime Higgins and Elco orders.

MTBSTC saw little point in any large PT boat, tracing excessive size to an overlarge torpedo, the Mark 8-3. Every extra pound of displacement carried its cost in speed, maneuverability, acceleration, and cruising radius. The center preferred the light, short-range, air-dropped-Mark 13 torpedo, four of which might fit in a 70-foot boat. Higgins would soon demonstrate that this could indeed be done.

In 1943, despite weight-saving measures, the speed problem recurred in the Higgins boat. The subchaser training center (SCTC) at Miami, responsible for shaking down Higgins boats, reported in mid-June that they averaged only 33 knots at war load. This was even worse than in 1942, although the boats performed somewhat better on their acceptance trials.

SCTC also criticized the Higgins on other grounds. Visibility from its bridge was poor—in particular, the turrets blocked the sides—and it was often difficult to pick up a target. The boat, moreover, was extremely wet; it was suggested that the chine be hardened to throw off water. The rudders had no feel, were overbalanced, nor did the wheel indicate when a straight course was being steered.[7] Mufflers were called for.[8] And the gasoline tanks, scattered from the bridge all the way aft, needed to be repositioned to lessen the danger of being hit. According to the training center, it made little difference to operating personnel whether they were blown up by 750 gallons of 100 octane or 3,000.

PT officers at SCTC complained that the Higgins carried too much nonessential equipment—heads and their plumbing, galley equipment, officers' quarters should all be dispensed with, as every pound added to the size and hence vulnerability of the boat and detracted from its speed. SCTC argued that boats rarely cruised for extended periods. MTBSTC replied that this equipment was indeed essential for some operations and should be installed when boats were built. Boats operating near a base, as in the Solomons, could have it removed in a forward area. Boats in the southwest Pacific often operated several hundred miles from their base and had to be self-sustaining. Moreover, they might cruise as much as a thousand miles when moving base.

Admiral Horne ordered the Board of Inspection and Survey to test a fully loaded Higgins boat in the open sea off Miami as soon as possible. In November 1943, two Elcos (PT 552 and 553) and two Higgins (PT 295 and 296) were compared. Their bottoms were carefully cleaned, their weights adjusted to near equality. One of each pair was powered by the W-8 version of the Packard, the other by the more powerful W-14. PT 296, with the W-8, displaced 106,500 pounds; PT 553, with the W-8, displaced 106,000. PT 295 displaced 105,100 pounds (94,300 light) and PT 552, 106,000 (94,500 light). The Higgins boats displayed superior acceleration and turning, the latter owing to larger rudders. Speeds were comparable: with the W-8, the Elco made 40.99 knots, the Higgins, 41.49; with the W-14, the Elco made 45.14, the Higgins, 43.90. At light displacement, the Elco W-14 boat made 45.83 knots, the Higgins, 44.91. The board concluded that engine performance made much more of a difference than any other design feature and suggested that the better acceleration of the Higgins could be traced to its propeller design.

The Elco was superior in visibility, and it rode better and was drier than the Higgins. Both seemed too large; the trials board wanted a smaller, longer-range boat with a true service speed of 40 knots, meaning a trial speed of 50. Most importantly, the Miami Beach trials proved that the Higgins was worth retaining in production.

Andrew Higgins himself did not doubt that a smaller, faster boat would be better. He had shown as much in designing the dream boat, and late in 1942, he built another boat of his own, the "Hellcat." He well understood the views of the MTB and SC training centers on size and speed, and PT officers who saw the Hellcat were suitably impressed. He argued that it was hard to ship the big PTs abroad owing to their size and weight, and that a smaller, faster boat, 48 to 50 knots, would be a more difficult target. It might even help cover landing operations, being carried to the amphibious objective aboard a transport (as, indeed, the contemporary LCS was). The VCNO, Admiral Horne, though he recognized that a new design might be valuable, thought that it would slow production. He was willing to buy the Higgins prototype but not to build it in quantity.

The new Higgins Hellcat, which became PT 564, was tested against a conventional 78-foot Higgins boat, PT 282, on 17 September 1943. Like the dream boat, the Hellcat was light, 80,000 pounds, for its heavy load, which was about 33,800 pounds (as opposed to 33,600 pounds for the 107,000-pound PT 282). With three Packards (W-8), it made 43.7 to 45.9 knots, compared with 39.6 to 40.9 knots for the larger boat (with W-14 engines). It handled well, turning faster and inside the larger boat. It did "hunt" back and forth while running straight ahead, probably because of the closing angle the spray strip at the chine made with the bottom (the bottom of the strip

The Higgins "Hellcat" (PT 564) epitomized the original PT philosophy, the elimination of any feature that seemed to be superfluous.

The Hellcat is seen here fitted with two experimental, fixed, remote-control machine guns, twin 0.50s. (Courtesy of Rear Admiral Barry K. Atkins, USN [Ret.])

was normal to the side and not faired in with the lines of the bottom).

Overall, the hull form, with its hollow V-sections, was quite similar to that of the early Higgins boats. However, the boat's chine was lifted, making for a slightly steeper bottom. Unlike the 78-footer, the Hellcat showed a tendency to pound.

Like the dream boat, it had a relatively light hull structure, the bottom of which would probably have to be redesigned, particularly in way of the engine room. Certainly this area worked noticeably during trials. Higgins also shrank his boat by omitting the galley and by squeezing down quarters; both modifications were perfectly acceptable to any officer who had seen Pacific service, as they bought high speed and, as importantly, fast turning for effective evasion. BuShips thought the pounding could be reduced simply by changing the slope of the bottom edge of the spray strip.

Elco also tried to develop a higher-performance version of its boat. Since PT 103 had been selected as the favored prototype for mass production, the company could hardly propose an alternative design. It could, on the other hand, seek to improve the performance of the existing boat. In effect, it rediscovered the steps used so successfully in World War I, fastening a series of six "Elcoplanes" to the bottom of a standard 80-foot Elco hull, PT 487. On 16 December, the boat made 55.95 knots lightly loaded, 53.62 knots fully loaded. It also proved extremely maneuverable.

The trials board was so enthusiastic that BuShips directed Elco to produce conversion kits for transport to operating squadrons. Early in January 1944, OpNav ordered a similar kit for Higgins boats. Later that month, MTBRon 29 tested Elcoplanes as it ran 1,250 nautical miles, at an average of 30 knots, from New York to SCTC. PT 558 and 559 did not have planes installed (but did have Thunderbolts on their sterns); PT 560–63, which were gunboats, had planes.

The squadron found that the Elcoplane had real drawbacks. The boats performed poorly in rough water. They were bow-heavy and tended to dig into heavy seas, making steering difficult. They accelerated more slowly, and their cruising radius was reduced by a quarter. Their oil consumption doubled. This should not have been altogether surprising, as the stepped planing boats of World War I had performed spectacularly in smooth water but very poorly in rough. Indeed, C&R had given birth to the U.S. PT in 1936 by emphasizing the advance in seakeeping achieved in foreign navies by adopting *stepless* (displacement) hulls.

Certainly some of the differences between planed and nonplaned boats could be attributed to differences in armament. The gunboats, carrying a heavy 40mm gun forward, were bound to be bow-heavy. They displaced about 8,000 pounds more because of their armament, and they carried an extra ton of spares and miscellaneous equipment. The Elcoplane was clearly sensitive to trim; MTBRon 29 found that the performance of the planed boats about equaled that of the unplaned with their bow guns removed and with about half their fuel aboard.

This sensitivity killed the Elcoplane. Whatever armament might be fixed in Washington or Melville or Miami, PT crews in forward areas tended to fill open deck spaces with guns and rockets. Boats would generally be overloaded, and sensitivity to trim or loading was clearly unacceptable.

By the end of the war, all the European boats had either been transferred to the Soviets or returned to the United States for reconditioning, pending movement to the Pacific. Many of those in the Pacific, concentrated at Samar, had seen hard service and were clearly not worth repairing. In September 1945, then, squadron commanders began to report which boats should be destroyed. The subsequent inquiry was a kind of final comparison between the Elco and Higgins designs, albeit one complicated by such factors as the degree of maintenance an individual squadron had been able to provide. The criterion was whether a boat required more than 1,500 man-hours of hull work to make it seaworthy, *not* to restore it to full combat capability. The squadron commanders initially recommended that 74 boats be destroyed. A special board surveyed the remaining 69 boats. The totals were 79 of 156 Elcos (50.6 percent) and 35 of 54 Higgins (65 percent).

The Elco was found to suffer from its relatively light construction, particularly when it had been loaded with heavier weapons. Decks were often soft and spongy, with leaky inner planking. Hull weaknesses revealed in 1941 still plagued the boat. The gunwales, apparently not strong enough to take the stress and working of a 65-ton boat, still sprung. This problem also applied to the Higgins boats. However, the heavier construction of the Higgins seemed better able to withstand the effects of the tropics and the recoil shock of heavy weapons. And, while the canopies of the Elcos tended to rot, the heavier Higgins structure seemed to remain fairly well preserved.

6

The Wartime PT

Elcos, which constituted the majority of wartime boats, were built in Bayonne, New Jersey, and commissioned at the New York Navy Yard, where some of the important armament modifications were developed. The squadrons trained at the MTBSTC at Melville, which also served as the PT research and development center. Higgins boats were commissioned at New Orleans, near the Higgins plant, and tested on Lake Ponchartrain. The SCTC at Miami was responsible for shakedown. It trained Higgins crews and, to a limited extent, duplicated the MTBSTC's role as a test and evaluation center.

Although the early BPTs (British PTs) were standard Elcos, later lend-lease boats derived from the standard British Vosper. Plans were circulated to potential bidders in June 1941, and the boats were built by Robert Jacobs, Harbor Boat (on the West Coast), Herreshoff, and Annapolis Yacht.

The initial U.S. war program, the "maximum effort program," was developed on the basis not of any tactical rationale but of an analysis of the capacity of America's shipbuilding industry. It was followed by a small craft program in FY 42.

From this point on, naval construction was planned according to the guidelines of continuing programs intended in part to make up for anticipated war losses, which were expected to increase as more boats reached combat areas, particularly in the southwest Pacific. Boats would also be discarded as they aged. The PT replacement program was set in March 1943 at ninety-six boats (eight MTBRons) per year, a figure confirmed in the fall of 1943 in the expectation that the boats would be increasingly important. Table 6-1 summarizes the wartime PT contracts.

Wartime PTs were usually built and commissioned for operation in twelve-boat MTBRons. Although some boats were transferred from one squadron to another to make up for losses, squadrons were usually homogeneous; modifications were generally squadron-wide. That is why construction contracts for these craft call for numbers that approximate the number of boats in an MTBRon.

The three original MTBRons consisted entirely of 77-foot Elcos:

MTBRon 1: PT 20–31, joined by boats assigned to the Philippines (31, 33, 35, 37, 39, 41–43), fought at Pearl Harbor, at Midway, and in the Aleutians. Many of these boats were later assigned to MTBRon 2. MTBRon 1 was initially formed to operate the experimental PTs, which were transferred to Britain during 1941.

MTBRon 2: PT 20–26, 28, 30, 32, 34, 36–40, 42–48, 59–61. Note that PT 59–61 were converted to gunboats in 1943 (see below). Squadron 2 operated the 70-foot Scott-Paines until their transfer to Britain in April and July 1941. Many of the squadron's Elco 77-footers were transferred to Squadron 1. With eleven Elcos, MTBRon 2 was assigned to the Panama sea frontier in December 1941, and then with six Elcos to the South Pacific late in 1942. The squadron was decommissioned on 11 November 1943. It was reconstituted—with PT 71–72, the prototype Higgins boats, and PT 199—for special (OSS) missions in the English Channel, carrying out twenty between May and October 1944.

MTBRon 3: In the Philippines, PT 31–35, 41 (the "expendables"); as reconstituted in the South Pacific, PT 21, 23, 26, 36–40, 45–48, 49–61 (in this form, MTBRon 3 was the first to arrive at Guadalcanal).

MTBRon 4 was the training squadron at Melville, initially including the first Huckins boats. However, PT 98–102 formed MTBRon 14 at the Panama sea frontier. The first two Higgins boats were also initially assigned to MTBRon 4, as were many others. PT 139–41 served in the training group at Melville. They had originally been assigned to MTBRon 8,

PT 462 was a typical midwar Higgins boat, armed with two 20mm guns aft and torpedo-launching racks rather than tubes. The bipod radar mast could fold onto the bracket.

which received PT 66–68 (ex-Melville) instead. In April 1946, MTBRon 4, which consisted of PT 613, 616, and 619–20, was transferred to the Operational Development Force.

Table 6-1. The Wartime PT Program

(a)	*Maximum Effort Program*	
	Elco	PT 139–96 (58) awarded 26 March 1942, directive of 19 January 1942
	Higgins	PT 197–254 (58) awarded 25 March 1942
	Huckins	PT 255–58 (4) awarded 16 May 1942
(b)	*FY 42 Small Craft Program*	
	Huckins	PT 259–64 (6) awarded 6 August 1942, directive of 5 June 1942
	Higgins	PT 265–313 (49) awarded 7 August 1942
	Elco	PT 314–61 (58) awarded 6 August 1942
	Elco	PT 362–67 (6) awarded 6 August 1942, subcontract to Harbor Boat on the West Coast
	Elco	PT 372–83 (12) awarded 15 October 1942; lend-lease for the Soviet Union (RPTs)
	Elco	PT 546–63 (18) awarded 15 October 1942; lend-lease for the Soviet Union (RPTs), but taken over by the U.S. Navy instead
(c)	*First Continuing Program (96 PTs)*	
	Higgins	PT 450–85 (36) awarded 27 April 1943, directive of 24 November 1942
	Elco	PT 486–545 (60) awarded 27 April 1943, directive of 15 March 1943
(d)	*FY 44 Second Continuing Program (96 PTs)*	
	Elco	PT 565–624 (60) awarded 23 March 1944, directive of 23 March 1943
	Higgins	PT 625–60 (36) awarded 6 May 1944
(e)	*FY 44 Third Continuing Program (48 PTs)*	
	Elco	PT 761–64 (4) awarded 2 March 1945, directive of 30 September 1944
	Elco	PT 775–90 (16) awarded 2 March 1945
	Higgins	PT 791–808 (18) awarded 30 January 1945[1]
(f)	*Lend-Lease (Vospers for Royal Navy)*	
	Jacob	PT 384–399 awarded 15 March 1943, directive of 20 January 1943
	Annapolis Yacht	PT 400–29 awarded 30 March 1943, directive of 19 March 1943
	Herreshoff	PT 430–49 awarded 14 May 1943[2]
(g)	*Lend-Lease for USSR*	
	Annapolis Yacht	PT 661–730 (60) awarded 31 March 1944, directive of 31 January 1944; Vospers
	Elco	PT 731–60 (30) awarded 13 March 1944[3]

1. PT 785–90 and PT 803–8 were canceled on 27 August 1945, PT 797–802 on 1 October 1945. Parts manufactured to build the canceled Higgins boats were reportedly used to build nine new boats for Argentina (LT 1–9), the first being commissioned in 1949. PT 791–96 were completed for U.S. service. PT 791–94 were stricken on 28 November 1945, PT 795–96 retained as noncommissioned small boats for experiments (reclassified 23 November 1945).
2. Earlier lend-lease boats for the Royal Navy were designated BPTs. BPT 1–10 were Elco 77-footers; BPT 11–20 were taken over by the U.S. Navy as PT 59–68. BPT 21–42 were Vospers built by Annapolis Yacht (21–33), Herreshoff (34–36), Jacob (37–42), Harbor Boat (43–48), and Annapolis Yacht (49–68). All later British boats were identified only by U.S. PT numbers.
3. PT 693–730 were completed after the end of the war, and PT 729–30 were retained by the U.S. Navy as uncommissioned small boats.

PT 103–14 formed MTBRon 5 in Panama and then, from the spring of 1943 on, in the South Pacific, joining earlier 77-foot Elcos (PT 62–65) transferred from Melville. PT 109–14 were transferred to MTBRon 2 on 22 September 1942. They were joined by earlier 77-foot Elcos (PT 62–65) in April 1943 and by later Elcos (PT 314–19) in March 1943. The squadron was broken up in February 1945. Some of its boats served in MTBRon 8 (PT 110, 113, 114), MTBRon 10 (PT 108) and MTBRon 18 (PT 103–05).

PT 115–26 initially formed MTBRon 6 in the southwest Pacific; they were later joined by 187 and 188. The squadron was decommissioned in May 1944, and its boats were distributed as replacements for other squadrons: PT 116 and 124–25 served in MTBRon 10, PT 126 in MTBRon 9, and PT 120–22 in MTBRon 8.

PT 127–38 formed MTBRon 7 in the southwest Pacific. The squadron was broken up in February 1945 as follows: PT 127 was transferred to MTBRon 12; PT 128 and 131–32 to MTBRon 21; PT 129–30 to MTBRon 8; PT 134 to MTBRon 25; and PT 137–38 to MTBRon 33.

PT 142–50 served in MTBRon 8 (with PT 188–89 after May 1944) in the southwest Pacific. PT 147–48 served in MTBRon 2 from January to June 1943.

PT 151–62 served in MTBRon 9 (with PT 126) in the South Pacific (southwest Pacific after May 1944). PT 318–19 joined this squadron in February 1945.

PT 163–74 served in MTBRon 10 in the South Pacific (southwest Pacific after April 1944). This squadron also operated PT 108, 116, and 124–25, replacing war losses including PT 165 and 173, lost when the tanker carrying them was torpedoed on 24 May 1943.

PT 175–86 constituted MTBRon 11 in the South Pacific (southwest Pacific from June 1944).

PT 187–96 served in MTBRon 12 in the Southwest Pacific (with PT 127, 145–46, 150–52 from the other squadrons listed above). The new boats of MTBRon 12 were the first to have torpedo launching racks instead of tubes (see below). PT 187–89 were transferred to MTBRon 6 in May 1943, before this squadron arrived in the southwest Pacific.

PT 73–84 formed MTBRon 13, which operated in the Aleutians, then the southwest Pacific.

PT 201–18 formed MTBRon 15 in the Mediterranean, operating under the tactical control of Royal Navy coastal forces. PT 201 and 203–17 were transferred to the Royal Navy in October 1944, and then PT 201, 204, 207–09, 211, 213, and 217 were transferred to Yugoslavia after British service. PT 202 and 218 were mined 16 August 1944. The Higgins boats in turn formed a pattern for reproduction by the Yugoslavs, who built ninety-six similar boats in 1951–60.

PT 219–24, 235, 241–42, and 295–301 served in

MTBRon 16 in the Aleutians and the southwest Pacific, after May 1944. In addition, PT 213–18 were originally included in MTBRon 16 but on 13 March 1943 were transferred to MTBRon 15 after only a few days' service. PT 295–96 were transferred to the Melville training center.

PT 225–34 formed MTBRon 17 in Hawaii, the Marshalls, and then the southwest Pacific, after April 1944.

PT 362–67, prefabricated by Elco and assembled on the West Coast, were assigned to MTBRon 18 in the southwest Pacific with PT 368–71. The latter had originally been ordered, in March 1942, for the Netherlands at Canadian Power Boat and were obtained under reverse lend-lease. They were 70-foot Scott-Paines, equivalent to early U.S. PT 9–19 and converted by installation of Elco fittings at Fyfe's Shipyard in Glenwood Landing, New York.

PT 235–44 formed MTBRon 19 in the South Pacific. It was broken up in May 1944, and its boats were transferred to make up losses in other squadrons. PT 236–38 and 240 went to MTBRon 20; PT 241–44 went to MTBRon 23.

PT 235–38 and 245–54 formed MTBRon 20 in the southwest and South Pacific after December 1944.

PT 320–31 made up MTBRon 21 (with PT 128, 131, and 132, transferred from MTBRon 7 in February 1945) in the southwest Pacific.

PT 302–13 formed MTBRon 22 in the Mediterranean. Originally, PT 265–76 were to have constituted MTBRon 22, but they were lend-leased to the Soviet Union instead. At the end of the Mediterranean campaign of April 1945, the squadron was shipped back to the United States for reconditioning and reassignment to the Pacific, but the war ended before it was ready.

PT 241–44 and 277–88 formed MTBRon 23 in the southwest and South Pacific after December 1944.

PT 332–43 and 106, the latter transferred from MTBRon 5 in February 1945, formed MTBRon 24 in the southwest Pacific.

PT 344–55 constituted MTBRon 25 in the southwest Pacific, along with PT 115 and 134, transferred from MTBRon 6 and 7 in May 1944 and February 1945.

PT 255–64 (Huckins) formed MTBRon 26 in Hawaii throughout the war.

PT 356–61 and 372–77 constituted MTBRon 27 in the southwest and South Pacific from December 1944.

PT 378–83 and 546–51 constituted MTBRon 28 in the southwest (from October 1944) and South Pacific.

PT 552–63 formed MTBRon 29 in the Mediterranean. PT 552–54, 556, and 560–63 were transferred to the Soviet Union in April 1945. PT 555 was mined on 24 August 1944, and 557–59 were transferred to Melville in November 1944.

PT 450–61 made up MTBRon 30 in the English Channel. It was returned to the United States for reconditioning and assignment to the Pacific, but the war ended first. PT 453–55 were transferred to MTBRon 31, Pacific Fleet, in May 1944, before shipment overseas.

PT 462–73 constituted MTBRon 31, Pacific Fleet (Solomons, Marianas, Okinawa).

PT 474–85 made up MTBRon 32, Pacific Fleet (Solomons, New Hebrides, Okinawa).

PT 488–97, and 137–38, transferred from MTBRon 7 in February 1945, formed MTBRon 33, southwest Pacific.

PT 498–509 constituted MTBRon 34 in the English Channel (498–504 and 506–8 were transferred to the Soviet Union in December 1944 and March 1945). PT 505 was assigned to Melville for training in December 1944. PT 509 was sunk in August 1944.

PT 510–21 made up MTBRon 35 in the English Channel. All boats were transferred to the Soviet Union in December 1944 and March and April 1945.

PT 522–32 constituted MTBRon 36 in the southwest Pacific. PT 533, which would have been the twelfth boat, was assigned to MTBRon 37 before it was placed in service.

PT 533–44 formed MTBRon 37, Pacific Fleet (Solomons, New Hebrides, Okinawa). PT 545, which had been intended for this squadron, was assigned instead to Melville for training.

PT 565–76 constituted MTBRon 38 in the southwest Pacific.

PT 575–88 made up MTBRon 39, Pacific Fleet. It arrived at Samar, in the Philippines, in July 1945 and saw no action.

PT 589–600 formed MTBRon 40, Pacific Fleet. Like MTBRon 39, it arrived at Samar only in the summer of 1945.

PT 601–12 constituted MTBRon 41. It was shaking down at Miami when the war ended.

PT 613–24 made up MTBRon 42, the only squadron commissioned after the end of the war (17 September 1945). It had been assigned to the Pacific Fleet but was never shipped. Construction of PT 623–24 was canceled on 1 October 1945. The four surviving experimental boats (PT 613, 616, 619–20) were ultimately transferred to the South Korean (ROK) navy.

PT 625–36 formed MTBRon 43. All were transferred to the Soviet Union under lend-lease in April 1945, together with the original MTBRons 44 and 45.

PT 761–72 were assigned to MTBRon 44 but canceled on 27 August 1944. The squadron was never commissioned.

PT 783–84 were assigned to MTBRon 45 but canceled on 27 August 1945. The squadron was never commissioned.

Tables 6-2a and 6-2b show the employment of U.S. PTs as of 1 January 1945.

The big Elco and Higgins boats fought up through the southwest and South Pacific to the Philippines and, in Europe, through the Mediterranean. They also served, less spectacularly, in the Aleutians and in the English Channel.

Particularly in the Pacific, their role shifted dramatically. The "expendables" in the Philippines in 1942 were classic torpedo boats, attempting to stop the Japanese transports and their escorts. In the Solomons, with the Third Fleet, they faced Japanese destroyers and cruisers, some participating in the infamous Tokyo Express, and so needed their torpedoes. They also had to attack and destroy coastal troop-carrying barges. In the southwest Pacific (Seventh Fleet), around New Guinea, barges predominated and larger ships were rare. There MTBRons concentrated on gun armament almost to the exclusion of torpedoes. As a result, MTBRons in the South Pacific sought a barge-busting gun that would not interfere with what they considered their primary torpedo battery, the 37mm cannon. The southwest Pacific MTBRons enthusiastically adopted the much heavier 40mm Bofors, accepting in many cases the removal of two torpedoes as compensation. The two theaters of PT operation in the Pacific merged in the Philippines in October 1944.

There, particularly at Leyte Gulf, PTs delivered classic torpedo attacks. Although few major surface targets were left by early 1945, the PTs were valued as the primary defense against sneak attacks that were expected to be made against the amphibious fleet invading Japan.

In Europe, the Mediterranean was the main PT theater; the boats there spent most of their time attacking heavily armed and armored shallow-draft barges. They also delivered small raiding parties and carried deception (beach-jumper unit) teams to support amphibious landings.

With their shallow draft, barges were poor torpedo targets. They had to be attacked by gun or rocket fire. Thus, PTs often functioned as gunboats; but they also served as torpedo boats to justify the retention of their torpedoes or, more importantly, to maintain the flexibility to convert back and forth between the two roles.

The roles did conflict. A torpedo boat needed stealth to approach its target, maneuverability and acceleration to escape after it had exposed itself by firing. Torpedo attacks were typically made in darkness, at low, quiet speed (approximately 8 to 10 knots) with engines muffled. If possible, the boat also retired slowly, on a radically different course. Maneuverability was, therefore, essential both at high and low speeds. Rapid acceleration was needed so that the boat could escape if it was detected.

Table 6-2a. PT Employment, 1 January 1945

Numbers of each type built and/or contracted for (total 808)	
Elco 70ft	11
Elco 71ft	49
Elco 80ft	358
Higgins 78ft	221
Huckins	18
Scott-Paine 70ft	4 (converted)
Vosper	136
Experimental	11
Status	
Commissioned USN (PT)	335
Reclassified USN	20
Stricken USN	71
Lend-Lease RN	65
Lend-Lease USSR	181
Undelivered USN	136
Stricken USN	
Surveyed, Obsolescent	12
Lost as Operational Casualty	17
War Loss (enemy action)	42
destroyed to prevent capture	18
shore batteries	7
planes	4
surface ships	7
mines	4
in transit	2
captures	0

The essential feature for torpedo attack, then, was minimum size, one reason the British chose the compact Vosper boat. The only advantages of greater unit size in a torpedo boat were sea-keeping and firepower (in the form of four rather than two torpedoes, and sometimes 21-inch ones, not 18-inch). Speed, maneuverability, and acceleration were all useful to a gunboat, but so was sheer size, particularly length to accommodate more weapons. Guns, moreover, would benefit from a larger, hence more stable platform.

The U.S. approach, almost always to combine the two roles and thus to accept relatively large boats, was by no means universal. The British chose a much smaller standard motor torpedo boat, the Vosper, and built solid motor gunboats during World War II. After the war, they tried to develop a small "convertible" boat but found that it would need a larger hull.

This evolution was reflected in PT armament. The officially approved battery was often radically altered in forward areas, particularly in the southwest Pacific, where barge-busting was so important. Extemporized weapons were sometimes adopted as standard, but many experiments were never officially recorded. PT armament remained controversial throughout the war.

At the outset, the heaviest item of PT armament was the Mark 18 torpedo tube, which had to be trained

Table 6-2b. PT Types Assigned (471 Total)[1]

	Elco 77ft	Elco 80ft	Higgins 78ft	Huckins	Experimental	Scott-Paine
Seventh Fleet	1	146[2]	78	—	—	2
Pacific Fleet	—	36[3]	39[2]	—	—	—
Hawaiian Sea Frontier	—	—	—	10	—	—
Eighth Fleet	—	—	11	—	—	—
Twelfth Fleet	—	23	9	—	—	—
Atlantic Fleet	—	13	5	8	1	—
Total	1	218	135	18	1	2

1. One PT unit is up for striking. Twenty-two PT units of the Twelfth Fleet are to be transferred to the USSR. PT 505 is in transit from the Twelfth Fleet for overhaul and temporary duty with MTBRon 4, Atlantic Fleet.
2. Twelve units under new construction.
3. Twenty-four units under new construction.

outboard to fire. Perhaps the most important wartime development was its replacement by a lightweight launching rack. Elco was the sole *wartime* maker of torpedo tubes, responsible for 820 of a total of 1,014 made between 1 July 1940 and 1 September 1943, when the program ended. The company subcontracted all of the work and encountered severe delays throughout the war; eventually PTs had to be delivered without tubes, then returned to the builders for installation.

At the same time, BuOrd encountered the severe shortage of 21-inch torpedoes predicted before the war. However, 22.5-inch aerial torpedoes (Mark 13) were relatively plentiful after the Battle of Midway demonstrated the problems associated with torpedo bombing. A 22.5-inch tube was designed for the Higgins-built PT 296–301, among others. By the time it had entered production, a much simpler side-launching rack had appeared, and only 92 of the 400 manufactured (out of 1,000 ordered) were ever installed.

The side rack could carry either the 21- or the 22.5-inch torpedo. The rack was similar to a pair of depth-charge releases and beginning in 1942, culminated a series of attempts, by the MTBSTC at Melville to find an alternative to the cumbersome tube. The rack was reportedly invented one night in February 1943 by Lieutenant George Sprugel and Lieutenant (j.g.) James Costigan of PT 188, who wanted a torpedo that could be rolled into the water rather than fired. At this time, BuOrd considered the tube essential to prevent the standard torpedo gyro from tumbling, but airborne torpedoes had to be able to slide into the water. An experimental model was built at the New York Navy Yard scrap shop, and PT 188 tested it at the Newport torpedo station with a Mark 13. A BuOrd officer witnessed the tests, and MTBRon 12 was the first to be equipped with the new system. By August 1943, the standard approved PT armament was four launch racks.

It is often reported that BuOrd had rejected rack launching on the ground that it would upset the gyro of the conventional 21-inch torpedo; but the bureau later described the rack as suitable for both 21- and 22.5-inch weapons, and PTs later carried 21-inch torpedoes in it.

When the lanyard was pulled, the torpedo gyro and motor fired. Once they were running at full speed, pins were pulled, and the torpedo rolled into the water. Much the same idea was later applied to the launching of lightweight ASW torpedoes.

Besides being lighter, 540 pounds as opposed to 1,450 for a Mark 18-1 tube—weight was also saved with the much lighter torpedo, 1,927 pounds instead of the Mark 8-3C or -3D's 3,026—the launch rack also saved considerable space and did not interfere with the temporary installation of guns. It therefore proved convenient when the Pacific PTs increased their gunboat role. Many boats were modified in forward areas.

Tubes were retained only for PTs destined for northern waters (MTBRons 13, 16, and 17 in the Aleutians and those for the Soviets), since the launching racks provided no weather protection for torpedoes.

Torpedo tubes were also manufactured under lend-lease for the British and Soviet governments. The British used a Mark 18-2. The Soviets used a Higgins-designed Mark 19, actuated by compressed air rather than black powder. Unlike the conventional Mark 18, Mark 19 was essentially a closed tube, loaded as a unit at a pier or from a tender. It was considered for use aboard American PTs in 1943, but it is not clear whether any were actually installed.

Both wartime classes were initially designed to carry four 21-inch torpedo tubes and two twin 0.50-caliber machine guns, the latter in adapted Bell aircraft mounts with hydraulic recoil absorbers. In addition, some Elco 77-foot boats had two single 0.30-caliber Lewis guns forward. As an alternative, others carried a 0.45-caliber Thompson submachine gun.

After the end of 1941, there was an alternative approved battery in which the after pair of tubes was replaced by eight 420-pound Mark 6 depth charges, either in individual release racks (depth-charge track type C) or in two four-charge stern tracks. The latter appeared in 77-foot Elcos only.

In these 3 July 1944 photographs taken at Puget Sound, PT 77, an early Higgins 78-footer, displays standard war modifications: torpedo-launching racks (with only mounting pads fitted for the after pair), a 40mm gun aft, a 20mm gun forward, and two free-swinging single machine guns abaft the torpedoes. Note the smoke generator aft and the pipe guard just forward of the 40mm gun that prevents it from firing into the boat; the free-swinging single guns have no such guard. The "ski pole" antenna forward of the radome is an IFF transponder.

PT 515, a typical late-war Elco, newly built on 11 April 1944. The gun forward is a 20mm Oerlikon with a ready-use ammunition box abaft. Aft, the boat has a standard army-type (unpowered) 40mm Bofors gun. Note the two torpedo-launching racks and the depth charges abaft them. The boxes aft probably held spare parts, which the Elco would carry to a forward area. The helmsman and captain can be seen in the cockpit.

Some of the 77-foot Elcos, and all of the others, carried a single 20mm Oerlikon (Mark 4) aft. In the short Elcos, the cannon was part of the alternative to the after tubes. This battery was standard through the end of 1942. In a few 77-footers, at least PT 20, 22, and 29, a single 0.30-caliber machine gun was installed early in 1943, when two additional torpedo tubes replaced the depth charges.

However, at the end of 1942, new Elco and Higgins boats were being completed with two 20mm guns (one Mark 14 forward and one Mark 10 aft), along with the usual two or four tubes and two twin 0.50s. Many earlier boats were fitted with the additional 20mm gun.

In mid-November 1942, Cominch (the CNO's organization) and the VCNO revised the ultimate PT battery to show single 20mm guns replacing the twin turret-mounted 0.50s in boats that would not be delayed by this change. In December 1942, BuOrd ordered the New York Navy Yard to manufacture twelve 20mm scarf mounts for MTBRon 10, the displaced 0.50-caliber guns would remain aboard until the reliability of the 20s was proven. One 20mm would be mounted in lieu of the starboard 0.50s in each boat. PT 172 conducted firing trials of a scarf ring 20mm mount (starboard turret) off Ambrose lightship on 21 January 1943, and early in February, installations were ordered for MTBRons 9 and 10. Few boats were altered, as the MTBSTC did not like the 20mm scarf ring. However, in April 1943, the manufacture of conversion kits (to turn a 0.50-caliber Mark 17 scarf-ring mount into a 20mm Mark 12 mount) began; a total of thirty kits were shipped to MTBRons 9 and 10, twenty-four to MTBRon 11, and twenty to MTBRon 12.

In February 1943, then, the approved but often not installed PT battery was changed to three 20mm guns—two single 20s in scarf rings replacing the earlier pair of twin 0.50s.

In March, PT 196 was fitted with an experimental lightweight Elco 20mm gun forward (515 pounds including one loaded magazine). MTBSTC liked the gun, but it could not be interchanged with other standard 20mm guns and thus did not enter production.

Combat experience showed that more forward-firing AA firepower was needed. Late in March, Cominch called for extra 20mm guns in each MTBRon. That in turn required a new lightweight mount. New York Navy Yard devised a Mark 13 mount, four of which were installed in MTBRon 18 boats. It was successful but had to be abandoned because it could not be interchanged with other 20mm mounts.

BuOrd developed a new standard, lightweight 20mm mount, the Mark 10, 675 pounds compared with 1,700 for the earlier Mark 4; New York Navy Yard devised a low-trunnion version, the Mark 14, suitable for installation in the bow of a PT. MTBRon 12 received ten Mark 14s and six Mark 10s before it departed for the southwest Pacific.

In April 1943, then, the planned battery was changed to a Mark 14 20mm gun forward, a Mark 10 aft, and two 20mm guns in scarf rings. Most boats, however, actually retained their two twin 0.50s, and by June 1943, with the failure of the 20mm scarf-ring mount, the approved standard was two 20mm fore and aft and two twin 0.50s. In the Elco boat, one 20mm was mounted forward and one aft; in the Higgins, one right aft and one amidships, leaving the forecastle open for another weapon.

In February 1944, MTBRon 17 boats (PT 225–34) received a second Mark 10 20mm gun aft on the centerline before being shipped to Majuro, in the Marshalls.

Many boats also carried 0.30-caliber machine guns. In July 1943, MTBRon 17 asked for two 0.30-caliber, air-cooled machine guns, with a pipe-stem mount fixed to each torpedo tube so the weapons could be set up to fire broadside or along the centerline. This became a standard installation, with the pipes set on the deck after torpedo tubes were replaced by launching racks. MTBRon 22 (Higgins 78-footers, PT 302–13) proposed a related barge-busting weapon in November 1943: a twin, water-cooled, 0.30-caliber machine gun on a shortened Lewis gun pedestal in the bow. The squadron argued that any larger gun would be difficult to handle. Water in the jackets would keep the barrels cool, even without any connection for fresh water. The twin 0.30 was actually tried aboard PT 293, but BuOrd rejected it in favor of an air-cooled weapon. In the interest of standardization, single 0.50-caliber guns replaced the single 0.30s late in the war.

PT 564, the Higgins Hellcat, was fitted around February 1944 with an experimental barge-busting armament: two twin 0.50-caliber machine guns fixed on deck and fired remotely from the cockpit.

The next step was to add heavier weapons:—the 40mm Bofors, the Elco Thunderbolt (multiple 20mm cannon), and the 37mm cannon. The 40mm was specially favored in the southwest Pacific, the 37mm in the South Pacific. The Elco Thunderbolt was tried in 1942–43, but installations began in earnest only at the end of the war.

Barge-busting, the forte of the 40mm gun, did not become important until mid-1943. However, in mid-December 1942, BuOrd suggested that the single 40mm be tested aboard a PT boat. Installation was authorized on 24 December. PT 174 was modified while under construction, her foredeck strengthened and a shielded, army-type (hand-operated) gun mounted. Neither MTBSTC nor BuOrd was enthusiastic. There was no structural problem, but the gun, placed in the area of the boat most subject to violent motion, would be difficult to handle in a rough sea. It was barrel-heavy, owing in part to its shield, and tended not to remain on target: it tried to train itself when the boat rolled. Even so, PT 174 served effectively in the South Pacific.

BuOrd, concluding that the gun would be usable if only it was moved to an area of less violent motion, the stern, requested experimental installations aft.

In 1943, Elco PTs in Morobe Inlet, New Guinea, show field modifications. The boat on the left has the field-installed SCR-517A (army air corps) radar and an extra shielded twin machine gun, probably 0.30 caliber, on the foredeck. The shrouded object on the righthand boat appears to be a multiple bazooka, perhaps appropriated from an airplane. There might also be a second shielded twin 0.30. (Courtesy of Rear Admiral Barry K. Atkins, USN [Ret.])

The gun replaced the 20mm cannon formerly mounted there. This position also allowed for greater angles of depression.

MTBRon 12 (PT 187–96), southwest Pacific) was selected in March 1943 for experimental armament installation. All the boats had scarf ring 20s in place of their 0.50s, and PT 187–92 had 20mm guns fore and aft (Mark 14 and Mark 10 mounts, respectively). PT 193–96 had one 40mm gun aft, replacing the after 20mm gun. The 40mm installation is sometimes attributed to Lieutenant Commander John Harlee, who commanded the squadron. He reported that the guns were very satisfactory. Elimination of the torpedo tubes saved so much weight that installation of the 40mm gun imposed almost no speed penalty; boats made 38.5 knots even being in the water for several weeks with their engines untuned.

MTBRon 12 arrived in the South Pacific in August 1943. By February 1944, its boats had the following: a 0.50-caliber machine gun (40mm aft) or a 37mm cannon on a low bow mount; a bow 20mm cannon; the usual pair of twin 0.50s; one 0.30-caliber machine gun on each side of the bridge; one or two single or twin 0.50s in low mounts on deck, either between the torpedoes or forward of them; and a 40mm or 20mm gun aft. Harlee considered the 0.30s effective against low-flying aircraft and barges at close range. One boat had 20mm cannon in its turrets plus the two twin 0.50s on deck in low mounts, a combination that seemed satisfactory, since the boat in question had destroyed barges and damaged aircraft without ever being hit. Harlee's officers generally agreed that this was a suitable battery, and that boats should have one 40mm gun aft and plenty of firepower in all directions.

In the southwest Pacific, the Seventh Fleet enthusiastically supported the 40mm gun. It was more accurate and powerful than the contemporary 37mm weapon and had greater range. Typically, it made a 12- to 14-inch hole in a barge, while the 37mm made

At Leyte in November 1944, massed PTs display a variety of field modifications. The boat in the center (PT 21?) has a gravity-fed, finned rocket launcher forward. Note also the helmets on the deckhouse and the tarpaulin covering the support onto which the mast can be folded. The ring under the tail of the forward rack-mounted torpedo was originally installed to take a conventional (trainable) torpedo tube. In both this boat and the one in the foreground, note the prominent aircraft-type star marking to prevent accidental attacks by friendly forces. A 37mm aircraft gun is visible on the boat alongside in the background. Note the free-swinging single machine gun, shrouded, on the boat in the foreground, just above the forward torpedo. A boat in the background shows depth charges abaft its torpedoes and a smoke generator athwartships, at the stern. The prominent boxes at the bases of the masts are the SO radar transmitter-receivers, fed by and feeding the cables visible beneath the box on the boat in the foreground.

MTBRon 9 boats at Leyte, December 1944. PT 150, in the foreground, clearly shows its mast brace as well as a pipe stem mount for an extra free-swinging machine gun on the cockpit bulkhead, just inboard of the torpedo.

a 3-inch hole. The smaller weapon was sometimes defeated by barge armor. The 40mm also suffered less from gun and ammunition failures. It was most effective when firing single shots (at about one shot per second); keeping it on target during automatic fire was a difficult task.

Selman S. Bowling, commander of MTBRon 21, which arrived shortly after Harlee's MTBRon 12, admired his 40mm guns so much that he obtained twelve from the Royal Australian Navy in early September 1943. In November 1943, the Seventh Fleet asked for thirty more single 40mm guns to complete the outfitting of its boats. At this time it rejected the 20mm Thunderbolt, which was not heavy enough to sink barges or to engage 75mm guns ashore. The fleet recommended a battery of two 40mm guns fore and aft and two twin 0.50s, arranged to permit simultaneous fire from both turrets to at least one side.

In June 1943, MTBRon 25, in the southwest Pacific, asked that four of its boats be fitted with 40mm guns aft. They were approved for PT 352–55.[1] In August, approving this installation, Cominch suggested a similar one for Third Fleet PTs so they could test the 40mm in the South Pacific, where barges were also now the primary targets. Thus MTBRon 27 had four PTs (374–77) fitted with 40mm guns.

In November 1943, six MTBRon 33 boats, PT 492–97, were assigned single 40mm guns aft and self-sealing tanks, In December, four MTBRon 28 boats, PT 548–51, were assigned single 40mm guns aft.

In October 1943, MTBRon 26 asked if it could mount 40mm guns aft in four boats, PT 255–56, 259, and 260. These were fitted, and the squadron later requested that all its boats be so fitted.

Boats for European waters were also modified. In November 1943, six MTBRon 22 boats were assigned single 40mm guns. They were also armored with 10-pound STS pending availability of self-sealing tanks. Rearmament of the remaining boats of the squadron, PT 302–7, was authorized the following March. In December 1943, the VCNO (VopNav) ordered single 40mm guns for MTBRons 30 (PT 456–61) and 34 (PT 504–9), both intended to operate in the English Channel. In January 1944, VopNav ordered single 40mm guns for two Mediterranean boats, PT 204 and 214 of MTBRon 15. Installations in the other boats—guns to be obtained locally from the army—were approved at the beginning of February. Similar installations were approved for MTBRon 35 early in January. In March 1944, OpNav authorized the single 40mm gun aft for eight MTBRon 29 boats in the Mediterranean, PT 552–55 and 560–63, excluding boats with Thunderbolts.

As of 1 March 1944, thirty-seven southwest Pacific boats had 40mm guns, fifty-eight had 37mm guns forward. The preferred battery was one 40mm aft, one 37mm forward, a 20mm gun forward to port, the usual two twin-turreted 0.50s, two twin 0.30-caliber machine guns in the bridge wings, and a twin 0.50 opposite the forward 20mm gun. Many boats also had twin 0.50s on low pedestals between the torpedoes, port and starboard, for a total of three such mounts. This very heavy battery was essential for barge-busting because the weight of the opening burst might well decide the fight. As a result, boats needed larger crews.

Boats were also expected to carry their four Mark 13 torpedoes in launching racks, with a depth charge between each pair of torpedoes.

By October 1944, all southwest Pacific boats except for the two 70-foot Scott-Paines had the 40mm guns.

The 40mm gun was much less popular in the South Pacific. Its weight cost speed and torpedoes, both of which the South Pacific MTBRons considered valuable insurance against Japanese destroyers. Thus they sought a *lightweight* barge-busting gun, while their southwest Pacific contemporaries were quite willing to adopt the heavy Bofors.

By May 1943, South Pacific PTs were clearly as much barge-busters as torpedo boats. Armored and well-armed barges were extremely frustrating targets. One PT commander reported that on 3–4 August 1943 he had to close within 100 yards, firing more than a thousand 0.50-caliber rounds and five hundred 20mm rounds in Hathorn Sound to damage only two of seventeen barges; only one sank. Even that required ten near-suicidal minutes close to Japanese shore batteries.

Since lightly protected PTs seemed impotent, the South Pacific commanders asked for a U.S. equivalent of the Japanese barges. On 14 August, for example, R. B. Kelly, commander of PTs at Lever Harbor, suggested that an LCM(3) be converted to a gunboat and carry a 40mm gun forward and two 20mm guns on each side amidships, the 0.50s being retained aft. Its engine would be muffled, and sandbags or armor would cover its engine, fuel tanks, and conn. Radio equipment would be limited to a walkie-talkie. One LCM actually was converted; its ramp was removed, a bow was welded on, and it was lightly armored and armed with one 3-inch/23 and two 37mm cannon. The boat, capable of 14 knots, made its maiden patrol on 18–19 January 1944.

The LCM, precursor of the Vietnamese river gunboat, was not particularly impressive. As an expedient, four LCIs were converted to antibarge gunboats, armed with a 3-inch/50 forward, a 40mm aft, four 20mm guns, and six to eight 0.50-caliber machine guns. The ships had limited armor protection and

Photographs of the field-modified "gunboats" are extremely rare. The boat on the right is PT 61, on 8 October 1943; note the two shielded 40mm guns and the complete absence of torpedo tubes. The position of two twin 0.50s proclaims it to be a 77-foot Elco. The boat on the left is PT 168. (Courtesy of Lieutenant Commander Edwin W. Polk, USN [Ret.])

could make 16 knots. Other LCIs were later converted to gunboats, but primarily for amphibious fire support.

The PGM was intended as the ultimate solution to this problem, and PGM 5, 7, and 8 arrived at Rendova in the spring of 1944. They could turn with a PT but, being much slower, were difficult to coordinate tactically.

Locally, the South Pacific force converted three 77-foot PTs, 59–61, into specialized motor gunboats. They were completed about the end of September 1943, with single shielded 40mm guns fore and aft. Their torpedo tubes were removed, and six to eight single 0.50-caliber machine guns were added along their sides. The single 20mm gun aft was retained, and all gun positions and the cockpit were armored with ⅝-inch (25-pound) STS. Also included were two bazookas for sinking barges after the automatic weapons had destroyed their firepower.

By April 1944, the first three gunboats were worn out. The tender *Argonne* converted PT 283–85 into replacement boats, with engine room armor, 40mm guns fore and aft, and a total of six twin 0.50-caliber machine guns, four on deck, all in shields. PT 283, soon lost, was by a converted PT 282.

On 25 October 1943, OpNav authorized the completion of PT 543 as an experimental gunboat with two 40mm guns and two twin 0.50s. PT 560–63 were also completed as gunboats, with similar armament.

The most suitable antibarge weapon available in the Solomons in mid-1943 was the army single-shot 37mm antitank gun, first fitted late in July 1943. It was clearly less than ideal with its limited traverse. By mid-August, three boats at Rendova (PT 116, 157, and 159) had had their 0.50s replaced by single-shot army 37mm antitank guns, some lashed to coconut logs, to counter the heavy batteries of Japanese barges in the New Georgia area. Another boat, probably PT 155, was similarly fitted at Tulagi and transferred to Rendova.[2] The jury-rigged 37mm was an obvious and immediate success; for the first time, barges had to flee the PTs.

An automatic weapon would be better, because its high rate of fire and tracers could compensate for the irregular motion of the boat. The two alternatives were the army 37mm AA gun, somewhat lighter than the Bofors, and the 37mm aircraft gun used in the P-39 fighter. The AA gun was much too heavy, but the aircraft gun was extremely successful. Again, mountings were devised by the local PT bases. The first guns were very successfully tested in September, and by mid-October several squadrons were

An MTBRon 8 boat off the coast of New Guinea in 1943 has a field-mounted 37mm antitank cannon forward and a field-installed army SCR-517A radar in a large "beehive" radome. (Courtesy of Rear Admiral Barry K. Atkins, USN [Ret.])

equipped with them. Most were mounted before the forward passage hatch to provide heavy firepower for a bow-on approach. The 37mm aircraft gun had a thirty-round magazine and therefore could be operated by one man rather than the two or three required to man the 37mm army AA guns or 40mm Bofors guns. However, standard practice, at least in the South Pacific, was to use a loader and a gunner. The loader spotted and monitored the number of rounds left in the magazine, which could easily be emptied in a fire fight. Because the gun had a light recoil, it could be used with a light, 213-pound mount. By February 1944, fifty-seven South Pacific boats had been fitted with it, and work was proceeding on all others.

At first, it was typically mounted with an armored shield in a low trunnion mount well forward, with a thirty-round, endless-belt magazine that could be unloaded easily without being removed from the gun. Unfortunately, the center of balance moved abaft the trunnion when the loaded magazine was in place, and guns had to be held down. They were fitted with counterweights, initially in the form of armored shields. Guns were unloaded and lashed down to a frame fixed on deck, except when the ship was on station or in contact with the enemy, because the lightly built magazine could not withstand vibration due to high speed or rough water. Another problem was warping, which caused the magazine to jam.

Later a magazine consisting of a curved track filled with shells was improvised, and so the crew was reduced to a single man. Many such guns were installed at forward bases on short improvised pedestals. Boats newly completed with 37mm guns had them on taller, factory-installed mounts similar to 20mm pedestals, which first became available about August 1944.

Late in the war, the medium-velocity M4 (2,000 feet/second) was replaced by the M9 (2,500 feet/second), an improved standard aircraft gun. MTBRon 21 had the first five boats fitted with the M9 in January and February 1945. By August 1945, when modification work was suspended, twelve Seventh Fleet boats had been fitted with the new gun, and enough had arrived at the main base at Samar to equip the others.

By the spring of 1944, the typical South Pacific boat was armed with two 20mm guns fore and aft, two twin 0.50s in the usual turrets, a twin 0.50 on the bow (starboard side forward), two single or twin 0.50s between the torpedoes port and starboard, one or two 0.30s in the bridge wings, and two 300-pound depth charges. Installation of 37mm guns was only beginning.

Barge-busting was generally undertaken at night. Although PTs had radar from 1942 on, effective fire required illumination. Tracers merely provided an enemy with a point of aim. The solution was a mortar that could fire its shell high above the battle area without showing a muzzle flash and blinding its gunner. MTBSTC tested an army 60mm mortar for PT use in November 1943. Installations aboard PTs, primarily for night illumination with starshell, began early in 1944. The standard allowance was ten high-explosive and twenty illuminating rounds. The successor, an army-type 81mm mortar, was tested late in 1944. Considered important for rocket-fire control at night, it was generally authorized in connection with the 5-inch spin-stabilized rocket (see below).

Thus fall 1943 and spring 1944 were periods of considerable flux in PT armament. In January 1944, BuShips proposed three alternatives for the ninety-six follow-on PTs (565–660), a Cominch/VopNav/BuOrd/BuShips conference having failed to standardize on any single armament.

In August 1944, the distinction between B and C began to break down. MTBRon 32 (PT 474–85) was assigned one 37mm forward *and* one 40mm aft (with 20mm as alternatives), plus two torpedo racks (four if the 20mm was mounted aft). MTBRon 31 (PT 453–55, 462–73) had two single 0.50s as well, plus two depth-charge racks and a 60mm mortar. Many other boats were fitted with 37mm guns and 60mm mortars at this time.

Table 6-3. BuShips Alternatives for Follow-on PTs, January 1944

	A	B	C
40mm (aft)	—	1	—
37mm SP (fwd)	—	—	1
20mm Mk 10 (aft)	1	—	1
20mm Mk 14 (fwd)	1	1	—
Twin 0.50	2	2	2
Single 0.50	2	2	2
Torpedo Racks	—	4[1]	4
Torpedo Tubes	4	—	—
DCT "C"	—	2[2]	2[2]
60mm Mortar	1	1	1

1. Two of four to be included in squadron spares.
2. In squadron spares, not normally installed.

The other major wartime PT weapon was the rocket. It offered high firepower without requiring deck reinforcement to compensate for recoil. The earliest rockets were the 2.36-inch sextuple bazookas tried aboard PTs in MTBRon 25 and in some PGMs. At Rendova, around March 1944, PT 47 tested the 4.5-inch fin-stabilized rocket launcher used by marine beach-jumpers; two racks, each holding twelve rockets, were installed on either side before the forward torpedoes. At maximum elevation, the rockets could reach one thousand yards. The general opinion in the South Pacific was that they were impressive for limited uses, but that their limited range and high trajectory were unsuited for most PT uses. However, others were installed both in New Georgia and in the Mediterranean. PT 336 was the first southwest Pacific boat so fitted. The Mediterranean installation, intended primarily for shore bombardment, was relatively unsuccessful because it could be outranged by shore batteries. In the Pacific, where shore batteries were less sophisticated, the rockets were more effective.

The next step was the 5-inch spin-stabilized rocket (SSR), which had longer range. Dr. L. A. Richards, a technical observer attached to the Seventh Fleet MTBRons, devised a launcher able to place a twelve-rocket salvo on target at up to 4,800 yards. Its success inspired the Seventh Fleet staff to propose an all-rocket PT that could deliver 4 tons of explosives at 5,000 yards. In November, the new approved PT battery included two eight-round Mark 20 5-inch rocket launchers, which were replaced within a few weeks by a sixteen-round Mark 50 5-inch launcher. MTBRon 33 received the first 5-inch SSRs (Mark 50 launchers) in March 1945; MTBRon 21 was the first older squadron to be refitted with them. PT 489 and 491 were the first boats in the Seventh Fleet armed with these weapons. Their high-capacity (HE) rockets had a range of 5,000 yards, but semi-armor piercing (SAP) rounds were rated at 11,000. The 5-inch SSR was used for the first time in combat by MTBRon 36 in the Tarakan and Borneo area. Sixteen rounds were fired at a range of about 800 yards at one floating and two grounded luggers. One was sunk, the bow was blown off another, and a 4-foot hole opened up in the third.

In April 1945, BuOrd circulated several proposals for power-driven rocket launchers to be installed in PT boats. The most promising was based on the existing Maxson quadruple 20mm mount. It was too heavy for existing boats, but the bureau persevered in hopes that it would be useful for other craft and for experimental PTs that could not accommodate the heavier Mark 102 rocket launcher then being fitted to landing ships. It became the Mark 107 of the postwar PTs.

In November, after Leyte Gulf, a new standard PT armament was announced. Each boat would have four Mark 1-1 side-launching racks, but only the two after racks would carry torpedoes. Each would also have the usual two twin 0.50s, one 40mm aft, one 20mm forward (a Mark 14 mount on the port side in the Elcos, a Mark 10 on the centerline amidships in Higgins boats), plus two single 0.50s in pipe stem mounts on the forward torpedo racks or in the cockpit. One 37mm gun on a Mark 1-1 mount would occupy the centerline forward, and each boat would carry two Mark 20 rocket launchers, each with eight 5-inch SSRs. Finally, each boat would be supplied with a 60mm mortar for illumination.

If the rockets were absent and a 20mm with a Mark 10 mount replaced the heavy 40mm gun aft, the boat could carry four torpedoes. The two extra

An Elco boat, PT 330, in the Philippines, early 1945. There is a 37mm cannon forward, a 20mm cannon partly obscured abaft it. The single free-swinging machine gun can be seen to port, abaft the radar mast; another, less visible, is installed between the torpedoes to starboard. No IFF antenna is visible; the short whip atop the bridge fed a UHF set for boat-to-air communication.

At Bastia, Corsica, on 19 February 1944, PT 211 reveals an early rocket battery (four quadruple launchers for finned rockets) as well as two free-swinging twin 0.30-caliber machine guns between the superstructure and the two 20mm cannon.

Newly completed in the spring of 1945, PT 588 carries the standard armament of that period. It has a long-barrel (M9) 37mm gun forward, together with a 20mm cannon and two launchers, one largely hidden, for 5-inch, spin-stabilized rockets. Note the spare magazines atop the cannon's ready-use ammunition box. Two torpedo racks and a depth charge are visible aft. The radar, the new SO-3, has a pair of IFF antennas on the mast and on the forward superstructure. The SO-3 required a resonance box, visible partway up the mast. The boat lacks the tall (HF) whip of earlier PTs.

PT 631 shows late-war features on a Higgins boat.

torpedoes would not be supplied initially to newly constructed PTs, an indication of the primacy of the gunboat role. Similarly, a pair of type C depth-charge tracks could be installed aft, but only their foundation pads were installed initially.

Several other weapons were tried during the war. The private-venture (Elco) 20mm Thunderbolt was evaluated in the period from 1942 to 1944. It was first proposed in mid-1942 as two 20mm and six 0.50-caliber guns in an armored aircraft-type carriage with stick control. At the time, Lieutenant Commander J. D. Bulkeley, who had fought through the Philippines campaign, had just taken over the new Squadron 7; he found the Thunderbolt attractive for its very high volume of fire, which might solve one of the main problems encountered in the Philippines, a lack of effective AA firepower. The weapon was clearly far superior to the single 20mm gun then being installed aft. Bulkeley hoped that recently authorized weight reductions in bridge armor, smoke generator, and torpedo launching racks would save enough weight to accommodate it.

Cominch found the Thunderbolt attractive but would not approve installation until the weight reductions had actually been effected. It did approve installation on one Squadron 7 boat, PT 138, for tests in action. The 20mm/0.50-caliber mount was tested at the AA training center at Price's Neck, Newport and, off Newport, aboard PT 160 in late November and early December 1942. BuOrd was instructed to develop a multibarrel machine gun and consider buying Thunderbolts.

BuOrd was unenthusiastic about the Thunderbolt, although it did prefer a version with four 20mm guns. The weapon required a strengthened deck and an additional 5-kw 24-volt DC generator. In January 1943, Cominch observed that considerable weight would become available as PTs converted from torpedo tubes to lightweight launching racks. That weight might be invested either in Thunderbolts or in 40mm guns, whose relative merits were unknown. The Thunderbolt, moreover, would be attractive even if it was not suited to PTs. Two hundred were ordered, their 0.50s eliminated for simplicity's sake.

Meanwhile, PT 138 tested the Thunderbolt in the southwest Pacific. The Seventh Fleet found it unsatisfactory. Vibration due to rough weather caused the trunnions to spread and the guns to drop, in most cases wrecking the mount. The sacrifice in ruggedness and watertight condition required to get the weapon into an 80-foot Elco seemed excessive.

Through much of 1943, the Thunderbolt received relatively low priority. However, it became more attractive as barge-busting became a prominent PT mission and as specialized gunboats were built.

Early in November 1943, OpNav ordered Squadron 29, which served in the Mediterranean, to be completed with experimental batteries for comparative trials:
- PT 552–55 had the conventional battery—four launching racks, two 20mm guns, and two twin 0.50s.
- PT 556–59 had one single 20mm gun forward (Mark 14), a Thunderbolt aft, the usual two twin 0.50s, and two launching racks.
- PT 560–63 were gunboats with two 40mm guns, two twin 0.50s, and armor around their cockpits. Since self-sealing tanks were not available, they received armor outboard of their fuel tanks.

On 21 November 1943, Cominch ordered the Elco Thunderbolt contract canceled upon delivery of the first fifty mounts. The service test in MTBRon 29 would continue, concentrating on the ruggedness of mounts.

Early in 1945, a Thunderbolt was experimentally mounted aboard PT 174, presumably on her reinforced foredeck. Firings were satisfactory, but the boat required an extra auxiliary generator and new batteries. Three more were assigned to Squadron 21, and more Thunderbolts were shipped to the Philippines. A few Squadron 10 boats may also have carried the weapon.

In spring 1944, PT 334 was experimentally fitted on her stern with an aircraft-type (M4) 75mm gun, the same one mounted in B-25s. This semiautomatic weapon weighed only two-thirds as much as a 40mm gun. It appears not to have been successful. OpNav rejected it, using a BuShips argument that it represented an excessive deck reaction load. It remained attractive, however; early in 1945, BuOrd rejected a proposed 75mm automatic gun only because it could not be developed in time for war use.

The existing 57mm and 75mm recoilless rifles were rejected at this time because of their back-blast problems. The recoilless rifle concept would be revived in the 1960s. A weapon of this type was reportedly developed for the post-Vietnam patrol boat (PB Mark 3). In 1945, the only new weapon in the making was a heavier aircraft machine gun of 0.60 caliber, which BuOrd had been developing for some time and which it expected ultimately to test aboard a PT boat.

PT armament remained controversial throughout the war, and in March 1945, a conference between representatives of Cominch, OpNav, BuOrd, and MTBSTC decided to hold extensive tests that May. The forces afloat—Seventh and Eighth fleets and commander in chief, Pacific (CinCPac)—were also asked for their views on various guns. The Seventh Fleet generally liked the existing armament. It considered the current 37mm M9 comparable to the

Newly completed in this December 1943 photograph, PT 556 illustrates the experimental installation of the Elco "Thunderbolt" quadruple 20mm cannon aft. Note that the radar mast has been folded down. Masts on Elcos folded back to the centerline at a 70° angle to clear the after 0.50-caliber turret. (Al Ross)

40mm in rate of fire, penetration, and muzzle velocity, objecting only to its limited elevation of 40°. It might have been preferable to have 40mm guns fore and aft, but space was not available and development of a new version of the 40mm would be too lengthy. A DP mount for the 37mm was then being worked on.

CinCPac had radically different views: it emphasized the torpedo boat role and considered guns strictly secondary, for AA protection only. The 40mm seemed too large and of little value for AA fire, and CinCPac also recommended against the installation of depth charges, smoke mortars, 37mm guns, or rockets. At this time, CinCPac controlled only forty-nine boats, ten of them the obsolete Huckins, and had none in combat. It is not clear whether its recommendations referred to requirements for the assault on Japan or whether they reflected earlier sentiments about PT operations.

A preliminary study suggested that if 37mm and 40mm guns were discarded, existing PTs could carry their four torpedoes, together with six 0.50-caliber guns (two twins, two singles), two twin 20mm guns, and one quadruple 20mm power mount. However, action was delayed, pending further service experience. Assuming the new battery was feasible, it could be installed in boats from MTBRon 41 and up (MTBRon 40 was being outfitted in April 1945).

In May, BuOrd approved a request by the prospective commanding officer of MTBRon 41 that six of his boats be equipped with the Elco 20mm quadruple mount (Mark 15-1) in place of the authorized single 40mm. BuShips objected that much more electrical power would be needed and that the entire new battery could be had only at the cost of two torpedoes.

By late July, tests at Melville had shown that the quadruple 20mm Mark 15-1 was not a satisfactory substitute for the manually operated 40mm, and the projected battery was abandoned.

Heavier batteries were accompanied by new electronics. The boats were designed with HF radios served by whip antennas and prominent radio direction-finding loops (in some 77-foot Elcos, one per squadron, but in most cases, one per boat). Later, VHF radios for short-range boat-to-boat and boat-to-air communication were added.

The original PTs had no radar, since there was no set light enough to be fitted. As a result, they were extremely vulnerable to air attack. After his experience in the Philippines, John Bulkeley urgently requested radars for his MTBRon 7 (PT 127–38). He received an adapted aircraft set, SCR-517A; a special PT set, SO, was soon developed.

Other wartime changes reflected operational experience. The controls, originally designed for one man, were revised for two-man operation, the throttles being moved away from the wheel. During a torpedo attack, one man typically steered while another manned the nearby torpedo director. Self-sealing tanks replaced the original aluminum gaso-

The standard early Elco cockpit was arranged for one-man control, with three lefthand throttles close to the wheel. The dials above the wheel show individual engine rpm and oil pressure; they are surrounded by switches for various lights. Note the intercom at the right. In later boats, for two-man control, the throttles were moved to the right and the light switches to the left, leaving the engine status dials in their own panel. A loudspeaker was fitted above the passageway to the right of the controls. The box in the foreground is a radar transmitter-receiver for the antenna atop the mast (which has been folded down in the direction of the reader). The flag bag is visible in the foreground, to the left.

PT 631 was a late-war Higgins. This is the cockpit on 1 March 1945. Behind and above the wheel is a rudder angle indicator, and above that a compass. The throttles and chokes are arranged to the right, and the dials, partly obscured, indicate engine-oil pressure. The shielded switch at the bottom of the panels on the right ignited the engines. The flexible cables led to red lights for night illumination of the instruments and controls. Switches on the lower panel controlled lights, including the searchlight; buttons at the top controlled buzzers, bells, and loudspeakers. The entranceway led down to the chart house, which at this stage was primarily a radar and radio room; it was often described, incorrectly, as a combat information center.

line tanks. Armor (STS) protection was installed around the steering station. Additional fuel was carried on deck in collapsible rubber tanks.

The effect of increased gun batteries was enormous. The Elco boat was designed for a displacement of 103,000 pounds (including 25,000 of ordnance), the Higgins, for 105,000. Both prototypes were badly overweight (fully loaded 106,000 and 126,000 pounds, respectively) and had to be rebuilt to make their speeds, though later boats did meet their speed requirements. The Elco PT 103 ran trials at 95,000 pounds, and design weight increased to 110,000 in PT 372 and PT 486. At the end of the war, PT 565, fully loaded, displaced 117,033 pounds (it ran trials at 110,000), and the rocket-equipped boat of 1945 was expected to displace 130,000 pounds, fully loaded, with 37mm and 40mm guns. PT 558, carrying the prototype 5-inch rocket battery, ran trials at 121,000 pounds.

By 1945, the standard ordnance and electronic load was 42,000 pounds, of which 30,000 pounds was ordnance, and further weapons and equipment were under consideration. Electronics in particular demanded more electrical power. The boats were designed with 2.5-kw auxiliary generators, supplementing motor generators fed by the engines at high speed.[3] By 1942, a 5-kw generator was standard. By January 1944, even that was considered barely adequate to carry the continuous load imposed by radar, IFF, VHF radio, TCS radio, and a fluxgate magnetic compass. By the end of the war, the standard PT had two 5.5-kw generators.

Adding weight and power was not too difficult, but because the hulls, in the interest of production,

The Elco "slipper," designed for field installation to improve PT performance, is installed aboard a new Elco PT in 1945. Note the three mounting struts and the mufflers above the slipper. PT 810, one of the four experimental postwar boats, had a similar installation, the angle of which could be varied by a pair of jacks. It extended 2.5 feet aft and was used to vary planing angle. PT 809, the postwar Elco boat, lacked any such fitting. (Norman Friedman)

could not be radically changed during the war, there was never any additional volume—and electronics maintenance and operations suffered. Volume was also necessary to accommodate additional men to operate and maintain the equipment and the added weapons; by 1945, the standard was three officers and fourteen enlisted men, up from a planned two and eight, respectively. According to many officers, working space was adequate but enlisted accommodations were much too cramped.

The net weight gain was partly due to the need to reinforce hulls taking extra loads. Because so many of the hull and armament changes had been improvised by forces afloat, they were often heavier than those incorporated in newly built boats. Even the best new boats displaced about 124,000 pounds fully loaded. This extra weight strained wooden hulls and could be expected to reduce their service life, which in 1945 was estimated at about two years, based on a survey of existing boats (see table 6-4).

Table 6-4. Age of MTBs Assigned to the Seventh Fleet, 1 January 1945

Age (months)	Number	Percentage
Over 30	4	1.9
24 to 30	54	26.0
18 to 24	81	38.9
12 to 18	58	27.9
Under 12	11	5.3
Over 24	—	27.9
Over 18	—	66.8
Over 12	—	94.7

Performance had to suffer, although the effect of the added weights was somewhat balanced by improvement in the Packard engine, whose power increased from 1,200 (W-8) to 1,500 (W-14) bhp, a similar percentage. The added power can be attributed in part to an increased supercharger ratio. In 1945, an 1,800-bhp version, W-50, was being studied. A heavier engine, 3,500 rather than 3,000 pounds, was scheduled for installation that December in the last Elco boat, PT 790. This engine, the 2,500-bhp, fourteen-cylinder W-100 (1M-3300), powered the four postwar PTs. Moreover, the tuning of standard engines was much improved, so that most operational boats could reach full power.

There was some question as to the effect of the added weights. PT officers argued that sea-keeping had not been seriously affected—one even felt that it had been improved. OpNav, however, suspected that losses of sea-keeping ability were masked because the boats had been operating in relatively sheltered waters. By the summer of 1945, conditions were clearly about to change. The fleet would have to operate in the more open waters north of the Philippines, and the PTs would, if anything, be even more important as a means of countering Japanese small craft

Photographed in October 1943, this Higgins boat, probably PT 268, has not yet been fitted with torpedo tubes (note the supporting pads). It has depth-charge tracks on the beam and right aft, single 20mm guns forward and aft. There is not yet any radar.

(including sneak-attack boats). The PTs had a unique capability as gunboats: although other light gunboats had been developed, only PTs were fast enough to penetrate far into enemy-held coastal areas under cover of darkness.

As for speed, PT officers maintained that it depended more on engine tuning and maintenance than on anything else. Successful planing required proper trim (that is, proper location of the longitudinal center of gravity, which did indeed depend on the location of added weights). Unless the boat was in good condition, it was restricted to the speed range below the 25-knot planing hump. Once the hump was passed, much higher speeds could be attained. A typical Elco in good condition could exceed 43 knots fully loaded; it could cruise at 30 knots on all three engines (radius about 400 nautical miles) or at about 6 knots on one engine, trailing the other two propellers (radius about 1,600 miles). Radius at full speed was about 300 miles.

The difficulty of getting over the resistance hump was reduced by installing smaller propellers, but then it became difficult to attain maximum speed.

The postwar PT program described in the next chapter was, in effect, the navy's verdict on the PTs.

They had proven extremely valuable in wartime, in the unexpectedly close quarters of the Solomons, the southwest Pacific, and to a lesser extent, the Mediterranean and English Channel. Had U.S. forces invaded Japan, PTs would have provided crucial cover against sneak-craft attacks—for the execution of which the Japanese were very well prepared. The question was whether it was worth preserving existing boats to meet future needs.

The first formal navy postwar plan of May 1945 showed four postwar MTBRons, all in the Pacific, plus four reserve boats, presumably the four postwar experimental ones. At this time, the navy had 409 PTs either on hand or under construction. A modified postwar plan (1A) was promulgated that December: it included only the four reserve PTs, which were retained through the later postwar plans. Their existence demonstrated a determination not to abandon useful wartime experience, but also a desire not to pay a heavy postwar price in maintenance and operation of craft that it seemed would be needed only in the event of a major war. The navy could not afford more, even though it often acknowledged the usefulness of the PTs in narrow waters, against an enemy dependent on the sea for reinforcement. That situation would clearly recur in Vietnam.

As for the numerous existing boats, those in the Pacific but not immediately burned in the Philippines were sold off for about $1,000 each (Huckins boats in Hawaii fetched only $150). The list at the end of this book shows that many were reclassified as small boats.

The two ex-European MTBRons, 22 and 30, were retained at New York. In 1948 they were scheduled for transfer to Turkey, but their condition was found to be too poor for transfer. In March 1947, Turkey received PT 203, 205, 212, and 216, which had served in the Royal Navy after September 1944. Italy bought twenty-eight surplus U.S.-built PTs lying at Palermo in 1947—nineteen Vospers (ex-U.S. BPT 37, 63–68, and PT 386, 388–94, 397–99), seven Higgins (ex-PT 88, 90, 92, 94, 206, 210, 215), and two Elco 77-footers (ex-PT 49, 57). They served initially as torpedo-recovery boats, but many were later converted to motor gun and torpedo boats. Two other ex-British Vospers, PT 385 and 387, were included in a bulk sale to Egypt in January 1947. A later Egyptian request for more PTs was rejected in 1948, as was a request by Denmark at that time.

Both PT 296 and the Hellcat, PT 564, were reserved for sale to foreign governments in 1947, but apparently the sales did not take place.

Some of the lend-lease boats, completed too late, were never transferred: four Higgins (PT 657–60) and thirty-eight Vospers (PT 693–730). In addition, the last Higgins boats (PT 791–96) never entered squadron service. Many of these craft were reclassified as small boats, some were later transferred abroad. Cuba bought PT 715–16, the Dominican Republic bought PT 709. Taiwan was later credited with a variety of ex-U.S. torpedo boats of uncertain origin. One source lists two Vospers, three or six Higgins, and remarkably, two Huckins boats.

The postwar transfer program began in 1948. Norway received PT 602–12 in December.

7

The Postwar PTs

The U.S. Navy disbanded its PT boat organization after VJ-day. That did not represent complete abandonment of the MTB concept. The official view was that, given sufficient preparation, the organization could be resurrected in an emergency. In peacetime, the U.S. Navy would rely on fast carrier and amphibious task forces to project its power overseas. Until the United States was heavily engaged in some particular area, large coastal craft such as PTs would have little role. They would be too small to deploy overseas on their own, too large to be carried aboard amphibious ships for combat deployment.[1] Similar arguments had been raised before 1941, yet the PTs had proven extremely important that December. After the war, then, the navy continued work on PTs: four new boats, PT 809–12, were built as experimental prototypes for possible future mass production. They in turn formed a bridge to the next U.S. program for small fast coastal craft, PT 810–11 becoming PTF 1–2 in 1962 (see chapter 10).

Admiral King authorized construction of four experimental PT boats on 13 February 1945. At this time, the most recent PT authorization, made on 30 September 1944, was for a total of forty-eight boats—thirty Elcos and eighteen Higgins. BuShips expected orders for more PTs and therefore saw no need to cancel any of these boats in favor of the experimental ones. Indeed, as of mid-February 1945, indications were that Admiral King wanted production stepped up to twenty-two per month by October. The bureau feared that this accelerated production would block any experimental PT program unless more builders were brought into the standard program. However, on 22 June, Admiral King decided not to request any more boats. On 5 July, he ordered two Elco and two Higgins boats replaced by experimental ones. By this time, both yards had assembled so much of the material for these last boats that BuShips could argue that modification of the existing contracts, as opposed to their cancellation, would cost about 60 to 70 percent of the contract cost. At Bayonne, for example, Elco had received 70 percent of the necessary material, and frames and engine girders were practically complete. Thus the bureau asked Admiral King to set up a separate PT experimental program, which he did at the beginning of August 1945. As a result, the four postwar PTs (809–12) survived the general cancellation of standard PTs later in 1945.

By 1945, development of the two principal wartime PT types had clearly reached a plateau. Something new was needed. It would have to be larger, given the steady trend towards heavier batteries and the prospect of more new weapons. For example, the existing, manually operated 40mm gun was limited because its operator could not compensate for the violent motion of a fast planing boat. The obvious cure was power operation with stabilization, but that would add weight. BuOrd was working on power-driven rocket launchers and on more multiple 0.50-caliber and 20mm gun mounts.

Existing PTs relied on stealth for their viability: they had to be able to sneak to within Mark 13 torpedo range. Because Japanese radar could not easily detect such small boats, stealth, in the Pacific, generally meant effective muffling, a low silhouette, and measures to reduce the boat's phosphorescent wake. However, advanced radar technology would become almost universally available after the war, and the minimum detection range of any PT would be considerably greater. The obvious solution was to return to longer-range torpedoes such as the hydrogen-per-

Newly completed, PT 810 runs trials in Chesapeake Bay. Although designed to carry torpedoes, none of the four prototype boats ever did. Instead, they operated as gunboats, with two 40mm and two twin 20mm guns and prototype 81mm mortars. In this photograph, the gun is being loaded to starboard, abeam the bridge. Surface-search radar has not yet been fitted, but the mast does show UHF dipoles as well as an HF whip. PT 810 later became PTF 1. Note its relatively spacious bridge, characteristic of the experimental boats.

oxide-fueled Mark 16, already being developed for U.S. submarines. The Mark 16 weighed considerably more than the wartime Mark 13, and its successful operation required a sophisticated fire control system, probably linked automatically to the PTs surface search radar.[2]

New weapons and equipment would require an increase in internal volume plus additional space for the concomitant increase in crew.

But any new PT had to be limited in size for transfer overseas aboard cargo ships and maintenance of a wartime standard of maneuverability and speed. PT tactics of stealth also limited size, and shallow draft was desirable if the PTs were to continue in their wartime shore bombardment role.

PT size was also limited by available power. It appeared that the existing 3,000-pound, 2,500-cubic-inch Packard V-12 engine had peaked at 1,800 bhp in its W-50 version, providing a total of 5,400 bhp for a three-shaft boat. The next planned step in engine development was a 3,500-pound, 3,300-cubic-inch Packard V-14 engine (IM3300 or W-100), which was expected to produce 2,500 shp.[3] It was similar to the existing V-12 except in being 15.5 inches longer. If, indeed, the new PT was to be a prototype for later mass production, an argument could be made in favor of a new engine with the potential for further development to match the inevitable weight growth for the next-generation boat.

BuShips also considered a gas turbine power plant in 1945–46. At this time, the characteristics for the new PT called not only for a maximum speed of 46 knots but also for an endurance of 500 nautical miles at 27 knots and "effective muffling" at speeds of 15 knots or less. Given a limiting displacement of 205,000 pounds, the bureau concluded that it would need to use diesels or gasoline engines to furnish at least half the power required for cruising endurance. However, even to reach 25 knots the boat would need 4,000-bhp diesels, and the combination of diesels and gas turbines would require too much internal volume. The alternative, a pair of 600-bhp Packards, would drive a PT at only 12 knots or so.

In addition, the boat would have to be furnished with controllable-pitch propellers so that the gasoline or diesel engines could operate economically both at cruising and at full speeds.

The combination diesel/gas turbine (CODAG) plant was, therefore, rejected in favor of a simpler all-gasoline plant using the near-term W-100 engine. As in Britain, though, the lightweight gas turbine was seen to be the key to future high-speed boats, and it was revived in the mid-1950s in the context of a new PT design. PT 812, one of the four prototypes, eventually became a gas turbine test hull.

The advent of a new PT represented an important opportunity to introduce other new technology, particularly in hull construction. The existing wooden PTs were credited with a combat life expectancy of no more than twenty-six months. BuShips had been much impressed by the durability of the aluminum PT prototype of 1941, the former PT 8. Aluminum also offered greater hull strength, which might translate into a shallower hull, that is, a lower silhouette. On a weight-for-weight basis, moreover, aluminum provided better fragment protection than steel. Limited armor protection could, therefore, be incorporated directly into an aluminum hull.

BuShips began work on a new PT design early in 1945, consulting with BuOrd to develop a suitable armament. However, the requirement for the new class was formally laid down only when the VCNO, Admiral Horne, signed a memorandum on 15 January. Although he expected detailed characteristics to be specified after a conference between experienced PT operators and representatives of BuShips and OpNav, he proposed a tentative version for discussion. Displacement would be limited, by the standard 75-ton crane used to unload the boat from cargo ships, to 165,000 pounds, and length to 100 feet. On the other hand, BuOrd wanted a 52,000-pound load (some suggested 55,000) of ordnance and electronics, considerably beyond wartime standards.[4] The requirement for more internal volume was stated as follows: the boat would need the largest possible conning station, with an adjacent combat information center (CIC),[5] and accommodations for three officers and sixteen enlisted men (there were two officers and eight enlisted men in the original PTs and up to three officers and fourteen enlisted men in the 1945 PTs).[6] The pressure these additional men exerted on the design was somewhat reduced because, although officers and men could be expected to live aboard the boat, it would rarely remain at sea for more than sixteen hours at a time. Thus the PT did not have to carry anything remotely like the standard of provisions common for larger ships and craft. The conning station, CIC, gun stations, and engine room would be armored as well as possible, using either conventional metal armor (STS) or the new glass or plastic armors. This would add considerable weight.

As a tentative armament, the VCNO suggested four Mark 13 torpedoes; two M9 37mm guns in two DP mounts (125 rounds each); four 0.50s in two twin scarf-ring mounts (1,000 rounds per barrel); four 20mm in two twin mounts (360 rounds per barrel); thirty-two 5-inch SSR barrels; one 81mm mortar (12 rounds: 8 illuminating, 4 HE); and two smoke-screen generators or smoke-pot equivalents. He claimed that this was actually less than what many boats were carrying, because one 37mm gun would replace a

THE POSTWAR PTs 179

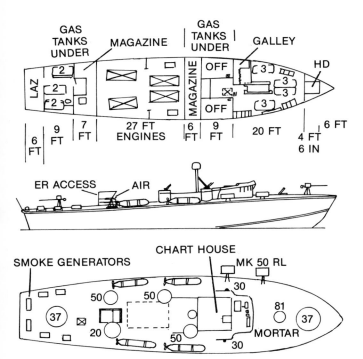

This is the 1945 PT boat sketched by BuShips code 516, which was responsible for PTs. The boat, 94 feet 6 inches long, is armed with 37mm guns fore and aft (large circles), two twin 0.50-caliber machine guns in the bridge wings, two 20mm cannon between the after torpedoes, an 81mm mortar abaft the forward 37mm gun, four Mark 50 rocket launchers forward, four Mark 13 torpedoes, and four 300-pound depth charges. There are positions for 0.30-caliber machine guns on the bridge coaming, inboard of the rocket launchers. The original sketch was not to scale, although lengths for internal components were supplied. (Norman Friedman)

heavier 40mm gun aft. The ammunition allowances were those of current PTs. As chapter 6 shows, armament was a matter of considerable controversy, leading to special trials at Melville; the tentative battery was hardly the final word.

Finally, the VCNO revived his wartime standard of 40 knots fully loaded, and radius of action was set at 600 nautical miles at 20 knots. Existing boats were credited with 400 miles at 30 knots. A month later the commander of Seventh Fleet MTBRons specially emphasized long range at high speed, 30 rather than 20 knots, because under the threat of enemy air attack boats often had to run to their patrol stations after dark. Time in transit had to be subtracted from total hours of darkness to give available patrol time. For example, a boat would need eight hours at 30 knots to travel back and forth to a station 120 nautical miles away (or 120 inside the range of enemy aircraft). Even given twelve hours of darkness, that would leave only four hours on station. A cruising range of 500 miles at 30 knots seemed worthwhile. The Seventh Fleet also wanted a long, slow-speed cruising range so that boats could be redeployed without having to wait for a tow or for ships large enough to carry them as deck cargo. The existing three-shaft machinery arrangement provided such a range, since the boat could operate on one engine. Multiple shafts were also an important safety feature, given the damage propellers often received when they struck logs and reefs; many boats came home on one engine. Gunfire often disabled one or two engines without destroying a boat.

At this time the tentative BuShips plan was to develop basic contract plans in-house, then to hold a builders' conference to consider the design and construction of a prototype. It would be best to build two boats of each design, so that if more than two development contracts were let (that is, more than the four planned boats), additional authorization would be needed. Admiral Cochrane, the bureau chief, rejected any extensive in-house preliminary design. Instead, he wanted to call an early design conference and ask the builders for suggestions and sketches to go into a new standardized design. The war ended before any such conference could be held, and this time BuShips did carry through a preliminary PT design.

The tentative VCNO characteristics were circulated throughout the fleet. Several officers doubted whether the existing boat had reached the limit of development. It was so close to the required endurance that better carburetor control rather than more fuel might solve the problem. A redesigned arrangement might allow room for additional electronics and crew.

Many officers called for better watertight integrity, meaning more compartments, and for two positive-action bilge pumps. Ventilation was considered grossly inadequate.

Maneuverability was, if anything, more important, and it represented a major argument against a larger boat, which would also be a larger target. At this time, although the Higgins boat seemed maneuverable enough, the Elco was still criticized for its substantially larger turning circle. An increasing scale of air attack made maneuverability more important, as a PT could often hope to dodge falling bombs and even kamikazes. At Mindoro, twelve kamikazes attacked Higgins boats, only one succeeding. This success was largely attributed to PT maneuverability, the result of big rudders.

There were definitely pressures for growth. A sufficient operational speed of 40 knots some weeks out of dry dock could only be ensured by raising the acceptance speed to 45 knots (with the boat perfectly tuned and aligned), which in turn would require more power. Electrical power was often inadequate; many favored a 110- rather than the existing 24-volt sup-

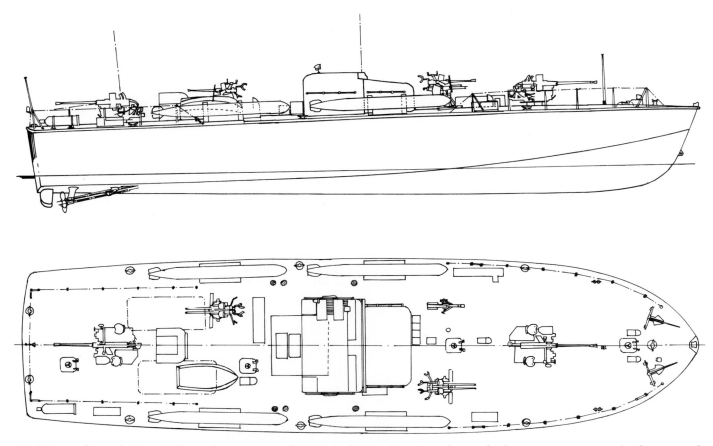

PT 809 as planned, May 1949, with a quartet of Mark 16 long-range torpedoes, which were never carried. The vertical dot-and-dash line is the space reserved for a radar mast. The dot-and-dash rectangles aft are removable plates over the machinery. Shaded circles are fuel-tank fillers and indicate the approximate position of tanks. (Norman Friedman)

ply. Electronic equipment needed its own special compartment. Moreover, the commanding officer had to be able to see either the radarscope or a repeater on the bridge.

Most officers wanted to keep the heavy 40mm gun. Some questioned the need for both 37mm and 40mm guns, asking that only one or the other be retained in the interest of simplifying ammunition supply.[7] Simplification would tend to favor the more powerful 40mm, which seemed to be the best AA and antibarge weapon. At least one officer wanted the 20mm gun eliminated as well, in favor of three twin 0.50s. In a related issue, it was deemed that a better internal communication system was needed, particularly for fire control during an air attack. The existing surface-search PT radar (SO series), used to detect snooping aircraft, the PT's worst enemy, was criticized for being unreliable.

Some were willing to surrender two torpedoes in favor of 5-inch spin-stabilized rockets, even though the VCNO had offered four torpedoes plus the rockets.

Torpedo control was still extremely primitive; the commander of Seventh Fleet MTBRons complained that there was no effective director and no automatic radar input for the fire control solution. Nor could his torpedoes be fired directly from the bridge. In the time elapsing between his order to fire and the actual moment of firing, moreover, the boat itself could turn off the required heading. That might not be important at short range, but it would become more and more important as longer-range torpedoes came into service.

By this time, the PT community was aware of the successes of diesel-powered German "E-boats" as well as the dangers of the gasoline fueling their own craft. Some officers asked for diesels, and many sought a better means of avoiding or fighting fuel-tank explosions and fires.

From the designer's point of view, there was one reason for optimism. The 75-ton limit referred to an empty boat being hoisted from a freighter; it could be considered equivalent to an 80-ton (179,000-pound) full load displacement. Existing PTs were carried in

PT 809 on 24 February 1951 off the Electric Boat plant at Groton, Connecticut. Note the framing on deck for torpedo-launching racks, which were never fitted.

22,000-pound steel cradles; most officers in the western Pacific preferred lifting pads built into the hull itself to save overall weight.

The proposed characteristics also went to MBTSTC, which took its own poll of thirty experienced PT officers. Although two-thirds favored the new characteristics, half saw an alternative solution to the overloading and congestion of existing boats; the U.S. Navy could abandon its existing policy and instead construct specialized boats, either MTBs or MGBs, as the British had already done. One-third wanted specialization anticipated in the new design.

Those favoring specialization wanted existing PTs divided into gunboats and torpedo boats, pending the appearance of a new class. Operating areas and squadrons would have both types, and one could easily be converted into the other if changing conditions required it. However, all the armament of a torpedo boat except its torpedoes would be limited to adequate AA protection, even though gunboats might carry torpedo racks and simple torpedo control equipment for emergency or supporting use.

The officers wanted lighter-weight, aircraft-type weapons and electronics to help reduce overloading; they also wanted a stronger hull, particularly on deck and amidships. They hoped a new boat would make a better gunboat (this reflected the contemporary combat emphasis on gunboat operation) because of greater steadiness and seaworthiness and a clearer deck. Even then it would suffer, being a larger target. The officers expected that, as a torpedo boat, its performance would be worse than that of existing PTs since it would be so much easier to detect. The skeptics knew that the existing PT rarely if ever approached its designed speed and range under combat conditions. The larger British Fairmile "D," slightly larger

Table 7-1. BuShips Combinations of W-50 and W-100

	BHP	Number of Engines	HP	Engine Shafts	Weight	Reliability	Figures of Merit[1] Availability	Resistance[2]	Risk
A	6,000	4	1,500	2	1	6	2	1	2
B	6,000	4	1,500	4	2	1	1	6	1
C	7,500	5	1,500	3(2-1-2)	6	4	3	3	3
D	7,500	3	2,500	3	3	5	5	4	5
E	9,000	6	1,500	3	7	3	4	5	4
F	10,000	4	2,500	4	5	2	6	7	6
G	10,000	4	2,500	2	4	7	7	2	7

1. Figures of merit were not weighted.
2. Comparative appendage resistance, ordered from higher to lower.

than the boat then envisioned and powered by four Packards, was not maneuverable and could reach 30 knots only under extremely favorable conditions. It could only be hoped that the heavier weapons and electronics aboard the new PT would make up for its reduced performance.

Given these considerations, many officers wanted the new boat developed exclusively as a gunboat, with existing PTs stripped of their non-torpedo weapons and thus restored to their original torpedo attack role. Some even wanted further work done on a 70-foot light PT like the Higgins Hellcat.

One-third of the officers specifically urged the adoption of diesels, if feasible.

Work proceeded on three related tracks: machinery, hull form, and armament. Ideally, all three would be integrated. However, BuOrd was never able to devote sufficient resources to developing weapons specifically for the new PT, and the design generally had to incorporate existing weapons. In some cases, they were space and weight reservations for planned future developments, most of which had to be canceled in 1946–47 (see below). In theory, too, hull form should have come first, since hull form and weight, largely ordnance, would determine required power. But the volume needed for machinery space would affect hull form, and BuShips, moreover, had to decide whether to continue development of the new W-100 engine. In mid-February 1945, the bureau was able to list a series of possible combinations of the near-term W-50 and the coming W-100 (see table 7-1).

Alternatives A, C, E and G would require the development of a combination angle-straight through-drive and a two-speed gear, and probably also a reduction gear. Reliability could not be assured; although BuShips did not say so at the time, that made it almost inevitable that a three- or four-shaft plant would be adopted (B, D, or F).

The choice between two and four propellers could be made on another basis. Existing PTs operated at a cruising speed of 20 to 25 knots on three engines about 90 percent of the time. Cruising on two of four engines, two would drag. That would mean either two dragging propellers, which might be designed to feather, or two engines that would have to be disconnected from their shafts. BuShips did not look forward to developing the necessary tandem coupling and considered gears a maintenance problem. If only two propellers were used, they would have to have relatively large diameters and so would increase draft aft, although they would reduce the number of underwater fittings and therefore appendage resistance. BuShips eventually chose four Packard W-100s on four shafts for a total of 10,000 bhp, more than twice the power of the standard wartime boats.

The first sketches, for a 99.5-foot boat, were dated 2 March. Arrangement generally followed that of the 78-foot Higgins. On 8 March, a complete sketch design was ready, showing a length of 95 feet overall, 88 on the waterline, and a displacement of 185,000 pounds. It was expected to achieve 45 knots at 8,000 bhp; 6,000 gallons of gasoline, in tanks both forward and aft of the machinery, would drive it 576 nautical miles at 20 knots without any allowance for fouling. Space would permit a crew of three officers and nineteen enlisted men, not the sixteen envisioned by the VCNO.

Armament would consist of the following:
- one DP, power-driven 40mm with 192 rounds (as opposed to the manually-driven 40mm in existing boats)
- one single DP 37mm with 125 rounds (instead of the single-purpose guns in existing boats)
- one power-driven, quadruple-mount 0.50 aft (a new weapon)
- two 0.50 twins with a total of 8,000 rounds and 0.50-caliber ammunition
- six torpedo racks (four torpedoes)
- two Mark 104 power-driven rocket launchers with thirty-two rockets
- one 81mm mortar with thirty rounds
- one smoke generator

By late April 1945, Preliminary Design had two alternative PT designs, I and II, both powered by four-shaft W-100s. Study II was armed with a battery proposed by BuOrd. Both had the same 94.5-foot hull, with lifting weights of, respectively, 166,000 and

171,000 pounds. At a sustained output of 8,000 bhp, study I could be expected to make 44 knots, study II, 43.5. Loads were 34,500 and 40,000 pounds, respectively.

BuOrd wanted a 40mm amidships with clear arcs of fire fore and aft, so study II showed a gun with its trunnions 8 foot 3 inches above deck. Preliminary Design thought it would be difficult to keep the gun attached to a wooden deck. It would suffer particularly badly from pounding.[8] The high gun, moreover, would raise the silhouette and displace the radar antenna aft, requiring that it be put higher. Preliminary Design suggested that the 40mm gun be shifted to the stern in place of the quadruple 0.50-caliber machine gun, which in turn would be replaced by two 0.50 twins forward.

The proposed weight seemed too high, and at an 18 May conference, Preliminary Design decided to limit both studies to a 165,000-pound lifting weight. BuOrd, agreeing that the 40mm gun should be astern rather than amidships, called for a new study II. Now it wanted the mortar eliminated and 20mm twins in place of the two rocket launchers. Quadruple 20mm guns would replace the 0.50-caliber mounts. It was understood at this time that OpNav would agree to the elimination of rockets and mortar.

Buord developed three preferred armament arrangements at this time. The most conventional showed a 37mm gun in the bow, a power-driven single 40mm or quadruple 20mm in the stern, and one twin 20mm and two twin 0.50-caliber mounts on each side in the waist. An 81mm mortar was mounted forward on the starboard side. Another version showed two quadruple 20mm guns; one was just abaft the 37mm, with the 81mm mortar on the opposite side forward, and the other was aft on the starboard side. Two twin 20mm guns were in the waist.

The two Preliminary Design studies were formally submitted to OpNav late in July. A wooden version would weigh slightly more than 165,000 pounds, an aluminum one about 145,000, both just within the weight limit. The difference of 20,000 pounds equaled 1 to 1.5 knots. Although BuShips strongly favored aluminum at this time, it was too early to switch to the alternative material in PT production. Experimental prototypes seemed to be an ideal way of gaining the necessary experience.

Each version of the prototype PT was expected to make 40 knots with the new 2,000-bhp W-100.

The two studies showed heavier electronics and ordnance loads—34,500 and 34,600 pounds, respectively, compared with 25,900 in existing PTs. These weights did not include any new electronic equipment. Alternative weapons could be substituted for those in the plans. For example, replacing a quadruple 20mm with a single 40mm would cost about 2,000 pounds, a quarter knot in speed. The wooden hull version of the design showed separate splinter protection, but Preliminary Design expected to be able to work such protection directly into the structure of the aluminum version, saving weight.

In accordance with the VCNO's requirements, the studies showed roomier bridges, 12 by 4 feet, with clear visibility to around 60° abaft the beam, where it was obstructed by the twin 0.50-caliber machine guns. A combination chart, radio, and radar room was placed immediately abaft the bridge with access directly to the bridge or to the officers' quarters below. The room was 13 by 9 feet, about 62 percent larger than the equivalent in the existing PTs. Berthing space and stores would be provided for four officers and eighteen enlisted men, spare berths (no stores) for two enlisted men.

The first hull lines were developed in mid-June, based on a combination of Elco and Higgins forms and on a review of the prewar work. The 1937–40 scale model tests had been extended to a twenty-hull test tank series in 1941–42; this made it possible to compare different proportions. Two quite different considerations had to be kept in mind for each hull. One was easy driving at maximum speed, which depended on planing. The other was endurance at 20 knots, which depended on the resistance of the hull in the displacement (conventional) mode. All models were assumed to be propelled by the four-shaft, 8,000-bhp W-100 plant.

The study began with a hull of 88 (waterline) by 19.25 feet and 185,000 pounds. The study varied dimensions and weights while holding *endurance at 20 knots* constant (hence varying fuel weight as hp at this speed varied). Hull lengths of 88, 85, 82, and 79 feet were studied. The fastest, an 82- by 20.66-foot boat, 192,450 pounds, became the basis for the next series.

This time hull length was held constant at 82 feet, but beam was varied at 20, 20.66, 21.5 and 23 feet, allowing 1,500 pounds of hull weight per foot of beam. Fuel weight was adjusted to keep the endurance constant. The comparison demonstrated clearly that beam had an enormous effect, but it did not show an optimum figure: the widest boat was 1.69 knots faster than the narrowest. Had displacement been held constant, the widest would have been 2.70 knots faster.

The next variable in the basic 82- by 22-foot, 195,500-pound boat was the lengthwise position of the center of gravity (corresponding to trim). With corresponding trim in parentheses, the alternatives were 56.5 percent of waterline length abaft the bow (0), 64 percent (2), and 71 percent (4). It was impossible to keep resistance constant; at 20 knots, with the 4° trim by the stern, it was extremely high. However, at constant displacement the greater trim was

Table 7-2. PT Alternatives, 31 May 1945

Comparative Table

	Vosper	Higgins	Elco	Study I	Study II
LOA (ft)	71	78	80	94.5	94.5
Full Load (lbs)	100,600	121,000	121,000	204,500	204,600
Lifting Weight (lbs)	92,000	106,000	106,000	166,000	167,000
BHP (emergency)	4,050	4,500	4,500	10,000	10,000
BHP (maximum)	3,375	4,050	4,050	8,000	8,000
Trial Speed (kts)	38.75	39	39	43.5	43.5
Fuel Capacity (gals)	2,960	2,850	3,000	7,000	7,000
Endurance at 20 nm	570	500	500	600	600
Engines	3	3	3	4	4
Shafts	3	3	3	4	4
Complement	2/8-10	3/14	3/14	4/18	4/18
Armament	—	1 40 dp	—	—	—
	—	1 37 sp	—	1 37 dp	
	1 20	1 20	—	—	1 x 4 20
	—	—	—	2 x 2 20	—
	2 x 2 0.50	2 x 2 0.50	2 x 2 0.50	4 x 2 0.50	
		2 x 1 0.50			
	2 torp tubes	4 torp racks	4 torp racks	6 torp racks	
	—	2 rl Mk 50	4 rl Mk 50		
	—	1 60mm mortar	1 81mm mortar		
	4-dc chute	—	—	—	—
	2 smoke racks	1 smoke generator	2 smoke generator	1 smoke generator	
Armament Weight (lbs)	7,710	12,534	12,815	15,293	16,234
Ammo					
40 mm	—	192	192	—	—
37mm	—	125	125	250	125
20mm	1,800	360	360	1,440	2,880
0.50	3,850	4,000	6,000	4,000	8,000
torpedoes	2	4	4	4	4
rockets	—	32	32	64	—
81mm	—	—	—	12	30
60mm	—	30	30	—	—
DC	4	—	—	—	—
smoke	8	—	—	—	—
Ammo Weight (lbs)	9,008	9,790	10,540	16,360	15,537
Total Weight (lbs)	16,718	22,324	23,355	31,653	31,771

extremely favorable for planing, so that the 4° boat enjoyed a 6.1-knot advantage over the 0° boat. With endurance held constant, the 2° boat required so much more fuel than the 0° boat that the latter was faster on the plane (41.4 as opposed to 39 knots).

The effect of greater beam was impressive, and thus it seemed likely that the first study had been misread: planing resistance was probably affected more by beam than by length. The first study was repeated with a constant beam of 19.25 feet and a varied length (400 pounds/foot). As in the case of the study on beam, there was no optimum length, but this time the longer hull was better; 88 feet was 0.6 knots faster than 82 feet. Regardless of this, greater beam was so important that it might be best to trade off length for a wider beam. The location of the center of gravity was important enough to validate the argument that the boat should be designed to change trim between the cruising and planing conditions. The change could be made by pumping fuel fore and aft.

The chief civilian in Preliminary Design, Mr. J. C. Niedermair, chose two alternatives, 88 by 19.25 feet and 84 by 21 feet. Lines for the long boat were started on 6 July. They generally followed those of the Elco, except aft. The chine heights represented a compromise between the Elco and Higgins hull forms, and the keel line generally resembled that of the Higgins. The transom was made considerably wider at the chine than either the Higgins or Elco boats, and the deck was a little wider.

The shorter hull was not developed in detail because design work on the prototype PT program was effectively suspended at the end of the war. The 94.5-foot (loa) sketch design became the basis for tentative characteristics issued in November 1945 and revised by OpNav in January 1946. At this time, Preliminary Design expected full-load displacement to be 204,000 pounds;[9] it hoped to achieve 46 knots on 10,000 bhp.

The idea of developing two separate classes of boat, both torpedo and gun, was rejected: the new PT, like its predecessor, would be a single hull with alternative armaments. It would have the aluminum hull proposed by BuShips for a combination of durability and protection. A BuShips proposal to build two wooden prototypes for comparison was rejected in view of the bureau's already extensive experience in wood shipbuilding. As the General Board later suggested, the bureau had only to maintain an active interest in wood and plastic construction so that it might be able to shift if that proved necessary. It was by no means clear, early in 1946, that the aluminum hull would be satisfactory. For example, at a General Board hearing in February 1946, a former PT operator, Captain S. S. Bowling, argued that wood would be easier to repair in forward areas. Thus, not only the hull form and general design but also the hull structure itself was experimental.

Maneuverability was intended to be comparable to that of the Higgins (in terms of ship lengths, not yards).

At this time the 37mm gun, no longer in production, was eliminated from the projected gun battery. The new PT would be armed with two power-operated, single 40mm guns controlled from a stabilized central station; two power-operated, quadruple 0.50-caliber machine guns, which could be replaced by twin 20mm mounts; one 81mm mortar; and four 4,000-pound long-range torpedoes, with automatic facilities for gyro setting. It was assumed that the 0.50-caliber and 20mm gun mounts would be adaptations of recent remote-control aircraft turrets. Ammunition allowances matched those standard in 1945: 192 rounds for each 40mm, 1,000 for each 0.50, and 10 illuminating rounds for the mortar. The total allowance for weapons and ammunition was to be 38,000 pounds, including 16,000 for torpedoes and 2,500–3,500 for torpedo fire control.

The characteristics required a trial speed of 46 knots at the standard full load, minus one-third fuel and including two torpedoes. Endurance was set at 500 nautical miles at 27 knots; this figure made provision for additional range to be attained with deck tanks capable of being jettisoned.

As requested in wartime, the electrical system would operate at 115 volts and be designed not only to carry the designed load but also to accommodate future developments. That entailed considerable weight. The standard wartime 5.5-kw unit, with its switchboard, weighed 345 pounds so the usual two-generator plant weighed 690. In February 1946, Preliminary Design assumed that the addition of a third generator totaling 1,035 pounds would suffice. However, the SCB asked for three 20-kw units, a total of at least 2,000 pounds out of a very limited total machinery weight. In the design, electronics was deliberately held down to the 1,826 pounds of wartime boats, despite a stated requirement for better radios and a new radar. The boat would take a crew of three officers and fifteen enlisted men.

The General Board slightly revised these characteristics in February; they were submitted by the CNO, Chester Nimitz, and approved by the secretary of the navy in May 1946. Nimitz's sole objection was to the General Board's comment that such craft could be expected to operate only in the tropics or in confined waters, where their limited sea-keeping would not be a great drawback. U.S. naval strategy was shifting to face a possible enemy, the Soviet Union, and the CNO could not really accept such a limitation. He also wanted continued effort to achieve further improvements in speed, endurance, and weaponry.

At about the same time, the projected armament was changed. Now the forward 40mm would be interchangeable with the Mark 107, an antishipping/bombardment rocket launcher then under development.[10] BuOrd abandoned the quadruple 0.50-caliber machine gun and was willing to settle for a pair of free-swinging, 20mm scarf-mount (Mark 28) guns, which might ultimately be replaced by a new 35mm.[11] Each Mark 28 had a thousand rounds.

BuOrd was still working on PT-related weapons, but by the late spring of 1946, voices complained that effort on these relatively low-priority projects was blocking much more important ones. For example, a new PT rocket launcher would require as many man-hours as any other new rocket launcher, such as the new ASW weapon Alfa. PT fire control work would displace efforts in ASW and missile fire control.

In October, Op-03 (operations) foresaw an additional role for the PT: ASW. PTs would not be fitted with sonar but rather would be used as pouncers for other ships, firing forward-thrown weapons like Hedgehogs. Op-03 also wanted a fully stabilized DP gun, at least 40mm, preferably something like a 3-inch. Both proposals were rejected—neither the 3-inch nor heavy ASW weapons could be accommodated in a fast lightweight hull.

The Preliminary Design studies had formed the basis for the General Board characteristics, but the characteristics were fairly general. In particular, discussions before the General Board showed that pure displacement hulls like those of the wartime German E-boats had important virtues. It seemed possible that at least one of the four prototypes would be a displacement boat, accepting lower speed in return for better sea-keeping, a significant goal in Admiral Nimitz's view. BuShips resisted the round-bottom hull for a time, but the Philadelphia Navy Yard was assigned to develop a round-hull design for comparison with conventional hard-chine hull forms.

BuShips issued design contracts and a circular of

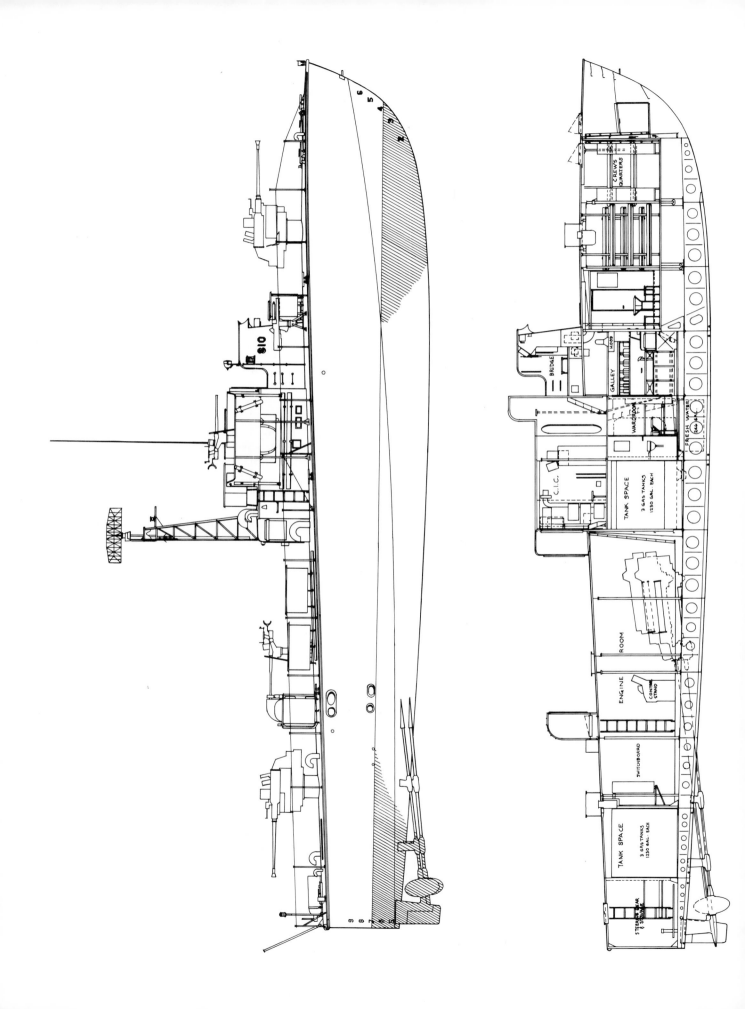

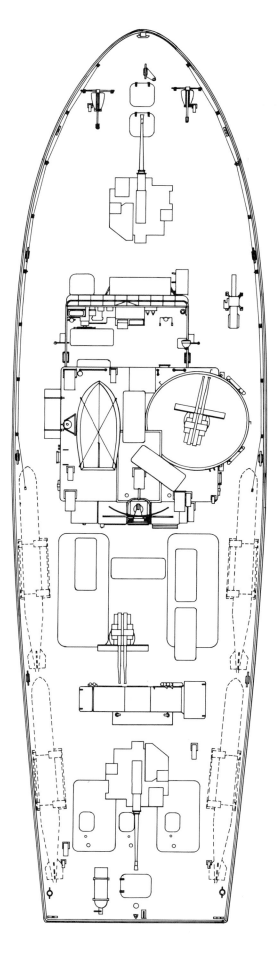

PT 810 as completed in November 1950. Without torpedo racks, it is armed with two Mark 3, Mod 4, 40mm guns with Mark 10 power drives, two Mark 24, Mod 6, twin 20mm guns, and one 81mm mortar. As PTF 1, it showed few external changes apart from the addition of a trim tab at the stern. It had SPS-5 radar and could carry BLR-1 ESM gear. Provision was made for four depth charges on tilt racks, two over the stern and two over the sides near the stern, or four Mark 16 torpedoes in Mark 2-1 and -2 dropping gear. In the plan view, dotted lines show how torpedoes would have been carried. (A. D. Baker III)

Table 7-3. PT Designs, 31 October 1947

	Bureau	Annapolis	Sparkman and Stephens	Elco	Philadelphia
LOA (ft)	94	94	89.5	98.6	105
LWL (ft)	88	88	85.2	93.3	100
BWL (ft)	19.5	19.4	22.1	20.7	16.0
B Max (ft)	24.8	24.9	24.1	25.8	21.0
Depth (ft)	11.07	10.94	11.0	11.6	11.8
Est Speed (kts)	45.9	47.0	52.0	48.8	42.0
Gasoline (gals)	7,000	7,000	7,380	7,000	6,000
LCG (%)	58.2	57.9	55.2	57.8	56.4
Hull (lbs)	38,213	41,105	38,555	38,400	35,998
Fittings (lbs)	24,506	15,044	16,604	18,799	20,280
Machinery (lbs)	38,385	35,189	36,023	33,471	34,613
Ordnance (lbs)	41,481	41,481	41,494	41,522	41,270
Stores (lbs)	9,510	8,980	8,980	8,980	9,123
⅔ Fuel (lbs)	28,560	28,560	30,100	28,560	24,480
Margin (lbs)	6,186	5,734	5,700	5,698	5,531
Trial (lbs)	186,841	176,093	177,456	175,430	171,295

Note: Speed estimates were based on 4,000 ehp, gasoline (gallons) on an endurance of 500 nautical miles at 20 knots. The weight of fittings included outfit and equipment; of stores, included complement and potable water. Machinery weight included machinery liquids. LCG is the location of the center of gravity as a proportion of the waterline length abaft the bow.

requirements to four potential builders on 7 June 1946; two of the four were each to build two boats. By this time, Higgins was no longer involved in the program, but the Philadelphia Navy Yard was added to the list. It was assigned to develop a round-bottom displacement design. The other firms were Annapolis Yacht, Elco, and the naval architectural firm of Sparkman and Stephens, working for Bath Iron Works. The bureau provided its preliminary design; the bidders would also be provided with data from ongoing full-scale and model basin experiments. Bidders wishing to depart from the BuShips plan had to remain within a length of 85 to 100 feet, a beam of 20 to 25, and a draft of not more than 7. Armament was now listed as two 40mm power mounts, two quadruple 0.50 caliber guns, one 81mm mortar, four racks for 21-inch torpedoes, and two rocket launchers.

Proposals arrived beginning in April 1947; they are shown in table 7-3. Annapolis Yacht and Elco offered modified versions of the BuShips preliminary design. Annapolis Yacht offered a bell bottom based on the form of the wartime Vospers it had built; this design carried all its gasoline aft. The structure of the Elco boat was designed by the Edo Aircraft Company. It showed an all-riveted hull built of 24-ST aluminum alloy and with ⅛-inch hull plating. All gasoline was carried below the waterline. The hull form incorporated a V-form with convex sections for a semiround bottom.

As in the 1945 hull studies, Sparkman and Stephens hoped to achieve a higher maximum speed (at some cost in cruising range) by using a shorter, beamier design. BuShips, fearing that the shorter hull would encounter difficulty getting over the hump, approved reduction gearing for it. Gearing was also expected to reduce cavitation, which in the past had markedly shortened the operating lifetime of PT propellers. Sparkman and Stephens also allocated more space to machinery; the engine room was 6 feet longer than that of the Annapolis Yacht design. The short boat was also unique in having its gasoline divided fore and aft. Opinions on this arrangement differed. Piping to a single gas tank would be simpler, but on the other hand it would be harder to destroy the fuel supply of a double tank—that is, if a single hit did not destroy the entire boat.

For comparison, Philadelphia Navy Yard offered a round-bottom (displacement) hull reminiscent of that of the German E-boat. Because of its greater length the crew could be divided fore and aft. The hull was completely welded.

On 22 April 1947, the award procedure was revised at a BuShips conference called to evaluate Annapolis Yacht Yard's proposal. One possibility was to defer any award until after the Sparkman and Stephens design, due on 28 April, had been submitted and evaluated. The Elco design was due in August, the Philadelphia design in September. The PT program was no longer particularly urgent, and it was not even clear whether the CNO still wanted the boats built.[12]

Nor was it clear whether two yards would each be assigned two PTs. Captain S. G. Nichols of BuShips' patrol boat section suggested that one vessel of each of the first two designs, from Annapolis and Bath, be awarded, and that the decision about the remaining pair of designs be deferred. Captain A. C. Mumma of

PT 810 is shown complete, with its SPS-5 surface-search radar, in August 1953. The two twin 20mm guns in all four postwar prototypes were mounted en echelon, so that both could fire on both broadsides, a decision reflecting preference for the arrangement of the wartime Elco.

Preliminary Design argued that the two later designs might prove better. He favored a round-bottom displacement hull, such as the boat Philadelphia was designing, and recalled wartime E-boat successes and sea-keeping qualities. The meeting concluded that no more than one vessel of any one design would be built.

PT 809 was built by Elco; 810 by Bath (to the Sparkman and Stephens design); 811 by John Trumpy of Annapolis (who had bought out Annapolis Yacht); and 812; the round-bottom boat, by the Philadelphia Navy Yard. Of the two long V-bottom boats, PT 811 enjoyed unusually good maneuverability at high speed, owing to a "turning fin" attached to the keel directly under the cockpit. The Bath boat, PT 810, had variable planing surfaces added at the stern to overcome excessive trim angle. They extended 2.5 feet abaft the transom and measurably improved sea-keeping.

All the boats other than PT 810 had four rudders, one abaft each propeller. PT 810 had only two rudders, abaft its outboard propellers, and proved much more difficult to handle. Larger rudders were installed in June 1952, but they did not completely solve the problem, which was attributed in part to additional underwater side area (triple supporting struts on each propeller shaft). PT 810 did have the best speed performance of the hard-chine boats.

Unlike the wartime E-boat on which it was broadly patterned, PT 812 did not have "Effekt" rudders; they added speed to the E-boat by channeling the propeller stream.

PT 810 had by far the largest silhouette. Although the other boats were considered satisfactory in this respect, all could be tracked at substantial ranges by existing surface-search radars. The minesweeper Peregrine tracked PT 811 (beam aspect) out to an average of 24,525 yards with her SPS-5(XN-1) radar. For the much longer PT 812, this range was 36,905 yards. Bow- or stern-on, PT 812 could be tracked out to about 30,000 yards. These figures confirmed the need for longer-range torpedoes, since by the mid-1950s many navies had radars capable of similar performance.

According to the final report of the Operational Development Force (OpDevFor), PT 812 was the stablest and easiest-riding boat with any sea running. In a smooth sea, a hard-chine boat could easily outrun it, but like the wartime E-boat, it could maintain speed in rough water. All of the boats could be expected to roll considerably with the sea broadside. Heading into the sea, the hard-chine boats were quite uncomfortable. PT 811 pounded constantly. Her bunks were untenable in sea state 3, her forward 40mm gun could not be manned at all, and manning the after 40mm was hazardous. The 20mm guns amidships could be manned, but motion rendered them inaccurate. PT 809 rode somewhat better, and the beamy PT 810 was considered the most stable of the hard-chine boats.

Table 7-4. Experimental PTs

	809 Elco	*810* Bath	*811* Trumpy	*812* Philadelphia, NY
LOA (ft-in)	98-9	89-5¼	95-0	105-0
LWL (ft-in)	—	—	—	100-0
Beam (ft-in)	26-1	24-1	24-11	22-5½
Draft (ft-in)				
(fwd)	3-0	3-0	3-4	3-5
(aft)	5-9	6-6	6-4	5-10
Displacement (lbs)	189,350	182,000	173,000	200,032
Trials (max speed, kts, warload)	46.7	44.3	47.7	38.2
BHP	—	9,300	—	11,040
Speed (kts)	—	46.5	—	44.75
Displacement (lbs)	—	170,000	—	175,000
Date	—	8.8.50	—	25.4.51
Cruising Range (nm)	Winter/summer	Winter	Summer	Summer
40 kts	464/383	488	497	—
27 kts	—/645	711	578	707
15 kts	—/692	854	732	—
Tanks (gals)	7,200	7,380	7,200	7,149
FL cap (115-145 octane, gals)	6,960	7,014	7,960	6,790
Burnable (gals)	6,720	6,774	6,720	6,550
Hull	⅛in 24 ST	61 ST Side 3⁄16 Bot 5⁄16	61 ST Side 3⁄16 Bot ¼ From frame 14 Fwd, rest ⅜	61 ST ¼in
Armament (no. of units)				
40mm	2	2	2	1
twin 20	2	2	2	2
81mm mort	1	1	1	1
Mk 107 rl	—	—	—	1
Fresh Water (gals)	250	246	260	300
Crew (off/enl)	4/16	3/15	4/18	4/18

Note: Displacement is wartime condition less ⅓ fuel oil (trial condition). All boats were powered by four Packard W-100 engines, and each had two Hobart 25-kw (120-volt) generators.

In a following or quartering sea, PT 811 rode more easily than PT 810, which had difficulty maintaining a steady course because of slewing and yawing. Again, as might be expected, boats moving through deep swells alternately planed on top of swells and then dug into the troughs in between. In PT 811, for which this effect was most marked, speed varied by as much as 10 knots, between planing and digging in. In extremely high following seas, the bow of PT 810 dug in considerably, and the boat took a great deal of water topside. PT 811 planed on top of the boat took a great deal of water topside. PT 811 planed on top of swells but slowed when its stern settled into troughs.

Owing to the combination of large sail area and shallow draft, PT 809 and 811 found maneuvering difficult in a wind greater than 10 to 15 knots. They had to run at 3 to 5 knots to maintain steerageway and could be expected to back into the wind, regardless of rudder, if wind speed exceeded 15 knots. PT 812 was less affected by wind. PT 810 was the worst. In a wind above 10 knots it could not be turned into the wind, unless the outboard engines were used and sufficient room was available to make headway. With only the inboard pair of engines in use, the boat had to accelerate to about 7 knots in order to turn, and that took considerable sea room.

Acceleration was roughly the same for all four boats, and 809 and 811 turned much more tightly than 810 and 812. In calm seas, OpDevFor concluded, 809 and 811 would evade more effectively, but this advantage would decline as the sea roughened.

Despite the successful prewar experience with PT 8, all four boats encountered structural problems. PT 809 was built with a light, riveted, aircraft-type hull (⅛-inch aluminum alloy). It was badly damaged by buffeting, and the after third of the underwater hull had to be replaced. The rest of the structure was of monocoque-girder design, so the stresses were evenly distributed among the longitudinals. Even so, the boat leaked badly after each repair and by 1954 had to have the forward part of its underwater hull replaced. Its ⅛-inch hull plating worked but enlarged

PT 811 at sea in May 1954, carrying an 81mm mortar on its port side, opposite a twin 20mm. It became PTF 2 in 1962.

some rivet holes, causing corrosion. Some rivets sheared. PT 810 showed heavier construction, partly welded and partly riveted: 3/16–3/8-inch aluminum on a monocoque-girder hull. Trumpy's PT 811 had a heavy-shelled, welded hull but encountered early troubles with its intercostal longitudinal members. The boat required additional structural members and internal bracing.

As might have been expected, PT 812 enjoyed a smoother ride (displacement rather than planing), and thus experienced fewer and less serious hull problems. Even so, it pitted badly while waiting five months in the water at Norfolk for a replacement supercharger. OpDevFor, assigned to test all four boats, concluded in 1954 that aluminum was at best a questionable material for the future.

The new engines were also a problem. By 1954, boats had all four operable only about 28 percent of the time. Unfortunately, their designs did not make for easy engine replacement. PT 810, the best boat, required twelve hours for an engine change, whereas four had sufficed for a wartime PT.

As in wartime PTs, the weapon position forward in the hard-chine boats was subject to severe pounding; PT 809 had a barrel support, which OpDevFor suggested for the other three boats. Nor did OpDevFor like the Mark 107 5-inch rocket launcher, which was actually installed only in PT 812 (and later replaced by a second 40mm gun). It suggested that the 40mm and 5-inch rocket functions be combined in a new 2.75-inch rocket launcher. Smaller rockets would be easier to handle on an unstable platform and might be carried in larger numbers. The 2.75-inch caliber was just then being introduced for air-to-air combat.

Apart form some projected rough-water tests, the evaluation of the four experimental PTs was complete by March 1954. Their future value depended very much on future U.S. involvement abroad. If the U.S. Navy confined itself to air strikes and deep-water ASW, PTs would be used primarily for base defense against sneak-attack craft, probably midget submarines or smaller vessels. If, however, the United States executed amphibious attacks, PTs would have other important roles, primarily deception to protect against amphibious attack (the beach-jumper unit, or BJU, role) and defense of the amphibious anchorage against attack by small enemy craft (fly catching).

OpDevFor reported two exercises: Atlantic Amphibious Exercise II-53 (27–31 March 1953) and Harbor Defense Exercise I (12–16 July 1953). The amphibious exercise included both a BJU phase and a post-landing antisneak attack (fly catching) phase. BJU operation was effective enough for the SCB to request a feasibility study in June 1956 for the conversion of PT 810–12). BuShips concluded the following March that all could be converted and that all would be satisfactory beach-jumper boats, capable of both types of countermeasures/deception (radar and communications). However, PT 810 and 811 had unreliable main engines and would lack the required cruising range of 1,500 nautical miles. PT 812 was being fitted with a new CODAG power plant, which was expected to overcome the problems of the earlier boats. Moreover, it would burn safer diesel fuel. BuShips felt that the other two boats should be converted only in an emergency, and then only if their gasoline engines were replaced.

PT 812 was the largest of the experimental boats and the only one with a round-bilge hull. The large shrouded object at the bow is probably an experimental Mark 107 rocket launcher. Astern is the 40mm mount, as yet without a gun.

In the fly-catcher role, with two boats patrolling off each side of the transport area, sneak attack from the beach was completely neutralized. With enough boats patrolling the entire perimeter in 3- to 4-mile sections, the sneak attack threat could be neutralized altogether. Alternatively, the boats could be used as "pouncers," vectored by slower patrolling craft, in this case, two patrolling LSMRs (large fire-support craft).

Their shallow draft allowed PTs to patrol very close inshore. PT 811 for example, operated only 700 yards from the beach. IFF was a problem, as the boats were well within the range of friendly fire. Patrol endurance was limited for example, to seventy-two hours 50 nautical miles from the beach at low loiter speed.

All of this still left one key issue unresolved. If they were to form an integral part of a future transoceanic amphibious task force, the PTs had to be transported in fighting configuration by the assault force, which was slowly being rebuilt for higher sustained speed, ultimately 20 knots. It would not suffice for the boats to move overseas as deck cargo, since there would be no time to unload and make them ready. Instead, any PT covering an amphibious operation would have to go either under its own power or in a well deck. In the latter, more logical case, it would displace one or more large landing craft.

PT 809–10 and 812 served as simulated sneak-attack craft in the harbor defense exercise. Their success was difficult to assess: defending forces claimed forty PT sinkings, but none of the PTs reported being challenged or visually attacked. Several of the detections had been reported by helicopters and blimps.

Finally, the Korean War illustrated the continued importance of what amounted to barge-busting or coastal interdiction. Although they relied primarily on railroads and roads, many of them coastal, the North Koreans also used coastal shipping to supply their forces—and to mine the harbors the United Nations navies used. Shallow-draft boats could expect to evade deep-draft UN warships. Aircraft could attack the warships in daylight, but after a few attacks on friendly craft, they came to require elaborate identification procedures that made coastal air attack almost impossible. The only shallow-draft craft initially available were motor minesweepers (YMS), which had to be detached for gunboat duty. They were badly needed to sweep mines laid by North Korean coastal craft that had evaded the blockade and, in any case, were not fast enough to catch many of the blockade-runners.

Naval raiding parties proved effective against the many crucial roads and rail lines running along the coast. Parties were generally landed by craft carried in the davits of destroyers or destroyer-transports. Similar parties had often been carried aboard PTs in World War II.

At this time, the United States retained only eight PTs—four of them in OpDevFor and four of them in reserve—in small-boat rather than combatant status. The commander of Naval Forces, Far East, requested PTs for Korea, but none were assigned. However, the four Elco OpDevFor PTs were transferred to the South Korean navy in January 1952. Even after the armistice, the South Koreans had a continuing need for coastal interdiction craft, as well as for a means of countering the postwar North Korean MTB fleet. This requirement led to post-war U.S. interest in motor gunboats (see chapter 9) and ultimately to the post-Vietnam CPIC (chapter 14).

All gasoline engines suffered from three major problems: the fire/explosion hazard, complexity requiring excessive maintenance, and short life at full power between overhauls. BuShips therefore considered the much simpler gas turbine the ideal PT power plant of the future. In April 1951, for example, it received a Fairchild Aircraft study of a 2,500-hp marine gas turbine installed in an 80-foot PT boat. In 1952, the bureau studied plants incorporating small auxiliary engines for low-speed cruising, in COGAG (gas-turbine cruise) or CODAG arrangements. The cruise engine would not be used at full power unless it could be provided with a controllable-pitch propeller. PT 812, with the longest hull, was the most attractive for conversion. Planning for the CODAG conversion of PT 812, using the existing British Metrovick Gatric G.2, began in mid-1952. The first report was completed that November.

Upon completion of the PT evaluation in March 1954, the SCB proposed that the lessons of these trials be reflected in a new prototype for future PT construction. That had, after all, been the original plan. Op-43 included four PTs in the tentative FY 57 shipbuilding program. BuShips believed that it could fix problems with the aluminum hull and that the PT 812 installation would provide useful operating experience. BuShips followed through in June, recommending at least two PTs for the FY 57 program.

Given the indication that a new PT might be ordered under the FY 57 program, BuShips began work in January 1955 on a suitable gas-turbine plant. Without any stated characteristics, the bureau assumed that a new PT would have to meet those originally framed in 1946. OpDevFor had found PT 810 and 812 most attractive, and the all-weather PT 812 was chosen as the best basis for further study. To accommodate the gas turbine, its hull had to be expanded in height and width in way of the machinery box.

As stated late in 1955, the requirement amounted to 10,000 shp on two, three, or four shafts. The available engines were the 1,600-shp Allison 510B, based on the T56 turboprop (development had been stopped, but it had a potential rating of 2,500 hp); the 900-hp GE XT 58 designed for helicopters (the lightest available engine, three of them could be mounted on a shaft for fuel economy at low speed); the 3,500-hp Packard MGT-10 under development; and the British 3,500-hp Bristol Preteus 1250, already installed in several British craft. Successful installation demanded light weight, about 2 pounds/shp; almost minimum space, about 0.5 cubic feet/shp; and an acceptable range of cruising speeds.[13] Any gas turbine would have the enormous advantage of eliminating the fire/explosion hazard of gasoline.

Gas turbines continued to be inefficient, so cruising range was still a problem. Endurance generally did not rise at low speeds because fuel consumption rose too fast. As a result, endurance would be about 40 percent less than in a boat with a gasoline engine. Even if the fuel tanks expanded to take advantage of savings in machinery weight, endurance would fall by about a quarter. Existing gasoline engines were still attractive; they too were very light aircraft engines, about 1.5 pounds/shp, and, at half power or below, they achieved 50 to 100 percent better specific fuel consumption (SFC). Worse, the gas turbine demanded much more volume for its air intakes and exhausts. In PT 812, for example, uptakes were expected to add 6,000 pounds, electrical installations 2,000, and the larger machinery space another 2,000, for a total weight growth of nearly five tons. On the other hand, Packard had closed down its gasoline engine plant, and the estimated replacement cost was $23 million.

The initial conclusion was that, the fuel rates of existing open-cycle (nonregenerative) gas turbines being so bad, a successful plant would have to incorporate several engines that could be cut out at cruising speed. They might be mounted in tandem with the cruising engines, or they might freewheel when the boat cruised. In either case, so much more power would be required that the plant would show no advantage unless its fuel rate could be halved.

The machinery study concluded that the Packard MGT-10 was most attractive for a short-range, 40-knot PT.

The other major line of new PT development was the hydrofoil. In February 1955, a hydrofoil PT was included among a series of possible warship designs presented to a senior BuShips advisory committee as part of the first study carried out by the navy's LRO group. The PT showed an overall length of 90 feet and an armament of four Mark 16-3A torpedoes, one 40mm gun, and one 20mm. A 4,800-shp gas turbine would drive it at 45 knots flying; hull-borne, it would cruise on twin diesels with a total rating of

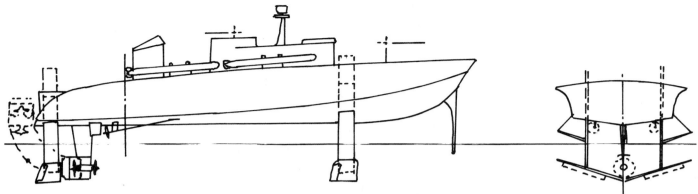

The hydrofoil PT boat, as sketched by Preliminary Design in 1955. It would have been powered by a 4,800-shp gas turbine, for 45 knots flying, and by twin diesels totaling 400 bhp, hull-borne. Dimensions were 90 (overall) x 28 feet, displacement, 78 tons. Armament consisted of one 40mm gun forward, one 20mm abaft the bridge, and four Mark 16-3A long-range torpedoes. Note the foil-control probe at the bow, used by the earliest submerged-foil craft. Note also that contraprops and rudder for foil-borne operation were retractable. (Norman Friedman)

Two of the four experimental PTs ended their days as PTFs. Here PTF 2, formerly PT 811, exercises with a Nasty, the next class of PTF, probably late in 1963. Externally, she shows little change over a decade after completion.

The planing and semidisplacement craft built under the 1945 PT program were inherently limited in sea-keeping; the old PT dream of maintaining high sustained speed in fairly rough water could not be achieved until the advent of hydrofoils. The Boeing *Tucumcari*, shown at launching, was a direct equivalent of wartime and postwar PTs. Her waterjet exhaust is visible in her transom, as are her main and forward foils. The water intakes for the jet are in the forward ends of the main strut tips, the boxy sections from which the foils themselves protrude. (Boeing photo)

400 bhp. The prototype would cost about $3 million, compared with $2.5 million for a conventional gas-turbine PT.

By 1956, the FY 57 program no longer showed any PT boat. Interest had shifted to inshore ASW (see chapter 8). PT 810 and 811 were withdrawn from service in August 1957. The other two experimental PTs served somewhat longer, but all four were stricken—reclassified as service craft—on 1 November 1959. PT 809, reclassified as a boat and renamed the *Guardian*, became a chase boat for the presidential yacht and later the drone recovery boat *Retriever;* it was still active in 1986. PT 812 was transferred to the army in 1959, then to South Korea in 1967, its gas turbines having been removed. It was discarded the following year.

PT 810–11 were reinstated as PTFs on 21 December 1962, beginning a new cycle of U.S. fast combatant boats.

8

Postwar ASW Craft

The small ASW craft described in chapter 3 had all been designed to deal with slow submarines incapable of diving deep.[1] Their speeds had been selected according to likely convoy speed, not submarine speed. Their sonars were relatively small, short-range, high-frequency "searchlights" that could easily be carried aboard small hulls. The associated weapons were both relatively simple and, apart from the sheer weight of depth charges, fairly light—again, having little impact on the ships carrying them and hence well suited to small hulls. For example, the smallest subchasers, such as the 83-foot Coast Guard boats, were armed with Mousetraps.

Several factors conspired to complicate matters after 1945. First, Mousetrap and the small sonars were already extremely limited. They were emergency measures, not satisfactory starting points for future development. The next step up in capability, the stabilized Hedgehog, demanded a much larger hull to absorb its weight and recoil.[2] Truly efficient submarine search required a larger scanning sonar. In 1945, moreover, submarine performance was changing radically. The Germans had just introduced fast diesel-electric submarines, types 21 and 23, which made 13 to 16 knots submerged, and after the end of the war it seemed inevitable that the Soviets, the next possible enemy, would soon duplicate them.

The immediate postwar U.S. reaction to a fast, deep-diving submarine was a trainable forward-thrown weapon such as the big Mark 15 Hedgehog or the much bigger Weapon Alfa. Unlike the Mousetrap or the wartime Mark 10 or 11 Hedgehog, which were fixed in train, this new weapon could follow a submarine moving faster than the attacking ship could swing. Ideally, the weapon would be directed by a big integrated sonar capable of tracking its target in three dimensions and feeding data into an electronic fire-control computer. Ideally, too, the future ASW ship, no matter how small, would have an advantage in speed over its submarine target, since the ship would have to bring its relatively short-range weapons to bear. The advent of the new submarines even affected traditional depth charges: the probability of killing a submarine with one pattern had decreased, and an ASW ship would have to carry more charges, more weight. By 1950 matters had grown worse, with the newer sonars, from SQS-4 onward, operating at lower frequencies; they required larger transducers, and hence larger domes and hulls, not to mention heavier transmitters and additional internal equipment.

The main hope was the homing torpedo. In theory, the kill probability of a single homing torpedo far exceeded that of a large depth charge or Hedgehog pattern, and at a small fraction of its weight. The homing torpedo would generate no recoil, and it might even tolerate a relatively poor fire-control solution. It would still require a good sonar, but that might be towed underwater as a variable depth sonar (VDS), thereby reducing impact on the hull.

The U.S. concept of what was required changed radically over time. In 1946, with the prospect of hundreds of Soviet-built type 21 clones vividly before it, the U.S. Navy's ASW establishment demanded a new generation of fast ASW ships equipped with big sonars, trainable weapons, and heavy homing torpedoes, principally the Mark 35. The effect of these requirements was to change the scale of ASW.

It might be said that existing ASW craft—SC, PCS,

Postwar inshore ASW was difficult because it was no longer possible to use simple, inexpensive craft. This is the ASW battery of the simplest wartime craft, PTC 62, an adapted, 63-foot aircraft rescue boat. The battery consists entirely of side-dropped depth charges. Low-cost inshore ASW became attractive again when the United States developed lightweight homing ASW torpedoes, such as the Mark 43/44/46 series. The weapons weighed little more than individual depth charges, yet they were several orders of magnitude more effective.

PC, PCE, DE, DD—formed a spectrum in size, speed, and to some extent, sea-keeping. The effect of the new submarines was understood to be size inflation; it pushed each class up one unit in size. Thus, the postwar destroyer escort was conceived as a 30-knot ship, and the converted *Fletcher*-class DDE was described as a destroyer escort. For that matter, the postwar destroyer *Mitscher* approximated a small light cruiser in size and capability. The postwar PC would come close to the wartime destroyer escort in size and capability. Initially, there was little interest in postwar equivalents of the small wooden SC and PCS and of the intermediate PCE.

The sheer cost, not to mention the political impossibility, of mass replacement of existing ASW units made a review of these ideas inevitable. In 1950, Admiral F. S. Low, who had commanded the Tenth Fleet during World War II, reviewed the ASW situation. He found that, despite fears to the contrary, the Soviets had not yet created a modern submarine fleet. They were clearly working in that direction, but a delay, due presumably to the need for postwar reconstruction, provided valuable breathing space. Admiral Low suggested that concentration on relatively inexpensive near-term measures would free funds for the research required to counter the massive Soviet submarine fleet once it did appear.

Suddenly existing ASW craft, such as the surviving PCs and PCEs, seemed far less obsolete. Moreover, it became possible to imagine replacing these craft with units of similarly limited performance—which might yet be valuable for some time, and which could be mass-produced.

As for the big trainable weapons, Low found that a much simpler weapon consisting of a pair of fixed Hedgehogs was about as efficient. Its much wider pattern compensated for the fact that the attacking ship could not swing fast enough. Unlike the trainable Hedgehog, the paired unit did not need an elaborate foundation or, for that matter, valuable centerline space.[3] It was installed aboard many destroyers and destroyer escorts and was part of the battery planned for new smaller craft.

Fortunately, lightweight homing torpedoes also reached the first stage of maturity about this time. They offered considerable firepower at a low cost in weight. For example, a single 250-pound Mark 43 could be considered equivalent to several nine-depth-charge patterns (3,060 pounds of 340-pound Mark 14s). Homing torpedoes began to appear in the batteries of planned U.S. ASW escorts in 1952. They made it possible to convert a minesweeper, with little available topside weight or space, to an efficient coastal ASW escort; it was easier to install a big sonar than to add the top weight of depth charges, tracks, and throwers.

No matter what the stated minimum requirements on a ship-by-ship basis, wartime ASW success would still require large numbers of escorts. As in the past, numbers were naturally equated with limited unit size and cost. Given the unhappy experience of the Battle of the Atlantic, it seemed prudent to prepare escort designs for future mass production, building a few peacetime prototypes. In Britain, this consideration led first to the second-rate frigate—the *Blackwood* class, which had half the ASW battery of the first rate—and then to proposals in 1950 for a third-rate frigate at two-thirds the cost of a second rate. The U.S. PC and PCE projects of 1951–53 might be equated with the British search for a worthwhile third rate frigate. In both cases the search was abandoned; too much capability had to be sacrificed, and other warship projects were so urgent that there was little point in building such mobilization prototypes.

Moreover, the basic attitude toward wartime mobilization changed. With the advent of the hydrogen bomb in 1952, U.S. national policy increasingly assumed that any central war against the Soviets would be fought largely by offensive strategic forces, and therefore that there would be no repeat of the Battle of the Atlantic. In such a context, the chief ASW roles were protection of the carrier striking forces and of the United States against Soviet missile-firing submarines, a role that initially justified the construction of the SOSUS system. The classical tactics of convoy ASW would still be practiced in the context of NATO's less apocalyptic strategy, but they could not justify special, austere mass-production designs.

Small-craft ASW was revived a few years later for harbor defense. As in the late 1930s, harbor defense would be a natural adjunct of the striking fleet strategy; without such defense, carrier task forces might be attacked in their bases. PT boats were proposed in the mid-1950s partly for harbor defense against sneak craft, and projected SC and PCS successors briefly took their place in U.S. naval building budgets. They in turn gave way to the first of the hydrofoils, the PC(H) *High Point,* the only U.S. ship described in this chapter that was actually built. But the *High Point* appeared just as U.S. strategy shifted again.

From 1955 on, the navy's long-range planners argued that Soviet possession of a substantial strategic arsenal would make strategic warfare almost impossible. Instead, the conflict between the two superpowers would be fought in the Third World, the Soviets using proxies to take over small countries, such as Vietnam, along the Eurasian periphery. The United States would respond on two levels. On one, it would arm those small allied states to resist aggression. That came to mean the construction of motor gunboats and the riverine navy developed for Vietnam. On another level, the navy would have to

The earliest postwar inshore ASW craft were Coast Guard cutters built to replace wartime 83-footers like the CG 83424, shown here on 16 September 1944. It is armed with a Mousetrap forward and depth-charge tracks on each beam and aft. The gun is a 20mm, the sonar, a QBE-series searchlight.

concentrate on carrier and amphibious groups capable of projecting sea power in support of the small allied states. Strategic strike would be transformed into strategic deterrence, and it would be essential for the navy to maintain its strategic striking arm at minimum cost to the limited-war forces that would actually have to fight Soviet proxies.

Coastal ASW lost its urgency. Few Soviet proxy states would be able to mount submarine offensives; many more of them would be armed with gun or motor torpedo boats. As for the big war, that would be fought, if at all, largely with forces already at sea, and then so quickly that most base defense would be of little importance. It also seems likely that, with the advent of SOSUS, striking forces such as naval aircraft and forward-deployed submarines could be expected to eliminate most Soviet submarines that might reach the open ocean.

These strategic projections, never quite made explicit, were acted on as money became tighter in the early 1960s. The inshore ASW mission was never formally abandoned; several classes, such as the PHM and the later patrol boat, were assigned alternative ASW armaments, but they were never equipped with them. Inshore and shallow-water ASW have not received much attention since the late 1950s.

The important exception was the Coast Guard, whose 95-foot patrol boat, conceived as a replacement for the wartime 83-footer, was optimized for harbor ASW protection with a modern hull sonar and a small forward-throwing weapon. The 95-footer was significant for navy development in two ways. First, it was a candidate for expansion as a replacement subchaser. Second, it evolved into an export patrol boat and ultimately into the standard U.S. export motor gunboat (see chapter 9).

In addition, the United States designed and bought small ASW craft for her allies, designs that show what she might have built for herself in a crisis.

The second (OpNav) ASW conference of April 1948 proposed a modern PC (SCB 51) for coastal escort and offshore patrol work. It had to be large enough to carry the new sonar and weapons, fast enough to catch submarines. It would displace about one thousand tons, with a length of 280 to 290 feet and a speed of at least 25 knots. Armament would consist of two

The chief inshore ASW craft after the war was the 95-foot Coast Guard cutter. The design—typified by the *Cape Carter* (WPB 95309), shown here in December 1969—formed the basis for motor gunboats exported to friendly navies. The *Cape Carter* was completed by the Coast Guard yard at Curtis Bay, Maryland, on 7 December 1953. As of 1965, she displaced 101.75 tons with a full load (78.77 tons with a light), could make 20 knots, and had a cruising range of 2,040 nautical miles at an economical speed of 8.9 knots. By the time this photograph was taken, cutters like the *Cape Carter* no longer carried ASW weapons or sensors. In the late 1950s, a typical battery included one Mousetrap (Mark 22) rocket launcher fired from the bridge and four type D depth-charge racks, controlled locally by a World War II–type QCU searchlight sonar. In peacetime, twenty-four Mousetrap rockets were carried but no depth charges. Cutters also carried a single 40mm gun with ninety-six rounds of ammunition.

3-inch/50 DP guns; two trainable Hedgehogs or, later, forward-thrown weapons limited in weight; depth charges in tracks and side throwers; and homing torpedoes. The ship would require the best available sea-keeping, maneuverability, and steaming radius at 20 knots on a displacement deliberately limited to make mass production possible.

Design work began early the following year. There were, as yet, no firm characteristics, and Preliminary Design sounded out various OpNav branches. In September 1949, Preliminary Design summarized the ship's features, as they were then understood. It would be armed with a pair of Mark 15 trainable Hedgehogs using rocket hoists from the magazine for quick reloading, one tube for a heavy Mark 35 homing torpedo, two depth-charge tracks, and eight to twelve depth-charge projectors. It would be equipped with the postwar integrated sonar, QHB plus a depth-determining set, probably SQG-1, preferably in a bulbous bow.[4] Preliminary Design sought a speed of 21 knots and an endurance of 5,000 nautical miles (later 7,000) at 18 knots. Successful attack would require tight turning, and early tentative characteristics showed a 300-yard turning circle, with emphasis on response to the rudder when first applied. All of this, it was hoped, could be obtained on a modified World War II destroyer escort hull. The Mark 15, quite massive, might cause problems.

Aside from these basic requirements, the battle efficiency of the new PC was understood to depend on its command and control arrangement. Much effort therefore went into designing an underwater battery (UB) plot, CIC, and sonar control room, all of which had to be combined and located on the same level as, and just abaft, the pilothouse. When actually applied to war-time destroyer escorts being rebuilt at this time under the SCB 63 program, this requirement was realized in a big bridge structure. Efficient sonar operation also demanded better conditions in sonar operator and ASW control stations; existing ones were noisy, inadequately lighted, poorly ventilated, and crowded.

OpDevFor reported that, in the face of fast submarines, a sonar search speed of 20 knots was both desirable and, with trained operators, practical; QHB could detect a submarine at 1,800 yards at that speed.[5] However, speeds well above 20 knots would be needed if the search ship were to regain her station after dropping behind to investigate a contact. Speed generally equaled greater size and cost.

In October, BuShips began work on a 6,900-bhp diesel plant, powerful enough to drive a wartime 300-foot destroyer escort at a sustained speed of 20 knots. A penciled note by one of the preliminary designers suggested that the CNO would surely ask for 23 on trial. As a result, sketch designs showed a variety of far more powerful steam plants by early 1950. For example, the early wartime 283.5-foot destroyer escort could be driven at 22 knots on 8,000 shp.

Early in February 1950, Captain McFadden, head of the ASW branch of OpNav's undersea warfare division, summarized his requirements: a small, cheap ship with a small crew, 1,200 to 1,300 tons, capable of 20 to 25 knots and 4,500 nautical miles at 15 knots, armed only with ASW weapons, and a tactical diameter of 350 yards. Commander in Chief, Atlantic Fleet, (CinCLant) wanted a beamier ship with its bridge well aft for dryness; CinCPac emphasized small size. In outline, this was not too different from the type 14 *Blackwood*-class frigate, design work on which the British had begun in mid-1949. Preliminary Design found its 15,000-shp single-screw power plant particularly interesting for application to SCB 51. The plant drove a 300- x 33-foot, 1,300-ton hull at 24 knots "deep and dirty."

By April 1950, Preliminary Design was trying to cut the size of what was now called a merchant convoy escort. Tentative characteristics were assembled from the notes of a February 1950 conference. They called for a 23-knot trial speed and an endurance of 4,500 nautical miles at 12 knots. By August, however, the minimum acceptable speed had increased to 25 knots. BuShips therefore developed a new 15,000-shp single-screw plant.

There would be no torpedo tube, and armament would consist of one Mark 15 (with 144 rounds); one twin 3-inch/50 aft, with 600 rounds per gun, 150 of them in ready service; four twin 20mm guns; two nine-charge depth-charge tracks; and eight depth-charge throwers (with five charges each). Complement would be 10 officers and 125 enlisted men.

Other important features were a low silhouette; quick turning, especially at the start of the turn; a sharp strong bow to ram submarines; the best possible design for sonar installation, with both scanning and depth-determining or attack sonar; towed sonar for long-range detection; an attack director such as a Mark 5 and an attack plotter; sonar countermeasures, towed and thrown; a surface antiperiscope radar; and good radio countermeasures (ESM). The conference also demanded the one-level control space described above.

Some unusual design concepts were considered. One had a very low bridge with a short forecastle and with the Mark 15 firing over the bridge. Another possibility was to push the bridge as far aft as possible, using a turtle-back forecastle with the Mark 15 in a well abaft it (as in torpedo tube installations in German destroyers of World War I). These radical reductions in top hamper were sought in hopes of cutting beam, which in turn would cut hull, machinery, and fuel weight and make the ship ride kindlier in a seaway. There would be a need, however, for a covered fore and aft passageway, which could be partially or completely buried under the main deck through the machinery spaces. Austerity for the purpose of mass production was so important that in February 1950 Preliminary Design considered eliminating the 3-inch/50 guns to leave only a 20mm battery.

By September 1950, the "PC" was growing much too large. SCB 51 was dropped in favor of a new study of an ocean escort, SCB 72, which evolved into the 20,000-shp *Dealey*-class ocean escort. (Alternative SCB 51 design data are given in table 8-1.) Work on the SCB 72 ocean escort began early in October 1950. Armament consisted of one Mark 17 projector firing sixty-one Hedgehog projectiles, with one reload in ready stowage and a hoist for rapid-reloading of six more, with a total of 366 additional charges; two twin 3-inch/50s; and eight depth-charge throwers with ninety-one charges each. The *Dealey* became the ancestor of modern U.S. frigates. Although she too was designed for mass production, interest in something smaller remained.

The first *Dealey*s were authorized as part of the FY 52 program. By mid-1951, the prospective FY 53 program included pairs of prototypes for a new PC (SCB 89) and PCE (SCB 90). Their common wartime ancestry seems to have been forgotten: the PC was imagined as comparable in size to the wartime PCE (180 feet long, 800 tons), the PCE to the wartime 220-foot minesweeper (about 1,000 tons). In both cases, slow speed, 17 knots on trial, 15 sustained, was accepted as a way of reducing ship size, hence cost. Both were assigned the wartime PC endurance of 3,000 nautical miles, but the PC was to have had its old endurance speed of 12 knots, the PCE, 15. The impact of the new sonars on ship size would be mitigated with a

Table 8-1. SCB 51 Alternatives

	240 Ft	260 Ft	283-6 Ft	283 Ft	320 Ft
Length (ft-in)	240	260	283-6	283	320
Beam (ft-in)	36	36	34	34	36
Depth (ft-in)	18	18-6	18-9	18-9	—
Draft, Full Load (ft-in)	10-11	—	10-3	—	11-3
Full Load (tons)	1,386	1,473	1,455	1,465	1,882
Standard (tons)	1,105	1,158	1,205	—	1,531
SHP	6,000 (4 1,500 diesel)	8,000 (steam)	8,000 (2 4,000 diesel)	8,000 (steam)	30,000 (steam)
Full Speed (95%, kts)	19.45	22	22	22	29.5
Sustained Speed (kts)	18.0	20	20.5	20.5	26.7
Endurance (nm/kts)	4,000/15	—	4,500/15	4,500/15	7,200/15
Armament	1 Mk 15	1 Mk 15	1 Mk 15	1 Mk 15	2 Mk 15
	1 Twin 3in/50	1 Twin 3in/50	1 Twin 3in/50	1 Twin 3in/50	2 Twin 3in/50
	2 Fixed TT	—	2 DC Tracks	2 DC Tracks	2 SGL 3in/50
	2 DC Tracks	2 DC Tracks	8 DC Projectors	10 Projectors	4 SGL TT
	4 DC Projectors	8 DC Projectors	2 SGL TT	—	2 DC Tracks
	Mk 63 GFCS	Mk 63 GFCS	Mk 63 GFCS	Mk 63 GFCS	6 DC Projectors
					Mk 56 GFCS

Note: DE 1, the short-hulled World War II DE that inspired many of the SCB 51 studies, is listed for comparison. In FY 52 dollars, she would cost $7.8 million, compared with $8.9 million for Ocean Escort II. In the ocean escort series, at the same length and power, 100 tons could be equated with 0.6 knots; at the same shp, 1 knot equaled 33 feet of length; at the same length, 1 knot equaled 2,250 shp at maximum speed; at the same speed, a twin-screw ship would require about 10 percent more shp than a single screw; at the same cruising speed, a single screw ship would have about 15 percent more endurance, owing to its greater efficiency. Full or trial speed was at full load.

new variable-depth sonar, SQS-9, then under development.

On paper, the two ships differed primarily in armament, their batteries corresponding approximately to those of the earlier PC and PCE. The PC was assigned one twin 40mm with no fire control, two homing torpedoes, and eighteen fast-sinking, magnetically fused Mark 14 depth charges with a nine-charge pattern. The ship would accommodate five officers and sixty-one enlisted men. The PCE was conceived to be about a third larger than the wartime PCE, its armament to consist of a new battery of six twin 40mm guns; two Hedgehogs or a new sixty-one-charge single Hedgehog (Mark 17) then under development; and one depth-charge track and four throwers. The PCE would accommodate seven officers and eighty-five enlisted men. There was some question as to where six twin 40mm guns could be installed; the wartime PCE had been armed only with one 3-inch/50 (equivalent to a twin 40mm) and three twin 40mm aft.

Estimated costs were, respectively, $5.3 and $6.9 million, based on the purchase of two ships of each type. The FY 53 program showed new DEs costing $10.1 million each; thus, the PCE represented an expected savings of about a third the cost of the DE, at a stiff price in speed and weaponry. Tentative characteristics emphasized better sea-keeping than that of the wartime PC and PCE, which might require fin stabilization. In addition, the PCE was to incorporate better conditions for sonar operation, such as air conditioning. Characteristics of the 1953 PC and PCE are given in table 8-2.

The SCB produced its first memorandum on the PCE mission and tasks on 24 May 1951, followed by preliminary characteristics on 12 September 1951. The original memorandum specified a 15-knot ship; the first preliminary characteristics added three Mark 32 lightweight homing torpedoes. Dimensions were now given as approximately 240 x 33 x 10 feet (1,000 tons), trial speed was increased to 17 knots, and estimated turning radius was 275 yards.

This postwar version showed improvement over the wartime PCE in the following features:

- ASW: two Mark 11s in place of one; Mark 32 torpedoes
- Sonar: hull-mounted scanning sonar; weight and space for VDS
- Habitability: more space per man; somewhat easier motion in a seaway
- Speed: 2 knots on trial, due to increased power and length; heavy seas, an additional margin could be expected because of added length and improved form
- Fuel economy: endurance speed 15 knots

Preliminary Design developed a series of alternative sketch designs. The optimum arrangement of all six twin 40mm guns was on a length of 246 feet; the minimum length to accommodate them was 216 feet. The other alternatives were lengths of 202 and 216 feet, mounting two twin and one quadruple 40mm. The short six-mount ship proved unacceptably cramped. Nor was it less expensive than the longer hull, since it required a forecastle for internal volume and a larger, more expensive power plant. The 246-foot hull required less power and had better fuel

Ocean Escort II	Ocean Escort III	Ocean Escort IV	DE 1	Ocean Escort
283-6	283-6	283-6	283-6	310-0
33-0	33-0	32-6	34-0	35-0
18-9	18-9	18-9	18-9	20-0
11-2	11-2	10-11	10-2	10-7
1,511	1,506	1,486	1,430	1,640
1,183	1,178	1,158	1,190	1,320
15,000	15,000	15,000	6,000	15,000
(steam)	(steam)	(steam)	(diesel)	(steam)
25.73	25.76	25.88	21.5	26.5 (29.2)
24.3	24.32	24.43	19.9	25.0 (27.3)
4,500/15	4,500/12	4,500/12	6,000	4,500 (2,800)
1 Mk 15	1 Mk 15	1 Mk 15	1 Mk 10	1 Mk 15
1 Twin 3in/50	1 Twin 3in/50	1 Twin 3in/50	3 3in/50	1 Twin 3in/50
4 Twin 20mm	4 Twin 20mm	4 Twin 20mm	1 Twin 40 mm	4 Twin 20mm
2 DC Tracks	2 DC Tracks	2 DC Tracks	5 SGL 20 mm	2 DC Tracks
8 DC Projectors	8 DC Projectors	8 DC Projectors	2 DC Tracks	8 DC Projectors
Mk 63 GFCS			8 DC Projectors	Mk 63 GFCS
			Mk 52 GFCS	

economy. It was also expected to be a better seakeeper owing to its greater length. This advantage was more than offset by much lower freeboard where the Hedgehog was installed. In view of the ship's ASW mission, it was considered extremely important to place the Hedgehog in the driest position. The 246-foot hull would also have the largest turning circle, a particular disadvantage against small submarines.

Preliminary Design preferred the three-mount alternatives. They provided a better arrangement on a shorter hull and still separated sonar and ship-control stations as far as possible from gunfire interference. They would be less expensive and would require fewer men. The 216-foot scheme, with its speed and seaworthiness, seemed preferable. These schemes are compared in table 8-3.

The SCB accepted the BuShips arguments but eliminated one of the two twin 40mm mounts, so that the January 1952 characteristics showed one twin and one quadruple 40mm, with two RF 3-inch/50s as an alternative. That fall, BuOrd and several senior officers pressed for adoption of the 3-inch/50 battery. Op-03 preferred the all-40mm battery but was outvoted. The only technical objection to the 3-inch guns was that the forward one would block visibility, and thus if it was provided with a weather shield, the pilothouse would have to be raised.

When work was suspended early in 1952, it seemed likely that topside weight could be reduced to give 215 x 30 x 10 feet, for 775 tons' standard displacement. When work was resumed, further analysis showed that a shorter ship would be badly cramped, both internally, in the berthing space, and externally, in the gun positions. Little weight would be saved.

The design showed no antiroll fins, because the United States had had no service experience with that device and therefore it was considered an unacceptable technical and maintenance risk in a mass-produced escort.

At about this time, the three twin 40mm guns were replaced by two single 3-inch/50s with weather shields; the pilothouse had to be raised 3.5 feet to maintain a reasonable line of sight over the forward gun. Later, in fall 1952, 7 feet had to be added to the length of the ship so that a trainable Mark 15 Hedgehog on a raised platform could replace the two Mark 11 Hedgehogs on the forecastle deck. The guns were to have received a Mark 63 fire-control system with a single director.

Space and weight were provided on the starboard side of deckhouse aft for eventual installation of an SQS-10 VDS. Studies were also made of the feasibility of a centerline installation on the bottom of the ship, but no actual design was available. A fixed sonar dome for the SQS-4 was provided at frame 37.

Second preliminary characteristics followed on 18 January 1952. They showed a somewhat smaller ship, 215 x 30 x 10 feet, or about 775 tons' standard displacement, with a 200-yard turning circle. The gun battery would be either one quadruple 40mm and one twin 40mm or two RF 3-inch/50s. Endurance was increased to the standard destroyer escort figure, 4,000 nautical miles at 15 knots.

Design work was suspended in January 1952 when the PCE and PC were removed from the FY 53 program. It soon resumed; as of March 1952, the ten-

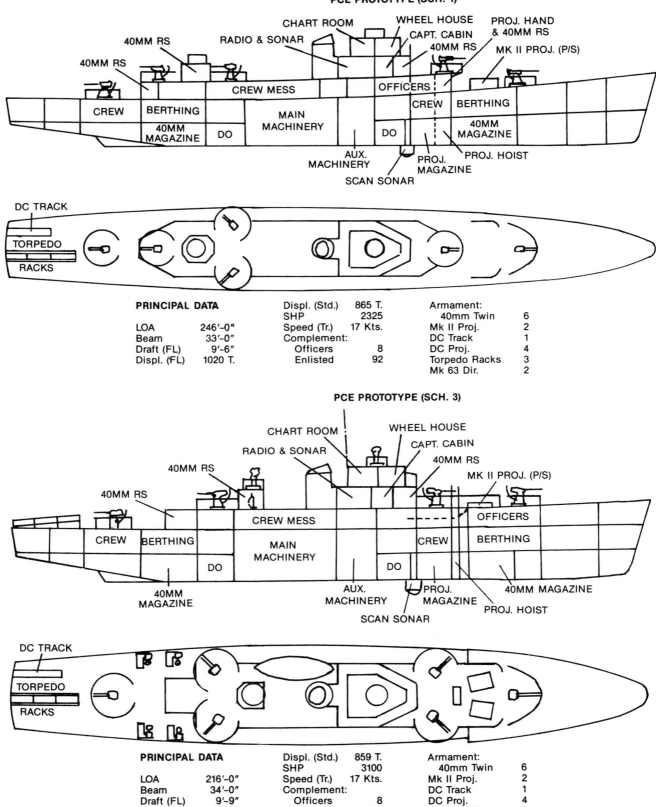

BuShips PCE designs, 1952.

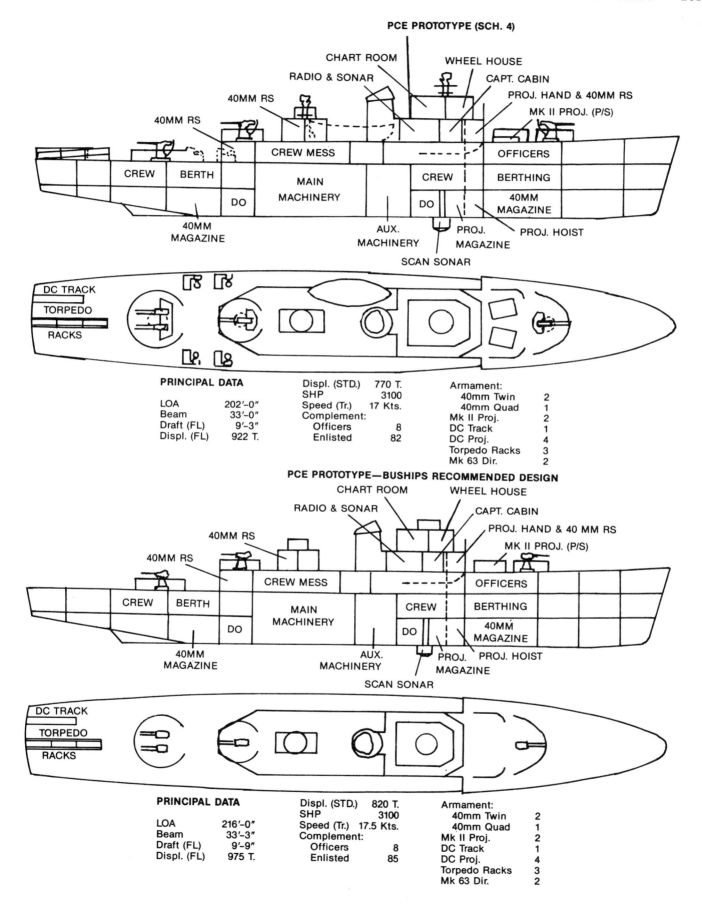

Table 8-2. The 1953 Escorts

	PCE 827–920	1953 PC	1953 PCE
LOA (ft-in)	184-6	—	223-0
LWL (ft-in)	180-0	175-0	217-0
Beam (ft-in)	33-0	28-0	33-6
Draft (ft-in)	8-7	8-0	9-9
Depth			
(fwd, ft-in)	—	—	27-6
(amids, ft-in)	14-0	15-0	17-6
(aft, ft-in)	—	—	19-3
Freeboard (ft-in)			
(fwd, ft-in)	—	—	17-9
(amids, ft-in)	5-5	7-0	7-9
(aft, ft-in)	—	—	9-6
Design Displacement (tons)	—	600	920
LS (tons)	595	—	—
Trial (tons)	795	—	880
FL (tons)	838	600	920
CB	—	0.536	0.449
CM	—	0.865	0.828
CP	—	0.620	0.542
CW	—	—	0.668
SHP	1,800	2,000	3,000
Trial Speed (kts)	—	16.0	18.1
Endurance	—	3,000/12	4,000/15
Armament	1 3in/50	1 40mm (twin)	2 3in/50
	3 40mm (twin)	—	—
	2 20mm	1 DC Track	2 H/H
	1 H/H (144)	4 DC Projectors	1 DC Track
	4 DC Projectors	(2-9 DC patterns)	4 DC Projectors
	2 DC Tracks	1 Mk 32 Torpedo-release	1 Mk 32 Torpedo
	(12 each)	track	launcher
Complement	5/5/84	5/60	8/90

tative FY 54 program included four PCEs, the two previous projects having been amalgamated. At this stage, the budget was still quite fluid; the initial wish list had an estimated cost of $3 billion, but it seemed likely that only $2 billion would be available. Based on priorities assigned at the time, two PCEs were to be included even if the ceiling fell to $1.5 billion. In fact, they were eliminated late in the budget process; the final PCE documents are dated December 1952.

The choice of a power plant had a considerable impact on the design as a whole. A twin-screw diesel plant was chosen in June 1952 after a comparison had been made. A single fixed-pitch propeller gave poor maneuverability when backing (a controllable-pitch or controllable-reversible-pitch propeller would have been considerably better); the single shaft would also be more difficult to arrange in the hull, since it would constrict the after magazine. The stern of a single-screw hull would be broader and flatter than that of a twin-screw hull, somewhat reducing speed in a seaway and causing more pounding aft.[6] The only important advantage of the single screw was that it might make mass production easier. Given the available range of diesels, however, it did not make production easy enough. Similar considerations would be raised in the case of the *Claud Jones*–class ocean escort a few years later. In 1954, the decision was based on the difficulty of cutting the large gears required for a single-screw plant.

In October 1952, approved characteristics showed somewhat larger dimensions, approximately 222 x 33 x 10 feet and 800 tons, and a larger complement, accommodations for eight officers and ninety enlisted men. The 40mm guns were abandoned altogether for two RF 3-inch/50s with weather shields. ASW weapons were listed as two Mark 11 Hedgehog projectors (total 144 charges for one projector, 288 for two); one stern depth-charge track; four side throwers (total 27 charges); and launching, handling, and stowage facilities for Mark 32 lightweight homing torpedoes (total 3).[7] Most of the ASW weapons requiring maneuverability on the part of the attacking ship, the characteristics included a turning circle of approximately 200 yards at full speed.

Characteristics of the preliminary design of December 1952, the final form of the 1953 PCE, are presented in table 8-4. Like the original PCE, this one had a forecastle for good sea-keeping. The ratio of length to freeboard at station 3 was about the same as for the existing PCE and considerably better than for the *Dealey*.

Two alternative sterns, a transom (as in the earlier

Table 8-3. PCE Designs, 1951

	PCE	BuShips Studies			BuShips Preferred Design
LOA (ft)	185	246	216	202	216
Beam (ft-in)	33-0	33-0	34-0	33-0	33-3
Draft (fl, ft-in)	9-5	9-6	9-9	9-3	9-9
Full Load (tons)	900	1,020	1,023	922	975
Standard Displacement (tons)	730	865	859	770	820
SHP	1,950	2,325	3,100	3,100	3,100
Complement	94	100	100	93	93
40mm (twin)	3	6	6	2	2
(quad)	—	—	—	1	1
(slow fire)	—	—	—	1	1
3-in/50	1	0	0	0	0
H/H	1	2	2	2	2
Freeboard (ft)	13½	12½	15¾	15¼	15¾
Tactical Diameter (relative)	0.90	1.33	1.00	0.94	1.00
Fuel Rate (at 15 kts, relative)	1.10	0.92	1.02	1.00	1.00
Cost (relative)	0.83	1.03	1.04	0.97	1.00

Note: Freeboard is at the Hedgehog. Cost is relative, that is, compared to the preferred BuShips design.

PCE) and a cruiser stern, were tested. In shp, the cruiser, with a reduced prismatic coefficient, was 7 percent better at the characteristic speed, 17 knots, and 10 percent better at the endurance speed of 15 knots. Because space inside was limited, a single all-moving rudder was used instead of the twin rudders of the transom stern, and turning was therefore slightly worse than in the existing PCE. In large part, this could be attributed to the larger length-to-beam ratio and the greater portion of the propeller disc area outside the influence of the hull. These tank test results were considered satisfactory because full-scale ships usually turned about 10 percent better than their models.

If the PCE is viewed as a second-rate, or mass-production, version of the contemporary DE, then it is fair to say that it later emerged in the form of the *Claud Jones* (DE 1033) class funded under the FY 56 and 57 programs. Although it was larger than the projected PCE, the *Claud Jones* showed little improvement except in speed. In effect, it demonstrated that the clear wartime distinctions between fast and slow escorts (DEs versus PCEs) lost a lot of their meaning after the war.

Work on the *Claud Jones* began with a June 1954 proposal by Rear Admiral W. K. Mendenhall, chairman of the SCB. He judged the *Dealey* too expensive and too complex for wartime mass production and sought a cheaper convoy escort. It would have a much slower sustained speed of 21 knots and would be armed with two slow-fire 3-inch/50s, one twelve-charge stern track, six throwers, and three homing torpedoes. Endurance would be very long, 10,000 nautical miles at 12 knots. This austere escort would take a crew of 10 officers and 143 enlisted men. Its design would emphasize sea-keeping.

Although the earliest versions of the design were simplified *Dealey*s, it is striking that in the end it resembled an enlarged PCE (SCB 90). Ultimately, it was armed with two RF 3-inch/50s, two Hedgehogs, and two triple homing torpedo tubes; endurance was reduced to 8,600 nautical miles at 12 knots. One might conclude, then, that SCB 90 failed because too much was attempted on too small a hull. However, once enough hull was added to support what was wanted, the result no longer seemed worth the trouble. Even before they were completed, the four *Claud Jones*'s were accounted failures, and the line of DE development turned toward more sophisticated ships that could not possibly be mass-produced: the *Bronstein*s, *Garcia*s, *Brooke*s, and *Knox*es.

The failure of the 1953 PCE did not eliminate the need for large numbers of relatively austere ASW craft. The U.S. Navy still had many war-built PCs and PCEs as well as many minecraft of similar size. After all, the PCE had been conceived as an escort version of the 180-foot minesweeper. In 1953, the SCB established a standard for fitting existing ocean and coastal sweepers for minor and local ASW, their primary sweep mission permitting. The existing minehunting sonar, UQS-1, had too short a range for ASW. However, the war-built 220-foot ocean sweepers had been built with searchlight sonars. The SCB suggested that the ships be fitted to fire Mouse-traps with forty-eight rockets, twenty-four of them in ready-service stowage, and one depth-charge track with a local release fitting. A tactical range recorder would suffice for fire control. For weight compensation, the sweeping floats, otters, and acoustic sweep gear could be removed.

The 180-foot steel sweepers (AM 136 class) were

Table 8-4. The 1953 PCE Design, December 1952

Dimensions (ft-in)	223-0 (loa) 217-0 (lwl) x 33-6 (dwl) x 9-3 (trial) 9-9 (full load)
Displacement (tons)	756 (standard) 880 (trial) 921 (full load)
Armament	2 x 3-inch/50 single-shielded RF (600 rounds total), 1 depth-charge track, 4 depth-charge projectors (27 depth charges total), 2 Mark 11 Hedgehogs (total 288 projectiles), torpedo stowage and launch (3 torpedoes Mark 32)
Machinery	2 shaft diesels; 3,000 shp at 18.7 knots trial; endurance 4,000 nautical miles at 15 knots; generators: 3 100-kw units, 2½-kw emergency diesel generator

	Light	Trial	Full Load
Displacement (tons)	705	880	921
GM (ft)	2.63	2.92	3.10
Max Righting Arm (ft)	0.97	1.41	1.54
Angle (degrees)	33.5	38.5	39.5
Range (degrees)	70.0	81.5	86.0
Weights (tons)			
Hull	308.53	—	—
Hull Fittings	90.97	—	—
Equipment and Outfit	45.45	—	—
Machinery	186.49	—	180.49
Armament	33.80	—	—
Margin	40.00	—	—
Light Ship	705.24	—	—
Ammunition	33.12	—	33.12
Complement	10.43	—	10.43
Stores	24.38	—	12.19
Potable Water	19.52	—	1.82
Diesel Oil (⅔)	74.58	111.67	—
Reserve Feed Water (⅔)	7.73	11.60	—
Fog Oil	5.00	5.00	—

Trial Condition[1]	CP 0.54 (original PCE: 0.60)
	CM 0.83 (0.93)
	CW 0.67 (0.75)

1. Comparison is with the wartime PCE (figures in parentheses): endurance at 15 knots, 4,000 (5,000); characteristics call for 4,000/15 in both cases. Trial speed is 18.0 knots (14.1), sustained, 17.3 (13.6), characteristic speed, 17 (16).

already close to PCE standards, with a single Hedgehog, Mark 10 or 11, directed by a searchlight sonar and an attack plotter. Depth-charge tracks could be added aft if floats and otters were removed. The smaller sweepers, a wartime YMS type, might be fitted with short, paired depth-charge tracks if floats, otters, and Dan buoys were removed as compensation. The tracks would be Mark 9-4, with four charges each.

Apart from the 180-footer, which already had an efficient ASW weapon in the Hedgehog, none of these proposals was particularly impressive. Mousetrap suffered because it was not stabilized. Depth charges were unlikely to catch a fast submarine. In December 1953, the assistant CNO (readiness) suggested that the new lightweight homing torpedo provided a solution. The new Mark 43-1 250-pound homing torpedo was credited with a single-shot kill probability of 59 percent in deep water, and with considerable capability even in shallow water. It could be substituted for depth charges at a substantial savings in weight, yet a minesweeper could still be converted on a quick and temporary basis.

Now the SCB project changed focus. With improvements in magnetic mines, the war-built steel sweepers were no longer viable. They could, however, be converted into convoy escorts equivalent to the abortive SCB 90 PCE. They could also be converted relatively inexpensively for U.S. allies. Second preliminary characteristics for this SCB 143 project of February 1955 showed all sweep gear removed—only the A coils of the existing degaussing equipment were retained. The new ASW battery would have consisted of a double Hedgehog with 96 rounds, two attacks in ready-service stowage, and another 228 rounds in the magazine; a new SQS-4 sonar; a Mark 2 homing torpedo launcher with six torpedoes; a single depth-charge rack with twelve charges; and a Fanfare noisemaker for protection against homing torpedoes. A simple attack plotter and tactical range recorder would suffice for fire control. The powerful electric plant of three 100-kw generators originally installed to power a magnetic-sweep coil sufficed for the new electronics and weapons. The PCE would also be armed with one slow-fire 3-inch/50 gun with 48 rounds ready service plus 300 in the magazine, two twin 40mm guns with 1,350 rounds per barrel, and two twin 20mm guns with 2,700 rounds per barrel. The converted PCE would carry a complement similar to the SCB 90's, nine officers and eighty-five enlisted men, with thirty days' worth of provisions.

In fact, the U.S. Navy never converted any 220-foot sweepers for its own use. Much later, under the military aid program, it did convert several with a single Hedgehog, one Mark 32 triple homing torpedo tube, and an SQS-17 sonar. Three of the ships went to South Korea, two to the Philippines, and three to Taiwan. Two, transferred to Peru in 1960, were less extensively refitted. Eighteen units transferred to Mexico with all ASW equipment removed serve as gunboats.

Postwar U.S.-built and -financed derivatives of the wartime ASW ships are described in chapter 3.

Interest in new navy-built harbor defense craft seems to have revived in fall 1956, as the FY 59 budget proposal was being prepared. The navy was then engaged in a strategic reappraisal by way of the LRO

Table 8-5. The 95-Foot Coast Guard Cutter

	95-ft CG	SC
LOA (ft-in)	95-0	110-10
LWL (ft-in)	90-0	107-6
Beam (extreme, ft-in)	19-10	17-11½
BWL (ft-in)	18-5	17-0
Draft (with skeg, ft-in)	4-1¼	4-5
Draft (max to DWL, ft-in)	5-7	5-8
Hull Depth (fwd, ft-in)	12-6¼	14-2
Hull Depth (amids, ft-in)	—	10-2¼
Hull Depth (aft, ft-in)	10-3¼	9-9¼
Freeboard (ft-in, fwd)	8-5	9-9
(amids)	—	5-9¼
(aft)	6-2	5-4¼
Displacement (tons)		
(to DWL)	78.1	108.0
(light)	80.7	—
(trial)	—	—
(full)	101.2	—
CB	0.403	0.469
CM	0.615	0.714
CP	0.665	0.657
CW	—	—
SHP	2,200	2,400
V (trial, kts)	20	22
Range (nm/kts)	1,950/10	1,500/12
Weights (tons)		
Hull	35.35	
Hull Fittings	10.35	
Machinery Wet	31.65	
Electric Plant	1.18	
Equipment and Outfit	2.19	
Light Ship	80.72	
Ammunition	2.61	
Stores	1.11	
Potable Water	5.00	
Fuel (95%)	10.15	
Complement	1.58	
(at 190 lbs each)		
Full Load	101.17	
GM (full)	5.5	
Armament	1 twin 20	1 40 single
	2 Mk 20	3 20
	4 2-charge DC rack	2 Mk 20
		10 DC single-release tracks
	—	2 single DC projectors
Complement (off/enl)	3/15	2/20

group; renewed interest in the security of its bases would have been a logical consequence of reappraisal. The only existing harbor defense craft was the Coast Guard's new 95-footer (see table 8-5), the prototype of which had appeared in 1953. It was armed with a Mousetrap and equipped with a small searchlight sonar; perhaps its most important feature was a seaworthy hull form.

The Coast Guard boat almost certainly was *not* conceived as a wartime ASW craft. Rather, it was the natural successor to the wartime wooden 83-footer. Built of unseasoned wood, the 83-footers badly needed replacement by the early 1950s. Once the requirement for replacement had been accepted, Coast Guard designers had to take into account their wartime mission of port security, particularly against swimmers and sneak-attack craft.

In fall 1956, the navy thought in terms of lineal successors to its two smallest ASW craft, the 110-foot SC and the extemporized 136-foot PCS. As with the PC/PCE of 1951, it was by no means clear that they represented distinct niches, and indeed both were originally lumped together under the designation SCB 183. The SC was later separated as SCB 184.

The most attractive prototype hulls were the 95-foot Coast Guard boat and the new *Cove*-class inshore minesweeper (MSI), in effect the modern successor to the wartime YMS from which the PCS had been derived in the first place. The Coast Guard boat was liked for its good sea-keeping but was criticized for cramped internal spaces.

The gun would probably be the single 40mm Bofors, supplemented, perhaps, by the twin aircraft-type 0.50-caliber machine gun. The 20mm Oerlikon was not considered.

The harbor defense boat would have to deal with three distinct underwater threats: the full-size submarine, which could lie on the bottom, the midget submarine, and the swimmer, who might straddle a manned torpedo. Suitable weapons existed for use against all three threats.

The new lightweight homing torpedoes were the obvious counter to full-size submarines. The existing Mark 43-1 was considered effective from 50 to 650 feet. It could be launched from the Mark 27X triple tube, which was 1,150 pounds, 110 inches long, and still under development. The next torpedo, Mark 44 (then designated EX-2), was already being tested. It was faster and could dive deeper; both Marks could be fired from the heavier, 1,800-pound Mark 32 triple tube. Alternatively, single Mark 32 tubes could be fixed at 45 degrees to the centerline. This was actually done in the hydrofoil PC *High Point*, Thai PCEs and PFs, and Philippine PCEs.

For the other two roles, the boat would have to use depth charges, two types of which were under development: a 50-pound ASW and a 5-pound anti-swimmer weapon. The 50-pounder was a modified Hedgehog projectile, 28 x 7.2 inches, with a 25-pound HBX explosive, sinking at 10 feet/second; it could be set from 25 to 155 feet in increments of 10 feet. Possible launching systems were roll off; the K-gun side launch with an explosive propellant; and the K-gun with a spring-loaded launcher. The cylindrical anti-swimmer depth charge could be set for 10 to 50 feet, normally 25, and would be dropped over the side by hand. It was 15 x 3 inches and carried 2 pounds of TNT.

The hydrofoil *High Point*, which employed sprint and drift tactics. It seemed to promise a viable small craft solution to coastal ASW problems, being able to use a large, low-frequency sonar effectively. Protrusions from the sonar dome are visible below the hull, directly under the mast, the bottom of which housed the retracted sonar strut. (Boeing)

At the end of September 1956, the chairman of the standing committee on the shipbuilding and conversion program asked for estimated costs for two craft with the following features:

(a) 90–100 feet; 20 knots; 1,000 nautical miles at 12 knots; two 0.50-caliber machine guns fore and aft, twin or single; lightweight shallow-water depth charges (SC).

(b) 130–40 feet; 20 knots; 2,500 nautical miles at 12 knots; sonar; lightweight shallow-water depth charges; ASW torpedoes; one single, manually operated 40mm gun; two twin 0.50-caliber machine guns (PCS).

These tentative ideas were developed further at an OpNav meeting late the following month, October, to develop characteristics for the then-planned FY 58 SC and PCS. Both were to operate close inshore, patrolling harbor and base approaches and coastal shipping lanes. The PCS would be large enough to function as a coastal convoy escort. Both were to maintain 20 knots in a force 2 sea. The basic difference between the two would be in range of operation from shore. The SC would generally operate within 15 nautical miles of its base, the PCS within 300. Tentative requirements for the SC emphasized quick-starting machinery so that the boat, at anchor in a harbor, could react to a submarine detected by a harbor defense sonar or other sensor. The PCS, patrolling offshore, would have no such requirement. Range was also expressed in provisions: five days' worth for the SC, twelve for the PCS.

Similarly, the SC would have only weapons suited to shallow water: lightweight and antiswimmer depth charges, the former thrown athwartships by spring-loaded launchers. The PCS would carry three EX-2 homing torpedoes (Mark 44) with two fixed launchers.

Both would have to be able to detect a coastal submarine of 250 tons at 1,000 yards at SC speeds, which suggested the need for a towed rather than a hull-mounted sonar. A list of required electronics showed a sonar similar to the SQS-17 (13 kc), which would later be exported to many Allied navies for installation on hulls from subchaser size up.

Both of the two main alternative sonars, the towed SQS-9 and the hull-borne SQS-17, presented problems. A table prepared at the time showed how sonar domes grew as sonar frequency was reduced; the smallest dome in the list, for 20 to 28 kc, corresponded to the largest World War II scanning sonar, the QHB—which had never been installed in anything smaller than a destroyer escort:

Table 8-6. Sonar Dome Sizes

Frequency (kc)	Length (in)	Width (in)	Depth (in)	
20–28	60	24	46	
12–14	100	40	54	SQS-4 (mod 3-4)
8	185	74	65	SQS-4 (mod 1)
5	270	108	66	SQS-23
13	75	30	54	SQS-17

Table 8-7. Increasing Ehp for Sonar Domes at Various Speeds

Frequency (kc)	10 Kts	13 Kts	17 Kts	20 Kts
20–28	12	26	57	93
12–14	24	51	115	187
8	52	114	255	415
5	76	166	372	605
95	—	307	—	905
(CG boat)		(at 12.65 kts)		(at 18.98 kts)

The figures for the Coast Guard boat showed *total* ehp required to reach the speeds listed. Actual required shp or bhp would be about twice the ehp figure, depending on propeller efficiency, and the dome ehp would have to be roughly added to the total for the rest of the hull. Thus, even to add a QHB to a Coast Guard boat would require about 10 percent more power at 19 knots. The smallest of the big sonars, the SQS-17, would probably add about 15 percent. It would also add about 3,000 pounds—2,000 for electronics, 1,000 for the transducer and dome. On the other hand, it could be expected to match the performance of the current and much more massive destroyer set, the SQS-4 (mod 4), with a nominal range of about 5,000 yards.

Like the SQS-17, the developmental SQS-9, two of which were expected to be available in January 1958, imposed a total weight of 3,000 pounds. It used a searchlight transducer in an 80- x 19-inch fish, towed alongside from a special boom on a 100-foot cable.[8] Tests showed that it could ping effectively at 20 knots. Typical operating depth was therefore 100 feet at 0 knots, 60 feet at 20 knots. Resistance at 20 knots was 56 ehp from the shape, 81 ehp from cable. Average range on a full-size submarine was 3,750 yards.

At this time, it was expected that the production version would use a 200-foot cable so that it could be streamed at a depth of 140 feet at 20 knots.

By 31 December 1956, the tentative FY 59 shipbuilding budget included both new types: a $1 million ($1.5 million lead ship) SC and a $1.5 million ($2 million lead ship) PCS. The following March, an SCB meeting heard that navy mobilization plans included 108 SCs and 102 PCSs, numbers that could be met only if the cost and complexity of both classes were held down.

It was by no means clear whether anything so expensive as the projected craft was justified. In mid-1957, the SC/PCS project officer wrote that he found it difficult to justify the PCS, and that the 95-foot Coast Guard boat could do both the SC and the PCS jobs. If something more powerful was really needed, that suggested a much larger hull, probably about 180 feet long and fast enough to keep up with modern submarines which ran at 25 to 26 knots.

Tentative FY 59 SC characteristics, issued in March 1957, showed an SQS-9 towed sonar, an attack plotter, and a Mark 32 torpedo launcher (three Mark 44 torpedoes), twelve lightweight depth charges in a stern track, and fifty small antiswimmer charges. Gun armament would be two single 0.50-caliber machine guns. The boat would be as simple as possible—commercial standards were acceptable—but it would also have to have minimum magnetic signature to evade destruction by magnetic mines an enemy submarine might lay in its shallow operating area. OpNav hoped to save money by coordinating the design with the Coast Guard for possible peacetime use. Preliminary Design looked at an enlarged 95-foot hull with the following dimensions, given in feet unless otherwise indicated: 105 LOA, 100 LWL x 20 x 12 hull depth x 8, 90 tons light, 115 tons fully loaded. Speed, 21.5 knots at 2,750 bhp, and endurance, 1,000 nautical miles at 12 knots, were as initially required. The characteristics showed twin screws; consideration was to be given to geared, gas-turbine drive. The crew was to be as small as possible, the SC carrying two officers and twenty-four enlisted men with provisions for five days.

As before, the PCS differed from the SC primarily in size and armament. Size in feet was 136 (loa), 132 (lwl) x 23 x 15 (hull depth) x 9 at 150/190 tons. Armament consisted of one twin 40mm with 904 rounds per barrel; two 0.50-caliber machine guns; two Mark 32 launchers with six Mark 44 torpedoes; one lightweight shallow-water depth-charge release rack carrying twenty-four charges; and fifty antiswimmer charges. It would be driven at 20 knots (3,000 bhp) and would accommodate three officers and thirty enlisted men.[9] Unlike the low-signature SC, the PCS would be a steel ship suitable for mass production.

Like the SC, the PCS might also be coordinated with the Coast Guard to produce a single mass-production design suitable for both services. The Coast Guard had already rejected the PCS itself, but it had also just completed a design study for a 165-foot WPC to replace the existing twelve 125- and eight 160-foot boats used for search and rescue. No such boat was built, and the requirement was later filled by the 210-foot *Active* class—which had a wartime PCS-like role, although no ASW equipment was installed. Even at this time, however, the Coast Guard expected to spend an average of $4 million for its new cutter, compared with an estimated $2.1 million ($2.8 million for the lead ship) for the 136-foot PCS.

In mid-February 1958, the SC and PCS were deleted from the FY 59 program and being considered for FY 60. In January 1958, BuShips reported that an FY 60 PCS would cost $2 million ($2.6 million for the lead ship). SCs, based on the 95-foot Coast Guard hull, would cost about $1.5 million each ($2 million for the lead ship).

OpNav (Op-31) considered both figures far too high. After all, neither design was fast enough to keep up with present or future submarines, so neither could reliably kill its targets. The only obvious way to reduce cost was to reduce the required speed. That would only have made matters worse, and an SCB proposal of 7 February 1958 that SC and PCS speeds be reduced from 20 to 15 knots was rejected.

Moreover, it was beginning to seem unlikely that any small conventional ASW boat could successfully operate a big, low-frequency sonar effective above 10 knots. Tests aboard the experimental PC *Weatherford* (EPC 618) showed that a small patrol ship at 10 knots generated background noise as intense as that of a 15-knot destroyer. Frequency and available sonar power would both be limited by the hull size of the patrol boat. Even a towed sonar would suffer from such limitations, as it, too, would have to use a relatively small transducer. BuShips therefore rejected conventional ASW craft as a waste of time and effort. They could not deal with the existing diesel submarine, much less the Soviet nuclear submarine, which was then expected in 1965.[10]

On 24 January 1958, Op-31 called for a new study of coastal and inshore ASW craft to be built under the FY 60 program and to include a new concept, a hydrofoil patrol boat.[11] BuShips reported early in April that, while there was no chance of either cutting the cost or improving the sonar performance of the conventional design, a hydrofoil, then designated PC(H), seemed attractive. As a result, the conventional PCS and SC were abandoned, but the prototype PC(H) was built under the FY 60 program as SCB 202.

A PC(H) feasibility study showed that the hydrofoil would be more expensive—in FY 60 dollars, $2.8 million, $3.7 million for the lead ship. But it would also be much more effective, with the low-frequency SQS-20 sonar. With 6,500 shp, it would fly at 45 knots on its foils and run at 12 knots while hull-borne; it would have an endurance of 1,500 nautical miles hull-borne plus 250 on foils. Armament would include four Mark 44 torpedoes and either a 105mm recoilless rifle or an aircraft-type Vulcan 20mm cannon.

The PC(H) made sense because it could exploit a new tactic already practiced by ASW helicopters with dunking sonars: sprint and drift. As in the case of the helicopters, self-noise was no problem at very low speed. A small boat, then, might be able to use a big, low-frequency sonar while making low, even bare steerageway, speed. It would have to ping and listen, then dash to a new pinging position or attack a distant contact. Low listening speed would create little noise; at this time, BuShips considered self-generated noise the principal limiting factor in sonar performance.

The phrase sprint and drift not yet born, BuShips called this a grasshopper operation. If the patrol boat were to achieve any significant net search speed, it would need a high dash speed, which in 1958 usually required a hydrofoil. The hydrofoil could be designed to retract its big sonar to avoid damage while taking off from the water to fly on its foils.

A lightweight low-frequency transducer might be lowered in as little as 10 seconds and retracted in as little as 20. The low hull-borne ship speed would even simplify signal processing by reducing the band width created by the Doppler effect. In sea states 2 to 4, an equipment weight of 8,000 pounds was expected to give a performance equivalent to that of the 50,000-pound SQS-23 then being installed in missile destroyers.

At this time, BuShips estimated that in sea state 2 the PC(H) could detect a submarine at periscope depth at 16,000 yards, compared with 5,000 for the conventional PC. In sea state 4, the hydrofoil would lose very little of its performance, achieving 13,000 yards, but the PC would be cut to 3,000. Ranges on submarines below the layer were much shorter, but even then the PC(H) could be expected to do about twice as well. Moreover, unlike the sonar coverage of conventional craft, the hydrofoil's was not limited by a baffle over the propellers (35 degrees for a destroyer, 80 for a PC at 20 knots). These figures did not take into account the effect of rolling on the conventional PC, which directed numerous pings into either the surface or the bottom.

The preliminary design of the hydrofoil ASW boat originally showed a retractable SQS-20 of 3.5 kc and a modified dipping helicopter sonar, AQS-10, to reach greater depths.[12] However, the AQS-10 performed poorly on trials as a helicopter sonar, and the design was completed only with space and weight for a future variable-depth sonar. It also included provision for the future installation of a magnetic anomaly detector (MAD) intended, as in ASW aircraft, to confirm that a sonar contact really was a submarine. The PC(H) was actually completed with an SQS-33 sonar, a modified SQS-20 that could incorporate a towed transducer sharing the same transmitter (but without any listening function).

The sonar was located amidships to minimize pitch and yaw motion when hull-borne. Its strut retracted upward, into the boat's mast, when the sonar was housed; this combination of mast and housing reduced air resistance when the boat flew.

As a platform, the hydrofoil seemed to have all the helicopter's advantages—high search speed, maneuverability, reduced vulnerability to torpedo attack—without the disadvantages of high noise level, limited endurance, and weather and weight restrictions.

The promise of hydrofoils was well known at this time. They were the only known means of moving a small boat at high speed through any substantial sea state. But as yet they had not realized their potential, at least in rough water. The U.S. Navy had been working toward a rough-water hydrofoil for over a decade, and the sprint-and-drift ASW boat project was attractive partly because it provided the occasion for building a rough-water hydrofoil prototype applicable to other missions as well.

The PC(H) was also the first U.S. warship to be designed primarily for gas turbine power. It flew on two British-built Proteus engines driving contra-props—one propeller at each end of each of two nacelles—through right-angle drives in the foil struts. An auxiliary Packard diesel was used for hull-borne operation.

The most radical and important feature of the PC(H) was its system of submerged (rough-water) hydrofoils coupled with an aircraft-type active-control autopilot. In configuration, the PC(H) duplicated the *Sea Legs*: flaps on the forward foils reacted to the height input of the autopilot, the after flaps to roll and pitch inputs. As in the *Sea Legs*, the system was fed by an acoustic height sensor. The foil and control systems of the gunboat *Tucumcari* and of the current *Pegasus*-class PHMs were all directly descended from those of the PC(H) prototype *High Point*. These systems, in turn, are responsible for the boats' effective rough-water performance.

Preliminary Design chose a canard configuration for the PC(H) because it permitted an arrangement with the machinery, and hence the hull's center of gravity, well aft. Power was transmitted over a very short path, and the turbine exhaust could be used to add thrust. Moving the machinery aft, moreover, left the most desirable part of the hull, the mid-length, for sonar, crew, attack, and control functions.

The optimum canard would have supported 25 percent of the weight of the boat forward, 75 percent on the main foils aft. The designers of the PC(H) had to settle for 30 percent forward, since the foil positions had to be set within limits—the bow foil as far forward as possible but still within hull waterline, the after foil at the forward machinery bulkhead, which served as its foundation. The proportion supported by the after foil could have been increased by moving it forward, but that would have interfered with crew's quarters.

The "square-cube" law and cavitation limited PC(H) size and speed. The weight each square foot of foil could lift depended on the speed at which the boat was driven. Preliminary Design considered 200 to 300 tons a practical limit; much later it would develop hydrofoils displacing as much as 1,500 tons. Lift per unit foil area was limited by maximum speed. Above 40 to 50 knots, the foil would begin to cavitate; flow over its upper surface would start to break down, causing a marked change in its lifting ability. A conventional foil shape, no longer effective, would have to give way to an ax-edge shape. Under supercavitating conditions, the flow over the top of the foil would break down completely, leaving a near vacuum. Water pressure below the foil would continue to be high. Experience limited Preliminary Design to a subcavitating foil, and thus to a maximum speed, in the PC(H), of about 45 knots. The forward foil was swept to delay the onset of cavitation and tapered for better lift distribution. The after foil was a simple rectangle because the "box plane" effect of the struts made it difficult to reduce drag.

Cavitation limited foil-lift coefficients and thus decided foil areas. Greater aspect ratio, that is, longer span and narrower chord, would have improved foil-lifting efficiency, but span was limited (to 20 feet forward, for an aspect ratio of 6.11, and to 31.5 feet aft, for 6.6) by hull width and structural strength.

Control was exerted by flaps. The alternative, varying the angle of the foil itself, was rejected because it would have reduced strength and increased drag at the intersection of the main foil and its strut.

The PC(H) was designed to fly in sea state 5 (10-foot waves). That translated into a combination of hull clearance above water and foil submergence below, which in turn determined strut length and affected boat weight and resistance, even takeoff conditions. In this case, the choice was 5-foot hull clearance and struts submerged to a depth equal to 1.2 times the chord of their foils.

For good all-weather performance, Preliminary Design chose a hull form similar to that of the 95-foot Coast Guard boat. Its flare was carried well aft to keep the main foils within the maximum hull beam. The narrow transom was chosen for good rough-water

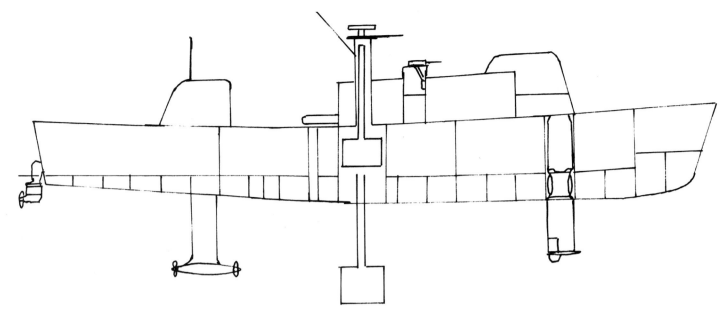

Inboard profile of the *High Point*, showing her sonar transducer in the extended and retracted positions. The struts of the after hydrofoils retracted into the exhaust-like stubs aft. (Norman Friedman)

flying performance. Waves striking the transom from aft would tend to pitch the boat into following seas; the wider the transom, the greater this force. The bow was given considerable deadrise to reduce impact loads in waves. Aside from these factors, the big foils would damp out roll and pitch when the boat was hull-borne.

Like the PTs, the PC(H) was a lightweight fast boat. It had to be built of aluminum (a steel hull would have weighed as much as 2.7 times as much). Preliminary Design hoped that its design could be based on experience with the aluminum PTs, despite the structural problems those craft had encountered.

For the non-ASW portion of the PC(H) battery, Preliminary Design considered both the 20mm Vulcan and the 106mm M46 recoilless rifle. It finally chose the PT boat's Mark 17 0.50-caliber gun, reducing the weapon's air drag by recessing it.

The PC(H) preliminary design is described in table 8-8.

As of 5 January 1959, before the PC(H) design had been completed, the approved FY 61–66 program included twenty-two PC(H)s at a unit cost of $2.8 million.

FY 60	61	62	63	64	65	66
1	1	—	2	6	6	6

The goal was to have thirty boats in service by 1972.

Only the FY 60 boat, the *High Point*, was built. Boeing was responsible for her detailed design, and she was built by Boeing at the J. M. Martinac yard in Seattle. Thus with her began Boeing's involvement in large submerged-foil hydrofoils, which continued through the PGH and PHM programs. Completed in September 1963, she achieved a flying speed of 51 knots at a displacement of 126 tons; range was 620 nautical miles on foils, 1,200 hull-borne.

Although the *High Point* had been conceived as a state-of-the-art hydrofoil, she required considerable work after completion; her problems led to the formation of a hydrofoil special trials unit at Bremerton in November 1966. Even after her defects had been corrected, it was found that she could not fly continuously for more than about six hours: as fuel burned, she was brought outside her flight envelope, that is, she became too light.

In the year after delivery, the *High Point* spent only 53 hours and 41 minutes foil-borne, of which only two hours were in rough water (sea state 4). Problems included erratic steering, cavitation damage to struts and foils, and saltwater contamination of lubricating oil in the transmission. In September 1964, therefore, the navy decided to refurbish her completely, and she was out of the water until June 1966. Her autopilot system was modified, a larger spade rudder was installed, and several test coatings were applied to various areas of her struts and foils. She was also much more extensively instrumented. Operational tests resumed on 22 October 1966; on 19 October, she successfully flew over 100 miles from Bremerton to Neah Bay in 3 hours and 19 minutes through swells averaging 6 feet. The forward foil periodically broached because of its short strut, and the boat slammed, but even so she was able to maintain speed.

The *High Point* used paired propellers in pods at

Table 8-8. The PC(H) Preliminary Design

Dimensions (ft-in)	110 (lwl) 115-9 (loa) × 31-1/34-0 (over foil guards) × 4-4 retracted/6-7 flying
Displacement	108 tons
Machinery	45-kw diesel generator, 40-kw gas-turbine generator
Performance	45 knots at 80 percent power; takes off at 25 knots at 2/3 available thrust, 1/2 hour rating; 12 knots hull-borne; range: 2,000 nautical miles hull-borne (12 knots); 500 miles foil-borne
Armament	1 twin 0.50-caliber machine gun (Mark 17 mount, 1,000 rounds), 2 modified twin Mark 32 TT (4 Mark 44 torpedoes), 1 shallow-draft rack (6 depth charges); in 1966: 1 twin 0.50-caliber machine gun, 1 81mm mortar Mark 2 (51 illuminating rounds), 1 40mm gun (160 rounds plus 16 ready service)

Weights (tons)	Designed	Built
Hull	30.00	29.72
Propulsion	12.44	15.13
Electric Plant	3.62	4.58
Command/Control	5.53	8.24
Auxiliaries	8.21	20.56
Outfit/Furnishings	9.44	5.90
Armament	1.77	1.61
Foils (retracted)	8.46	—
Margin	4.00	—
Light Ship (retracted)[1]	83.61	85.74
Loads	—	26.55
Full Load	—	112.29

1. Inclined (15 July 1963), light ship displacement was 93.94 tons.

the intersection of her after struts and foils. It was discovered that, although the forward propellers did not suffer cavitation damage, their wake ruined flow over the pod. The after propellers cavitated badly and had to be replaced after about 40 hours of flight. For subsequent hydrofoils, Boeing shifted to waterjets.

In May 1966, NavSea awarded Boeing a contract to study advanced strut/foil/propulsion combinations that might overcome deficiencies in the *High Point*; alternative propulsion systems included waterjets. The study, completed in September, formed the basis for a planned Mod 1 configuration in which the boat would be steered by rotating the forward strut rather than by flaps and spade rudder. Existing propellers would be replaced by two pushers on each pod into which air could be injected to overcome cavitation. New struts and foils would be constructed out of a stronger material, perhaps the HY-130 NavSea was then developing. As of mid-1967, the new configuration was scheduled for installation in 1968, and the *High Point* was to return to service early in 1969. However, the Mod 1 conversion was delayed until a major refit during which the boat received Harpoon missiles.

The *High Point* served as a test platform for a variety of hydrofoil applications. She towed sonar bodies at a depth of 400 feet and at a speed of up to 42 knots. Mark 44 torpedoes were fired successfully at 45 knots in March 1968. Between September 1971 and July 1973, she was converted at Puget Sound to test and evaluate the Harpoon cannister system planned for the *Pegasus*-class PHM; a quadruple cannister was mounted on her centerline at her stern. At the same time, her struts and foils were modified and the forward foil was made steerable. She was temporarily transferred to the Coast Guard in April 1975 for operational tests.

The *High Point* was scheduled for decommissioning in 1978, but congressional action kept her in service. On 30 September 1979, she was stricken but retained as "equipment," her four torpedo tubes removed. Later she was used to test a U.S.-Canadian high-speed towed array (HYTOSS), and in 1984–85, she was on loan to Boeing.

The *High Point* could be seen as the culmination of a decade of BuShips/ONR research in subcavitating, fully submerged hydrofoils, all of it conducted at a relatively low level of effort. By 1960, with a massive Soviet nuclear submarine fleet in the making, interest in fast oceangoing craft to maintain what was seen as the necessary edge over submarine speeds was growing. Several new ideas, some of them not completely developed, might permit the construction of hydrofoils big enough to fly at high speed in the open sea: the new, fully automatic sea-keeping control system, the powerful marine gas turbine,[13] and fully cavitating and ventilated foils. In 1958, Grumman's ONR-funded XCH-6, the *Sea Wings*, tested the first supercavitating foils and propellers, which could produce a speed exceeding 60 knots.[14]

In February 1960, the ASW panel of the president's Science Advisory Committee pressed for major new investment in fast open-ocean hydrofoils, which seemed the most likely candidates for the anti–nuclear

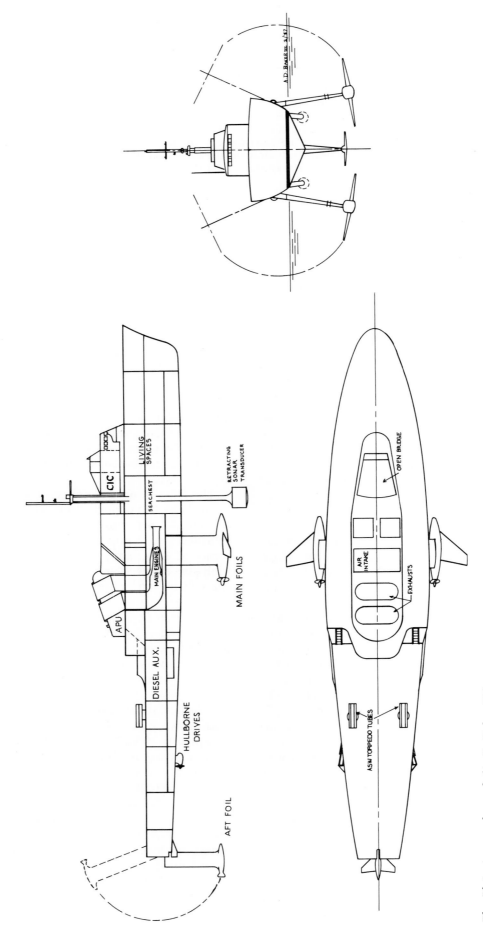

The *Plainview* as planned. (A. D. Baker III)

Table 8-9. The *Plainview*

Dimensions (ft-in)	212 (loa) 205-1¾ (lwl) x 40-5 (hull) 82-8 (foils retracted) 70 (foils extended) x 6-3 (foils retracted, hull-borne); the forward foils carried 90 percent of the total weight of the ship
Displacement (tons)	265 light ship; 290 normal takeoff; 328 fully loaded
Machinery	2 GE LM1500 gas turbines, total 29,000 shp (foil-borne); two GM V12-71 diesels, total 1,000 bhp (hull-borne); two GM V8-71 diesel generators (100 kw each)
Performance	Maximum speed 50 knots; cruising speed foil-borne, 42; takeoff speed, 33; cruising speed hull-borne, 12; maximum sea state foil-borne, 5; estimated endurance (1965): 500 nautical miles at 50 knots on foils, 7,000 miles at 12 knots hull-borne; in service, the *Plainview* was able to fly at 25 knots, about 10 below her minimum design speed

submarine role. Given the obsolescence of World War II–built destroyers and escorts, the navy would have to begin mass replacement about 1963. A conventional, though gas turbine-powered, destroyer, the Seahawk, was tentatively chosen; it was abandoned because of the rising cost of the war in Vietnam.

Since 1950, the navy had invested only about $200,000 to $300,000 per year in basic hydrofoil research. Moreover, NASA, the successor to NACA, had recently reduced the national hydrofoil research base by closing its hydrodynamics division. The president's panel recommended that the navy immediately begin studies of a 250- to 350-ton, 60- to 80-knot hydrofoil to serve as the forerunner of a class of relatively inexpensive, very-high-speed warships. The minimum speed of 60 knots was chosen to restore what the panel considered the conventional two-to-one speed advantage over future submarines. It hoped that the test craft could be completed in 1963, so that operational experience would be available in time to affect the ASW modernization program. This craft might be able to use conventional, subcavitating foils. Some papers written at this time described the big hydrofoil as a DD(H), a hydrofoil destroyer.

The panel also pressed for the immediate design of an 80-ton, 80-knot hydrofoil using supercavitating or superventilating foils.

The panel saw both the *High Point* and the Maritime Administration's *Denison* as useful test vehicles and recommended that the latter be fitted with an SQS-20 sonar for testing two-hydrofoil tactics with the PC(H).

The ASW panel estimated that its program would cost $40 million over three years, including $18 million for the big hydrofoil and $13 million for the fast, supercavitating craft. On 6 April 1960, BuShips asked for emergency research and development funds to explore the upper limits of hydrofoil size and assess the use of such craft, particularly in ASW. The bureau was given $11,472,000. It scaled down the ASW panel recommendations to one large hydrofoil, which became the *Plainview*, and a 16-ton test vehicle, which became Fresh-1. It also set a long-range goal: a hydrofoil with a 65-ton payload (total displacement greater than 500 tons), capable of taking off at 30 knots and flying at up to 90 (120,000 shp), large enough to fly in sea state 6, and with an endurance of 500 nautical miles on foils, 7,000 hull-borne. The sub-cavitating *Plainview* came close to meeting the range requirements, which were 500 miles on foils and 6,500 hull-borne, but its payload was 40 tons, its maximum speed 50 knots, and it was limited to sea state 5 and below. As of September 1962, BuShips estimated that the next major design-capability plateaus would be 300 to 700 tons at 60 knots in FY 65 and 500 to 900 tons at 90 knots in FY 67. In June 1963, the ultimate goal was set for FY 69 and redefined: about 1,000 tons' displacement, a speed of up to 100 knots, and the demonstration of an optimum ASW system.[15]

At this stage, then, BuShips emphasized high speeds requiring supercavitating foils. Reviewers of its proposals asked whether lower foil-borne speeds of 30 to 45 knots might not also be valuable.

The big hydrofoil, which became the *Plainview*, was conceived from the first for ultimate conversion to supercavitating (very high speed) foils. Thus, her hull had to be able to withstand wave impacts at speeds as great as 70–80 knots, and she had to carry enough engines to drive her at those speeds while remaining light enough to plane with the lower-powered (subcavitating) power plant. After a design competition, Grumman was awarded the AG(EH) design and construction contract, but the navy could not reach agreement on a satisfactory construction price, and Lockheed won the detail design and construction contract. When foil-borne, the *Plainview* was powered by two GE LM1500s, and she had sufficient internal space to take another pair. She was armed with two lightweight, triple torpedo tubes (Mark 32) and designed to carry a big sonar like that of the *High Point*. In 1972, she was fitted with a single canister to launch the NATO Sea Sparrow vertically. Details are given in table 8-9.

The *Plainview* was the largest naval hydrofoil ever built. Authorized under the FY 62 program, she was originally scheduled for delivery in July 1965. She was launched 28 June 1965, but construction proved more difficult than anticipated—the hydraulic system caused particular trouble—and she did not fly until 21 March 1968. She was delivered on 1 March

The high speed of hydrofoils seemed promising for open-ocean work. The large experimental *Plainview* was conceived as a step toward a deep-water escort, sometimes designated DE(H). Like the *Glover*, she had an experimental auxiliary designation. She was to have been fitted with more engines and supercavitating foils for speeds of up to 80 knots, but the supercavitating program died before her completion.

In flight in 1964, the *High Point* displays the propeller she uses when hull-borne. It is neatly folded up against her transom, under the two gas-turbine exhausts. The winch above was installed for experiments with a high-speed towed sonar, which would have complemented her big hull unit. She appears to be fitted with torpedo tubes on her starboard side only. The significance of her two big cradles is unknown.

Experience with the *High Point* demonstrated that rough-weather hydrofoils could be built. The Boeing *Tucumcari*, shown in March 1971, used much the same control system. She displays her waterjet, with its protective cover raised, at the lower edge of her transom, with the gas turbine exhaust uncovered above it and a guard protecting both. One of the two rudders she used in the hullborne mode is visible below her hull, aft.

1969. The *Plainview* was stricken on 30 September 1978.

In mid-1965, with the *Plainview* under construction and badly delayed, the hydrofoil program was redirected to evaluate the capabilities of existing subcavitating craft. The high-speed goal was abandoned. This was partly the result of difficulties experienced in the *High Point*, but it also reflected a general decline in ASW research funding due to the costs of the Vietnam War.

Even so, the success of the *Plainview* encouraged NavShips to consider larger hydrofoils. An advanced ship concepts team was formed in 1970, and later the CNO, Admiral Elmo Zumwalt, formed an advanced ship types and combatant craft office within OpNav. The high-speed hydrofoil concept of the early 1960s was revived, and for a time, reconfiguration of the *Plainview* was again contemplated. Two notional types were developed: the developmental big hydrofoil (DBH), essentially a hydrofoil frigate, and the developmental fast hydrofoil (DFH). The DBH would carry an 80- to 160-ton payload at 40 knots or more on a total displacement of less than 1,500 tons, and it would have an endurance of at least 75 hours at 25 knots to minimize the need for underway refueling. For a time in 1971–72, it appeared that research and development funds would be used to build a DBH in FY 72. That plan was abandoned, although feasibility design work on a 750-ton DBH continued as late as 1974.

Zumwalt's special office also sponsored work on big surface-effect ships that superseded the DBH—and that died a few years later.

9

Small Combatants for Counterinsurgency

From 1945 on, U.S. strategy developed on two distinct levels, with corresponding naval implications, to counter a dual Soviet threat. On the one hand, the Soviets might attack the West directly. The naval counter was a blue-water fleet built around carrier and amphibious striking forces. Although U.S. small combatants would have no more than a limited role in such a fleet, NATO small combatants could help deny the Soviets access to the open sea, on and below the surface, at such chokepoints as the Danish straits, the Sicilian narrows, and the Dardanelles, and in Norwegian coastal waters. The United States therefore assisted local NATO powers by supplying or financing small fast combatant boats, subchasers, and minelayers.[1] This policy had another, unintended, consequence. The U.S. funding that helped support postwar small combatant design in Europe automatically provided the U.S. Navy access to that technology some years later. The U.S. *Nasty*-class PTF was a direct result.

The other Soviet threat was political subversion leading to revolution, which might be supplemented by invasion, not by the Soviets themselves, but by their proxies. This threat existed in both Europe and the Third World, particularly in Asia. The United States tended to arm countries facing Soviet-inspired insurrection in the expectation that their own armies and navies would fight the wars. Thus, American intervention in the Greek and Chinese civil wars (1944–49 and 1946–50), in the Indochina war (1946–54), and in the Philippine Huk rebellion (1948–56) was limited to supplying advisors and materiel. Only in the case of Korea (1950–53) did the United States directly intervene during the post-1945 decade.

Foreign assistance was often given in the form of specialized small combatants. In many countries, rivers and shallow coastal waters were the principal highways, and it was imperative to military success that they be controlled. These craft fall into three broad categories. First, through the early 1950s, the United States supplied World War II–type amphibious and patrol craft, in the case of the Philippine PGMs, reproducing their design after the war. Second, in Indochina, the French radically modified existing U.S. craft and introduced some new ones of their own. The French craft formed the pattern for the later U.S. riverine navy. Third, the United States developed a small combatant specifically for export, a motor gunboat (PGM) based on the Coast Guard's new 95-footer, described in chapter 8.

Craft were initially transferred through special single-country programs, then through the Mutual Defense Assistance Program, later through the Military Assistance Program (MAP).

In 1946, the U.S. Navy had as candidates two shallow-draft gunboats: PGMs converted from 173-foot steel PCs, and converted infantry landing craft (LCI), particularly the sophisticated LSSL. The wartime wooden PGMs (SC conversions), all disposed immediately after the war, were not available.

Six PGMs went to Greece in 1947, six others to China in 1948. Of the latter, three ended in the PRC

Many small combatants supplied to Vietnam by the United States were slightly modified versions of earlier, French-converted American amphibious craft. This monitor, converted from a U.S. LCM(6) at San Francisco, was being prepared for shipment to Vietnam on 20 May 1965. The turret contains a 40mm cannon and a 0.50-caliber machine gun. The well deck is built to take an 81mm mortar. The LCMs in the background, on the right, have been armored for Vietnamese service, in line with designs developed by the French more than a decade before. U.S.-converted monitors can be recognized by their sharper and more shiplike bows. They were different from the monitors developed for use by the U.S. Navy a few years later.

Of necessity, U.S. foreign aid began with existing small craft such as this Burmese 83-footer. Note that, unlike the Coast Guard boat, it has a funnel abaft the bridge and a 40mm cannon in place of the Mousetrap typically carried forward. Although at one time the standard battery was two 40mm guns, this boat has a 20mm Oerlikon aft. (Burmese navy)

navy and two were torpedoed in January 1955 by PRC MTBs, one being sunk and the other damaged beyond repair. PGM 31, the only unit then remaining in U.S. service, was transferred to nationalist China in 1954.

PGM 33–38, modified versions of the converted World War II subchasers (PGM 1 class), were supplied to the Philippines, then fighting a communist insurgency, under the FY 52 program. They were armed with two 40mm guns (total 48 ready service rounds and 2,016 rounds in the magazine), one 60mm mortar (72 HE and 72 illuminating shells), and four twin 0.50-caliber machine guns (3,000 rounds per barrel).

The LSSLs were the most heavily armed of the wartime gunboats, in effect the end result of one solution to the barge-busting problem treated in chapter 6. American advisors to the Vietnamese navy later characterized them as the best solution to several Third World problems, but there was never any attempt to duplicate them.

Although most of the 130 LSSLs built were transferred abroad, the only transfers fitting this category were made to France for combat in Indochina: six under the FY 51 program (LSSL 2, 4, 9–10, 26, and 80) and two more in February 1954 (LSSL 105 and 111). LSSL 2 was sold to Vietnam in November 1954, and LSSL 4 was retransferred to the new South Vietnamese navy in September 1955. LSSL 9–10 and 26 were returned to the United States but later turned over to Vietnam. In March 1956, France transferred LSSL 105 to Vietnam and LSSL 80 and 111 to Japan. Finally, the Japanese navy retransferred LSSL 96, 101, and 129 to Vietnam (FY 65).

The LSSL program will be described in greater detail in a subsequent volume of this series, on the amphibious force.

The one war the United States fought at this time left little scope for U.S. small combatants. The North Koreans received Soviet torpedo boats, but they limited their naval operations to minelaying by small coastal craft such as sampans. As in the Pacific, then, the potential role of U.S. small combatants was primarily the equivalent of World War II barge-busting (coastal interdiction). It was limited further in that the North Koreans depended more on road and rail communication than on coastal waters. Small parties of special forces were occasionally inserted behind enemy lines, but generally by fast landing craft slung from destroyer davits; motor whaleboats were sometimes used for coastal antisampan work. The four surviving Elco PTs remained in the United States with the Operational Development Force until they were transferred to the Korean navy in January 1952.

However, the Far East naval command did want PTs. The Soviets had so many at their disposal (ONI estimated in mid-1951 that the Soviet fleet included

265 MTBs and 232 motor ASW boats) that it was difficult to imagine that any future war would *not* include such craft. In June 1951, Commander John Harlee, the World War II MTB squadron commander, gave advice based on Korean War experience. U.S. PTs would be useful for coastal interdiction, and they would certainly be needed to protect Korean harbors should the North Koreans use MTBs (which they did not yet possess) of their own. Harlee proposed that at least two squadrons be procured at once.

He wanted something closer to a motor gunboat than to a classic PT, a boat powered by diesel engines to obviate the explosion risk posed by 100 octane PT fuel. A reduction in trial speed, perhaps to 30 knots—which in any case was often the performance of operational wartime PTs—could be accepted in return for the safety of diesel fuel. Boats would still need substantial speed for transit and to close targets rapidly enough to prevent their escape. Harlee favored wood construction because he saw a need for large numbers of such boats in case of a major war, when aluminum would probably be in short supply. Nothing was done at the time, for evaluation of experimental PTs had not been completed.

The French experience in Indochina is important to this story for three reasons. First, the tactics developed, particularly those of the riverine assault divisions (*divisions navales d'assaut*, or *dinassaut*) formed the basis for much later U.S. thinking. Second, French operations demonstrated the urgent and special requirement for river minesweepers as the Vietnamese enemy soon came to appreciate how important the rivers were. Third, in two quite distinct ways the boats themselves deeply influenced U.S. craft used a decade later. American experts working in Vietnam, who developed the concepts for the new craft, found ex-French craft in Vietnamese service and at the same time had access to extensive French reports on the lessons of riverine warfare using such craft. Remarkably, some of the major French craft and their U.S. successors were built on the same hulls, World War II–designed medium landing craft (LCMs).

Although the French had ruled Indochina for many years, they had not developed specialized riverine craft before 1946. The French Marine Brigade of the Far East, conceived during World War II to fight the Japanese in China, initially had no amphibious craft. Arriving in Indochina in the fall of 1945, it had to improvise. The Royal Navy transferred to the brigade an ex-Japanese 200-ton motor junk, which was used to open routes to Mytho, Vinh-Long, and Cantho. French marines armed a variety of native craft, including 30-ton junks, larger river boats, and existing shallow-draft landing craft, many of them U.S. designed and built, which formed the basis of squadrons based at Cantho and Mytho.

At the beginning of 1946, the French riverine force consisted of two armed junks, four armed and armored 200-ton, 6-knot, former river rice freighters (self-propelled barges built for the Gressier company), one small recovered launch, two ex-Japanese landing craft in poor condition, three ex-Japanese small landing craft equivalent to the U.S. LCVP, one motor river ferry, one ex-British motor barge, and fifteen ex-British landing craft (LCA). The French immediately began to buy surplus landing craft—LCAs, LCVPs, and LCMs—at Singapore and Manila. The ex-Gressier boats, the *Lave*, *Foudre*, *Tonnante*, and *Devastation*, became the first French river monitors; they could transport a company with its materiel. Junks armed with one 75mm cannon, one 20mm cannon, and two machine guns patrolled the interior rivers of the country. The small prewar naval launch and local and native craft were found to be too slow, hence too much at the mercy of the strong currents and tides and too apt to anchor and thus invite attack.

Of the existing landing craft, the LCT (later redesignated LCU in U.S. service) was valued for its capacity of 250 men, but it was too large to navigate many rivers. The ex-British LCM was criticized for its limited freeboard, that is, the limited protection it afforded personnel on board. The ex-U.S. LCM was more powerful, better armed with its two heavy machine guns, and better protected. It became the basis for French and later U.S. riverine forces. Of the smaller craft, the British provided the LCA and LCS, the latter similar in size but armed with a 20mm cannon and two 0.30-caliber machine guns. The French liked the LCA for its protection against 0.50-caliber fire and for its quiet engines (two Ford V-8 gasoline engines). This silence was a consequence of the design origin of the LCA, which had been conceived for the purpose of transporting commando raiding parties against defended shores. The French obtained the LCA without weapons and armed it with machine guns and mortars. However, it was slow, 6 knots, and had a limited radius of action, 8 hours at 6 knots. Moreover, the engines proved fragile, their water-cooling system ill adapted to river conditions. The ex-U.S. LCVP, which the French would obtain in large numbers, was criticized for its lack of protection and its loud engine, which the French called its lack of discretion. However, the diesel-powered LCVP enjoyed a much greater radius of action, 80 nautical miles, and a higher speed of 7 knots. The French tried, and failed, to silence the LCVP.

All of these craft were considerably modified in Indochina. The LCM was armed with two Japanese-manufactured Lewis guns, the LCA with one on each beam, plus a 0.50-caliber gun aft. Ex-Japanese 2-inch

This Vietnamese navy monitor was typical of craft left by the French. Note the canopy and the machine gun turret aft.

mortars were installed. The ex-British LCS had been designed to mount a 20mm cannon, but that proved impossible and a 0.30-caliber machine gun had to be used instead; a twin 0.50 was mounted in the turret, and a small Japanese mortar was also carried. The official French history observes that a 20mm cannon "would have had a considerable moral effect on the enemy." The LCS suffered because without heavy armament it offered no advantage to compensate for its lack of troop capacity. Most LCVPs carried a 20mm cannon in the well deck, a 0.30-caliber machine gun on each side aft, and a heavy machine gun right aft.

Armament increased as the war progressed. In September 1948, the standard battery of an ex-U.S. LCT was two 20mm cannon and mortars; the larger ex-British LCT carried one 40mm cannon and four 20mm guns along with mortars, and an LCI carried a 75mm gun, a 40mm cannon, two 20mm guns, one 0.50, and mortars. Motor launches, generally ex-British HDMLs, carried two 20mm and two 0.50s. Forty- and 70-ton native craft could be expected to carry two 25mm, two 13.2mm, and three 7.7mm ex-Japanese machine guns as well as mortars; one also carried a 75mm gun. The LCM was typically armed with one 20mm and two 0.50s or 20s; the LCA, with one 0.50 and two 0.30s; and the LCVP, with one 20mm and two 0.30s.

Craft were also armored, the LCVP against 0.30-caliber fire. Unfortunately, although the well deck and the weapon positions could be lightly armored, little could be done for the helmsman apart from lowering his position.

Perhaps more importantly, landing craft designed for short passages between transports and shore had to be modified for lengthier river operations, their personnel protected against sun and rain.

By the end of 1946, then, the French could conclude that they needed something about the size of an LCA but with more powerful and still silent engines using a better cooling system (circulating pure fresh water through the engine and exchanging heat with filtered river water, for example). This craft would

Table 9-1. French Amphibious Forces in Indochina, 1946–49

	March 1946	End 1946	Late 1947	Mid-1948	Mid-1949
LST	2	3	2	2	2
LCI	8	6	9	9	10
LCG	—	—	—	—	1
LCT (250 tons)	—	7	4	6	9
LCT (120 tons)	—	—	5	5	5
LCM	6	28	28	29	37
LCVP	9	32	32	34	48
LCA	26	24	23	21	20
LCS	2	—	—	—	—
VP	2	8	11	11	13
MFV	—	6	5	6	6
Junks	5	—	2	2	1
River Barges	6	—	5	6	4
LC (Japanese)	5	—	2	—	—
Coast Guard	—	—	3	3	3

Notes: In March 1946, the French had two ex-Japanese patrol craft (VP) as well as sixteen civilian launches (steam and motor). In addition to the craft listed, thirteen ex-Japanese landing craft were awaiting armament. At the end of 1946, the French operated eleven armed civilian craft and eighteen miscellaneous craft. VP designated a harbor (port) patrol launch, in most cases an ex-British HDML. The variation in MFV numbers reflects refits at Singapore. As of mid-1948, there were six DNAs: 1, 3, and 5 in the north, 2, 4, and 6 in the Mekong. Except for DNA 5, each was led by an LCI and included one LCT and lesser craft—one LCM, two LCVPs, and one VP in the Mekong Delta, two LCMs and one or two LCAs in the north (with an additional repair LCM in DNA 5). There were also sector forces and river forces directly protecting Saigon.

These figures are somewhat misleading, in that total numbers of amphibious ships were rather larger. The French began with seven LSTs (9001–7), followed by an eighth (9088) transferred in November 1951; ultimately, they had twelve. Not all served in Vietnam. All the LCTs listed were ex-British. Ultimately, there were eighteen LCTs of about (U.S.) LSM size—LCT(3) (9083–4) and LCT(4) (9060–72, 9080–82).

have a hull better adapted to higher speed—12 knots was sought—and a well deck large enough to take a jeep or a 105mm howitzer. It would be armed with a 20mm cannon and one or two 0.50s in a high turret capable of firing over the full 360 degrees. Habitability would be much improved, the crew being provided with collapsible bunks, and the boat would have horizontal as well as vertical protection.

Such a craft, derived directly from the LCA, would suffice for the full spectrum of riverine missions: transport, surprise assault, and patrol.

A much smaller armored sampan with a single machine gun would be required to patrol narrow streams.

These requirements presaged those the U.S. Navy would develop two decades later. They presented problems. Speed required power, and silencing was difficult to achieve. Heavy weapons and armor increased draft and made it difficult to match even the low speed of the LCA. The French never did produce their modified LCA. Rather, they soon came to depend heavily on U.S.-built landing craft.

Table 9-1 shows the growth of French amphibious forces in Indochina up to mid-1949. At this time, the United States was unwilling to provide direct assistance, although the French were permitted to buy surplus U.S. warships such as PCs, of which three were bought for Vietnamese customs service in 1949 and then rearmed, and LCIs (L9040–52). The French bought six SCs at Long Beach in 1948, a seventh at Manila in April 1948, and an eighth, the ex-smuggler *Bluebird*, in Saigon early in 1949. All were owned by the Saigon Customs Office, but they supplemented local French naval forces.[2]

On 20 January 1950, the JCS decided that, "within the general area of China," French Indochina would receive highest priority for aid under the new Mutual Defense Assistance Program established the previous year. At this time, too, the Viet Minh first began large-scale as opposed to guerrilla attacks on the French. President Truman approved $15 million in aid on 10 March on the basis of an evaluation that communist victory in Indochina would lead to communist domination of the rest of Southeast Asia and possibly of areas further west.

The Mutual Defense Assistance Program furnished the small amphibious craft that became the basis of French riverine operations, replacing or supplementing the small World War II–supplied craft with which the French had begun the war.[3] According to the official French history of the naval war in Indochina, apart from French-built second-class escorts, six minesweepers, and all but one of the LSTs, from 1951 on the mutual assistance program provided most of the small combatants, riverine craft, and naval aircraft for the Indochina War.

For FY 51, the French urgently requested thirty-six LCVPs (36-foot landing craft), six river craft with a shallow draft and a speed of over 12 knots, fourteen harbor tugs, thirteen subchasers, and a troop ship.

Table 9-2. MDAP Assistance to France During the Indochina War, 1950–54

	FY 50	FY 51	FY 52	FY 53	FY 54	Total
LSM	—	—	—	—	9	9 (0)
PC	2	6	—	—	—	8
SC	—	1	3	—	—	4 (1)
AMS	—	—	—	6	—	6
ARL	—	1	—	—	—	1
YOG	—	1	—	—	—	1
LSD	—	—	1	—	—	1
LST	—	—	1	—	—	1
LSSL	6	—	—	1	2[1]	9 (7)
LSIL	—	3	1	7	—	11 (4)
LCU	2	5	5	5	3[1]	15 (4⅓)
LCM	—	40	14	36	40[2]	130 (86)
LCVP	36	30	59	75	45[3]	245 (167)
YTL	—	14	—	3	—	17 (14)
Launches	—	26	1	—	—	27
LCR	—	—	—	50	—	50 (0)
Assault Boats (M2)	—	24	—	48	—	72 (24)
River Craft	—	—	—	—	70[1]	70 (0)

1. For the Vietnamese navy.
2. Half for the Vietnamese navy. All were delivered in 1954–55.
3. Thirty for the Vietnamese navy. Fifteen were delivered in 1954 and thirty in 1955.

Notes: This table does not include two supplementary LSMs received in 1954. In each entry, figures in parentheses reflect deliveries as of 1 January 1954 when they differ from the totals on order; the figures tend not to include FY 54 and some FY 53 ships and craft. Of the LCU entries, two-thirds of one boat was lost in passage (LCUs were often transported in sections). The FY 53 craft were delivered in 1954; four were for river use, and one was a workshop. The FY 55 craft were delivered in 1955.

The launches listed here were 40-foot Coast Guard utility boats. Presumably, the French 83-footers were not obtained through the mutual assistance program. LCRs are navy rubber boats; the M2s were army boats. YTLs were small tugs for use as river minesweepers.

By March 1955, the French were surveying surplus craft for return to the United States: one LSM, three LSSLs, twelve LCMs (ten of which had already been returned), eight LCVPs (three already returned), eight Coast Guard utility boats (five already returned), and three YTLs.

For delivery before 1 October 1951 (FY 52), they wanted, in addition, seventy-five more LCVPs, two LSDs, seven LCTs, three LCIs, fifty-five LCMs, three small tugs, one gasoline barge, and an escort carrier. The United States offered twelve LCVPs and six LSSLs, a total of $2 million, plus forty naval aircraft for $6 million. However, the North Korean invasion of South Korea in June 1950 considerably increased navy funds, and aid to Indochina in FY 50 increased to $31 million, including $13.5 million for naval supplies and aircraft. Note that, although the LSSL and three LCIs (9021–9033) arrived at once, the LCVP did not arrive in Vietnam until early 1951. Transfers to France under the mutual assistance program are summarized in table 9-2.

In the fall of 1950, the United States reevaluated the requirements of Southeast Asian assistance, making Indochina its first priority, followed by Thailand, then the Philippines, Indonesia, and Malaya. French requirements mainly for spare parts, maintenance facilities, and training, were estimated at $298 million, including $50 million for the navy. Navy members of the evaluation group emphasized the need for maritime patrol aircraft (ten PB4Y Privateers with high-definition radar) and recommended small riverine and coastal craft, including LSSLs and fast, shallow-draft boats. This U.S. aid allowed the French to form two additional riverine assault divisions to reinforce the two in Tonkin and the four in the Mekong Delta.

In 1951, the United States provided eight PCs, one LST, thirty LCMs, thirty-six LCVPs, and twenty-six 40-foot Coast Guard utility boats armed with one 0.50-caliber machine gun forward and one 20mm cannon aft.[4] Small harbor tugs (YTL) were transferred to serve as river minesweepers, supplementing two existing ex-British motor fishing vessels (MFV) and a force of motor minesweepers (YMS) and specially equipped LCVPs. The YTLs used the Herrison Sweep, similar to the chain drag sweep the United States would adopt over a decade later, to destroy the control wires of command-detonated mines laid on river bottoms.

By the end of 1951, the French navy, including its forces outside Vietnam, had one LCC (landing craft, control), fifty-three LCM(3)s and (6)s, four LCM(1)s, twenty LCAs (British), seventy-three LCVPs, one LCPR (ramped landing craft, a predecessor of the LCVP), and one LCPL (a small landing craft, also a predecessor of the LCVP). The LCVP figure included French-built steel LCVPs called EAs. At the beginning of 1954, this force had grown to 106 LCM(1)s, (3)s, and (6)s; 130 LCVPs and EAs; 2 LCPRs; and 3 LCPLs.

In addition, from 1951 on, the French army received

The French learned, and the U.S. Navy had to relearn, that command-detonated river mines were a major threat. This Vietnamese *commandement* was sunk by a mine, then salvaged by a U.S. derrick. (Commander Thomas B. Wilson, USN [Ret.])

its own LCMs for river transport. It formed two river transport companies with four platoons apiece; each platoon had eight LCMs for a total of sixty-four.

In 1953, the French decided to form a large Vietnamese national army that could free the limited number of French troops in Indochina for mobile operations, including several large-scale landings. They urgently requested additional U.S. assistance and between November 1953 and March 1954 received six LSMs, two LSSLs, three LCUs, fifty-five LCMs (twenty-five modified for river work), forty-five LCVPs, and thirty-seven (of seventy ordered under the mutual assistance program) armored river craft of French design built in Japan. Additional LSMs were lent until the end of 1954 to make up a total of eleven. The Vietnamese forces, listed in table 9-3, included a Vietnamese navy.

In addition, in May 1954, the United States approved the emergency procurement of 300 15-foot "Wizard" plastic boats powered by fourteen HP outboards to patrol the upper reaches of Indochinese rivers. The first shipment, 138 boats, arrived in July 1954, too late to affect the fortunes of war. Armed with a single 0.30-caliber machine gun, a Wizard could make 21.9 knots with two passengers, or about 17.5 with six.

It is not clear whether all of these craft were delivered. In all, the Mutual Defense Assistance Program provided the French with 438 naval craft between 1950 and 1954, 330 of them by September 1953. At the end of the war, the French left their surviving small craft behind, to be taken over by the Vietnamese and Cambodian navies.

The French modified their U.S.-supplied amphibious craft at their navy yard in Saigon, fitting them with armor, additional weapons, and in many cases, the amenities required for extended periods of operation on the Vietnamese rivers. A French LCI was typically armed with a 75mm cannon in the bow, a 40mm abaft the superstructure, and a 20mm aft; others had a 40mm in the bow and sometimes another just forward of the pilothouse. The LCI was, then, both a gunboat and a transport that could carry 250 troops. By way of contrast, the LSSL, built on the same hull, carried no troops but was armed with two twin director-controlled 40mm guns and one 3-inch/50.

LCTs, later redesignated LCU by the U.S. Navy, were used almost entirely for transport; they were typically armed with four 20mm cannon fore and aft and one 40mm forward of the superstructure. Some had an 81mm mortar.

Like their U.S. successors, the French considered the 50-foot Mark 3 or the 56-foot Mark 6 LCM the basis of the riverine force. Very maneuverable thanks to its twin screws, it received the most extensive modifications and hence could deliver a considerable weight of fire for its tonnage. Troop-carrying LCM(6)s were armed with three shielded 20mm cannon aft and could transport a 25-ton tank or 120 men. Their holds were covered over and their sides armored. The shorter LCM(3), with one 20mm gun and two 0.50-caliber machine guns, could carry a 16-ton vehicle or 100 men.

At the end of 1951, the French introduced a monitor version of the LCM, more heavily armored, with 12mm plating effective against the standard enemy 13.2mm machine gun at 50 meters, and with a ship bow to escort valuable craft and force dangerous passages. An armored turret in the bow had one 40mm and sometimes also one 20mm cannon; three 20mm cannon installed aft, around a newly built-up superstructure, carried an 81mm mortar. Some had light tank turrets forward with 37mm guns. A few monitors were fitted with rockets, others with flame throwers, but they never entered combat. By the fall of 1953, ten were in service, all in North Vietnam, and the French planned to convert another ten. Given their success, the naval commander in North Vietnam called for the conversion of a much more powerful gunboat, an LCT with a 5.5-inch gun and an armored conning tower.

This could not be done, but the French did decide to arm their LCTs in North Vietnam with the maximum battery that did not reduce their carrying capacity. One possibility was to mount two army-type wheeled 25-pounders aboard each boat. This idea was rejected because the guns could not easily be trained. One ex-British unit, formerly the *Pierre Idrac* (LCT 9069), was armed with two 75mm cannon and one 120mm mortar in the well deck, two 40mm cannon in the bow, and four 20mm guns aft. Other LCTs were similarly armed except for the 75s, the installation of which required too much labor. At this time, too, LCIs operating in South Vietnam received a 120mm mortar in place of their 75mm cannon. The LSIL, a newer version of the LCI, was modified so that 120mm mortars could be installed if required. This happened when three LSILs in use in South Vietnam were moved north to replace three older LCIs.

Table 9-3. French Amphibious Forces in Indochina, 1950–54

	Dec 1950	April 1953	July 1954
LSSL	—	3	7
LSIL	—	4	10
LCI	13	9	8
LCG	1	1	1
LCT(4)	10	13	13
LCT(6)	4	4	2
LCT(AT)	1	1	1
YTL	—	14	14
VP	16	7	5
VP (armored)	—	12	10
CGUB	—	17	27
MON	—	10	12
CDT	—	9	4
LCM	51	78	104
LCA	13	—	—
EA	—	55	49
LCVP	13	17	63
LCVP (armored)	78	13	10
River Patrol	8	7	5
Armored Barges	2	1	—
MFV	6	—	—
Coast Guard	3	—	—
M/S Launches	4	—	—
Vietnamese Navy			
LSIL	—	—	1
MON	—	—	2
CDT	—	1	2
LCM (armored)	—	—	13
LCM	—	4	2
LCVP (armored)	—	—	12
LCVP	—	8	4
LCU	—	—	2

Notes: This table shows how heavily the United States participated in the French naval effort in Indochina; by the end, virtually all the craft were modified versions of standard U.S. landing craft. In this list, MON is an LCM converted to a monitor; CDT is an LCM converted to a command craft (*commandement*). CGUB is the 40-foot Coast Guard utility boat. Note the appearance of mine-sweeping launches (m/s launches). EA was the *engin d'assaut*, the French-built metal LCVP. LCT(AT) was an LCT converted to a repair craft (*atelier*). The VP included a former civilian launch named the *Cypris*. Altogether, in December 1950, the French had 229 river craft, of which 143 were in service, 17 were unserviceable, 59 were in reserve, and 9 (7 armored LCVPs and 2 LCMs) were being armed. These lists are somewhat misleading, as they do not include similar craft used at sea or aboard larger ships. In April 1953, that amounted to three LSSLs, fourteen CGUBs, and sixteen LCVPs aboard larger craft. In 1954, the French listed two LCUs, nineteen LCMs, and twenty-five LCVPs aboard seagoing ships assigned to Indochina.

At this time, the French navy experimented successfully with army quadruple grenade launchers, and at least some were installed.

New armor plating was fabricated for the existing landing craft, and special engine armor enclosures were mass-produced for LCMs, LCIs, and later for LCTs.

In 1952, the French converted an LCM(6) into a

A U.S. monitor conversion for the Vietnamese navy, ready for shipment from San Francisco on 20 May 1965, shows how the slab-sided underwater hull of the standard LCM was modified to form a more shiplike bow. The weapons aft are a 20mm cannon on the centerline and two machine guns.

A Vietnamese navy *commandement* in 1964. Note the telescopic crow's-nest aft, the 20mm turret in the bow, and the bow guard.

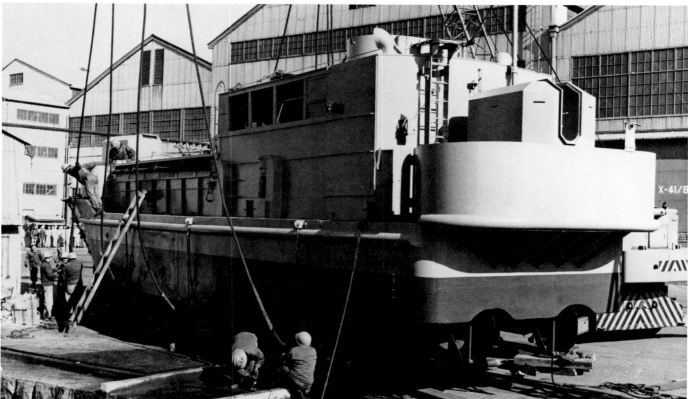
The U.S. version of the French-designed *commandement*, the LCMC, is shown at San Francisco ready for launch on 25 January 1965 and departing for trials two days later. Armament consists of 20mm cannon in the turrets and an 81mm mortar in the well deck, covered here by an awning. Note the raised crow's-nest in the photo of the completed boat.

Although they disliked its loudness, the French found the U.S. LCVP reliable. They armored and used it as a minesweeper and gunboat, the Vietnamese later following suit. This Vietnamese LCVP was photographed in 1964 armed with a 20mm gun forward and 0.30-caliber machine guns amidships and aft. The canopy protected the open well deck from fragments and grenades.

command ship (*commandement*) to replace LCIs in dangerous areas or very shallow water. It entered service in the Transbassac area, based at Cantho, and when it proved successful, a second was built for the Saigon river patrol. As table 9-3 shows, more were soon converted. These boats had massive superstructures. They were armed with 20mm cannon and 0.50- and 0.30-caliber machine guns. By May 1953, the navy yard at Saigon could convert an LCM into a *commandement* in about six weeks or into a monitor in about ten.

The LCMs, monitors, and command boats supplied to Vietnam under the mutual assistance program were similar to the French-modified craft.

LCVPs were covered over, armored with quarter-inch plate, and armed with a 20mm cannon forward and five 0.50-caliber machine guns—two each side, one in the stern. These five were later reduced to two aft.

The French also built a steel version of the U.S.-designed LCVP which they called an EA (*engin d'assaut*). EAs were used for minesweeping, escort, and patrol; they were typically armed with one 20mm gun forward and three 0.30-caliber machine guns aft.

The first EA, 1–35, were built at Cherbourg in 1950, officially entering service on 12 September 1950. They were usually armed with one 20mm cannon, one 0.30-caliber machine gun, and two rifle grenade launchers, and they carried seven men. The LCVPs were used primarily for logistics, the EAs as minesweepers.

The French further modified the basic LCVP design, installing a ship-type bow and considerably increasing deadrise to create the V-bottomed "STCAN"—which inspired the U.S. ASPB of the Vietnam War.[5] The Japanese-built boats delivered in 1953–54 under the mutual assistance program were almost certainly STCANs. The STCAN, the only entirely new craft of the riverine assault divisions, was intended for liaison, raiding, scouting, and rapid sweeping. The French criticized it as too slow at 10 knots, too noisy, and difficult to handle in choppy water. Both in French and in Vietnamese service, the STCAN gained a reputation for resisting mine damage.

The French army operated a smaller, 8-meter standard boat called an FOM (*forces outre-mer*), armed with one heavy and several light machine guns and capable of 7 knots. It generally operated with the

From the American point of view, the most significant development in the LCVP story was the French STCAN, shown here in Vietnamese service. It was built of steel, had a shiplike bow, and was typically armed with one 0.50-caliber machine gun in a bow turret and three 0.30s, in the sponsons and after turret. The boat impressed U.S. observers, who were inspired to write the specification that in turn produced the U.S. ASPB.

navy. U.S. documents tend to lump the FOM with the STCAN, forty-five of which the Vietnamese ultimately received.

By 1953, the French wanted to equip their LCMs and LCVPs with Canadian Iroquois flame throwers, predecessors of the U.S. "Zippos". After successful tests early in 1954, the U.S. Navy suggested that the American M-4 flame thrower would be easier to use. However, the war ended before craft equipped with flame throwers could enter combat. At this time, too, the French sought additional protection for their riverine craft, as the Americans would do a decade later. French boats were too heavily loaded to accept much additional weight, so ballistic nylon was shipped to Vietnam beginning around April 1954. It also arrived too late to see combat, and its possibilities seem largely to have been forgotten until the late 1960s, when it was valued as an antifragment measure.

STCANs, EAs, and FOMs were supplemented under the assistance program by U.S. Coast Guard 40-foot utility boats, about seven of which survived as Vietnamese navy *vedettes*, or launches. The Vietnamese also had at least fifteen 44-foot "interceptor" vedettes and one 54-foot "surveillance" boat. All were generally described as French in origin, but at least the 40-footers were ex-American.

The commander of French naval forces in the Far East submitted his notes on lessons learned in March 1955. Apart from EAs, FOMs, and barges, he had relied on ex-British and -U.S. amphibious craft, considerably modified. He liked the LSSL gunboat, but considered its draft excessive at times, its protection of 8 to 10mm plating light, and its projection into the air too great to allow passage under many canal bridges. The old LCI, replaced after 1952 by the U.S. LSIL, was valued for its low superstructure; it could pass under the Doumer bridge by folding down its mast.

The ex-British LCT(4) was "the most remarkable of the craft used in Indochina" because of its simplicity, maneuverability, compartmentation, shallow draft, and low superstructure. It could carry a battalion, a 105mm battery, or half an amphibious squadron, and it could be heavily armed without losing its transport capacity. Unfortunately, none were received after 1950. The U.S. LCT(6), smaller and slower, was used only as a transport.

LCMs formed the basis of French riverine warfare. They existed either in armored form, with 8mm plating and two to four 20mm cannon, or as monitors with 12mm armor and telescopic crow's-nests that were extremely valuable among the dikes of North

Vietnam, or as *commandements*, which had special crow's-nests and lighter armor.

The LCVP was dismissed, as before, for its loud engine and its limited capacity and power. The French-built FOM and STCAN were considered superior, the FOM using the quieter French Renault engine, the STCAN the louder U.S. Gray Marine diesel.

Finally, the British-built VP (HDML in British service) was praised for its robustness and silence. It was, however, too vulnerable for "front line" riverine service. The Coast Guard boat was fast but noisy and rated as unsafe with its uncovered propellers and open (floodable) deck. Tried on the rivers, it was ultimately reserved for coastal surveillance. The Wizard, which arrived only after the cease-fire, was admired for its "brilliant qualities" of speed and maneuver, but it was too small and fragile with a plastic hull and outboard gasoline engine.

Overall, the boats, although they tended to be too slow and their Gray Marine engine too loud, were acceptable. Flat bottoms and deep draft made the ex-U.S. landing craft too vulnerable to river mines.

The greatest handicap to the French force was a lack of heavy armored assault vessels, a type successful in earlier river campaigns such as those of the American Civil War. A specially built craft would have been best, but it was out of the question given the limits of French industrial and fiscal resources. The LCM monitor was not powerful enough, which left the LCT, with a 5.1- or even 5.9-inch gun in place of the 40mm and with thicker armor, 100mm rather than 12mm. The only such LCT converted, LCT 9069, proved successful in 1953–54.

As for the smaller craft, the French wanted a small, cheap, easily produced craft capable of 20 knots and of towing a sweep. It should have two silent (nongasoline) engines, a strong, flat-bottomed hull suitable for beaching, a draft of a meter (3.28 feet) or less, and a tunnel stern. And, ideally, it would be armed with one 20mm gun or one 57mm recoilless rifle and two machine guns, it would be adequately armored, and it would have a crew of three. By way of comparison, the STCAN had a crew of eight and was armed with one 0.50- and three 0.30-caliber machine guns.

Given the lack of effective armor, automatic weapons were the best safeguard against successful attack. The report considered the balance between long- and short-range weapons satisfactory. Although no order of merit for different weapons could be established, the report did note that twin 40mm cannon were extremely effective in neutralizing enemy units at short range. The 120mm mortar was very capable, especially at short range. The 20mm machine gun was superior to the 20mm Oerlikon, which was harder to protect and more subject to stoppage, but both were effective, particularly with incendiary bullets. Finally, the old 75mm gun, the M1897, had rendered excellent service aboard the LCI and LCT. Single 40mm guns that were not director-controlled, unlike the twin mounts on the LSSLs, were still valued as counterambush weapons. As for more exotic types, recoilless rifles had been tried only aboard army FOMs and LVTs, the navy preferring to avoid the risks of their back blast. Flame throwers were tried against ambushes but abandoned as unsafe.

The heaviest weapon, the 120mm mortar, was considered inconvenient, being a muzzle loader and slow and dangerous to fire. Its ammunition presented a grave fire hazard. It would have been better, the French concluded, to use their prewar 5.1- or 5.5-inch destroyer gun, the single U.S. 5-inch gun, or the German 5.9-inch gun then in French service. The ideal gunboat battery would have been one or two such navy guns, two 120mm mortars, two twin 40mm Bofors guns, and six to eight 20mm Oerlikons, as well as grenade launchers and hand-held automatic weapons.

As the U.S. Navy would discover a decade later, armor was a problem because no existing plate could keep out the shaped-charge projectiles fired by 57mm and heavier recoilless rifles. In some cases, craft were protected by a double layer of armor, with or without cement in between, the outer layer to trigger the shaped charge. By 1954, the Viet Minh had 105mm artillery, and it would have taken 10 to 15 cm (3.9 to 5.9 inches) of armor to resist such shells; that was entirely beyond the carrying capacity of the French craft. Even the armor fitted had, by its top weight, destroyed the vessel's sea-keeping qualities.

These boats formed the basis for the *dinassauts* (DNA), or naval assault divisions, which, as previously mentioned, evolved from the French Far East Naval Brigade formed in December 1944. The brigade joined French marines with suitable small craft. It was used extensively during the French reoccupation of Indochina in 1945–46. Early in 1946, it was broken down into two separate elements: the First River Flotilla had thirty small craft for Tonkin, in the North; and the Second River Flotilla had about sixty small craft for the South, including the Mekong Delta. At the beginning of 1947, the brigade was reorganized as the Naval Amphibious Force, Indochina, directly under the Far East French naval commander. It comprised a northern group, based in Tonkin and at Danang, with five attached commandos or marine units, and a southern group in the Mekong Delta with two attached commandos. The combination of commandos and craft soon became the DNA.

Each consisted of a command and fire support ship, usually an LCIL; a transport (LCT); landing and support ships, usually two LCMs and four LCVPs;

and one patrol and liaison craft, usually a harbor patrol boat such as a STCAN. The DNA brought these craft together with French marines.

DNAs functioned independently during most of the war, but by its end they sometimes had to operate together with transports from other services. They were most effective in the interconnected waters of the Mekong and Red River deltas. Along the coast, in Annam, the rivers were shallow and craft could move from one to another only by way of the sea. The heavily armed and armored ex-landing craft were too top heavy for sea swells.

The war effectively ended with the fall of Dien Bien Phu on 7 May 1954; the French and the Viet Minh concluded an armistice on 20 July. French riverine craft were transferred to the new Cambodian and South Vietnamese navies. As the French withdrew, the United States took over responsibility for supporting and advising the South Vietnamese military through a small military assistance advisory group.

In 1952, the French planned a Vietnamese navy consisting of two DNAs, a river flotilla, and minesweepers. The first unit, the Cantho DNA, had been commissioned on 10 April 1953 with three LCMs and two LCVPs. A second Vietnamese DNA, similarly configured, was formed in June. Two more Vietnamese DNAs were formed during 1954–55, a fifth in 1956. Each ultimately consisted of six LCMs, five LCVPs, and five shallow draft outboard motorboats backed by five LSILs, two LSSLs, and five LCUs, one a floating boat-repair craft. These DNAs became the Vietnamese river assault groups. Unlike the DNAs, the river assault groups had no integral army troops and were subordinate to a district army commander, who tended to use them primarily for logistics rather than as offensive assets.[6]

Unlike its French predecessor, the new Vietnamese navy did not emphasize riverine warfare. What growth it experienced between 1955 and 1960 was concentrated on oceangoing ships, which were more prestigious. The only new craft that proved useful on the rivers, the LSSLs, were bought as coastal gunboats. The development of riverine forces in Vietnam only resumed with increased U.S. aid after 1960.

From 1953 on, the United States supplied specially built gunboats to many of her Third World allies. The boats had two distinct missions. First, they had to control rivers and coastal waters, which in many countries were the principal means of communication. Second, in Asia they had to counter armed junks, of which potential enemies could mobilize large numbers for invasion. A large number of relatively low-performance gunboats would serve this second function.

An evolving need to counter the new Soviet-bloc torpedo and missile boats created a third mission requiring much higher performance and therefore too great a degree of sophistication for local navies.

The Coast Guard's new 95-foot, steel-hulled patrol boat replaced the wartime wooden SC as the preferred U.S. export PGM after 1953. Four units were specially built in 1953 for transfer to Thailand in 1954, four more went to Iran, one in FY 54, two in FY 56, and one in FY 57, and one went to Saudia Arabia in FY 59. All had a single 40mm gun aft (rather than the twin 20mm that was standard in Coast Guard boats), a Mousetrap (Mark 20) forward, and four tilt racks (type B), each holding two depth charges. The vessels came with searchlight sonars typical of World War II practice.

The Coast Guard also transferred completed 95-footers from stock: two to Ethiopia (ex-CG 95304 and 95310 in FY 56 and FY 57), two to Haiti (ex-CG 95315 and one other in 1956), and nine "C" types to South Korea (ex WPB 95323, 95325, 95327, 95329–31, 95333–35 in September 1968). The latter were transferred instead of newly constructed boats because they were so urgently needed; Coast Guard boats were built to replace them.

After FY 58, new boats built specifically for export were designated in the navy PGM series. The first were PGM 39–42, for the Philippines, and PGM 43–46, for Burma. Construction paused in FY 59, then PGM 51–52 were built for Burma (FY 60), PGM 53–54 for Ethiopia (FY 60), PGM 55–57 for Indonesia (FY 60), PGM 58 for Ethiopia (FY 61), and PGM 59–70 for Vietnam (FY 62).

This PGM was built in three versions: the PGM 39 class, for the Philippines and Indonesia, the PGM 43 class for Burma, and the PGM 53 class for Ethiopia. PGM 39 was the original Coast Guard hull lengthened to 101 feet for cargo, a distilling plant, and air conditioning. It had hot water for showers but no heating plant. It was powered by four 550-bhp Cummins NVMS-1200 diesels for a speed of 21 knots on trial. As in the Coast Guard–built craft, armament consisted of one 40mm gun, one Mark 22 Mousetrap (with twenty-four projectiles), and four depth-charge racks (six charges) with a wartime-type searchlight sonar. However, the Philippine navy used its boats only as gunboats, removing the sonar and Mousetrap and mounting one 40mm and three 20mm guns. Eventually that battery was reduced to two single 20mm guns. The three Indonesian boats were powered by Mercedes MB 820 Bb diesels, which produced the following: 1950 bhp, 17 knots on trial, 15 knots sustained, and 1,000 nautical miles at 12 knots (as opposed to 1,500 miles at 14 knots in the Cummins version). They were initially armed with two 0.50-caliber machine guns.

The U.S. Navy modified many LCMs for Vietnam according to a special MDAP configuration. These three boats, modified but not yet armed, were photographed in San Diego in June 1965. Note the lightly armored enclosure with three gun houses aft. The ramp forward is also lightly armored. U.S. forces used these boats until ATCs became available, and at least some were modified to fire large bursts of rockets.

Apart from specialized river craft, the main U.S. small combatant export of the 1960s was the PGM based on the 95-foot Coast Guard cutter. This is PGM 116 before its transfer to Thailand in 1969, armed with a single 40mm gun forward, a 20mm aft, and two 0.50-caliber machine guns.

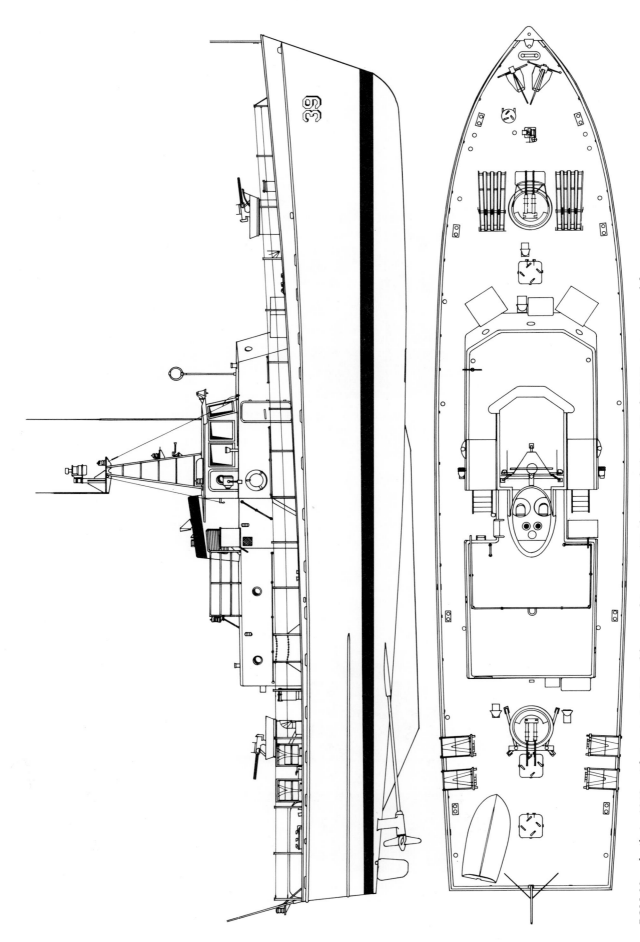

PGM 39 as built in 1960, with two twin 0.50-caliber machine guns (Mark 17 Mod 1), two Mark 20 Mousetraps, and four double depth-charge tilt racks. (A. D. Baker III)

The PGM 43 class was an austere version of PGM 39, using the original 95-foot hull, retaining the cargo space and distilling plant, and eliminating air conditioning and hot water. With lower-powered engines—Detroit Diesel 6-110 Tandems of 1,330 bhp—the class was limited to 16 knots on trial, 15 sustained. The boats cost $455,000 in 1961. Although the Burmese boats were delivered with two 0.50-caliber machine guns and ASW weapons, they were rearmed as gunboats with two single 40mm guns. Remarkably, this increase in firepower actually reduced top weight by 2,693 pounds, because ammunition stowage was brought down from 10,088 to 7,395 pounds.

The PGM 53 class closely resembled the 95-foot Coast Guard boat previously supplied. With 2,200-bhp Cummins VT-12-M Tandem engines, the boats achieved a trial speed of 20 knots, eighteen sustained, and an endurance of 1,400 nautical miles at 12 knots. Each carried a single 20mm gun, two depth-charge racks, and a Mousetrap, and cost $625,000 in 1961. The Ethiopian boats were eventually refitted with one 20mm gun forward and one 40mm aft.

PGM 59 was the first of a new series of 101-foot boats conceived as austere repeats of the PGM (PGM 43 class) for Vietnam. Eight such 101-footers replaced a single planned fleet minesweeper in the FY 62 Mutual Defense Assistance Program. However, the U.S. military advisory group in Vietnam wanted a faster boat and suggested Mercedes-Benz MB 820 Bb diesels. Armament was also upgraded from the originally proposed two 0.50-caliber MGs, four depth-charge racks, and two Mousetraps, in which form costs amounted to $510,000 (FY 63). The advisory group asked for more guns, the threat being much more on the water's surface than beneath it, and the boats were completed with a single 40mm forward, two twin 20mm guns aft, and two single 0.50-caliber guns beside the bridge. PGM 69 had two twin 0.50-caliber guns on Mark 17 mounts instead of the two singles. All ASW weapons were deleted.

The motor gunboat program was reviewed in 1959. Several alternatives were rejected in favor of further construction of the basic Coast Guard design (in the following list, asterisks indicate craft built locally, to U.S. plans): PGM 71 for Thailand (FY 64), PGM 72–74 for Vietnam (FY 64), PGM 75–76 for Ecuador (FY 64), PGM 77 for the Dominican Republic (FY 64), PGM 78 for Peru (FY 64), PGM 79 for Thailand (FY 66), PGM 80–83 and 91 for Vietnam (FY 65), PGM 102 for Liberia (FY 65), PGM 103 for Iran (FY 66), PGM 104–6 for Turkey (FY 64), PGM 107 for Thailand (FY 66), PGM 108 for Turkey (FY 66), PGM 109–10 for Brazil (FY 66),* PGM 111 for Peru,* PGM 112 for Iran (FY 67), PGM 113–17 for Thailand (two FY 67, three FY 68), PGM 118–21 for Brazil,* PGM 122 for Iran (FY 68), and PGM 123–24 for Thailand (FY 69). One modified boat without a U.S. number, the *Cirujano Videla*,* was built as a hospital craft for the Chilean civil assistance program in 1964, and the same basic hull design was also used as the basis for a U.S. deep-water torpedo retriever.

All of these boats were powered by pairs of GM quad diesels, each consisting of four 6V71 engines clutched to a single shaft. The change was particularly unfortunate for Vietnam, as it meant that two quite different machinery plants had to be maintained simultaneously. The U.S. military advisory group asked to have the Mercedes engine reinstated, but that could not be done because of U.S. congressional pressure against the use of foreign-made equipment. With two GM quads (2,000 bhp), a PGM 71 could make 18 knots with an endurance at that speed of 630 nautical miles. It cost about $460,000 in 1964 and could carry 30 tons of cargo. The standard armament in 1966 duplicated that of the Vietnamese gunboats: one single 40mm gun, two twin 20mm, and two single 0.50-caliber machine guns.

Turkey and Iran received an ASW version, equipped with the large SQS-17A medium-frequency scanning sonar. To accommodate it, the positions of the 40mm and 20mm guns had to be switched, the 40mm going aft. The Turkish boats were fitted with a double Mousetrap (Mark 22, with twenty-four rather than twelve rockets). The arrangement, if feasible, was awkward, since the sonar was quite large relative to the boat. In effect, the craft duplicated the FY 60 SC rejected in 1958. It was also expensive. In 1965, for example, the U.S. mission to Turkey estimated that although five PGMs could be bought for the price of one PC 1638, then being built for the Turkish navy, the ratio would fall to three to one for boats equipped with SQS-17A sonars.

Finally, the Brazilian-built boats were armed with an over-and-under 0.50-caliber machine gun and an 81mm mortar forward, and two single 0.50-caliber machine guns aft.

The PGM was not altogether popular. Many small Allied navies regarded it as inferior to the 173-foot PC, a natural, if unrealistic comparison. Others felt that too much had been attempted on too small a hull. For a small naval craft, the PGM had great metacentric height: it rolled too easily and rapidly. Some military advisors abroad reported that boats rolled even if a slight wave hit them. The only cure was to fit bilge keels, at a cost in speed.[7]

The PGM was also relatively large. However, as the record of the Mutual Defense Assistance Program shows, the U.S. Navy had no satisfactory prototype for anything much smaller. The French received the standard 40-foot Coast Guard patrol boats. They were also transferred to other nations: fifteen to the Phil-

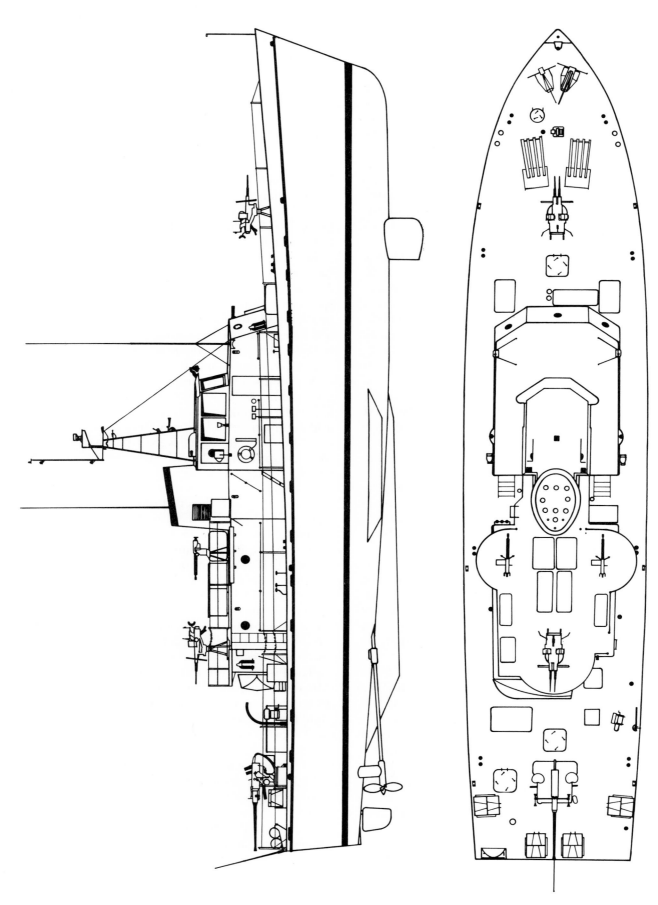

The PGM 71 class delivered to the Turkish navy. (A. D. Baker III)

ippines in 1952, six to Iran in June 1953, one to Thailand, two to Ecuador, two to Saudi Arabia in November 1957 (C-3233–34), two to Pakistan in April 1959, four to Taiwan, ten to the Netherlands, two to Colombia in 1961 (FY 60), two to El Salvador in August 1962 (40523 and 40428), and two to Guatemala (40409 and 40487, also reported as 40520–21) in August 1963. The boats were much too heavy and slow for satisfactory riverine use.

A few Mark 4 steel LCPLs, intended originally to control the waves of landing craft, were transferred, although it is not clear whether they were considered patrol boats. Three went to Italy in 1953 (C-79723-5), five to Spain in October 1958, four to nationalist China in May 1960 (C-60612, C-87270, C-87556, C-87578), and three to the Philippines. One LCPL patrol boat was later transferred to Thailand in November 1968, another, C-4391, to Greece in February 1973. LCPLs were used for riverine patrol during the Vietnam War.

Private American yards also built a few patrol craft to their own designs. Korody Marine produced two 66-footers for Ceylon in 1956, and Trumpy constructed ten 40-footers for Cuba (SV 1–10) in 1953–56.[8]

The absence of satisfactory naval craft explains both the need to extemporize coastal and riverine craft for the war in Vietnam and the use of specialized commercial craft, built by Bertram and Sewart, from about 1963 on. These two firms constituted the U.S. technology base in mass-produced fast small craft on the eve of the Vietnam War, and Sewart translated its prewar expertise into large orders for the Swift program (see Chapter 12).[9]

The 31-foot, deep-V Bertram was unusual in incorporating a series of keels running along its bottom. It could be driven at up to 44 knots by twin 275-bhp Interceptor or twin 280-hp Chrysler M413 gasoline engines, and it was credited with good seagoing performance, sea state 3 or worse.

The first four U.S. Navy 31-foot Bertrams, C-5928–31, were bought on an emergency basis by the naval base at Guantanamo Bay to answer the threat of attack by Cuban forces. Ordered in November 1962, they were delivered a month later to replace slower LCPLs. C-5931 later served in the Canal Zone. In October 1964, a 38-foot Bertram, C-13384, was bought specifically for service at Key West to intercept Cubans who might approach U.S. waters. These early purchases encouraged further official interest in the boat, but the U.S. Navy has never bought it in any numbers.

Six 25-foot (yard numbers 251-532 through 537) and four 31-foot (311–63, 365–67) Bertram sports conversions were ordered for Venezuela in December 1963. The Venezuelans rejected them, and they were stored at Norfolk. One of each length was sent to Panama City for evaluation, and the others were shipped to SEAL boat support unit 1, on the West Coast, in February and March 1966. Additional boats were ordered: three for Iran in May 1965, three more (31-543 through 545) in September 1966; two for the U.S. Air Force in February 1967 (31NS672–73); three for Iran in April 1968 (31NS681–83); one for the U.S. Air Force in July 1968 (31NS684); and two for Iran in November 1969 (31NS701–2). Similar Bertrams were exported to Egypt (twenty in 1973), Haiti (two), Jordan (four in 1974), and Venezuela (seventeen for the national guard). Iran reportedly received a total of over forty Enforcers.

Except for the first boats, which had Chrysler engines with outdrives, these boats were powered by paired GM 4-53 diesels. Details varied; some craft were cabin cruisers, Bertram Express, and others were Bertram Enforcers, designed as patrol craft. All had a common hull.

Apart from these official naval purchases, Bertrams have enjoyed long and successful sales to police forces in the United States and abroad.

Located on the Gulf of Mexico, Sewart concentrated on boats to support offshore drilling rigs. They had to maintain fairly high speed in moderately rough water. Sewart built its first aluminum boat, the 26-foot *Claire* (yard number 406), in March 1955. She achieved 18 knots with a 110-bhp gasoline engine. In 1959, for service on Lake Maracaibo, Venezuela, the company introduced the civilian prototype of what became the Swift—the *Semarca 27*, yard number 1039, and the *Semarca 29*, yard number 1079. The *Semarca 29*, the first Swift to be powered by twin 478-bhp GM V-12 diesels, attained 37 miles per hour on trials. Sewart claimed that it had solved the problems of aluminum construction, that such hulls were actually superior to steel hulls.

Sewart received its first contract under the Mutual Defense Assistance Program in September 1963 for four 40-foot boats for Korea (yard numbers 1249, 1325–27); shipped in February and March 1964, they were available for inspection in the Far East. Two more were ordered for Ethiopia in November 1964 (yard numbers 1392–93); they were shipped in February 1965). Another 40-footer was ordered for Iran on 3 January 1966 (yard number 1572) and shipped in June 1966; two more were ordered for Ethiopia in November 1966 (40NS671–72) and completed in June 1967; two were ordered for Paraguay in June 1967 (40NS673–74) and completed in November 1967; and five were ordered for Iran in May 1968 (40NS681–85) and completed in July and August 1968. In 1969, another ten 40-footers were ordered: four for Paraguay (40NS701–3 and 40NS7010) and six for Iran (40NS704–9).[10] Four of the Iranian 40-footers were

The postwar navy did not have many small gunboats available for export, but it did have numerous small ASW craft to send to U.S. allies. These PCs—the Korean *Chirisan* (704), *Kumkangsan* (702), and *Samkasan* (703)—were stripped of their Mousetraps and depth charges and rearmed at Mare Island. They are shown on 16 June 1950, not yet fitted with their 20mm and 40mm guns.

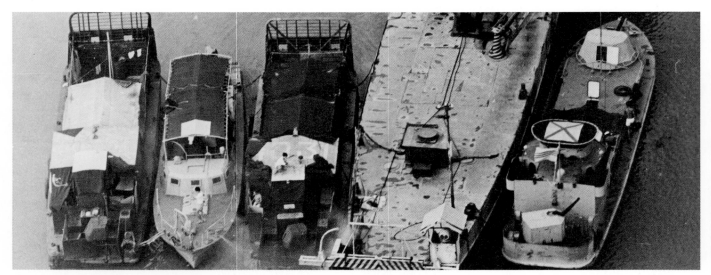
The fruits of military assistance: South Vietnamese small craft of the Fourth Naval Zone alongside their tender, a cut-down LST, January 1966. The three converted LCMs show the boxy machine-gun turrets characteristic of boats supplied under MDAP or converted in Vietnam. The boat between the two troop carriers on the left is a launch (*vedette*), probably a 54-foot "surveillance" craft, almost certainly of French origin. The canopy conceals its armament, a 0.50-caliber machine-gun turret atop the bridge and a larger 20mm turret aft. The boat on the right appears to be a new U.S.-converted monitor.

Much of the MDAP program was designed to build up Allied ASW forces. The Korean *Kumseong* was the former U.S. PCS 1445, a sonar school ship, with a Hedgehog forward. Transferred in 1952, she was returned and discarded about 1970. The MDAP program as a whole encountered some difficulties because U.S. views on appropriate national priorities (such as ASW) were not always shared by local navies, which might be more concerned with defense against coastal infiltration (as in Korea) or anti-ship capability.

transferred to the Sudan in 1978. One slightly larger 43-footer was ordered for Venezuela in July 1965, followed by three more in July 1969 (43PB701–3), which were completed in December 1969.

Sewart also built an 85-foot crew and offshore supply boat. From 1964 on, the company sold several military versions—patrol boats and torpedo and missile retrievers. In June 1964 (FY 64), BuWeps ordered two weapon retrievers, C-14252 and C-14253, for Key West and Point Mugu, followed by four in May 1965 (FY 65), 85TR651–54. A patrol boat version with a Sewart hull number of 1589, the *Holland Bay*, was ordered for Jamaica in July 1966 and delivered the following March (FY 67). Five more, 85NS671–75, were ordered that December, two for the Dominican Republic (the *Procion* and *Bellatrix*), one for Guatemala (the *Utatlan*), one for Jamaica (the *Discovery Bay*), and one for Nicaragua. Jamaica received a third 85-footer, the *Manatee Bay*, in July 1967. In May 1967, the air force ordered two as drone targets, and four more (85NS681–84) were ordered in June 1968, one for the Dominican Republic (the *Capella*), one for Uruguay, one, a missile retriever, for the air force, and another of the same for Point Mugu. Point Mugu ordered another missile retriever, 85NS691, in August 1968. This was the boat Sewart offered the navy in a gas turbine version as an alternative PTF and which, therefore, led indirectly to the Osprey. Two more 85-footers, 85NS721–22, were bought under the FY 72 program for the Dominican Republic (the *Aldabaran*) and Guatemala (the *Osorio Saravia*), and Honduras later bought an 85-footer. The navy revised the Sewart design in 1976, and two prototypes were built by Tacoma Boat.

Later Sewart boats are described in appendix E.

10

Motor Gunboats and PTFs

The U.S. Navy's interest in fast patrol craft revived at the end of the 1950s, as Soviet fast torpedo and then missile boats proliferated. The Soviets had boats and active production lines because they alone, of the major powers, had maintained large small-attack-boat fleets after World War II, in keeping with their coastal defense orientation.

Allied navies needed something faster than the PGM to face such boats, supplied by the Soviets first to its satellites and then, from 1956 on, to such client states as Egypt. Soviet transfers to China and North Korea began as early as 1952, and the Chinese transferred torpedo boats to North Vietnam in 1958.[1] That year, the U.S. Navy reviewed the mutual defense program requirements for a future PGM.

Soviet exports supported Khrushchev's new policy of "peaceful coexistence," which meant, in effect, carrying on a struggle with the West in the Third World rather than through a major nuclear confrontation. Motor torpedo and missile boats, and later, submarines, were particularly valuable because they countered the U.S. carrier forces that would otherwise easily dominate the coast of any small, hostile country. Missile boats, with their long reach, presented a particular threat to carriers. Torpedo and gun boats could be expected to break up amphibious operations.

During the Cuban missile crisis, Soviet-supplied torpedo boats threatened U.S. surface operations. In the early 1960s, the Pacific Fleet nervously watched North Korean PTs near South Korean shores and, operating further south, North Vietnamese boats. During the Cyprus crisis of February 1964, the U.S. amphibious task group had to keep moving because it was always within striking range of Egyptian PT boats based at Alexandria. The North Vietnamese torpedo boat engagement with U.S. destroyers in the Tonkin Gulf seemed only to demonstrate what had long been feared, that Soviet-supplied fast attack craft would contest free use of the sea near communist-held coasts.

In 1959, the CNO, Arleigh Burke, asked for studies of a small fast patrol boat (PTF). He appears to have been motivated, at least at first, by the need to protect amphibious operations from motor torpedo or motor gun boat counterattack. PTFs would also be useful in guerrilla warfare, which seemed likely to be the primary means by which the Soviets would pressure the West in the future.

Fast combatant development diverged. The PTF was ultimately valued primarily for its ability to support clandestine guerrilla operations. Something bigger, the *Asheville*-class PGM, or at least something more expensive like the PGH was needed to deal with Soviet-type fast combatant boats. As it turned out, both were bought, for the new Kennedy Administration considered limited war, in both the Caribbean and Vietnam, its greatest challenge.

The U.S. fast gunboat/PTF program began in an SCB review of the export gunboat program. In September 1958, having read requests from U.S. missions to allied navies, the SCB asked BuShips to develop a rugged, simple, 20- to 30-knot gunboat of less than 500 tons with an endurance of 1,000 to 2,000 nautical miles at 10 knots. A request for a hydrofoil alternative reflects the SCB's interest in the high-speed anti-PT mission. The following March, the SCB extended the speed range to 40 knots and indicated that an endurance of 1,000 miles was acceptable. It

An "Osprey"-type fast patrol boat takes shape at Sewart Seacraft in Berwick, Louisiana, 24 October 1967. The object to one side, in the frontal view, is a Napier Deltic engine, the standard power plant of the U.S. PTFs. Behind the Osprey is a Swift, the boat that, much more than any other, propelled Sewart into the patrol craft business. The Osprey hull form, so different from that of the *Nasty*, was derived from a much smaller fast craft designed by Sewart a few years before.

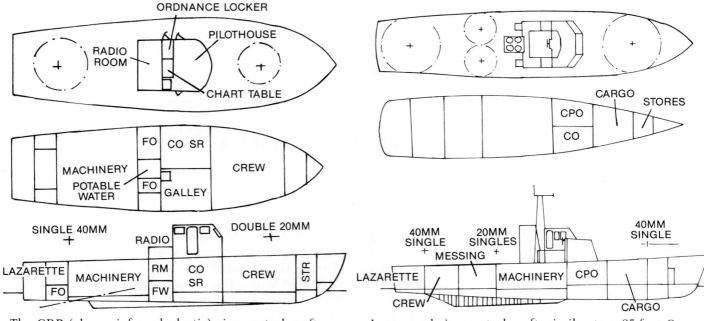

The GRP (glass-reinforced plastic) river patrol craft, as proposed in August 1959. Dimensions in feet-inches were 61-0 (waterline) x 18-6 x 9-1 (depth) x 5-0 (draft). Displacement was 36 tons. The boat would have achieved 20 knots on 1,000 bhp. Armament consisted of a twin 20mm forward and a single 40mm aft. (Norman Friedman)

A proposed river patrol craft, similar to a 95-foot Coast Guard cutter, August 1959. Dimensions were 90 (waterline) x 20 x 11 (depth) x 6 (draft) feet. Displacement was 105 tons. It would have made 25 knots on 3,600 bhp, and endurance was 1,000 nautical miles. This design included provision for 5 tons of cargo. Armament consisted of 40mm guns fore and aft and two single 20mm guns abaft the superstructure. (Norman Friedman)

preferred a wooden hull, which would be relatively easy to maintain.

In August 1959, BuShips replied with a series of "river patrol boats," most of which were really PTFs. In the list of alternatives (table 10-1), scheme 1 was the existing PGM. Schemes 3 and 4 were the preferred but expensive fast boats, 30- to 40-knot hard-chine craft based on the existing PT 809 design. Scheme 3 had diesel, 4, gas turbine, propulsion. Scheme 5 was a hydrofoil for speeds over 40 knots. Scheme 2 was a compromise, a 25-knotter, the fastest that could be derived from the existing slow PGM. It was still much too slow to deal with fast Soviet-bloc MTBs.

Steel was too heavy for the fast boats, and BuShips rejected wood on the ground that it could not take pounding loads at high speed. Wartime PTs had deteriorated badly in service, and the bureau blamed their wooden construction for a variety of structural failures. Similarly, ONI reported structural failures in the bottom of Soviet-built motor torpedo boats. Reinforced plastic was attractive for its light weight and resistance to corrosion, but as of 1959, it had not yet been used for anything as large as the boats contemplated. It would soon be used for the 52-foot LCSR.

That left aluminum, which had certainly been tested in the experimental PTs.

Even a simple hydrofoil with fixed surface-piercing foils would outperform a conventional planing hull in rough weather. In calm weather, above approximately 30 knots, it would require about half as much power for the same speed. Given the MAP requirement, there could be no question of using the kind of complex submerged foils envisioned for the contemporary PC(H). BuShips dismissed the argument that a hydrofoil was particularly vulnerable to damage when hitting underwater objects; fixed foils were quite sturdy. Given the hydrofoil's potential to achieve very high speed, it was reasonable to pare its hull weight to the minimum required for the support of one 40mm and one 20mm gun and to enclose its bridge for reduced wind resistance.

As in the past, efficient but light engines were the key to high performance, most critically in the hydrofoil. The PT boat's 2,500-bhp Packard was rejected as unavailable, the plant having long since closed down. For speeds up to 35 knots, Preliminary Design suggested multiple 1,300-bhp Mercedes diesels. For higher speeds, it proposed the British Proteus gas turbines already used in "Brave"-class patrol boats.

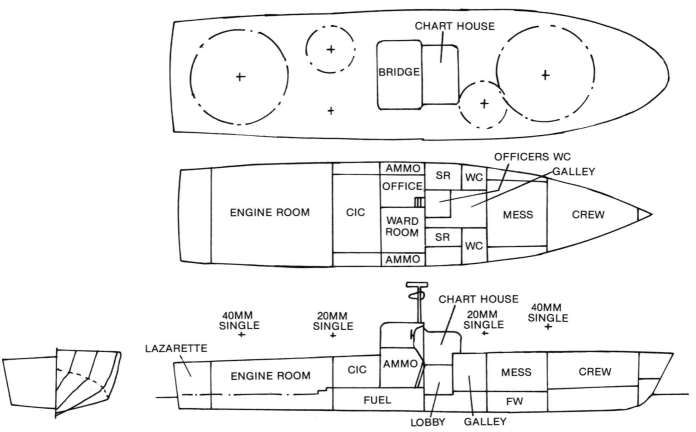

A proposed river patrol craft, similar to PT 809, 1 May 1959. Dimensions in feet-inches were 93-4 (waterline) x 25-9 x 11-7 (depth) x 5-10 (draft with gas turbine engines). Displacement was 95 tons or, with a diesel power plant and 6-foot draft, 100 tons. The gas turbine alternative produced 5,000 shp and incorporated a 500-bhp cruise diesel for a maximum speed of 38 knots and an endurance of 1,000 nautical miles at 10 knots (240 at maximum speed). The diesel plant produced 3,900 bhp for 29 knots; endurance was 1,000 miles at 10 knots, 525 at 29. Armament consisted of two 40mm and two single 20mm guns plus a 106mm recoilless rifle, indicated on the drawing by a cross to starboard, abaft the superstructure. (Norman Friedman)

The best domestic diesel, the Cummins, was much too heavy—the maximum of four would drive a boat at only 23.5 knots. Packard was working on a new 900-bhp diesel, but it had not yet entered service and so could not be considered reliable.

Fast patrol boat alternatives developed at this time are described in table 10-2.

Meanwhile, in May 1959, the SCB asked BuShips to evaluate the new Norwegian *Nasty*-class torpedo boat. This private-venture design was developed in 1956 by analysis of earlier fast planing and displacement hulls. The prototype was completed in the fall of 1957, and the Norwegians ordered twelve the following spring after successful trials. They went on to build eight more for their own navy, six for Greece, two for Turkey—and fourteen for the U.S. Navy. They also produced components for six boats built by John Trumpy of Annapolis.

The hull had an unusually wide beam, about a third of the length, and a V-bottom with a hard chine aft and a relatively round bottom forward to reduce slamming and increase rough-water speed. Even so, operating experience in the Baltic clearly showed that, like any other hard-chine boat, the *Nasty* rode badly in rough water. The Swedes preferred a larger boat based on wartime German practice. According to the BuShips report, a Norwegian officer who witnessed exercises involving both the *Nasty* and an ex-U.S. Elco of the Norwegian navy said that in rough water (sea state unspecified) both boats were reduced to about 15 knots. The Elco did not turn and maneuver as well as the *Nasty*.

All *Nasty*s were powered by Britain's Napier Deltic 18-cylinder diesel engine, developed largely as a reaction to British wartime dependence on the U.S. Packard. The boats could sustain 41 knots (46 maximum) in smooth water and provided an endurance of 600 nautical miles at 25 knots or 450 at 41. In Norwegian service, a *Nasty* carried two 40mm guns and four 21-inch torpedoes.

The bureau found that the *Nasty* was the most attractive of existing European fast attack boats. In

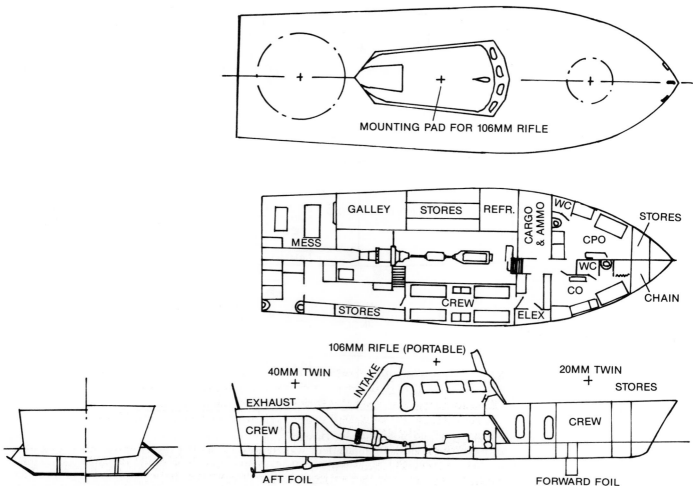

Hydrofoil river patrol craft, dated 30 January 1959. Dimensions in feet-inches were 80-0 (waterline) x 27-9 x 11-0 (depth) x 3-0 (flying), 6-0 (hull-borne). Displacement was 70 tons with 5 tons of cargo. Maximum speed was estimated as 45 knots on 3,000 shp (gas turbine), and endurance at 2,100 shp (39 knots) was 380 nautical miles. Hull-borne, the craft would run at 10 knots on a 300-bhp diesel for an endurance of 1,000 miles. Armament consisted of a twin 20mm forward, a twin 40mm aft, and a portable 106mm recoilless rifle atop the pilothouse. (Norman Friedman)

July, having read the bureau's report, Admiral Burke asked for the cost and characteristics of possible PTs. In April 1960, he ordered a squadron of fast patrol boats (PTF) bought under the FY 62 program, to be operational in 1963. They became SCB 220. The size of the squadron would be determined by cost per boat and economies of scale, but OpNav envisaged four to eight boats rather than the twelve of wartime MTB squadrons. The *Nasty* was attractive as a prototype for U.S. development, just as Power Boats' MTB had been in 1939.

Along with the PTFs, work was being done on a fast swimmer reconnaissance boat (LCSR, SCB 221). It was important both as a possible prototype for further combatant craft and as an experiment in large plastic-hull construction.[2] Characteristics written at the time suggest that the PTF and LCSR were conceived as complementary craft for amphibious operations in the context of limited warfare.

As in 1939, the chief obstacle to buying a foreign boat was political. In May 1960, Admiral Burke asked BuShips to report on the acquisition procedure, and on 2 June, he ordered the bureau to obtain one *Nasty* for evaluation. The Buy-American Act required presidential confirmation that such a foreign purchase was essential to U.S. defense; on 13 June, the letters for presidential authorization were completed, then hand-carried to the CNO the next day. They remained in the Office of the General Counsel until 14 July, awaiting a legal opinion on the Buy-American Act. The legal obstacles apparently prompted Admiral Burke to abandon the planned purchase on 23 August.

That hardly eliminated the need for a new PTF by 1963. The *Nasty* was still attractive: in August, the SCB cited its simplicity and reliability as valuable guides to any U.S. design. Modernization of the three surviving experimental PTs (810–12) was to be considered a way of providing initial numbers in min-

Table 10-1. River Patrol Boat Alternatives, 1959

	Displacement		Planing		Hydrofoil
	1	2	3	4	5
Hull	Plastic	Steel	Aluminum	Aluminum	Aluminum
LOA (ft)	65	95	95	95	85
LBP (ft-in)	61	90	93-4	93-4	80
Beam (ft-in)	18-6	20	25-9	25-9	27-9
Depth (ft-in)	9-1	11	11-7	11-7	11-0
Draft (ft-in)	5	6	6	5-10	6
					3 (flying)
Displacement (tons)	36	105	100	95	75
Trial V (kts)	20	25	29	38	43
Endurance					
(nm at 10 kts)	1,000	1,000	1,000	1,000	1,000
(nm/kts)	—	—	525/29	240/38	380/39
Machinery	2 500D	4 900D	3 1300D	2 3000GT	1 3000GT
				1 500D	1 300D
Armament (mm)	1 40	2 40	2 40	2 40	1 x 2 40
	1 x 2 20	2 20	2 20	2 20	1 x 2 20
		———————————— 1 Flamethrower ————————————			
		———————————— 1 106 Recoilless rifle ————————————			
Cargo	None	———————————— 5 tons ————————————			
		——— Commercial navigation radar, portable underwater-object-locator equipment ———			
Lead/Follow Cost (in millions of $)	0.7/0.475	1.4/1.2	3.3/2.8	3.8/3.3	3.2/2.7

Note: An increase in endurance to 2,000 miles would increase full load displacements by approximately 10 percent and decrease trial speed by about 2 knots. Use of a steel hull in schemes 3, 4, or 5 would increase the displacements by about 15 percent and decrease trial speeds by about 5 knots. Trial condition includes 2/3 fuel. Schemes 3 and 4 were similar to the PT 809 hull.

imum time;[3] after all, the timetable of the FY 62 program required cost data by 1 September and initial design proposals by early October 1960.

The SCB issued tentative PTF characteristics on 19 August 1960. They resembled those written for the experimental PTs of 1946: length was limited to 80 to 100 feet, beam to about 25, and draft over propellers to 6 (50 to 80 tons). The boats would be loaded either as deck cargo or in the well decks of LSDs. The latter made them suitable for operation with the fast amphibious force. The unhappy experience with the experimental PTs and a desire for stealth showed in a demand for a wooden hull. Gasoline engines were rejected because of the fire hazard, but speed was to be 45 knots, endurance 500 nautical miles at 25 knots (draft characteristics issued in September showed a speed of 40 knots with half fuel). Machinery would be chosen partly for ease of maintenance, using a tender or a base, and reliability under adverse conditions. Although high speed would not be required in rough weather, the boats would have to operate in seas up to state 5.

Armament would be the most powerful possible on a PTF hull of this size; it would include two torpedo racks, which could also be used to lay mines, one-man RF guns, a long-range flame thrower, and a Redeye missile launcher with a blast deflector. Tentative missions entailed the classical PT antiship role, minelaying in sensitive areas, the support and defense of amphibious operations, beach-jumping, reconnaissance, and the support of guerrillas or other small units behind enemy lines.

The PTF would carry four officers, eighteen enlisted men, seven days' worth of water and refrigerated stores, and a ten-day supply of dry stores.

In September 1960, BuShips made a quick feasibility study of installing new engines in the three PTs to achieve 45 knots and 500 nautical miles at 25 knots. At this time, each was armed with four torpedo racks and two 20mm cannon. Two single 40mm guns could replace the torpedoes at a cost of two tons and one knot. Model tests and full-scale data are summarized in table 10-3. The alternatives included two, three, and four shafts as well as all-diesel, all-gas turbine, and combined plants. BuShips suggested as most feasible the following:

- For PT 810, a three-shaft plant, with two Proteus gas turbines on the outboard shafts driving through reverse-reduction "V" gears and one Deltic diesel on the centerline with a controllable-pitch propeller. This combination could achieve 45 knots for 15 minutes. With 10.5 tons of endurance fuel and the centerline diesel uprated to 2,700 bhp for continuous operation, it could meet the 25-knot requirement. The existing 2,400-bhp Deltic could make a sustained speed of 23.5 knots, and 500 nautical miles would cost 9.7 tons.
- For PT 811, a similar plant driving it at 50 knots for 15 minutes and sustaining 45 knots. It would require 9.9 tons of endurance fuel if the center-

Table 10-2. Fast Patrol Boats, 1959

	PT 809	809 Conv	809 Conv	95-ft Conv	Hydrofoil
Hull	Aluminum	Aluminum	Aluminum	Steel	Aluminum
LBP (ft)	93	93	93	90	80
FL (tons)	93	95	88	100	70
Trial V (kts)	46	30.5	39.5	27	45
Endurance (nm at 10 kts)	650	1,000	1,000	1,000	1,000
Machinery	4 2,500 Gasoline	3 1,300 Mercedes	2 3,000 Proteus 1 500 GM diesel	4 900 Packard diesel	1 3,000 Proteus 1 300 Packard
Draft (ft)	5.8	5.8	5.6	6.1	6.0

Note: A steel hull would add 15 tons, equivalent to a loss of about 5 knots. Cargo weight of 5 tons was included, as well as total ammunition requirements. Ballistic protection was not included; it would add about 6 tons, or cost 1.5 knots.

line diesel were uprated to a continuous output of 2,600 bhp. The 2,400-bhp Deltic would achieve 24 knots and require 9.7 tons of endurance fuel.

The three-shaft plants were also chosen because they would be simpler and cheaper than new four-shaft machinery. In neither case would BuShips have used the existing shaft lines.

- For PT 812, with its much longer, narrower hull, much more power was needed to make the required high speed. With the Proteus/Deltic combination, its maximum 15-minute speed would be only 41 knots, its endurance speed 22 knots. The most feasible four-shaft arrangement would be two Proteus turbines inboard and two Deltics with CRP propellers outboard. This would drive the boat at 42 knots maximum, and it would require 12.1 tons of endurance fuel with two 1,600-bhp diesels.

In each case, the weight of the diesel plus its fuel was less than that of a gas turbine plus its fuel; hence, the use of the Deltic as the cruise plant in a CODAG arrangement. In each case, too, endurance fuel included 0.1 ton for a ship service load of 15 kw (two 30-kw diesel generators).

The estimated cost to activate, reengine, rearm, and reequip each boat was $1,250,000. Even then the boats would be unsatisfactory, too delicate for rough service, with aluminum hulls subject to corrosion since they had been designed before aluminum welding had been perfected. It seemed best to shift to a new PTF design.

Once again, the key considerations were propulsion and armament. As in the PT reconstruction studies, a CODAG turbine was most attractive, although a two-shaft 6,000- to 8,000-bhp diesel plant was also considered. There being no suitable domestic engines, the new British Proteus looked like a good candidate.[4] It produced 3,100 shp, 3,500 bhp for a short time. The 25-knot endurance speed could be attained by a cruise/maneuvering diesel clutched to each shaft and declutched when the gas turbine was cut in to achieve the 45-knot dash speed.

At this time, the 106mm recoilless rifle was to have been replaced by a 120mm model under development, and it seemed likely that a single pintle mount would be suitable for a wide range of weapons: the 7.62mm and 0.50-caliber machine guns and both recoilless rifles.

Preliminary Design submitted proposals for new-construction PTFs late in October (see table 10-4). The gas turbine–powered scheme A, which met the full SCB requirements except in cruising at 20 rather than the required 25 knots, would cost $2.4 million per boat ($3.5 million for the lead boat); the SCB staff, however, wanted to limit cost to $1 million ($1.5 million for the lead boat). Preliminary Design proposed the $1.4 million ($1.9 million lead ship) scheme B, which it compared with the *Nasty*. As an alternative, aluminum could be substituted for wood to meet the 45-knot speed requirement (scheme C). Finally, scheme D, powered by three diesels, would cost $200,000 more than either B or C and would not meet the 45-knot requirement. Preliminary Design offered the new British *Ferocity* as a foreign alternative to the *Nasty*. Note that the *Nasty* did meet the basic SCB requirements in being wooden and powered by two non-gasoline engines (Deltic diesels).

These craft were convertible gun or torpedo boats. As gunboats, they would be armed with a single 40mm gun forward and twin 20mm en echelon aft. As torpedo boats, they would have one twin 20mm forward, one twin 20mm aft flanked by torpedo racks. Each would carry four officers and eighteen crewmen, but at minimum habitability standards. The listed speeds did not include the effect of the 4 percent future-development margin usually imposed by the SCB, which would cost about 1.5 knots.

Although the SCB wanted a wooden hull, BuShips preferred aluminum, which would save about 8 tons and thereby gain about 3 knots. It claimed that most

Table 10-3. Model Tests of PT Performance for Modernization, September 1960

	15 Kts	25 Kts	35 Kts	45 Kts	55 Kts
Best Hull, Bare	812 (0)	812 (0)	812 (0)	811 (0)	811 (0)
(% inc. in EHP	809 (41.4)	809 (13.0)	811 (2.6)	809 (1.4)	809 (0.2)
over best hull)	811 (63.3)	811 (23.4)	809 (2.9)	810 (3.9)	810 (2.9)
	810 (66.1)	810 (32.7)	810 (12.7)	812 (17.2)	812 (18.0)

Original Data from Standardization Trials

	25 Kts BHP	45 Kts BHP
809 (185,600 lbs)	2,500	8,250
810 (171,900 lbs)	2,700	10,000
811 (182,700 lbs)	2,600	8,400
812 (185,750 lbs)	3,200	14,000 (est)

Table 10-4. PTF Sketch Designs, October 1960

	A	B	C	Nasty	D	Ferocity
Length (ft)	89	81	81	80	89	91
Beam (ft)	25	24.5	24.5	24.5	25	24
Draft (ft)	6	6	6	6	6	6
Displacement (tons)	85	82	74	73	93	95
Machinery	2 GT 2 Diesels	2 Diesels	2 Diesels	2 Diesels	3 Diesels	2 GT 2 Diesels
SHP	9,400	6,200	6,200	6,200	9,300	8,500[2]
Speed (kts)	46 (1/2 fo)	42	45	45	44 (1/2 fo)	45 (at 80 tons' disp.)
Endurance (nm/kts)	500/20	500/25	500/25	600/25	500/25	400/42
Hull	Wood	Wood	Aluminum	Wood	Wood	Wood
Complement (off/enl)	4/18	4/18	4/18	22 Total	4/18	14 Total
Armament	2 Torps 1 20mm 1 Flame thrower 1 Redeye 1 Smoke generator	1 40mm 2 20mm 1 Smoke generator or 2 20mm 2 Torps	— 4 21-in Torps —	— —	1 40mm 2 20mm 1 Smoke generator or 2 20mm 2 Torps 1 Smoke generator	1 40mm 4 21-in —
Cost (millions of $)	2.4	1.4	1.4	1[1]	1.4	1.6

1. Delivered in USA, built in Norway with USN electronics and minor modifications.
2. Built in the United Kingdom, delivered with USN electronics and minor modifications.

of the problems with the experimental PTs had been solved and that aluminum hulls would be at least as easy to repair as wartime-type double-diagonal mahogany. For example, the bureau pointed out that welding techniques had been improved, so that a weld retained 95 percent of the parent metal strength, as opposed to a loss of 35 to 40 percent in the experimental PTs. This was by no means a universal view. A survey of European navies in December 1961 showed a preference for wood, except for navies that had to operate in ice and therefore preferred steel. Wood was stronger, needed no insulation, and could be coated with fiberglass to eliminate its one major drawback, moisture absorption.

Given the long hiatus in U.S. PT development, BuShips suggested that the 1963 squadron be limited to four boats, three modernized, one new, for minimum initial cost and evaluation before production. The bureau also advocated a simple, inexpensive battery that could act as a weight and space reserve for future lightweight and flexible weapons.

To give an example of the armament considered at this time, in January 1961, personnel at the SCB working level discussed the following:

- two Mark 37 torpedoes, two racks
- one twin 20mm, one single 40mm gun
- four 7.62mm machine guns
- two 40mm grenade launchers
- one 81mm mortar, mount Mark 2
- twelve Redeyes
- two 75mm recoilless rifles
- two 0.50-caliber machine guns

In 1959, BuShips had argued that no conventional small displacement or planing boat could match a hydrofoil in rough water (sea state 4 or more). Now BuShips sent the SCB a series of alternative patrol boats (table 10-6), including two hydrofoils, one based on the new PC(H) described in chapter 8, a 95-foot round-bottom boat, and a 70-foot hard-chine boat based on a new projected 70-foot missile-range patrol boat for rough water.

Table 10-5. Armament Alternatives for PT 810–12

	(1)	(2)	(3)
40mm Single Mount	1	1	1
7.62mm Machine Gun	4	4	4
Redeye Missile (SAM)[1]	12	12	12
40mm Grenade Launcher	2	2	2
0.50-cal Machine Gun	4	4	4
Torpedo Racks	4	4	4
81mm Mortar	1	1	1
Mk 37 Torpedo	4	—	—
106 Recoilless Rifle	—	1	1
0.30-cal Machine Gun	—	—	3

1. Three per shipping box.

BuShips warned that, despite its good fuel economy, the Deltic was extremely complicated and so could be difficult to maintain. That proved accurate when boats were heavily used in Vietnam. The bureau liked the British Proteus gas turbine—which was already at sea in *Brave*-class fast patrol boats and on order for the PC(H)—but argued that the U.S. Navy should not order any foreign-built engine in large numbers. The alternative American-built Curtis-Wright and Solar gas turbines lacked sea experience.

For NATO or SEATO, the *Nasty* was the least expensive of all available fast planing boats, providing foreign-built equipment and engines were acceptable. The bureau was loath to buy a *Nasty* solely for evaluation, as it believed that existing U.S. technology could achieve better performance. With the CNO's tight schedule, it turned out to be more important that the *Nasty* was already in production, hence immediately available.

Work on PTF characteristics continued through late April 1961. At that time, there was no PTF in the FY 62 program, and none was planned for FY 63. However, the PTF was identified with counterinsurgency, and that became increasingly important and urgent as the situation in Vietnam deteriorated. The LRO group suggested that the PTF be included in the FY 63 program and that the first boats be bought during 1962. LRO emphasized MAP considerations, particularly simplicity and easy maintenance.

According to the anonymous BuShips minute-taker in an SCB meeting,

> At this point there occurred what might be called "a revolting development." Capt Barnes, who had experience in PTs during WW II, asked if he could have the floor for a few minutes. He then launched into a discussion of realistic requirements and pointed out that during WW II there were only two instances where PTs were required in conventional PT roles.... Some 500 [sic] of this type craft had been built, and it was his personal opinion that a great sum of money had gone down the drain.... We are about to make the same mistake all over again.... Initially it was our intention to furnish the deep-water capability and to expect our allies to come through with their own craft for inland and coastal operations. However, nothing has happened and the U.S. must pick up the ball. He suggested that it is time to look at the entire picture again in terms of what is actually required. During this discussion I inferred from a side remark that the Navy may be investigating means whereby PTs may be purchased abroad....

Captain Barnes emphasized that the boat would have to fight a primitive kind of war requiring size rather than speed. Like Commander Harlee a decade before, he doubted the value of very high speed and therefore of a boat comparable to the *Nasty*. It would be more worthwhile to guarantee 30 knots after prolonged operations. After all, the World War II Higgins had been good for only 28 knots after such operations. A boat would have to flee an area only if it had been defeated.

LRO admitted that it wanted a long-range coastal (not riverine) motor gunboat much more than a true PT. Its next priority was a river patrol boat, and only after that a PT. PTF work would continue as an alternative to the motor gunboat, but the SCB members doubted that any PTF would be built.

The shift to work on a slower, more heavily armed and less complex motor gunboat, the PGM, was formalized a few days later. However, interest in PTFs revived in April 1962 as the Soviets transferred motor torpedo boats to Cuba. There was some well-founded concern that they would be followed by missile-firing *Osa*s, which could attack U.S. warships operating in the Caribbean. Now the PTF was seen primarily as an anti-PT weapon.

The available alternatives were a fast, 30- to 35-knot motor gunboat with a 3-inch gun (which became the *Asheville*) or a hard-chine fast attack boat like the *Nasty* or the British *Ferocity* or *Brave*. The quickest alternative, available within two months if it was pushed, was clearly to reactivate the two gasoline-powered PTs, 810 and 811, without giving them new engines. PT 812 would take one and a half to two years because it would receive a new engine and the hull would be repaired. PT 810–11 were reactivated at Philadelphia in October 1962 as PTF 1–2. They were shipped to Subic Bay in January 1963. Each was armed with two single 40mm and two twin 20mm guns.

Table 10-6. BuShips Patrol Boat Designs, 1959

	PCH-1	Hydrofoil	Round Bottom	V-Bottom
LOA x B (ft)	116 x 32	85 x 26	95 x 21	70 x 20
Draft (ft)				
Foils Up	6	5	6	9[1]
Flying	8	6		
Maximum Speed (kts)	—	55	50	45
Sustained Speed (kts)	45	50	45	40
	2 Proteus	2 CW GTs[2]	3 CW GTs	2 Solar
	1 Packard Diesel	—	—	10 MV
Maximum SHP	—	8,000	12,000	2,000
Sustained SHP	5,800	7,000	10,500	—
Fuel (gals)	6,450	7,000	10,500	3,150
Endurance	2,000/12	740/15	740/15	630/12
(nm/kts)	500/45	600/50	600/45	540/40
Armament	1 0.50-cal	1 3-in/50	1 3-in/50	2 20mm
	4 Mk 44/46	1 40mm	1 40mm	1 Mk 44/46
	SQS20 (VDS)	6 Mk 44/46	6 Mk 44/46	—
Full Load	108	75	85	35
Complement	13	20	20	12
Hull	Aluminum	Aluminum or reinforced plastic		
Cost (millions of $)[3]	4.7	3	3.5	1

1. Foils nonretracting.
2. Curtis-Wright gas turbines.
3. All 1959 prices.

The only other near-term possibility was to buy Norwegian *Nasty*-class boats already under construction under a partial U.S. subsidy. They were being completed without their planned armament and electronics. Objections included the cost in foreign exchange (gold outflow) and likely delays due to the heavy workload in Norwegian yards. However, to use the Norwegian plans, to which the U.S. Navy had rights under the subsidy contract, would have required redrafting for English rather than metric units, which could take as long as the completion of a fresh contract design for the new fast motor gunboat.

Two Norwegian *Nasty*-class boats, the just-completed *Skrei* and the partly completed *Hvass*, were bought under the FY 63 program and became PTF 3–4. Acquired in December 1962, they went to the West Coast the following August, then to Subic Bay. They soon moved to Vietnam to support SEAL and other semi-guerrilla operations, having been requested by the CIA as early as January 1963. In December 1963, Secretary of Defense Robert McNamara ordered PTF 1–2 transferred to Subic Bay and four more PTFs for Vietnam. He estimated that the United States needed a total of ten to fifteen more *Nastys*, including eight for Cuban areas.

Senate staff members questioned the need for such foreign purchases. The navy officially replied that there was no existing U.S. alternative, but that no purchases were planned beyond the eight boats for the Cuban area; the new hydrofoil gunboat, the PGH, would replace them.

The four Vietnam boats were acquired under the FY 64 program on 1 March 1964, becoming PTF 5–8. They were in service at Subic Bay by mid-1964. The acquisition of McNamara's eight additional boats, PTF 9–16, also in FY 64, was announced on 2 September 1964. The U.S. boats differed from their Norwegian counterparts in being equipped as single-purpose gunboats without provision for torpedoes and in achieving longer range by carrying extra fuel in compartments otherwise reserved for berths.

On 26 January 1965, PTF 3–9 and 14–16 were leased to the South Vietnamese navy for SEAL operations along the North Vietnamese coast. They typically carried twelve-man Vietnamese SEAL teams. The boats, except for six sunk—PTF 4 in 1965, PTF 8–9, 14, and 15–16 in 1966—were returned in 1970.

In July 1964, after the Tonkin Gulf incident, McNamara suggested that the United States buy four more *Nastys*, under the FY 65 program, but that was not possible. Building *Nastys* abroad could be justified only by the urgency of the escalating war in Vietnam. Otherwise the legal requirement to buy American had to be satisfied. In the case of the PTF, that was done in two ways. First, additional *Nastys* were all assembled in the United States by Trumpy, albeit from Norwegian parts. Second, the navy bought

One of the first *Nasty*-class PTFs at Pearl Harbor. These craft were bought specifically to take agents into North Vietnam, a concept formally suggested in September, 1962. The first *Nasty*s were placed in service at Little Creek in May 1963; they arrived at San Diego in late August and at Pearl Harbor in September. At that time, the Pacific Fleet demanded that their armament be increased by two 3.5-inch (super bazooka) rocket launchers and two or more flame throwers, none of them visible in this photograph. A PTF could carry a ten-man SEAL detachment.

PTF 3, the first *Nasty*, in Pearl Harbor, September 1963.

PTF-1, formerly PT 810, in 1963. Two 0.50-caliber machine guns were added in Vietnam. The PTF conversion included special provision for silencing.

an alternative PTF, the *Osprey*, from a U.S. firm, Sewart Seacraft, which had designed and built the famous Swift boats used in Vietnam.

Trumpy completed six boats under the FY 67 program: PTF 17–20 in 1968, and PTF 21–22 in 1970.

At this time, Sewart was already exporting a patrol boat version of its 85-foot aluminum commercial cruiser. Some time in 1966, Sewart proposed a combined gas turbine/diesel version of the 85-footer built for Jamaica, which was expected to achieve 40 knots. The navy rejected it, but Sewart did build the *Osprey*, a 95-foot aluminum patrol boat using the same Deltic engines as the *Nasty* and a similar planing hull. The broad-beamed planing hull form of the *Osprey* was patterned after that of the *Eagle*, an experimental gas turbine–propelled 53-footer that Sewart had built some years earlier.[5] The *Osprey* and the *Nasty* shared the same armament. Four, PTF 23–26, were purchased under the FY 67 program and delivered in 1968. Although all were sent out to Subic Bay, they saw little service because their direct-drive propellers eroded much too quickly at high speed. In 1978, PTF 26 was disarmed and converted to a gas-turbine trials boat with 5,000-shp Garrett 990 engines.[6]

The PTFs were originally armed with single 40mm guns fore and aft plus two single 20mm (1,920 rounds). In most cases, the forward 40mm was replaced by the standard Vietnam War combination 0.50-caliber machine gun and 81mm mortar. There were 4,000 rounds of 0.50-caliber ammunition; 16 rounds of ready-use 81mm HE and 8 illuminating, and in the magazine, 48 rounds of HE and 24 illuminating. The single 40mm gun aft typically had 256 ready-use rounds and 832 in the magazine. Space and weight were reserved for nine Redeyes, but these weapons were never carried.

By the end of 1969, the approved ultimate battery was two single 40mm and two single 20mm guns. Of the surviving ships, PTF 3 and PTF 19–23 all had two 40mm guns, and others such as PTF 18 were later refitted. PTF 5–7 were unique in that their ultimate battery included no 20mm guns at all.

The trunnion-mounted (Mark 2) 81mm PTF mortar, which looked like a conventional naval gun, first appeared aboard the four experimental postwar PTs; it was rejected at the time. All PTF weapons were aimed by eye, and the mortar could place illumination shells beyond a target ship without creating

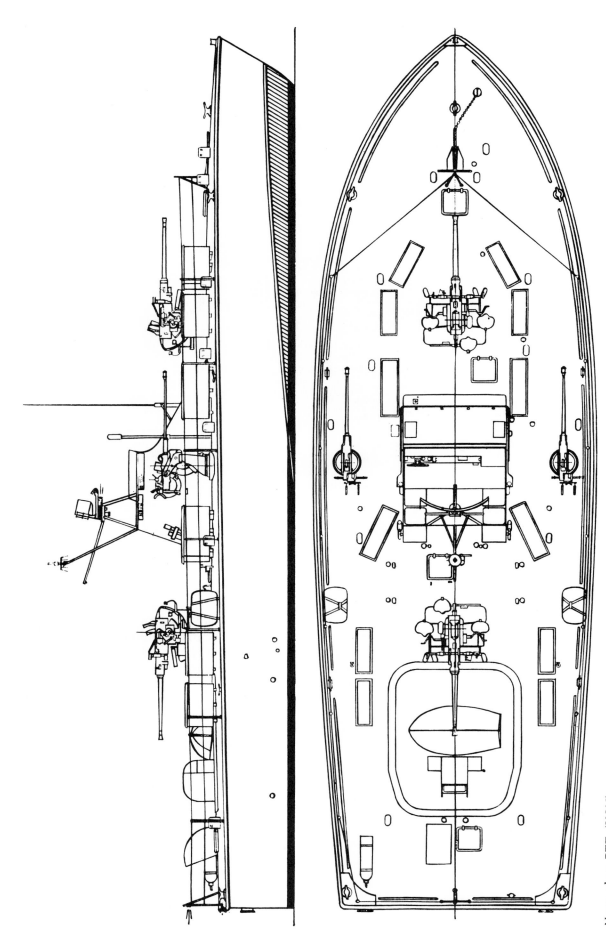

Nasty class PTF. (USN)

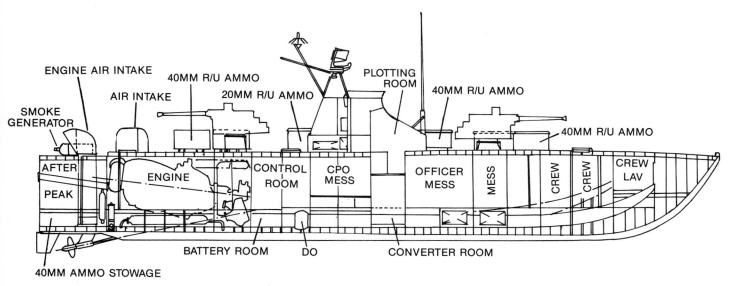

Inboard profile of an early PTF, as delivered. The large boxes around the two 40mm guns contain ready-service ammunition; the smaller boxes are filled with 20mm ammunition. (Norman Friedman)

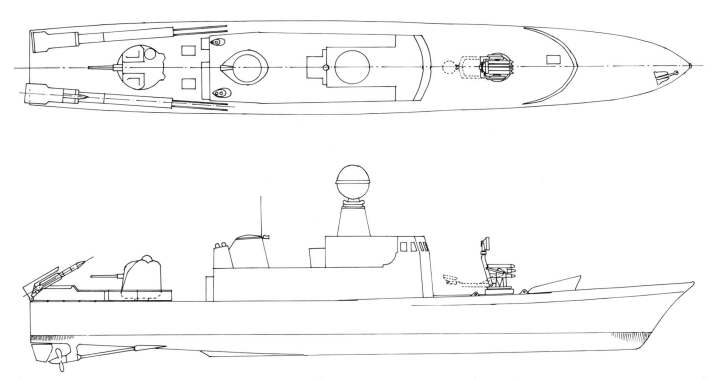

There were several alternative schemes for arming *Asheville*-class fast gunboats with missiles. This appears to be one of several undated studies of fast attack craft armed with a surface-to-surface version of the Tartar AA missile, developed by General Dynamics for the West German navy. Other craft studied included the *Nasty* and a German fast attack boat. The missile forward is Sea Mauler, a projected point-defense weapon initially developed for the army. It was canceled in December 1964; this drawing probably dates from earlier in that year. The Bullpup-Tartar launch system was later adapted to the Standard interim surface-to-surface missile. (Norman Friedman)

PTF 23, the prototype Osprey.

At Little Creek in 1973, PTF 6 has the later PTF arrangement, a combination 81mm mortar/0.50-caliber machine gun replacing the forward 40mm gun. Initially, the 40mm guns were removed to provide weight and internal space for additional fuel, but ultimately, mortars were installed to compensate somewhat for the loss of firepower. Because of the PTFs covert role, they were formally transferred to the Vietnamese navy on five-year lease. Although the boats were in the charge of American officers while in Danang, only South Vietnamese personnel operated them when they went North. Later *Nasty*s (PTF 9–16) differed internally from the first boats: the last eight Norwegian-built craft had silenced engines (the noise level reduced by 50 percent) and larger fuel tanks for longer range, 900 nautical miles at 35 knots. Note PTF 6's new radar and bridge windshield.

an aim point by muzzle flash or flame, as in the case of any flat-trajectory weapon. A mortar could keep enemy heads down on a beach where SEALs were landing. In Vietnam, it had the additional advantage of firing a low-velocity incendiary (White Phosphorus) shell that could destroy a wooden junk at 250 to 300 yards. At higher velocity, it passed through without exploding.

As of 1964, BuOrd was achieving a range of 4,000 yards, compared with 2,350 for the standard army weapon. Much greater ranges seemed achievable.

PTF 13 remained in the United States as a training and experimental boat. It fired Redeye missiles against Firebree drones in November and December 1966, the first operational boat to destroy a drone with that weapon. In August and September 1967, she evaluated a proposed new armament outfit consisting of a power-driven 40mm gun, a Hispano-Suiza 20mm, which doubled the rate of fire and the reliability of the existing World War II weapon, and an Oerlikon 81mm mortar, which tripled the range of the existing weapon. Results were impressive, but the new suit was not adopted. PTF 13 also evaluated an experimental twin Oerlikon rocket launcher.

As BuShips had observed in 1959, the *Nasty* was limited in rough-water performance. By 1964, the strike warfare division of OpNav had been studying hydrofoils for two years, seeing them as the ideal successors to conventional planing boats; the hydrofoils added a rough-water, high-speed capability at a cost per boat about twice that of a *Nasty*. As a result of initial studies made in May 1962, the division asked BuShips to make a feasibility study of a 50-ton boat with a 5-ton payload that would move faster during cruising and combat than a planing boat. It would also have a low-speed, hull-borne mode for island hopping or on-station endurance, and it would be small enough to be hoisted aboard larger ships for long-distance deployment. This gunboat would have the best possible armament, but it would be small and simple for MAP and carry minimal electronic equipment.

BuShips claimed that it could assure combat superiority up to sea state 3. In sea state 1, the hydrofoil would enjoy a 5-knot advantage, in 6-foot waves, a 25-knot advantage. For example, its bow acceleration (pitching motion) would average only one-eighth that characteristic of a planing boat (4 G).

The requirement for a counterinsurgency hydrofoil can be traced back to a report by the joint DoD/CIA counterinsurgency committee in December 1962. However, the Kennedy Administration's pressure for a hydrofoil gunboat was a direct reaction to the Cuban Missile Crisis: the *Nasty* seemed best for covert operations, partly because its wooden hull was a poor radar target, but its margin over a Soviet-type torpedo or missile boat was thin at best.

OpNav announced a requirement for a hydrofoil gunboat in 1963. BuShips delivered a feasibility study in May 1963, showing a performance considerably better than that of a planing boat.

In February 1964, McNamara requested that the program be reshuffled so that two hydrofoil PTFs (PTH) could be bought with FY 64 funds. Afterwards, he was identified with the program. Secretary of the Navy Paul Nitze ordered an urgent study of tradeoffs between the PGH and PTF, and that April, OpNav's strike warfare division suggested a force level of forty-eight craft: sixteen PTFs and thirty-two hydrofoil PGMs (PGHs). It was not, however, formally entered into the navy program. Similarly, a program change proposal submitted in May showed a force level of twenty hydrofoil PGMs, the first two to be delivered in FY 66 and funded under FY 64.

Secretary McNamara signed the program change proposal on 18 March 1964; on 30 May, he authorized procurement of two PGHs. At that time, plans called for two FY 64 prototypes, followed by production craft from FY 66 on. But during the summer of 1964 the comptroller of the navy deleted the FY 66 boats on the ground that the FY 64 prototypes had to be evaluated before further boats could be ordered. Production, therefore, tentatively slipped to FY 67. In June 1964, senior admirals on the CNO advisory board commented that, despite McNamara's personal interest in the PGH, it would be well to keep in mind that they were still experimental craft. The navy would best avoid buying any more, at least until its two experimental hydrofoils, PCH and AGEH, had been fully tested and operational concepts proven.

As a result, only the two prototype PGHs, the *Tucumcari* and *Flagstaff*, were built. Being the lineal descendants of the PTF, they are described here before the large, fast PGMs (*Asheville* class) that actually preceded them in time.

The SCB began work on a hydrofoil PGH in January 1964. Missions included countering enemy PTs and missile boats, intelligence reconnaissance and covert operations, support of air-surface surveillance, and support of U.S. and indigenous counterinsurgency forces, including guerrillas and SEALs.

In cold war, boats would be deployed as follows:
- Western Pacific: primarily Vietnam counterinsurgency, clandestine intelligence or sabotage, and support contingency operations
- Caribbean: close-in surveillance, intelligence, support of counterinsurgency, and clandestine operations
- Mediterranean: support contingencies
- U.S. coasts: support missile-range operations, surveillance, and harassment of Soviet intelligence-gathering trawlers

The small gunboats were initially armed with a single 40mm gun forward, two twin 0.50s, and an 81mm mortar aft. The *Tucumcari* tested a twin Emerson Electric EX-73 (84MX) mount, originally developed for the armored troop carrier. She carried two 20mm Mark 16 guns, with ammunition beneath the mount. Although the EX-73 could be controlled remotely, the version in the *Tucumcari* was locally controlled. A mortar was originally carried in its position.

The PGH was always primarily a PT-killer, its weapons chosen accordingly. BuOrd considered the existing 40mm gun ineffective and, at 8,500 pounds in its hand-worked version, too heavy. It preferred missiles. The only ones available were the French SS-11, comparable to a 5-inch shell, and SS-12, comparable to an 8-inch shell. Either could be expected to sink any PT it hit. The army had bought the SS-11 as an antitank weapon, but BuOrd considered its fuse defective and expensive to replace.

Both missiles were guided visually, the operator transmitting commands along a single wire. That could confer spectacular accuracy, 3 mils at 3,000 yards, but it required some form of illumination at night, which in turn required a mortar. BuOrd expected to mount four missiles on each side of the PGH. The French offered four SS-11s at about 4,000 pounds, and BuOrd expected to do better. Later it proposed four SS-12s on one mount and four more in a ready-service magazine, for a weight of 5,600 pounds plus 4,000 for a two-man director. Maximum ranges were 3,000 yards for the SS-11, 6,000 for SS-12. Cost per round was $1,600 and $4,000, respectively.

Op-03 backed the missile, suggesting that Congress would never buy a PGH unless it was armed with a missile or a homing torpedo; guns were passé. Those at the flag level of the SCB disagreed. The new missiles had never been tried at sea. The SCB working level did decide to propose operational evaluation, which the SS-12 adherents considered the kiss of death. The missile was never adopted, although early tests at Point Mugu were extremely encouraging.

As a result, preliminary characteristics issued in June showed something much closer to the PTF battery: one 40mm gun, with space and weight for replacement by a suitable missile in the future, two 0.50-caliber machine guns, and an 81mm mortar for night illumination. A British study of 1955 was cited to show that the 40mm was the minimum weapon required to inflict serious damage on an enemy patrol boat. Within the 10,000 pounds available, it would eventually be possible to fit a launcher and eight SS-11/12 missiles. Weight was also available for twenty Redeyes, six in night configuration.

The 40mm would have to be hand-operated, because the lightest U.S. director, the Mark 87, weighed 5,000 pounds and the power-operated gun, 4,200. On the basis of British and U.S. experience, the 40mm was considered a single-shot weapon, although its accuracy was poor. The only other weapon considered at this time, the recoilless rifle, was rejected for its dangerous back blast.

The characteristics called for both maximum seaworthiness and minimum complexity to make the boat suitable for MAP. The two requirements had to conflict, since good sea-keeping required the complex submerged foil system. Pressed, the SCB chose sea-keeping, rejecting a naval version of the principal

Table 10-7. The PGH Designs, 1965

	Grumman Flagstaff	Boeing Tucumcari
Length (oa, ft-in)	74-0	71-0
Beam (ft-in)	21-0	19-6
Draft (foils extended, ft-in)	13-5	13-11
Draft (foils retracted, ft-in)	4-3	4-5
Full Load (tons)	57	58
Machinery		
On Foils	RR Tyne	Bristol Proteus
Bhp	3,150	3,100
	Super-cavitating propeller	Water jet
Hull-borne	GM 250 bhp(2)	GM 160 bhp(1)
Configuration	Conventional	Canard
Weight on Forward Foil (%)	70	31
	Prop in nacelle at intersection of after strut and foil	Anhedral foils to facilitate turning

Note: The *Flagstaff* was completed at a displacement of 67.5 tons, with a range of 560 nautical miles on foils, or 2,000 hull-borne. In 1970–71, she displaced 72.5 tons and could achieve 400 miles on foils, 1,500 hull-borne. Her maximum continuous speed was 48 knots, her maximum speed 51 knots. She could fly in seas worse than sea state 4.

commercial hydrofoil, the Italian Supermar surface-piercing boat, in favor of the submerged-foil type. It expected the 27- to 50-ton PGH to maintain 42 knots in sea state 4, while a 450-ton fast gunboat, the SCB 229 PGM, would have to reduce speed to 26 to 28 knots. This performance was justified with the argument that the PGH would encounter seas up to sea state 4 80 to 85 percent of the time and seas up to state 3 70 to 75 percent of the time.

It was difficult to decide on a particular maximum speed. Soviet PTs were rated up to 50 knots, U.S. PTFs up to 47—in smooth water. The PGH could afford a marginally slower smooth-water speed because it could outrun such craft in the moderately rough conditions that were more likely to exist. The preliminary requirement was 54 knots in smooth water with 60 percent fuel. Endurance was initially specified as 500 nautical miles at 48 knots foil-borne; later, in June 1964, it was increased to 700 foil-borne.

Like a PTF, the PGH had several secondary missions. It was expected to support beach jumpers and UDTs. The standard beach-jumper detachment was one to four men, and their equipment would not exceed 8,000 pounds. UDTs needed troop space, demolition materials, and 120 kw of generating power.

Preliminary Design began a detailed feasibility study for a 57- to 59.4-ton PGH after the March SCB meeting. As in the case of the PTF, the choice of an engine dominated the design. Only three engines were available in the required 3,000-shp range: the Pratt & Whitney FT-12 (3,000 shp continuous/3,400 shp for takeoff), a Lycoming (1,500/1,625 hp, but not yet approved for service use), and the British Proteus (2,800/3,400). The advantage of the Lycoming was that its exhaust and drive shaft were at opposite ends. The FT-12 had both at the same end, but was slightly lighter, though more expensive, than the equivalent pair of Lycomings. The Proteus might be difficult to import under buy-American legislation.

All of these engines were heavier than expected, and propulsion weights grew as PCH data were assessed. Thus a 51-knot burst speed (at a 15-minute rating) had to be accepted where 54 had been sought. The boat was expected to cruise at 45 to 48 knots. More generally, the addition of one ton could be expected to cost a half to three-quarters of a knot.

The BuShips feasibility design is described in table 10-8.

Table 10-8. BuShips PGH Feasibility Study, 1964

Length	80 ft
Beam	22 ft
Draft	4 ft (no foil)/13 (at rest)/4 ft 5 in (foil-borne)
Displacement	57.5 tons (full load)
Machinery	3,000 hp (gas turbines) = 48 kts continuous 8–12 kts hull-borne
Endurance	500 nm/48 kts, 2,000 nm/8 kts
Complement	1 off/12 enl
Payload	10,000 lbs of weapons or ammunition; fuel adds 200 nm; can carry 22 troops and equipment, or beach-jumper detachment with equipment
Cost	Estimate $4.4 million (lead ship), 3.5 million (follow ship)

The bureau wanted to use its feasibility design as the basis for an in-house contract design. Several other alternatives were considered: an existing boat, probably a derivative of the Supramar PT-50, could be adapted, a design competition could be held, or two prototypes could be built and flown off. Requests

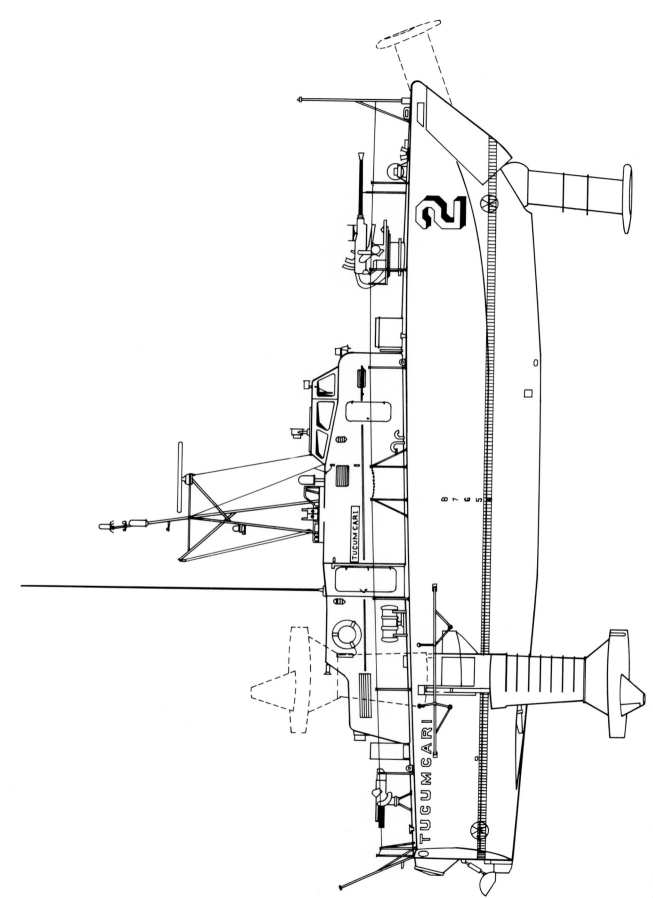

The *Tucumcari* as completed. Dashed lines indicate the position of retracted planes. (A. D. Baker III)

Perhaps the most spectacular achievement of the hydrofoil gunboat program was the installation of an army-type 152mm turret aboard the *Flagstaff*.

for proposals were sent to seven firms in July 1965. This procedure was radically different from the usual one, in that contractors were responsible only for meeting SCB characteristics without any detailed preliminary design to guide them. Only Boeing and Grumman, both of which were already designing and building hydrofoils, responded. Their proposals were quite different, although estimated costs were similar. BuShips decided to build both and test them competitively—an aircraft-style procurement choice. PGH 1, the *Flagstaff*, and PGH 2, the *Tucumcari*, were ordered in 1966.

The winning Boeing and Grumman designs are described in table 10-7. Boeing followed the subcavitating canard foil arrangement of the earlier PCH *High Point* for its *Tucumcari*. The main foils were given anhedral (negative angle) to reduce their tendency to ventilate in a banked turn. Propulsion, foil- or hull-borne, was by waterjet, with inlets at the strut-foil junction. Control was achieved with flaps in the main foils.

Both hydrofoil gunboats were deployed to Vietnam in August 1969 aboard the LSD *Gunston Hall*. They participated in the coastal interdiction operation, Market Time, based at Danang, rotating duty every five days, one on ready alert and the other on standby or carrying out maintenance. Patrols gen-

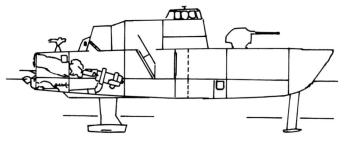

Inboard profile of the *Tucumcari* showing the gas turbine aft, exhausting through the transom and driving a waterjet pump. Both hull- and foil-borne waterlines are indicated. Armament is a 40mm gun forward and an 81mm mortar aft. (Norman Friedman)

erally lasted six to ten hours. The boats would use high speed to reach distant contacts. They typically covered 190 to 270 nautical miles per patrol (at sea states up to 5), and upon returning to the United States in February 1970, they had covered 14,843 miles at an average of 43.5 knots. The main lesson was that the new electronic autopilot, which made the rough-water hydrofoil practicable, was indeed reliable and suitable for forward operations—an important conclusion for the PHM program that followed (see chapter 15).

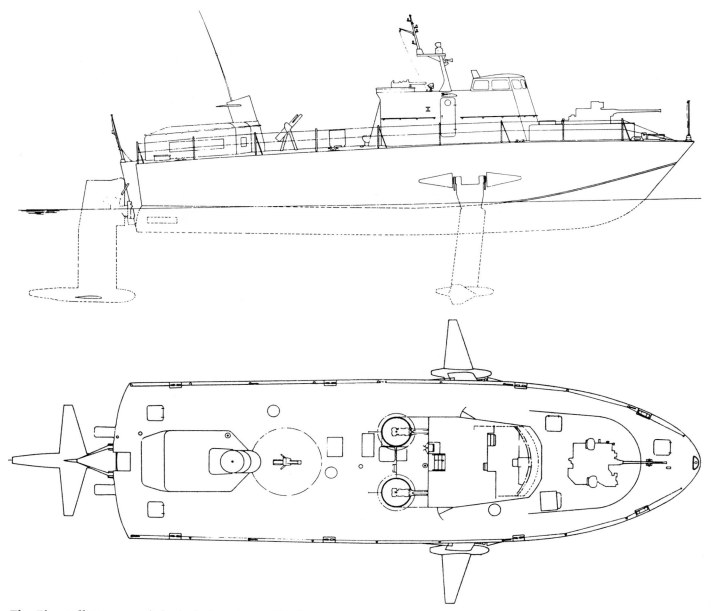

The *Flagstaff*, Grumman's hydrofoil gunboat. (USN)

The *Tucumcari* was deployed to Europe in 1971 as a demonstrator for NATO's PHM program. In November 1972, while operating at night with amphibious forces, she ran aground off Puerto Rico and had to be stricken. Her hull was sent to the Naval Ship Research and Development Center for fire and structural tests.

In 1964 Boeing, the Italian government, and Carlo Rodriguez of Messina, a builder of commercial hydrofoils, formed the Italian company Alinavi to develop and market advanced marine systems. Alinavi in turn was taken over by Cantieri Navali Riuniti. The Italian navy awarded the company a design contract for the missile hydrofoil *Sparviero*, a modified *Tucumcari*. The first was delivered in 1974, and six more were ordered in 1977. They are armed with a single OTO-Melara 76mm gun and two Otomat missiles. Two planned units were canceled.

Grumman had already built a surface-piercing hydrofoil, with the main foils forward, for the Maritime Administration. The company adopted much the same split-foil configuration, with submerged foils, for the *Flagstaff*. Foil-borne, she was propelled by a supercavitating propeller driven through a right-angle

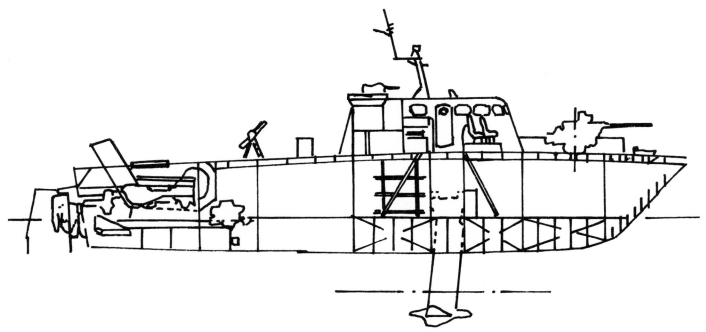

Inboard profile of the *Flagstaff*. The lower portion of the after foil strut is not visible; at its lower end, the boat carried a propeller that was gear-driven by the engines aft. (Norman Friedman)

drive, and hull-borne, by two Buehler waterjets, each powered by a GM diesel. Control was achieved by varying the incidence angle of the foils.

Perhaps the most impressive feature of the hydrofoils was their steadiness and stability at high speed, which invited tests of relatively heavy weapons, the most spectacular being the 152mm gun/launcher that had been used in the Sheridan light tank. To save weight, a plastic bubble replaced the armored top of the Sheridan turret. Three months of installation in the *Flagstaff* began in October 1970, followed by tests off San Diego for compatibility and off San Clemente for accuracy. Finally a demonstration firing was held in June 1971. The full installation weighed 7,700 pounds, not including any special fire control system. It could fire a solid shot or an antipersonnel cannister carrying ten thousand flechettes. In all, thirty-eight rounds were fired, twenty of them when the ship was foil-borne, including sixteen at full charge. The tests clearly showed that a large weapon could be carried effectively by a small hydrofoil.

The *Flagstaff* was evaluated by the Coast Guard between November 1974 and February 1975 and was transferred to the Coast Guard in October 1976. Maintenance was expensive, and she was stricken the following year.

Grumman later developed a Flagstaff Mark II. Israel, the only export customer, planned to build a prototype named the *Shimrit* at Lantana Boatyard, Florida, under subcontract and fourteen more at Haifa.

Orders were placed in 1978, but due to a deteriorating economic situation, Israel had to cancel twelve boats in 1982. The first three entered service in 1982, 1983, and 1985, of which the latter two were built in Israel. They are armed with a twin BMARC 30mm gun, two single 0.50-caliber machine guns, four Harpoons in two twin cannisters, and two Gabriel missiles.

In May 1961, the SCB asked for a new round of PGM cost and feasibility studies, citing developments in the Caribbean and Southeast Asia, that is, Cuba and Vietnam. This PGM, which might be operated by the U.S. Navy, and which therefore might be fairly sophisticated, was intended primarily for surveillance, blockade, operations against other small craft in coastal waters, and limited support of troops ashore. The SCB distinguished it from the PTF, which it considered too limited in endurance and armament payload and possibly too complex in hull and engineering design to meet all the requirements for a PGM. A new PGM would therefore have to be developed to supplement the fast patrol boat. The Coast Guard 95-footer and the export PGMs could be accepted as interim designs.

Tentative characteristics included a length of 95 to 125 feet, a maximum draft for coastal operations of 8 feet, a speed, using a *non*-gasoline engine, of 30 knots, and an endurance of 1,500 nautical miles at 17 knots. The U.S. Navy had no favored small craft weapons, but the BuWeps had developed a suitable "weapons shopping list," including 3-inch/50 and 40mm guns, nonrecoiling rifles, rocket launchers, mortars, and Redeye hand-held AA missiles. European navies had a variety of attractive weapons. For example, the new Norwegian *Storm*-class gunboat was armed with a new Bofors low-angle 3-inch/50 controlled by an integrated fire-control system designed in the Netherlands for small craft. In the spring of 1962, a joint CNO/BuWeps/BuShips/BuPers (Bureau of Personnel) group visited the Netherlands to consider the Dutch system for the U.S. PGM then under design.

The U.S. Navy was also interested in a small antiship weapon. The only available one was the French SS-11, whose value was open to question and whose performance hardly equaled that of the Soviet SS-N-2 soon to enter service in Cuba.

Tentatively, however, the new PGM would be armed with America's single slow-firing (Mark 26) 3-inch/50, as that weapon was clearly powerful enough to deal with Soviet-bloc MTBs.

BuShips' immediate reply was limited to modified versions of the existing PGMs. Although they were only modestly armed, they had the weight to take on more weapons. At a cost in speed of 1 knot, PGM 43 could be armed with a single 40mm gun, two 81mm mortars, four 40mm grenade launchers, four 0.50-caliber machine guns, and twelve Redeyes, at a cost of $690,000 (FY 63).

The more powerful Cummins Diesel VT-12-M Tandem Uprated, of 320 bhp, could be installed in this modified boat to achieve a speed of 22/20 knots and an endurance of 1,250 nautical miles at 12 knots. Putting new engines in the PGM 39 class would increase its speed to 21 knots on trial, 19 sustained, and would give it an endurance of 1,000 miles at 12 knots.

The preferred alternative was a new PGM design, which would meet the SCB requirements for speed, endurance, and firepower. It became SCB 229, the *Asheville* class. LRO supported the fast PGM as a PT and anti-PT. The group included sixteen craft in the 1961 version of its report, but admitted that this was no more than a vague estimate. Other commands also liked the PGM. For example, in November 1961, the CNO recommended the new PGM as the optimum anti-MTB boat for Korea, which wanted one, and nationalist China, which wanted ten, in preference to the scheme C PTF CinCPacFleet was then suggesting. At this time, two PGMs were included in the tentative FY 63 program.

A representative told the SCB that it was the LRO's conviction that in the future, there would be "a place for small, relatively inexpensive, lightly manned coastal patrol craft aimed primarily at possible requirements in support of limited wars and primarily used in the western Pacific areas." LRO medium-term plans showed about sixteen such boats.

Vietnam was on the LRO's mind; the PGM had a definite role. It could patrol off the coasts of countries supporting unfriendly guerrillas or subversives, keeping track of coastal shipping, blockading gunrunners, and if need be, opposing enemy craft smaller than destroyers. It could also support minor amphibious operations ashore. The LRO described the PGM as suitable for destroyer-type missions in waters where destroyers could not go or could not be risked.

Given adequate endurance and sea-keeping, the PGM would be useful outside limited-war areas (that is, outside Southeast Asia), for example, in "a blockade of shipping [in] some Latin American area."

The United States could set an important example with the new PGM. Many local navies did not care for the small gunboats; they wanted something much more impressive, something the U.S. Navy would operate itself, however impractical. Including the new PGM—the existing one—in the U.S. Navy would enhance its prestige and thus help encourage local navies to buy it. That put a premium on high performance, even though it might make maintenance more difficult.

DoD initially rejected the navy's proposal to buy two PGMs in FY 63. However, in March 1962, Secretary of Defense McNamara specifically called for a navy program of small combatants to deal with Cuban-based covert aggression in South America. A navy reclama to the initial DoD action was received with interest by Congress. The PGM might also be needed for Vietnam. The LRO argued that if the DoD awoke to the problem of sublimited war, a crash program of existing designs would be needed. The PGM would be cheap enough to make such a program palatable. It urgently asked the fleets for their own PGM force goals so that a revised program could be submitted to DoD not later than April 1962.

For contingency readiness and cold war in the Caribbean, South Atlantic, and Indian oceans, CinCLant wanted PGMs: four for the South Atlantic and four for the Caribbean, including two for Panama, before the end of 1965. These eight ships were to be presented in the FY 64 budget as additional forces for limited or unconventional warfare, not as competitive candidates for general-purpose forces.

For the longer term, CinCLantFleet wanted two PGMs per year, from 1965 to 1971, assuming that suitable lightweight weapons could be found. MAP requirements would depend on whether an ASW version could be developed.

The Pacific Fleet badly wanted a fast shallow-draft boat, particularly for Southeast Asia. MAP requirements projected through 1971 called for ten for South Vietnam, four for Cambodia, ten for nationalist China,

Visiting the Washington Navy Yard in the 1970s, the *Asheville*-class gunboat *Green Bay* displays a World War II weapon system in a hydrodynamically advanced gas turbine hull. Her enclosed RF 3-inch/50 gun was controlled by a Mark 63 director in the shroud atop her bridge. The radar dish on the gun mount was part of the unstabilized system. (A. D. Baker III)

and three for South Korea. CinCPacFleet considered the PGM less useful for the U.S. Navy and felt that a total of sixteen—eight in the Pacific—would suffice, bought at the rate of one per year in FY 63–70.

As of August 1962, the tentative FY 64–66 program showed twenty-two follow-on PGMs for a planned total of twenty-four ships, in addition to fourteen PTFs and twenty-two PGHs. In fact, only seventeen were built, two each in FY 63 and FY 64 (PGM 84–85, 86–87), three in FY 65 (PGM 88–90), and ten in FY 66 (PGM 90–101).[7] That did not end official interest in PGMs. The Five-Year Defense Program for FY 70–75, approved in July 1967, showed eight PGMs in FY 69. The navy program objectives at that time showed eight more in FY 70 and in FY 71. The proposed FY 69 budget later showed two PGMs, which would probably have been smaller than the *Ashevilles*.

The big *Ashevilles* were so clearly more capable than the PGMs of the Mutual Defense Assistance Program that they were redesignated gunboats (PG) on 1 April 1967, less than a year after completion of the name ship. They retained their old hull numbers and thus duplicated some of the gunboat numbers assigned in World War II to reverse lend-lease "Flower"-class corvettes.

Design work on the *Asheville* began in June 1961. Within the stated limits (a hull not too much larger than that of the export PGM), BuShips could achieve the desired speed but not the desired range. For example, sufficiently powerful diesels weighed so much that the hull could not store sufficient fuel. BuShips submitted three alternatives ranging in length from 95 to 115 feet overall.

On the existing 95-foot hull it could install a reasonable armament, including a 40mm gun; Mercedes diesels could drive the boat at 30 knots at a cost of $1 million each. At this time, BuShips considered the Mercedes 3,000 bhp unit the only satisfactory foreign engine; it is surprising in retrospect that the Deltic was not taken into account. In any case, the 40mm was unimpressive as an MTB-killer, so in scheme 2 Preliminary Design tried to mount a 3-inch/50 gun on a slightly beamier 95-foot hull. It could get the speed but not the required long, low-speed endurance. Scheme 3, lengthened to 115 feet to restore good hull proportions, had a third shaft. It was rated at 27 knots but did not quite meet the required range. The SCB wanted both the 3-inch/50 and high speed, an unworkable arrangement in a small boat. Nor could the usual PGM goals of simplicity, ease of maintenance, and low cost be achieved. Moreover, it was by no means clear that the estimates reflected in scheme 3 were realistic; the PGM might turn out to be much larger.

The firepower represented by the 3-inch/50 was important. Manual control was not acceptable because it would limit effective range to that of a 40mm gun; hence the complexity of a computing radar fire-control system, the World War II Mark 63. The 3-inch/50 was also attractive on account of its proximity-fused ammunition, which might do damage through an air burst even if it passed over a target. As in the PGH, the main alternative was a recoilless rifle. BuWeps rejected it because it trained and elevated too slowly and its back blast was powerful enough to blow the pilothouse off the boat. The danger area would extend as much as 200 feet back. Nor did the rifle have a high sustained rate of fire.

Now that the SCB's wishes were clear, Preliminary Design developed a much more thorough 30-knot scheme 4. For the first time in this project, it considered gas turbines. The combination of high maximum speed and long endurance made a CODAG power plant attractive; in this case, it would consist of two Proteus gas turbines on outboard shafts and two 900-bhp Curtis-Wright 3-D diesels geared to the centerline shaft. All would have controllable-pitch propellers for efficient functioning at both cruising and high speeds, which meant the gas turbines had to be set fairly far forward in the hull.

The study of alternative hull forms showed that greater length could improve sea-keeping and speed performance. On a waterline length of 150 feet (161 feet LOA, 225 tons), it was easy to maintain 28 knots and reach a burst speed of 30. The designers tried to keep the 3-inch/50 forward, but that, it was decided, would throw the balance off. They considered moving the gun aft to balance the gas turbines forward. Arrangement turned out to be the major flaw in the design.

Scheme 4 showed a total armament and ammunition load of about 20 tons: one 3-inch/50, one twin 0.50-caliber machine gun, and a Mark 63 fire-control system.

As for cost, the BuShips studies showed that it would not be too expensive to achieve 22 knots and a bit more range in a conventional PGM, with 40mm and single 0.50-caliber guns, 81mm mortars, and Redeye missiles. Cost rose sharply with either gas turbines or lightweight diesels: $1.2 million for scheme 2, $1.5 million for scheme 3 or 4.

The next step was to rearrange the machinery so that the single 3-inch gun could be moved forward again. In scheme 5, each of two shafts could be driven by either a diesel or diesel plus gas turbine, with controllable pitch propellers for efficiency at either power level. The Proteus gas turbines would be located aft to exhaust through the transom. This arrangement was hydrodynamically superior, since in scheme 4 the center (gas turbine) shaft would windmill at cruising speed.

The PGM could be driven at 30 knots on half fuel or at 28.5 knots fully loaded (continuous rating); endurance at 16 knots would be 1,700 nautical miles (400 at 28). The armament weight of 20 tons allowed for one 3-inch/50, one twin and two single 0.50-caliber machine guns, two 81mm mortars, twelve Redeyes, and four 40mm grenade launchers. However, some versions of the design showed only one 3-inch/50 with one SS-11 or -12 launcher *or* a twin 40mm gun.

The estimated cost was $2 million ($2.1 million for the lead ship, in FY 63) compared with $1.7 million for scheme 4.

The bureau noted that scheme 5 was extremely complex, that "the design as dictated by performance considerations does not fulfill the requirement for simplicity, low cost, ease of maintenance, and operation by indigenous personnel of Southeast Asian countries. The cost and complexity of Scheme 5 are justified only if the stated performance satisfies a firm requirement of the U.S. Navy."

The various schemes developed up to this point are listed in table 10-9.

At an SCB meeting in February, the BuShips representatives explained their design as the optimum version possible with the performance OpNav wanted. The controlling requirements were seaworthiness and high maximum speed, long endurance at moderate speed, and the 3-inch/50 gun. Given these factors, the long hull and the sophisticated propulsion plant were inevitable. The board reacted by eliminating the requirement that the ship be operable by non-U.S. personnel. Maximum draft was increased from 7 to 8 feet, and required endurance at 28 knots was reduced from 400 to 325 nautical miles.

Scheme 5 was reflected in characteristics approved on 5 March 1962. The stated mission was very much like that later listed for the PTF: attacks on enemy coastal shipping and perimeter defense of amphibious shipping. Blockade/surveillance and support of paramilitary and guerrilla operations were added in May 1962. The new PGM would be 165 feet long with

Table 10-9. Fast PGM Designs, October 1961

	PGM 39	1	2	3	4	5
LOA (ft)	101.25	95.0	95.0	115.0	161.0	161.0
Beam (ft)	21.1	19.8	22.0	24.2	25.0	24.0
Draft (ft)	6.9	6.3	6.5	8.1	7.0	7.0
Full Load (tons)	117.7	107.3	124.1	209.9	225.0	225.0
Max Speed (kts)	17	30	28	27	30	30
Endurance (nm/kts)	1,000/12	1,150/12	600/12	1,300/17	1,700/16	1,700/16
Armament	2 0.50			———— 1 3-in/50 ————		
	Mousetrap			———— 1 0.50 twin ————		
	4 DC			———— 2 0.50 single ————		
	—	—		———— 2 81mm mortar ————		

Note: PGM 39 data are included for comparison. Maximum speed is with half fuel on board. Scheme 1 has approximately the same dimensions as PGM 43 with about 5 tons more of displacement.

draft limited to 8 feet, a displacement of 230 tons, a maximum speed of 30.5 knots, a cruising speed of at least 16 knots, and an endurance of 1,700 nautical miles at 16 knots.

Estimated cost was $1.9 million ($2 million for the lead ship). Gibbs & Cox was the design agent.

BuShips considered this PGM too slow and continued to develop alternative machinery arrangements. The key was a more powerful gas turbine. Soon it became clear that two U.S. aircraft engines, GE and Pratt & Whitney 13,000-bhp units, would be available for FY 63. They promised spectacular performance, up to 39 knots. A slightly longer hull of 166 feet, displacing 251 tons, would improve seakeeping. Very high speed would make the PGM a plausible counter to communist-bloc torpedo boats in, for example, Cuba, Korea, China, the Baltic, and the Adriatic, and the PGM would also face escorts and destroyers with a reasonable chance of success. It would carry a much heavier armament—beside the single slow-firing 3-inch/50, two single 40mm guns, two 81mm mortars, two 0.50-caliber machine guns, two 40mm grenade launchers, and twelve Redeye missiles for self-defense.

BuShips could also make the case that its proposed PGM followed the general trend in European navies typified by the new Swedish *Spica* gun/torpedo boat.

In late May 1962, the first version of the high-speed plant showed three shafts, the centerline one driven by a GE MS420 or a Pratt & Whitney FT4A with a supercavitating propeller. Cruise diesels would drive the two wing shafts, which had controllable-pitch propellers. A full preliminary design was due in July. There was not nearly enough time to carry through more than one design. Early in June, BuShips decided that, rather than pursue the lower-speed alternative, scheme 4b, it would order Gibbs & Cox to center the preliminary design around the high-powered plant (scheme 7b).

At this stage, the bureau revealed its decision to the SCB. Scheme 7b was exciting, but it was also about $300,000 more expensive than 4b ($1 million more for the lead ship). The officer at SCB in charge of the design felt sure that the increased cost would eliminate it. On 13 June, a presentation to the SCB received a "courteous but cold" reception. The BuShips machinery team began to seek some less expensive though still powerful alternative.

One possibility, scheme 7c, was to have only two shafts, with the gas turbine driving both shafts through gears. It would be superior hydrodynamically, since there would be no trailing shaft at cruising power. However, the gearing would be expensive. Other alternatives seemed equally bleak, and a BuShips meeting broke up on 15 June with only two weeks before the preliminary design was to be completed.

Under pressure, BuShips chose 7b for the preliminary design so long as it could be held within the total FY 63 allocation of $4.1 million, which was to have bought two ships. Failing that, a low-powered, three-shaft alternative, with a Proteus on each of two outboard shafts and a 3,000-bhp diesel on the centerline, would be developed. The SCB reluctantly accepted these rules. However, it was clear that the radical departure BuShips advocated would have to be submitted to the CNO.

The resulting preliminary design was completed in July 1962. Overall length was 166.2 feet (154 LWL), beam 23.8, and draft amidships 5 (maximum 8); fully loaded, the PGM would displace 251 tons (190.3 light). Unlike a PTF, this was an all-weather boat. It therefore had a round bottom, although a sharp chine and hollow buttocks had been worked in aft to achieve enough dynamic lift to keep it from squatting at high speed.

In August, then, Preliminary Design prepared a comparative presentation for the CNO. The three shafts of the original design were abandoned in favor of the somewhat more complex scheme 7c, so that both "low"- and "high"-power schemes had two shafts. The low-power plant consisted of two Bristol-Siddely

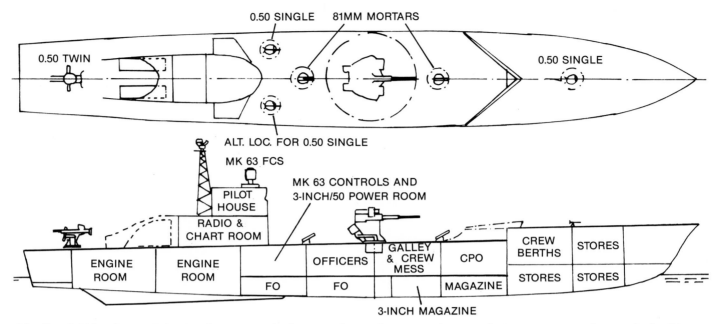

The fast PGM, scheme 5, 19 October 1961. The big gun forward is a 3-inch/50, with 81mm mortars fore and aft of it on the centerline, a 0.50-caliber gun on the centerline forward, and a 0.50 on either beam forward of the bridge. The twin mount aft is also a 0.50. (Norman Friedman)

Proteus engines, 3,400 bhp each, and two Curtiss-Wright CW12V142A cruise diesels, 750 bhp each, for a maximum speed of 31 knots and a continuous speed of 28. Cost was estimated at $2.3 million ($3.4 million for the lead ship at FY 63 costs). The alternative high-power plant used a single GE MS240 or Pratt & Whitney FT3C gas turbine (13,000 bhp) geared to both shafts with the same two cruise diesels to achieve an estimated 40 knots of continuous speed. It would also achieve greater cruising range, 1,900 rather than 1,700 nautical miles at 16 knots. The cost would rise to $2.7 million ($3.9 million for the lead ship).

BuShips argued that the low-power alternative should be discarded because it used foreign engines, whose purchase would upset the U.S. balance of payments, and because the larger, faster, more seaworthy alternative would be more responsive to real needs, even though it exceeded the characteristics. It was technically outstanding. It used American engines. It would be a prestige ship.

It was true that buying two such ships would violate the cost ceiling established by OpNav, but a single, extremely capable ship could be bought within that limit.

On 17 August 1962, the chief of BuShips presented both alternatives to the CNO, Admiral George W. Anderson, who chose the faster alternative for the entire class. The characteristics were revised to call for a maximum speed of 40 knots carrying half the fuel load. Later, when less optimistic tank test results came in, this figure was cut to 37 knots.

The first approved characteristics showed a single, slow-firing (Mark 26) 3-inch/50, a single 40 mm gun, two single 0.50-caliber machine guns, two 81mm mortars, two 40mm grenade launchers, and twelve Redeyes, the latter subject to successful test and evaluation, within a total of 20 tons, four times that allowed for in the PGH. Armament weight became more critical as the design evolved towards very high speed, and the elimination of various weapons was suggested. The 81mm mortar was initially saved because its illumination capability was too important in counterinsurgency warfare. BuWeps proposed eliminating the single 40mm gun in December 1961.

For the FY 64 boats, the SCB wanted to consider alternative weapons. In August 1962, it asked BuShips and BuWeps to consider both domestic and available foreign weapons that might provide effective surface fire out to 6,000 yards; effective AA or antimissile fire to at least 3,000 yards, which meant a lightweight missile; a lightweight surface homing torpedo; and a new gun. The U.S. RF 3-inch/50 was relatively heavy and cumbersome; its Mark 63 fire-control system would be ineffective at high speed. BuWeps suggested the new Bofors 3-inch/50 controlled by a Dutch M20-series integrated radar/fire-control system.[8] The Swedish gun fired faster and required fewer personnel, but BuWeps noted that it could elevate only to 30 degrees, as opposed to 65 for the U.S. gun, and therefore would provide no AA fire. That would have been acceptable if there had been an effective lightweight AA missile to take its

The newly completed *Defiance* at Sturgeon Bay, Wisconsin, 21 August 1969.

place. An October SCB meeting rejected the gun but found the M22 itself attractive. With its AA potential, it could restore the single 40mm gun that had previously been eliminated from the characteristics.

The Dutch fire-control system was ruled out for FY 63 PGMs because it would take too long to adapt. In July 1963, the SCB approved it for the FY 64 ships. By September, however, it was clear that the change would delay the FY 64 ships a year, and the change was shifted to the FY 65 series. In fact, only the two FY 64 boats, the *Antelope* (PGM 86) and the *Ready* (PGM 87), had the Mark 87 track-while-scan radar fire-control system developed from the Dutch system. The Mark 87 could compensate for ship motion much better than the World War II Mark 63 of the other units, and it was credited with increasing effective gun range from 2,000 to 3,500 yards out to 5,000 to 6,000. The approved characteristics were modified in April 1963. The gun would be an RF (Mark 34) 3-inch/50 rather than the slow-firing Mark 26. Fiscal compensation would include elimination of the two 81mm mortars originally planned. Trial speed was reduced from 40 to 37 knots, but endurance speed rose from the 28 knots of the original slow PGM to the 35-knots sustained speed of the fast one. All boats were completed with an enclosed Mark 34 gun forward, a single 40mm aft, and two twin 0.50-caliber machine guns.

Since Soviet-supplied missile boats were, in theory, the PGM's main prey, their Styx (SS-N-2) missiles were the major threat it faced. Existing 3-inch/50 and 40mm guns were clearly inadequate, and shipboard Redeye did not materialize.[9] In 1967–68, the Sidewinder was proposed as a point-defense system, but it was rejected in favor of an offensive surface-to-surface weapon, which materialized later as the Standard ARM.

The *Asheville*s were powered by one GE LM1500 gas turbine of 13,300 shp for a maximum speed of 37 to 40 knots; they achieved about 38 in service. However, most were limited to about 32 knots because their stainless steel propellers suffered from the cavitation pitting common among very fast ships. The cruise engines were two 725-bhp Cummins diesels, good for 16 knots. Rated endurance was 3,000 nautical miles at 12 knots and 490 at 37.5. The FY 66 series (PG 92–101) had more fuel and an improved turbine gear box and machinery arrangement. They were therefore sometimes considered a separate class.

Because the PGs were by far the smallest U.S. oceangoing surface warships, they had a rougher ride than the destroyers and frigates with which most surface officers were familiar. They developed a reputation as poor sea-keepers, one reason being that few officers were familiar with the much rougher rides of other small combatant warships.

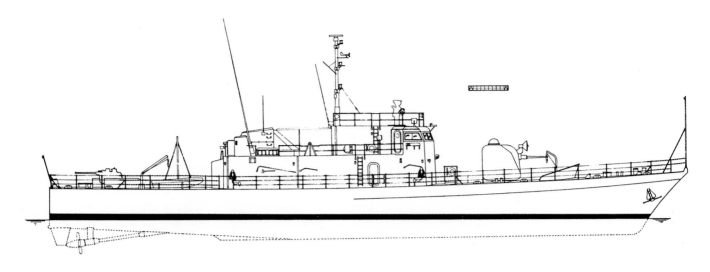
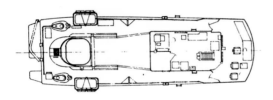
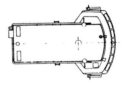
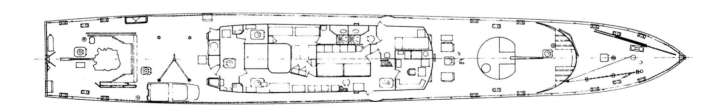
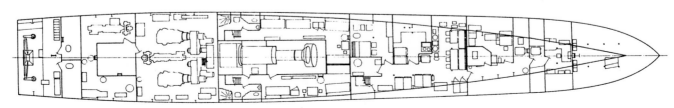

The *Asheville* class, as built. (USN)

The *Ready* was equipped with a gun weapon system better adapted to the kind of rough motion characteristic of a fast gunboat, the Dutch Signaal radar "egg" (Mark 87 fire-control system). Mark 87 was a more austere form of the Mark 92 used in the *Pegasus*-class missile hydrofoil (PHM) and the *Perry*-class patrol frigate. No gun mount radar dish was needed. The rest of the battery was standard: a single 40mm gun right aft and two single 0.50s.

They were little affected by waves up to 8 feet, but over that they rolled and pitched heavily, and crews riding out 10-foot waves for over seventy-two hours could become badly fatigued. Speed would generally be cut from 37 to about 20 knots. In sea state 5, a PG could roll 65 degrees each way. There was no question, though, of its seaworthiness: it could transit open oceans under its own power and ride out major storms without severe damage. In many places, moreover, waves were not so very high. In the Mediterranean, for example, waves were less than 5 feet 70 percent of the time, and in only 8 percent did they rise higher than 8 feet. Several of the later *Asheville*s were used to test stabilization systems, including fins, for example, in the *Chehalis*, and a box keel. All were restricted from cold-weather operations, as topside ice would have made them unstable.

In Vietnam, the big PGs found little use for their high speed, but they were valued for their sea-keeping and heavy gun. In the Market Time coastal blockade, they could remain on station longer than small craft and were less expensive to run than destroyers. The usual duty cycle was seven or fourteen days on, three off. Their shallow draft made them useful river gunboats, in which role they were used extensively, but being lightly built, they were subject to damage. The *Antelope* was badly mauled in Vietnam riverine operations. The modern *Pegasus*-class missile hydrofoils described in chapter 15 are their direct successors.

The PGs were considered as alternatives to the PTFs for UDT and SEAL support, as they were roomier and more comfortable.

While the PGMs were being built, the Soviet naval posture changed radically. In the Mediterranean, the USSR used fast surface ships, "tattletales," to trail U.S. carrier task forces; in wartime, they would call in missile strikes by more distant ships. In 1971, Admiral Elmo Zumwalt began his tour as CNO with Project Sixty, a quick look at new ways of using U.S. warships. One of its conclusions was that the United States should begin to trail Soviet naval formations in the Mediterranean. The *Asheville*s were ideal—small enough to be expendable, fast enough to keep up in most Mediterranean seas, and, as radar targets, very small. The *Surprise* and *Beacon* were used for this mission. Their main defect was that, by modern standards, they were barely armed.

The *Douglas* was fitted to fire the standard Arm antiship missile (the interim surface-to-surface missile) as an emergency measure until the Harpoon was ready. Each box aft consisted of two sections: a launcher holding one missile and a reload forward of it. Ranges up to 40 nautical miles could be achieved, actual performance depending on the extent to which the target used its radar. An active-seeker version of standard was also developed but never deployed.

The next step, then, was to provide them with a surface-to-surface missile, a version of the standard AA weapon (the ISSM, or interim surface-to-surface missile) adapted from the General Dynamics Tartar-Bullpup originally developed for the West German navy. The *Benicia* conducted test firings in the spring of 1971. The two Mark 87 ships, the *Antelope* and *Ready*, were each fitted with two launch cells aft plus reloads in boxes on deck. Their 40mm guns were removed, and they lost 3 to 4 knots in speed. The *Grand Rapids* and *Douglas*, which lacked the Mark 87, were able to take an improved Standard ARM missile, which did not require such an elaborate fire-control system.

As of 1973, three had been transferred, three operated as Mediterranean tattletales, and three more were based at Little Creek with other U.S. amphibious and coastal force craft. The other seven, ironically, performed classic gunboat duty, patrolling the Marianas trust territory from Guam. Their high speed must have been an asset over so vast an area.

Like their PGM predecessors, they were most useful to U.S. allies in direct confrontation with enemy fast patrol boat fleets. South Korea received the *Benicia* in 1971, Turkey, the *Defiance* and *Surprise* in 1973. The *Tacoma* and *Welch*, the last active U.S. units, were used to train Saudi personnel to operate the PGGs before being transferred to Colombia in 1983. The *Asheville* was transferred to the Massachusetts Maritime Academy in 1976, followed by the *Ready* and *Marathon* in 1977 and 1978. The *Antelope* and *Crockett* were stricken in 1977 and transferred to the Environmental Protection Agency. Two ships, the *Chehalis* and *Grand Rapids*, were disarmed and converted for the Mine Defense Laboratory at Panama City into high-speed tugs for helicopter-towed, mine-countermeasures gear. They became the *Athena I* and *II*. The *Douglas* was to have become the *Athena III* in FY 83, but funds were not available and she was discarded instead in December 1984. Of the remaining units, the *Beacon, Canon, Gallup,* and *Green Bay* were stricken on 31 January 1977; the *Canon* and *Gallup* were restored to the navy list on 17 July 1981, then stricken again at the end of 1984. As of this writing, all are in storage, two at Bremerton and two at Little Creek; the *Beacon* and *Green Bay* had been scheduled for transfer to Colombia, the *Gallup* to Taiwan in 1981, but none of these transfers occurred.

Tacoma Boatbuilding built four modified *Asheville*s as the PSMM-5 class for South Korea. Korea Tacoma built four more for the South Korean navy and another four for Indonesia (PSK Mark 5, or the *Mandau* class); four more planned Indonesian units were not built. The Taiwanese *Lung Chiang/Suikiang* classes are further variations on the PSMM Mark 5 design. As of 1986, four had been completed and at least four more were planned. Finally, the U.S. Navy totally revised the basic *Asheville* design as the Saudi navy PGG class, nine of which were completed by Peterson in 1980–82.

Gunboat design work continued along two lines after 1962. The first was a series of hydrofoil PGMs, early data for which are listed in table 10-10. The

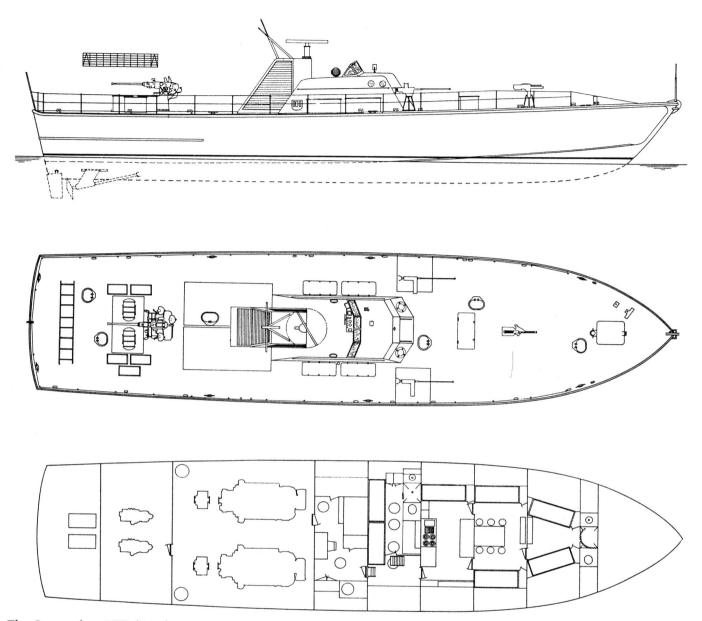

The *Osprey*-class PTF. (USN)

other was a smaller fast patrol boat or PGM. In the fall of 1963, Preliminary Design was working on a 130-footer, which was conceived as a minimum boat suitable as either a PGM or an ASW boat and which also would have significant counterinsurgency possibilities. It is compared with a contemporary 80-foot PTF design in table 10-11. None of these craft was built, as the Vietnam War consumed counterinsurgency funds from 1965 on.

No account of the fast attack boats would be complete without the SWAB (shallow-water attack boat) conceived as an anti-*Komar* measure by the Naval Weapons Center at China Lake. In December 1962, BuWeps asked China Lake to adapt its Sidewinder air-to-air missile as a surface-to-surface weapon for small ships. It was clearly far from optimum, and late in December, John Boyle of the aviation ordnance department proposed an alternative, a very fast, small, 30- to 50-foot fiberglass boat, carrying only a driver and a weapon-system operator strapped in semireclining positions to minimize the effect of pounding at high speed. Launched off shore by an LSD, it would use infantry-type weapons such as the SS-11 to ambush *Komar*s or *Osa*s emerging from harbor. Boyle considered it attractive because it was so much less expensive than a conventional missile boat.

The Naval Weapons Center began work using its

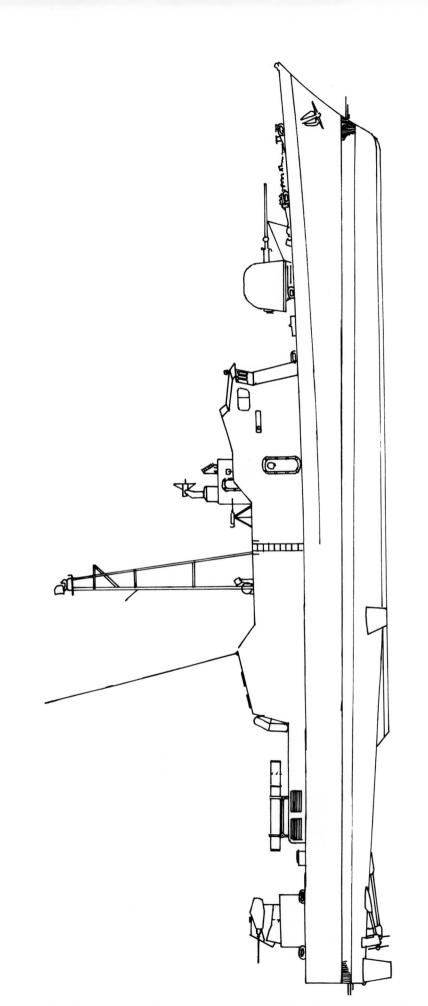

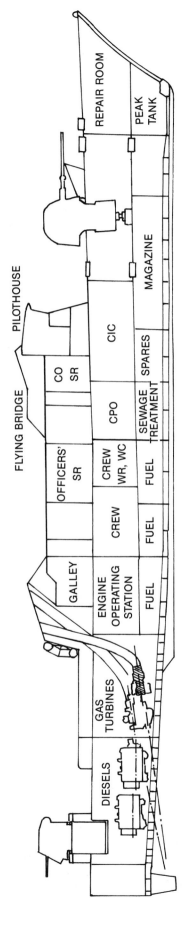

Tacoma Boat developed the *Asheville* into the successful PSMM Mark 5 shown here, armed with an OTO-Melara 76mm gun forward, a twin 30mm gun aft, and an unidentified missile system. CODAG machinery (one 1,200-bhp diesel and one 3,800-shp gas turbine on each of three shafts) was intended to drive her at up to 38 knots (18 knots at cruising speed, on two diesels, for a range of 2,000 nautical miles). Dimensions were 165 (overall) x 23.9 (molded) x 13.58 (molded depth) feet. (Norman Friedman)

Table 10-10. Hydrofoil PGMs, October 1962

	PGM	PGM(H)	PGM(K)
Displacement (tons, full load)	230	150	142
LOA (ft)	165	139	95
Beam (ft)	23	35	38
Draft (ft)			
At Rest	8	5.2	7
Full Speed	8	7.0	8
Armament	——— 3-in single auto, 40mm single, 2 0.50, 2 81mm single ———		
Machinery	——— 14,000-hp gas turbines, 2 750-bhp diesels ———		
Speed (kts)	40	50	50
Cruise Speed (kts)	16	15	16
Range (nm)			
High Speed	325	500	470
Cruise	1,700	2,700	1,500

Table 10-11. The 130-Ft PGM, November 1963

	80-ft FPB	130-ft MGB
LOA (ft-in)	80-0	130-0
LBP (ft-in)	75-0	120-0
Beam (ft-in)	22-0	20-10
Draft (ft-in)	6-3	8-10
Displacement	132,600 lbs	157.5 tons
Machinery	——— 2 3,200-hp gas turbines ———	
	2 150 Diesel	2 75 Diesel
Max Continuous Speed (2/3 fuel, kts)	45	30
Cruise Speed (kts)	10	12
Endurance	200/45	300/30
(nm/kts)	1,400/10	1,000/12
Fuel Oil[1] (gals)	3,720	7,460
Armament	1 20mm	1 3-in/50 RFSM
	1 40mm	1 twin 50, 2 single 50 machine gun
Payload		
Armament	6,200 lbs	10.5 tons
Ammunition	3,800 lbs	7.0 tons
Electronics	2,000 lbs	1.5 tons
FCS	—	4.1 tons
Total	12,000 lbs	23.1 tons
Complement (off/CPO/enl)	2/4/12	3/4/16

1. Including 700 gallons for 192 hours of ship-service-generator operation.

own exploratory development funds in October 1963, and its special-warfare attack boat was completed in January 1965. It was a modified standard 31-foot Bertram boat with a modified superstructure and with radar-absorbing material arranged to make head-on detection difficult. It was initially armed with two standard Aero 7 2.75-inch folding-fin aircraft rocket (FFAR) pods alongside its pilothouse. Initial testing of the pods showed excessive dispersion, and they were replaced in July 1965 by two 57mm nonrecoiling rifle pods with seven weapons each, two 0.50-caliber aircraft-type machine guns, primarily for spotting, and one 20mm Mark 3 cannon. To prevent damage to the boat, the 57mm guns were limited by cutout cams to a maximum range of 2,500 yards. Even so, their back blast did rip sections of fiberglass cloth and radar-absorbent material from the engine-room cover.

As might have been expected, all of this was too much for the small Bertram hull and its two 325-bhp Chryslers. The designers hoped for 30 knots, 45 knots burst, but the SWAB, carrying standard ammunition and 400 gallons of fuel, labored to plane and could not plane at all when manned by four men. The boat's maximum speed with battle load was only 18 to 22 knots. Nor was it particularly maneuverable, though it had to maneuver to bring the 57mm cannon onto a target.

The gun pods restricted visibility to 25 degrees on either side of dead ahead.

Boat Support Unit 1, evaluating the SWAB, suggested a new engine—a gas turbine or diesel—rear-

Table 10-12. Fast Attack Craft

	Nasty	Osprey	PGM 71	PGM 84 Asheville	PGH 1 Flagstaff	PGH 2 Tucumcari	LCSR
Length (ft-in, oa)	80-4	94-8	101-1	164-6	74-5.5	71-9	52-4.5
Beam (ft-in)	24-7	23-2	21-1.75	23-10.75	21-5	19-6	14-9.25
Draft (ft-in, loaded)	3-10.25	7-0	—	—	13-6 (foils down) 4-3 (foils up)	13-11 4-5	5-6
Engines	2 Napier Deltics T-18-37	2 Napier Deltics T-18-37K	2 Quad GMs	1 GE LM1500 12,500 shp 2 Cummins VT12-875M	1 Rolls Royce Tyne Mk 621 2 GM C6V-53N	1 Proteus 1273 3,200 shp 1 GM C6V-53N	2 Saturn T-1000
Power (bhp, each engine)	3,100	3,100	1,000	1,400	—	—	1,000
Speed (kts, full load)	—	36.8[1]	17	—	—	—	35 (max)/30 (continuous)
Fuel	5,800 gals	65,527 lbs	4,927 gals (15.3 T)	—	3,682 gals	3,639 gals	1,300 gals
Endurance (nm/kts)	—	1,004/32.6	1,000/12	—	—	—	200/35
Complement	17	19	15	29	13	13	8 (enl)
Weights (lbs)[2]							
Hull	47,870	—	107,184	65.76	40,245	28,515	20,141
Propulsion	48,710	35,870	48,138	47.16	19,180	12,028	6,145
Electric	6,230	—	7,459	8.79	7,526	6,742.4	2,217
Command/Control	1,830	—	3,450	8.24	29,512	6,697.6	2,814
Auxiliaries	3,620	—	19,779	20.77	7,743	6,585.6	713
Outfit	6,680	—	26,163	23.10	6,315	5,129.6	—
Armament	9,560	—	10,998	14.18	—	15,254.4	—
Margin	800	—	4,458	—	3,624	5,488	714
FO	40,780	—	—	—	117,596	88,995.2	36,410
LS	125,300	150,930	227,629	188.01	151,309	25,160.2	53,630
FL	181,240	245,000	315,056	242.44	23,100	25,400	9,048

1. The Deltic cruised at 2,500 bhp at 1,800 rpm (3,100 bhp at 2,100 rpm); cruising speed on trials was 32.6 knots at 232,640 pounds' displacement. Full load in this condition was 245,638 pounds, light displacement was 155,920.
2. Weights for the *Asheville* are in tons. They apply to the *Benicia* (PG 96).

ranging the armament for better visibility, and enlarging the conn to receive two men.

The project died by the end of 1965. The thoroughness of its demise is demonstrated by the decline of its classification. The original China Lake report of late 1965 was so secret that reproduction was prohibited, but less than a year later a crisp photograph of the SWAB was included in an article on counterinsurgency craft (which it was not) published in the *Naval Engineers' Journal*.

11

Vietnam: Beginnings

Vietnam was very much the war both Khrushchev and U.S. long-range planners envisioned in the mid-1950s: one fought by a Soviet proxy or ally on the Eurasian periphery, which the United States could not deter by threatening a nuclear attack, naval or otherwise. It took on particular significance for the United States because the North Vietnamese/Viet Cong attack on South Vietnam seemed to be a possible prototype for Soviet-backed aggression in many other areas, including Latin America.

Vietnam was also special because the United States had helped establish South Vietnam in 1954, at the end of the Indochina war, preventing the victorious Ho Chi Minh from seizing the whole of the former French colony. The Viet Minh victory had been seen at the time as the high-water mark of post-1945 communist expansion; any further expansion into South Vietnam would be seen as the beginning of a new and threatening cycle.

In 1961, the Kennedy Administration was keenly aware of the potential of communist-led revolutionary warfare. In the Congo, in Laos, and in Bolivia, revolutionaries had shown that they did not need direct Soviet support. Nor was it clear whether the Soviets could control local proxies, and thus whether the U.S. nuclear deterrent could have even an indirect effect on them. Guerrilla warfare, perhaps of a decentralized form, was possible anywhere in the world.

A few weeks after entering office, President Kennedy directed Secretary of Defense McNamara to increase emphasis on the development of counterinsurgency forces. The Joint Chiefs of Staff (JCS) did not immediately react to this directive; for example, Army Special Forces were still designed primarily to conduct guerrilla warfare and sabotage in support of a major war against the Soviet Union. The navy had no equivalent force; it formally established a counter-guerrilla organization, the SEALs, on 1 January 1962.[1] More than any other U.S. Navy organization in Vietnam, the SEALs were conceived to conduct counterinsurgency or sublimited warfare. Their specialized craft are described in chapter 14.

President Kennedy reemphasized his concern with counterinsurgency in a special budget message to Congress on 28 March 1961. Secretary McNamara, who would be personally identified with many of the navy's Vietnam programs, pressed the services for specific initiatives. In response, the JCS ordered quarterly reports on the current status of counterinsurgency projects. Even so, through the end of 1961, the administration felt that the military was not doing enough. One service response was to revise the FY 63 budget (see chapter 9 for some of the navy revisions).

By 1964, it was official U.S. doctrine that low-intensity insurgency would continue to be a primary threat for the period envisioned in long-range planning, fifteen years.

Because of such interest on the part of the administration, much of the technological development was carried out under the auspices of the secretary of defense's own advanced development organization, DoD's Advanced Research Projects Agency (ARPA), which had been established in 1958 to coordinate missile work in the wake of Sputnik.[2] The administration's belief that counterinsurgency was much more than a military problem was reflected in the extensive early involvement of the CIA.

At the outset, as in past administrations, U.S.

Of all U.S. Navy small craft, the LCPL came closest to filling the requirements of riverine and coastal warfare. All four active units were sent to Vietnam in 1964 to help secure the river approaches to that country. This LCPL, 36PL6447, was built specifically for foreign transfer under MAP by the Miami Beach Yacht Company. The Vietnamese navy initially relied on LCPLs of its own for port security in South Vietnam (Operation Stable Door). Some had shark teeth painted on their bows, and all had the searchlights on the bow machine gun.

The river patrol craft (RPC) was the first fruit of U.S. attempts to develop specialized craft for river warfare in Vietnam. This is C-13456, an FY 64 boat under construction at Birchfield Boiler, Tacoma, Washington, on 14 July 1965. The large box forward protects a gun mount. Note that the screws protrude from the hull aft, and that the boat shows considerable draft, both unfortunate features in riverine operations.

national strategy was to support the local friendly government without directly involving U.S. ground troops. Naval forces were a somewhat different case, because by their mere presence in an unstable area, they could exert considerable influence on friend and foe.

The U.S. government perceived North Vietnamese pressure on the South as a Soviet-backed proxy war. It seems more likely, in retrospect, that the North Vietnamese were bent on national reunification from the first, and that they used Khrushchev's policy to obtain the material support they needed. The war itself was initially fought by Viet Cong guerrillas in the South. As it intensified, they needed arms, which the North Vietnamese supplied. Ultimately, North Vietnamese troops were fed into the battle.

The North Vietnamese and the Viet Cong had never recognized the legitimacy of the 1954 settlement of the war against the French, which split French Indochina into North and South Vietnam, Cambodia, and Laos. From 1954 to 1959, the Viet Cong concentrated on building a political base in the South through such measures as the assassination of provincial offi-

cials. That is why the South Vietnamese navy spent so little of its limited resources on riverine forces. Note, however, that the Vietnamese river assault groups did have to fight dissident religious sects and the Binh Xuyen forces in 1955–56.

The United States initially sought to reinforce South Vietnam with advisors and specialized weapons. This effort led to tests of a very wide variety of small craft and to the design and construction of specialized river patrol craft (RPC).

As the South Vietnamese added to their defeats, the United States started exerting pressure on North Vietnam to abandon its war, on the theory that only outside assistance was maintaining the Viet Cong. U.S. naval forces were initially employed, then, to threaten the North and so to deter it from active involvement. For example, the Pacific Fleet amphibious force, by its presence in nearby waters, could threaten to land either U.S. or South Vietnamese troops on the North Vietnamese coast and tie down North Vietnamese forces.[3] However, U.S. policy was, generally, to make this threat without actually carrying it out. Policy makers feared that invading the

C-13371, a FY 64 RPC, is shown at full speed on trials off Tacoma, Washington. It was built by Peterson.

An undated photograph of a South Vietnamese RPC. The walkie-talkie (visible in use in the forward gun tub) would have been for communication with troops; the high-frequency whip was for boat-to-boat communication. Note the frame for a canvas canopy, amidships.

North would drag China into the war, just as the military defeat of North Korea in 1950 had lured China into the Korean War. That reality did not make the threat any less important.

As the war escalated, so did the U.S. presence. In the Tonkin Gulf, American destroyers demonstrated U.S. concern. When the North Vietnamese attacked them in August 1964, the United States retaliated, signaling the beginning of active, as opposed to passive, American intervention in the war in the South.

Pressure was also exerted directly on the North by bombardment and by agents and saboteurs who landed along the North Vietnamese coast to operate in the enemy's rear. This was part of a program, begun in February 1964, designated Operation Plan 34A.[4] U.S. PTFs were turned over to the South Vietnamese navy for the purpose of carrying out the plan. They were as much a part of naval deterrent/pressure as the amphibious force, which later actually landed marines for operations in the South, or carrier air strikes.

For the purposes of this book, the extent of direct U.S. intervention defines phases of the war. President Kennedy announced an expansion of MAP aid to Vietnam in May 1961. For the next three years, relevant U.S. combatant craft design was intended primarily for operation by the Vietnamese, although the U.S. Navy did carry out some limited blockade operations from December 1961 to August 1962 (see chapter 12). As the American role expanded, the U.S. Military Advisory Group (MAAG) was supplemented in February 1961 by the U.S. Military Assistance Command, Vietnam (MACV).

President Kennedy's special military advisor, General Maxwell Taylor, visited Vietnam in October 1962 and recommended increased aid ("Project Beefup"). Even so, the military situation continued to deteriorate through 1963.

Initially, the Viet Cong had been able to rely on weapons cached in the South in 1954 or captured from South Vietnamese forces. Late in 1963, it appeared to the North Vietnamese that the government in South Vietnam had been so badly undermined that a major escalation would swiftly defeat it. Escalation required troops and large quantities of weapons. Soon new Chinese-made 7.62mm rifles and light machine guns were being captured in South Vietnam. The obvious conclusion was that they were being brought in by sea, as U.S. advisors had assumed from the beginning. However, South Vietnamese patrols found no weapons smugglers.

A team of eight U.S. naval officers headed by Captain Philip H. Bucklew and meeting in Saigon in January 1964 concluded that demoralized South Vietnamese forces had failed to prevent infiltration; U.S. forces were needed. This might be considered the genesis of Operation Market Time, but nothing happened for over a year, until concrete evidence of arms smuggling came to light. The Bucklew committee argued further that arms could come in through the Cambodian and Laotian borders and the Mekong Delta. It wanted a mobile patrol force along the border and a Mekong force. The Mekong antismuggling force became Operation Game Warden (see chapter 12). The border patrol came only in 1968, when the strategy of the naval war was radically revised.

In May 1964, MAAG was absorbed by MACV to form a single U.S. military command in the country; the MAAG navy section became the Naval Advisory Group, Vietnam. The Johnson Administration still wanted to avoid most direct involvement in the war, and major U.S. naval investment in Vietnam was concentrated in the development of ports such as Da Nang, which were needed to accept increasing quantities of U.S. materiel. This was true even after the passage of the Tonkin Gulf Resolution in August 1964; major direct U.S. involvement began only with Operation Market Time the following year.

Thus, until 1965, the United States developed special craft to fight the war, but combat was almost entirely in the hands of the Vietnamese. After 1966, the U.S. Navy operated a large combat force of specialized riverine craft, unlike anything it had had since the American Civil War. The coastal interdiction operation, Market Time, played an intermediate role between advice and direct participation.

U.S. strategy was to split the war into two parts: isolation of the South, to cut off the source of arms and, perhaps, men flowing to the Viet Cong, and an in-country war against the Viet Cong themselves. Isolation could not be total; the North Vietnamese were able to use the port of Sihanoukville, in neutral Cambodia, and the Ho Chi Minh Trail in neutral Laos.

Even so, the arms interdiction effort, both along the Vietnamese coast and in the rivers of the Mekong Delta, was a major part of the U.S. naval effort in Vietnam. Only very late in the war did U.S. forces operate in strength on the Cambodian border, as part of the new Sealords strategy.

Much of what happened in Vietnam, where water was a major means of transportation, involved riverine or coastal warfare. The Mekong Delta, where about a third of the population lived, was crisscrossed by numerous rivers, canals, and streams. It was difficult country for troops, and until the arrival of the Riverine Force, it was a virtually uncontested Viet Cong sanctuary. Another third of the population lived along the coast, served by coastal waterways and small rivers that were not interconnected. Another third lived in greater Saigon, dependent upon the main shipping channel, the Long Tau River, for supplies. The Long Tau in turn ran through the swampy Rung Sat Special Zone (RSSZ), traditionally termed the Forest of Assassins.

Operations were often described in terms of the corps tactical zones into which South Vietnam was divided, north to south, from 1957 on. The demilitarized zone (DMZ) that divided the two Vietnams was the northern border of I Corps. The DMZ was eventually the defensive responsibility of the marines, and after 1968, the U.S. Navy maintained patrols on the Cua Viet and Perfume rivers nearby. The corps included the port of Da Nang and the old imperial capital of Hue. South and west of I Corps, II Corps was headquartered at Pleiku and included the ports of Qui Nhon and Nha Trang. Around Saigon, III Corps included the port of Vung Tau. It originally contained the Mekong Delta, but in 1962 the delta was split off as IV Corps and its naval headquarters became An Thoi. Each corps had a corresponding naval zone.

For the first period of the war, U.S. assistance focused on the South Vietnamese navy. It was divided in 1957 into two main forces, the Sea Force and the River Force, supplemented, after 1960, by the paramilitary Junk Force.

The Sea Force consisted initially of only one 173-foot PC and two 110-foot wooden subchasers. Four more PCs were transferred in 1956, a sixth in 1960. A 180-foot PCE was transferred in 1961, followed by two more in 1966 and 1970 as well as five 180-foot minesweepers (MSF) used as large patrol craft: two in 1962, two in 1964, and one in 1970. From 1961 to 1964, the Vietnamese navy also received twelve PGMs, two LSMs, three LSTs, one YOG, and twelve minesweeping motor launches (MLMS). Thus the Sea Force accounted for most South Vietnamese naval growth between 1955 and 1960.

The River Force inherited about one hundred landing craft from the French, and it had about the same strength in 1960. U.S. assistance roughly doubled this force, so that in 1963 the Vietnamese navy included 208 river craft.

U.S. assistance initially provided sufficient craft for a sixth riverine assault group; each group consisted of one command boat, one monitor, five armored LCMs, eight armored LCVPs, and four STCANs. The Vietnamese had LSSL gunboats as well as a gunboat version of the closely related LSIL troop carrier, the only classes for which the later U.S. riverine force had no direct equivalent. By 1965, each of nine riverine groups consisted of one LCM(6) command boat, one LCM(6) monitor for 40mm fire support, five LCM(6)s for armored troop/cargo transport, six LCVPs for armored troop/cargo transport and minesweeping, and six STCANs.[5] The United States transferred six of the larger LCM(8)s in December 1964; they were assigned to Group 27 at My Tho. Each group could lift, support, and land a battalion of about five hundred men; LSSLs or LSILs could be attached when additional fire support was needed.

In 1961, the Vietnamese asked the United States for a specialized monitor that could escort river convoys and counterattack ambushers; in December 1961, the SCB discussed a converted LCM with a spoon bow plus armor and armament. It became the prototype for the later U.S. monitor described in chapter 13.

President Kennedy's initial aid program of May 1961 included sixteen LCVPs and sixteen LCMs for the riverine assault groups. Converted LCMs (command boats and monitors) began to arrive in 1963. Resembling their French predecessors, they were lightly protected with quarter-inch STS (⅛-inch STS on pilothouses and guntubs) and, unlike standard U.S. landing craft, had raised deckhouses aft with covered machine-gun mounts.

Ten LCMC command boats were transferred in December 1963, another four in 1966 (two in April, one in July, one in September). The LCMCs were originally armed with 20mm guns fore and aft and an 81mm mortar. Later, two 7.62mm Mark 21 machine guns and, in some cases, two 0.50-caliber machine guns were added.

The first seven U.S.-converted monitors were transferred in December 1963, seven more in March and June 1965, eight more in April, July, and August 1966. The monitors were armed with one Mark 52 turret forward carrying one 40mm gun and one 0.50-caliber machine gun, an 81mm mortar in a deep well amidships, and two 20mm guns in single turrets on either side. Later, a Mark 19 40mm grenade launcher on a Mark 64 mount was added.

The U.S.-modified, armored LCM was denoted an LCM-combat (LCM-C). Transfers included at least twelve in December 1963; eight in December 1965; twenty-four in 1966. These craft incorporated a raised and enclosed poop and were armed with three 20mm cannon and two 0.50-caliber machine guns (one 20mm in a turret on the centerline aft, with single 20mm and 0.50-caliber guns on each side of it, forward, under a wooden canopy). It is not clear whether any of the sixty-eight on the Vietnamese navy list had been converted by the French or by the Vietnamese themselves.

LCVPs were armed with one 20mm cannon and three 0.30-caliber machine guns.

The later U.S. river assault squadrons, modeled on the riverine groups, consisted of converted LCM(6)s plus a specially designed patrol boat, the ASPB, which itself was largely inspired by the STCAN. The concept of the U.S. riverine monitors and command boats owed a great deal to craft built or rebuilt for transfer to Vietnam under MAP. For example, the standard 40mm Mark 52 bow turret was initially developed for the MAP program, although the protection of the U.S. craft reflected new technology. The U.S. riverine force had to borrow Vietnamese LCM-Cs until its own armored troop carriers (ATC) arrived.

In addition to the assault groups, the River Force included the river transport escort group, primarily to protect convoys to Saigon, with four LCMs, eight LCVPs, and eighteen STCANs, and a logistic force of seven LCUs, one of them a mobile repair barge.

On 1 June 1963, the Vietnamese instituted a special Mekong river patrol consisting of a Sea Force LSIL supported by two river craft, either LCVP or STCAN. These craft could be heard two miles away, and they accomplished little.

The Kennedy aid program also included 14-foot swimmer support boats to improve South Vietnamese army mobility on the rivers. In November 1961, the United States assisted the navy yard at Saigon in building twenty.

The paramilitary Junk Force was the first major creation of the U.S. advisors, who had advocated such a coastal patrol force at least since 1956. It was formed on 12 April 1960, and by May 1961, the Vietnamese had 80 junks (four divisions) on counterinfiltration patrol at the seventeenth parallel (the DMZ). In December 1960, the navy MAAG group advocated a force of 420 junks. President Kennedy's initial Vietnamese assistance program of November 1961 included 63 motorized command junks and 420 sailing junks for coastal patrol; 3 PCEs were transferred to augment them. In February 1962, Secretary McNamara and CinCPac decided to increase the Junk Force to 644 craft (28 junk groups). In theory, each junk group included 3 motorized junks and 20 31-foot sail junks.

It was also decided at this conference to provide 136 (later 144) LCVPs to the South Vietnamese civil guard and 220 more swimmer support boats and 12 MSMLs in 1963.

In May and June 1962, the United States contracted with five South Vietnamese builders for 500 sailing junks. The navy yard at Saigon was building 24 U.S.-designed command junks, and the South Vietnamese navy contracted for 12 more.[6] Although the Viet Cong harassed the South Vietnamese builders, all of the U.S.-contracted junks were complete by May 1963 for a total junk or coastal force strength of 632 craft.

The Junk Force was absorbed into the regular Vietnamese navy in July 1965.

These forces formed the basic pattern the U.S. Navy in Vietnam would follow. The DERs (radar picket destroyer escorts) and ocean minesweepers of the Market Time patrol could be considered an extension of the Ocean Force. The Mobile Riverine Assault Force was certainly patterned loosely on the river assault group. The inshore Swifts were bought to supplement the Junk Force boats.

The first efforts to develop craft specially for South Vietnam predated the Kennedy Administration. As a result of the visit of South Vietnamese Defense Minister Thuan to Washington in April 1960, two each of two types of shallow-water craft were sent to Vietnam in December 1960 and February 1961 for evaluation: a "swamp buggy" (a 19-foot rescue boat propelled by an airscrew at 15 to 20 knots) and a water-jet boat. They were, in theory, direct descendants of the Wizards bought for the French in 1954. Both had been rejected by April 1961. The swamp buggy was too noisy and the water-jet boat was fouled by vegetation. Three more airboats were shipped in May and June 1961, but they also proved too noisy.

From the perspective of combatant craft design, then, intense U.S. Navy interest in counterinsurgency warfare can be dated to the presidential directive of 1960. BuShips established code 404, the coordinator for unconventional warfare equipment, on 4 June 1961. The SCB formed a corresponding counterinsurgency watercraft panel in 1961. The problem was clearly urgent, so the panel and the SCB sought to adapt existing navy and commercial craft rather than develop new types. Standard commercial boats were attractive because they could be obtained quickly, often from stock. However, very often they failed to meet the difficult requirements imposed by Southeast Asia.

On 5 July 1961, the SCB specified requirements for three classes of craft: a shallow-water boat, a transport/logistics boat, and a pair of patrol boats, one for offshore work and one for rivers. On 27 July, CinCPacFleet listed his needs for a limited war. The list foreshadowed what the U.S. Navy would develop:

- Coastal and waterway surveillance craft—seaworthy, simple, quiet, dependable, armed, lightly armored. The coastal surveillance requirement was later filled by the Swift (PCF). The river surveillance requirement led first to the river patrol craft (RPC) and then to the plastic river patrol boat (PBR).
- Coastal river and swamp mobility—troop-carrying vehicles, fast, rugged, quiet. Some would have to be amphibious. In a sense, the mobility requirement could be associated with the creation of the riverine navy. The two main approaches to the swamp mobility problem were the Hovercraft and the Marsh screw amphibian, neither of which was entirely successful.

The obvious candidate for coastal surveillance was

The LSCR, the first large U.S. fiberglass naval boat, was considered but rejected as a possible Vietnam patrol craft. It also proved unsuited to its designed swimmer-support mission, largely owing to problems with the Solar gas turbine engine. This LCSR is on trial off Bellingham, Washington, soon after completion.

the PGM, already in production for MAP. With its 6-foot draft it could enter river approaches, but it could not penetrate a river deeply. The ASW version then being bought for Turkey and Iran would be worse, given the extra draft of its SQS-17 sonar dome.[7] Even so, in June 1961, a CinCPacFleet/CinCPac team recommended that the United States supply eight PGMs of the 16-knot PGM 43 class to Vietnam under the FY 62 MAP program in place of the planned fleet minesweeper (MSF) as a means of improving patrol coverage. As recounted in chapter 9, these PGMs received new engines, Mercedes-Benz diesels, and their armament was changed to reflect a gunboat mission. The boats became the PGM 59 class.

At this point, the navy stated a new requirement for a reasonably fast, well-armed patrol craft, seaworthy enough for offshore patrol but with a light enough draft to permit operations in coastal waters. It would be armed with one single 40mm gun, two twin 20mm, and one twin 0.50-caliber machine gun.

As of October 1961, the principal candidate was a more heavily armed version of the new plastic LCSR swimmer-reconnaissance boat, SCB 221. Like most other navy boats, the LCSR was relatively heavily loaded, in this case with a full twenty-two-man UDT team, 4,000 pounds of equipment, and 2,000 pounds of armament for a total of 10,000 pounds, the figure later associated with the PGH armament alone. It could also carry 10,000 pounds of fuel on a total displacement of only about 50,000 pounds. The design was limited by requirements for shallow draft to approach enemy-held beaches and, like the design of an LCM-6 landing craft, for stowage in the well deck of an LSD or LPD. Wet-well stowage also required that the LCSR hull incorporate a large skeg to support it when the LSD well was pumped dry. It was later discovered that the skeg acted as a fin when the boat turned at high speed, almost tipping it over; LSDs had to carry LCSRs in special cradles.

At the same time, the LCSR was designed for a high sustained speed of 30 knots in moderately rough weather, sea state 3, 32 knots in a calm. With fiberglass-reinforced plastic construction—the LCSR was the largest plastic planing hull—it promised the sort of maintenance advantages that would be important in counterinsurgency warfare, where local forces would have to use American-supplied equipment. Note that the 10,000-pound payload was about the same as that projected three years later for the PGH.

All of these figures made the LCSR attractive. However, because it was designed to be launched close offshore, the LCSR did not need great endurance or, for that matter, living quarters of any kind. It was designed to run at maximum continuous speed for eight hours, or for 200 nautical miles at 30 knots. Lightly loaded, the LCSR was expected to achieve

40 knots. As in the much larger PGM, such high speed in a limited weight was achieved by gas turbine, in this case two 1,000-shp Solar units. They had been downrated from 1,100 shp to achieve what BuShips hoped would be a 2,000-hour overhaul cycle. Although the bureau considered supercavitating propellers, it chose conventional ones to avoid any unpleasant surprises in a boat intended for service use and to maintain the speed and loading capacity that had been planned. In fact, like the skeg, the Solar engine caused considerable trouble and contributed to the overall failure of the LCSR.

The boat was arranged with a well deck aft for the swimmers and a cabin in which they could be seated with scuba gear already on their backs. The engine room was located amidships, in part to leave the stern free for swimmers.

As a patrol boat, the LCSR hull would have a solid transom with the swimmer gate sealed. The next most powerful engine was the T-55 turboprop, which was nearly the same size as the Solar but promised 1,300 to 1,500 shp. That would add 6 to 8 knots. Anything more powerful would require a new hull. For example, a similar boat powered by two 3,000-shp Proteus would be 65 to 75 feet long with an 18-foot beam and a 6-foot draft. It would probably make 50 knots in calm water. However, some alternative power plant, probably a CODOG (combined diesel or gas turbine), would be needed to operate efficiently at low as well as high speed or to exceed the eight-hour endurance. Diesels were rejected because they could not combine very high speed in a heavily loaded boat with fuel economy.

The issue of low-speed operation, to loiter off a coast, for example, in turn brought up the question of hull form. The LCSR would begin to plane at 16 knots, but its hard-chine form could hardly be described as sea kindly. A requirement for sustained low-speed cruising might make a round bottom attractive.

Further design work was concentrated on a new dual-speed power plant. In an all-diesel version, the engine room was moved aft to accommodate two diesels with V-drives and the swimmer cabin was moved forward, connecting with the swimmer well by a cramped passage between the diesels. Displacement would increase to 57,000 pounds, and speed would fall to 26 or 27 knots (210 nautical miles at 26 knots). The alternative was CODOG, in which a diesel would replace the swimmer well aft. Weight would rise to 55,000 pounds, the limiting displacement, and in addition to the eight-hour dash, the diesel would drive the boat for 50 hours at 10 knots.

The LCSR patrol boat project was dropped in December 1961, leaving only the big, deep-draft PGM.

The only alternative fast coastal boat, the PTF, which was delicate and expensive, was revived only the following April. Early in 1964, the CNO and CinCPacFleet asked the SCB's watercraft panel to evaluate a commercial Gulf Coast crew boat, the Sewart Swift. It was initially bought in small numbers, outside the normal navy procurement chain, and possibly by the CIA rather than the navy itself. Three had been bought in 1964 and were based in Vietnam, supporting OpPlan 34A, well before the Swift had been chosen for the Market Time blockade. They were all in South Vietnamese hands by March 1965.

The Mine Defense Laboratory in Panama City, which was responsible for much small craft work, leased a 51-foot Sewart boat for evaluation and later bought it. Built in 1963, it was named the *Mustang*.[8]

As for river patrol, the Vietnamese themselves wanted something faster than the boats in the river assault groups. The design of the river patrol craft (RPC) began when the Thai navy, which had no river craft, asked ARPA for PTs on short order; the agency overhauled six small landing craft (LCVPs) while a more satisfactory design was being developed. The Thai/MAAG river patrol design was forwarded to BuShips, which developed what it considered a superior version based on discussions with representatives of CinCPac, CinCPacFleet, and MAAG. In mid-August 1961, a preliminary design was forwarded to MAAG with the following characteristics: 44 x 12 x 2.5 feet, steel, twenty-four troops plus a crew of six, two high-speed diesels, a maximum speed of 18 miles per hour, a cruising speed of 15, a radius of 100/10 knots, and two 0.50-caliber guns as well as lighter weapons.

At this time, February 1962, the Vietnamese also wanted a smaller boat to carry cargo and personnel in a protected enclosure. The existing larger LCM, the LCU, and even the small LCVP were efficient cargo and personnel carriers, but they were considered too slow. The SCB tentatively sought 15 knots and a shallow draft of 3.5 feet in a new river craft personnel boat (RCVP). Since the boat would not have to be carried in standard davits, it could be longer than the existing 36-foot LCVP; a BuShips study, conducted with the assistance of MAAG, Vietnam, showed a length of 40 feet. The boat would generally resemble the LCVP (one wag on the SCB called it a "sports model LCVP"), but it would be built of welded steel and the ramp would be lower because the boat would not have to pass through 6-foot surf. Displacement would be 30,000 pounds fully loaded (20,000 light), and 375 gallons of fuel would drive the boat 180 nautical miles. Weights included 2,000 pounds of armament and 4,200 pounds of general cargo in a well sized to take a jeep. By way of comparison, the 29,500-pound (loaded) LCVP could carry 8,000 pounds of cargo.

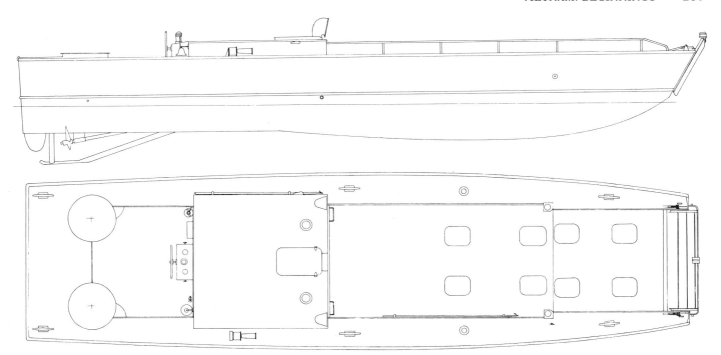

The RCVP, a riverine version of the standard amphibious landing craft (LCVP), designed but apparently not built for the Vietnam War. These drawings were made from undated contract plans. Overall length in feet-inches would have been 39-11 (37-6 on the waterline), maximum beam 10-9¾, maximum draft 3-5, for a full-load displacement of 29,900 pounds (light load 20,815). Machinery would have been two Cummins 8V-265W engines with 1.5:1 reduction gears (270 hp each at 2,800 rpm) or two 6M6-71 (64HN9) (225 hp each at 2,100 rpm), for estimated speeds of, respectively, 16.5 and 14.5 knots. Speed without cargo would have been 18 knots. A fuel capacity of 310 gallons would have sufficed for an endurance of 180 nautical miles. Armament would have been four machine guns: two 0.30-caliber in the tubs aft, and two 0.30- or 0.50-caliber on pedestals. In addition to armament, the boat would have carried 4,200 pounds of cargo and twenty-four men (175 pounds each), or 2,100 of cargo and twelve men, or 867 of cargo, six men, and a jeep. (Alan Raven)

The 9-knot LCVP used a single diesel; this RCVP would be powered by two standard LCVP engines (275 bhp each) to make 14 knots or, later on with better engines, more. The hull would have to be W-shaped aft, to accommodate twin screws in tunnels, and skegs would protect the propellers. A typical armament might be two 0.30-caliber and two 0.50-caliber machine guns, as well as miscellaneous hand-held weapons such as a portable flame thrower, a Redeye missile, or a 7.62mm machine gun.

The RCVP could not be armored, since it would take 1-inch (40-pound) plate to stop even a rifle bullet. However, its twin-engine power plant, internal subdivision, and all-steel construction would provide some protection.

The SCB specifically recommended against a larger boat, since the combination of additional speed and reduced draft was not yet possible. Slow LCUs might be used for logistics.

The SCB hoped that RCVPs could be built in the navy yard at Saigon. None appears to have been built.[9]

The designers found it attractive to base the RPC on the same hull as that of the RCVP. The RPC's was smaller than the one originally proposed. In effect, the design traded troop space for armor and a heavier armament. There was no need for the bow ramp of the RCVP; instead the RPC had a blunt, though not bluff, bow. The RPC did retain the W-stern and the twin engines—a standard MAP item—of its sister craft. At the SCB meetings, the unusual W-section hull was criticized because it might gather bottom mines. On the other hand, a more conventional V-stern would draw significantly more water. As it was, the RPC failed partly because it drew too much water.

As in the case of the RCVP, all-steel construction, with eighth-inch steel over the pilothouse, would provide a measure of protection. The RPC was designed with considerable reserve buoyancy, but it had less freeboard than the U.S. Navy's armored amphibious-assault boat.

The boat would be armed with a completely enclosed 20mm gun forward and a completely enclosed 0.50-caliber machine gun aft, plus two 0.30-caliber and two 7.62mm machine guns in spool mounts abeam the completely enclosed pilothouse; total armament weight would be about 5,000 pounds.

Table 11-1. The RPC

	Preliminary Design			As Built					
	August 1961	February 1962	March 1962	1964	RCVP 1962	RCVP(K)	LCVP	LCPL Mk 4	Coast Guard Utility
LOA (ft-in)	44-0	40-0	39-11	35-9	39-11	36-0	36-0	35-9.5	40-3
Beam (ft-in)	12-0	10-10	10-8	10-4	10-10	11-0	11-0	11-2.5	11-4
Draft (ft-in)	2-6	3-6	3-6	3-9	3-5	—	3-5	3-6	3-3
Displacement (light, tons)	—	30,000	30,700	29,220	29,500	25,000	26,600	21,300	22,400
Engines (no./bhp)	—	—	2/225	2/225	2/225	2/300	1/225	1/250	2/180
Max Speed (kts)	18*	16	13	—	15	26	9	16	22
Cruise Speed (kts)	15*	—	—	15	—	—	—	—	—
Fuel (gals)	—	—	—	350	—	—	—	160	—
Endurance (nm/kts)	100/10	180/15	180/13	—	180/14	115/26	110/9	140/16	283
Crew	6	—	6	6	—	3	3	3	—
Troops	24	18	12	—	(4,200 lbs cargo)	(6,200 lbs cargo)	(8,100 lbs cargo)	17	20
Armament	2 0.50	1 20mm 1 0.50 2 0.30	1 20mm 1 0.50 2 0.30	2 twin 0.50 2 0.30	—	—	—	—	—

Notes: Starred figures are in miles per hour in the original. They have not been translated into knots. The RPC design allowed for 5,000 pounds of weapons. As of 1962, the planned 20mm mount could also accept the 81mm mortar Mark 2 as an alternative. Armament was also listed as one twin Mark 17 and one single Mark 26 0.50-caliber machine gun, plus two single 0.30-caliber. The LCVP listed was the standard type. The MAP conversion (quarter-inch STS protection) added about 3,000 pounds, which had to be subtracted from cargo weight. A single 20mm cannon or 81mm mortar could be installed in the cargo well, at a further cost in cargo weight. As of March 1962, a new LCVP cost $13,500, but activation, conversion, and replacement of an LCVP for MAP cost $24,000, excluding electronics and weapons. The RCVP was expected to cost $20,000 ($40,000 for the lead ship), based on its contract plans. At this time the RPC was expected to cost $25,000 ($45,000 for the lead ship), and the LCPL Mark 4, $30,000, including government-furnished material but not armor or armament. The alternative wooden hydrokeel, modified-air-cushion RCVP listed above (K) would probably cost $40,000 ($150,000 lead ship). Its underbody differed from that of a standard LCVP in that it had two sidewall keels and a centerline keel forming twin tunnels for an air supply. Air was injected behind hinged flaps at the bow to form an air bubble on which the boat rode. A hydrokeel LCVP was successfully tested, but it never went into mass production. No price is shown for the Coast Guard utility boat because it was no longer in production.

Table 11-2. Small Boat Alternatives, 1962

	SSB	Zodiac Mark 3	20-Ft Sampan	Boston Whaler	Army Assault Boat
LOA (ft-in)	14-6	15-4	20-0	13-3⅓	16-5
B (max, ft-in)	6-10	4-4	4-2	5-2	5-4
Light Displacement (tons)	410	320	300	275	320
Full Load (tons)	1,500	2,200	1,500	1,500	3,375 (or 15 troops)
Max Speed (kts)	16[1]	7	7	9.4	7.5
HP	40	40	6	40	25
Engine	Gasoline	Gasoline	Gasoline	Gasoline	Gasoline
Endurance (gal/hr)	4	4	1	4	3
Hull	Reinforced fiberglass, foam core	Inflatable rubber-coated fabric	Wood or reinforced fiberglass	Reinforced fiberglass, foam core	Reinforced fiberglass
Cost w/o Engine ($)	1,300	650	—	500	381

1. Has been tested with a 75-hp engine, 8 gallons/hour. Engine costs: 6 hp, $240; 25 hp, $292; 40 hp, $460 with a long shaft or $445 with a short shaft; 75 hp, $895.
2. On trial, 9 June 1961, the 10-foot boat at 800 pounds made 21.8 miles per hour; it made 6.7 at 1,230 pounds. The 13-foot boat made 20.5 knots at 750 pounds, 17.5 at 1,200 pounds, 15 at 1,600 pounds, and 24 at 350 pounds.

An 81mm mortar could replace the machine gun aft.

Normally, the RPC would carry twelve troops and a crew of six—one each for four weapon positions, a coxswain, and an engineer or relief coxswain or officer-in-charge. The troop guns would provide a significant part of the RPC's total firepower. Weight was reserved for up to eighteen troops.

The SCB proposed the existing LCPL Mark 4 as an interim river patrol boat pending production of the RPC. BuShips estimated that an RPC would cost $25,000 ($45,000 for the lead boat) as opposed to $30,000 for the LCPL, which included government-furnished equipment but not armor or armament.

Table 11-1 compares the preliminary RPC characteristics with the RPC as redesigned and then as built, with two existing alternatives, the small personnel boat (LCPL) and the Coast Guard 40-foot utility boat. The table also describes the RCVP and an alternative, faster RCVP(K) based on the new hydrokeel LCVP, an air-cushion (surface-effect) craft.

The RPC was the only major failure of the U.S. Vietnam small-craft program. Even before construction, it was criticized in a report from the Naval Ordnance Laboratory for its noise, vulnerability to weeds, mine-attracting tunnel stern, readily identifiable silhouette, low speed, and limited armament. Having been designed for the relatively small Vietnamese, it was too crowded. Finally, its tunnel stern encountered terrible problems with air bleed. Although they were largely solved after much effort, the craft was never particularly popular.

A total of thirty-four RPCs were built. Ten came from Peterson Boatbuilding of Sturgeon Bay, Wisconsin, under the FY 64 program: six ordered on 24 May 1964 (C 13371–76, completed between 8 January and 11 March 1965), and four on 30 June 1964 (C 13378–81, completed between the same dates). Twenty-four were ordered from Birchfield Boiler of Tacoma, Washington, on 28 January 1965 under the FY 65 program (C 13456–79, completed between July 1965 and February 1966). Each boat was powered by two Gray Marine 64HN9 diesels. All were originally delivered to Vietnam, but six, C 13474–79, were transferred to Thailand in March 1967.

The other requirement was for a small, shallow-draft boat. In May 1962, the SCB described a variety of alternatives, including the existing swimmer support boat and the semirigid Zodiac (see table 11-2). These troop utility boats were needed specifically to transport guerrilla teams in shallow and restricted waters. They had to be rugged, capable of carrying up to six men with their equipment (1,500 pounds), operated by one man, and powered by outboard motors. Dimensions were initially limited to 13 to 15 feet in length, 4 to 7 in beam, and 2 in draft, and light weight was not to exceed 400 pounds. Rigid hull boats were to be able to mount the 0.30-caliber Mark 21-4 or the 7.62mm M60 machine gun. Minimum acceptable performance was 15 knots fully loaded with a 40-hp outboard engine and an endurance of 50 nautical miles at 15 knots. Rigid-hull boats were to absorb light machine-gun fire without sinking. Solid-core hulls had to be strong enough to accept a 75-hp motor; the rubber-hulled, pneumatic boats had to accept engines of up to 45 hp.

In keeping with their guerrilla mission, boats had to be adaptable to air dropping and towing, and they had to be able to tow each other. They had to be light enough for their crews to paddle or pole them while they were loaded.

The navy began to buy large numbers of very small boats for the Vietnamese. Procurement was fast because the need was so urgent. There were three categories:

- Boats with a very shallow draft, for example, for swimmer support. The chief of BuShips evaluated two small commercial boats he had seen in Panama City; they became the first swimmer support boats. They were shipped to MAAG on 4 August for tests. On 5 September, MAAG asked OSD (the Office of the Secretary of Defense) to provide twelve 14-foot boats, not to weigh more than 60 pounds each but to carry a load of 500 (two or three men); they had to be suitable for paddling. By October 1961, ARPA was in the process of shipping thirty very small boats to Vietnam. Late in 1961, 222 swimmer support boats (*dong nai*) were shipped to Vietnam to replace the native sampan. They were derived

	SSB Mark I	SSB Mark II
	10-0	13-7.5
	6-0	6-10
	200	345
	1,600	1,600
	—	—
	35	20
	Gasoline	—
	—	—
	Styrofoam block covered with one layer woven roving, one layer plain weave	—
	7,500	—

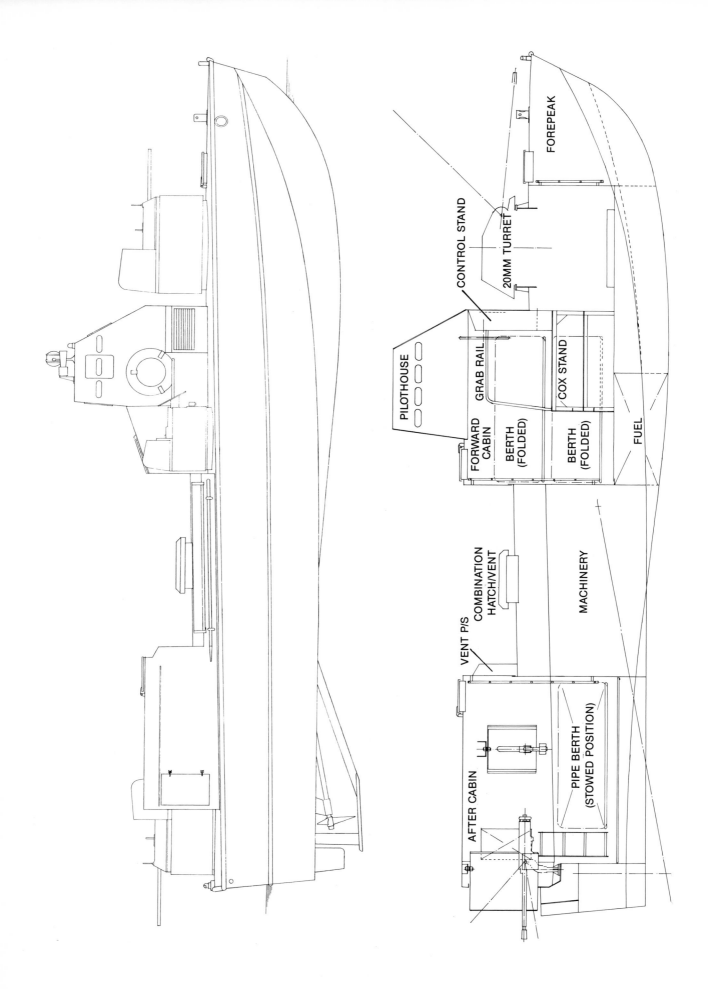

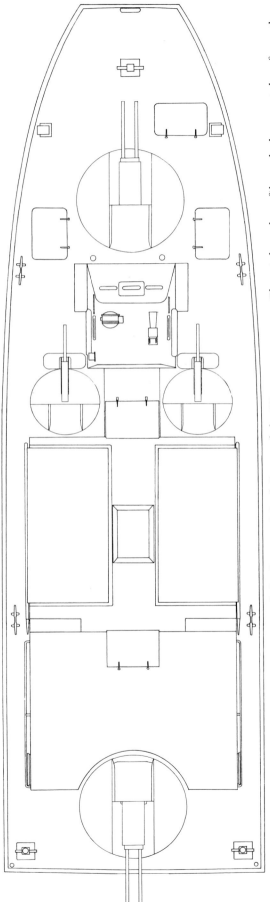

The 36-foot steel river patrol boat (RPC) was one of the few outright failures of the Vietnam program. The outboard profile and plan are taken from the official plan issued in January 1964 and revised that September. In June 1964, the after mount was changed from a single 0.50-caliber gun to a twin Mark 17. The two 0.30-caliber guns were in Mark 21 mounts. All shields were nonstandard, designed by the builder, and made from 5.1-pound MS plate. The inboard profile represents an earlier stage of the design: a 20mm cannon forward and the 0.30-caliber guns on pintle mounts in the after deckhouse (rather than on scarf rings abaft the pilothouse, as in the craft built). In this version the 0.50-caliber gun aft is a single pintle mount. The associated plan view shows the 20mm gun mounted off center to port, so the forward turret was probably a forerunner of Mark 48, carrying a coaxial 40mm grenade launcher. (Alan Raven, Norman Friedman)

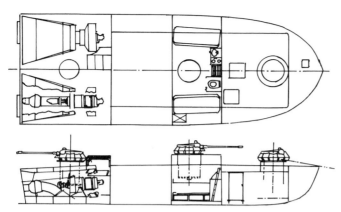

With the beginning of direct U.S. involvement in Vietnam, numerous contracts for counterinsurgency studies were let. United Aircraft proposed an inland-water warfare patrol boat, a stripped-down 40-foot armored gunboat version of which is shown here. The forward cupola was the conning tower, and the main battery consisted of two 20mm guns with 250 rounds each in stabilized armored cupolas. Armor was concentrated around the magazines and gas turbine engines (driving water jets), and there was space for cargo between the two gun mounts. Displacement amounted to 25,000 pounds, the crew totaled four, and speed was 35 knots, endurance, 200 miles. The drawing is undated, but it was published by the U.S. Navy in July 1966. It was probably submitted (and rejected) early in 1965. The study began in 1964. (Norman Friedman)

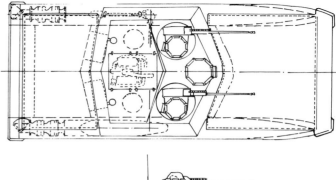
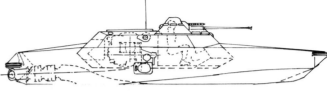

FMC, the manufacturer of armored personnel carriers and amphibious tracked vehicles, proposed an air-cushion armored river patrol craft (ARPC). It employed standard gun mounts—in the case illustrated, a pair of 20mm/0.30-caliber cupolas. Total height was 6 feet 4 inches, length slightly over 30 feet. Sponson assemblies would have been built of aluminum and filled with foam. The two engines were mounted under an armored hatch amidships. Propulsion was to have been achieved by water jet; details are not available. (Norman Friedman)

from the two Panama City boats, being 14 x 7 feet, 345 pounds light (with a payload of 1,200), and driven at 20 knots by a 40-hp outboard. The *dong nai*, built of shaped styrofoam planks encased in two layers of resin-impregnated fiberglass, were unsinkable. Their medium deadrise hulls were highly maneuverable. Similar boats were later built by the Vietnamese, and a modified version, SSB Mark IV, was manufactured for the SEALs. These craft were succeeded by commercially designed Boston whalers, also unsinkable and often used both as utility boats and to transport officers within Vietnam. Small boat alternatives are summarized in table 11-2.

- Boats to transport people and things in quantity on rivers. In June 1961, BuShips was authorized to assist the Vietnamese in obtaining 700 small sampans with minimum dimensions of 17 x 4 feet. They were built in Vietnam with basic materials supplied through MAP, and technical assistance from BuShips.[10] On 1 August, CinCPac advised the U.S. MAAG in Vietnam to urge the Vietnamese to proceed. BuShips forwarded preliminary designs for a six-man wood or plastic sampan on 14 August, and a molded plastic sampan was shipped from Saipan to Vietnam later in the month. The preliminary design for a 10-foot, two-man plywood sampan was forwarded at the end of August.
- Patrol boats, including conventional river patrol craft and air-dropped, semirigid Zodiacs for the marines. BuShips was authorized on 30 June 1961 to buy 125 British-built Zodiac Mark 3s powered by 40-hp outboards. Procurement began in London early in July, and the boats were delivered in October and November.

By about December 1962, SCB's watercraft panel had formulated a program to include the following:

- A compact boat, which could be air dropped, for escape and evasion. Six such "instant" boats were built under a March 1962 contract, five going to Thailand (OSD/CDTC) and one to the USN SEAL teams in St. Thomas. The instant boat consisted of a dacron cover that could be filled in about thirty minutes with a rigid, monocellular plastic foam. The air-drop kit weighed only about 50 pounds, and the boat was 14 feet x 3 feet x 5 inches with a total depth of 10 inches. It could withstand a 16-foot free fall onto concrete, was designed to take a 5.75-hp outboard, and could carry a 500-pound load. The alternative folding boat would be stacked in an airplane, dropped by parachute in bundles, and then discarded. It

had similar dimensions and weighed about 70 pounds.
- A small, fast, silent reconnaissance boat. Work on silencing outboard motors led eventually to the quiet fast boat (QFB) used by the SEALs some years later in Vietnam.
- A fast, 55-knot boat for intelligence, clandestine attack, interdiction, or special missions. The SEALs never obtained anything this fast, as the PTF was limited to about 45 knots in smooth water. Inside Vietnam, it was limited to about 35 knots.
- A 30 to 40-foot, high-speed boat for coastal anti-smuggling blockade. This requirement was filled by the 50-foot Swift, capable of 30 knots.
- Shallow-draft river and coastal patrol boats, particularly the RPC and RCVP, for support, escort, attack, and logistics.
- A lightweight amphibian for reconnaissance, escort, attack, logistics in shallow water, and operation in rice paddies, swamps, and off-road terrains. It became the marsh screw amphibian, based on a commercial roto-sleigh; three amphibians were funded jointly by the navy and ARPA. As of early 1964, they were being tested in snow, swamp, sand, and water. Considered very promising, they later figured in plans for the Riverine Force. At this time, the aluminum marsh screw weighed 2,500 pounds; it was 12 feet 8 inches long and 8 feet wide with a maximum height of 5 feet 9 inches. A 140-hp engine drove it at 12 miles per hour in water and at 8 on land, including mud and sand. Endurance in water was four hours at full power.

ARPA and the navy also tested another amphibian, the air roll (LVA-1), built by Ingersoll's Kalamazoo division. Eight air rolls were ordered. They had two tracks, each consisting of a series of free-running tires or rollers on an endless belt. They weighed 4,000 pounds, had dimensions in feet-inches of 14-8 x 8-0 x 6-10, and were expected to attain 10 miles per hour loaded, 15 empty. ARPA and the army ordered the Canadian-built Jiger, a small, six-wheeled, two-man amphibian scout car.

Apart from the big riverine craft, this agenda shaped the U.S. small-craft contribution to the Vietnam War. Table 11-3 summarizes naval programs for Vietnam as of mid-1962.

By early 1964, related navy programs included the following:
- The development of a pump-jet propulsor for very shallow water by David Berkeley. Pump jets required minimal hull appendages, which resulted in faster craft. Positive nozzle steering made for excellent maneuverability and sharp turns. Pump jets propelled the standard PBR.

Table 11-3. Special Projects for Vietnam as of Mid-1962

Project	Number	Funding
Air Cat	1	MAP*
Air Boy	1	MAP*
Sampan	1	MAP*
Skimmer (Danville air boat)	1	MAP*
Coast Guard 40-Ft	2	MAP*
Swimmer Support Boat (Swamp boat)	20	MAP*
SSB	190	MAP*
LCM	16	MAP*
LCVP	136	Army
Junks (including 220 engines)	524	MAP 1962
Folding Boat	2	MAP*
Instant Boat	6	MAP*
Roto Craft	3	ARPA
Lepel Water Scooter	1	MAP*
Zodiac Swamp Boat	125	MAP*
50-Ft Utility Boat/Minesweeping Launch	12	MAP 1962

Note: Starred items indicate funds had already been committed.

- The evaluation of the Rule (shallow-water) propeller by the David Taylor Model Basin.
- The development of a Hovercraft, the Bell SKMR-1 hydroskimmer, for a feasibility test. Air-cushion vehicles were later used experimentally in Vietnam.
- The purchase of two "aqua skimmers," which were placed with UDT personnel at St. Thomas for evaluation.

Work also proceeded on new lightweight armor. Four main types were available. A new face-hardened steel was considered twice as effective, that is, half as heavy, as World War II plating. In tests, quarter-inch plate stopped thirteen of fifteen 0.30-caliber bullets fired from a range of 15 yards. New ceramic armors were even lighter, although metal plating was more effective against oblique hits. A 6-pound/square foot thickness of boron carbide was equivalent to but much more expensive than a 9-pound/square foot thickness of aluminum trioxide ceramic; the equivalent face-hardened quarter-inch steel plate weighed 10 pounds/square foot. Because ceramic could shatter, it was normally backed by laminated fiberglass or metal. Ceramic was not considered effective against heavier machine guns such as the standard Soviet-bloc 14.5mm weapon.

Transparent armor of 17 pounds/square foot would stop the 0.30-caliber armor-piercing round. Composite armor consisted of aluminum trioxide balls or cylinders embedded in aluminum and backed by metal. Finally, ballistic nylon was typically used in body armor. It was later applied to boats in Vietnam in the form of antifragment canopies. These new armors were actually employed in Vietnam.

Appendix D describes special small-craft weapons developed and tested at this time.

The interior of a program 5 ATC shows shock-mounted seats. They had collapsible basket-weave legs to absorb shock and were mounted on a rigid platform with its own collapsible legs (barely visible behind the angled crosspieces). Seat belts kept troops from flying when the craft was jolted by a mine. These features were developed by the David Taylor Model Basin, which had already worked for some years on minesweeper shockproofing.

Fiberglass offered the advantages of a light hull and ease of production in large numbers. The first large U.S. Navy fiberglass boat was the 52-foot LCSR; this hull is shown at United Boatbuilders, Bellingham, Washington. (Later the company adapted its production technique to the 31-foot PBR, changing its name to Uniflite.) Ridges were molded into the hull for strength, and later in the production process a rectangle was cut in the transom to allow swimmer access. (USN)

All of these programs promised new types of craft at some later date. For immediate use, the navy had to rely on existing commercial craft, which meant the Bertram and the 50-foot Sewart Swift.

There was also the small commercial Boston whaler. Several whalers were bought early in 1964 for evaluation by the commander of amphibious forces in the Pacific and by the U.S. Navy in Panama. They became standard personnel boats in Vietnam and were also used by the SEALs. The Boston whaler was both stable and "unsinkable" because of the inherent buoyancy of its plastic, foam-filled hull material. It weighed 250 pounds, could take a crew of about ten, and could carry 1,500 pounds or 500 pounds at 24 miles per hour. Dimensions were 16 feet 7 inches x 6 foot 2 inches x 1 foot. The Boston whaler also formed the basis for the SEALs' QFB. Like the RPC, the Boston whaler had a W- (trimaran) hull.

Two other skimmers similar to the Boston whaler were also used in Vietnam: the Kenner ski barge (KSB) and the Miami surfer. The Boston whaler was the most durable; the KSB, which the army bought because it was cheaper than the whaler, suffered because of a thin hull.

In December 1964, the CNO announced a new study of cold war patrol craft. The naval advisory group (NAGP) of MACV reported the following February. It was particularly concerned that the U.S. Navy avoid excessive sophistication in craft that would, after all, have to be operated by local sailors. It found the PTF, which the Vietnamese had specifically requested, expensive and far too fragile. That was only one example of the more widespread tendency of small allied navies to request relatively large ships for their prestige value without reference to concrete naval requirements. The Vietnamese, for instance, wanted APDs, whose steam plants they could not support.

The advisors emphasized the need to control maintenance costs, which over the lifetime of a ship or boat far exceeded initial cost. Existing World War II craft, in Vietnam and elsewhere, already presented severe maintenance problems because of age and a lack of spare parts. The hulls of Vietnamese LSMs and PCs had weakened over the years. The U.S. Navy had solved a similar bloc-obsolescence problem by extensive reconstruction under the FRAM (fleet rehabilitation and modernization) program, but a similar effort would be expensive for Vietnam. NAGP reported that material inspections of fourteen Vietnamese navy ships were scheduled for February 1965; it expected that replacement would be shown to be more desirable than rehabilitation. If all World War II ships needed to be replaced within ten years, as expected, new craft would have to be designed.

NAGP suggested two complementary craft, a small gunboat and a small river and coastal patrol boat. Plans to test the new RPC and the interim LCPL for river patrol were arrested by the suspicion that an entirely new coastal craft would have to be designed.

The larger gunboat might best be based on the World War II–built LSSL/LSILs. Advisors considered them the most valuable Sea Force ships with their simple and reliable machinery (twin-screw diesels, useful for maneuverability on rivers and in strong currents and powerful enough to retract from the ship's groundings or beachings), their relatively high speed of about 15 knots, their silence (they were considered the quietest craft in-country), their shallow draft, and their steel construction (for maintenance and for protection against small-arms fire). They could be deliberately grounded to produce more stable gun platforms, and they had simple fire-control systems. A counterinsurgency boat built along these lines should be armed with 3-inch/50 or lightweight 5-inch/54 guns, 40mm guns, and mortars. Its simple electronics would have to include a high-definition radar capable of tracking wooden boats to maximum gun range, about 6 nautical miles. Modernity could well be restricted to communications, which would have to be compatible with other equipment in-country.

A future MAP gunboat would probably need some small troop lift because inventories of appropriate small landing craft, particularly LCMs, were declining. It would carry small craft such as plastic assault boats, a loudspeaker for civil action and population control, night vision equipment, and searchlights.

The gunboat might be made more palatable to its recipients if it had a new gun such as the lightweight 5-inch/54 and if its speed were increased to 18 knots; the advisors clearly were unaware that the only MAP gunboat in prospect was the 40-knot PGM.

They decried the lack of appropriate standard craft in the size range for river or coastal warfare (30 to 50 feet in length). To be sure, there had been sporadic attempts to obtain commercial craft, primarily using 31-foot Bertrams, but no concerted effort.

NAGP had in mind a quiet, simple, reliable, sturdy boat with a range of 400 to 500 nautical miles for coastal patrol, a speed of 20 to 25 knots, a nonwood hull to avoid rot, and screws and propellers protected against grounding. It is clear from the context that NAGP had not yet met the Swift.

12

Vietnam: Market Time and Game Warden

For the convenience of the reader about to start the next few chapters, it is best to summarize later developments in Vietnam. During 1965–66, the navy implemented two of the Bucklew recommendations, forming Task Force (TF) 115 for Operation Market Time, the coastal blockade against arms smuggling, and TF 116 for Operation Game Warden, the extension of Market Time into the seaward reaches of the Mekong Delta. By this time, the U.S. Navy in Vietnam was clearly an operational rather than an advisory entity. However, there was no overall naval commander. The post of commander of Naval Forces, Vietnam (ComNavForV) was established on 1 April 1966 and given authority over TFs 115 and 116. Meanwhile, MACV evaluated proposals for a U.S. equivalent of the old French DNA, and operation in the Mekong Delta of the Mobile Riverine Force, including TF 117, its naval component, was approved in Washington. The force entered combat in April 1967.

There was still no mobile naval patrol of the Cambodian and Laotian borders, and the North Vietnamese were able to ship arms into South Vietnam via the Cambodian port of Sihanoukville. Cambodia was theoretically neutral, and the U.S. government was loath to expand the limited war it was trying to fight. Until late 1968, it had neither the logistical nor the operational capacity to seal the Cambodian border, and with allied Vietnamese troops reluctant to undertake an aggressive patrol of river banks to close it, the Viet Cong were able to evade the Riverine Force by falling back on small waterways.

The Tet Offensive of 1968 radically changed conditions. The Viet Cong, evading sea and river patrols, had managed to build up the necessary stocks of weapons, demonstrating that something beyond the existing strategy was needed. The North Vietnamese opened a new theater of naval war after the city of Hue was cut off from land communications. In February 1968, Task Force Clearwater was formed from TF 116 and 117 craft to keep open the remaining supply route, the Perfume River. It also patrolled the Cua Viet River, just south of the DMZ.

Although U.S. counterattacks effectively destroyed much of the Viet Cong, in the United States the effect of the offensive was to demonstrate what appeared to be the weakness of the American position. Richard Nixon won the 1968 election partly on a pledge to end the war on favorable grounds. By that time, it was clear that U.S. forces would soon have to turn over major responsibilities to the Vietnamese. On 31 December 1968, President-elect Nixon introduced the term Vietnamization to describe this process.

Vice Admiral Elmo Zumwalt became ComNavForV on 30 September 1968. He decided to integrate his forces to stop infiltration of the Mekong Delta and to pacify it; to gain the initiative in the increasingly troubled Rung Sat special zone; and then to accelerate the turnover of his command to the Vietnamese. The offensive strategy, Sealords (Southeast Asia lake, ocean, river, and delta strategy), included an inland naval patrol of the Cambodian border northeast of the Gulf of Thailand to the inland "Parrot's Beak." Sealords was executed by the new Brown-Water Task Force (TF 194), which combined elements of all three existing task forces: the unarmored Swifts of TF 115, the fast "plastics" (PBRs) of TF 116, and the slow, heavy riverine craft of TF 117, all supported by navy helicopter gunships (Seawolves), fixed-wing OV-10 Broncos (Black Ponies), and SEALs.

TF 194 established a series of interdiction barriers: Operation Search Turn (2 November 1968, on the Rach Gia, Long Xuyen, and Ca San canals in the upper delta); Operation Foul Deck (16 November 1968,

---

An early river patrol boat, PBR Mark I. The Mark II version had a larger superstructure and lacked the armored ring around the forward gun mount. The plates of vertical armor protect the coxswain and machinery.

on the Rach Giang Thanh and Vinh Te canals at the Cambodian border); Operation Giant Slingshot (6 December 1968, on the Vam Co Dong and Vam Co Tay rivers astride the Parrot's Beak); and Operation Barrier Reef (2 January 1969, on the La Grange-Ong Lon Canal). Giant Slingshot was so far upriver that it employed floating advanced tactical support bases (ATSB) built up from army (*Ammi*) pontoons.

The net effect of these barrier operations was to reduce the Viet Cong from large shipments of fifteen to twenty sampan loads to small shipments of two to three sampan loads. Even the small shipments became extremely risky, making it difficult for the Viet Cong to replace wastage in the Mekong Delta.

Sealords also envisaged the pacification of the Ca Mau Peninsula and the U Minh Forest, the areas of the Mekong Delta controlled by the Viet Cong. Swifts raided it in October 1968; the Viet Cong erected fortifications that were destroyed in December by riverine assault craft engaged in Operation Silver Mace. Through the first half of 1969, Sealords depended on forces attacking from the sea. In May, CTF 115 proposed a mobile advanced tactical support base on the Cua Lon River. It materialized as Sea Float, protected by its attached riverine craft and helicopter gunships. The fast river current precluded swimmer attack. The base's components were built at Nha Be and carried to the mouth of the Cua Lon River by LSDs. The first pontoons were towed up the river on 25 June 1969 by YFUs. A Sea Float annex was later established six miles away. These bases established a permanent navy presence in the area and so made for effective pacification.

The other major Sealords operation was to secure the Rung Sat Special Zone (RSSZ), particularly the Long Tau shipping channel to Saigon. After a lull in attacks on merchant ships, due in part to the clearing of a broad strip on either side of the channel, attacks began to increase early in 1969. The sanctuary of the Viet Cong sapper unit responsible was known; it was not far from Saigon. A strike in June badly damaged the unit, and pressure on the Viet Cong continued through the rest of the year.

From a tactical point of view, perhaps the most important feature of Sealords was the use of small craft in ambush along rivers and canals. These waterborne guard posts used electronic sensors emplaced on shore to detect the nearby enemy, and they could use their speed and associated helicopter-borne troops to pursue him. The waterborne posts radically changed the nature of river warfare. Radomes, for example, became shrapnel hazards and were often removed, since radar was no longer particularly valuable as a surveillance sensor.

By early 1970, Zumwalt could reasonably claim that he had achieved his operational objectives. Vietnamization was the logical next step. He hoped to turn over virtually all U.S. operational responsibilities by 30 June 1970, the end of FY 70, and all support functions and bases by 30 June 1972. The first unit, River Assault Division 91, transferred its craft to Vietnamese River Assault and Interdiction Divisions (RAIDs) 70 and 71 on 1 February 1969. All U.S. operational craft except the SEAL's were handed over by the end of 1970.

The account that follows frequently refers to program 4 and program 5. Direct U.S. naval contributions to the Vietnam War were broken down into annual programs of which these two, FY 67 and FY 68, respectively, were the most important. The RPCs were bought under programs 1 and 2 (FY 64 and FY 65), the earliest Swifts under program 2 (FY 65), and the first PBRs under program 3 (FY 66). The later riverine aid package for Cambodia was called program 6 (FY 73), at least unofficially. Note that the riverine craft of programs 4 and 5 were frequently described as Marks IV and V, although there were never references to Marks I through III.

U.S. involvement in the coastal blockade began in December 1961, when U.S. ocean minesweepers (MSO) joined Vietnamese barrier patrols near the North-South border. The MSOs did not actually stop suspicious junk traffic, but they vectored Vietnamese ships to it. Similar patrols by U.S. destroyer escorts began in the Gulf of Thailand in February 1962. Few arms carriers were intercepted, and the patrols ended on 1 August. From a tactical point of view they were significant, though, in being precursors of the large Market Time operation.

Market Time itself began in 1965, as the Viet Cong armed for what they expected would be the decisive phase of their war. Over 100 tons of munitions were found aboard a North Vietnamese coastal ship forced to beach in Vung Ro Bay, South Vietnam, on 16 February 1965, and the JCS approved a U.S. Navy coastal patrol on 16 March. The navy resumed blockade operations with ten Seventh Fleet ships (TF 71, the Vietnam Patrol Force) on 11 March, President Johnson formally approving it on the fifteenth. In August 1965, responsibility shifted from Seventh Fleet to the specially formed TF 115, the Coastal Surveillance Force. CTF 115 was also responsible for port security within Vietnam (Operation Stable Door).

Market Time was targeted against two distinct types of craft, coastal junks, which could lose themselves among thousands of civilian craft, and trawlers or larger vessels, which had to operate further out to sea. On any given day, about five thousand junks were at sea, fishing or transporting cargo. Initially, the United States and South Vietnam's Navy

The 50-foot Swift was bought off the shelf to enforce Market Time. Its commercial origins show in the shape of the side windows. This boat was attached to Coastal Division 102. The lookout position is built onto the radar mast, the bumper stowed beneath one of the windows. This photograph was released in February 1967.

Sea Force took joint responsibility for the larger craft, while the Coastal Junk Force was assigned to the smaller coastal craft.

Thus, TF 115 initially consisted of seven radar picket destroyer escorts (DER), two MSOs, two LSTs to provide radar coverage of the entrances to the Mekong River, and six PTFs (*Nastys*), based at Danang, to operate near the North–South border. The PTFs were soon reassigned to other duties.[1] The large ships were joined by seventeen 82-foot Coast Guard cutters (WPB).[2] The surveillance area was divided into nine patrol zones, each controlled by a DER or MSO; the 82-footers were used for barrier patrols along both the North–South border and along the South Vietnamese–Cambodian border. Coastal surveillance centers were set up at Da Nang, Qui Nhon, Nha Trang, Vung Tau, and An Thoi.

The outer barrier ships, working beyond the 12-mile limit, ultimately included large *Barnegat*-class U.S. Coast Guard cutters (WHEC) and *Asheville*-class gunboats. The latter were valued for their seaworthiness, speed, and firepower, which could neutralize a suspect trawler. However, note that of seventeen trawlers attempting to infiltrate before 1970, only two were detected initially by shipboard radars. One was spotted by a patrolling Swift, all of the others by patrol aircraft. Market Time often worked by means of a seaborne equivalent of ground-controlled air interception.

By 1 April 1965, the NAGP in Saigon was pressing for an inshore U.S. patrol to supplement the offshore ships. It considered the Coastal Junk Force neither effective nor sufficiently aggressive. Of its 530 junks, only half were motorized and none had radar. However, there was no completely suitable U.S. Navy patrol boat, although a few U.S. commercial craft had been bought by Venezuela. Nor was there any suitable Coast Guard boat available in sufficient numbers. The existing 82-foot cutter was somewhat larger and slower than desired, and the 40-foot utility boat was too small.

For the Vietnam inshore patrol the navy bought a commercial boat, the 50-foot Sewart (later Teledyne Sewart) Swift, which had been designed to support offshore oil rigs in the Gulf of Mexico. That mission required high speed in moderately rough water at shallow draft, the two main requirements of an inshore patrol boat. Most importantly, the Swift was in production and therefore could be obtained quickly in relatively large numbers.

The Swift program demonstrated both the benefits and the pitfalls of "off-the-shelf" procurement. The great benefit was speed of production, which met an urgent requirement. The NAGP proposal of seventeen 82-footers and fifty-four Swifts was dated 1 April 1965. On 14 April, Secretary of Defense McNamara authorized Secretary of the Navy Paul Nitze immediately to procure twenty Swifts to counter

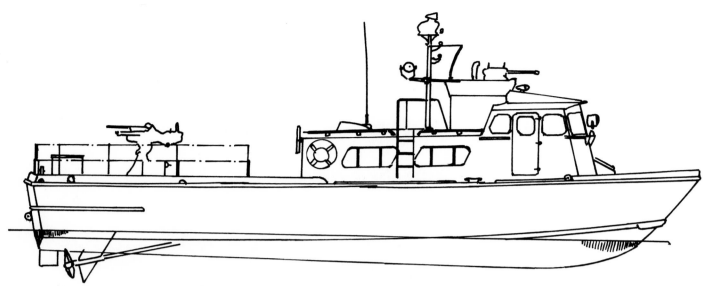

The Swift Mark I. (Norman Friedman)

infiltration. The FY 65 contract, for sixteen Swifts, 50NS651–6516, was signed on 20 May, the first boat was delivered on 9 July, and the first boat arrived in Vietnam on 30 October.

An FY 66 contract for thirty-eight more Swifts, 50NS661–6638, followed on 23 August. Then Market Time was reviewed in September 1965, and the force level raised to twenty-six Coast Guard boats and eighty-four Swifts. As a result, the navy ordered fifty more Swifts under FY 66 on 1 December 1965, 50NS6639–6688, for a total of 104 Mark Is, including twenty for training in the United States. All eighty-four operational boats were on patrol by November 1966. Of the 104 Mark Is, four were transferred to the Philippines after serving in the U.S. Navy, and two more were transferred upon completion. The first eight improved Swift Mark IIs were ordered for the Philippine navy on 16 August 1967 under FY 68: 50NS681–688, yard numbers 1445, 1451, 1453, 1573, 1609–12. These were followed by twenty-two more on 31 December 1967, 50NS689–830. They included two Thai and six Vietnamese boats.

Finally, thirteen Mark 3s (50NS691–6913) were ordered under the FY 69 program—three for the Philippines, five for Thailand, and five to make up for attrition in Vietnam—and twenty (50NS721–7220) under FY 72, all for Cambodia. Other Swifts were ordered directly by foreign governments: 4 for Brazil (with at least six more built locally), 10 Mark IIs for Thailand, and 6 Mark I or IIs for Zaire, for a total of 193. In addition, the Philippines and South Vietnam each built a prototype ferro-cement Swift as part of a U.S.-sponsored program for inexpensive mass production of such craft.

Secretary Nitze asked his assistant, then–Captain Elmo Zumwalt, to evaluate the Bell Hovercraft (SRN-5, later PACV) as an alternative. Two days later Nitze asked whether any existing navy boats, such as the *Nasty* and the 63-foot air-sea rescue boat, could serve as interim patrol boats. Patrol craft were needed so badly that he suggested buying privately owned Swifts.

Zumwalt broke Market Time into three categories: offshore in deep water; river/waterway; and overlapping areas, including the difficult, unusual terrain of the Mekong Delta and the shallow waters of the Gulf of Thailand near the Cambodian border. In the latter, from Cat Lo to An Thoi, the sea floor sloped so gradually that even Swifts could not come in close enough to maintain their patrols. No conventional craft could stand patrol duty 10 nautical miles offshore and still closely approach a shallow beach.

The Swift was clearly more effective than the *Nasty*, which could not cruise at less than 12 knots. However, because its habitability was limited, the Swift could not be used to cruise offshore. It would have to function, Zumwalt thought, more as an interceptor. The Hovercraft was more sophisticated and therefore more expensive, but it was clearly the solution to the very shallow area near the Cambodian border. Conventional boats seemed more cost-effective elsewhere. Zumwalt therefore recommended buying the twenty Swifts plus three Hovercraft specifically for use in the Mekong Delta. He reserved the possibility of buying ten more of the latter.

All of these boats would operate with U.S. crews and Vietnamese navy boarding parties.

Three Swifts were already in Vietnam, based at Danang. They had been bought (probably by the CIA, since they do not appear on navy records) for Oper-

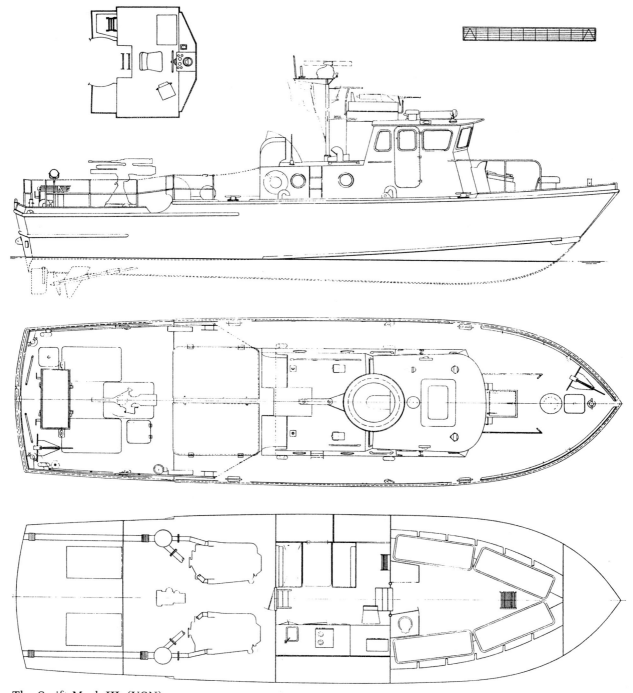

The Swift Mark III. (USN)

ation Plan 34A, and by April 1965, they were being used for border surveillance. By early 1965, all were manned by the South Vietnamese navy. Their builder, Sewart, was a relatively small company, and quick procurement of the secretary's twenty Swifts would require rescheduling. It appeared that by delaying delivery of four 85-foot missile-range boats being built for BuWeps, one boat being built for Venezuela, and one being built for General Motors, and by working two shifts, Sewart could deliver one boat on 1 June, three on 1 July, and two in each of the following months for a total of twenty by March 1966. In fact, Sewart did much better.

The 82-foot Coast Guard boat already existed; first deliveries were made in June 1965. A second group, units 18–26, was planned in September and October

1965 and delivered in January and February 1966. Comparative prices of the Market Time craft are shown in table 12-1, and the entire navy small craft program, as it existed in April 1966, is shown in table 12-2.

Table 12-1. Comparative Costs of Market Time Craft, March 1966

	Costs (thousands of $)	Total Procurement
PGM	4,900	17
PGH	3,850	2
PTF (*Nasty*)	1,400	14
PCF	187	104
PBR	75	120
PACV	350	3
LCPL XI	100	16

Table 12-2. USN/USCG Small Craft as of 12 April 1966

	Programmed Through FY 66	Active	Reserve Ready for Issue	Reserve Not Ready for Issue	Now in South Vietnam
PTF	16	12	—	—	11
PGM	2[1]	0	0	0	0
PCF (Swift)	104[2]	63[3]	0	0	25
ACV	3	3[4]	0	0	0
PBR	160	78[5]	0	0	20
ASPB	0[6]	0	0	0	0
LCPL (1)	114[7]	114	27	8	0
LCPL-4/11	256[8,9]	236[5]	18	5	4
LCM-3	50[9]	50	30	46	100
LCM-6	523[9]	475	35	89	217
LCM-8	103[9]	75[2]	0	7	25
LCU	94[9]	92	0	1	24
WPB (CG)	26	26	0	0	26

1. Seven building.
2. Eighty-nine delivered through FY 66.
3. Thirteen in Subic, seventeen en route, eight training.
4. En route to Vietnam.
5. Twelve en route to Vietnam, twenty-four awaiting shipment, twenty-two training.
6. In design, forty planned for delivery by January 1967.
7. Wood hull, small, slow.
8. Steel or plastic hull, larger, faster.
9. Includes active and new construction; planned assignments from ready-for-issue craft not included.

The pitfall of quick, off-the-shelf procurement was that the Swift had never been designed as a patrol boat, and certainly not for conditions as they existed in Vietnam. It was neither quite large enough nor quite strong enough for the arduous service it undertook. For example, a boat trying to pick up a man overboard was torn in half by surf off Danang. Others almost flooded in very heavy seas. The aluminum structure itself corroded, forming a magnesium layer; in some cases, a finger could be put through a quarter-inch plate. Structural material problems were cured when Alcoa changed the processing of the material itself. By 1967, there were several proposals for larger replacements. They resulted in the postwar 65-foot patrol boat Mark III. However, the Swifts fought the offshore and some of the river war in Vietnam. Once they had been bought, once they had proven largely successful, there was little point in replacing them.

Swifts in navy service were designated PCF (patrol craft, inshore). Their commercial origins are reflected in their "nonstandard" (NS) hull numbers.

NAGP force-level figures were based on the performance of the first three Operation Plan 34A Swifts. Their Decca 202 radar could detect a junk at 2½ nautical miles. In theory, then, a single Swift could sweep a 5-mile width at 20 knots. That being the performance expected of five motor junks at 10 knots, the NAGP equated one Swift to ten motor junks. However, each Swift would have to inspect all of its radar contacts, whereas the junks could split up inspections. The NAGP therefore considered a five-to-one ratio more realistic; 600 motor junks could be equated to 120 Swifts. The oldest motorized junks might be phased out when about 50 Swifts were available in country.

The NAGP insisted on U.S. crews because the Vietnamese navy could not be expected to use Swifts aggressively enough. (The NAGP cited "the unwillingness of the VNN to patrol effectively with junks now, and the known reluctance of the VNN to accept new ideas and operational concepts in any timely manner.") The advisory group wanted the Swifts manned by U.S. personnel until the Vietnamese navy showed adequate ability to perform. It had grown so rapidly that its experienced personnel were overwhelmed by the need to train new recruits. It might, therefore, find the Swift, with its sophisticated engine, difficult to operate and maintain.

As of 1 January 1965, the Vietnamese had about 330 motor junks and needed another 270, equivalent to 54 Swifts. These approximate figures were based on 50-percent usage of the junks. The program was submitted to Washington around May 1965, and it was sold to Secretary of Defense McNamara during his mid-June 1965 visit to Vietnam. At that time, the force was set at fifty-four Swifts plus seventeen Coast Guard 82-footers. McNamara approved the program before he left Vietnam, but further details were later settled by the NAGP under Captain Landis.

The Landis study considerably enlarged Market Time, increasing the offshore force from nine to fourteen ships, the Coast Guard 82-footers from seventeen to twenty-six, and the Swifts from fifty-four to eighty-four. A fourth LST was added, and patrol plane coverage doubled. The same report recommended the formation of a river patrol force of 120 craft operating from LST tenders, their patrols extending about 25 miles upriver. The river patrol operation became Game Warden.

As of March 1966, Market Time employed fourteen radar pickets (DER) as well as sufficient Seventh

The Swift was often criticized for its relatively fragile hull. This boat, firing its 81mm mortar into enemy positions on the Gulf of Thailand, probably during the Tet Offensive, shows a battered hull and a row of rubber tires atop the deckhouse, which acted as bumpers.

Fleet minesweepers to maintain eleven 17-knot (DER) equivalents simultaneously on station. (An ocean sweeper could maintain 14 knots, a coastal sweeper [MSC], 12.)

There were also twenty-six Coast Guard 82-footers, seventeen continuously under way on assigned barrier and inshore patrol stations, and twenty-five Swifts, with fifty more scheduled to arrive within eight to ten months. Three air-cushion vehicles (Hovercraft) were expected the following month, April. Support required one repair ship (ARL), one repair barge (YR), and two small cargo ships (AKL). Another repair ship was due in July 1966, and a self-propelled barracks ship (APL) built on an LST hull was scheduled to arrive later in March 1966. Four LSTs, each carrying two armed helicopters, were expected in October 1966, the first thirty fast river patrol boats (PBR) in March and April 1966.[3] Note that the PBRs were all carried under Game Warden (TF 116), although from a planning point of view thirty of them were part of Market Time.

The seaborne patrols were supported by seven Neptunes (SP-2) at Tan Son Nhut, by a combination of P-3A Orions at Sangley Point in the Philippines, and by P-5 Marlin seaplanes in Vietnam, the latter supported by a seaplane tender. Two patrol planes were always airborne, maintaining designated patrols.

Patrol planes flew on northern and southern offshore tracks, with one continuously airborne on each. In addition, patrols were flown over most of the coastline twice each day by army OH-1 Mohawks that had side-looking radar for nighttime contacts.

The 82-footers patrolled barriers at the Seventeenth Parallel and the Cambodian border. The DERs and minesweepers patrolled the "B-Ring," a band extending 10 to 20 nautical miles off the Vietnamese coast. The Vietnamese navy generally patrolled the A-Ring closer in, using an average of 10 patrol ships and 190 junks. There was some hope of expanding the Vietnamese contribution to 16 Sea Force ships and 250 junks daily; U.S. advisors had been attached to the twenty-eight coastal groups (formerly, "junk divisions").

One LST was to be stationed 5 miles off the mouth of the Bassac, the Co Chieu, and the Cua Tieu rivers. Each LST controlled two army-manned helicopters and ten river patrol boats (PBR), which worked the shallow waters off the Delta land mass and as far as 25 nautical miles up-river, linking Market Time with the Game Warden river patrol.

The entire Market Time operation was controlled by CTF 115 in Saigon, with orders and data relayed to coastal surveillance centers at Danang, Qui Nhon, Nha Trang, Vung Tau, and An Thoi. The centers were responsible for coordination with the Vietnamese navy.

Note that, with the exception of the Mekong and Bassac rivers, U.S. Special Forces were responsible

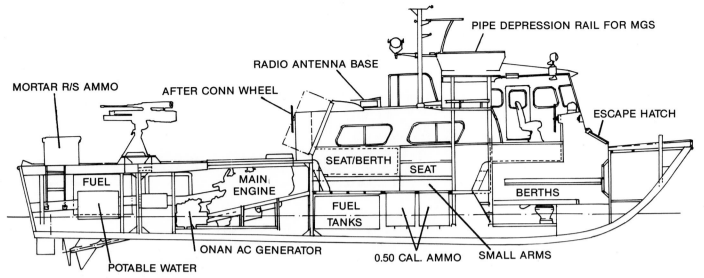

An inboard profile of the Swift Mark I. (Norman Friedman)

for securing the Vietnamese land and river borders against infiltration.

Swifts were unloaded from freighters at Subic Bay in the Philippines, then brought to Vietnam by LSD, either in the well deck or on the superdeck, eight at a time. The first twenty—ten at Danang, ten at An Thoi—were organized almost as boat pools. Captain Landis, however, decided that the boats should be split into divisions, each commanded by a lieutenant. Divisions were numbered 101 through 105; the air-cushion PACVs were Division 107. Initially, there were to have been fourteen to nineteen boats per division, but when the United States pulled out of Vietnam, the largest division included twenty-one or twenty-two boats. Each boat was commanded by a lieutenant j.g., and the ratio of crews to boats was three to two. The Landis report suggested patrols limited, by the strain on a crew, to twelve hours, but in fact, Swifts could generally remain at sea for twenty-four hours at a time.

Swifts, compared with the original Sewart crew boats, were little altered. Each had a crew of six, one officer and five enlisted men, and was armed with a single 0.50-caliber machine gun aft, where the deck was reinforced. The passenger compartment was stressed for 5,000 pounds of cargo or twenty passengers, with a 5 x 5-foot access hatch. There were four compartments—one watertight compartment fore and another aft of the engine compartment, and one collision bulkhead forward. Maximum range was 600 nautical miles (400 normal), and the boat was rated at 32 knots.

The U.S. Navy modified the Sewart crew boat by gutting the main cabin, removing the bench-type seats, and cutting a hole in the pilothouse roof for a twin, ring-mounted 0.50-caliber machine gun. A single 0.50-caliber machine gun and later a combination 81mm mortar/0.50-caliber machine gun were mounted aft, where the deck was strengthened. The major field modifications were the relocation of the radar console from the main cabin to the pilothouse and the installation of a remote radio microphone and receiver in the pilothouse. Some problems remained. The Swifts never had a boat-to-air radio; they relied on a relay through their bases. Their main (VRC-46) radio transceivers were located in such a way that frequency could be changed only from the cabin, not from the pilothouse. In addition, it was impossible to see into the engine compartment without opening the hatch cover on deck.

Market Time tactics were based on a mixture of interception and patrol. Because a central plot of all major traffic was maintained at Cam Ranh Bay, Swifts and Coast Guard boats would generally know in advance about any fast trawler approaching them. Thus the high speed of the PCF, about 25 knots in practice, was important not only for transit but also for giving chase once a suspect vessel attempted to evade. (Since evasion was itself evidence that a vessel might belong to the enemy, it was rare.) Coastal boats rarely, if ever, being faster than 15 knots, PCFs would attempt to transit at about 2,000 rpm (25 knots) but would limit themselves to 1,700 to 1,800 rpm on patrol, and would often cut their speed further, to about 1,200 rpm or less, to save fuel, particularly at night. Sometimes, in a calm, they would drift on station. Speed was also governed by noise. Many in the Swift fleet believed that small coastal craft escaped at night because they were masked by the noise of the Swift

Table 12-3. Vietnam-Era Inshore and River-Patrol Craft

	RPC	PBR I	PBR II	PCF I	PCF II
Length (ft-in)	35-9	31-0	31-11.5	50-1.5	51-3.75
Beam (ft-in)	10-4	10-7	11-7.5	13-1	13-7
Draft (ft-in)	—	1-10.5	—	3-6	—
Engines	2 64HN9	———— 2 6V53N ————		——— 2 GM 7122-7000/3000 ———	
Power (bhp)	225	———— 216 ————		425	—
Fuel (gal)	350	160	160	800	828
Weight (lbs)					
Hull	13,510	3,660	5,653	13,590	14,418.9
Propulsion	7,899	4,600	5,241	12,220	13,343.3
Electric	250	600	749	970	2,202.4
C2	248	410	373	500	761.5
Auxiliary	1,017	130	219	1,950	1,432
Outfit	845	760	1,547	1,750	2,000
Armament	1,902	1,640	895	5,210	—
Margin	513	240	294	724	773.9
Fuel Oil	2,874	1,260	1,114	5,477	5,565.1
Potable Water (with lubricating oil and coolant)	499	—	—	—	—
LS	26,184	12,040	14,971	36,913	34,934
FL	32,548	14,590	17,946	47,047	45,093.8

Note: Average Swift loads in Vietnam were as follows:

Light Ship	11,988	Small Arms	50
Ammunition	1,001	Complement	1,075
Stores	60	Lube Oil	35
Water	33	Full Load	15,356

itself. Although noisy when cruising, Swifts normally slowed to their quietest speeds when intercepting radar contacts.

In service, the engines performed poorly in the heat and deteriorated when they operated for long periods at low speeds, for example, 500 to 700 rpm in heavy seas, or 800 to 1,000 on patrol. The boats tended not to make their designed speed of 28 knots (2,000 rpm). Over time, they did worse as their propeller shafts and blades bent from frequent groundings. These problems were aggravated by overloading and by the bottom fouling endemic to tropical waters. Average service speeds as low as 23 knots were reported.

For interception based on the central traffic plot, it would have been ideal to keep very fast craft at their advanced bases awaiting calls; this would have held maintenance to practicable levels. In 1969, NavSea argued that none of the existing craft, which then included PTFs and *Asheville*s as well as Swifts, could adapt to such techniques; none could maintain sufficient speed in rough water, and sea state 3 or worse would be encountered about 30 to 40 percent of a typical month. Cloud cover would be frequent enough to preclude air surveillance and, therefore, air tracking. As a result, conventional small craft held at advanced bases could intercept their targets only about 20 percent of the time. NavSea argued at the time that the only solution was a boat capable of very high speed in rough water, a hydrofoil.

The commercial crew boat was not intended to operate in very rough water, and neither was the Swift derived from it. The Swift's operating limits, set by the crew's endurance, were 6-foot waves with a 4- to 6-second period, at which point arms and legs could be broken. With its bridge so far forward and its bow so low, the Swift tended to dig into a sea and take green water through broken pilothouse windows. A Swift nearly sank when it flooded that way in January 1967. In a heavy sea, green water and spray could even reach the open gun tub atop the superstructure. All openings had to be secured, which made for poor ventilation in the earliest Swifts. Boats were later modified with stronger pilothouse windows and with blowers for internal ventilation.

In persistent monsoon weather, Coast Guard 82-footers, which were much better sea-keepers, were assigned to supplement the Swifts. However, even that was not always enough. In 1967, one boat squadron commander called for a new heavy-weather patrol boat to supplement the Swifts—in I Corps during January, the Swifts were able to go to sea for only three days, the 82-footers for only twelve.

The Mark II Swift was an attempt to cure the worst of the sea-keeping problems; it embodied a series of previously approved sea-keeping boat alterations.

Swift Mark II solved some of the problems of the original boat. Note the new shallow forecastle, the bridge windows squared off and slanted inward, and the smaller ports in the superstructure's side. This boat was photographed off Little Creek in July 1968.

Compared with the Mark I, it showed less sheer because of a bow raised 22 inches, a broken deck line, a pilothouse set about 3 feet further back from the bow, and an overall length increased about 1 foot to 51 feet 3¾ inches. The original 3-kw Onan generator was replaced by a 6-kw unit. Officers in Vietnam proposed a more radical improvement, a new prefabricated 5-foot bow with a 32-inch bulwark, but nothing this extreme could be introduced into the Market Time boats because they were busy.

In Market Time operations, Swifts generally carried a Vietnamese police or naval petty officer who examined boats they encountered. If the boat was larger than the Swift, the Market Time craft would generally stand off 10 or 20 yards so that it could not be boarded and taken over. When coming alongside, the Swift was at general quarters, one crewman aft armed with a shotgun and another in the machine gun tub, atop the pilothouse, with an M-16 rifle. At such times the coxswain would operate the boat from the aft control station, where he had unobstructed visibility for coming alongside.

Searchlights were the main sensor for night work. However, the ones fitted to the Swifts were not powerful enough to illuminate a possibly unfriendly craft outside the Swift's gun range. They merely presented a point of aim. The searchlights were, moreover, entirely dependent on AC power from the auxiliary generator. Ironically, overhauled boats with 6-kw water-cooled generators were less reliable than the original boats with air-cooled 3-kw units, because sufficient spares were not bought and impellers for the sea-water coolant pumps frequently broke down.

Market Time operations revealed another important limitation. The PCF had no open space onto which a junk's cargo could be offloaded while the junk was searched. That, more than any obvious factor such as armament, limited the efficacy of the search. Ironically, the Swift's radar, which had made it so attractive in the first place, was deemed badly in need of replacement by 1967.

There were also problems inherent in the crew boat design. The Swift was unarmored, so until 1969 it was kept out of Vietnamese rivers.[4] Its noise made it vulnerable to ambush. Its aluminum hull was subject to corrosion and to damage when coming alongside. However, it suffered nothing remotely like the problems of the experimental aluminum PTs, and in this sense, it justified BuShips' faith in aluminum construction. Perhaps worst of all, the Swift was so uncomfortable that patrols could not be extended beyond the usual twenty-four hours, no matter that

An official model of the Swift shows how far below water its propellers and rudders protruded.

The Swift was a big boat—too big, it was said, to operate easily on rivers. This one was photographed on the Duong Keo River, Ca Mau Peninsula, April 1969.

the boat was able to replenish at sea. For extended operations, crews had to be exchanged at sea when the boat refueled.

Because the PCFs arrived before shore facilities were complete, they had to depend partly on sea-based support. One LST could support two PCFs and an 82-footer. The LST carried spare crews, spare parts, maintenance personnel, and fuel. In January 1966, DERs off the west coast of Ca Mau Peninsula began to support a Swift with two crews. They rendezvoused each morning, taking about thirty minutes for turnover and fueling. That March an ocean minesweeper was tried for a similar experiment, but it had no berths for spare crewmen, who had to live ashore. The LST was clearly the best of the lot, and throughout the war, a series of LSTs and LSDs were used as PCF mother ships.

A larger boat could keep the sea better and for a longer period—but at a much higher price. The 82-footer had a five- to six-day endurance, but no dash capability, and it cost about three times as much as a PCF. Moreover, its deep draft of 6½ feet kept it well offshore, in waters 10 to 12 feet deep. Even so, experience with the Swifts made it obvious that they were too small for Market Time.

Swifts were also used to drop small groups on the coast for special operations, part of the crews' routine twenty-four-hour patrol. The rate of such activity varied from section to section, but was on the order of one per month to two per week. In these operations, the men used their 81mm mortars for harassment and interdiction fire, fire support, or reconnaissance fire. Generally, Swifts carried about a thousand rounds of 0.50-caliber and twenty of 81mm ammunition over the rated allowance.

As mentioned, aluminum Swift hulls corroded much more quickly than expected; by 1967, hull thickness in some older boats had been halved in places. A major PCF overhaul program was established at a planning conference held at Cam Ranh Bay on 8–9 September 1967. The repair cycle began in December, with twenty-four craft overhauled in six-month increments, four each at Subic and Sasebo. Meanwhile, new contracts for Mark IIs were let with Sewart. More importantly, enough Swifts were in service for the navy to consider a more satisfactory alternative. The Swift was good enough for most officers to look to Sewart for its replacement, and the company had 65- and 85-foot boats in production for commercial customers and for some foreign navies.

In 1967, a survey of PCF commanders showed that the craft and its capabilities had generally improved. One officer merely wanted a longer bow to meet short seas about 4 feet in height. Another cautioned against buying a second-generation craft designed to perform several missions, none of them well. The commander of Amphibious Force, Pacific, touring Southeast Asia in November 1966, wanted to add 10 feet of length to obtain more fuel and an 81mm mortar (not yet fitted to Swifts) for better direct fire.

A PCF squadron commander wanted to mix the Swifts, inshore, with larger craft, 85-foot, 30-knot Sewarts, for offshore work in rough seas and during the monsoon. Another senior officer proposed the Sewart 65-footer as an interim solution but agreed that an 85- or 95-foot heavy-weather boat was needed.

Finally, ComNavForV suggested a larger patrol boat with a ten-day patrol endurance. It would patrol at 12 to 14 knots and dash at 30. As in Vietnam, a shallow draft of 4.5 feet was needed for effective coastal work. However, in the ideal Market Time boat, it would be combined with good sea-keeping and maneuverability, at least up to the standard of the Coast Guard 82-footer. This kind of compromise was well beyond what could be accomplished on short order, but the requirement reappeared in the postwar CPIC project.

For the moment, a requirement for nine longer-endurance craft was established, and the acquisition of twelve Sewart 85-footers was planned as a long-term program. These boats were tentatively identified as heavy-weather patrol craft (HWX). A contract for these boats, dated March 1968, appears to have been canceled shortly thereafter.

Intense U.S. Navy interest in air-cushion vehicles predated the Vietnam War. In 1964, OpNav directed the Operational Test and Evaluation Force (OpTevFor) to study their use in various naval missions. Three, including the British SRN-5, were investigated, and in January 1966, OpTevFor recommended specially designed air-cushion vehicles for counterinsurgency, mine countermeasures, and amphibious assault operations. The latter, representing their "greatest potential," were particularly attractive because an air-cushion vehicle could so easily make the transition from land to water, traversing minefields on both and thus eliminating the need for inshore sweeping prior to landing. Similar qualities, for example, the ability to pair gunships and troop carriers, made air-cushion vehicles attractive for the semi-amphibious Vietnamese river war.

The gunship would cruise at 40 to 50 knots, with a maximum speed of 60 to 70. It could be protected by light ceramic armor and armed with 0.50-caliber machine guns for suppressive fire, with more exotic weapons to destroy foxholes and fortifications. Possibilities included remotely fired 105 or 106mm recoilless rifles and mortars, 2.75-inch rocket pods, aircraft-type 20mm Vulcan (or 7.62mm minigun) pods, and even the helicopter-type remote-control 40mm

One of three navy air-cushion vehicles (PACV) in Vietnam runs across Cau Hai Bay near Hue, August 1968. They were the last operational U.S. Navy hovercraft until the advent of the air-cushion landing craft (LCAC) about a decade and a half later.

grenade launcher. The accompanying troop carrier would be similarly equipped but would carry either troops or cargo (at a minimum, one army rifle platoon, comprising one officer and forty-three enlisted men).

As in the case of more conventional craft, any near-term application depended on the existence of a suitable commercial craft, in this case the British 7-ton, 70-knot SRN-5 Hovercraft. It was initially designed for operations in sheltered waters, particularly between the English mainland and the Isle of Wight, and was being imported by Bell Aircraft. Bell could deliver a navalized form fairly rapidly. Admiral Zumwalt's report showed that, if the navy decided to proceed on 1 May 1965, it could have two craft, 004 and 008, at Norfolk on 1 August, and a third, 015, at the Bell plant at Buffalo on 1 November. One could be delivered each month after that. In fact, only six, three navy, three army, were sent to Vietnam.

Bell modified three SRN-5s, which it designated SK-5s, under a BuShip contract in 1965.[5] They were intended to carry a crew of three plus an armed three-man boarding party. In Vietnam, though, the crew was generally five men, an officer in charge, boatswain's mate, gunner's mate, engineman, and radarman.

A single GE LM-100 (1,050 shp) burning JP-4 or JP-5 jet fuel drove both the 7-foot lift fan and a 9-foot controllable-pitch (air) propeller. Shorter but broader than a Swift, an SRN-5 could make over 60 knots in a calm, 35 knots in 5-foot waves. Its radius of action at 50 knots, with 30-minute reserves, was 125 nautical miles, and at 40 knots, 150. In 1966, it was reported that the SK-5 could make 41 knots in 1- to 2-foot waves and a 10-knot wind, and often made 45 to 47 knots in sea state 0.

The SK-5 maneuvered by means of a pair of rudders behind the propeller. Tactical diameter was about 500 yards when the craft was down wind at 20 knots, about the same up wind at 30 knots.

Designed range was 200 nautical miles at 15,665 pounds in calm water with 295 to 310 gallons of usable fuel. Range decreased with increasing weight because that required more lift power. For example, at 14.8 percent overload (17,975 pounds), the air-cushion vehicle needed 580 rather than 280 shp to lift an inch. In sea state 2, operating range was about 120 miles with a 15-minute reserve; much more fuel was needed in a high sea or a high wind.

The navy designated the SK-5 a patrol air-cushion vehicle (PACV).

Armament comprised a single, later a twin, heavy 0.50-caliber machine gun in a scarf ring atop the cabin, with two boxes of ammunition and 300 rounds each); two M60 light machine guns in the side windows with 500 rounds each; and two remotely fired M60s in panniers aft, fixed in train, with 500 rounds each, and boresighted to a range of 500 to 1,000 yards astern. A crewman in the bow could fire an M79 lightweight (single-shot) grenade launcher. Extra 0.50-caliber and 7.62mm ammunition was carried in the cabin deck. The 0.50-caliber gun had a night sight (crew-served weapon sight, or CSWS) that could detect a target at about 1,000 meters in moonlight and an infrared searchlight for use with the infrared sight. There was also a searchlight on the radar mast, above and behind the gunner.

At night, the SK-5 depended on radar surveillance to sort out the numerous junks, sampans, and fish traps in patrol areas. It carried a Decca 202 radar, which could detect an LST at 24 nautical miles, a medium to large junk at 7 to 10, and a small junk or sampan at 1 to 5.

Fixed armor totaled 1,670 pounds in removable 2- x 3- or 3- x 3-foot sections of 23-pound, half-inch homogeneous rolled steel, including removable plates over the engine. The windows were shatterproof but not bulletproof. Before Operation Quai Vat, the plating on PACV-1 was partly replaced by equivalent 10-pound high-hard steel.

Early in 1966, the three PACVs were tested by Boat Support Unit 1 at Coronado, California. In March, it recommended many improvements along with an extensive development program. The PACVs then went to Vietnam as PACV Division 107, arriving on 1 May 1966. Under evaluation by the Naval Research and Development Unit, Vietnam, they spent about four weeks in Market Time, then about two weeks in Game Warden. They operated from LSD 26 with a section of PBRs in the Mekong and Bassac river estuaries, "an extremely hostile combat environment." They also supported U.S. and Vietnamese Special Forces in IV Corps, in the Plain of Reeds, which was far too shallow for conventional craft.

The PACVs were first used in the Mekong Delta in Operation Quai Vat (Vietnamese for "monster").[6] They departed Vietnam on 15 December 1966.

Although extremely noisy—reportedly they could be heard up to 7 miles away under certain conditions—the PACVs were so fast that they could hope to surprise and catch small Viet Cong sampans moving in the large shallow bay inside the mouth of the Perfume River. Typically, two boats worked as a team, an anchored one vectoring the moving one. They could also be used to land and extract troops, carrying twelve to fifteen and placing them on the beach next to a treeline. It became clear that troops landed this way avoided the usual hazard of wading ashore under fire, and so this type of operation led directly to the later air-cushion landing craft (LCAC).

The PACVs were limited in not being able to climb the steep, 5- to 6-foot river banks common in the delta or to cross the tree lines that often bounded them. It was difficult to turn these craft sharply on narrow rivers. Another problem arose from tides, which in the flat delta region might travel 50 to 60 miles inland. Rivers normally flowing 4 to 5 knots downstream would reverse at high tide. One PACV chased the enemy over a 3-foot river bank at high tide and into a rice paddy; by the time the chase had ended and the tide receded, the PACV was blocked by a bank 4 or 5 feet high.

During Operation Quai Vat, the PACVs suffered from ingested debris that made small holes in their skirt but had no effect on their performance. Even after a collision in which the entire bow section of a PACV's plenum was lost and its lift skirts were severely damaged, the boat was able to return under its own power.

It was found that the heavy spray the PACV produced could cut visibility, a problem further aggravated by the 2- to 3-degree trim by the stern the operators had to adopt to overcome the boat's inherent tendency to plow in. When drifting with its engine shut down, a PACV used its auxiliary power unit. That, however, had an inadequate muffler and could be heard at 200 to 400 yards. After four hours, crews experienced severe fatigue in the hot, cramped cabin, which had totally inadequate ventilation. The Naval Research and Development Unit also wanted the craft silenced more effectively.

Modifications suggested in March were clearly needed. These included more armor and installation of twin 0.50-caliber machine guns on each side. More significantly, the Naval Research and Development Unit concluded that the PACVs were ill suited to both Market Time and Game Warden because of their noise level and limited range and endurance. But they were far superior to any vehicles in the Plain of Reeds for search and destroy, quick strike, and armed reconnaissance missions. If redeployed to Vietnam, they would be most useful for

- Patrol and interdiction, along the west coast of Cau Mau Peninsula, from the Cambodian border to the southern tip, based at Rach Gia or on an LSD.
- As part of a PACV/helicopter task unit, centrally based on the Mekong Delta for search and destroy, patrol and interdiction, and quick reaction strike along rivers and inundated areas.
- For Medevac, search and rescue, and support of civic action programs, especially during rainy season in the delta.
- Augmenting Game Warden and Market Time in areas inaccessible to assigned patrol craft.

The three PACVs returned to Vietnam in 1967, arriving at Vung Tau on 4 December. Modifications

The origin of the name Monster for the PACV is evident in this photograph of 1966. PACVs originally had only a single machine gun in the tub atop the pilothouse.

included the installation of auxiliary fuel tanks, additional radios, new segmented skirts, and air-puff ports for better directional control at low speed. Armor was added, and the rear-firing fixed M60s were removed in favor of a single gun that fired through the center-bow window. Later, each craft was fitted with a pair of Mark 19 40mm long-range grenade launchers.

The redeployed PACVs were initially assigned to the Mobile Riverine Force in the delta, beginning operations on 29 December, when two of them landed infantry squads. One craft had to be airlifted home after mechanical problems developed. Eventually, all three were assigned to I Corps, which had large, salty sand flats and lagoon areas like Cau Hai Bay. They were based at Hue and employed in Market Time.

The PACVs served in Task Force Clearwater (Hue River Security Group) between July 1968 and May 1969, before being withdrawn to the United States. Encountering excellent operating conditions—few tree lines, many wet rice paddies, and large inland lagoons, they were able to range over 90 percent of their potential operating area. However, they suffered severe maintenance problems as they advanced in age; maintenance requirements approximated those of a helicopter rather than a small boat.

Bell also produced modified SK-5s for the army; three were completed by February 1968 and airlifted to Bien Hoa in May. Two were designated assault air-cushion vehicles (AACV), the third as a transport air-cushion vehicle (TACV).

From late 1966 on, TF 115 was also responsible for Operation Stable Door, that is, security for the four major ports—Vung Tau, Cam Ranh Bay, Qui Nhon, and Nha Trang. By July 1967, each harbor-entry control post had an SPS-53A radar for surface surveillance.

Stable Door initially employed LCPLs and 16-foot Boston whalers, which had M60 machine guns installed on their bows in March 1967. By January 1967, a total of twenty boats was on station. They were supplemented by 45-foot picket boats. These craft had been built during the Korean War and then laid up when U.S. harbor defense units were disbanded in the late 1950s. In Vietnam, they were armed with a single 0.50-caliber or twin 7.62mm machine gun. See appendix D for details.

The River Patrol Force (TF 116), which carried out Game Warden, complemented Operation Market Time. Its defensive role was to secure the Long Tau channel to Saigon through the RSSZ. Its offensive role was to deny the Viet Cong free passage across the rivers of the Mekong Delta, to wipe out the ene-

PBR 66 (Mark I) of River Squadron 543 at speed off Nha Be. Note the canopy and the vertical armor athwart the steering station. Ultimately, canopies were made of ballistic nylon to keep out fragments such as those created when shells and rockets hit the big radome. (Arnold A. Adams)

my's tax collection system along the rivers, and to help destroy his infrastructure. Operations began late in 1965 with four 36-foot LCPL Mark IV landing craft operating from the navy yard at Saigon. They were to secure the RSSZ and test the concept of waterborne interdiction patrols to complement Market Time. The LCPL, designed to control waves of amphibious boats, was the U.S. Navy boat that came closest at the time to the requirements of riverine warfare, although clearly it was not entirely satisfactory.

TF 116 was formally established in December 1965 to carry out Operation Game Warden.

The RPC having proven a failure, representatives of the major U.S. commands met in Saigon in September 1965 to decide on a new-generation boat. They believed that about 120 patrol craft were urgently needed for work in the delta and the RSSZ. BuShips was ordered to procure a new type, then designated PCR. It asked for bids for a 25- to 30-knot craft with a draft of only 18 inches when stopped, 9 when cruising. Boats would be able to operate in seas up to sea state 3. Given the urgency of the requirement, the bureau had to accept a trial speed of 25 knots.

Early in 1965, Willis Slane, president of Hatteras Yacht, suggested converting one of his stock civilian cruiser hulls into a shallow-water patrol boat. He proposed a 50-footer, but the Swift had already been chosen. The navy asked instead for something smaller to operate in the delta against small sampans. Slane offered his standard 28-footer, newly placed in production, modified to take water-jet pumps so that it could operate in particularly shallow water. Navy representatives, enthusiastic, called for a prototype. Slane offered to convert a hull at his own cost within two weeks. Although the hull itself was a stock item, its machinery was not; it was powered by special turbocharged diesels, and its jet pumps came from the turbo-power division of Indiana Gear Works. The hull was a hydrodynamic demonstrator: it was fitted with a simple wooden deck and a speedboat-type windshield.

On trials, Slane achieved 33 knots, and the pump jets resulted in extraordinary maneuverability at high speed. The boat was able to do end-for-end turns and crash stops. Several weeks after these successful trials, Slane, in poor health, died.

The navy was impressed, and BuShips asked for bids for Hatteras-style hulls with waterjet propulsion. On 29 November 1965, it awarded a contract for 120 boats based on an existing 31-foot plastic hull to United Boatbuilders of Bellingham, Washington,

Table 12-4. Comparison of River Patrol Boats

	PBR	LCPL XI (modified)
Dimensions (ft)	31 x 10.5 x 2.5	36 x 13 x approx. 4
Displacement (tons)	7	11
Material	Fiberglass	Fiberglass
Engines	2 diesel/water jets	1 diesel/propeller
Speed (kts)	25.7 max., 23 sust.	16/15 sust.
Weapons	1 twin 0.50 forward	Single 0.50 forward and aft
	1 x 0.30 aft	2 x 0.30 amidships
Electronics	Radar, radios, fathometer	Radar, radios

the low bidder. Delivery was for the following April. United already manufactured a line of commercial boats, which it called Uniflites, ranging in length from 17 to 40 feet. Its experience in fiberglass construction included the 52-foot LCSR, the largest U.S. naval fiberglass hull to date. The navy patrol boat, then, was an adapted 31-foot Uniflite, displacing about 50 percent more than its civilian counterpart.

Because of this large contract, United Boatbuilders was renamed Uniflite and the preliminary designer of the BuShips PBR (patrol boat, river) became its chief engineer. Table 12-4 is a comparison of the LCPL and the PBR.

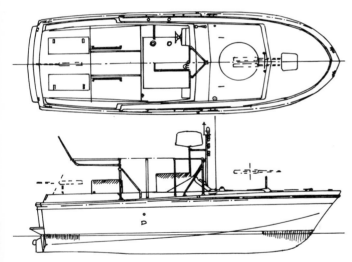

PBR Mark I. Vertical armor shields have been shaded; they are shown as heavy lines in the plan view. (Norman Friedman)

The boats had been redesignated PBRs in October, but they were generally called plastics for their hull material. They shared only an outer hull with their commercial forebears. They had an entirely new, navy-designed superstructure, a new internal arrangement, and a Jacuzzi water-jet pump propulsion; one navy internal report described the boats as a large-scale experiment in jet pumps. The two General Motors 220-bhp truck diesels were used to run generators in other warship classes.

The choice of a fiberglass (GRP) hull was particularly fortunate, because it did not rot or corrode in the humidity and heat of Southeast Asia. Nor when it was hit—which was often, given its service on narrow canals and rivers—did it create a spray of fragments. Many of the shaped-charge warheads that struck the hull did not even trigger, and it tended to absorb shocks such as mine explosions. The hull was also strong enough to tolerate beaching. As it had demonstrated in trials, the water jet conferred exceptional maneuverability at medium and high speeds; the boat could stop or turn 180 degrees in its own length. It drew so little water that it could pass through virtually all Vietnamese waterways.

The PBR typically carried a complement of four: a boat captain, forward and rear gunners, and a coxswain. In addition, one boat of every patrol carried a patrol captain, and by 1968 most also carried a South Vietnamese regional policeman. Like the RPC, the PBR could carry troops—up to twenty Vietnamese or forty SEALs for short periods.

PBRs were designed to carry a twin 0.50-caliber machine gun (Mark 36 mount) forward and a single 0.30-caliber weapon aft, but the latter was replaced by a single 0.50 before the craft appeared in Vietnam. PBRs also generally had an M60 light machine gun (7.62mm) amidships. From 1 January 1967 on, boats were fitted with a Mark 18 40mm grenade launcher aft, carried either piggyback on the after 0.50 or on a separate pintle mount. They also had hand-held M79s. The somewhat larger Mark II typically had a lighter, lower Mark 56 twin mount forward and was completed with a single 0.50s aft. Most sections also had an army-type (later navy Mark 4) 60mm mortar, which could only be fired at certain angles because the boat was not sufficiently stressed for the recoil. About half the craft of one boat section replaced one of the forward 0.50-caliber guns with a 20mm cannon. At least some boats experimented with a 90mm recoilless rifle.

Most boats carried hand-held antitank rockets (LAWs), although they were rarely fired, and then usually against bunkers and buildings. Their back blast was considered too dangerous in a fire fight.

At least one PBR section and several SEAL teams

An early PBR Mark I displays a heavily armored forward machine-gun position.

tried the GE Minigun, a 7.62mm version of the Vulcan. It was liked for its high rate of fire and because it was easy to aim and hold on target. The muzzle flash was a problem at night, however, and the recoil of the gun, mounted aft, could swing a PBR off course at low speed. There was no problem at or above normal cruising speed.

None of these weapons could fire very accurately from a moving, rolling boat. However, the plastics often operated in "free-fire zones," in which accuracy was not nearly as important as volume of fire. The 0.50-caliber machine gun had a considerable effective range, 1,900 yards, but it could not penetrate the bunkers or paddy dikes used by Viet Cong ambushers, and its flat trajectory prevented it from hitting an enemy behind a river bank or in a spider hole. The grenade launchers, effective to 300 yards, were considered more useful for harassment and interdiction fire. The hand-held M79 was used primarily to counter ambushers. It was extremely versatile owing to the number of different types of ammunition it could fire, but it was only a single-shot weapon (an experienced man could get four rounds in the air at once). The RF Mark 18 could jam, especially if the fabric belt linking its rounds had been used more than twice.

Unlike Mark Is, Mark II PBRs had their M-60s mounted so that they could easily be switched from one side to the other.

Ammunition stowage varied widely, boats carrying 1,300 to 3,000 rounds of 0.50 caliber, usually 2,000 to 2,800, and 800 to 2,000 of 7.62mm (M60). The boats carried 20 to 120 40mm grenades, depending on whether they were armed with the Mark 18, and up to 40 rounds of mortar ammunition. Crewmembers were typically supplied with such small arms as M16s and even AK47s. All sections complained of insufficient ammunition, and many devised new and larger ammunition cans.

Unlike the Swift, which operated under fairly benign conditions, the plastic was armored against ambushes mounted in the narrow waterways it frequented.[7] Most plastics had splatter or fragment shields at the coxswain flat and over the engine, mounted in positions decided by the boat or section commander. At the coxswain's flat, one plate was always mounted to port, the other either on the centerline or all the way to starboard. Engine shields were either at the sides or about 3 feet apart, equally spaced around the centerline and facing outboard, or at about 45 degrees to the centerline. Guns were protected against 0.30-caliber rounds by quarter-inch, dual-hardness plate.

This armor was intended to resist small arms (sniper) fire and fragments, since shaped-charge (Heat) rounds tended not to do much direct damage. When they were triggered by the hull, the only sources of damage were the hot jet the round produced and the fragments of its case. However, the radome and recognition lights could trigger Heat rounds and thus spray the coxswain's flat, which was open from above, with fragments. The remedy was the ballistic-nylon, fragment-suppression canopy, which was also fitted to riverine assault craft (ASPB and ATC). It could defeat fragments produced by the standard Viet Cong hand grenade and even a dud 81mm mortar round at close range.

Viet Cong sniper fire was so ineffective that PBR crews generally preferred to fire their weapons from the open, where arcs of fire were less restricted.

Viet Cong ambush tactics were determined by the range of weapons. At first these were relatively heavy 57mm and then 75mm recoilless rifles, which had to be emplaced in bunkers because they could not easily be moved. As the war progressed, they were replaced by lighter rocket launchers that did not have to be emplaced in bunkers. The standard RPG 2 (B40) penetrated better than the 57mm but was lighter than the 75mm rifle. It was most effective within 100 yards, so throughout 1968, ambushes were generally mounted between islands and river banks or in canals, where the waterway might be 30 to 100 yards wide. As the later RPG 7 (B41), effective at 500 yards, became plentiful, wider rivers grew more dangerous. Ambushes involved ten to twenty positions spread over 100 to 300 meters. The usual mix was four or five semiautomatic or automatic weapons to one recoilless rifle or RPG.

The PBR would escape at full power, then return firing at full speed. Guns were swung over 30-degree arcs to cover the target, which was often invisible.

Relatively few PBRs were lost, considering the kind of close combat in which they had to engage. What happened on 8 February 1967 to the first, PBR 13, exemplified their experience. PBR 13 approached a sampan near Phu Vinh. The three men in the sampan went overboard. When one was tossed a life ring, he threw a grenade. It started an intense fire in the engine compartment, which in turn cooked off ammunition stowed aft. The boat was gutted from the stern to the forward bulkhead of the coxswain's flat. It had to be stripped and destroyed. Three of its four crewmen, who jumped ship, were saved by covering fire from an accompanying PBR.

Each boat had two army VRC-46 radios. In the PBR Mark I, they were at ankle height in the coxswain's flat without remote speakers or controls. As a result, unless the call light was continuously monitored, messages were missed. Only a few boats had hand sets. In Mark II boats, remote speakers were set at ear level in the cox flat, but frequency and volume could only be changed below decks. The relative location of the two speakers, moreover, caused feedback howl, which not only obscured messages but also traveled a considerable distance on a quiet night.

The Raytheon 1900N radar had been intended as the primary PBR night sensor. It was effective to 2,000 to 3,000 yards; boats generally tried to stay within 2,000 for mutual support and optimum radar and visual cover. Experience in Vietnam showed that the Raytheon 1900N was unreliable (a section rarely had 50 percent of its radars working), that it was often useful only as a navigational radar, and that it could not detect small obstructions nearby. Visual navigation generally sufficed at night.

The other night sensor was the battery-powered starlight scope, which could see out to about 200 yards with a quarter moon. It was used by the forward gunner. It had no indicator to show that it was working properly. According to a 1969 report from the Naval Ordnance Laboratory, "as patrols start before dark, it is not unusual for the crews to find that they have no night detection capabilities." Crews also considered the scope heavy and fatiguing to use.

PBRs rarely, if ever, attained their designed speed. The navy asked for 30 knots but had to settle for 25, fully loaded, on trials in the United States. In Vietnam, however, the PBRs did not perform nearly as well. It was discovered that the jet pumps were operating on a steep portion of their efficiency curves, so that even a slight loss of engine speed, induced, for example, by the heat in Vietnam, sharply reduced their efficiency and therefore a boat's speed. Some boats could not even plane.

The Jacuzzi water pumps suffered badly in service. Their internal coatings came off in the brackish water. As parts wore out, replacement was difficult because many of the spares did not match properly. Existing parts often had to be partly or totally rebuilt instead. The sandy bottom of I Corps was particularly troublesome: sand sucked into the pumps wore out their impeller blades. The drive shaft to the pump was also a problem. The angle on the universal joint was very large, far beyond the allowance specified, and it had to be lubricated after every twelve-hour mission. Many boat sections carried grease guns for this purpose.

Engines suffered as they overheated because their sea-water intakes were clogged, for example, by plastic bags, nylon line, and palm leaves. Engine lifetime was estimated at seven hundred hours.

By 1968, speed had plummeted. The average speed of twenty-one Mark I plastics examined in the delta was 14.6 knots (the worst, 12.6, the best, 19.5). Besides water-jet problems, low speed could be attributed partly to overloading and waterlogged bow flotation compartments.

Operational experience showed that the PBR engine was overloaded in service. For example, one reason the radar was not used as often as it might have been was that it represented a severe drain on the boat's electrical system. Late in the war, a more powerful engine, the 8V53 instead of the 6V53, which would

An official model of a PBR Mark II shows the twin water-jet propulsion and steering system that allowed the boat to operate in shallow water. Note the arrangement of flat vertical armor plating around the controls and engine room hatch. Mark II differed from Mark I in both superstructure arrangement and hull form.

have weighed about 150 pounds more, was proposed, but it was never fitted.

Dissatisfaction with the Mark I led to the procurement of an improved Mark II, which NavSea hoped would make 27.5 knots fully loaded, with a greater margin to increase payload without paying a penalty in speed. The Mark II had a new hull form based on the lines of the *Nasty*, the rights to which the United States had acquired. Its silhouette was slightly lower, and it had an aluminum gunwale so that it would not be damaged by junks coming alongside. However, the fore end of the keel remained a tender spot, and forces afloat evaluated an aluminum cap for the hull. The Mark II also incorporated flotation material.

In response to complaints about Mark I performance, Mark II had a better water-jet pump, the Jacuzzi Mark IV, more powerful engines, and enlarged mufflers. Mark II was somewhat disappointing in service, however, showing poorer sea-keeping and speed performance than the Mark I.

Speed was often unimportant, since a PBR on patrol in a narrow river could not run very fast anyway. But because speed could enable the PBR to escape from an ambush or capture a fleeing enemy, it was valued. Hence the interest in solving the engine and water-jet problems.

The speed crisis was finally solved in mid-1968, as the supply of spare water-jet parts finally caught up with demand. (There was another spares crisis from March to May 1969.) Mark IV pumps were adapted to the Mark I boat and existing boats were dried out, their hulls resealed with gel coating to eliminate water seepage. Many Mark Is, rated at 28.5 knots at 14,600 pounds, could now exceed 30 knots, and late in 1968, the commander of River Patrol Flotilla 5 actually had to restrict his division commanders to 27 knots to avoid structural damage.

Both Marks were loud. A Mark I could be heard two miles away on a quiet night, the quieter Mark II over a mile away. Some sections therefore installed silent, low-speed outboard motors; others drifted onto the Viet Cong at night. The problem in Mark I was that the fiberglass hull widened as it aged, so that gaps developed between the deck and the engine-hatch cover. The cover was further damaged because it was used as a platform for firing 60mm mortars. Silencing material installed on the hatch cover came loose, and once it began to clog the engine air intakes it was removed altogether. By 1968, new engine-room covers and sound-absorbing material were planned for installation in Mark Is. As of 1968, the PBR silencing program was the most extensive to date.

The initially planned figure of 120 boats was derived from experience on the Saigon River, which suggested that bases, which could send patrols out about 20 nautical miles, should be 40 miles apart on major waterways. The PBRs would have to cover about 400

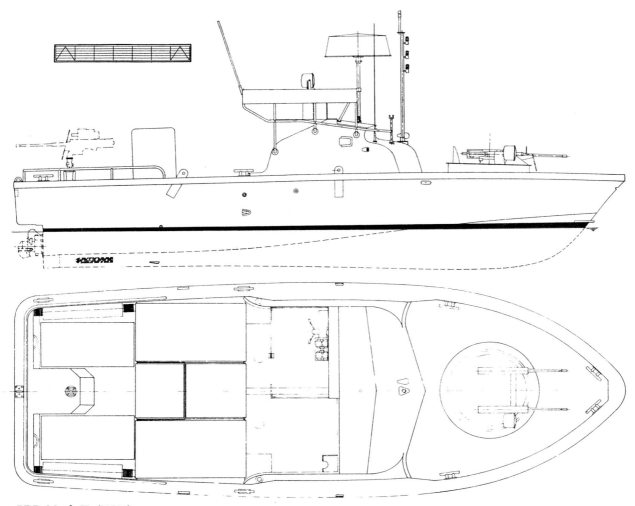

PBR Mark II. (USN)

miles of major rivers, and they would be based at Can Tho, My Tho, Sa Dec, Vinh Long, Chau Doc, and Tau Chan. Throughout the war, boats were organized in ten-boat sections operating in two-boat patrols.

Each of three river-mouth patrols, with ten PBRs and two UH-1B helicopters, would be supported by an LST and an LSD (for repairs), with another LST, LSD, and helicopter section as a rotating reserve. Reportedly, World War II LSTs had been considered for river-mouth control as early as 1962. The first such tender was the *Jennings County* (LST 846), which served from May 1966 on. Others were the *Harnett County* (LST 821), *Hunterdon County* (LST 838), and *Garrett County* (LST 786). They were redesignated AGPs but retained their original LST hull numbers.

The LST cargo hatch was enlarged to 13 x 32 feet, so that a PBR or a helicopter could be lowered onto the tank deck for repair and maintenance, and a 10-ton crane was installed on the starboard side just forward of the deckhouse to lift the patrol boats. A 50-foot-square section of the main deck was strengthened as a helicopter deck. The CIC was enlarged as a PBR command center, and allowance was made for JP-5 fuel and 2.75-inch-rocket stowage. Transient housing facilities on board were increased to accommodate 8 officers and 112 enlisted men. Finally, improved surface-search radar was fitted, as the LST had a secondary surveillance role.

The major problem was weather, which could be rough off a river mouth; boats found it difficult to transit from their LSTs to river estuaries in 3- to 4-foot seas. In July 1966, for example, River Patrol Section 512 reported that it could operate only about half the time. PBR support therefore gradually moved up rivers into calmer water. Once that happened, the LSTs had to move every night so that mortar-aiming stakes planted during the day by Viet Cong would be useless.

In addition, two special barges, an APL and a YRBM, were outfitted as mobile PBR bases, providing berthing and messing for the crews as well as command and control and support for the boats.

TF 116 was broken down into two task groups, the Delta River Patrol Group and the Rung Sat Patrol Group, in February 1966. One year later, TF 116 was assigned to support the new Mobile Riverine Force, TF 117, on a reciprocal basis. The Delta River Patrol Group was further divided in January 1968 into separate commands to operate on the Bassac, Co Chien, and My Tho rivers. The initial distribution of PBRs between the delta and the RSSZ (two to one), attributed to fears that the Viet Cong might close the vital shipping channel to Saigon, gradually changed. At first, each task group was to have about the same number of PBRs and attached attack helicopters. By mid-1968, distribution was approximately six to one, owing to the additional demands of the Upper Delta River Patrol Group and TF Clearwater.

Game Warden also had attached SEALs, which sometimes used its PBRs. A SEAL platoon was assigned in February 1967, and as of mid-1968, each of TF 116's four task groups had a SEAL unit and a SEAL-support unit.

In 1967, the size of the Game Warden force increased rapidly, to 155 (30 in the RSSZ). For the first time, a section of PBRs was sea-lifted to operate in I Corps, near Da Nang, and others operated in II Corps in the spring of 1968. In June 1968, a new Upper Delta River Patrol Group was formed to operate near the Cambodian border. In February 1967, total authorized PBR strength was increased to 250. In mid-1968, rather than activating them as the twenty-fifth PBR section, the last ten boats were earmarked as a battle damage and overhaul pool. By that time, more than a dozen boats had been stricken for being either unsalvageable or beyond repair. The ten boats earmarked for the pool had to be diverted to TF Clearwater, which had two PBR sections by late 1968.

The first 16 PBRs were shipped to Coronado for training. The hundred operational boats went directly to Vietnam, the first 11 arriving on 22 March 1966. Patrols began on 10 April 1966 in the RSSZ. PBRs arrived in the Mekong Delta in May. Beyond the 120 programmed for 1966, 60 more were requested for 1967.

The program initially consisted of 120 plastics: 1 for shock tests (it was stricken), 19 for training and attrition, 30 for the river mouths, 10 for the RSSZ (supplemented by 20 LCPLs), and 60 for river patrols. These were supplemented by Vietnamese river assault groups, Vietnamese national police, Vietnamese army (regional and popular forces) boat companies, U.S. Army Mohawk reconnaissance aircraft, and U.S. Air Force strike planes. In fact, all 30 RSSZ craft were PBRs, the LCPLs having been diverted to harbor defense. Soon another 10 PBRs had been requested for delivery during 1967, that is, under the FY 66 program. Another 30 were requested for other areas, and 40 more, for a total of 160, were approved late in March 1966.

In all, then, 160 Mark Is were built under the FY 66 program, 31RP661–66120, 31PR66121–66160. An additional boat, 31PR66161, was shipped to NorDiv, the small boat design organization. A prototype, 31PR671, and 80 Mark IIs, 31RP672–6781, were built under FY 67.

The program 5 (FY 68) boats were to have been bought competitively, but the program was accelerated and it was more expedient to give the contract to a single source, United Boatbuilders. Awarded on 1 January 1968, this FY 68 contract covered 145 boats: 39 for the army (31RP681–6839, designated J-7803 through J-7841), and 106 for the navy (31RP6840 through 68145). Twenty-seven of those from the navy contract were built for Vietnam, one for Brazil.

More plastics were built under later programs: in FY 69, 31RP691 through 6937 (eighteen for Vietnam, plus one bare hull for the Vietnamese navy); in FY 70, 31RP701 through 7023 (including three for the army, which became J-7842 through 7844); in FY 71, 31RP711 through 7130 (ten for Thailand and twenty for U.S. riverine squadrons, initially earmarked for Cambodia); in FY 72, 31RP721 through 7210 (eight for the United States and two for the Philippines); in FY 73, 31RP731 through 7337 (twenty-two for Cambodia, five for Thailand); in FY 76, two for Thailand; and in FY 77, five for Thailand. Other plastics were exported to Burma (six in 1978); Israel, which built some for herself as the *Yatush* class for a total of at least twenty-eight boats by 1986; and Thailand, which reportedly obtained a total of thirty-seven. An unknown number of extra bare hulls was bought to replace damaged ones. Note that the last of the FY 73 boats had Turbo Intercooled 6V53 engines.

In a program paralleling the initial PBR program, ninety plastic LCPL Mark 11s were ordered to replace steel (Mark 4) craft in the fleet amphibious force. The first sixteen, funded under MAP, were armed, armored, and equipped for Game Warden for delivery in April and May 1966. At this time, four Mark 4s were patrolling the RSSZ. However, few LCPLs ultimately participated in Game Warden. Some did operate with TF Clearwater.

By September 1968, TF 116 had 197 PBRs out of an authorized strength of 250; by October, there were 220. PBRs were organized in sections of ten, one on each river. In 1968, it also contributed the following to TF Clearwater: PBRs and a section of eight LCPLs on the Cua Viet River, just south of the DMZ, and the previously mentioned section of three PACVs on the bays at the mouth of the Perfume River.

By this time, there were seven floating bases: mobile

A PBR Mark II with modified superstructure, operating out of Little Creek in December 1973. Craft of this type have been exported widely since the end of the Vietnam War.

bases 1 and 2; a barge, YRBM-18; and LSTs on the My Tho, Ham Luong, Co Chieu, and Bassac rivers. Each had its own boats and its own detachment of two UH-1B Seawolf helicopters. PBRs typically patrolled in pairs within sight of each other and could call in helicopters for quick reaction strikes or air cover. Unlike the Swifts, the PBRs had direct boat-to-air (Seawolf) communications, and both boats and Seawolves communicated with the operation centers at the bases to avoid delay when air support was needed.

In addition to the PBRs, LCPLs with Xenon searchlights borrowed from the marines patrolled the Cua Viet River. The lights had two modes, white light and infrared. The LCPL could use the latter to see movements within 100 yards or more. Patrolling at idling speed, the LCPL was almost impossible to detect against background noise. LCPL crews could call in air strikes or even blind hostile personnel with white light. The Cua Viet patrol differed from the others in that five LCPLs and three PBRs were assigned, together, to less than seven miles of river.

Game Warden was at best a limited tactical success. PBRs patrolling rivers and canals could not stop the Viet Cong from crossing or from using the smallest streams and waterways for transport. In 1969, therefore, as part of Sealords, the navy switched to the alternative strategy of waterborne guard posts (WBGPs); boats now waited in ambush along the canals and waterways, using special electronic sensors ashore to detect enemy movement. Although Sealords never had enough troops to back it up, the strategy was quite effective. As might have been expected, it cost a lot in numbers of craft per mile of waterway, so that many additional small craft, such as the army's Kenner ski barges and the STABs described in chapter 14, had to be pressed into service.

WBGP ambush operations required in a boat the smallest possible silhouette, as well as both the ability to start engines quickly and to dash, either in escaping superior enemy forces or in closing on Viet Cong watercraft. The small Boston whalers were particularly effective because they were easy to hide.

Many a PBR crew on WBGP duty found that they had to remove radars. The sets were noisy, they drained batteries that were needed for a quick start, and their radomes were a shrapnel hazard.

The Vietnam interdiction experience, and for that matter, the experience of the Cuban quarantine, colored broader U.S. naval perceptions. In October 1965, inspired by the success of the Hughes' new towed-array passive sonar, LRO suggested a low-cost ocean reconnaissance ship (PR) built specifically for the surveillance and control of surface traffic.[8] It would also be useful for augmenting convoy escorts and performing local ASW surveillance. It would need a

A PBR Mark II in its element, early in 1969, with three crew members returning after a search using a captured Viet Cong sampan. There are two M60 light machine guns just abaft the main superstructure and a starlight scope mounted above the canopy.

Plastics were so light they could be airlifted by helicopter. This Mark II was delivered to the Cai Cai Canal on the Mekong Delta, 1969.

Table 12-5. Schemes for Ocean Reconnaissance Ships, 31 March 1966

	A	B	C	D	E	May 1967
LBP (ft)	317.5	311	311	315	—	374
LOA (ft)	—	318	318	322.5	320	—
Beam (ft)	35.2	34.6	—	36.0	35.5	41.6
Depth (ft)	23.45	23.1	—	24.0	24.0	27.0
Draft (ft)	11	—	—	—	—	11.42
FL (tons)[1]	1,650	1,565	1,585	1,560	—	2,357
	—	1,628	1,647	1,612	1,804	
Machinery	2 x 3200D	2 x 3200D	1 x 3200D	4 x 2000	1 x 16000	1 x 13,500GT
	—	—	1 x 25000GT	RGT	RGT w/CRP	2 x 3000D
SSDG (kw)	2 x 300	—	—	—	—	—
Emergency (kw)	300	—	—	—	—	—
Trial Speed (kts)	20	20.5	27.5	21.4	24.0	25.0
Sustained Speed (kts)	18+	19.5	26.3	20.6	23.1	—
			17.0 (diesel)			
Endurance (nm/kts)	————————————————— 9,000/12 —————————————————					4,500/16 (towing)
Complement (off/enl)	10/150	———————————— 10/125 ————————————				18/158
Cost (lead/follow, millions of $)	—	16.7/14.3	17.8/15.3	19.6/16.9	19.6/16.9	—
R&D for RGT	—	—	—	22	30-35	—
Weight (tons)						
Hull	537	503	503	—	—	833
Propulsion	128	128	144	—	—	190
Electric	59	59	59	—	—	109
C2	48	40	40	—	—	57
Auxiliary	174	167	167	—	—	238
Outfit	139	128	128	—	—	185
Armament	38	37	37	—	—	50
Margin	112 (6%)	106	106	—	—	160 (9.6%)
Light Ship	1,235	1,168	1,184	—	—	1,822

Note: In machinery lists, D indicates a diesel, GT a gas turbine, and RGT a regenerative gas turbine.
1. Second series of weights shows the effect of requiring additional eight-hour, high-speed endurance.

light helicopter for surface surveillance; the same helicopter supplemented by lightweight torpedo tubes would suffice for ASW attack. LRO hoped to build reconnaissance ships in FY 70–72.

Preliminary Design offered scheme A (see table 12-5). It would be powered by two diesels geared to one shaft for cruising economically over a wide range at up to 16 knots. Armament would consist of one lightweight 3-inch/50 (a projected gun, controlled by the new Mark 87 director), two 0.50-caliber machine guns, two Mark 32 triple torpedo tubes (six torpedoes), and a six-round antiship missile launcher (eighteen surface-to-surface and twenty point-defense missiles; this was space and weight reservation only, since no missile had as yet been chosen). The Mark 87 director would double as an air-search radar, and the major surveillance sensor was a towed array backed up by an active, short-range, hull-mounted SQS-36 sonar. Stowage and maintenance were provided for a lightweight helicopter as well as starting and fueling space for a UH-2A. There would be sixty sonobuoys and six helicopter-launched torpedoes. The projected complement was 10 officers and 150 enlisted men.

The design was austere, with no margin for future growth. Still LRO considered it far too large, suggesting that a regenerative gas turbine might save space and that joint weapons-handling and stowage facilities might reduce manning. In February, LRO proposed the elimination of surface-to-surface missiles and, as compensation, the substitution of a lightweight 5-inch/54 gun for the proposed 3-inch/50. The BPDMS (basic point-defense missile system), using Sea Sparrow, would be retained with an eight-missile box launcher plus twelve reloads.

LRO also wanted alternative power plants explored, for a maximum speed of 20 to 22 or 25 to 27 knots at eight hours in addition to the desired basic endurance. Such burst speed entailed either an all-gas-turbine plant or at least a gas-turbine boost plant. Endurance speed would be at least 16 knots, with a long range of 9,000 nautical miles at 12 knots.

LRO listed draft characteristics in April 1966. The

Swifts at a river patrol base. In the nearer nest the boats are all Mark IIs, but those at the ends of the more distant nest are Mark Is. The differences between Mark I and Mark II are clearly visible: the Mark II bridge is farther aft, it is higher, and the Mark II deckline is broken aft. Several Mark II's show tape crossed over their after pilothouse windows, presumably to protect against shattering.

The Coast Guard contributed 82-footers to Market Time. The *Point Kennedy* is shown during a langing at the mouth of the Co Chien River in Operation Deckhouse V, May 1967. Modifications for service in Vietnam included installation of the 81mm mortar/0.50-caliber gun forward, the four 0.50-caliber machine guns aft, and the small radome (probably a radio direction-finder) atop the pilothouse, just abaft the searchlight. The 82-footers were relatively slow (18 knots maximum in service), but they had considerable endurance (1000nm at 12 knots) and were considered good sea-boats. All twenty-six in U.S. service were turned over to the Vietnamese Navy, the *Point Kennedy* becoming HQ 713, *Nguyen van Ngan*, in May 1970. She was seized by the North Vietnamese in 1975. (USN)

River security includes mine countermeasures. This 57-foot MSB, modified for riverine operations, swept the channel of the Long Tau River below Saigon in November 1967. Her major improvements were light armor and the addition of light weapons (M60 machine guns and a 20mm cannon, aft). Like other U.S. small craft, the MSB was designed for transport overseas aboard larger ships, in this case in the well deck of an LSD. Twenty years later, surviving MSBs were transported to the Persian Gulf in similar well decks, moving much faster than the relatively slow ocean sweepers.

mission of the PR would be to assist in patrol, surveillance, and control of ocean and coastal areas, and to conduct special reconnaissance missions. Operating individually, the PR would go on extended ocean patrols, but it would also have to be able to enter restricted waters and minor ports throughout the world. Individually or in conjunction with aircraft and additional ships, it would trail merchant and other craft and, if necessary, visit and search them. It would therefore assist in blockade or quarantine.

The PR was not a pure patrol ship. LRO wanted it to be able to deal with anything smaller than a destroyer and even to control and support small surface ASW craft. As a result, the PR was close to a frigate in size and capability.

This resemblance was emphasized by analysis of the role of the towed-array ship in convoy protection. In July, the PF (patrol frigate) designator, then being applied to small frigates built for Iran and Thailand under MAP, was applied to the towed-array ship. Design requirements now included SEAL support, increasingly important in Vietnam. Characteristics changed to show 100 rocket-assisted rounds out of 400 for the 5-inch gun, and the Mark 32 tubes were reduced to fixed twins, as in the contemporary *Knox*-class frigate, with eight reloads. The reloads could also be used by the helicopter. Required performance was altered to a sustained speed of 18 knots with the array streamed. Oddly, the desired endurance (with the array streamed) of 4,500 nautical miles at 16 knots did not quite correspond with that required of other U.S. warships. Machinery was to be arranged to permit economical patrol at 5 to 10 knots.

The patrol craft ancestry of the project was revealed in a requirement that the helicopter hangar be adaptable for use in beach-jumpers, SEALs, or UDTs.

Table 12-5 lists several attempts to reduce the size of the ship. In scheme B, reserved space for surface-to-surface missiles was eliminated. In scheme C, one diesel was replaced with a 25,000-shp simple-cycle gas turbine to provide a high, short-term speed. One propeller shaft had to be added because otherwise tandem propellers might be needed to absorb the plant's considerable power. Scheme D employed four regenerative gas turbines for greater endurance. Finally, scheme E of early 1967 had a single gas turbine, which showed poor fuel economy at 12 knots. All of the gas-turbine designs indicated a prominent air intake atop the bridge.

Preliminary Design preferred scheme C and presented a developed version in March 1967. It was armed with one lightweight 5-inch/54, a BPDMS launcher, aft, with twenty missiles, two 0.50-caliber machine guns, and two twin Mark 32 torpedo launchers with twelve torpedoes; the boat would carry two 6,000-pound helicopters and be able to land and launch UH-2. It was also anything but LRO's small, cheap super-Swift, and it died.

13

Vietnam: The Riverine Force

The U.S. riverine force, TF 117, brought the U.S. Navy into combat throughout the heavily populated Mekong Delta. The force was conceived following the formation of the coastal and riverine arms interdiction patrol described in the previous chapter, although a U.S. riverine force had often been proposed earlier; from 1961 on, at least six U.S. studies had called for a riverine force to fight the Viet Cong in the delta.

NAGP submitted its riverine force recommendations in January 1966. Each U.S. river assault squadron (RAS), like the Vietnamese river assault group, would be able to lift and support a single infantry battalion (combat experience showed that two RASs could support three battalions). A RAS would consist of two command boats (CCB, converted LCM); four monitors (converted LCM); nineteen armored troop/cargo transports (ATC, converted LCM), each carrying a platoon of about forty men; twelve ASPBs (fire support/minesweeping/patrol boats patterned on the French-designed STCAN); and a refueler converted from an LCM.[1] These numbers, but not the basic types, later changed, so that the RAS included twenty-six ATCs, five monitors, and sixteen ASPBs.

NAGP considered the RAS ill suited to the shallow but extensive waterways of the coastal plain; only the proposed ASPB would be useful there, and then only in limited areas because of its draft.

MACV proposed the brigade-size Mobile Afloat Force consisting initially, in FY 67, of two RASs. Unlike the Vietnamese river assault group, the RAS would incorporate its own troops, moving along a river system conducting independent operations from a mobile riverine base. As of January and February 1966, it was expected that the first RAS would become operational in the last quarter of that calendar year. It would be joined by a second RAS during FY 67, to achieve full brigade lift. The force would deploy for thirty days at a time to make four- to five-day search-and-destroy missions aboard the sort of craft used by the assault groups. This proposal was reviewed by a conference in Saigon on 18 February.

Under FY 68's program 5 another two RASs, of improved craft, were sent to Vietnam to double the size of the Mobile Riverine Force. Three of the RASs operated from the mobile riverine base, supporting four battalions. A fourth RAS was available for other purposes.

By way of contrast, the Vietnamese river assault group, which consisted of about twenty boats (most of them converted LCMs akin to those of the RAS) had no organic infantry, although it could carry one infantry battalion. Its secondary mission was river patrol. By 1967, thirteen such groups were deployed around III and IV Corps. At that time, utilization depended upon the U.S. commanders on the ground. Thus the assault groups of III Corps were used for assault, whereas those of IV Corps were used for routine patrol. U.S. naval advisors were trying to change this pattern. They reported fair success with the Seventh Division but still had not convinced the ninth and twenty-first, which were reluctant to use anything but helicopters.

In February 1966, a conference in Saigon called for the Mekong Delta Mobile Assault Force consisting of a main force (two assault groups) and a salvage force. U.S. assault-group boats would be modified

ASPB 50AB6711, newly completed with the notorious stern mortar well, 17 August 1967. After several boats swamped because of insufficient freeboard, the well was ordered decked over. Note the recognition markings on the canopy over the main superstructure, the national star and "A-91-1," which showed that the boat was assigned to River Assault Division 91. Many boats in Vietnam had a pair of single shielded machine guns mounted on tripods alongside the stern well.

versions of craft already in use by the Vietnamese, albeit with increased armor and improved armament. Compatible with MAP, they would be able ultimately to replace Vietnamese assault-group craft. That was not possible until the U.S. Navy had turned over its own river craft to the Vietnamese: the U.S. riverine force consumed all available American production.

Each river assault group was to have consisted of fifty-two LCM(6) ATCs, five LCM(6) command and control boats (CCB), ten LCM(6) monitors, two LCM(8) monitors (which were never, in fact, built), thirty-two ASPBs (a new type), sixty Marsh screws (a new type), and two LCM(6) refuelers. They in turn would be supported by five mobile barracks ships (APB) operating up-river, resupplied by two LSTs. The riverine force would also be supported by two repair ships, ARLs, on LST hulls. Costs and details of the projected force are given in table 13-1. The total cost was $107.3 million, not including the planned development of second-generation craft. This was high, and several efforts were made to scale it down. For example, a 14 March 1966 version developed by CinCPac deleted the LCM(8) monitors and cut the marsh screws to twelve per river assault group.

The entire force would have required 1,460 naval personnel, 460 of whom were immediately approved.

The main objection to the up-river concept was that the APB was too vulnerable. It also drew too much water to cross over to the Bassac River. Some in CinCPac's office suggested, therefore, that it be replaced by an LST 1156, which was more readily available, more capable, and could berth 392 troops, 200 more in cots on the tank deck. The LST could be protected against fire by carrying side-loaded pontoon causeways. However, the post–World War II LSTs were considered too valuable to tie up this way, and the riverine force was ultimately built around APBs and other barracks craft.

A CinCPac officer studying the basic riverine concept attached a note to his copy of the proposal complaining that *implementation* requirements were being firmed up even while the search for alternative *concepts* proceeded. No alternative existed if the force was to be in place by the end of 1966. For example, development and procurement of optimum craft were being studied, but interim solutions (LCM conversions) would have to be accepted to meet the December 1966 deployment deadline.

Nonetheless, on 8 April Secretary of the Navy Paul Nitze asked the CNO and the chief of the marine corps for an alternative operational concept. OpNav developed a mixed river assault group/amphibious force concept based on Operation Jackstay, the first inshore amphibious operation of the war. It rejected up-river floating barracks (APB) in favor of sea-based forces working up-river and across coastal strips. Estimated cost fell to $24.6 million, compared with $107.6 million for navy elements of the proposed MACV package.

CinCPac had not yet commented on either the original concept or the alternative. The marines were reluctant to comment; they were already heavily committed on the ground in I Corps. Some decision was needed immediately if any delta force were to be in position by the first quarter of 1967.

The navy's position was simplified because the basic RAS idea had already been accepted; in fact, 460 men for two RASs had been authorized. Craft development could thus proceed, despite disagreements over the tactics to be used. The use of army landing craft was initially projected, but ultimately the craft were converted from among units in navy stocks.

The program was so urgent that, in the end, it was submitted without CinCPac approval to Secretary of Defense McNamara for consideration by the JCS.

Secretaries McNamara and Nitze feared that the mobile riverine base (the barracks ships) would be too vulnerable. They also considered the original plan too ambitious as a starter. A scaled-down RAS was tentatively approved. Figures below give the cost of two RASs:

	Unit Cost (thousands of $)	Total
30 (vice 52) LCM-6 ATCs	230	6,900
9 LCM-6 MONs	225	2,025
24 (vice 32) ASPBs	400	9,600
4 LCM-6 CCBs	240	960
1 LCU Refueler	150	150
50 Marsh Screws	060	3,000
4 LCM-6 Refuelers	100	400
TOTAL		23,035

The figures below give the makeup of the two river assault groups:

MY THO based	LSD based
20 ATCs	10 ATCs
5 MONs	4 MONs
12 ASPBs	12 ASPBs
2 CCBs	2 CCBs
25 Marsh Screws	25 Marsh Screws
1 LCU Refueler	
4 LCM-6 Refuelers	

No barracks ships would be refitted. Instead, existing amphibious ships would be used as bases: one LSD as an RAS base, one LPH or LPD for helicopter support, and two LSTs for river support as well as to berth additional troops and support units. When the LPH/LPD was not available, an additional LSD and LST would be required for troops and equipment. Some helicopters might have to be based ashore.

Riverine operations were possible because of the presence of mobile logistical bases like the self-propelled barracks ship *Benewah* (APB 35), shown here in 1968. Alongside are three ASPBs and five armored troop carriers (ATCs), two with helicopter flight decks and three with protective canopies. Note the heavy armament installed aboard the *Benewah*, including quadruple 40mm guns fore and aft and two single, slow-firing 3in/50s abaft the helicopter platform.

Table 13-1. The Proposed Riverine Force, February 1966

Main Force (millions of $)

5 APBs	34.951
2 LST Resupplies	0.583
2 ARLs	10.486

RAG (millions of $, each)

52 LCM(6) ATCs	7.743
5 LCM(6) CCBs	1.938
10 LCM(6) Mons	4.140
2 LCM(8) Mons	1.775
32 ASPBs (new)	8.883
60 Marsh Screws (new)	3.311
2 LCM(6) Refuelers	0.100

Salvage (millions of $)

2 ANs	8.218
2 Light-Lift Craft (LCU)	2.624
6 LCMs (fire and pusher)	0.771
2 YTBs	1.455
2 Heavy-Lift Craft (ex-German)	7.500
3 100-Ton Pontoon Dry-docks	0.300

R&D, Test and Evaluation (millions of $)

Weapons	9.998
Armor	2.524

Personnel (including pipeline)

177 officers, 2,362 enlisted

The projected riverine salvage force was also scaled down (costs given in thousands of dollars):

1 LCU weight lifter	1,137	1,137
4 LCM-6 pusher tug/fire fighters	100	400
TOTAL		1,537
GRAND TOTAL		24,572

OpNav proposed that U.S. river assault groups begin operations with trained amphibious troops, marines, rather than army forces, and in Go Cong Province rather than the entire delta. The terrain problems in Go Cong was considered similar to that of the delta, so the lessons learned there would be fully applicable to the whole of IV Corps.

As OpNav envisaged it, the force would operate in four phases:

1. As a battalion landing team conducting five- to ten-day Jackstay-type operations in the immediate future. As in Jackstay, the initial thrust would be from the sea inward.

2. As a RAF comparable in size and composition to the phase 1 team. The RAF would be specially trained and equipped for riverine operations from

In the Mekong in 1968, program 4 monitor M-91-1 beaches to provide a stable mortar-firing platform in support of the Ninth Infantry Division. A crewman can be seen muzzle-loading the mortar in the monitor's well deck. The crew has stuffed the space between the bar armor and its solid backing with boxes, a common practice after the Viet Cong introduced shaped charges that could penetrate the armor system. Reportedly, crews could not be convinced that the empty space between trigger (bars) and backing was what protected them, and that any filling actually decreased protection. Bar armor around the conning station has been folded down to improve visibility.

an afloat base off the Cui Soirap River estuary to continue phase 1–type operations.

3. As a brigade introduced into an assigned tactical area of responsibility ashore. An RAF would be in supporting operations to clear/destroy Viet Cong and initiate "revolutionary development."

4. Still as a brigade, but in littoral areas with the initial tactical area of responsibility transferred to other forces as revolutionary development reached an accelerated level.

Secretary McNamara chose a middle route: the size of the final RAF was scaled down from MACV's first proposals, but it was based afloat on the delta and included the full complement of river craft. Plans were approved in July.

River Assault Flotilla 1 (TF 117) consisted of River Assault Squadrons 9 and 11 (Task Groups [TG] 117.1 and 117.2), each consisting of two river assault divisions (91, 92, 111, and 112, respectively). The approved RAS organization was considerably more powerful than that envisaged by the CNO the previous April:

LCM(6) ATC	26
LCM(6) MON	5
ASPB	16
LCM(6) CCB	2
LCM(6) Refueler	1

Marsh screws were still insufficiently developed to be included.

The attached army force, the Ninth Infantry Division, provided 105mm howitzers on "ammi" barges, which the force could tow and which could be anchored in place.

River Support Squadron 7 (TG 117.3) comprised two (later four) self-propelled barracks ships (APB); one (later two) non-self-propelled barracks ship (APL); one berthing, messing and repair barge (YRBM, ex-YFNB); one (later two) repair ships (ARL); two support ships (LST 1156 class); two heavy salvage craft (LHC); two light salvage craft (LLC); two tug boats (YTB); and two harbor clearance teams. The barracks ships *Benewah* and *Colleton* were recommissioned for Vietnam use on 28 January 1967, followed by the *Mercer* and *Nueces* in 1968; all were decommissioned between 1969 and 1971 as U.S. forces disengaged.

The *Benewah* was configured for flotilla/brigade command and control, and the *Colleton* was equipped for the RAS and battalion commanders. At the end of 1967, the *Colleton* was provided with hospital facilities. In Vietnam, the standard APB complement of 12 officers and 186 enlisted men was supplemented by 900 troops and boat crews; armament was two 3-inch/50s, two quadruple 40mm guns, and eight 0.50- and ten 0.30-caliber machine guns.

Monitor M-91-2, newly completed. Note the stretcher stowed on the after side of the superstructure and the pilothouse protective plates folded down for better visibility. The cross-shaped well in the extreme bow housed an anchor. Aft, the large, superfiring 20mm turret (Mark 51) has not yet been covered. The smaller turrets are Mark 50s with 0.50-caliber machine guns; the forward turret (Mark 52) carries a 40mm cannon and a 0.50-caliber machine gun. None of these turrets was armored; all were provided with plastic weather covers, but those tended to distort in the hot, humid conditions of South Vietnam. All of the LCM conversions were fitted with the hydraulic minesweeping (chain drag) winch that can be seen on the fantail.

For repair and maintenance, the program 4 riverine force used the converted LST *Askari* (ARL 30), which had been decommissioned since World War II. The corresponding program 5 ship was the *Sphinx* (ARL 24). The supporting LST 1156–class ship was chosen for its size, to provide sufficient storage space without excessive draft; it carried a ten-day supply of ammunition and emergency rations as well as the brigade detachment of four UH-1B helicopters. A second LST shuttled weekly from Vung Tau to the mobile riverine base.

Program 5 support ships resembled the original series but had minor improvements. The two APBs (the *Mercer* and *Nueces*) duplicated the *Colleton*, except that two 3-inch/50 guns and the assault boat handling/stowage were deleted, while the helicopter platform and hospital space were enlarged. APL 26 duplicated program 4's APL 55 but had better, air conditioned living facilities (for 75 officers, 18 petty officers, and 292 enlisted men), came with more electrical power (a total of five 100-kw generators), and was rearranged internally to provide more offices and shops. A helicopter platform was fitted, and the ship was armed with 0.50-caliber machine guns and 81mm mortars.

Improvements in the program 5 YRBM included more and better living facilities, machinery to make the craft self-supporting (this included three 100-kw generators), officers, a helicopter platform, 0.50-caliber machine guns, and 81mm mortars.

Late in 1966, River Flotilla 1 began training with the Ninth Division in Vietnam borrowing boats from the Vietnamese navy. RAS 9 began combat operations on 16 February 1967, the first U.S. river assaults since the Civil War. The first three converted LCMs (ATCs 91-1, 91-2, and 91-3) arrived at Vung Tau on 8 March, the first two entering combat on 18 March. Five more ATCs arrived on 30 March, another five on 31 March. All were outfitted locally by the USS *Askari*. The first support ship, the *Kamper County* (LST 854), arrived on 14 April.

From then on, the force built up quickly. As of 30 April, one CCB, two monitors, and twenty-six ATCs were available in Vietnam. By 31 May, the entire planned force had arrived except for two CCBs and four monitors, which arrived early in June. By September 1968, TF 117 operated 103 ATCs, 17 monitors, 6 CCBs, 4 refuelers, and 31 ASPBs in Vietnam. Table 13-2 shows total numbers of boats converted under programs 4, 5, and 6.

A program 4 command and control boat (CCB) shows a communications hut built in the well deck that, in the contemporary monitor, was occupied by an 81mm mortar. Note the open hatches; the module was not air-conditioned. The decking over the high turrets aft is plastic weather covering, not armor.

Table 13-2. LCM(6) Conversions

	Program 4	Program 5	Program 6
ATC	52	64	8
Refuel	2	2	4
CCB	4	8	1
Monitor	10	14	5
LCMM	—	10	—
HSSC	2	—	—
Total	68	98	18

Note: There were two types of program 5 monitors, 105 mm and Zippo, and some ATCs had helicopter pads.

The Cambodians ultimately received twenty ATCs (TC 2–22, omitting TC 13), the last in September 1973; two CCBs; and seven monitors. The program 6 listed here was ordered late in January 1972, and the greater numbers indicate a supplemental program.

Boats received unofficial hull numbers referring to their divisions. For example, C-91-1 was the first command boat (C) within River Division 91. Similarly, a monitor might be M-91-2, an ATC, T-91-5, and an ASPB, A-111-5.

The program 4 riverine force had been scaled down and did not include allowances for training or losses. Nor did its size allow for drastic changes in the kind of operation the river force came to mount. By June 1967, the commander of River Flotilla 1 was complaining to his superior, the commander of Pacific Fleet amphibious forces, that his boats were being stretched too far. He had not yet received his ASPBs, but that was only a small part of his problem. The main part was that assault groups were using the flexible ATC for many new roles.

Each ATC could carry a platoon. A battalion consisted of nine rifle platoons (three companies) and one reconnaissance platoon, so it required ten ATCs for transport. Each battalion also needed three 81mm mortar boats and one battalion aid station. One boat served as a medical clearing station, two as artillery battery fire-direction centers, and at least two for logistic supply. The total ATC requirement, then, for a two-battalion attack, was thirty-three; the nominal strength of the force was fifty-two boats.

Available strength was much less. One river assault division of thirteen boats had to be assigned to defend the mobile riverine base, allowing thirty-nine for operations. Because of the very success of the riverine concept, those boats were worked hard, and at best their availability rate was only 75 percent, leaving an average of twenty-nine (four below the requirement) available at any one time.[2]

A program 4 ATC discharges its troops over the bow ramp, July 1967. Troops in the well deck could be killed by rounds passing through the nearly unarmored ramp, a major flaw. Moreover, troops could come only over the bow ramp, which slowed their exit; specifications for later ATCs showed a desire for alternative routes from the troop deck.

And there were other demands for ATCs. The army wanted them for blocking forces and the escort and supply of barge-mounted artillery-fire support bases. Pending the arrival of the new ASPB, ATCs were also needed for patrol, blocking, and minesweeping, the latter a particularly hazardous assignment.

Meanwhile, the riverine force was penetrating almost untouched Viet Cong sanctuaries. Its boats had demonstrated their ability to withstand concentrated recoilless rifle and heavy automatic fire at close range, and they had survived water mine attacks, but surely some would be lost. This point was made clear when the Viet Cong changed their mine warfare tactics. In the past, Vietnamese convoys and river assault groups had been led by relatively expendable small craft such as LCVPs and STCANs. The Viet Cong had fallen into the habit of ignoring the minesweepers and concentrating on the valuable troop carriers, *commandements*, and monitors that followed. The U.S. ATC minesweeper, however, was worth attacking, and the Viet Cong revised their tactics accordingly. As a result, by June 1967, two ATCs were laid up for repairs, extensive in one case. Both ships had been caught while sweeping ahead of a force.

Given the shortage of craft, any loss of an ATC, monitor, or CCB would have an immediate and severe effect on TF 117's operations. River Flotilla 1 suggested an immediate new-construction program to be completed by late 1967: twelve ATCs and six mon-

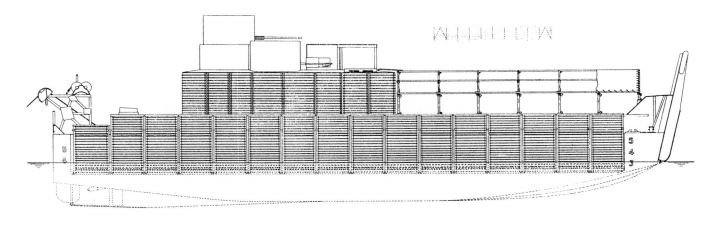
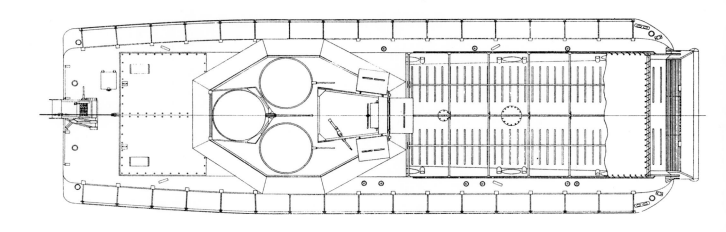
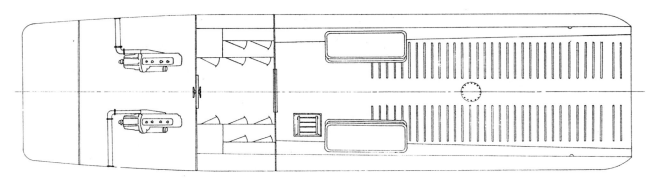

Program 4 ATC. (USN)

Program 4 armored troop carrier T-112-3 of River Assault Division 112, May 1967. The canopy protects the troop space from fragments.

itors, half of them to be based at Subic Bay in the Philippines, half in the United States at Mare Island. By August, the proposal had been changed to two CCBs, four monitors, and eight (later six) ASPBs.

The following attrition and training boats were included in program 5 (FY 68) as additions to two new RASs, 12 and 15: twelve ATCs, two CCBs, four monitors, and six ASPBs. The total program 5 force, then, included sixty-four ATCs, eight CCBs, fourteen monitors, two refuelers, and thirty-eight ASPBs. The overall figures were approved in October 1967.

There was also a program 5 river minesweeping force of ten LCMMs (converted LCM), ten MSRs (modified ASPB), fourteen minesweeping drones (MSD), and fourteen modified (refurbished) minesweeping boats (MSB).

Program 5 required additional salvage and support: two light lifting craft (LLC, a modified LCU), two modified barracks ships (APB), two modified berthing barges (YRBM), and two modified, non-self-propelled barracks ships (APL). These figures were later changed, leaving only one APL and one YRBM; the LLCs were deleted in February 1968.

Program 5 funds were also intended to pay for other Vietnam programs: nine (abortive) heavy-weather patrol craft for Market Time, sixty-six PBR Mark IIs and two LCPL Mark XIs for Game Warden, and even two Boston whalers. The total estimated cost, including these other craft, was $101.9 million, $99 million of it in FY 68 money. This had to be cut by $2.5 million to fit the FY 69 budget. At the same time, the total of FY 68 PBRs was increased. That combination killed the heavy-weather patrol boat, which in any case was not well enough defined.

The last production LCM conversions were carried out for Cambodia in 1973 under program 6.

MACV submitted initial design requirements for all four types of riverine craft in January 1966, and BuShips immediately began work on the new ASPB and the ATC, CCB, and monitor converted from landing craft. The latter had to be used because the program was so urgent. As in 1962, the only capacious, shallow-draft vessels in U.S. Navy service were landing craft. The LCU was somewhat too large, the LCVP too small. The LCM(6), a lengthened version of the original LCM(3) developed specifically for the army, formed the basis for the U.S. riverine force. The larger LCM(8), considered for monitor conversion, was apparently judged too valuable for that. Some preliminary work was done to develop a temporary command/control shelter for it, but ultimately the LCM(8) did not participate in the Vietnam riverine war.

Program 4 encountered special difficulties, because it was carried out just as the BuShips small-boat design section moved to Norfolk and lost most of its personnel. Much of the burden of preliminary design fell on what would otherwise have been a management group in Washington, headed by Richard R. Hartley of PMS 300. The Long Beach Naval Shipyard

A CCB tied up to a river bank, 18 October 1967. Note the tins stowed between armor and bar armor, the life ring stowed on the superstructure, and the anchor forward, with bitts abaft it.

(assisted by FMC Corporation), which had been responsible for the MAP LCMs, produced the working drawings and acted as lead yard.

Program 4 and 5 LCM(6)s were converted to a common basic design, differing only in the space from the main bulkhead forward. The monitor and the CCB had identical ship-shaped closed bows, but the CCB had a command and control module in its cargo well, while the monitor had an 81mm mortar. As conceived by NAGP, each had identical communications equipment in the wheelhouse (one PRC-25 and two VRC-46 transceivers), and the CCB had three more VRC-46s in its command module.

From the first, planning for the riverine force included a second-generation series of craft specially designed for the purpose. They are described later in this chapter, but the reader should keep in mind that their design characteristics were laid down at the same time as the design of the LCM(6) conversions.

Although both the army and the navy maintained stocks of landing craft, the former had drawn down its stocks starting in the late 1950s and transferred beaching craft to the navy because it preferred amphibians, considering them less vulnerable as they crossed the beach. Many of the riverine craft were converted from ex-army LCM(6)s. Table 13-3 shows the navy landing craft inventory in mid-1964 and the LCM/LCU inventory in March 1966.

The weapons of the riverine force are described in appendix D.

NAGP wanted to arm its ATCs with a pair of 0.50-caliber machine guns with associated 40mm area-fire grenade launchers to port and starboard, just forward of the pilothouse, and with two 20mm cannon, one each side forward of the machine gun turrets. The ATC would also have had four M60 machine guns on pintle mounts, one on each outboard armored bulkhead of the wheelhouse, one on each side of the cargo well.

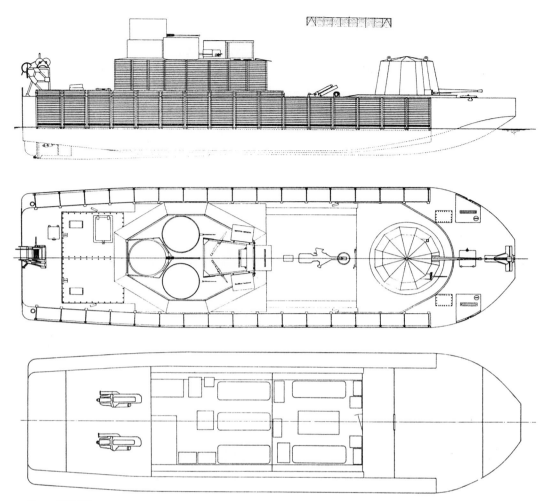

Program 4 monitor. (USN)

Table 13-3. U.S. Navy Landing Craft Inventory
As of Mid-1964[1]

	In Service	Storage
LCVP	1,022	254
LCPR	62	—
LCPL	391	56
LCM(8)	58	24
LCM(3)	175	105
LCM(6)	396	194

LCM/LCU Inventory, February 1966[2]

	LCM(3)	LCM(6)	LCM(8)	LCU
Army Depot Assets[3]	—	72	0	11
Navy Depot Assets				
Issuable	40	74	0	0
Reparable	65	206	12	3
Total	75	280	12	3
FY 67 budget	—	15	32	3

1. Ninety LCM(6)s on order for FY 65.
2. Total of 124 LCM(8)s in use and in storage.
3. Army assets as of 15 February 1966, navy assets as of 6 January 1966.

NAGP's CCB and monitor both had Mark 52 40mm/0.50-caliber turrets forward; one 20mm cannon on the centerline aft; and two 0.50-caliber machine guns with associated area-fire 40mm grenade launchers in armored turrets port and starboard forward of the wheelhouse, or two twin 0.50-caliber machine guns with grenade launchers. The CCB also had two M60 machine guns on pintle mounts on each side of the wheelhouse. The monitor had an 81mm mortar in a well, and four rather than two pintle mounts.

Note that the ATC would incorporate a chain-drag mine sweep. ATCs were later actually employed as minesweepers while the riverine force awaited delivery of its ASPBs.

The Mark 48 combination turrets NAGP wanted were not available in time, and the program 4 LCMs were armed with single 20mm and 0.50-caliber guns. Moreover, all three types were identical aft of the well deck for simplicity of construction and maintenance.

LCM conversion characteristics were first tentatively set in February 1966. As of July, their approved armament was as given in table 13-4.

Table 13-4. LCM Conversion Armament

	ATC	CCB	Monitor
40mm Cannon/0.50 (Mk 52 turret)	—	1	1
20mm Machine Gun (Mk 51 turret)	1	1	1
0.50 Machine Gun (Mk 50 turret)	2	2	2
7.62mm Machine Gun (Mk 21) (pintle mount)	4	2	4
81mm Mortar (Mk 2 mount)	—	—	1
Mk 18 Grenade Launcher (pintle mount)	2	2	2
M79 Grenade Launcher	2	3	3
M1 Rifle (7.62mm conversion)	5	8	8
12 Gauge	1	1	1
0.38 Pistol	7	11	11

The extemporized armament of table 13-4 was based on that of the MAP boats supplied to the Vietnamese. So many of the gun mounts were new to American sailors that two LCUs had to be converted to train gunners in the United States.

In August 1967, with program 4 boats in service and program 5 boats under conversion, NavOrd described improvements which could be carried out in Vietnam. The Mark 50 (0.50-caliber) turret could be converted to a 20mm Mark 51; a 20mm cannon in a Mark 51 mount could be replaced by a long-range (Mark 19) grenade launcher. The Mark 21 7.62mm machine gun could be replaced by an M60, and the Mark 18 short-range grenade launcher converted to a long-range Mark 19 or to a short-range Mark 20. This report induced River Flotilla 1 to propose a series of retrofits of existing boats to coincide with changes in the new program 5 craft in October 1967.

Program 5 ATCs were to have the two pintle-mounted 7.62mm machine guns relocated from their pilothouses to their well decks (these ATCs had six well-deck pintles in all). The top weight saved would go into converting their two Mark 50 turrets (0.50-caliber machine guns) to Mark 51 20mm cannon; the Mark 51 would be converted to a Mark 63 carrying a Mark 19 machine-gun grenade launcher. All converted 0.30-caliber Mark 21 machine guns would be replaced by M60s. The pintles would carry four M60s and two Mark 18/20 grenade launchers or two 0.50-caliber machine guns.

River Flotilla 1 wanted all existing ATCs retrofitted to this standard, but no refit appears to have been done. The command was dissatisfied with the existing Mark 50/51, as it had been unable to depress fast enough to hit nearby targets in a recent fire fight; all weapons had to be able to shoot at targets 70 to 100 feet away.

For the CCB and monitor, the new Mark 48 turret already in use aboard ASPBs would be installed at OpNav's suggestion. Standard Vietcong weapons, RPG-2 and -7 and 57mm recoilless rifles, had already hit ASPB turrets with little effect, and one ASPB that had suffered hull damage after being mined had to have the structure under its Mark 48 replaced. Nonetheless, damage was minimal: a bent train drive handle struck by the operator's knee and feed mechanisms torn from their supports. The Mark 48, then, would resist existing weapons as well as the earlier Mark 50/51, and it was more weather tight. Its operator enjoyed better vision, he was seated, and he was protected against most fragments if his legs were not dangling beneath the armor skirt of the mount.

The river flotilla suggested that the CCB's Mark 52 turret forward be replaced by a Mark 48-0 (a 20mm cannon and Mark 19 grenade launcher) as far forward as possible. The Mark 50 and 51 turrets of earlier CCBs would be replaced by two centerline Mark 48s, one slightly higher for a 360° arc of fire. Existing CCBs would be refitted like the ATCs, with 20mm cannon and a grenade launcher aft and pintles moved to the well deck. The 0.50-caliber guns would be removed from their Mark 52 turrets. In fact, Mark 48 production was too slow because the necessary supply of half-inch dual-hardness steel was insufficient. Program 5 CCBs had a Mark 48 forward and two 20mm cannon and one long-range grenade launcher aft. The earlier CCBs were refitted as proposed.

The flotilla's monitor backfit plans were similar to those for the ATC and CCB, except that the 81mm mortar was eliminated and pintles for Mark 18 or 20 grenade launchers were added in the well deck. In this case, the Mark 19 grenade launcher was installed, but the mortar and the two 0.50-caliber machine guns remained.

None of the program 4 weapons could smash the mud bunkers that were a major riverine force target. The 40mm gun could penetrate, but the bunker could be reused after being hit. Moreover, the 40mm had so great a range that a boat using it risked hitting friendly forces inadvertently; special permission was required before firing. The 81mm mortars on monitors were conceived as indirect-fire (high-trajectory) weapons to be used from beached boats, but that tactic was soon discarded and the 81 reverted to a short-range, direct-fire role.

The bunker problem was reminiscent of barge-busting in World War II. As in the case of the PTs versus the barges, the riverine force began to experiment with more powerful weapons almost as soon as it arrived in Vietnam. In May 1967, a 4.2-inch mortar was successfully tested from an ATC. In August,

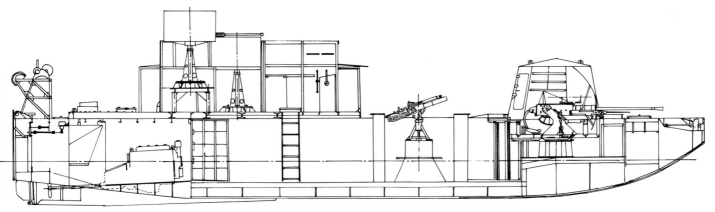

Program 4 monitor, inboard profile. (Norman Friedman)

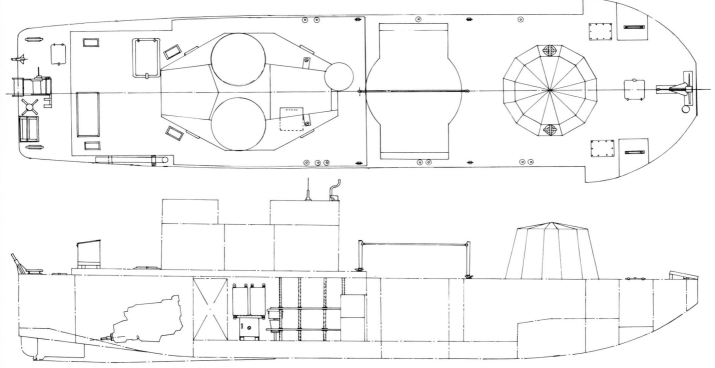

Program 4 CCB/monitor arrangement. (Norman Friedman)

a 90mm recoilless rifle, a light antitank rocket, and a back-pack flame thrower were all tested as potential antibunker weapons for the monitor. The recoilless rifle could be fired safely from the monitor's bow mount and probably from the ATC. However, a more powerful 106mm rifle had sufficient back blast to peel nonskid paint from the ship's deck. It had to be rejected. A light antitank weapon (LAW) could be fired from between the two 0.50-caliber mounts, but it could not destroy a bunker.

In August 1967, River Flotilla 1 proposed replacing the Mark 52 turret with the army 90mm gun (M56) or with the army Flame Cupola (tank flame thrower). Alternatively, three or four 106mm recoilless rifles could be mounted on the outside of the Mark 52 turret, which could be modified for one-man control. In this case, the mortar would have been closed over to avoid blast. This line of development led to the adoption of the 105mm howitzer, in a turret taken from the standard marine corps LVTH-6 amphibian, in eight program 5 monitors. The howitzer was also badly needed for indirect support. Early program 5 plans called for all fourteen monitors to have been armed with 105mm howitzers, but only ten were available, and OpNav had to accept a weapon it did not like, the flame thrower.

The riverine force found the flame thrower extremely promising. Tests with the back-pack unit,

A program 5 monitor shows its 105mm howitzer turret forward and two Mark 48 turrets aft. Note the sandbag protection around the pilothouse and the drums carried behind the bar armor on the main deck.

A new Zippo at Long Beach displays two flame turrets, 1968.

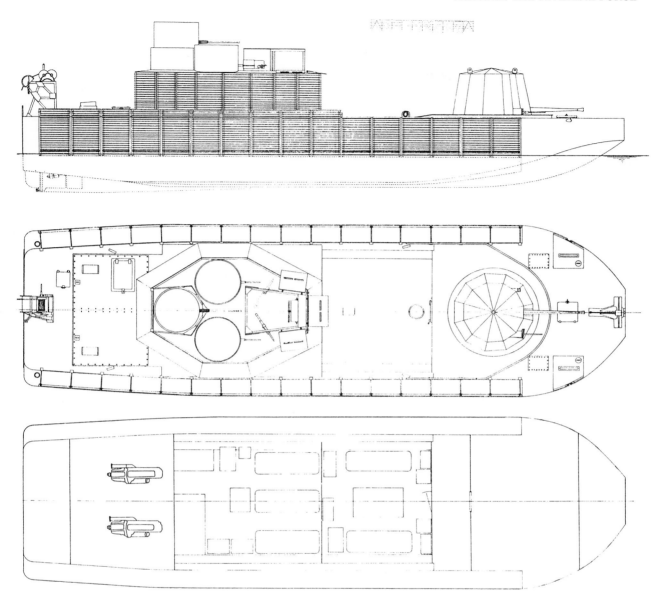

Program 4 CCB. (USN)

which had insufficient fuel and a range of only 40 feet, encouraged the riverine force to look for something more powerful. On 4 October, the force tried a big army flame thrower, the M132A1, the flame-throwing version of the M113 armored personnel carrier; the weapon was "shoehorned" into an ATC of River Division 92. The M132 weighed 23,000 pounds, and the ATC had a capacity of only 24,000, so the boat's canopy and mine protection seats had to be removed. This combination proved quite successful; there was no blowback even when it fired into the wind. The first test shots were fired into the hostile shore at Kien Hoa, and the experimental flame thrower entered combat on 5 October. The riverine command considered its psychological effect so great that it asked the Ninth Division to attach up to four M132A1s to the riverine force indefinitely.

The force requested six flame monitors. The M132A1 was clearly too heavy; instead, its M10-8 flame thrower could be removed and installed in modular fashion aboard a monitor, replacing the 81mm mortar in the well deck, with a flame turret replacing the existing 40mm/0.50-caliber turret. Alternatively, the entire assembly might be installed aboard an ATC. The ASPB was considered too small for a proposed flame conversion.

In M132A1 form, the flame thrower had a range of 150 to 170 meters, and 200 gallons of fuel sufficed for 32 seconds of continuous fire.

In October, River Flotilla 1 reported that its flame

throwers effectively penetrated bunkers and spider holes, asphyxiating the enemy within. The effective range of 200 to 300 yards was more than adequate, since the average battle range was less than 50 yards, with targets bordering the Vietnamese canals at distances of only 5 to 25 yards. OpNav disliked the flame thrower, but the weapon was incorporated in six of fourteen program 5 monitors delivered in May and June 1968. Four went to Vietnam, two being retained in the United States. Conversion entailed ripping out much of the LCM well deck to stow compressed air and napalm bottles below the waterline.

These "Zippos" had two M10-8 flame throwers with sufficient fuel (1,350 gallons) for a 225-second flame. Range at 90 percent fuel capacity was 140 meters, and 160 meters could be exceeded when 90 percent of the fuel had been expended, so that the propelling gas was working on a lighter liquid mass.

In addition, two program 4 monitors, one of them built on an LCM(3) hull, were converted locally. They retained their 81mm mortar and their 0.50-caliber and 20mm turrets aft; the ex-LCM(3) also retained its Mark 52 turret forward. Both had the modified flame system Mark 1 Mod 0.

Six LCM-6s were converted to "Zippo rechargers," one of the boats normally assigned to each. Four LCM-6s had army M4A2 flame thrower tank trucks parked in their well decks; two had E49R3 napalm refueling units, for which the bow ramp was cut down and sealed. All were provided with canopies over the well deck, special fire-fighting equipment (foam nozzles in the tank area), and watertight bulkheads between tanks and the engine area.

Besides their 105mm howitzers or flame throwers, the fourteen program 5 monitors were armed with two Mark 48 gun mounts aft (each carrying a 20mm cannon, the high one also carrying a Mark 19 grenade launcher). The craft also had two 7.62mm M60 machine guns on Mark 26 pintle mounts. The 81mm mortar and well deck were eliminated, and the radio was relocated.

Table 13-5 shows program 5 ammunition loads.

Operationally, there was some question, at least early on, as to just how effective riverine weapons were against enemy troops, particularly those behind banks or trees. On 4 December 1967 and 8 May 1968, riverine divisions accidentally fired on single companies of U.S. infantry ashore. In one case, U.S. troops were in a fairly dense treeline of nipa and coconut palms and were fired on by a full riverine division using weapons including a flame thrower. In the second case, troops were in an open treeline with virtually no underbrush; this time no flame thrower was used. In both situations, the tide was full and the boats fired level. In neither situation was any soldier scratched. One company commander said in an interview that the experience was "traumatic" but "did make one question our effectiveness." Interviews with Viet Cong defectors about this time confirmed this lack of success in achieving kills.

Table 13-5. Program 5 Ammunition Loads

	ATC	CCB	Monitor	ASPB
105mm	—	—	350 RS	—
81mm	—	—	—	40-60
20mm	2,000 RS	2,000 RS	2,000 RS	4,350
	2,000 Mag	3,200 Mag	3,200 Mag	—
Mk 63	1,000 RS	4,400	—	—
	2,000 Mag	—	—	—
7.62mm	5,000 RS	—	3,000 RS	—
0.50	1,000 RS	—	1,000 RS	4,260
Mk 18/20	250 RS	—	100 RS	—
Mk 19	—	360 RS	360 RS	2,108[1]
	—	360 Mag	360 Mag	—

Note: RS indicates ready-service rounds, Mag, rounds stowed in magazines.
  1. Including 100 illuminating rounds.

As for taking out bunkers, by 1968 the Zippo seemed the only effective weapon at hand. The Viet Cong had acknowledged that fact by moving their bunkers about 30 yards back from river banks. Even so, the Zippo was not considered entirely satisfactory, for a burned-out bunker could be reoccupied. The Naval Research and Development Unit, Vietnam, produced an alternative, the "douche." Two ATCs were fitted with a 12V71 diesel, vacuum tank, and centrifugal pump in their well decks. They could discharge 2,000 gallons per minute from two nozzles and direct the stream up to 200 meters to clear foliage and trees and, more importantly, erode bunkers. One of the two ATCs was further modified as a douche-dredge: its water could be diverted to scoop away mud before a modified bulldozer scoop.

LCM conversions were carried out at a number of yards, locations determined by the availability of craft in stock. Contracts were awarded by a mixture of competition and invitation to bid. Spreading the contracts over yards ensured that many boats would be completed early. The first two monitors were competed for, the others bought by formal advertisement. The first two CCBs were negotiated, other two advertised. All ATCs were bought by formal advertising. Program 4 conversions were distributed among two naval and twelve private yards. The entire program was completed in ten months, including prototype construction and tests. Program 5 was even faster, being finished within fourteen weeks. Program 6 took less than nine months; all boats were converted at Seattle and San Diego.

The LCM conversion design was based on sufficient armor to defeat both conventional projectiles and shaped charges. NAGP proposed dual-hardness steel-2, which it believed would defeat a 0.50 caliber armor-piercing bullet at 20 meters. Armor was to be spread over the outer skin of the engine room, from 1½ feet below the waterline to the main deck; over the gun turrets; on the interior wing walls of the cargo well, from 1½ feet below the waterline; and as an armor bulwark above the main deck, aft of the cargo well. The machinery would be covered by a quarter-inch STS deck. To defeat shaped charges, sixteenth-inch steel plate would be installed 15 inches outboard of the armor bulwark and 12 inches outboard of the wheelhouse. The monitor and the CCB were to have spaced, sixteenth-inch armor 10 inches outside the Mark 52 shield forward, turning with the mount. In addition, the monitor's mortar well would be lined with armor.

Republic Steel was unable to provide enough dual-hardness steel quickly enough. Instead, equivalent 1-inch (nominal 0.9-inch) high-hardness plate had to be used to resist 0.50-caliber armor-piercing rounds at 20 meters, and compensation had to be made for the extra weight.[3] The boats were blistered like the old battleships to bring them back out of the water. The foot-square blister ran their full length and was filled with foam. Armor was mounted on top; foam blocks filled the intervening space.

Instead of the proposed sheet-steel armor, heavier armor bars were installed to trigger delayed-action and point-detonating recoilless rifle rounds through 75mm and against 57mm shaped charges; their hot jets would splatter against the high-hardness armor. This protection initially extended to all areas except the turrets, the bow ramp (in the ATC), and the above-shoulder-level portion of the cox flat. Bar armor was later extended over the top of the pilothouse. Gun turrets had only high-hard armor, effective only against 0.50- and lesser-caliber projectiles. Bars had to be heavier than sheet-armor triggers to get sufficient rigidity.

However, bar armor was much more effective than sheet armor. The standard North Vietnamese weapons were detonated by a small nose charge that fired into the base of the shaped-charge explosive cone. If the charge was knocked off center, it fired into the side of the cone and so ruined the shaped-charge effect. A shaped-charge round tended to distort upon hitting bar armor off center. Thus, the spaced-armor effect was important only for the small percentage of rounds hitting precisely between adjacent bars; for the others, the main effect of the bars was to cause either duds or ineffective explosions. The lighter-sheet spaced armor of the ASPB had no such beneficial effect.

The Viet Cong adopted more powerful weapons. The 57mm shaped charge was replaced by the 80mm Soviet RPG-2 and -7, lighter weapons that could penetrate four times as much armor.[4] Existing riverine craft were suddenly vulnerable to armor penetration and exposed to a greatly increased scale of attack since the enemy could carry so much more ammunition. But there was no hope of adding sufficient armor plate; the LCM(6) conversions were already badly overloaded. The ATC carried 6,800 pounds of armor, but the program 5 CCB carried 17,860 and its 105mm monitor companion carried 19,740.

In July 1967, a NavSea officer in Vietnam described attacks on ATCs in which shaped charges had penetrated armor plate, bulkheads, voids, and inner bulkheads to wind up in the magazine—in some cases, the distance on angular shots from bar-trigger shield to high-hard plate was 31 inches. As for the new RPG-2 round, a NavSea officer commented that "if one of the turrets takes a hit from this it will be all she wrote. Would sure like to see a solution on this one."

In October 1968, for example, a flame monitor was hit by a B-40 that detonated on the trigger bar, penetrated the styrofoam and internal armor plate, and finally spalled against the flame fuel tanks without penetrating. This was the first instance of a Heat round making a direct hit on a flame fuel system.

Crews reacted to the new threat by filling the empty space between the trigger bars and the armor plate with sandbags, watercans, and empty ammunition and ration containers. That completely negated the armor system, guaranteeing that all RPG shaped charges would detonate; without the filler many charges, striking air after passing between the armor bars, would be spent.

The South Vietnamese doubled the thickness of high-hardness armor around the coxswain's flats of many boats, but that too was useless; it was later estimated that 5 feet of sand or 10 inches of high-hard armor (at an 18-inch standoff) was needed to defeat the RPG jet.

U.S. boat developers were keenly aware of this problem, and in 1968 they tried a variety of new armor schemes. None was implemented, because the LCM conversion program effectively ended in FY 68 and the second-generation prototypes were not placed in production. Many boats were, however, modified to mitigate the effects of the hot jet the shaped charge produced. It was found that 30-ply ballistic nylon blankets developed to suppress fragments could be extremely effective, and installation began in January 1969. A blanket in the ATC troop well, it was discovered, could cut fragments by 99 percent. The blanket generally had to be about 18 inches inboard of the armor that the jet could be expected to fragment.

As in the large armored ships of the past, boat captains had to give up protection for better vision, standing outside their protected pilothouses and leaning over a 0.50-caliber turret. Most used cases of

The extended rear bar armor of a new Zippo is shown at Long Beach, 1968. Note that the styrofoam sponson is constructed from modules, each replaceable if damaged.

Bar, or standoff, armor was essential to the success of the numerous LCM conversions. It is shown here aboard a program 5 flame thrower ("Zippo") monitor newly converted at Long Beach, 1968.

This program 5 LCM minesweeper, at Wilmington, California, on 29 August 1968, has a cutdown bow door (sealed) and an armored pilothouse, not yet equipped with bar armor. Two bases for minesweeping davits can be seen in the well deck; one is being worked on. A pintle for a 0.50-caliber machine gun is visible in the bow.

C-rations for protection. Other crewmen could not reach a wounded captain without exposing themselves; because of this, at least one commanding officer bled to death in a fire fight.

Aside from penetrating hits, spaces such as the mortar pit in the monitor, the communications module of the CCB, and the troop well of the ATC were vulnerable to mortar hits when boats were anchored. Unless covered by ballistic nylon canopies, they were also exposed to shrapnel from hits on the forward turret or the after superstructure. Many program 5 ATCs benefitted because the helicopter pad covering the troop well was made of high-hard steel.

The ATC presented a particular dilemma in this respect. Most ATCs were fitted with shock-absorbing seats to protect troops from mine explosion. Troops using them had to sit up during long transits, and because the seats made it difficult for the men to scramble ashore quickly, they were exposed to extra mortar and shrapnel hits. By 1968, the seats were used only under direct orders; otherwise, they were removed so troops could lie down in transit.

Still, mines were unquestionably a problem; no riverine boat had been designed specifically to resist their explosion. The first severe damage was suffered by a boat in the Xang Canal near Dong Tam: one engine was completely destroyed; 18 feet of the blister on the starboard side was crushed; the bottom plating and framing on that side were distorted; the engine rails were sprung about $1\frac{1}{4}$ inches both horizontally and vertically; and the 0.50-caliber mount and turret on the starboard side went overboard.

The riverine force required its own special minesweepers. Mines were an ideal guerrilla weapon, homemade and command-detonated under their targets. The only effective countermeasures were the chain drag and the mechanical, type 0 sweep, which tore up the command wires. Boats normally patrolled in pairs, one running along each bank of a river about 5 to 15 yards off shore, trailing chain-drag gear that consisted of about 30 fathoms of cable, a 52-pound counterweight, a swivel, and 10 feet of 1-inch anchor chain with at least twenty-two steel strips. The strips protruded at different angles to ensure that some would constantly drag through the muddy river bottom, snagging or cutting command wires. Because the sweepers had to stay close to the banks to cover as much as possible of the river's width, they were easy ambush targets.

By March 1967, there were only two river minesweepers, borrowed Vietnamese RPCs, to supplement twelve U.S. MSBs. The latter were both

Departing for practice sweeps off Long Beach in September 1968, an LCMM reveals one of two Mark 51 turrets, with the bulwark cut down abeam to open the field of fire. Note the sweep davit on the port side amidships and the type O floats aft.

expensive and relatively old. Three MSBs had already been lost. The commander of minecraft, Pacific, proposed a cheaper alternative, an LCM minesweeper. The other alternatives were a sweeper version of the new ASPB and a minesweeping drone (MSD), the prototype of which would be ready for evaluation in the United States in May 1967. Four deployed to Vietnam in June, and others were later used there.

The sweeper version of the ASPB, designated MSR, was bought in program 5, but LCMs designated LCMM were converted as interim solutions to the problem. In November 1966, anticipating this requirement, PMS 84 asked the Naval Ship Engineering Center, the newly organized design arm of the new Naval Ships Systems Command, to prepare drawings. Six LCMs were quickly converted, armed with three 0.50-caliber machine guns (one in the bow, two amidships), one 7.62mm M60 machine gun, and two M79 hand-held grenade launchers. Their bows were sealed and the ramps cut down to accommodate the bow gun.

In March 1967, then, the commander of minecraft, Pacific, called for a total force consisting of six MSBs, plus six at Subic or Sasebo; six MSRs (modified ASPB), plus two for training and four for attrition; ten LCMMs (modified LCM), plus two for training and four for attrition; and eight MSDs, plus two for training and eight for attrition.

The program 5 LCMM, redesignated MSM, was basically an ATC with sweep gear armed with two Mark 51s (20mm cannon), one 0.50-caliber machine gun forward, and two M79 hand-held grenade launchers. Their bows were sealed and rounded, and they were fitted with minesweeping davits and canopies. The earlier LCMMs were transferred to the Vietnamese when the program 5 craft arrived in June 1968.

Some ATCs were also used as stopgap sweepers. Early riverine operations showed that they and the MSMs were two slow and vulnerable. ASPBs were used instead.

The other side of mine protection was the special shock-absorbent seating and flooring for the ATC developed by the Naval Ship Research and Development Center after 1966. It was first installed in mid-1967, in ATC-112-11. Chairs and matting for thirty-two sitting and eight standing men took up about two-thirds of the well deck and could be removed in about two hours. Shock seats and flooring were standard in the program 5 ATCs.

The other major program 4 LCM innovation was the ATC helicopter platform, initially conceived for medical evacuation and requested personally by Commander W. C. Wells in June 1967. A small helicopter was photographed at Tan Son Nhut Airfield, and its scaled dimensions were applied to ATC drawings. The portable platform was built aboard the *Askari*, completed on 2 July, and installed on the fourth, when a light helicopter landed and took off four times. There had been some concern that the helicopter's downwash would blow the ATC away, but the craft's coxswain did not even have to use his

The MSR was a simplified ASPB adapted for river minesweeping. Note the special bow frame and open gun mount aft. This photograph was taken in Long Beach Harbor, January 1969.

A new program 5 ATC, 1968, shows a flat armored helicopter deck designed to resist 81mm mortar hits. A ladder leads up to the deck. The cross-shaped objects aft on the bar armor are for stowing life rings.

rudder; the first approach turned into the first landing. A UH-1D made three landings and takeoffs on 5 July. This particular platform could be mounted on any ATC, and several ATCs equipped with it were used as battalion aid stations. The removable platform, which incorporated armor, became standard in program 5 ATCs.

The program 5 craft incorporated several other improvements, most of which reflected requests summarized by the Pacific Fleet staff in October 1967. In all three LCM types, the pilothouse was to be enlarged to make room for four personnel, and a small protected space containing a collapsible chair for mine shock was to be built above and behind it for the boat captain or task unit or task group commander to use during an engagement. Some space would be made by relocating the VRC-46 radio from the pilothouse to the well deck; that would also make for better weather protection.

Battle experience showed in several requests for improved protection: flexible engine-fuel lines that would withstand shock, bar-trigger armor for gun mounts, and armor sufficient to stop the RPG-2 and -7 (which could not be provided). The summary also suggested a variety of counterambush weapons: Claymorettes, fuel/air explosives, and portable flame throwers.

The CCB command module was air-conditioned and lengthened by moving the forward bulkhead to contain an additional VRC-46 for the army and an additional PRC-25. One PRC-25 was converted to a GRC-125 by the addition of a power pack. CinCPacFleet proposed that a data link compatible with existing VHF (FM) and HF (AM) radios be installed to provide the MRF command ship with the identification and location of the CCB.

More bar-trigger armor was added around the pilothouse and the aft deckhouse.

The program 5 boats had more powerful electrical systems: ATCs and monitors had 2-kw (24V) rather than 1-kw generators to power their new turrets; note that the program 4 CCBs had 8-kw generators. All but eight ATCs had 24V DC starters in place of the program 4 12V DC starters.

Perhaps the most fitting conclusion to the LCM(6) story is a message that River Flotilla 1 sent to the Naval Ship Systems command in July 1967. All the program 4 conversions had been delivered, and they had been tested in combat. The flotilla liked them and cited their outstanding performance:

- Their soundness was demonstrated by their ability to withstand near-miss mining without damage.
- Their blisters absorbed the mining shock and reduced the subsequent damage when mines made direct hits.
- Their bar armor/ballistic shield combination had defeated the 57mm recoilless rifle round on numerous occasions.
- After initial problems had been solved, the 20mm gun (Mark 16-4) was proving adequate in combat.
- The value of the 40mm gun as a bunker-buster had been demonstrated in action on 15 May and 19 July.
- The boats proved "maneuverable well beyond initial expectation."
- The flexibility of the craft seemed virtually unlimited: the ATC served as troop carrier, salvage boat, resupply boat, and miniature hospital; it had just been fitted with a locally produced portable helicopter deck and platform for the UH-1D helicopter.

There were, to be sure, several criticisms. All of the converted LCMs were slow. However, as long as the ATC was required to carry not only troops but also the organic vehicles of the reinforced infantry platoon, such as the M113 armored personnel carrier, it had to have a big bow ramp and therefore a bluff bow. When it was armored, the bow ramp was too heavy for its winch. ATCs rarely if ever carried vehicles, and it was rarely used in service; troops usually went over the side.

The coxswains often had inadequate vision, requiring hand signals in order to steer.

As for weapons, the main complaint was the absence of firepower forward. The combination of bow ramp and helicopter pad blocked the bow sector—which was important as a boat beached. The projected shallow-water craft of the early 1970s would have had a heavy weapon forward. Some operators disliked the 20mm cannon primarily because it was supported by an unreliable electrical system. Some wanted to substitute a hydraulically operated 20mm mount; others would have been happy with a twin, 0.50-caliber, hand-operated machine gun.

As for the monitors, opinion on the Zippo was mixed. Some considered it useless, but one officer recalled that in November 1969 his patrol group was ambushed almost every night on Cambodia's Vinh Te Canal until a Zippo was assigned to antiambush duties. The douche was considered extremely effective. As for the howitzer, its beehive (flechette) rounds were quite effective, but many disliked it for its low rate of fire and limited magazine capacity. The 40mm gun was preferred for a fire fight. The next step up in size, the 155mm howitzer, was attractive, but there was some fear of its shrapnel ricocheting. The monitor exceeded the lifting capacity of repair ships (ARL) in Vietnam by 10 to 20 tons.

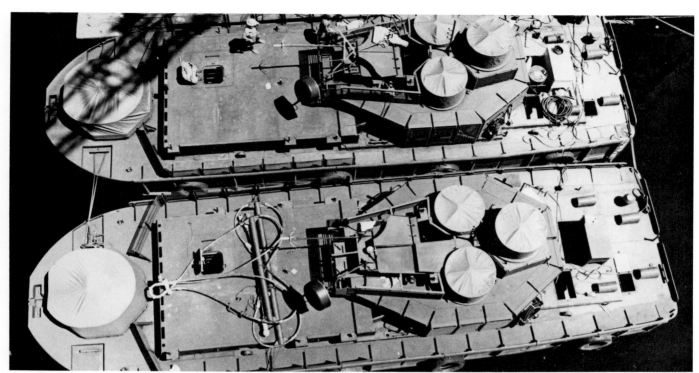

Two program 5 command boats (CCB) being prepared for shipment to Vietnam at Wilmington, California, 24 August 1968. Note their flat, air-conditioned command modules and the new bulwark arrangement forward. Whip antennas have not yet been fitted, though the stubs are visible, and the radar and UHF radio mast, with its AS-390 antenna for ship-to-air communication, have been folded down. The position right forward was occupied by a Mark 48 turret carrying a 20mm cannon and a Mark 19 grenade launcher, and the three turrets aft carried two 20mm cannon and one long-range grenade launcher (Mark 19). None of these weapons had as yet been fitted. There is no minesweeping winch: program 5 included specialist river sweepers.

Table 13-6. CCB and ATC

	Mk 1	Mk 2	CCB Mk 2
Length (ft-in)	56-1.5	49-9	49-9
Beam (ft-in)	17-6	24-0	24-0
Draft (ft-in)	4-0	3-3	3-3
Displacement (lbs)	155,000	97,023	78,670
Engines	2 Gray Marine 64HN9	2 Avco-Lycoming	TF-1260
Power (shp)	225	920 (800 shp continuous)	
Speed (kts)	8.5	20	23
Endurance (nm/kts)	—	250/15	250/15
Armament	—	2 Mk 48-2 (twin 0.50 or 20mm, Mk 19 40mm GL)	
Tactical Diameter (yds/kts)	—	35/6	
Cost (thousands of $)	215 (375 for CCB Mk 1)	1,500	—
Weights (lbs)			
Hull	—	46,359.1	48,266.9
Propulsion	—	6,017.1	6,017.1
Electric	—	3,313.6	3,313.6
C2	—	889.1	889.1
Auxiliary	—	1,585.9	1,521.2
Outfit	—	5,237.5	2,999.9
Weapons	—	4,293.9	4,293.9
Module	—	—	5,800
Margin	—	—	1,462
Light Ship	125,000	69,050	74,563.7
Full Load	155,000	97,138	93,235.6

Finally, there was a widespread belief that the CCB was superfluous, that a suitable module in an ATC could do as well.

These ideas were reflected in both the Mark 2 ASPB/ATC/CCB program (see below) and in the post-1970 small craft development program described in chapter 16.

The assault support patrol boat (ASPB) was the only entirely new craft of the riverine navy. In 1965, NAGP conceived it as an American successor to the French STCAN, sharing that boat's shallow draft (maximum 3 feet 6 inches) and antimine hull form and construction ($5/16$-inch steel with large scantlings). The ASPB would be used for escort, fire support, and mine sweeping during a U.S. river assault. It would, therefore, be extensively armored above the waterline: like the LCM conversions, it would have dual-hardness steel-2 to defeat 0.50-caliber armor-piercing bullets at 20-meter range, and it would have quarter-inch STS over its engine room. The ASPB would also have the same radios as the LCM, two VRC-46s and one PRC-25.

NAGP envisioned a boat 36 to 39 feet long and up to 13 feet in beam, although a greater beam might be acceptable to keep the draft to a maximum of 3 feet 6 inches.

Armament would comprise a combination 20mm (or twin 0.50-caliber)/area-grenade-launcher turret forward; a combination 7.62mm M60 machine gun/area grenade launcher aft; two M60 machine guns at the radio operator's position port and starboard; a 60mm (M19) mortar on a bipod firing from a bedplate at the stern; and a LAW rocket launcher for use against bunkers or armor. The crew would also have two M14 rifles, five revolvers, and a shotgun. NAGP also specified night-vision equipment: a manned weapons sight on the forward mount, an infrared weapons sight on the after centerline mount, and a spotlight bright enough to illuminate the entire length of a 40-foot sampan at 1,000 yards.

The ASPB would also have a chain-drag sweep.

Its crew would consist of a coxswain/boat captain, a radio operator, and three gunners.

The only existing boat even approaching these characteristics was the unsatisfactory RPC developed for the Vietnamese navy. MACV considered it deficient because of its lightweight and easily damaged hull, flat bottom, and lack of adequate protective armor, especially in vicinity of engines. Indeed, local U.S. military advisors wanted RPC production stopped in favor of the ASPB. However, there was some question, early in 1966, as to whether any new boat could be developed quickly enough to enter service with the riverine force by the end of the year. In February 1966, that meant twelve ASPBs had to be available by the end of the year.

CinCPacFleet was skeptical; he asked MACV for a fallback involving either continued production of the RPC or interim substitution of another craft, such as the LCPL Mark XI. MACV refused to accept any more RPCs, even if ASPBs could not be provided in time. However, other suitably modified craft might be satisfactory. The LCPL could be armored, or monitors used (although that would restrict the river assault groups to deeper water), or a few Vietnamese RPCs diverted to the U.S. Navy.

BuShips began design work in February, basing

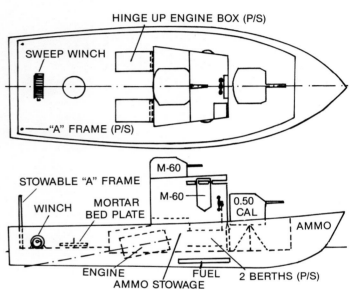

The ASPB proposed in early 1966 by the Military Assistance Command in Vietnam. It is 36 to 39 feet overall, 13 feet in beam, with a maximum draft of 3.5 feet, and has a rugged hull similar to that of the successful STCAN, with $5/16$-inch plating and I-beam framing. To defeat 0.50-caliber armor-piercing bullets at ranges of 20 meters or more, the hull above water was to have been built of dual-hardness steel, as was the deckhouse enclosing the coxswain position, radio operator, and after gunner, the after rotating turret, and the forward 20mm turret. The forward turret was to have carried either a lightweight 20mm gun or one or two heavy-barrel 0.50-caliber machine guns as well as a long-range 40mm grenade launcher. The superfiring turret would carry an M60 light machine gun and a second long-range grenade launcher; two more M60s would be mounted on pintles in deckhouse ports. The bedplate platform aft would support a 60mm mortar M19 on a bipod mount, and the boat would carry a small LAW to deal with bunkers or armor. As for the crew, it would carry two 7.62mm M14 rifles in quick-release stowage brackets, behind armor, one at the forward gun position and one at the coxswain's position; two M79 single-shot grenade launchers; five Smith and Wesson pistols; and a shotgun. A winch was provided aft to stream a chain drag sweep. In the drawing, note the well deck extending for most of the length of the boat, outlined in the plan view and indicated by dashed lines in the elevation; note also the stowable A-frame aft, for sweeping. The ASPBs actually built were larger and had stowable sweep davits aft. (Norman Friedman)

ASPBs under construction at Gunderson Brothers Engineering show shallow V-bottoms and propeller tunnels. The boat on the right, 50AB6811, was the first to have the new lines that precluded the need for sponsons. Note the massive sponsons fitted to the hull on the left, even though it is less advanced than 6811.

its work initially on NAGP's proposal. MACV presented more detailed characteristics in a 1 March message. The most important ASPB characteristics were (1) speed in excess of the speed of the LCM(8); (2) fire power for gunfire support and protection; (3) armor to protect the crew and vital boat machinery; (4) minimum draft and maximum maneuverability, a turning circle of 50 feet at 6 knots, to allow flexibility in canal assault-support and blocking operations; and (5) mine-resistant hull design. Quiet operation would be important for canal-blocking operations.

The hull would have to withstand the explosion of a 55-kg (TNT equivalent) mine—the one most commonly encountered—2 meters below the keel and 3 meters to one side without rupturing or being crippled. Styrofoam or styrofoam-substitute cofferdams within the hull would limit damage from above-water hits, and the boat would have enough compartments to prevent sinking when any single one was flooded. It would make maximum use of styrofoam for flotation and fragment protection.

The crew's battle stations would be armored against 7.62mm armor-piercing bullets at 75 meters. In addition, bar armor would be fitted to protect the crew and vital boat machinery from recoilless rifle projectiles. The coxswain's armored position would be large enough to incorporate the radio operator's position behind.

The boat's bow would be reinforced for high-speed beaching and to prevent pentration by sharpened logs. A skeg would protect the propeller. French experience with the STCAN was reflected in a requirement that the boat have sufficient sea-keeping capabilities to operate safely at river mouths, which could produce 6-foot swells and 4-foot chop in high winds.

Maximum length was set at 38 feet, maximum beam at 13.

Speed, which NAGP had not mentioned, was set at 18 knots, range at 200 nautical miles at 10 knots.

Armament was reduced by one M60 light machine gun: the amidships weapon would be arranged so that it could be shifted from side to side.

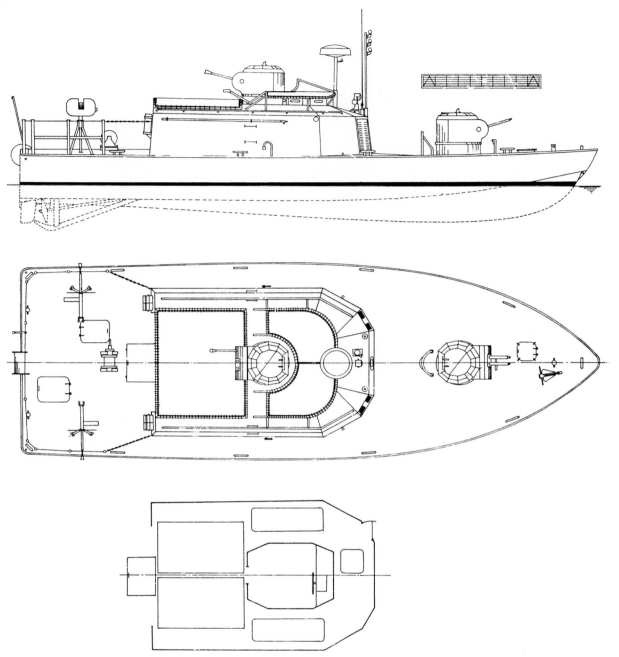

ASPB Mark I, program 5. Note that the mortar well aft has been covered over. The forward Mark 48 turrets are farther forward here than in the earlier ASPB. (USN)

To NAGP, the ASPB was a relatively simple boat adapted from the existing STCAN. BuShips found otherwise. It tried to stay within the 38-foot limit, but the combination of weapons and armor weighed far too much. In June 1966, the bureau completed the design of a 15-knot steel 50-footer. Even then, the limiting draft had to be relaxed to 3 feet 9 inches (from 3 feet 6 inches). Table 13-7 describes a series of 38-foot alternatives. The 38-foot boat could be expected to squat more than the 50-footer, so that its operating draft might be as much as a foot more than the static draft shown in the table. Moreover, providing it with a tunnel stern to reduce draft to the required figure would cut speed drastically.

As table 13-7 shows, heavy aluminum construction would have helped, but at a cost in advanced-base facilities. If such construction was considered acceptable, it could help the 50-footer even more,

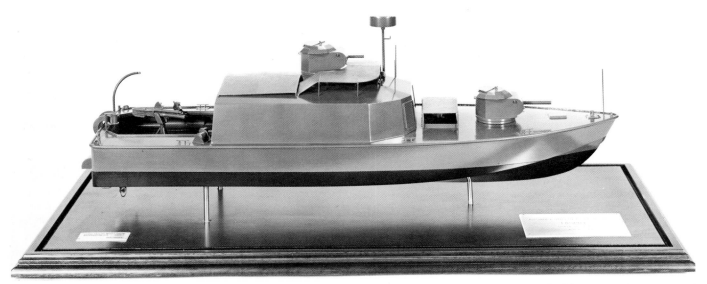

The official model of the ASPB shows its hard-chine, high-speed hull form and the massive deckhouse with its spaced armor.

saving enough weight (9,000 pounds) to add about 3 knots for the specified maximum speed of 18 knots. In fact, aluminum was rejected as too risky and expensive, and the ASPB was built with an aluminum superstructure and a steel hull. It never made the designed speed because it was too heavy.

Table 13-8 compares several 44- and 50-foot alternatives with the 38-footers. Only the largest boats could make anything like the required speed, and comparison with the data tables (table 13-9) shows that even the 50-footer ended badly overweight. Even gross and presumably unacceptable reductions in protection would not have saved the shorter designs.

The ASPB did differ from MACV's proposal in having space for a six- to eight-man assault team of troops in addition to its crew of five.

BuShips made the ASPB extremely tough. Like the LCMs, the ASPB was protected against 57mm shaped-charge recoilless-rifle shells and 0.50-caliber AP bullets—in this case over the coxswain's flat (superstructure), the assault team area aft, and the engine room. The ASPB had a new type of protection. Instead of bar armor, it used a thin, quarter-inch aluminum trigger plate at a 22-inch standoff (44 inches around the coxswain's flat). Weight was so critical that ⅝-inch dual-hardness armor had to be used behind the trigger plate.

This brittle plate could shatter when struck by delayed-fuse shells, which failed to detonate on the aluminum trigger plate. This kind of catastrophic failure, first experienced in July 1968, caused many casualties. It was solved after July 1969 by installing fragment suppression immediately behind the armor plate to stop the large, flat pieces before they rotated into a cutting position.

There was also lightweight armor over the fantail to shield personnel operating the sweep gear and the fantail machine guns. This armor protected against 0.30-caliber armor-piercing bullets at pointblank range and fragments from directional mines (Claymores). It also protected against the blast and fragments produced by 57mm recoilless rifle shells and RPG-2 and -7 shaped charges hitting the armored superstructure amidships. Additionally, the squat of the ASPB at speed produced a protective wall of water around the fantail.

The hull plating was ⁷⁄₃₂-inch steel, but from the gunwale to 2 inches above the lower-side longitudinal, it was ⅛-inch steel. The spray rail was ¼-inch steel, the chine ⅜-inch. Bulkheads were 14-gauge steel, and ³⁄₁₆-inch steel isolated the fuel compartment. The weather deck was 12-gauge steel, but the internal coxswain's flat (deck) was ¾-inch plywood.

As might have been expected from its origins, the ASPB was designed specially to overcome the shock effects of underwater explosions.

ASPBs were armed with two Mark 48 combination turrets: one forward with a 20mm cannon or twin 0.50-caliber machine gun plus a Mark 19 grenade launcher, and one above and abaft the pilothouse carrying a single 20mm cannon and another grenade launcher. ASPBs also had two single 0.30- to 0.50-caliber machine guns, or one 81mm mortar, or a piggyback mortar with an 0.50-caliber machine gun in their after cockpits. Modified boats had the after cockpit decked over flush; two crewmen could fire small arms from the engine cover.

Other ASPB features included a chain-drag sweep and an underwater exhaust for silencing.

The first thirty-six boats and one test platform

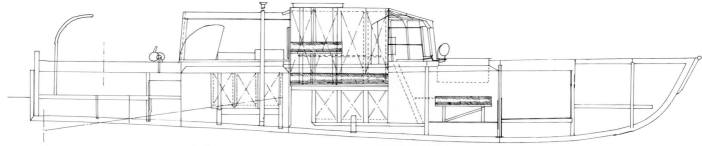

An inboard profile of an early ASPB shows the patches of internal armor protecting the engines, magazines, and coxswain position; armor is represented by crossed dot-and-dash lines. Dense diagonal hatching indicates the three shock mats, two for gunners, one for the pilothouse as a whole. The upper gunner's mat may not have been fitted in the earliest craft. Aft, note the minesweeping winch and the davit, fitted on the port side only. The vertical line is the centerline of the mortar in the well deck. Note that later ASPBs, at least those following 50AB6811, had their forward gun mounts well forward of the superstructure. (Norman Friedman)

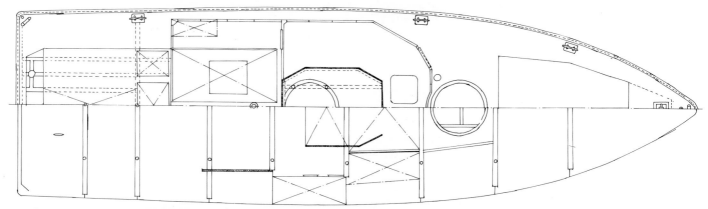

A plan view of an early ASPB showing the armor arrangement (hatched lines). The drawing's upper portion displays the main deck and the lower, the boat's interior. Note the open cockpit forward, characteristic of early ASPBs, and the well deck aft. The large circles are foundations for Mark 48 gun mounts. (Norman Friedman).

were ordered from Gunderson Brothers Engineering Corporation, Portland, Oregon, on 25 October 1966: 50AB671–6737 (FY 67); 50AB675 was the test article. The first was completed in May 1967. These boats were followed by fifty program 5 craft ordered on 24 January 1968, 50AB681–6850 (FY 68). Fourteen program 5 boats were delivered directly to the South Vietnamese navy upon completion. Gunderson also built ten very similar river minesweepers (MSR) under program 5: 50MS681–6810, the first of which was completed on 25 November 1968.

The first two ASPBs arrived at Vung Tau aboard the SS *Ohio* on 20 September 1967. They first saw action in Operation Coronado 5, 30 September–1 October 1967, in the Mo Cay and Huong My districts of Kien Hoa Province. They were fast enough to sweep for TU 117.2.1 during its transit, then return and sweep for 117.1.2 during its transit of the Song Ben Tre; River Flotilla 1 suggested that their higher speed produced a better overall force speed of advance and improved the chances of surprise.

Gunderson had erred in lofting the ASPB hull, and boats were completed with inadequate freeboard and metacentric height. As a result, they sometimes swamped in the wakes of other boats—or in their own wakes, if they stopped short or turned sharply. The remedy was blistering with styrofoam-filled sponsons, and the mortar well aft had to be decked over. The sponsons had the incidental advantage of adding protection against hull hits by rockets.

In March 1968, the sinking of five ASPBs (only two by enemy fire) in less than thirty days prompted the commander of River Flotilla 1 temporarily to curtail ASPB operations to the minimum required for combat. On 2 March, A-91-1 (50AB676) was swamped by the wake of two passing ASPBs; it sank in less than one minute at the junction of the Can Tho River and the Ba Long Stream. The boat was recovered on 6 March. This was not to say that the ASPB was unduly vulnerable to enemy fire. On 1 March, A-112-8 (50AB6737) and M-112-1 were hit by several B40 rockets in the Can Tho River while covering the sal-

Table 13-7. 38-Foot ASPB Tradeoffs

Scheme	Description	Displacement (tons)	Beam (ft)	Without Tunnel Stern Draft (ft-in)	Without Tunnel Stern Speed (kts)	With Tunnel Stern Draft (ft-in)	With Tunnel Stern Speed (kts)	Armor	Engines
1-A	Heavy steel hull, Heavy armor	52,200	14	4-9	11.5	3-6	8.5	50 AP 57mm RR	Twin 8-V71
1-B	Heavy steel hull, Light armor	48,200	14	4-6	13.0	3-6	10.0	30 AP	Twin 8-V71
1-C	Heavy steel hull, Light armor, Small engine	45,200	14	4-4	11.5	3-6	8.5	30 AP	Twin 6-71
2-A	Light steel hull, Heavy armor	45,200	14	4-4	13.5	3-6	10.5	50 AP 57mm RR	Twin 8-V71
2-B	Light steel hull, Light armor	41,200	14	4-2	15.0	3-6	12.0	30 AP	Twin 8-V71
2-C	Light steel hull, Light armor, Small engine	38,200	13	4-0	13.5	3-6	10.5	30 AP	Twin 6-71
3-A	Heavy alum hull, Heavy armor	45,700	14	4-4	13.5	3-6	10.5	50 AP 57mm RR	Twin 8-V71
3-B	Heavy alum hull, Light armor	41,700	14	4-2	14.5	3-6	11.5	30 AP	Twin 8-V71
3-C	Heavy alum hull, Light armor, Small engine	38,700	13	4-0	13.5	3-6	10.5	30 AP	Twin 6-71
4-A	Light alum hull, Heavy armor	41,200	14	4-2	15.0	3-6	12.0	50 AP 57mm RR	Twin 8-V71
4-B	Light alum hull, Light armor	37,200	13	4-0	15.5	3-6	12.5	30 AP	Twin 8-V71
4-C	Light alum hull, Light armor, Small engine	34,200	13	3-10	15.0	3-6	12.0	30 AP	Twin 6-71

Note: All versions had propulsion, armament, ammunition, and accommodations identical to those of the 50-foot ASPB. The heavy hull version had special hull structure and shock mounting features similar to those of the 50-foot boat. The light hull version had neither. Aluminum hulled boats would require additional repair personnel, equipment, and material at the RAS bases.

Table 13-8. ASPB Alternatives, 30 June 1966

	50-Foot				44-Foot				38-Foot			
	Heavy Steel	Light Steel	Heavy Alum	Light Alum	Heavy Steel	Light Steel	Heavy Alum	Light Alum	Heavy Steel	Light Steel	Heavy Alum	Light Alum
Hull (lbs)	32,000	22,000	23,000	17,000	28,500	20,000	21,000	15,500	25,000	18,000	18,500	14,000
Loads (lbs)	10,000	10,000	10,000	10,000	10,000	10,000	10,000	10,000	10,000	10,000	10,000	10,000
Propulsion (lbs)	8,500	8,500	8,500	8,500	8,500	8,500	8,500	8,500	8,500	8,500	8,500	8,500
Full Load (lbs)	58,200	48,200	49,200	43,200	55,700	46,200	47,200	41,700	52,200	45,200	45,700	41,200
Speed (kts)	15	17	16.5	18.5	13	15.5	15.0	17.0	11.5	13.5	14.5	15.0
Reduced Armor (lbs)	−4,000	−4,000	−4,000	−4,000	−4,000	−4,000	−4,000	−4,000	−4,000	−4,000	−4,000	−4,000
Full Load (lbs)	54,200	44,200	45,200	39,200	51,700	42,200	43,200	37,700	48,200	41,200	41,700	37,200
Speed (kts)	15.5	18.0	18.0	20.0	14	16.5	16.0	18.0	13.0	15.0	14.5	15.5
With 12V71 Engine (greater power, lbs)	Too heavy	+6,000	+6,000	+6,000	—	—	—	—	—	—	—	—
Full Load (lbs)	—	50,200	51,200	45,200	—	—	—	—	—	—	—	—
Speed (kts)	—	20	19	22	—	—	—	—	—	—	—	—
With 6-71 Engine (less power) and Reduced Armor (lbs)	—	—	—	—	—	—	—	—	−3,000	−3,000	−3,000	−3,000
Full Load (lbs)	—	—	—	—	—	—	—	—	45,200	38,200	38,700	34,200
Speed (kts)	—	—	—	—	—	—	—	—	11.5	13.5	13.5	15.0

In service in 1968, this ASPB carries an additional weapon, the grenade-launching Mark 19 machine gun. It is mounted alongside the superstructure, probably because in that position it could deal with sudden ambushes. By the end of the Vietnam War, NavOrd was developing multiple grenade launchers for the same purpose. The object just abaft the bow turret is an open hatch.

Many ASPBs were fitted with free-swinging machine guns aft, initially in the mortar well itself. The port gun is clearly visible, a starboard gun less so. This ASPB of TF 117 was photographed in December 1967. The boat in the foreground is an ATC; note the pintle-mounted machine gun protruding from under its canopy. Note also the sweep wire the ASPB is towing.

Table 13-9. ASPB

	ASPB Mk 1	Sewart Mk 2	Sikorsky Mk 2
Material	Steel	Alum	GRP
Length (ft-in)	50-1.5	50-0	49-8
Beam (ft-in)	15-2.5	21-4	19-8
Draft (ft-in)	4-3	2-9	3-5
Displacement (lbs)	76,814	90,310	74,537
Engine	2 GM diesels 12V71	2 Avco/Lycoming TF-1416	3 United Aircraft (Canada) ST 6-570
Propulsion	Propeller	Water jet	Water jet
Power	430	1,050	565
Max Continuous	—	950	443
Speed (kts)	16	25	25
Fuel (gals)	650	2,000	—
Endurance (nm/kts)	130/10	250	250/15
Armament	1 Mk 48-0 (20mm, 40mm GL) 1 Mk 48-1/2 (twin 0.50 or 7.62, 40mm GL) 1 81mm Mortar	3 x 20mm 1 x 40mm 1 x 81mm	1 x 105mm Howitzer (M102) Twin 20mm (bow) 1 Mk 48 (twin 7.62mm M60C or M2HB 0.50 MG and Mk 19 40mm GL)
Cost (thousands of $)	325	2,700	2,600
Weight (lbs)			
Hull	38,020	39,566	34,730.8
Propulsion	13,383	10,665	8,830.8
Electric	1,149	4,866	1,239.0
C2	1,320	2,569	1,113.0
Auxiliary	2,115	5,262	810.8
Outfit	3,574	4,121	2,271.5
Armament	5,054	5,094	8,065.0
Margin	—	721	1,141.2
Fuel Oil	—	13,500	11,700
Loads	12,199	17,446	16,235
Light Ship	64,615	72,864	58,202
Full Load	76,814	90,310	74,537

vage of A-112-4 (50AB6721), which had sunk on 27 February. A-112-8 took a direct B40 hit in her engine cover and had to be towed back. A-112-4 was recovered and towed to Can Tho on 4 March.

On 14 March 1968, A-92-7 (50AB6731) sank on the Sam Giang River after a rocket hit it directly on the stern below the waterline. Salvage was not feasible, so the boat was stripped and destroyed in place by an explosive ordnance disposal team. That same day in I Corps, T-112-7, sweeping the Cua Viet River two miles northeast of Dong Ha in Quang Tri Province, was flipped over by a large water mine estimated at 900 pounds. The boat was a total loss, and six men were killed, one seriously wounded.

Eleven river assault craft were damaged on 4 April. A-92-1 (50AB6710) and A-92-4 (50AB6719), which suffered the most, were hit by a total of six RPG-7s. These were direct hits on 0.50-caliber mounts.[5]

The alarming rate of ASPB sinkings led to the removal of engine-compartment armor from river assault force boats to reduce top weight. Only later, in program 5 boats beginning with 50AB6811, were the ASPB hull lines corrected; program 5 ASPBs also had their mortar wells decked over. Even then, the ASPB had a hair-raising ride, tending to heel outboard rather than inboard in a turn.

The ASPBs were known as delta destroyers because of their speed and firepower. In retrospect, the speed was almost irrelevant, as formations were only as fast as their slowest components, the 6-knot ATCs.

In October 1967, looking toward a second series of ASPBs to be built under FY 68's program 5, River Flotilla 1 proposed improvements. Unlike the LCMs, ASPBs had hardly begun to enter combat at this time. The flotilla wanted better firing arcs, to be secured by knocking down the radar mast in combat and removing the depression limit on the number 2 (high) Mark 48. All Mark 48 mounts would mount 20mm cannon.

BuShips made other improvements in program 5 craft: better shock resistance, redesigned engine compartment louvers for better cooling, reduced noise levels, more armor for crewmen, relocated (more protected) sweep controls, better hydraulics, and a smaller turning circle through steering and rudder redesign. The program 5 contract was awarded to

Newly completed and still unarmed, ASPB 50AB6816 makes three-quarter speed on its trial run, 17 October 1968. Note the decked-over mortar well and the two machine-gun tripods fitted above it. The high bow wave and low freeboard were the features that caused so much trouble earlier in the year.

Gunderson Brothers to avoid delay; the company would use its existing production line.

Program 5 also included a river sweeper (MSR) version of the ASPB. The ASPB was relatively expensive at $325,000, but half of the cost was armament. The MSR was simplified by replacing each of the expensive Mark 48 turrets. Forward, MSRs had a twin 0.50-caliber mount, an armored Mark 56. The pilothouse Mark 48 was replaced by a pair of single 0.50-caliber machine guns (Mark 55 mounts). The 81mm mortar aft was eliminated to make space for a chain drag and an O-type sweep, and the MSR carried a bow sweep as well. Perhaps more importantly, it could control a minesweeping drone (MSD). Estimated cost fell to $200,000.

By 1969, ASPBs were used primarily as minesweepers in the Riverine Force, even though they had been conceived, in effect, as floating tanks.

In its mature form (1968), the riverine force generally needed about seven to ten days to evaluate intelligence, plan an operation, and develop an operations order. Movement would begin very early, so that the troops would arrive at first light at an objective up to 70 kilometers away. Three ATCs carried each of two companies and used recognition lights of the same color (one, two, or three for each boat).

The formation was always led by a pair of chain-sweeping ASPBs. Typically these were followed in line ahead by a monitor, a company of troops in three ATCs, three more troop-carrying ATCs, a helicopter pad ATC, and a monitor bringing up the rear. Another helicopter pad ATC and the command boat would occupy the center of the formation, with a monitor or a Zippo separating them from the forward company of troops.

Speed was generally limited to about 5 knots to avoid scattering the force. Boats kept a one- to two-length spacing.

The entire force communicated by CCB to a joint (army-navy) tactical operations center (JTOC) aboard the riverine flotilla flagship at the mobile riverine base. The JTOC, in effect the CIC of the operation, combined brigade and task force commands like a miniature amphibious flagship (AGC). Normally, an air force liaison officer was attached to the JTOC to help call in air strikes. Navy and army commanders often established another CIC in a helicopter, which flew above the objective area and bypassed the JTOC.

The CCB, manned by both army and navy personnel, handled the routing of communications and decisions at the objective, functioning as a miniature JTOC. Operation was complicated because the services maintained their own radio nets, crossing only at command centers. For example, an army request for navy action would typically be communicated to the army section of the CCB, thence to the army section of the JTOC or the helicopter CIC. Message traffic could be considerable, and the JTOC was small, which caused intense noise and some confusion. Despite these drawbacks, the system worked quickly and well.

In 1968, a research team from the NSRDC compared these riverine force operations with those of the 199th Light Infantry Brigade, which worked

The official model of the Sikorsky ASPB displays triple water jets and a minesweeping cable reel aft.

adjacent to the RSSZ with Vietnam's River Assault Group 27. Here the army was in full control, using helicopters and a few LCM(8)s to supplement the group. The latter typically had one monitor, six to eight RPCs, four to six LCM(6)s, and perhaps two LCM(8)s. Plans were always minimal, and strikes were made nearly every day at an area chosen the day before and a spot chosen the morning of the operation. The brigade tried to use the latest intelligence; it usually reacted in two hours (once in twenty minutes) to a report of Viet Cong activity.

Whenever possible, the 199th flew in troops by helicopter. Otherwise troops were flown to the nearest available landing zone and moved by boat the rest of the way. The boats followed the troops as they swept, to extract them quickly or to transport them across large streams in their path. At the end of an operation, troops were removed by boat.

Because the assault groups used unmodified LCM(6)s, and because they moved only over short distances, each LCM(6) could transport an entire company; up to 175 men could move aboard a single LCM(8). Troops were immediately landed when resistance was met; unlike the riverine force, the river assault groups never intentionally triggered an ambush to force the Viet Cong to fight.

The river assault group commander and his U.S. advisor felt that they could profit from higher speed; they believed that the assault group could maintain formation at 12 to 15 knots. They also wanted some smaller craft such as Boston whalers to take troops across small streams too shallow for the assault group itself.

Program 5 included research and development funds for second-generation riverine craft. The new ASPB, ATC, and CCB would overcome the inherent limitations of the existing converted LCM(6), which was criticized for its deep draft (over 4 feet), slow speed (8 knots or less), vulnerability (it was protected only against 0.30-caliber bullets, not at all against mines), and lack of maneuverability and sea-keeping qualities.

The demand for increased speed was somewhat controversial. Riverine force commanders, for example, considered their 8 knots about the most they could safely use in shallow water with little warning of obstructions and grounding. High speed, moreover, would generate high stern waves, which would make following difficult. On the other hand, as noted, river assault group commanders would have welcomed craft capable of 12 to 15 knots. Some argued that craft designed to run up fast-moving rivers required a speed of 12 to 15 knots merely to make good a speed of 8 against the current and, often, the tide.

All the second-generation craft would have to be transportable aboard standard amphibious and commercial ships, which imposed length (50 feet) and beam and weight limits.

Official interest in second-generation riverine craft can be traced to an OSD (director of defense research and engineering) memorandum of 1 July 1966; on 30 September 1966, the newly established Naval Materiel Command established a naval inshore warfare project, PM-12. The director of the project in turn issued a specific operational requirement (SOR) for a family of riverine war boats (assault, troop transport, and command), SOR 38-05. However, SOR 38-05 was later limited to a second-generation ASPB, and a second SOR, 38-08, was issued for a second-generation ATC and CCB on 30 January 1967. All

The official model of the Sewart ATC. There was a parallel CCB with a command and control module that could fit into the troop well forward.

were merged in SOR 38-16 on 30 April 1968. As of January 1967, it was hoped that craft developed with FY 67 funds could complete their operational evaluations by January 1968.[6]

SOR 38-05, for the ASPB, called for a speed of 25 knots and an endurance of 250 nautical miles at a cruising speed of 15 to 20 knots. It recalled the length and draft limits of the original ASPB specification: length 36 feet (although up to 50 would be acceptable), draft no more than 3 feet 6 inches. Maximum height was set at 8 feet 6 inches above water. At a maximum speed of 25 knots, the ASPB would maximize firepower to a range of 6,000 yards (experience showed that the 81mm mortar could reach 3,650). The boat's design would take river currents, obstacles, and drag due to the chain sweep into account. Like existing river craft, these new boats would be protected against 0.50-caliber armor-piercing bullets at 20 yards.

SOR 38-08 described the Mark 2 ATC and CCB. The ATC would carry at least fifteen and up to thirty troops, compared with forty in the converted LCM(6). The design was to emphasize rapid embarkation and disembarkation, over the bow or the side or via a ramp. The ATC would carry inflatable rubber boats or plastic assault boats. The CCB would carry a staff of five to coordinate boat/base, boat group/boat group, fire support, boat/troop, boat/helicopter, and boat/close-air-support communications.

Although, ideally, the ATC/CCB would duplicate the 25-knot performance of the ASPB, the SOR allowed for a reduction to 20 knots. Minimum sustained speed was set at 15 knots for an endurance of at least 200 nautical miles. Overall length was set at about 50

Sewart's unsuccessful ATC Mark II. It has not yet been armed. Note the long bar armor screens around the superstructure.

feet, and as in the ASPB, maximum height at 8 feet 6 inches.

As the ASPB story showed, these boats were extremely cramped. High speed, that is, high power, required gas turbines. Moreover, installation had to be relatively sophisticated, since both SORs included the specific silencing requirements that reflected early combat experience.[7]

The emphases of SOR 30-08 were later changed to mold these craft to the exigencies of combat experience. The ATC's load was increased to sixty fully equipped troops for a two-day mission. Given this increased load, maximum speed was cut to 15 knots at combat load (15,000 pounds) and 12 knots at maximum load (20,000 pounds); 18 knots was considered desirable. Cruise performance therefore fell to 200 nautical miles at 12 knots. The SOR even allowed for a further reduction to 12 knots, if that was needed for sufficient protection and troop capacity.

Required protection was increased to defeat the 75mm recoilless rifle and the RPG-2. Maximum height above the waterline grew to 9 feet when masts were retracted or folded. Combat experience showed in a requirement that boats be able to engage hard-baked earthen bunkers from 50 to 2,000 meters away. For close fire support, they would have to provide indirect fire to 4,000 meters, direct fire to 2,000, and flame to 100. The boats would also have to be able to destroy bunkers.

The final version of the SOR covered five boats (the ASPB, ATC, CCB, monitor, and RUC, or Marsh screw) and four special modules, which would convert the ATC into virtually any riverine force element: medical station, helicopter support station, or monitor. The helicopter support version would also act as a refueler and ammunition boat.

The monitor module was the responsibility of the Naval Ordnance Laboratory. The module could carry a 152mm turret, as in the Sheridan light tank, firing HE, Heat, Beehive, or conventional ammunition at four rounds per minute out to 10,000 yards. Alternatively, the module could carry an M10-8 flame thrower or a 106mm recoilless rifle, firing ten to thirty rounds per minute to 14,000 yards. The heavy-gun version would also carry one 0.50-caliber and 7.62mm (M73) machine gun. Design problems included weight (the limit for all modules was 20,000 pounds) and recoil, which was quite considerable for the 152mm gun. Evaluation was scheduled for July 1969 through March 1970, but the ATC Mark 2 program died before the prototype module could be completed.

Owing to the urgency of the riverine requirement, the usual contract definition process was cut short, and a request for proposals for the ATC and CCB was issued in April 1967. Proposals arrived late in May. Sewart was chosen on 1 August 1967, and two prototypes, one ATC and one CCB, were ordered on 4 August 1967. Because neither was ever accepted, neither ever received a navy number; the ATC was yard number 1619. It was completed 6 March 1968; the CCB was completed on 5 March 1968.

The Sewart ATC/CCB prototypes failed their oper-

The official model of the Sewart ASPB Mark II.

ational evaluations, which were completed in January 1969. Although the ATC prototype achieved speeds as high as 26.2 knots at the Sewart plant, at the required continuous rating and fully loaded it made only 13.2 knots; a minimum of 16 had been set by the circular of requirements. Sewart's proposal had promised up to 21 knots. Endurance was 170 nautical miles, compared with 250 in the circular, and turning diameter, 75 rather than 50 feet. Draft was 3 feet 9½ inches, compared with the required 3 feet 6 inches and Sewart's proposed 3 feet 3 inches. The Sewart boat was also too noisy, its coxswain had insufficient visibility, and its hydrodrive was evaluated as subject to damage in obstacle-strewn water.

There was no covered access to the Mark 48 mounts, and they exposed their crews to hostile fire.

The Sewart ATC could carry forty-four troops or an M113 or a 105mm howitzer and its prime mover. The M113 was a very tight fit and took too long to unload. The boat could not really be loaded in an LSD well deck because its screws were not protected enough against damage in rough water when entering and leaving. Nor could the relatively sophisticated ATC operate, as required, with minimum logistical support.

Ultimately, the ATC Mark 2 was much better than the converted LCM, but it was still not good enough.

ASPB proposals were received at the end of June

The Sikorsky ASPB Mark II was built around a massive 105mm turret with a Mark 48 turret in the bow. Note the bar armor around the superstructure.

Sewart's ASPB Mark II prototype was armed with three Mark 48 turrets. Its bunker-buster was an 81mm mortar in a well right forward. The turrets carried 40mm guns in mounts adapted from those used in air force C-130 gunships.

The hull of the Sewart Mark II prototype is shown under construction in Berwick, Louisiana, 8 November 1968.

1967 from United Aircraft (Sikorsky), Teledyne-Sewart, and Atlantic Research. United Aircraft's ASPB Mark 2 prototype was formally ordered on 12 June 1968 (50AB691), the Sewart prototype on 28 June (50AB692). Both passed their operational evaluations, and both were declared surplus on 25 August 1980. The Sikorsky boat was appreciated for its bunker-busting 105mm howitzer, but it received poor technical ratings and encountered some serious problems.

Finally, on 25 July 1968, Chrysler received a sole-source contract for ten of its Marsh screws, which were designated riverine utility craft (RUC). As envisioned throughout the history of the riverine force, these craft were intended for reconnaissance/scouting/special operations, rescue/evacuation, and supply in marginal terrain such as swamps. Powered by a 280-bhp Chrysler marine engine, they were expected to attain 15 knots in water, 20 in mud at an endurance of five hours, and to traverse a 2-foot vertical obstacle or a 60 percent grade. They would also be able to operate in water of any depth, so they could cross deep streams between swamps. Hoisting weight was 7,200 pounds for a 2,000- to 3,000-pound payload or seven combat-loaded troops, and dimensions were 18 x 11 x 7 feet. The prototype carried 1,050 pounds or seven to eight men; its dimensions (feet-inches) were 13-8 x 8-2 x 5-6, and it attained 10 to 14 knots in water and 8 to 12 knots in a marsh. Range was 125 nautical miles. Prototypes were delivered in the summer and early fall of 1969. By 1971 only five were left, four of them operated by the marines.

Details of the other craft and of their program 4 predecessors are given in tables 13-6 and 13-9.

None of the Mark II craft was ever produced in quantity. Initially, the navy hoped to produce new riverine craft to replace existing ones, so that the latter could be turned over to the Vietnamese navy without eliminating American riverine capability. In August 1968, a navy program change proposal called for sufficient craft for two new RASs in each of FY 70 and FY 71, leaving one RAS for the Pacific and one for the Atlantic after the war. Prices would have been quite high. The Mark II ASPB cost $800,000, compared with $341,000 for a program 4 monitor. The Mark II ATC/CCB cost $750,000 ($172,000 for a program 4 ATC, $300,000 for a refueler, or $470,000 for a CCB). The RUC was expected to cost $50,000. All of the Mark II costs are given in FY 69 prices.

The planned production schedule is shown in table 13-10. Most of the surviving program 4 and 5 craft were turned over to the South Vietnamese navy, falling into North Vietnamese hands in 1975: 293 PBRs, 107 Swifts, 84 ASPBs, 9 CCBs, 22 program 4 monitors, 42 program 5 monitors, about 100 ATCs, 8 MSMs, and 8 MSRs. The U.S. Navy retained a few riverine craft as a nucleus for training. In 1978, the navy operated 20 PBRs, 5 Swifts, 2 ASPBs, and 1 CCB.

Table 13-10. RAS Craft-Production Schedule Proposed for FY 70–71

ASPB Mk II	4/69 Service approval of prototype (prototype delivery 12/31/68)
	7/69 Production contract
	3/70 Production copies available 4 per month
ATC/CCB Mk II	12/69 Production copies available at monthly rate of 4/4/4/5/6/6/6
RUC	3/69 Service approval of prototype
	1/70 Production copies available at monthly rate of 5/5/5/5/4/4
Boats for 2 RASs	Produced in a continued two-year contract with no break between either
Manufacturers	ATC/CCB Mk II: Teledyne/Sewart Seacraft
	ASPB Mk II: Teledyne/Sewart and United Aircraft/Sikorsky
	RUC: Chrysler

14

Vietnam: SEALs and STABs

The SEALs were established by the same CNO directive that set up the watercraft panel described in chapter 11, to conduct sabotage, demolition, and other clandestine activities in and from restricted waters, rivers, and canals. The SEALs are counterparts to army and air force special forces, the "special" referring to small-unit, almost guerrilla, operations behind enemy lines or in enemy-dominated territory. In this sense, special forces seemed to be the best U.S. vehicle for fighting guerrillas who expected to hide themselves among the Vietnamese people, and the SEALs were among the first U.S. military men actually to see combat. During the Vietnam War, SEAL activities included intelligence gathering and attacks deep in Viet Cong–held territory.

SEALs trained many foreign counterparts such as the Vietnamese LDNN, and they evaluated special swimmer-support craft for both their own use and that of their counterparts.[1] The LDNN in turn carried out, among other things, Operation Plan 34A attacks on North Vietnam using U.S.-supplied PTFs to launch their swimmer support boats.

At first, the SEALs drew many of their craft from the established underwater demolition teams (UDTs), which had been created during World War II to reconnoiter beaches and clear the way for amphibious landings. During the Korean War, UDTs participated in behind-the-lines attacks on enemy targets, together with British SAS troops and American army troops. After the war, the UDT mission included maritime sabotage and attacks on ships in enemy harbors, to which they could be delivered by specially modified submarines. The SEALs were therefore initially criticized for their rather limited or special counterinsurgency mission. Ironically, they later absorbed the UDTs and assumed the general maritime sabotage role in addition to many other special-warfare missions.

The UDTs themselves developed special swimmer-delivery boats, including the gas turbine LCSR, which the SEALs evaluated for their own purposes. By about 1960, the UDTs were using the 36-foot LCPL Mark IV, a rampless craft intended primarily to control waves of amphibious boats. One LCPL was converted in 1960 as a UDT carrier, the Mark 2 LCP(R) (R for reconnaissance). The faster and unsuccessful 52-foot LCSR was intended as its successor.

In offshore SEAL operations, the LCPR or LCSR or PTF would drop smaller swimmer-support boats or plastic Boston whalers close to shore.[2] The smaller boats in turn would covertly deliver their SEAL passengers.[3] As the war in Vietnam developed, the SEALs operated largely on inland waterways, being delivered directly by larger boats that could support them with automatic fire.

SEALs were organized in two teams, one for each U.S. coast. Team 1, based at Coronado, was responsible for weapons development. Team 2, at Norfolk, was responsible for boat development. Detachments from each served in Vietnam. Their boats were supported and operated by boat support units. Boat Support Unit 1 was established on 1 February 1964 at Coronado as part of the Naval Operations Support Group, Pacific. It was intended to organize, train, and provide personnel and small combat craft for special warfare. It also tested and modified several of the SEAL craft. Its size reflected the increasing importance of special operations in Vietnam. It was formed with 24 officers and 140 enlisted men, but by 1967 it included 36 officers and 310 enlisted men. It had achieved commissioned status on 18 February 1966 on account of its unprecedented growth.

A bow view of the official model of the Sea Fox, the current SEAL delivery craft, shows its specially rounded forecastle, its well deck aft, and its maximum battery of two 7.62mm M60 machine guns and one 0.50-caliber heavy machine gun aft, to starboard. The standing figure and the portable radio indicate scale.

SEALs served in Vietnam throughout the war. First they trained the Vietnamese. Detachments from both teams were deployed to hard-core Viet Cong areas of the RSSZ and the Mekong Delta.[4] This was different from what was initially conceived, swimmer insertion of SEALs. The small counterguerrilla units were delivered and extracted by helicopter and boat. The SEALs, without special craft of their own, used Vietnamese river assault craft, RPCs, and some lightly armed LCPL Mk IVs and LCM(6)s. As the Viet Cong began to receive heavier weapons in the RSSZ, these standard craft became extremely vulnerable. An LCM was heavily damaged, and the SEALs began to seek specialized craft of their own.

Boat Support Unit 1 established an advanced base at Danang in February 1964 to support the Operation Plan 34A boats, including the first PTFs. At Coronado, it evaluated or helped develop such craft as the Marsh screw amphibian, the PACV, the shallow-water attack boat (SWAB), and the 52-foot LCSR. It assisted Sewart in Swift detail design work and contributed to the development of Swift boat doctrine, training crews from September 1965 onward.

It was Boat Support Unit 1 that operated seven LCSRs, which had been bought for UDT 1, and evaluated them as unsatisfactory for combat operations. They were dangerously noisy, their gas turbines took in saltwater, and clutch problems had not been resolved. As of 1966, they could not yet operate for over 350 hours before overhaul. The unit's verdict, then, was that the LCSRs should not be deployed to Vietnam, and that they could best be used to support SEAL and UDT training off southern California and anti-PT/Komar (missile boat) training for ships deploying to the western Pacific.

Even had its engine been reliable, the LCSR would have been difficult to deploy. Too large for the standard (Welin) davit, it had been designed to fit within the well deck of an LSD. It originally had a big skeg to protect the propellers, but the skeg caused serious problems and had to be removed. Any LSD carrying an LCSR overseas therefore had to carry a special cradle in its well deck, itself a severe limitation on well-deck capacity for other amphibious craft. Work began on an alternative 36-foot (Welin-size) LCSR(L). The LCSR(L) is important here because it was a candidate craft for SEAL insertion.

Two boats were built by Harbor Boat under the FY 68 program, one (36SR681) to a navy design and the other (36SR682) to a commercial design. The former arrived at Coronado for evaluation on 25 December 1969. One of the two designs was to have been selected for further production, but that never happened.[5]

Boat Support Unit 1 was assigned to SEAL support in Vietnam in December 1966. At that time, CinCPac established two SEAL support-craft programs: an interim program of modified existing craft and a program of specialized craft to be developed over the following year. Unit 1 modified the interim craft in less than two weeks at Coronado—to two LCM(6)s and four LCPL Mark IVs—under the codename Project Zulu. The LCM were considered heavy SEAL support craft (HSSC), the LCPLs, medium SEAL support craft (MSSC). All were in combat in Vietnam by March 1967.

Neither boat was entirely satisfactory. The LCM had considerable capacity and could be silenced, but it was very slow. Viet Cong agents often watched it being loaded and could predict the range of its possible destinations. The LCPL was faster, though not fast enough, and its capacity and firepower were both limited.

A follow-on series of craft was to have consisted of three types: a new light SEAL support craft (LSSC), a new MSSC to replace the LCPL, and an HSSC, a modified LCM. The interim craft were wearing out by December 1967, and OpNav stated requirements for the LSSC and a replacement MSSC on 14 December 1967. At a SEAL support-boat conference on 6 February 1968, it was agreed that sixteen LSSCs and eight MSSCs would be provided within 90 to 120 days. At that time, it appeared that only Goodyear Aerospace and Aerojet General could provide boats quickly enough. All LSSCs, as it turned out, were built by Grafton Boats.

Like the STABs (see below), the new LSSCs and MSSCs were to be armored against 0.30-caliber bullets at a range of 100 yards (that is, at a velocity of 2550 feet/second).

Trials of a prototype LSSC were successful, and sixteen were bought. Because the FY 69 and FY 70 programs initially included no funds for these craft, a modified stock LCPL XI was proposed as the MSSC. It failed its tests, and in December 1968, a newly designed commercial boat had to be bought instead, using funds originally assigned to ASPB Mark 2 procurement.

The LSSC was descended directly from a 26-foot shallow-draft commercial fiberglass (unsinkable) trimaran hull converted by SEAL Team 2 at Norfolk before it deployed to Vietnam. This SEAL-team assault boat (STAB) had room for ten weapon stations—nine about the gunwales and one amidships—for four standard SEAL weapons: the Stoner 5.56mm machine gun, the M60 7.62mm machine gun, the 0.50-caliber machine gun (from two mounts on the sides or the amidships tripod only), and the 40mm Honeywell grenade launcher. In addition, the 7.62mm minigun (mini-Vulcan) could be mounted on the amidships tripod. The boat was armored against 0.30-caliber

A Grafton LSSC being tested, demonstrating its ability to turn within its length because of jet-pump propulsion. At this time, the boat was armed only with two M60 light machine guns.

bullets from stem to stern, and two 110-bhp Mercury gasoline engines could drive it at over 30 knots while carrying a full boat crew of three and full ammunition.

This STAB proved extremely useful in combat, launching squad-size operations along river banks in the delta and supporting them with its heavy firepower. It appears that further boats were converted, although precise numbers are not available. They lasted through late 1968. Their outboards were reliable, but their flat-bottom hulls were torn apart and they had to be replaced.

The official monthly summary of naval operations in Vietnam applied the STAB designation to an armed 20-foot Boston whaler used to land and recover SEAL units and powered by a pair of 50-hp outboards (40 knots). The summary listed an armament of one 0.50-caliber machine gun, one M60, and one Mk 18 grenade launcher; a 57mm recoilless rifle could also be carried.

The prototype of a new 24-foot aluminum LSSC to replace this STAB was tested from 11 to 21 June 1968. It combined the capabilities of earlier SEAL support craft and the STABs. It had a new V-form hull and Ford Interceptor (Thunderbird) engines driving pump-jets, and the SEAL Team 2 command historian described it as more stable and quieter than the earlier STAB. It also enjoyed greater range, firepower, communications capability, and weather protection than the earlier boat. It had a crew of three plus seven troops, could make 26 knots fully loaded, and had a shallow draft of only 29 inches fully loaded (9 inches when planing).

Trials showed that the boat needed more bottom area to carry its required load of 2,250 pounds of personnel, 1,200 pounds of armor, and 800 pounds of cargo. However, it was limited to dimensions in feet-inches of 23-6 x 9-6 (beam), partly because of a requirement that it fit within the cargo hold of a C-130 airplane. The solution to the lack of bottom area

LSSCs in service in Vietnam were heavily armed. This one carries two light machine guns and two grenade-launching machine guns.

was to add a 12-inch trim tab that could be extended for additional planing area without lengthening the hull itself.

Typical armament for the new STAB was two M60 light machine guns and one 40mm grenade launcher. The cockpit was armored against 0.30-caliber machine guns, and tests conducted in July 1968 included firing 0.30- and 0.50-caliber armor-piercing rounds and an RPG-2 against the craft's gas tank.

Ten of these LSSCs were built immediately. They were so badly needed that they were airlifted to Vietnam, the first ones shipped 23 July 1968. One was assigned to each of ten SEAL detachments (six from Team 1, four from Team 2). By early 1969, with six detachments from each team in Vietnam, twelve LSSCs were operational. In all, sixteen were ordered from the Grafton Boat Company of Grafton, Illinois, on 6 May 1968, as 24UB681–6816 (LSSC 1 through 16), the last being completed on 21 November 1968. Helicopters often transported LSSCs to otherwise inaccessible rivers and canals. On 15 November 1969, LSSC 10 was destroyed by being dropped from a Chinook helicopter carrying it at 6,000 feet.

As early as July 1968, CTF 116 (Game Warden) proposed a new operational concept, a special interdiction force of quiet fast boats (QFBs). At this time, the only existing QFB was a Boston whaler with a special 9.9-hp silent motor in addition to its 105-hp outboard. The larger LSSC could carry small numbers of troops fast enough to react to intelligence such as that obtained by capturing Viet Cong tax collectors. CTF 116 did consider the initial and maintenance cost of the LSSC prohibitive. However, the new operational concept was extremely attractive, and it was supported by the naval force commander in Vietnam, Admiral Zumwalt, who was impressed by LSSC operations in An Xuyan Province.

He envisaged high-speed strikes, by what he called STABs, into areas not accessible to PBRs or PCFs, their boats inserted by helicopter where necessary. They would also react to real time intelligence or contact by friendly forces so as to interdict enemy waterborne traffic and forces near waterways. The STABs would be organized in six-boat sections from an operational base at Binh Thuy, but they might also operate from advance bases when STABs operations continued in an area far removed from the operating base. STABs would not conduct routine patrols and would not be manned by SEALs.

The STABs were urgently needed—Zumwalt did not want procurement delayed to FY 71—so they had to have an existing hull. The STAB mission required a fast, silent, shallow-draft, highly maneuverable armored boat that could be transported by helicopter and that had a draft of 2.5 feet or less, a beam of 10 feet or less, a length of 25 feet or less, and a speed of 35 knots. The new LSSC was ideal.

As a result, a requirement for thirty more LSSCs was set to meet a total requirement of twelve for SEAL support and twenty-four for STABs (with six in-country backup for either). The situation was aggravated by the failure of the MSSC conversion (LCPL). Details are given in table 14-1.

A total of twenty-two of these boats was ordered from Grafton on 3 July 1969 (24UB701–7022). They were completed between September and October 1969. All became STABs.

Table 14-1. LSSC

	Mark 1	Mark 2
Number Built	16	22 (70UB701–7022)
Length (ft-in)	24-6	26-2
Beam (ft-in)	9-6	10-4
Draft (ft-in)	1-10	3-9
Displacement (lbs)	11,900	14,900
Engines	Ford Interceptor	Ford Mercruiser III
Hp	300	325
Range (nm)	150	190
Cost ($)	39,732	76,000

Note: Mark 2 was lengthened to restore buoyancy (boot on stern).

STABs were considerably modified LSSCs, lengthened by 2 feet for extra buoyancy and lined with ceramic armor. Inboard of the ceramic armor were ballistic nylon flak blankets to stop fragments formed when the ceramic armor was hit. Styrofoam hull filling made the STAB virtually unsinkable. Gasoline fuel was carried in two 150-gallon, foam-filled, self-sealing rubber bladders under the floor plate of the cockpit, below the waterline. The foam filling was apparently an effective preventive against the usual explosion threat, since bladders were penetrated by machine-gun bullets on several occasions (and once by an RPG warhead) without fire or explosion.[6]

Unlike the LSSC, the STAB was powered by two 350-bhp Chevrolet gasoline engines driving propellers, for a maximum speed of about 40 to 45 knots and acceleration to that speed in 15 seconds. They occupied separate watertight and nearly soundproof compartments. As a result, the slap of water on a STAB's sides could be heard before its engines. The very low STAB silhouette (the boat's radars were removed, and it had no canopies) also made it difficult to detect.

Each STAB had four pintle mounts, and normally carried two M60s on the two forward mounts and either two M60 or one M60 and one grenade launcher aft. Heavy, 0.50-caliber machine guns were substituted for the forward M60s if boats expected to fire through bamboo. Boat crews normally carried starlight scopes on patrol.

As boats, the STABs were not altogether satisfactory. Many problems arose with the maintenance of engines, outdrives, fuel bladders, and hulls, and it was estimated that they would need a major hull overhaul every twelve to eighteen months. The hulls weakened noticeably in their eight months of Vietnam operation. Several boats showed warping and cracking in the after part of the hull where the weight and movement of the outdrives "worked" the aluminum. Other STAB hulls "dished" forward after the longitudinals and stringers had weakened or torn loose from their welds. Several rattled when planing because structural members had come loose. Nor was the propeller outdrive entirely successful; it extended 2 feet 7 inches below the bottom of the keel, giving the STAB a deeper draft (3 feet 9 inches) than a PBR and keeping it out of several tentatively selected areas of operation.[7]

In August 1969, the STABs were organized as a strike assault-boat squadron, STABRon 20, at Mare Island's Naval Inshore Operations Training Center (NIOTC). The STAB was so new that NIOTC could not provide handling/maneuvering, engine maintenance, or tactical training. Instead, STABRon personnel went through the standard PBR and riverine assault craft courses. Twenty STABs were shipped to Vietnam; they arrived at Dong Tam early in January 1970. The other two were retained at NIOTC.

The typical STAB crew comprised four men: a coxswain, a boat engineer, a gunner's mate, and a line handler, the gunner or coxswain serving as boat captain. A patrol officer was generally assigned to every three or four STABs on patrol.

The STABs arrived during Sealords and, as part of Operation Barrier Reef, were assigned to night waterborne guard post (WBGP) ambush positions along the Grand Canal. Boats were spaced about a kilometer apart on the north side of the east-west canal, using electronic sensors and night vision devices to warn of the approach of the enemy. They protected themselves with Claymore mines. Some STABs were used in daytime to screen and search sampans in the canal (part of the resource control program). The former commander of STABRon 20, Lieutenant Commander John K. Ferguson, suggests that this essentially static employment was a waste of fast strike boats.[8]

By early 1970, certainly, U.S. strategy in Vietnam emphasized barriers within which pacification might proceed rather than strikes into enemy-controlled territory—which would only temporarily dislocate his activities. In this sense, the STAB was victim of a changing war.[9]

Other STAB operations at this time included a shallow ten-boat penetration into Cambodia on 9 May 1970.

A strike assault boat (STAB) at speed near the Cambodian border, June 1970.

STABs prepare for a night patrol of the Grand Canal, June 1970. The boat in the foreground is 24UB7017, with 7011 behind it. The machine guns are M60s, and the boxy, shrouded objects are grenade-launching machine guns.

Table 14-2. SEAL Support Craft

	Seablazer	*LSSC*	*MSSC*
Length (ft-in)	24-0	23-6	36-2½
Beam (ft-in)	9-3	9-5	11-9
Draft (ft-in)	2-6	2-5	3-4
Displacement (lbs)	13,640	11,938	34,365
Engines	2 Mercruiser	2 Ford	2 Mercruiser III
	—	300 Hp	325 Hp
	Gasoline	Gasoline	Gasoline
	—	210 Gals	300 Gals
		(93 octane)	
Speed (kts)	33	35	30
Endurance (nm/kts)	150/30	132/33	182/25
Complement (off/enl)	—	152/30	207/9
Armament	—	2	4/6
	—	M60 MG	1-7.62 Minigun
	—	0.50 MG*	2-0.50 MG
	—	Mk 18 GL*	2 M60
	—	Mk 19 GL*	Mk 4-0 60mm mortar*
	—	Mk 20 GL*	Mk 18 GL*
	—	—	Mk 19 GL*
	—	—	Mk 20 GL*

Note: Starred entries were optional. The LSSC could carry up to five weapons in any combination.

In August, the squadron was split into two divisions, one of which was reassigned to a SEAL platoon; they continued to man WBGPs at night. The other division, reassigned to Nha Be, provided support in the RSSZ. Night operations were initially limited to troop insertion and extraction, but later the STABs inserted troops and then remained in nearby WBGP positions.

The squadron was deactivated late in October 1970, and all its personnel and boats were returned to the United States. The latter were placed in storage at San Diego. Their engines were removed, and many (or all) of them were eventually sunk as targets.

Atlantic Research, which had developed the MSSC, proposed the Seablazer, an LSSC replacement, about 1970. One prototype was evaluated but none bought. The Seablazer is interesting here as an indication of the state of the art at that time. It had an aluminum hull and two-shaft inboard-outboard propulsion. Details are given in table 14-2.

The LSSCs all disappeared from service after the Vietnam War.

In December 1968, following the failure of the LCPL conversion, NavShips was ordered to procure ten 36-foot GRP MSSCs from Atlantic Research for $140,000 each (see table 14-3). They were provided with five Mark 16 pintle mounts stressed for 0.50-caliber machine guns, and they were also armed with a 7.62mm Minigun (Gatling). Other equipment included a Raytheon 1900ND radar, one URC-58 HF radio, one VRC-46 VHF radio, and one RRC-77 VHF radio. Unlike the LSSC, the MSSC was propeller-driven, by two 325-bhp Mercruiser gasoline engines (30 knots) drawing 3 feet 4 inches with its stern drive down and 2 feet 5 inches with it raised. An MSSC typically carried a five-man crew plus a sixteen-man SEAL platoon.

These MSSCs were supplemented by existing LCPLs.

The HSSC was a converted LCM. The number converted during the Vietnam War is not known; certainly the SEALs used LCMs extensively in 1966–67, and as recounted above, Boat Support Unit 1 converted at least two. The HSSC differed from the ATC because the SEALs demanded a heavy, forward-firing weapon. Typical armament, late in the war, included a 106mm recoilless rifle mounted on a new deck above the tank deck, an 81mm mortar, a Minigun, six to eight 0.50-caliber machine guns, and two Mark 18 or 19 40mm grenade launchers. These weapons called for a large crew, typically eight to ten men, and the HSSC also carried a fourteen-man SEAL platoon. HSSCs had helicopter platforms, both for medical evacuation and for SEAL insertion. They were much slower than MSSCs but also much better armed and armored.

In 1971, ComNavForV urgently requested ten new conversions, three LCM(3)s and seven LCM(6)s, which were transferred to the Vietnamese navy in January 1972. Nine had come from the boat pool at Subic Bay. These HSSCs were silenced by muffling their underwater exhausts, which could not be heard at a low rpm. The cost was a speed even lower than that of the standard LCM(6), 6 knots maximum (4 sustained), compared with 9 knots maximum. HSSCs could launch and control smaller insertion craft, such as Boston whalers or sampans, and could act as communications relays for SEAL operations. These HSSCs were armed with one 81mm mortar, one pintle-mounted M60 light machine gun, six pintle-

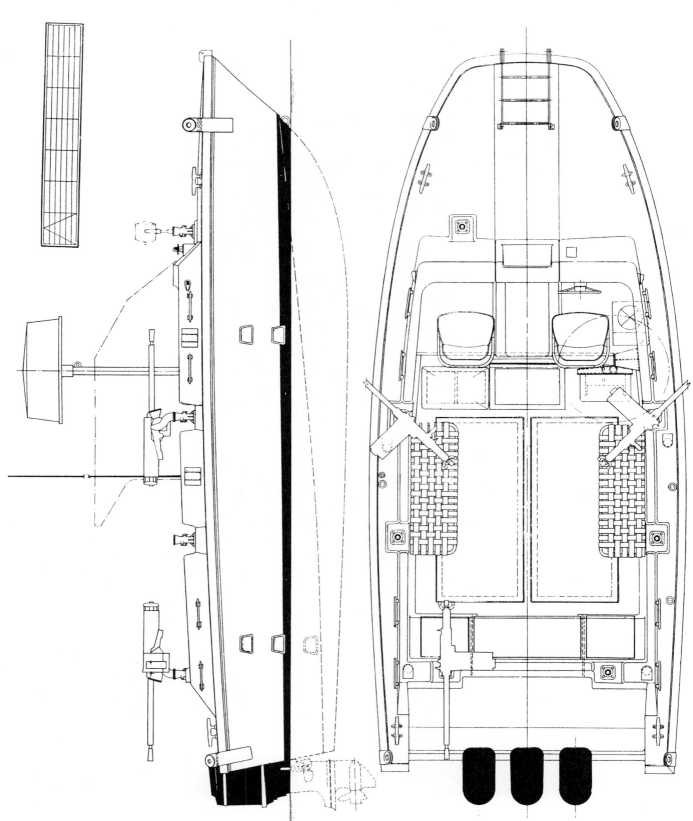

The abortive 24-foot LASSC ("River Fox"). (USN)

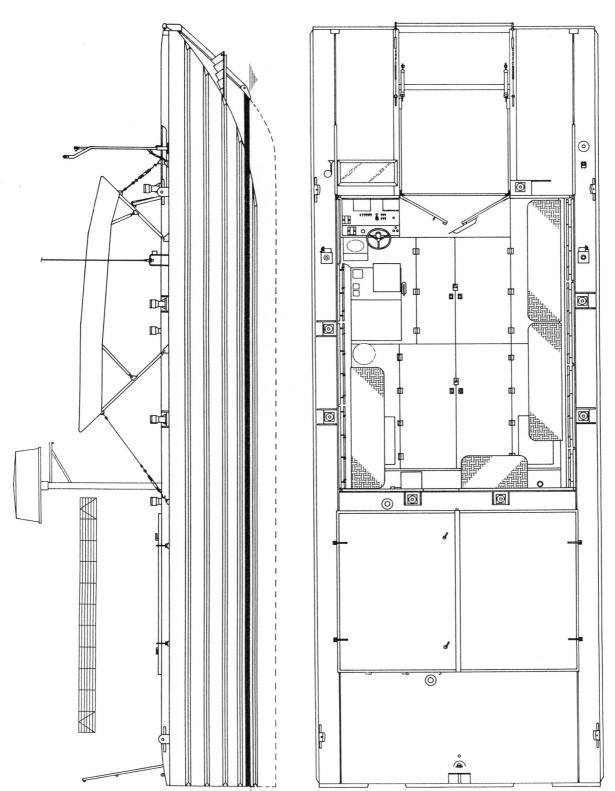

The 36-foot "mini-ATC." (USN)

Never officially identified, these are almost certainly HSSCs (heavy SEAL-support craft) at Long Beach Naval Shipyard. Note the heavy, cut-down bow doors and the slab of armor aft, over the machinery spaces. In Vietnam, these craft were provided with helicopter pads over the troop well; 106mm recoilless rifles were often mounted on the pads, lighter weapons in the well deck.

The "mini-ATC" reflected lessons learned in Vietnam. Note that, unlike the LCM-derived wartime beaching craft, its ramp is located aft, enabling it to back onto the beach. (John L. Shepherd)

The Boston whaler was the standard SEAL boat. Here SEALs riding their assault boat demonstrate ambush tactics, January 1967.

After Vietnam, the SEALs took responsibility for U.S. small combatants. New craft, such as this 65-foot Patrol Boat Mark III (Sea Specter), were built partly to replace the existing Vietnam-era craft (in this case, the Swift) and partly to provide very specific capabilities for the SEALs. The platform folded up at the boat's waterline was for swimmers. Two machine guns and an 81mm mortar are visible on deck, as well as a 20mm cannon above the folded platform abeam the pilothouse. This boat was photographed in August 1978; PBs are now more heavily armed, with a 40mm or 25mm gun forward.

Table 14-3. SEAL Craft

	MSSC	Sea Fox	LASSC (River Fox)	Mini-ATC
Length (ft-in)	36-3	35-6	23-1½	36-0
Beam (ft-in)	11-9	9-9	9-9⅝	12-9
Draft (static, ft-in)	3-4	—	—	3-6
Height (ft-in)	6-9	—	—	—
Above Waterline (ft-in)	5-2	7-0	5-6	5-11
Full Load (tons)	—	25,000	10,900	29,500
Light Load (tons)	—	19,600	6,200	22,000
Engines (each)	2 Mercruiser 325 Hp	2 GM 6V-92 TA 450 Bhp	3 Outboard 115/120/135 Bhp	2 GM 6V53T1 280 Bhp Water-jet
Fuel (gals)	280 Gasoline	325 Diesel	130 Gasoline	500 Diesel
Speed (kts)	30 Full Load (SS1) 25 (SS2) 15 (SS3)	—	—	—
Range (nm/kts)	150/25 Plus 2-hour loiter	—	—	—
Tactical Diameter (yds/speed)	50/full	—	—	—
Audible at (yds/kts)	200/10	—	—	—
Pintles	5	4	5	7
Load (lbs)	3,400 (14 men)	2,500 (10)	3,335	4,400 (15)
Weights (lbs)	(estimate of 6 Dec 69)	—	(estimate of 2 April 73)	(as built)
Hull	10,047.6	—	2,952	9,251.2
Propulsion	3,803.8	—	1,299	8,086.4
Electrical	636.0	—	783	349.4
Communications/Control	485.0	—	106	486.1
Auxiliary	455.3	—	212	1,657.6
Outfit	723.6	—	509	710.1
Armament	137.0	—	126	414.4
Margin	325.8	—	239	
	16,614.1	—	6,226	20,955.2
Draft (light, ft)	1.18	—	1.33	1.05
Weights (lbs)				
Ammunition	647	—	—	—
Stores	100	—	—	—
Potable Water	94	—	—	—
100% Fuel Oil	1,240	—	—	—
Cargo	3,150	—	—	—
Crew	660	—	—	—
Miscellaneous	800	—	—	—
Full Load	23,805	—	—	—
Draft (ft)	1.58	—	—	—
GM (ft)	7.91	—	—	—

0.50-caliber machine guns, and two Mark 19 40mm grenade-launcher machine guns to provide fire support and illumination to the SEAL raiders. HSSCs could block the escape of enemy troops the SEALs were ambushing, and they could act as small platforms for medical evacuation helicopters. The boats were heavily armored, even against RPG-7s, to extract a heavily engaged SEAL unit in relative safety. Finally, they had sufficient carrying capacity to support SEALs in remote operating areas, serving as independent, self-sustaining mobile operating bases. Electronic equipment included the usual Raytheon 1900ND and one VRC-46 radio.

All of this carrying capacity and protection cost speed and required a relatively deep draft, 5.6 feet.

Finally, SEAL units employed Boston whalers. About 1968, for example, each SEAL team had an allowance of eight 16-foot Boston whalers. When Market Time (CTF 115) was assigned seven 16-footers (four on hand), the boat pool at Danang had twelve and Game Warden (CTF 116) was assigned ten, which it had.

The LSSC and MSSC were conceived as interim elements of a larger program; their longer-term successors, the LASSC (River Fox) and the mini-ATC, appeared only at or after the end of the war. The

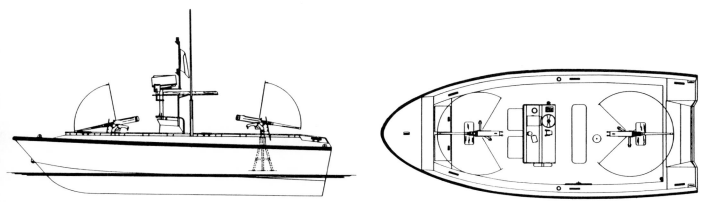

The 26-foot "Stinger," in effect the successor of the River Fox. It has a deep-V aluminum hull and is powered by two 155-bhp commercial gasoline outboard engines. Dimensions in feet-inches are 26-0 (overall) x 10-8 (molded) x 3-0 (fully loaded). Displacement is 5,000 pounds fully loaded. The Stinger carries a crew of four and is armed with two 0.50-caliber machine guns. As of 1986, there were two commercial prototypes, but the boat was not yet in navy service. (USN)

LASSC design was completed in 1973, and specifications were written for an FY 73 prototype, but none was built. In 1986, Boston Whaler demonstrated a private-venture LASSC derivative, the "Stinger." It is not clear whether more Stingers will be built.

The MSSC was replaced by the fast mini-ATC designed by Sewart. It had an aluminum hull and ceramic armor, radar, quiet engines for 28 knots, a shallow draft of only 1 foot at speed, and space for up to seven pintle-mounted weapons. The troop-carrying well deck was covered by a ballistic nylon canopy. As in the original ATC, there was a ramp. Teledyne-Sewart built seventeen under the FY 72 program (36AT721–7217), followed by eight from Marinette under the FY 76 program (36AT761–768) and fourteen from Tacoma Boat under the FY 78 program (36AT781–7814), of which ten (36AT781–782, 787–7814) were for the Philippines.

The characteristics of the LASSC and mini-ATC reflected SEAL involvement in riverine warfare. In the mid-1970s, after American withdrawal from Vietnam, the SEAL mission changed and interest shifted back toward insertion of forces from the sea. This shift is symbolized by the Seafox described in chapter 16.

15

The PHM

The six U.S. *Pegasus*-class missile hydrofoils (PHM) are the current equivalents of a long line of antiship fast attack craft that can be traced back to the PT boats and ultimately to the *Asheville* class. They are also the direct results of the success of the PC(H) *High Point* and of the two hydrofoil gunboats (PGH).[1] In the late 1960s, it became clear that missiles could provide a small hull with enormous antiship firepower. The hydrofoil was a natural candidate for an antiship missile platform, since it combined small size (and hence, presumably, limited cost) with large-ship sea-keeping and sustained rough-water speed. It could loiter in ambush, then run out to attack or run off after attacking. This combination of prolonged, hull-borne loiter and high flying speed also made the PHM attractive for blockades like Market Time, where it could intercept contacts or be used for offshore patrol.

The PHM was clearly too small to support itself. It required either local bases—trailers are used to support the PHMs at Key West—or specialized tenders comparable to the PT tenders (AGP) of the past. The *Graham County* (LST 1176) was adapted as a tender (AGP 1176) to support *Asheville*-class tattletales in the Mediterranean, and the *Wood County* (LST 1178) was scheduled for a similar conversion (AGHS) to support PHMs. It was canceled in 1977 because the LST's propulsion plant was in extremely bad condition; the cost of plant replacement plus conversion was excessive. Rather more fancifully, a contractor conducted a study of a possible AOR tender conversion.

However, as in the past, it was not always clear that the strategic requirements of the U.S. Navy (as opposed to those of U.S. allies) were best met by a small fast missile boat. The same limited size that restricted manning also restricted the size and weight of the weapon system the PHM could carry. Perhaps the most imaginative proposed solution to this problem was a team of PHMs carrying different weapon packages and functioning, in effect, as a multihull equivalent of a conventional warship. This idea was seriously considered several times during the history of the program, but it never came close to adoption.

The missile hydrofoil was initially attractive, in the late 1960s and early 1970s, as a means of countering a newly active Soviet surface fleet, particularly in the Mediterranean. Being more weatherly, for example, it would have been a more effective tattletale than an *Asheville*.[2] However, that was a limited role, and U.S. practice has generally been to emphasize general-purpose ships more adaptable to the full spectrum of naval operations. As a result, PHM production stopped at six ships, and the PHM community sought alternative roles and configurations. The unique capabilities of the PHM have made it attractive for Caribbean patrol operations in which the mere presence of a fast boat is much more important than, say, its electronic sophistication or total weapons load.

The history of the PHM can be broken into four phases. The first was a PM(H) design study based on experience with the *High Point*, PC(H)-1. This design was sometimes designated PCH-2. It was followed by PXH, which would have used alternative modules for different missions. Here the *X* emphasized the range of alternatives achievable with the same hull. PXH evolved into a navy-designed missile hydrofoil,

The *Aquila* runs at high speed, her forward foil visible under the bow. At the time of this photograph, she had only two of her eight Harpoon cannisters, but her chaff rocket launcher is on board, barely visible at the edge of the deckhouse, nearly abeam the lattice mast. It was suggested earlier in the program that the PHMs carry eight reloads for a total of sixteen missiles. The ships were designed to carry two single, free-swinging 20mm guns abaft the bridge, but these weapons were never mounted. The small horizontal waveguide forward of and below the egg-shaped radome housing the *Aquila*'s Mark 92 fire control and search radar is an SPS-63 navigational radar. (Boeing)

a PHM, which became a NATO project. Meanwhile Boeing, which had designed both the PC(H) and the successful *Tucumcari* (PGH 2), proposed a somewhat larger PHM to the West German navy. This design was adopted as the NATO PHM—which, in the end, was bought only by the U.S. Navy, because of its great unit cost. The Boeing hydrofoil configuration was also incorporated in some commercial craft, in the offshore British patrol vessel *Speedy* and in a proposed series of Indonesian hydrofoil patrol boats.

In May 1967, the NavShips research directorate asked the ship-concept design division to investigate a hydrofoil to protect coastal installations, surface shipping, or amphibious operations against fast-attack boats. An unusually stable weapons platform with superior maneuverability and high speed even in sea state 5 was required—that is, a hydrofoil. The craft was to enter service in 1975. Given so short a lead time, NavSec had to work from an existing prototype. It sought to adapt the *High Point* as a missile boat, armed with one 40mm gun and the existing German-type Tartar antiship missile. The latter, installed in the *Asheville*s, had been tested in 1966.[3]

Preliminary Design reported this feasibility study on 18 August 1967. At the same time, it worked out an optimized missile hydrofoil boat (PCH 2), reporting what amounted to a base-line configuration in April 1968. NavShips 03 issued a draft SOR on 27 May 1968. The Pomona division of General Dynamics was working on a lightweight, six-weapon, antiship Standard missile package that might later be upgraded to surface-to-air use.

Meanwhile, work proceeded at Newport's Naval Underwater Research and Engineering Station on an advanced ASW hydrofoil weapon system for the AG(EH) *Plainview*. The system was intended for rough-weather screening, area ASW search, and hunter-killer operations. At-sea tests were scheduled for mid-1973. This line of development would have led to a very large escort hydrofoil, sometimes denoted DE(H) and at other times described as a deepwater or developmental big hydrofoil.

By November 1968, Ships 03 was considering a 100- to 150-ton missile hydrofoil or PM(H), formerly PC(H)-2. Three such craft would be bought with FY 71 research and development funds, followed by seven squadrons, probably of six boats each, from FY 73 on. At this time, the cost per PM(H) was estimated at $10 to $15 million, and the small crew, which reduced life-cycle cost, was a major selling point. Tentative FY 71 and 72 programs still included eight PGMs per year and six PGHs in FY 71, eight in FY 72; some argued that more PGHs should be bought for squadrons with eight or twelve boats each.

As of July 1968, NavShips conceived the PM(H) as a single hull design supporting three base-line alternatives. The result of the feasibility study of April 1968 is shown in table 15-1. The proposed boat was to have been armed with a trainable, single Standard missile launcher, three to five reload Standard missiles effective out to 30,000 yards, a single 40mm gun (as in the PGH), and two twin 0.50-caliber machine guns. Projected missions included defense and attack against Soviet-built Osa- and Komar-type missile boats, protection of amphibious forces against fast patrol-boat attack, covert operations, reconnaissance and surveillance (as in Market Time), interdiction of increasingly aggressive Soviet intelligence-collection ships, and Market Time–like patrol.

Experience with the earlier hydrofoils suggested that this one, with its crew amidships in the area of least motion, would be able to sustain 45 knots in sea state 5 with an 8-foot hull clearance. Like the PC(H) and *Tucumcari*, the design showed a canard foil arrangement. It had twin gas turbines for foil-borne propulsion and a single diesel for hull-borne travel. The foils were sized to avoid cavitation at full load and at design speed. The bow foil would retract into a centerline well, and the stern foils were split to allow retraction. Ship and engine controls were concentrated in the pilothouse, with tactical command exercised from a CIC.

The hull form was based on that of the PC(H)-1, with some minor changes—a hard chine aft and a foot more width at the transom waterline—to reduce squat during takeoff, landing, and wave impact. The flare of the bow was slightly increased to permit the ship to knife through wave crests and throw spray well clear. The hull had about 2 feet less beam than that of the PCH-1 and 1 foot less depth; hull weight was traded for payload. The forward sheer line had to be changed to provide a level platform for the 40mm gun.

Estimated cost of production craft, excluding concept development costs, was $10.2 million; the prototype was expected to cost $18 million.

The PM(H) could be viewed as an alternative to further construction of the PGH design. No one knew how hydrofoils would perform in a realistic environment, and it could be argued that an advanced hydrofoil in service was more important than an optimum hydrofoil later. An internal NavSec study of March 1969 argued that to choose the large PM(H) over a PGH follow-on would be to accept a five-year delay in acquiring realistic service experience. The PGH could be justified by what was predicted to be its much superior Market Time performance.

At this time, Market Time conditions were described as sea state 3 or worse 30 to 40 percent of the time on the Vietnamese coast, with $^8/_{10}$ cloud cover 35 to

40 percent of the time. Interim evaluations by Op-34 showed that, despite coordinated air, surface, and shore surveillance, successful responses to requests for interdiction and identification of coastal traffic were low in 1967–68, perhaps as low as 20 percent. This problem could be attributed to the mismatch between PCF performance and the need to operate in rough coastal waters. Availability of individual craft was also low.

NavSec argued that a smaller number of hydrofoils could cover the same area in rougher weather. A hydrofoil might cost up to twice as much as a 55- to 60-ton displacement-hull patrol boat, but it would cost less to operate than an *Asheville* or a PTF. A hydrofoil would bring additional advantages: better availability, 70 as opposed to 40 percent; improved all-weather performance, 93 compared with 60 percent of the time; reduced maintenance costs; and good mission keeping, with twice as many missions carried out by two-thirds as many boats.

Table 15-1. PM(H), April 1968

Length (oa, ft-in)	115-6
Length (wl, ft-in)	110-0
Beam (ft-in)	29-0
Draft (down, ft-in)	19-3
Draft (up, ft-in)	4-6 (keel)
Draft (up, ft-in)	6-0 (forward foil)
Displacement (tons)	125 (foils down)
Displacement (tons)	122 (foils up)
Speed (kts)	
Takeoff	25–30
Flying	45
Hull-Borne	12
Endurance (nm/kts)	500 (foil) or 2,000/12 (hull)[1]
Armament	1 40mm (352)
	2 x 2 0.50 (8,000)
	6 SM-1 Missiles
Engines	2 P&W FT-12A (3,100 shp)
	Water jets
	1 Waukesha L1616 DSIN Diesel (600 bhp)

Weights (lbs)	PCH-1	PCH-2
Hull	69,780	58,441
Propulsion	35,677	21,384
Electric	11,827	13,860
C2	19,376	8,433
Auxiliary	48,922	48,313
Outfit	14,313	13,201
Armament	3,763	14,547
Margin	—	13,890
Light Ship	203,658	192,069
Ammunition	3,153	14,526
Full Load[2]	117.5	125

1. The design incorporated reserve tanks that could extend hull-borne endurance by 925 nautical miles, foil-borne endurance by 180. The PM(H) was designed to carry a crew of fifteen on a fourteen-day patrol.
2. In pounds, it was 263,126 (PCH-1) and 280,000 (PCH-2).

The secretary of defense did not finally reject the PGH alternative until July 1970. NavShips continued to develop the larger missile hydrofoil design.

The PM(H) study showed that existing engines, two 3,000-shp Proteus or FT-12 gas turbines, would limit any 45-knot hydrofoil to 125 to 150 tons, which in turn would provide about 18 tons for payload. The only immediately available propulsion alternative was four 2,000-shp AVCO TF-35s. According to a study of February 1969, the PXH would cruise about 600 nautical miles on foils and, like the PM(H), would have a hull-borne endurance of about 2,000 miles at 12 knots. The modular payload would fit within a 40- x 15- x 9-foot box, and it would be supplemented by a secondary armament consisting of either one 20mm Vulcan cannon or Redeye missiles and two twin 0.50-caliber machine guns in Mark 48 turrets.

Typical modular payloads were six antiship missiles in fixed launchers (Harpoon or standard); or fifteen Sparrows in fixed launchers; or an electronic warfare suit; or space for fourteen SEALs and their equipment; or equipment for inshore undersea warfare or for ASW outer screening. The PXH might also support beach-jumpers, who were responsible for deception in support of an amphibious landing.

By this time, the standard PXH briefing included the team operation concept described above. The PXH might help defend the fleet by tactical deception or early warning.[4]

By late 1969, 3,500-shp gas turbines were on the horizon, and the PXH was slightly larger than the PM(H), at 120 x 30 feet and with a 150-ton full-load displacement. The U.S. Navy developed five alternative configurations (NATO later added a sixth):

PXH-A	Coastal ASW (using sonobuoys and torpedoes, with a high-speed, two-mode sonar to be developed later)[5]
PXH-E	EW (demountable deckhouse)
PXH-G	Gun (an OTO-Melara 76mm, with a lightweight 3-inch/50 as a fallback)
PXH-M	Missile (six Standards or Harpoons in fixed launchers; fifteen Sparrows as an alternative)
PXH-S	Special operations (with a deck-mounted module for SEALs)
PXH-N	NATO (gun plus missile)

The PXH merged with a NATO fast-patrol boat project first proposed in mid-1969 by commander in chief, Allied Forces, Southern Europe, as a means of combatting a perceived fast-patrol boat threat in the Mediterranean. In November 1969, operations and technical personnel from eleven NATO nations met

An official artist's concept of the PHM, dated 1972, shows the smaller NavSec design rather than the larger Boeing design actually built.

at Brussels to devise a joint program. They expanded the scope of the project to include all of western Europe, that is, the northern and Baltic seas. The hydrofoil was chosen as the only candidate that would guarantee a significant speed advantage over large Soviet missile ships in rough weather. The consensus was that the NATO boat would need at least four surface to surface missiles plus a secondary gun for surface warfare. It would enter service in the mid-1970s. In fact, the craft was built only by the United States, as the PHM.[6]

The new CNO, Admiral Zumwalt, made the big hydrofoil an important element of his Project 60 program.

The United States chaired the first NATO PHM meetings, of subdivisions of an information exchange group, at the request of NATO's naval armaments group. When interest and enthusiasm appeared to lag somewhat, the United States offered to chair the next meetings of what became exploratory group 2. An offer to demonstrate the PGH *Tucumcari* in NATO waters was eagerly accepted.

By mid-1970, then, the navy could offer the secretary of defense several options. He could buy more PGHs, which might be too small to be useful. He could choose the five-mission PXH, or limit it to a single mission. He had to decide whether to continue advanced development on a large, 1,000-ton hydrofoil. And, finally, he had to decide whether to lead the NATO fast patrol boat program. A draft decision paper written but apparently not submitted in July 1970 proposed an immediate decision in favor of the PXH, leading to delivery of three prototypes to the special hydrofoil trials unit in May 1974 and delivery of the first unit to the fleet in August 1976. Ultimately, the force should consist of about fifty boats. The alternative, three improved PGHs per year for a total of seventy-one to seventy-five boats, was supported by the Italian navy. However, the internal U.S. analysis rejected the idea—the boats would be of marginal use outside the Mediterranean because of their low-powered 3,500-shp engine and short range.[7]

The United States chose to lead the NATO fast patrol boat program, offering a 140-ton PXH in August 1970. In November, the United States formally offered a 142-ton hydrofoil with a maximum speed of 48

Newly completed, the *Hercules* turns at speed in September 1983. Her Harpoon missile cannisters have not yet been fitted, but she has her 3-inch gun and her triple chaff rocket launcher just abaft the base of the fire-control radar; the launcher is controlled by an SLR-20 ESM system. Note the frames for the Harpoon cannisters. The circular housing of an athwartships propeller for low-speed maneuvering can be seen forward.

knots at sea state 0 and taking off at 27 knots. Endurance would be 860 nautical miles at 47 knots, or 1,500 at 12 knots hull-borne. The power plant was now two Rolls Royce Proteus 1273 gas turbines, each of 3,200 shp, and two P&W ST-6A gas turbines for hull-borne operation (500 shp each).

The U.S. versions showed a 20mm Phalanx anti-missile gun with 18,000 rounds. In the NATO version, six Standard missiles in three box launchers were added, and the Phalanx was replaced by a 76mm OTO-Melara DP gun with 160 rounds, controlled by a Mark 87 G/MFCS.

The U.S. gun version showed one lightweight 3-inch/50 aft, with fifty rounds of ready-service ammunition plus fifty rounds in a magazine, controlled by a Mark 87; a 20mm Phalanx close-in defensive gun forward; and two 0.50-caliber machine guns. In the U.S. missile version, six Standard missiles replaced the 3-inch/50.

In August 1970, based on a forty-two-ship purchase, the estimated cost was only $3.2 million per boat ($6.9 million for the lead ship).

NATO's project group 6, formed to run the program, first met in November 1970. At that time, the United States offered to sponsor the program. Admiral Zumwalt authorized U.S. delegates to the second meeting, March 1971, to announce that the initial U.S. purchase would consist of eight PHMs. POM-73, the projected U.S. five year program for FY 73–77, showed thirty PHMs: two in FY 73, eight in FY 74, ten in FY 75, and five each in FY 76 and 77. These boats would most logically be organized in six-boat squadrons. It was expected that advanced design and long-lead items such as turbines and water-jet pumps would be paid for under the FY 72 budget.[8]

The open-ended phase of the program, in which all NATO nations could participate, terminated in June 1971. Only those nations willing to commit resources could participate in phase 2, which began with a meeting in Brussels in October 1971. The U.S. position was that interested, friendly, non-NATO nations such as Sweden, Israel, and Australia might be permitted to join later. The committed NATO participants were Canada, Italy, the United Kingdom, and West Germany; Denmark was undecided.

NATO considered the PHM useful primarily to augment surface forces in the narrow seas and coastal areas. The United States, on the other hand, saw the PHM as an offensive craft attacking surface forces, a surveillance platform, a screening ship, and a vehicle for special forces. Another important difference between the U.S. and NATO missions was that American forces operated far from home, where they would have to depend on mobile logistical support. The U.S. Navy was also expected to place greater emphasis on AAW (antiair warfare), and it might want an ASW capability at a later date. The program itself did not include ASW, because work on high-speed towed sonars had not yet been completed.

In peacetime, the U.S. PHM would be used primarily for surveillance: patrol of areas such as straits, exits through restricted waters, amphibious objectives, selected sea lanes, anchorages, ports, and islands; and tattletale trailing in support of task forces or blockades. Since the Soviets dispersed their missile ships, relying on long missile range to concentrate fire at the target, the fleet in the Mediterranean needed numerous tattletales, typically two PHMs for every one to three Soviet ships. The PHM would suffer from limited endurance, but it could fly to and from its station at high speed, for example, 42 knots, to make up for time lost in replenishment. For area surveillance, six PHMs would typically be required to maintain two on station.

As for wartime, commander in chief, Allied Forces, Southern Europe, estimated that he would need sixty NATO fast patrol boats: thirty-six in the Mediterranean and twenty-four in the Black Sea. His war scenario forecast a high-speed Soviet exit from the Black Sea, which might be intercepted by a squadron of six PHMs at Suda Bay, Crete. They would be ordered to stations near islands at the narrow part of the Aegean Sea to carry out ESM and controlled radar surveillance of the presumed track of the Soviet force. Their mission profile, then, would show a flight out to their patrol station (200 nautical miles at 40 knots), hull-borne loiter on station (24 hours at 10 knots), a foil-borne engagement for up to four hours, and a return to port at 40 knots.

The West German navy had special requirements. In the target-rich Baltic, its PHM would have to operate with an integrated AGIS-162 auto-engage/information combat system, controlled from shore posts by Link-11 data links. The German version of the PHM would, therefore, need a much larger CIC and a much larger crew than the other versions. Estimates showed a total payload (weapons and sensors) weight of 45.5, as opposed to 32.7 metric tons for the NATO base line.

NavSec prepared a new PHM design for the March 1971 NATO meeting. Where the PXH had been a multimission ship with interchangeable modules, this was a single hull accommodating alternative national weapon systems. It had to be much larger (122 x 25 x 13 feet in depth) than the PXH, and therefore it needed more fuel and power, the latter in the form of four TF-35 gas turbines. Foil configuration (total area and subcavitating cross-section) defined the maximum flying weight of the boat, 162.7 tons.

NavSec presented other PHM alternatives as heavy as 175 tons, which would have made a sustained speed of 49 knots under normal continuous power, a maximum continuous speed of 50, and a burst speed, under maximum intermittent power, of 54. These alternatives were not pursued.

Table 15-2. Hydrofoil Designs, 1971

	PXH	PMH	Boeing
LOA (ft)	115.5	122	119.0
LWL (ft)	110.0	110.0	108.3
Beam (ft)	29.0	25.0	32.7
Draft (foils down, ft)	19.3	17.7	22.5
Draft (foils up, ft)	6.0	5.2	9.0
Hull Depth (ft)	12.0	13.0	±16.0
Full Load (tons)	140	162	230
Fuel Oil (tons)	32.3	43.2	36.4
Endurance on Foils (nm/kts)	860/47	750/48	400+/50
Endurance on Hull (nm/kts)	1,500/12	1,500/15	1,100/10
Foil Engines	2 Proteus	4 TF 35	1 LM 2500
Hull Engines	2 ST6A	—	2 MTU
Foil Speed (kts)	48	—	52.5
C3 (tons)	5.7	5.4	9.9
Armament and Ammunition (tons)	18.5	16.7	24.6

The PHM differed from earlier U.S. hydrofoils in using the same engines hull- and foil-borne, cruising at 15 knots on its hull. However, the separate hull-borne engine had to be restored after a June 1971 NATO meeting; NavSec proposed an AVCO TF-14 gas turbine with a controllable-pitch propeller.

The PHM beam was derived from that of the PC(H)—which, however, used the relatively large Proteus in a 25-foot-wide engine room. TF-35s were small by comparison, and all four would fit in a 23-foot-wide engine room. In both cases, the hull widened forward of the engine room to provide deck space. The boat's length was fixed by adding items that had to be placed in line. The internal layout was particularly driven by berthing. Berths could be placed side by side, but the design could not easily accept more beam, since increasing the span of the after foil between struts would lead to structural design problems. Moreover, too short a boat would be a poor sea-keeper.

The PHM deckhouse was moved further aft than the PXH's to make better use of deck space for underway replenishment and to incorporate the engine air intakes for streamlining.

By late 1971, the PHM had grown once again. The base-line displacement was 160 tons, including 40 tons of fuel for a foil-borne endurance of 750 nautical miles (1,500 hull-borne), and total armament weight was 21 tons. This gradual growth was important because, unlike a conventional ship, a hydrofoil had relatively little margin of acceptable added weight. Its foils could add only so much lift, after which it could not fly at all. Thus, the weight margin designed into the ship in the first place was crucial for any future growth (many conventional ships outgrow the future-growth margins designed into them).

Meanwhile, the choice among alternative antiship missiles had been cut to the Exocet (two prototypes of which had recently flown), the Standard semi-active (in service), and the Harpoon (expected to enter service in 1975).[9]

The Harpoon was recommended for its range and good hit probability, estimated at one out of three even in the worst case. At this time, it was argued that providing four weapons would ensure one hit, but that to add more would be uneconomical on account of declining marginal effectiveness.

The principal choices for the secondary battery were OTO-Melara 76mm cannon (400 rounds, of which 80 could be ready service); Oerlikon 35mm cannon (3,000 rounds, of which 820 could be ready service); or the Sea Sparrow missile (four cell launchers). All of these were thought to be effective against a fast attack boat, and the 76mm, with its penetrating power, would be best against a larger ship such as a destroyer. The Sea Sparrow was rated as fairly effective but unacceptable because an all-missile PHM might be helpless in the face of a destroyer capable of jamming its missile-control electronics. Therefore, the 76mm gun and the best available U.S. fire-control system, the Mark 92, were recommended. Sea Sparrow would be best against missiles, but its four weapons would soon be exhausted. However, on its limited displacement, the PHM could not fit both the 76mm anti-destroyer gun and the effective antimissile gun, the Phalanx.[10] NavShips deemed it impractical for the PHM to carry three weapon systems, considering deck space and center-of-gravity limitations.

Five weapon suits were developed at this time. Suit A was the combination recommended by NavShips, described above. C was a base line presented to NATO, the heaviest combination. E was a fallback in case the Harpoon failed to meet its expected 1975 operational date. B and D were configurations of six Harpoons, which required superstructure design modification and violated commonality. Weight and space limits made it impossible for the boat to carry six Exocet or Standard missiles.

As of October 1971, configurations favored by Allied navies were as follows:

Italy	76mm cannon, Vega fire control, and four Otomat missiles (slightly heavier than Harpoons, arranged two forward and two aft).
United Kingdom	Four Exocets (two forward, two aft), leaving space aft for possible ASW equipment. The Royal Navy considered three Proteus engines, but that would have required considerable redesign.
West Germany	Four Exocet missiles aft. This boat showed a major reconfiguration to accommodate the LM 1500 engine and its elaborate combat system in a lengthened deckhouse.

The U.S. Navy chose sole-source procurement rather than another design competition (proposed by several member nations) to save six to nine months and eliminate the costs of source selection. The selection committee was chaired by a U.S. flag officer, and the United States announced that it would buy the first two boats. The contract award was initially expected in January 1972, but in October 1971, the United States announced that Boeing would receive the contract. It was awarded that November, and feasibility design was completed in March 1972. Letters of intent were received from Italy in April and from Germany in May 1972, all other participants having either dropped out or reverted to observer status. The Italians favored a 168-ton base-line boat armed with six Otomat missiles and showed interest in building PHMs at home for outside customers, such as Venezuela. The third-party market for patrol boats was expected to be substantial.

NavSec completed a feasibility study of a 150- to 160-ton base line PHM in May 1972. It expected Boeing to develop a similar design; the NavSec baseline would be used for comparison. The Boeing design and the *Pegasus*-class PHM turned out to be substantially larger, and the NavSec base line formed the end of the third stage of the PHM story.

NavSec developed a series of Marks:

Mark 1	Baseline, April 1971.
Mark 2	U.S. Navy version, 170 tons, with 35mm Oerlikon guns, six Harpoons, and a Mark 92 fire-control system.
Mark 3	West German version.
Mark 4	U.S. Navy version, 170 tons, with one four-cell Sea Sparrow forward, six Harpoons, and Mark 92 fire control.
Mark 8-0	With a 76mm cannon, four Harpoons, and Mark 92 fire

control. Because these weights displaced only 158.1 tons, the boat had to take on extra fuel to increase foil-borne range to 750 nautical miles. In Mark 8-1, the data link, electro-optical fire control, and centralized information-display equipment were removed.

Mark 10 — 150 tons, length reduced to 99 feet, four Harpoon missiles, and one Emerlec EX 74-0 twin 30mm turret, as in the CPIC (see chapter 16).

Another design was evolving alongside the NavSec PHM: a Boeing missile hydrofoil. By 1970, Boeing had accumulated extensive design experience with the PC(H) and PGH-2. It knew that West Germany badly needed a new generation of fast attack boats and had already arranged to buy twenty missile boats from France for delivery in 1973. Late in October 1970, Boeing requested State Department permission to offer Germany, as an alternative to the French package, a contract for ten missile hydrofoils at an estimated unit cost of $12.5 million. Five of the ten boats would have omitted the German automated combat system. Boeing planned to offer the same design to the U.S. Navy.

The Boeing model 928-70 design was considerably heavier but also shorter than the PXH; at 230 tons and with a length of 108 feet, carrying an OTO-Melara gun and four Exocets, it was capable of 50 knots (52.5 knots on a temperate day). Unlike the PXH, it would be able to take the massive German combat system with its automated data links. The Boeing model would be propelled entirely by gas turbines—one LM 1500 for foil-borne propulsion, one LM 100 for hull-borne, and two AIResearch Garret 83153 turbogenerators.[11] Hull lines were based on those of the successful *Tucumcari*, hydraulics on those of the 747 airliner.

Compared with the NavSec PHM, the Boeing design reduced length about 6.5 feet by stacking auxiliary machinery vertically. It had a deep-V hull form forward, with a high-deadrise planing surface aft. Continuing this deadrise aft moved the center of buoyancy up for better stability and also raised the center of gravity.

The Boeing PHM showed more structural weight than the NavSec design because of its two-level pilothouse, its deeper and hence stronger hull girder, its use of built-up rather than integral tankage, and its 10 percent margin on hull drag.

Boeing pressed hard to sell its PHM to the U.S. Navy, knowing that a purchase would encourage the Germans. NavSec argued, however, that the Boeing boat, with its relatively short range, large payload, and high cost (probably 25 to 35 percent more than that of the 160-ton PHM), was too large for the U.S. Navy. Moreover, the navy would be unable to exert full control over a design so heavily influenced by German requirements.

Boeing defined its craft in December 1970 and presented a preliminary report the following June. It argued that purchase of its PHM would support the one major active U.S. hydrofoil group. Its *Tucumcari* had already won a competitive runoff, and its new PHM was primarily an expanded *Tucumcari*. Surely there was little point in yet another destructive competition. Boeing emphasized that sales to Germany would bring substantial foreign earnings.

NavSec argued that there was little for Boeing to choose from between the 160-tonner and the 230-tonner, since the company, the only viable contender at the time, would receive the design and construction contract in any case. Boeing had, after all, several smaller hydrofoil designs in hand.

This argument was pressed through February 1972, as NavSec continued to develop its internal baseline design with progressively more powerful engines. The new TF40 gas turbine made a 186-ton PHM practicable.

Compared with a 170-ton NATO PHM, the Boeing PHM was faster, 50 versus 48 knots; carried a heavier payload, including the Link-11; and used a fully proven foil engine, the LM2500. The TF-35 had not yet been tested at sea. On the other hand, the Boeing boat would have to depend on a single foil engine, whereas three of four TF-35s could keep a PHM foil-borne at 40 knots. The small TF-35 could be repaired on board using modular spares. The Boeing design had a significantly shorter range, both foil-borne and hull-borne. This problem could be solved, but only at the expense of military payload, which would fall below that of the NavSec design.

The Boeing design was selected, partly because it was so clearly preferred by the West Germans, the major foreign partner. Certainly this larger PHM had a greater margin for additions, and it was probably more rugged than the smaller NavSec design. Ironically, later on, as the United States began to back out of the program, the Germans also abandoned it.

Ironically, too, in December 1972, Boeing found it possible to double the Harpoon battery at virtually no cost in added weight, because the Harpoon launcher turned out to be much lighter than expected.

The navy request for FY 72 money was approved, and the January 1971 budget submitted to Congress included $5.1 million for advanced PHM procurement. The full thirty-ship program first appeared in May 1971 in POM-73, an internal navy program. OSD

BOEING NATO PATROL MISSILE HYDROFOIL (PHM)

Length	131.2 feet	40.0 m
Beam	28.2 feet	8.6 m
Draft		
Hullborne (foils retracted)	6.2 feet	1.9 m
Hullborne (foils extended)	23.2 feet	7.1 m
Foilborne (nominal)	8.8 feet	2.7 m
Displacement	231 long tons	235 metric tons
Speed		
Hullborne	12 knots	
Foilborne	in excess of 40 knots	
Propulsion		
Hullborne	2 diesels with 2 waterjets	
Foilborne	1 gas turbine with waterjet	
Crew	21	

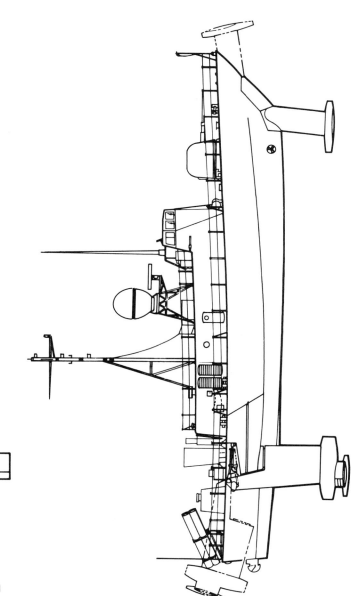

The *Pegasus* class. (Boeing)

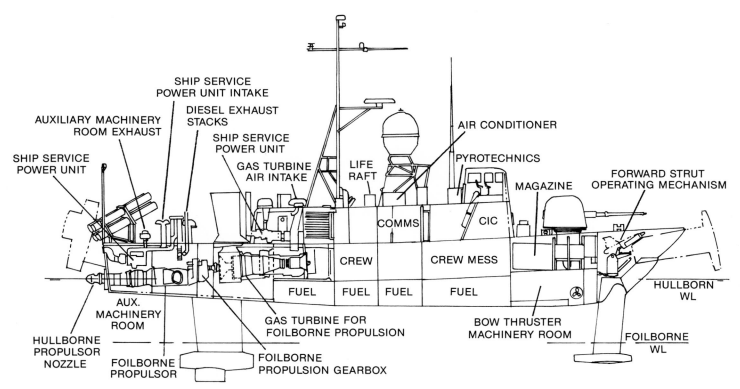

The *Pegasus* inboard profile. (Norman Friedman)

proposed that this funding not be approved until the Defense Systems Acquisition Review Council (DSARC) reviewed it that September. The navy POM was, however, accepted "for planning purposes." DSARC was delayed, and OSD again challenged inclusion of the PHMs in the navy's proposed POM-74. This time the deputy secretary of defense agreed, and the program was cut to eight ships pending the review of DSARC.

The review was held on 26 September 1972, and afterwards the deputy secretary authorized the navy to plan for eight ships. Further production would depend upon "full justification and consideration of alternatives." However, in January 1973, mid-range plans such as the POM continued to show the thirty PHMs.

Two of these *Pegasus*-class PHMs were included in the FY 73 research and development budget, and four more in the FY 75 shipbuilding (SCN) budget. Boeing now encountered severe cost overruns, and on 15 October, the navy decided to suspend work on PHM-2 until more funds could be obtained. DSARC met on 31 October to review production problems. At this time, the five-year shipbuilding plan (FY 75–79) still showed twenty-three production PHMs plus the two research and development boats, but the FY 76–80 plan showed a final reduction to four ships plus the two other units.

By early 1974, when Boeing discovered problems with hull fabrication, the program was in deep trouble. The PHM's aluminum welds required more reworking than had been expected, to the point that man-hours expended bore little if any relation to actual progress. At one point, it was discovered that two and a half times as many feet of welding than was estimated was needed, and that working with each foot of weld required one and a half times the estimated number of man-hours. Steel welding was also taking considerably longer than planned.

Originally, PHM-1 had been intended as an unarmed test boat, with the first combat system, using a Dutch-supplied Mark 94 fire control, in PHM-2. As delays built up, the prototype combat system was shifted to the lead boat. The second boat was laid down on 30 May 1974, but costs continued to rise and its construction was stopped in 1976, at 20 percent completion, to provide funds to finish the lead ship, the *Pegasus*.

PHMs beyond the six already funded were officially canceled in February 1976 because of escalating costs. On 6 April 1977, Secretary of Defense Harold Brown terminated the entire program apart from the

On Puget Sound, the PHM *Aquila* turns with a Boeing jetfoil built as a fast troop carrier for the Indonesian navy. Although their hulls are radically different, they have a similar system of foil control.

lead ship, citing cost and the limited potential role of the class. Congress disagreed, reportedly because of pressure from Boeing, and refused to rescind funding; construction of the remaining five ships was officially approved in August 1977. The PHM 2 (the *Hercules*) had to be laid down anew, on 13 September 1980. Given the rising cost of the program and the uncertainty of American support, the Germans—who had in effect shaped the PHM—withdrew from the NATO hydrofoil project.

Boeing sought other applications for its fully submerged, automatically controlled hydrofoil system, building several smaller but similar jetfoil passenger boats and designing a naval patrol variant of the jetfoil. As with airplanes, the design of the wings and controls was far more critical than the design of the fuselage or, in this case, the hull, which was out of the water most of the time. One such ship served in the Royal Navy as HMS *Speedy*, ordered in June 1978 after the 1977 evaluation of a jetfoil. Indonesia bought a jetfoil, the *Bima Samudera I*, for evaluation in 1981; it also signed a contract for two each of two versions of the basic design, the troop transport and gunboat, to be built in Indonesia. These first four were actually built in Seattle and fitted out in Indonesia. The Indonesians, who at one time planned up to forty-seven, have an option for six more. However, since delivery of the basic hydrofoil hulls and propulsion plants, little additional work has been done on the craft.

The U.S. Navy was left with the six *Pegasus*-class missile hydrofoils, primarily being used today for patrol in the Caribbean. With interest in that area increasing in recent years, there has been some pressure for further construction of fast gun/missile combatants. Boeing, for example, actively promoted construction of repeat PHMs.

16

Post-Vietnam Small Combatants

The post-Vietnam small combatant program can most conveniently be broken down into two distinct periods. In the first, lasting until 1975, the Norfolk Division (NorDiv) of the Naval Ship Engineering Center developed four new classes, reflecting the experience and concerns of the Vietnam War. U.S. naval priorities changed radically after the United States left Vietnam, and the craft developed during the second period, notably the Sea Fox, the patrol boats Mark 3 and 4, and the SWCM, reflect those new priorities.

The preceding chapters show the great variety of small craft extemporized for Vietnam. As early as 1967, there was an attempt to simplify the list of craft by amalgamating the monitor and ASPB. As the war progressed, pressure increased to simplify the bewildering variety of small craft and their attendant logistical problems.

A tentative advanced development objective (ADO 38-XX, naval inshore small craft) was written around October 1969 for the production of new craft by about 1980. In 1969, second-generation Vietnamese riverine craft were being tested. The ADO called for higher and quieter speed, better payloads, more effective sensors and weapons, and hulls relatively insensitive to overloading. With such hulls, the modular approach begun in the ATC/CCB could be extended (see chapter 12), since module weights and centers of gravity would not have to be specified precisely. Vietnam experience also showed in a requirement that craft performance not be unduly limited through a wide range of temperature and humidity. As in all such documents, the ADO called for simpler, smaller, but more lethal boats, easier to maintain and man.

The Naval Special Warfare Craft Program (S38-20) was formally established when ADO 38-20X, begun in January 1970, was officially issued that July; it was later renamed the Naval Inshore Warfare Craft Program. Its objective was to develop four new types of craft by about 1980:

- The coastal patrol and interdiction craft (CPIC), which would have been the heavy weather patrol craft Market Time wanted. Its design was shaped largely by the South Korean need to prevent infiltration from the North.
- The coastal patrol craft (CPC), in effect the successor of the Swift (PCF), and the Coast Guard 82-footer in Market Time.

The CPIC and CPC would complement each other in coastal waters; the CPC was to have been the more austere craft, primarily for patrol and reconnaissance. The faster CPIC was conceived as more powerfully armed, with offensive striking capability. As the program evolved, the CPIC was reoriented towards South Korean requirements and the CPC was considerably upgraded.

- The shallow-water attack craft, medium (SWAM), replacing the entire ASPB/ATC/CCB/monitor family as well as the MSSC and LCSR(L).
- The shallow-water attack craft, light (SWAL), replacing the PBR, QFB, STAB, and LSSC.[1]

As in Vietnam, the SWAM and SWAL would complement each other. The variety of possible SWAM missions required modular design.

Initial requirements for the four craft are listed in table 16-1.

Only the CPIC itself was built and tested, in 1974. For the future, it is probably most significant for its double-chine planing hull form, a great advance over previous types and an indicator of future possibili-

The 65-foot Sea Specter, functional successor of the Vietnam-era Swift. Photographed from ahead, the boat shows a 20mm cannon and two machine guns on open tripods forward, with what appears to be a combination 81mm mortar/40mm grenade launcher on the heavy-weapons platform abeam the pilothouse. The aerial view shows the new one-man 40mm gun forward, with 20mm guns abeam the pilothouse and aft and two machine guns aft. The swimmer platform is folded up against the transom. Note the asymmetrical superstructure in both views. (Giorgio Arra)

Table 16-1. ADO 38-20 Requirements, July 1970

	CPIC Coastal Interdiction	CPC Coastal Patrol	SWAL Shallow Water Attack Light	SWAM Shallow Water Attack Medium
Maximum Draft (ft)	6 (6.5)	6 (4.5)	2.5 (1.5)	5 (3)
Length (ft)	65–95 (99.875)	60–70 (60–80)	17–27 (17–36)	36–50 (65)
Beam (ft)	(18.53125)	—	—	(22; 18 desired)
Maximum Speed (kts)	45 Smooth (43.5) 35 SS3 (38 SS3)	25 (60) 10 —	30–35 (40) — (30SS1-2)	15–20 (30–40) — (25SS2)
Range (nm/kts)	350–450/45 (456)	400–500/25 (500/25)	100–50/max (150/40)	200–300/max (250–300)
Designed to Survive in Sea State	5	4	3	4
Designed to Fight in Sea State	4	3	2	3
Endurance at Sea without Support (hrs)	(60)	(72–96 maximum) (24–36 nominal)	—	(12–72)
Transportability	(deck cargo; well deck)	(deck cargo; well deck)	(trailer, C-130, CH-47)	(well deck; desired C-5A)
Typical Opponents	Ships, craft	Light craft	Unbunkered troops, light craft	Bunkered troops, tanks, helos, heavily armored craft

Note: Figures in parentheses are the final requirements. There were also specific silencing requirements for both inboard spaces and outboard radiated noise.

ties. The 72-ton CPIC performed so well in rough water as to encourage the hope that the relatively simple, robust displacement/planing hull would still compete effectively with the much more expensive, complex hydrofoil and surface-effect craft.[2] NorDiv later developed 130-foot (576-ton) and 250-foot (1,000-ton) CPIC-type designs under the Advanced Naval Vehicles Concepts Evaluation (ANVCE) Program. Not only was the CPIC itself the best U.S. sea-keeping planing hull to date, but elaborate tests make it possible to optimize a future planing hull that will halve the accelerations at center of gravity that the CPIC experienced. The boat also introduced a novel, lightweight, welded aluminum hull, a type that will cut the structural weight fraction from the usual 28–35 percent to 21–26 percent.

The CPIC was conceived at the end of Market Time, just as the Koreans became particularly concerned with seaborne infiltration from the North. (Note the number of coastal patrol boats, Coast Guard 95-footers and Sewart 65-footers, that they obtained in the late 1960s. These are listed in chapters 8 and 10.) By 1969, the Koreans wanted better performance than those craft could give, and their needs fitted the long-range requirement embodied in the CPIC.

The Naval Ship Research and Development Center (NSRDC) began a study of Korean seaborne infiltration in November 1969, and an OpNav/NavShips/NavOrd/NSRDC group visited the special warfare group, Pacific, and the joint U.S. military advisory group in Korea to discuss adaptation of the CPIC to Korean needs in February and March 1970. That resulted in a mission analysis, which was completed in April, when the request for proposals for a CPIC feasibility design was issued. In March 1971, an OpNav/NavShips/NavOrd/OSD conference in Hawaii heard CinCPacFleet call for an accelerated CPIC program for Korea, and a month later commander, Naval Forces, Korea, issued an "operational concept for countering the seaborne infiltration threat." ARPA provided $500,000 at this time, a direct result of the Hawaii conference. In May 1971, Deputy Secretary of Defense David Packard ordered the navy to develop a fast patrol boat.

The Koreans' ultimate goal was sixteen boats. The program was, therefore, shifted toward joint development and production, with half the boats to be built in Korea, and from long-range to short-range technology.

CPIC design requirements reflected the anti-infiltration emphasis: the boat would spend much of its time loitering, awaiting infiltrators it would have to catch or sink. Table 16-2 describes its expected mission profile. Note, too, that although CPIC had an impressive performance, its design emphasized sea-keeping *and weapons* instead of the more traditional

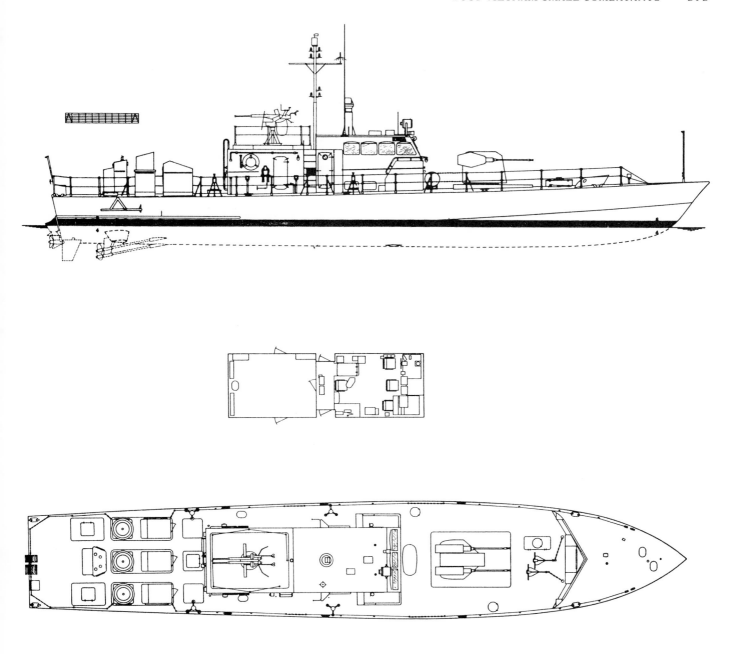

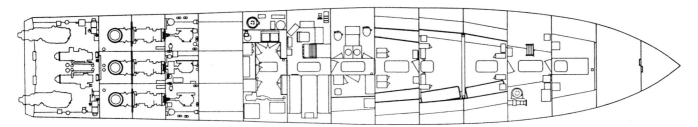

CPIC, the experimental fast patrol boat. This drawing displays a twin 20mm cannon on the 01 level aft, but the single boat completed was armed only with the twin 30mm gun. There are four pintle mounts along the sides, which can carry two twin 7.62mm machine guns and two 40mm grenade launchers. As an alternative, the boat was designed to mount a second twin 30mm mount in place of the 20mm gun, with a second director, a Mark 93 fire-control system. (USN)

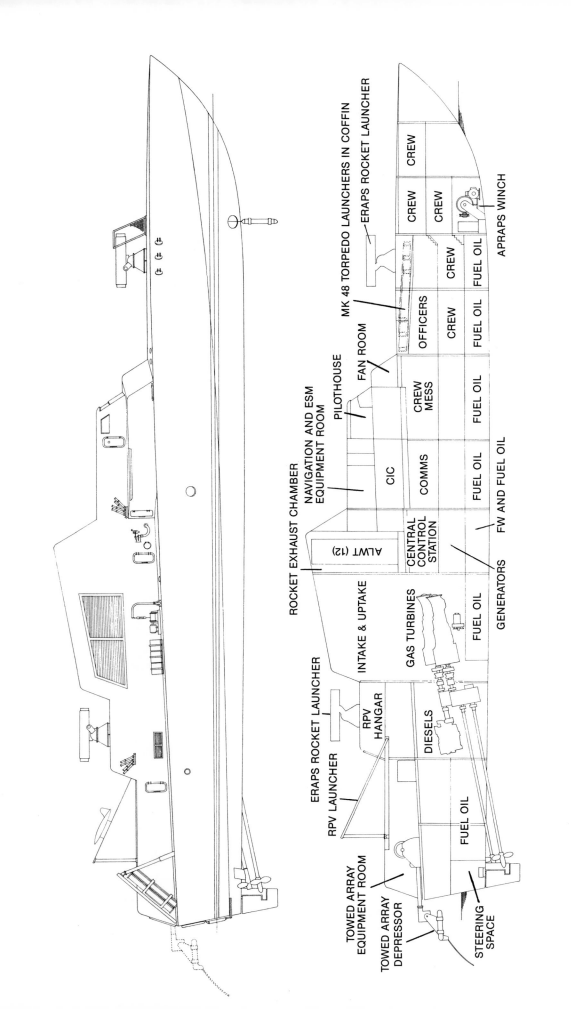

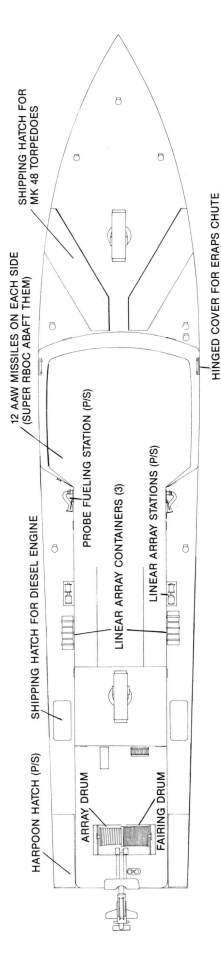

The 76-meter (250-foot) "Spirit of '76" was the largest of a series of planing craft developed under the advanced naval vehicles program, using data gathered during CPIC design and trials. This design was developed at NorDiv, and the drawings were made from plans dated 1 December 1976. Full-load displacement would have been about 1,100 tons; the 76 meters referred to overall length. As befitted a craft, the Spirit of '76 incorporated numerous new weapon and sensor systems, including many that had only been proposed. For ASW, the ship would have used a combination of expendable linear arrays launched amidships, a towed array aft, and reliable acoustic path (RAP) sensors: rocket launchers forward and aft for each of six RAP buoys; chutes to drop the expendable buoys from the sides of the superstructure forward; and the active RAP dipping body below the launcher. High speed would have allowed the boat to close a distant contact, attacking with six Mark 48 torpedoes carried in tubes forward and twelve ASW standoff weapons in the stack-like structure amidships. Now, a decade later, the latter are about to materialize as vertically launched ASROCs. The other major offensive weapon was the Harpoon, in quadruple cannisters aft, targeted with the aid of the RPV shown on its launcher. Space was provided to stow twelve RPVs. For air defense, the ship would have had twelve "AAW standoff weapons" under a hinged cover on each side. For close-in protection, it would have carried a Super-RBOC (rapid-blooming chaff). Plan views show space for two radars: a target acquisition system, TAS Mark 20, presumably conceived as a follow-on to the TAS Mark 23 used with NATO's Sea Sparrow; and APS-116, the sea-search radar of the S-3A Viking ASW airplane. It appears that the antennas would have been set into the roof of the pilothouse. Note the portable screen before the forward buoy launcher, for reducing air resistance at high speed. There would have been three LM2500 gas turbines, with two diesels on the outboard shafts for cruising, in a CODAG arrangement. (Alan Raven)

concentration on sheer speed. The CPC emerged the faster boat (prospectively) by far.

Table 16-2. CPIC Mission Profile and Requirements, May 1971

	Speed (kts)	%
Flank	45	5.0
High	35–40	5.0
Transit	30	10.0
Cruise	8–14	15.0
Loiter	5	58.3
Draft	0	6.7

Note: Requirements included speeds of 45 knots minimum in calm, 40 minimum in sea state 3, and 8 to 14 cruise, and an endurance of 350 nautical miles at 45 knots. The boat had to turn in a 400-foot (maximum) radius at 40 knots and stop within three boat lengths of 40 knots.

By May 1971, interest on the part of nine "priority" nations (Korea, Thailand, Taiwan, Iran, Greece, Vietnam, Turkey, Cambodia, and Israel) led the CNO, Admiral Zumwalt, to reprogram funds well beyond the $500,000 originally planned for FY 71.[2] The following is given in millions of dollars:

FY	71	72	73	74	75	76
	3.0	5.6	5.5	4.6	2.9	2.0

At this time, the CPIC was important enough to be the first project under DoD's "new incentives" program; the boat had rapid development authorization and priority immediately below that accorded the hydrofoils. This priority was reflected in the reprogramming of $1.6 million of shortstop money for a program prototype and four test craft, with additional MAP money ($5 million) for the Korean boat. The figures above include both research and development and MAP money.

Coproduction was never quite practical, because no satisfactory "memorandum of understanding" could be written and U.S. costs, owing to inflation, were escalating in 1973–74. Having brought the CPIC into existence, as it were, the Koreans later killed it by withdrawing funding.

Three alternative designs were evaluated: the double-chine by Atlantic Hydrofoils, a more conventional NorDiv hard-chine hull, and a catamaran by Atlantic Research Corporation. Atlantic Hydrofoils won the competition to develop both the preliminary and contract plans, and detail design was done by the builder, Tacoma Boatbuilding. NorDiv continued its own CPIC hull design so as to build its own expertise for the design evaluations.

Atlantic Hydrofoils began with a 77-foot hull. By February 1971, it had been lengthened to 95 feet to allow for more payload at a higher speed. It was ultimately lengthened to 100 feet. The length-to-beam ratio and beam loading had to be correspondingly increased to limit vertical acceleration.

When the program began in 1971, the most recent complete data on planing hull performance dated back to a series of 1949 tests of the YP 110 (ex–PT 8). The requirements of smooth-water planing hulls were well understood, but there was not sufficient data to design such a hull for good sea-keeping at low to medium speeds (up to 25 knots). It was by no means clear that a planing boat could maintain good sea-keeping at high speed (above 30 knots).

The primary design consideration was to reduce impact acceleration at high speed. That was accomplished by using a narrow chine with high beam loading. At high speed, the hull planed on the narrower lower chine, but the upper chine broadened the hull enough to provide good stability at the displacement waterline. Spray rails ensured clean flow breakaway to minimize wetted surface (friction drag) at all speeds. Trim was controlled by transferring fuel between forward and aft fuel tanks. Single active fins between the upper and lower chines reduced rolling motion. The hull had a flared, deep-V entrance to cut through waves.

Resistance was further reduced by cooling the machinery through the skin rather than through radiators fed by hull ducts.

The CPIC mission required three speed ranges: very high speed for pursuit, medium speed for patrol, and very low speed for loiter. Atlantic Hydrofoils provided, in effect, three power plants. For high speed, there were three gas turbines, originally 1,250-shp AVCO Lycoming TF 14s, but 2,000-shp TF 25As in the final lengthened version; each turbine drove a fixed-pitch propeller. For very low speed, two diesels with outdrives were provided for endurance. The center turbine was used for medium speed, the outer propellers windmilling with slipped clutches.

The CPIC achieved 43.5 knots in calm water and made 37 to 38 knots in 4- to 6-foot waves, 35 knots in 5- to 9-foot swells. These figures fell somewhat below those originally desired, but they were still extremely impressive.

Compared with a conventional planing hull, the CPIC had reduced beam and depth. Fabrication cost was cut by using longitudinal framing stiffened every 80 inches by transverse bulkheads or deep webs. This was equivalent to two and a half times the hull frame spacing of a conventional boat, such as an *Osprey*. The result was a radically lighter hull, its weight fraction of 24 percent approaching the optimum achievable in a planing hull. The hull was also expected to experience much higher stresses. These departures were possible partly because the CPIC program was handled, not as a conventional production project, but rather as advanced development. Thus, no material or design specifications were provided to the contractor. Instead, he developed his

CPIC was the only element of the post-Vietnam small craft development program to be built. It is shown in August 1974 before any weapons had been installed. It received the planned twin 30mm cannon forward, but a second position, above the deckhouse, was never filled, reportedly because the boat lacked sufficient stability. A periscope atop the pilothouse was part of the fire-control system. Note how shallow the hull is.

CPIC runs in company with an "Osprey" (PTF-24) and the hydrofoil *Flagstaff*, August 1974, off the coast of southern California.

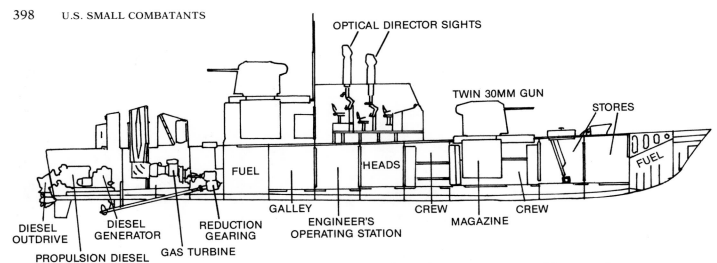

CPIC inboard profile with the alternative projected armament of two twin 30mm cannon. The second mount was not fitted because of limited available top weight. Note how shallow the CPIC hull is. (Norman Friedman)

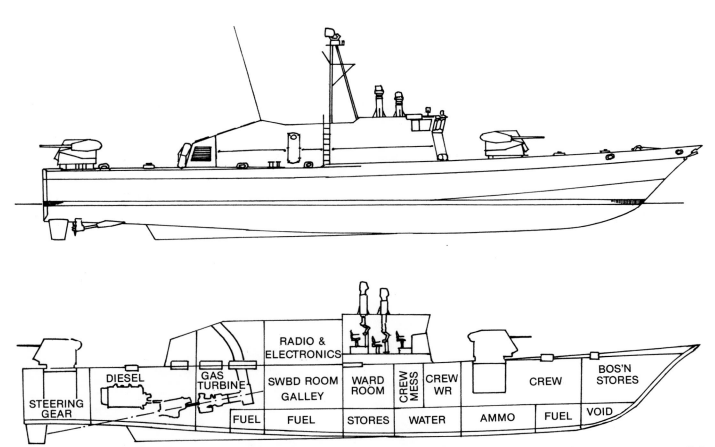

Tacoma Boatbuilding's PCMM was broadly equivalent to the more conventional CPIC developed by NorDiv as a parallel project. A diesel version, the "Sea Dolphin" class, appears to have been built for the Korean navy by Korea Tacoma; none of the boats was exported. This PCMM Mark 5 was designed to achieve 44 knots with CODOG propulsion (one 1,100-bhp diesel and one 4,450-shp gas turbine on each of two shafts). Fuel capacity was 7,000 gallons, endurance, 1,800 nautical miles at 10 knots. Dimensions were 116 (overall) x 23 (molded) x 13 (depth amidships) feet. Design displacement totaled 125 tons. The maximum complement consisted of three officers, three petty officers, and twelve crewmen. (Norman Friedman)

Table 16-3. Post-Vietnam Small Combatants

	CPIC	PB MK III	PB MK IV	PB MK I	SWCM
LOA (ft-in)	99-10.5	64-10.75	68-5	65-6	82-6
Beam (ft-in)	18-6.375	18-0.75	18-0.75	18-4	35-0
Draft (ft-in)	6-6	—	3-5.825	5-0	8-4
Full-Load Displacement (lbs)[1]	158,000	78,000	99,000	71,000	251,800
Engines	3 TF 25A	3 8V71TI	3 12V71TI	2 12V71TI	2 16V149TIB
SHP (each)	2,000	600	650	—	1,800
Speed (kts)	43.5	27	—	—	—
Fuel (gals)	—	1,800	1,800	1,200	—
Range (nm)	456	—	—	—	—
Weights[2]					
Hull	18.23	24,574	28,624	26,030	33.95
Propulsion	10.17	19,045	22,243	16,192	28.25
Electric	4.18	3,700	6,160	3,026	3.62
Communications/Control	1.81	928	1,232	482	3.50
Auxiliary	4.89	4,267	6,249	3,139	5.81
Outfit	5.90	6,933	6,518	5,540	6.81
Armament	4.09	1,858	5,002	1,443	3.48
Margin	2.47	—	—	—	2.54
Light Ship	51.74	61,305	76,070	55,852	87.96

1. Late in 1976, CPIC full-load displacement was listed as 182,800 pounds.
2. Weight data are estimates in tons for the CPIC (95-foot version of 28 March 1972) and for the SWCM. Other weights are in pounds.

own load and stress criteria, which NorDiv had to approve.

Because the CPIC represented so extreme a departure from standard practice, NorDiv decided to carry out a full-scale evaluation to determine the stresses and assess the structural adequacy of the design. Structural trials were held in the Strait of Juan de Fuca from late January to early February 1974.

Finally, the CPIC was the occasion for a new small-boat weapon development, an Emerson Electric twin-gun turret (EX 74) carrying two GE 30mm EX28 machine guns. The CPIC was to have been armed with two turrets, one forward and one atop the pilothouse. Both would be controlled by a new Kollmorgen remote optical sight (EX 35) and a Honeywell radar-gun fire-control system (EX 93). All equipment was delivered for land testing in September 1972, whereupon the machine gun was rejected and replaced by a Hispano-Suiza weapon. In September 1974, one modified mount was installed aboard the prototype CPIC, which completed a successful operational evaluation in February 1975. It could engage targets at 5,000 yards in split (radar range/optical track) mode, or 3,000 yards in radar-only (track-while-scan) mode. None of the other craft in the program progressed far enough for such detailed ordnance work. The Emerson gun turret has since found wide acceptance in foreign navies, but it has never been adopted by the U.S. Navy.

The prototype CPIC was delivered in May 1973. It was tested for eighteen months on the West Coast, particularly in the rough waters of the Strait of Juan de Fuca, before being delivered to the South Korean government in June 1975 for test and evaluation. The South Koreans returned it after making the tests, and it was ultimately laid up at San Diego, where its hulk could still be seen in 1986. A second CPIC hull was delivered to NSRDC (Carderrock, Maryland) for tests of its unusual structure. Other hulls planned a few years earlier never materialized.

Tacoma Boat did incorporate many CPIC features, albeit with the NorDiv hull, into its PCMM, which the associated Korean Tacoma yard built.

As the CPIC shifted from an advanced development project to a near-term operational boat, the CPC was upgraded. By this time, the primary characteristics of the boat were very high speed and relatively large deck area. Length was limited to 80 feet for air transport.

A request for proposals was issued in December 1972; eight proposals were being evaluated when the program shifted and the CPC was canceled. NorDiv favored those submitted by Hydronautics, Rohr Industries, which was then heavily involved in the SES program, and Teledyne Sewart Seacraft. Details of the Sewart proposal have not survived. All three successful proposals were unconventional twin-hulled, partially air-supported craft.

Rohr proposed an 80-foot, 200,000-pound "surface effect and stepped planing catamaran" with a 35-foot beam. At high speed, it would rely on its captured air bubble. Rohr considered its stepped hulls unusually tolerant of variations in loading; they would also result in efficient cruising off-bubble or in a partial air-cushion mode. In rough seas, their fine lines would make it easy to enter waves.

Rohr chose a hybrid because the wave drag of the

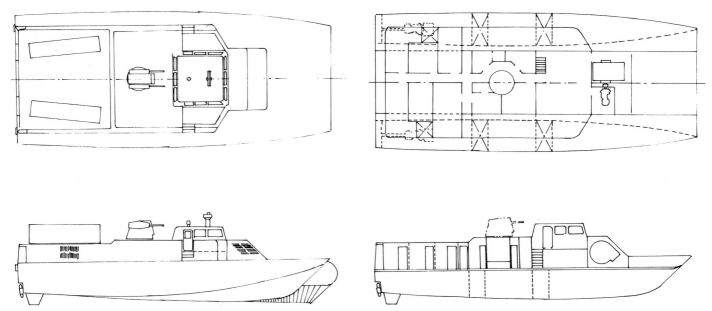

The Hydronautics proposal for CPC-X, armed with a twin 30mm cannon and with surface-to-surface missiles aft. Dimensions are 80 x 35 feet. (Norman Friedman)

air bubble itself created an undesirable hump between 20 and 40 knots. The hump might have been reduced by changing the length-to-beam ratio of the bubble, but in effect, that was fixed by the length limitation. In the critical speed region, bubble pressure could be reduced and more reliance placed on planing and buoyancy for a net reduction in drag. At higher speeds, the bubble itself had the minimum drag, and the cushion supported about 90 percent of the craft's weight, the remainder being divided between planing lift in the catamaran sides and buoyancy.

Hydronautics also chose a hybrid, consisting of semiplaning catamaran hulls with flexible air seals between them. Below the hump, the boat would operate as a conventional catamaran with the seals lifted. Above the hump, part of the total weight of the craft would be supported by air pressure in the space between the hulls.

NorDiv pursued a parallel hybrid, a "ski cat" whose hydrofoils provided partial support analogous to that given by the air bubble in the other craft: each of two catamaran hulls was supported by a single submerged foil with a high-aspect ratio and low drag just aft of the longitudinal center of gravity. It was a Grunberg system, with a hydroski at each bow and a broad, submerged horizontal plate at the stern of each hull to damp pitching. NorDiv had a third-scale manned model built and tested. The rigid catamaran could support a long hydrofoil, which could be designed for speeds as great as 60 knots without cavitation. In addition, because the foil had a low induced drag (that is, very high lift-to-drag ratio), it could

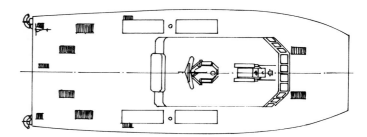

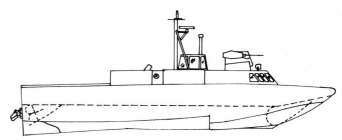

The Rohr proposal for CPC-X. Dimensions are 80 x 35 feet. (Norman Friedman)

accommodate considerable increases in weight without a disproportionate increase in drag.

For FY 74, the program was redirected from the CPC to the SWAM. NorDiv considered two alternatives, an air-supported boat and a pure planing boat. The planing hull, cheaper, simpler, and quieter, was

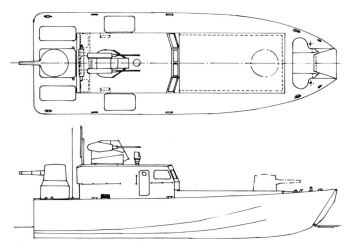

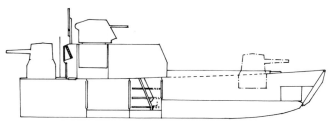

SWAM-X inboard profile. Note the access between the bridge area and the well deck. (Norman Friedman)

The SWAM-X notional base-line design with a large covered well deck forward. No propulsion system is shown, because water jets and propellers were considered throughout the life of the program. The weapon aft is a 20mm Gatling gun in a projected turret; the dashed lines forward show an alternative location, and there are also two pintle mounts in the bow. Dimensions are 63 x 22 feet. (Norman Friedman)

preferred. A request for proposals was released in September 1975, but it had to be withdrawn that December, as the entire program ended.

As in earlier riverine programs, the SWAM suffered from a limitation on draft; beam and length were also limited, by the demand that the craft be transportable by air, which in turn restricted total height (although sea-keeping demanded a minimum freeboard). As in the ASPB, the only way out was to increase length.

The modular approach for the abortive ATC/CCB was adopted for SWAM. Required mission variants were troop carrying, CCB, fire support (ASPB), advanced base support (logistics), medical aid and evacuation, minelaying and countermeasures, and logistical support. The basic SWAM had a deep well with a removable cover and bow door forward of its pilothouse, two pintle mounts forward of the well, a turreted Vulcan cannon aft, and a CPIC-type twin 30mm gun atop the pilothouse. A second Vulcan could be mounted at the forward end of the well. The mission module could be fitted by the builder or in a forward theater, and base-line boats thus adapted to varying circumstances.

In the troop carrier version, the module space was filled with forty-four shock-absorbing seats. Otherwise, the space was to be filled with one 20- x 8- x 8-foot container module. The fire-support version, for example, carried the 105mm howitzer of the program 5 monitors. An advanced base support (defense) version showed a Vulcan forward and two quadruple Harpoon missile launchers aft. In the mine-warfare version, the Vulcan aft was replaced by a standard triple sweep winch, and a Vulcan was mounted forward. There was even a SEAL-support version in which the container carried a swimmer delivery vehicle. Consideration was also given to a close-coupled trailer that would ride its rooster tail.

The unusually wide range of required speeds—40 knots for interdiction and escaping ambushes, very low speed for loiter and slow patrol—corresponded to the wide range of craft the SWAM was intended to replace. The program ended before any definite choice of propulsion could be made; both water-jet and tunnel hulls with propellers survived to that point.

SWAM was the least advanced of the four projects. A draft request for proposals was submitted to NavSea in November 1972, but money was not available and it was canceled in April 1973. Sketches of NorDiv feasibility studies show a low superstructure and modular weapons such as the twin TOW launcher. Propulsion would have been similar to that of LASSC, with three outboard engines that could be changed quickly—a valuable feature for a boat operating in remote areas.

With the demise of the integrated S38-20 program, further small combatant development split in two. On the one hand, there was a family of new SEAL craft, the Sea Fox (special warfare craft, light, or SWCL) and the new surface-effect SWCM (special warfare craft, medium), unofficially called the Sea Viking. On the other, there was a new patrol boat that could replace the existing Swift and also carry out some of the duties envisioned for the abortive CPC.

The new SEAL craft were needed because the SEAL mission changed abruptly from the support of riverine and jungle warfare to the earlier one of inserting small teams by sea. New SEAL boats had to emphasize sea-keeping and covertness.

The Sea Fox was designed specifically for over-the-horizon support of SEAL teams, taking them from well out to sea into shallow water. It had a special trailer for transport by C-130, which required remov-

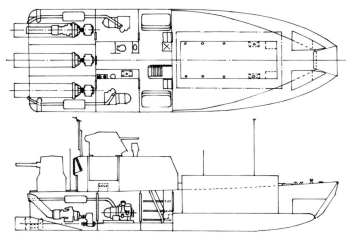

SWAM-X command and control variant, showing gas turbines driving propellers in partial tunnels. The engines forward of the turbines are ship service diesels. (Norman Friedman)

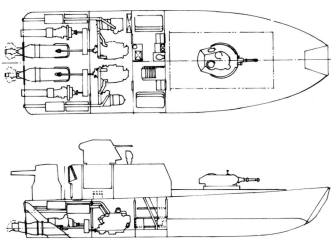

SWAM-X fire support variant, showing CODOG engines driving water jets. (Norman Friedman)

able or, better, telescoping and hinged structures. Including the trailer, it could be no more than about 7 feet high; maximum width was about 9 feet 9 inches. Covert activity required quieting, that is, heavy silencers and special antiradar hull materials. The boat needed elaborate communications gear and had to be able to load and offload swimmers; for the latter, its transom had a hinged platform that dropped down so that a man could run out of the cockpit into the sea. The boat also had to carry a heavy payload of SEALs, yet it was to be operated by one man. Seakeeping required a heavy hull structure and a deep-V form with considerable deck camber to shed water. Operation in heavy seas also prohibited the use of outboard air scoops, so the engines had to operate from inboard scoops without the benefit of ram air effect. Finally, operation in shallow water required the use of tunnel propellers.

The Sea Fox was designed at Norfolk to meet a 1976 operational requirement, and the prototype (36PB771) was built by Uniflite under the FY 77 program. Completed in November 1977, it was retired to the amphibious museum at Little Creek in February 1984. Uniflite built another seven (36PB801–807) in FY 80, one in FY 81 (36PB811), and sixteen in FY 82 (36PB821–8216), completing the last in March 1984. As of 1987, the official figure for total Sea Fox production is about forty-five units, a few of which have been transferred to Egypt. Because it was the closest thing to an easily transportable navy fast patrol boat, the Sea Fox was used as a picket for interdiction off Lebanon, a function far from its design role. Not surprisingly, its performance was relatively poor.

NorDiv did some design work on the larger SWCM as early as 1977–78. However, the program lay dormant until 1982, when Congress appropriated money for immediate construction of what was then labeled a PBM (patrol boat, medium). As in the case of the CPIC, NorDiv conducted a design competition. SWCM was conceived as a mother ship capable of launching smaller insertion craft or landing a platoon of troops directly onto a beach. High burst and continuous speeds were required, with low speed for transit (that is, long endurance). At least two SWCMs would have to fit in the standard amphibious wet well, without cradles, and the craft had to meet the usual SEAL requirements for conducting covert missions and avoiding detection.

NorDiv developed two in-house designs as yardsticks for the competition, a planing hull and a surface-effect ship (SES). The competition semifinalists were Bell Aerospace Textron, which built the LCAC, Rohr Marine, the group that had designed the abortive 3KSES, Swiftships, and Uniflite. Rohr Marine was chosen, and the design option contract, which included long-lead material and a thousand-hour engine test, was awarded on 8 May 1984. The company received the construction contract on 2 November 1984.

The winning SWCM was a 116-ton, 82- x 35-foot surface-effect ship carrying a payload compartment on its centerline aft. It is suitable for troops, insertion craft, and bulk cargo. The NavSea evaluators found the surface-effect hull far less expensive than a comparable planing hull in both initial and life-cycle costs. While planing hulls, for example, have to employ a combination of gas turbines and diesels to meet the burst/cruise requirements, the surface-effect

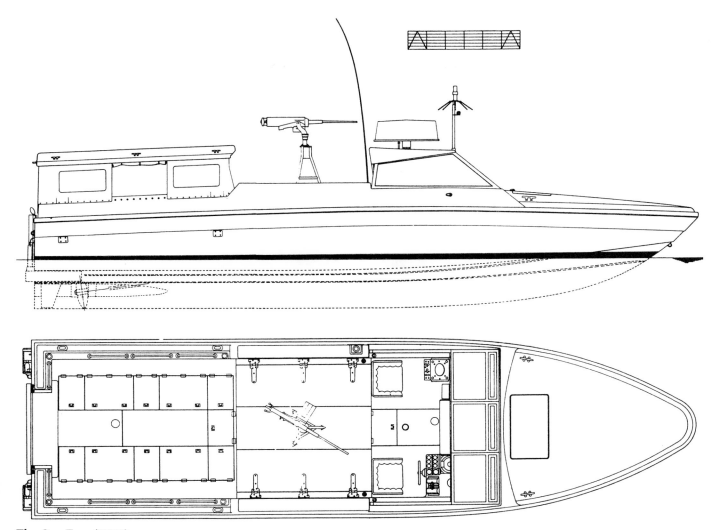

The Sea Fox. (USN)

SWCM, though it has separate fan and propulsion diesels, does not have to carry an idle power plant at any speed. Planing hulls are also limited in stability because of the requirement that two be fitted (in their case, side by side) in a well deck; thus, they may not meet operating conditions when iced. The surface-effect ship, inherently broad beamed, can easily accommodate a large cargo compartment on the centerline. The ship's twin hulls eliminate the need for a well-deck cradle.

The design did encounter one major problem. Model tests showed that the propeller might surface and thus lose thrust. Rohr Marine hoped to solve this by immersing the propeller more deeply at sea, using an Arneson drive. Not even that sufficed. Well into the project, a deeper propeller requiring a wedge-shaped skeg was provided, at the cost of increased craft height (2.5 feet), which in turn reduced flexibility. The boat could still be transported by wet well, but now it could be moved into and out of the smaller well decks only in calm water.

To some of the boat's congressional sponsors, the *M* in the original designation PBM meant missile rather than multimission. The missile requirement soon evolved into a space and weight reservation. The current design shows a station for a man firing a hand-held defensive Stinger, the direct descendant of the Redeye of the PTF program. The gun was initially to have been GE's GAU-25 Gatling (EX-35) in a stabilized mount. When the price of the weapon increased, it was canceled in favor of Hughes' 25mm chain gun. As this is written, it appears that the GE weapon may be revived. Sketches published in 1985 show alternate armaments, including Harpoon and Penguin antiship missiles.

Plans call for nine for each coast plus one for train-

The official model of the Sea Fox displays the well deck and the swimmer gate in the transom. Note the collapsible struts supporting the roof aft; the boat was designed to be road- and air-transportable, which limited overall height. The view from aft also reveals prominent propeller tunnels.

A Sea Fox, unarmed, in 1980. It carries a rubber boat for swimmer retrieval and side curtains for the swimmer compartment. Although not designed for the task, Sea Foxes were used by the U.S. Navy as patrol boats in Lebanon in 1982–83.

The Norwegian Penguin missile system was tested aboard Mark III patrol boats. Up to four weapons were installed, with a director on a short tower on the open platform abeam the bridge. (Giorgio Arra)

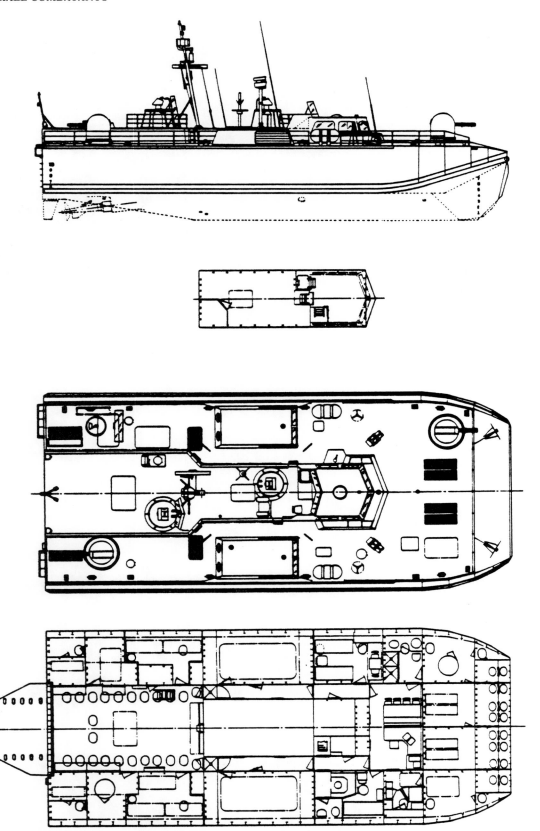

The 80-foot SWCM now under construction. Its future is troubled—the contractor, Rohr, filed for bankruptcy, and as of early 1987 no contract for completion or for a successor boat had been let. Here armament consists of two 25mm Gatling guns (EX 90 mounts), two 0.50-caliber machine guns, one Stinger defensive missile-launching station, and one Mark 34 decoy launcher.

POST-VIETNAM SMALL COMBATANTS 407

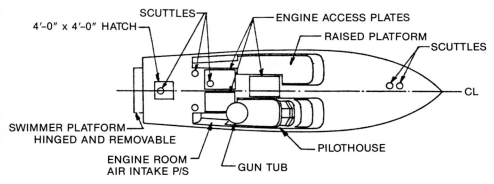

BARE BOAT—NO ARMAMENT SHOWN

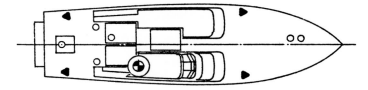

ARMAMENT OPTION NO. 1

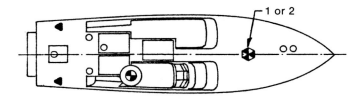

ARMAMENT OPTION NO. 2

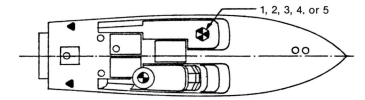

ARMAMENT OPTION NO. 3

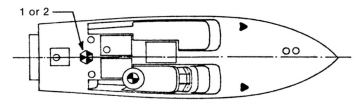

ARMAMENT OPTION NO. 4

TYPE OF ARMAMENT

▲ = **One of the following:**
.50 cal. MG
7.62mm MG
40mm Grenade Launcher
20mm MG

⊕ = **One of the following:**
Twin .50 cal MG
Twin 20mm MG
Twin 7.62mm MG

⬢ = **One of the following**
1. Single 30mm MT
2. Twin 20mm MT
3. 106 Recoilless Rifle (port only)
4. Torpedo Launcher (port only)
5. Rocket Launcher (port only)

Alternative armament options for the Mark 3 patrol boat. (USN)

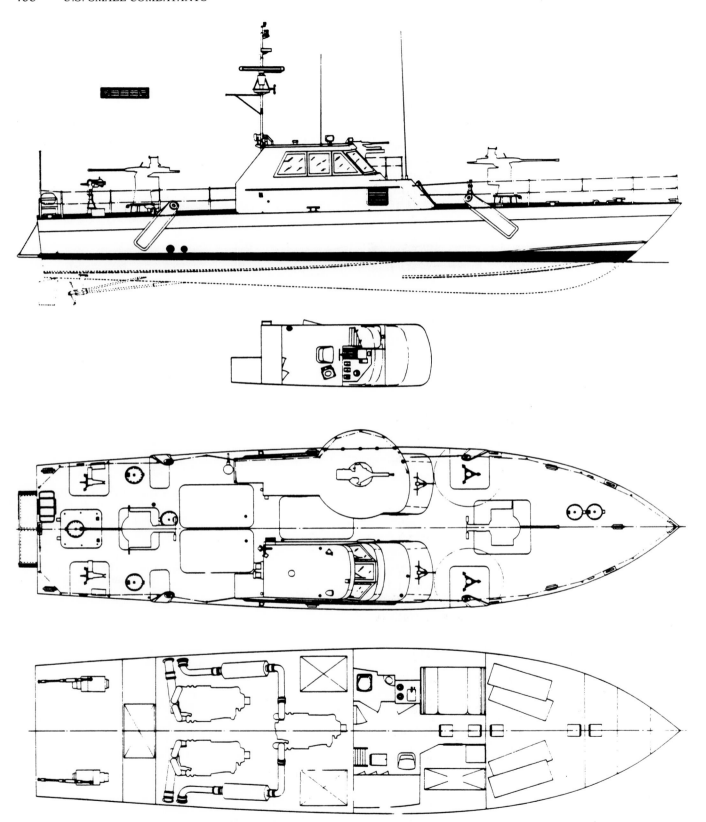

Patrol Boat Mark 4, the 68-foot derivative of the Mark 3 with a less cramped machinery compartment. Armament consists of two 25mm chain guns on the centerline, two 40mm grenade-launching machine guns (Mark 19 Mod 3), and one combination 81mm mortar/0.50-caliber machine gun. (USN)

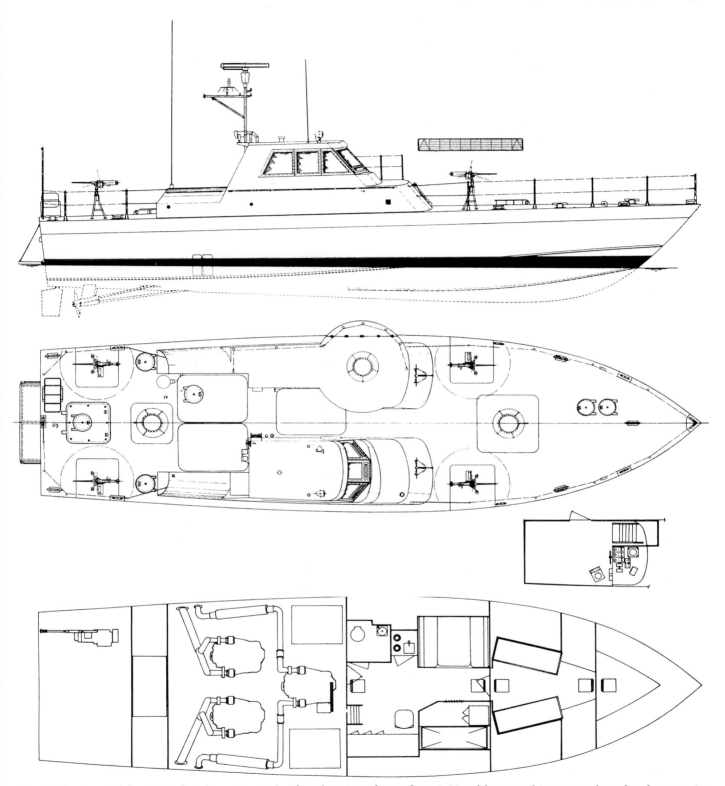

The 65-foot patrol boat Mark 3 (Sea Specter). This drawing shows four 0.50-caliber machine guns, but the three main weapon positions are empty. The main position opposite the bridge can accommodate a twin torpedo tube, a 20mm gun, or an 81mm mortar. With minor modifications, it can take a larger weapon such as a 40mm gun. The two centerline positions were designed to support 20mm guns or 81mm mortars. In 1986, the standard battery is two 7.62mm machine guns on the open bridge, port and starboard, and two 0.50-caliber machine guns fore and aft; the forward one may be replaced by a 40mm gun or a 25mm chain gun. (USN)

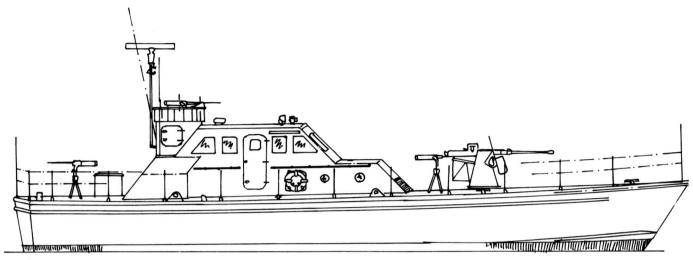

The 65-foot Patrol Boat Mark III sketched by NorDiv. The mount for two 0.50-caliber machine guns atop the pilothouse, reminiscent of the one in earlier craft, appeared only on the prototype boat. (Norman Friedman)

ing. However, the program has been badly delayed and costs have risen. Rohr Marine went bankrupt in the summer of 1986, and the future of the SWCM is in doubt, especially as the incomplete hull structure was some 9 tons over design weight when work stopped. Congress appropriated funds for a second SWCM in FY 87 and directed the navy to recontract and complete the first unit.

The other major post-Vietnam program was the 65-foot patrol boat. As recounted in appendix E, Swiftships built a series of 65-footers, in effect enlarged Swifts. Those built for the navy, for transfer to the Philippines, were called patrol boats Mark I. The same hull was developed as a personnel boat, a utility boat, a torpedo retriever, and an aircraft rescue boat. The two FY 72 boats were redesigned as Mark I Mod 1s to test Mark III engines and several hull features. The vessels were later disarmed to serve as pilot boats. No patrol boats Mark II were built.

The 65-foot Mark III was a very different boat, conceived as a sort of mini-CPIC at a much lower cost.[3] Designed by NorDiv, it veered radically from other boats in being asymmetrical; most of its superstructure was on the starboard side, and there was an open gun platform to port. That provided three rather than the usual two heavy-gun positions; top heaviness was avoided by using a superfiring position. It also left an open space for a back-blast weapon such as a big recoilless rifle or a missile launcher. In service, 65-footers tended not to be heavily armed, and features adopted specifically to provide space for absent weapons were often ridiculed by their operators.

The prototype had a Swift-type gun tub atop its pilothouse that was later removed; the design sketch showed the tub, a 20mm cannon forward, and machine guns on deck pintles.

A Mark III test-fired Norwegian Penguin II missiles in 1981–82. The four missiles were mounted aft, and their optronic director was placed on a platform abeam the pilothouse. The missiles, heavy for the boat, were not adopted for service use.

More recently, Mark IIIs have been armed with a modernized World War II weapon, a 40mm gun with power ammunition feed, and as of this writing, they are to receive the 25mm Mark 88 mounting. A Mark III will be used to test GE's EX-35.

The Mark III hull design was also unusual, with a heavily cambered hull and a carefully curved superstructure for minimum radar cross-section. A critic from NorDiv later suggested that such elaborate measures gained little if weapons were mounted, since they themselves inevitably had large radar cross-sections.

The Mark III has three engines—compared with two in a Mark I—improving the boat's chance of returning after receiving battle damage. The centerline engine is silenced for low, quiet speed, and there is a single rudder. The craft can, therefore, be steered with the outboard throttles. The Mark III is somewhat cramped, being powered by three 375-bhp 8V71TI engines. Overstressed to produce 600 bhp, the engines create terrible maintenance problems.

Mark III production began under the FY 73 program (65PB731–738 at Peterson), followed by ten boats in FY 75 (65PB751–7510 at Peterson) and eight in FY 77 (65PB771–774 for the Philippines, 775–778

A Sea Specter tested the prototype Hughes 25mm Bushmaster or chain gun. The weapon further forward is a 20mm cannon, and two 0.50-caliber machine guns are visible right aft, turned to fire broadside. (Hughes Helicopters)

for the U.S. Navy). In 1986, the U.S. Navy operated seventeen Mark IIIs, thirteen in special boat units and four in the reserves.

Three Mark IVs, which are 3 feet longer to make room for larger 650-shp 12V71TI engines, were bought in FY 85 for Canal Zone service. They realize the armament potential of the design, with 25mm Mark 88 cannon fore and aft, an 81mm mortar/0.50-caliber machine gun abeam the pilothouse, two 40mm Mark 19 grenade launchers aft, and two pintles just abaft the forward 20mm cannon for two 7.62mm machine guns.

Certainly, the United States is not as well equipped for sublimited or inshore warfare as it was in, say, 1970. However, the Mark IV demonstrates a continuing interest in and capacity for building an emergency force of small combatants. In that sense, the United States is much better equipped than it was in 1960—or 1950.

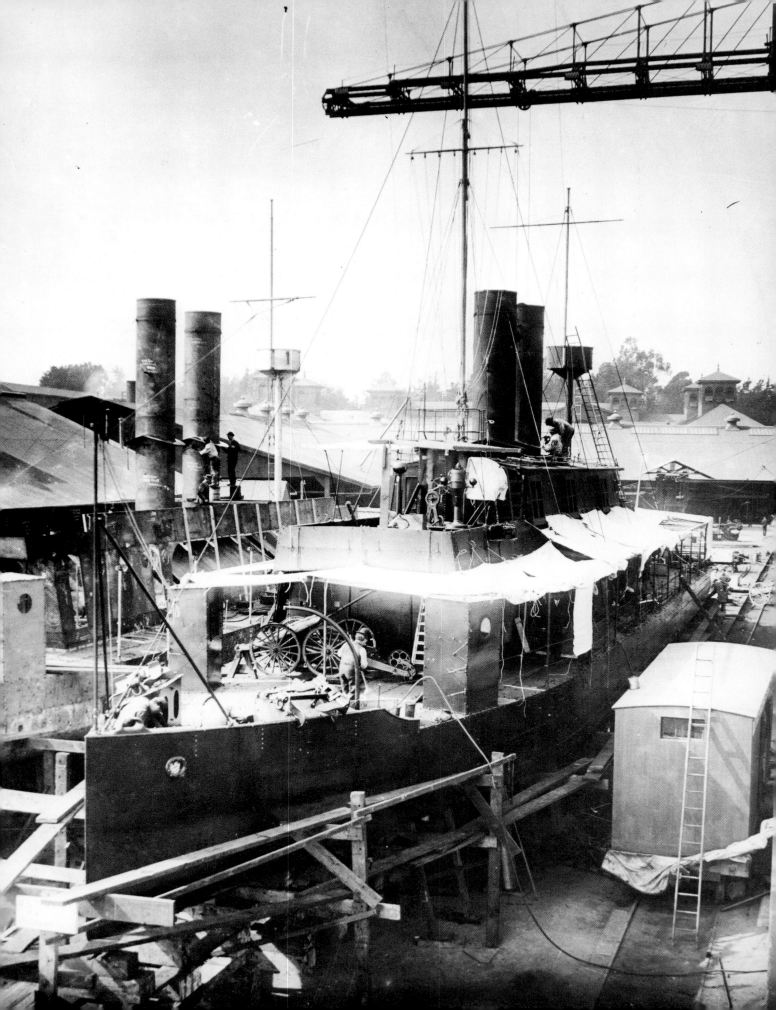

A

Gunboats

Gunboats were never a particularly homogeneous group. Many of the early ones and two of the later ones, the *Erie* and the *Charleston*, were actually small cruisers; they are described in some detail in the appropriate companion volume of this series.[1] This appendix is limited to modern oceangoing gunboats of the *Nashville*, *Wilmington*, *Annapolis*, *Wheeling*, *Dubuque*, *Sacramento*, and *Asheville* classes, and to river gunboats of the *Monocacy*, *Wake*, *Panay*, and *Luzon* classes.

Modern gunboat (PG) designators were assigned as part of the hull number system in July 1921; the new category was defined officially as including all the gunboats and third-class cruisers then on the navy list. Thus PG 1–22 were old gunboats previously numbered as such; PG 23–24 were previously unnumbered gunboats; PG 27–36 were old third-class cruisers (as was the *Topeka*, purchased during the Spanish-American War); and PG 37–42 were ex-Spanish gunboats. PG 43–48 were river gunboats for Chinese service, later renumbered in a separate PR series. The *Fulton* became PG 49 after having been built as a submarine tender; she subsequently became a submarine rescue ship. The last prewar gunboats, PG 50–51, were really small cruisers, the *Erie* and the *Charleston*.

Several PG numbers were not used: 25–26 were almost certainly reserved for the old cruisers *Cincinatti* and *Raleigh*, which were placed on the sale list before they could be redesignated, and it appears that 72–84 were mistakenly reserved for the thirteen *Owasco*-class Coast Guard cutters following a February 1942 directive for BuShips to proceed with their construction (together with seventy-three miscellaneous auxiliaries). These designations were never officially entered. PG 72 was assigned in April 1943 to the ex-yacht *Nourmahal*, acquired by the Coast Guard in 1942 and then transferred to the navy.

The largest yachts taken over during World War II were included in the PG series with the numbers 52–61, 72, and 85. They included PG 61, the *Dauntless*, Admiral King's wartime flagship, and PG 56, the *Williamsburg*, which served as a presidential yacht (AGC 369) from November 1945 through June 1953. She was stricken in 1962 and turned over to the National Science Foundation as an oceanographic research ship. Her hulk still rests in Washington, D.C.

Modern U.S. gunboat operations were concentrated in two areas, the Caribbean and the Far East. In the Caribbean, the Special Service Squadron was sometimes called the State Department's Navy because so many local ambassadors called upon it. The squadron became important after the fleet shifted from the Atlantic to the Pacific in 1919, taking so many of its smaller warships with it. On the Yangtze River of China, gunboats protected U.S. trade from the depredations of warlords, particularly after the Revolution of 1910. The two theaters were very different, and as a result, Yangtze gunboats presented difficult and unusual design problems.[2]

As in the case of the cruiser, gunboat development closely followed changes in U.S. naval and national strategy. Before the Spanish-American War, gunboats supported American citizens abroad, but they had little or no police function. Because the United States did not have foreign bases, her gunboats often transited the open sea and therefore had to be fairly large. On the other hand, the essence of the gunboat, as opposed to the cruiser, was limited cost and size. The first modern U.S. gunboat, the *Petrel*, was essentially a slow, 11.8-knot cruiser armed with cruiser weapons, four 6-inch guns in sponsons. Her two suc-

The river gunboat *Palos* being constructed at Mare Island, 28 April 1913. She was disassembled and shipped to China, where she was rebuilt for service on the Yangtze River. To the left, her sister ship *Monocacy* is being built. Note the wheeled field artillery pieces on her foredeck.

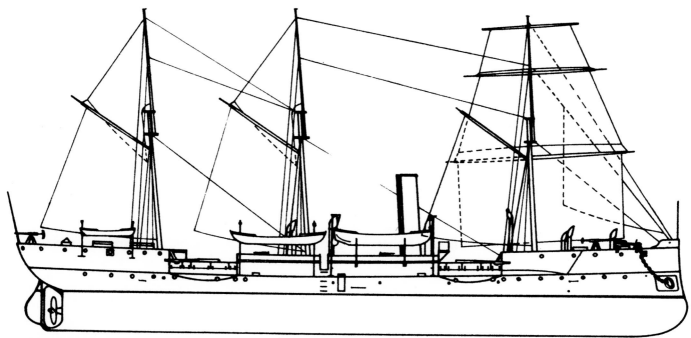

The *Petrel*. (A. D. Baker III)

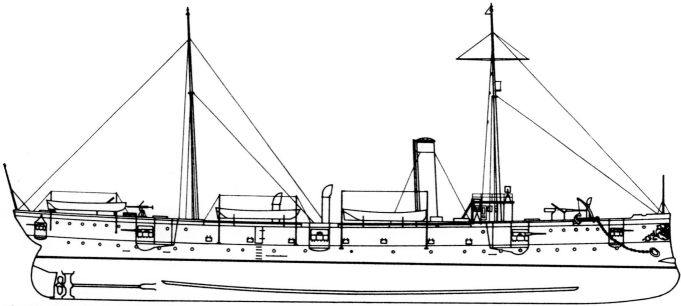

The *Machias*. (A. D. Baker III)

cessors, the *Castine* and the *Machias*, were larger and faster and therefore closer to cruisers in concept, but they were armed with eight 4-inch guns. They did not have sufficient stability, and after completion they had to be ballasted and lengthened by 14 feet.

When Secretary of the Navy Benjamin Tracy adopted a battle fleet strategy in 1889, he ceased the construction of gunboats, which were too specialized for a fleet expected to exert U.S. power wherever it was needed abroad. That was not entirely practical, and construction of specialized gunboats resumed with his successor's 1893 (FY 94) program. It included three shallow-draft, lightly armored 1,200-tonners built specifically for service in the Far East.

Gunboat operation in the Far East required long range, shallow draft (Chinese coastal rivers were shallow), and roomy, well-ventilated quarters. It was generally accepted, at least at first, that a gunboat

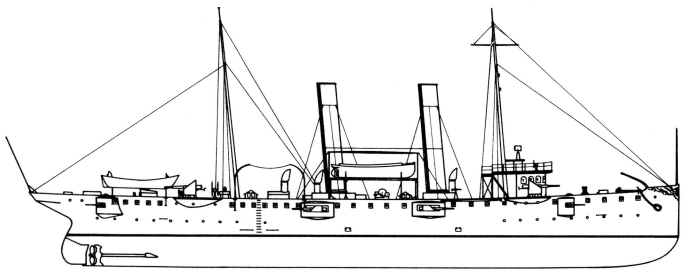

The *Nashville*. (A. D. Baker III)

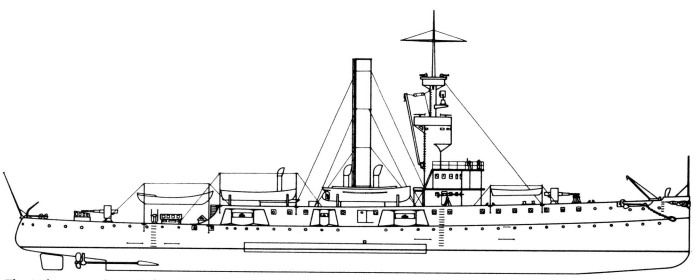

The *Wilmington*. (A. D. Baker III)

would often have to transport troops, so it had to be able to carry considerably more than its nominal complement. All of this had to be achieved on a small displacement, since every ton or every dollar spent on gunboats had to be extracted from the main fleet, the war-fighting force.

The *Nashville*, the first of the new gunboats, had unusual quadruple-expansion engines, one cylinder of which could be disconnected so that she could cruise on the other three cylinders. Unlike her predecessors, she had a flush deck for more enclosed volume (to accommodate troops, for example). C&R proposed composite construction, wood on steel frame, but the Congressional Authorizing Act specified a steel hull.[3] The program initially consisted of two *Nashville*s and one specialized Chinese river and coastal gunboat, but in fact, two of the latter and only one *Nashville* were built. Her design descent from earlier cruiser/gunboat showed in the provision of a bow torpedo tube.

The other two ships, the *Wilmington* and the *Helena*, were the first specialized China gunboats. They had a limited draft of 9 feet, which would allow them to operate in river mouths but not up-river, twin screws, and large rudders for maneuverability in narrow waters. Their unique military masts were intended to provide a commanding view over the 50-foot banks and dikes of the Yangtze. Unlike the *Nashville*, they had shallow hulls with substantial forecastles extending about three-quarters of their length

The *Annapolis*. (A. D. Baker III)

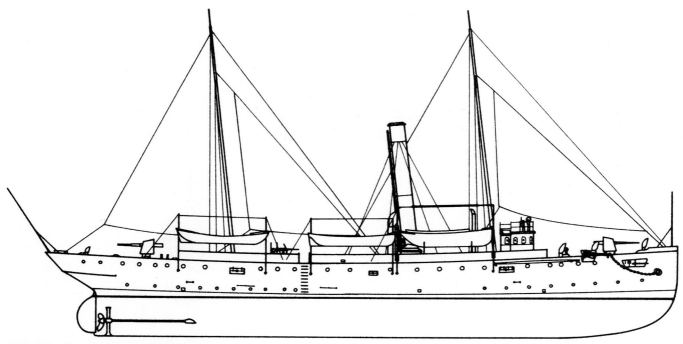

The *Wheeling*. (A. D. Baker III)

not only for crews but also for refugees and special landing parties. Their unusually large reserve buoyancy, 209 percent, was a measure of the need for internal volume. The forecastle did not extend all the way aft; this minimized hull weight and also reduced loading at the stern, which was cut away (and hence not very buoyant) to improve water flow over the propellers, particularly in shallow water.

All three ships were large enough for $5/16$-inch, watertight protective decks with $3/8$-inch slopes. They

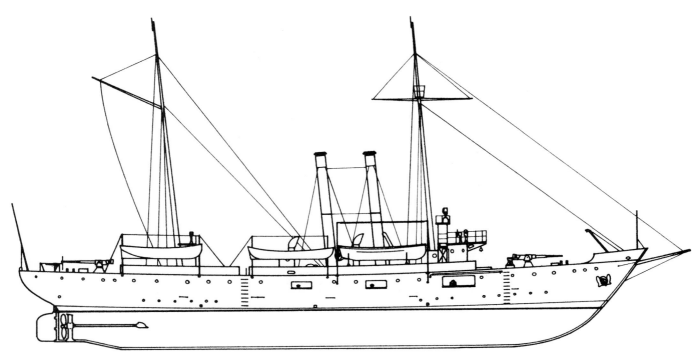

The *Dubuque*. (A. D. Baker III)

were armed with eight 4-inch/40 guns. The China gunboats were unique in having short belts amidships, in this case 1 inch think.

Six more gunboats were authorized in 1895 (FY 96). All had composite hulls (steel above the waterline, wood below), displaced about 1,000 tons, were entirely unarmored, and mounted six 4-inch/40 guns. The Navy Department offered bidders two alternative designs, in effect alternative approaches to balancing speed and endurance: a single-screw gunboat with full sail power and a twin-screw steamer with steadying sails only. Each had to make 12 knots under steam. The nonsail steamer carried 120 tons more coal, against which the barkentine-rigged gunboat required 15 additional tons of spars as well as more length to carry sails, that is, greater hull weight. The four barkentines, *Annapolis* class gunboats, were built on the East Coast; Union Iron Works built their two virtually sailless equivalents, the *Wheeling* class, on the West Coast.

The last two gunboats of this period were the slightly larger *Dubuque* and *Paducah*, authorized in 1902, specifically designed for police and surveying duties in the Caribbean. They were characterized by an unusual bow, high and rounded, and two tall thin funnels, but otherwise they generally resembled the *Wheeling* class.

U.S. gunboat construction reached a high point after the United States seized the Philippines and became the preeminent power in the Caribbean in 1898: the navy built six *Denver*-class peace cruisers, which were actually large oceangoing gunboats and were later redesignated as such. The U.S. position in the Philippines increased U.S. economic involvement in China, thus paving the way for the construction of specialized Yangtze River gunboats.

The *Denver*s were designed primarily for the new American sphere of influence in the Caribbean. They were effective, but large and expensive to operate. In the fall of 1909 the General Board conceived a less expensive oceangoing gunboat, its size, cost, and complement halved by eliminating the requirement to carry troops and by a drastic reduction in gun battery, from ten 5-inch to three 4-inch guns. Authorized in 1911 (FY 12), the ship became the *Sacramento* (PG 19). As might be expected, the design emphasized good sea-keeping, with high freeboard and an unusually great coal capacity for her size, 420 tons. All three of her 4-inch guns were mounted on the centerline.

The *Sacramento* was originally to have been substantially larger, but Cramp, the only bidder, was unable to come within the stipulated $500,000.[4] Secretary of the Navy Meyer asked whether the austere Cramp design would meet navy requirements. The General Board offered to cut speed to 12 knots and endurance to 4,000 nautical miles at 10 knots to stay within the cost limits, and the bureaus developed a new and even more austere design.

The board hoped to do better with the next gunboat, which it sought under the FY 14 program. This time it offered to accept much less endurance, 2,000 nautical miles at 10 knots, in exchange for another 4-inch gun. No gunboat was included in the 1914 program, so in April 1913, the General Board proposed new characteristics for an FY 1915 gunboat, this time to include six 5-inch guns. BuOrd was unenthusiastic; the standard 5-inch/51, weighing about two and a half times as much as the 4-inch gun of the *Sacramento*, would impose considerable growth. The board suggested instead the obsolete 5-inch/40, which did not weigh too much more than the 4-inch gun, then retreated to suggest using the production 4-inch. The 1915 characteristics also showed twin rather than single screws and a complement of 200 rather than 150 men. No gunboats were included in the 1915 program, so in May 1914, the General Board repeated the same characteristics for what it hoped would be the FY 1916 gunboat, except with six 4-inch/50 guns.

At this time, the board hoped for four gunboats in FY 16. The United States was becoming increasingly involved in Mexico's affairs, and ships were badly needed in Mexican and Central American waters. Although thirty ships were officially listed as gunboats, few were available for general service. Seven were assigned to local naval militias, three just having been withdrawn from that service.

Eventually, the board had to settle for a slightly modified *Sacramento*. In October 1915, her commander recommended that any repeat ship be lengthened by about 15 feet to accommodate two more boilers for a 25 percent excess capacity. One boiler could be overhauled while the ship steamed, which would permit it to operate much longer on a distant station. The additional 75 tons would also provide space for a fourth 4-inch gun. As an indicator of the conditions of gunboat operation off Mexico, the captain of the *Sacramento* complained that his standard 2-kw radio was not nearly powerful enough. It had been necessary to station a gunboat at Tuxpan to relay between Vera Cruz and Tampico, and his ship had been unable to reach the *Marietta* at Progresso. He wanted steaming endurance doubled to 8,000 nautical miles at 10 knots.

In October 1915, the General Board adopted these suggestions in its gunboat 1917 characteristics, except that it retained the three-gun battery. BuEng argued that four boilers would be impractical, since each would have to be extremely small and hence could not burn oil efficiently. The design was therefore developed with three somewhat larger boilers. It was authorized for construction under the FY 17 program as the *Asheville*. One sister, the *Tulsa*, built to identical characteristics, followed in 1918 (FY 19).

The board did try to shift back to a more impressive armament; in November 1917, it asked how difficult it would be to substitute two, preferably three, 5-inch/51 for the three 4-inch guns, and to increase speed from 12 to 16 knots. These improvements were certainly possible, but C&R discouraged them. It could fit only two 5-inch/51s into the existing hull, since a third gun, right aft, would be too close to the rudder head and could not be supported to absorb its recoil force. Ammunition stowage would have to be reduced from 350 to 250 rounds per gun.

Both ships were built, at the Charleston Navy Yard, as slightly improved *Sacramento*s.

The U.S. naval presence in the Yangtze basin of China dates from 1854, when British and American envoys demanded the right to trade there. The Chinese acceded after losing the Second Opium War, and after 1860 Western trade, protected by gunboats, penetrated the Yangtze. Ships of the U.S. East India Squadron first steamed up the Yangtze in the spring of 1861, and from then on the U.S. Navy maintained a China station. It had no specialized shallow-draft gunboats, although two Civil War shallow-draft paddle-wheelers, the *Ashuelot* and *Monocacy*, were used on the river for many years.[5]

Gunboat protection became more important as anarchy reared in China, leading ultimately to the rise of the warlords after 1910. From the point of view of gunboat design, what mattered was how far up the river trade had to be protected. The Yangtze could be divided into four regions: from Shanghai to Hankow (600 miles); from Hankow to Ichang (370 miles); from Ichang to Chungking (430 miles); and beyond Chungking to Suifu (300 miles). The fourth region was never patrolled, for only small craft could penetrate the rocks leading up to it. The middle stretch, Hankow to Ichang, was considered treacherous with its rapids and deep gorges. Pirates or guerrillas could fire at river traffic from the cliffs; U.S. river gunboats built in the 1920s were armed with AA guns specifically intended for returning their fire.

This division of the river explains why the *Wilmington*s, with their 9-foot draft, were described as river gunboats. They were intended for the lower river. Further up, a gunboat had to be limited to about half that draft, which the new special craft built after 1911 had.

The United States first obtained shallow-draft gunboats as a result of her victory over Spain in 1898. War prizes included a variety of small gunboats, notably a 620-tonner, the *Elcano* (1884); two 350-tonners, the *Quiros* and *Villalobos* (1895–96, length 138-0, draft 9-3); six 243-tonners, the *Arayal, Belusan, Callao, Pampanga, Paragua,* and *Samar* (1888, length 115-0, draft 7-8); three 170-tonners, the *Albay, Mar-*

The *Palos* at Chungking, around 1935, with a Chinese sampan alongside. (Vice Admiral T. G. W. Settle, USN [Ret.])

iveles, and *Mindoro* (1885–86, length 100-0, draft 7-8); one 162-tonner, the *Panay* (1885); three 150-tonners, the *Calamianes*, *Leyte*, and *Manileno* (1888 and 1885); two 100-tonners, the *Alvarado* and *Sandoval* (1895); and the 83-ton *Mindanao*. The launch dates noted after each name suggest that many of these ships, even those fit for U.S. service, had to be replaced within a few years.

The three largest gunboats were considered the best suited to Chinese river service, and by 1904, two of them were operating there. The usefulness of boats in the Philippines was governed by their draft, 7 feet being the critical figure for bars and shoals. Many rivers were navigable 20 to 30 miles upstream of their bars, at depths of 10 to 60 feet. Thus the *Mariveles*, which could be trimmed down to draw less than 7 feet, was very useful. The *Pampanga*s, which were better sea boats, rendered more service than the larger *Villalobos* because of their lighter draft. A still lesser draft, about 3 feet, was needed to operate on the lakes of Mindanao and the rivers feeding them. Similar characteristics would have to be given gunboats operating on the interior rivers of China.

The General Board first asked for specialized river gunboats in October 1903, when it endorsed an Asiatic Fleet proposal for two to work the interior rivers of China. As of the following September, there were only two American gunboats on the Yangtze, one of which, the *Elcano*, badly needed major repairs. The two specially built gunboats required new boilers. Even with them, however, the boats would be too large and would draw far too much water. Having seen foreign shallow-draft gunboats, the captain of the *Elcano* advocated a U.S. equivalent, limited in dimensions to 140 x 28 x 3 feet fully loaded, so that it could negotiate the twists and turns of the upper river. It would have a flat bottom and a spoon bow, and it would be armed with two 6-pounders (57mm) and four heavy machine guns (1-pounder pom-poms)—this in spite of the fact that U.S. business interests did not extend above Hankow, that is, into the middle stretch of the river.

In December 1909, answering a query from the secretary of the navy, the General Board proposed building an improved *Helena*, perhaps with better sea-keeping qualities, and two river gunboats. At this time, the Asiatic station operated the two *Wilmington*s, the 620-ton *Elcano*, the 400-ton *Quiros* and *Villalobos*, five 250-tonners, and two 170-tonners—and no light-draft gunboats built specifically for river service. The board envisioned an active force of two large *Helena*s for rivers, two light-draft gunboats for the upper rivers and shallow lakes of China's interior, and four small gunboats of the *Samar* and *Quiros* type for cruising in the Philippines and on Chinese rivers. These would be backed by two small reserve gunboats of 200 to 400 tons.

Since 1898, the navy had had the authority to build a small gunboat (PG 16) for the Great Lakes, but no such ship had been built in view of the long-standing demilitarization of the U.S.-Canadian border. The FY 12 naval bill transferred this appropriation to a new shallow-draft gunboat for China. The same act authorized the *Sacramento*, which might be loosely construed as an improved *Helena*.

The navy bought plans from the British firm Yarrow, which had designed and built all the British river gunboats and several French ones. The *Monocacy* and *Palos* were built at Mare Island, disassembled, and re-erected at Shanghai. They served into the late 1930s.

The two British-designed gunboats were criticized for being slow. Their engine foundations were too light, so that they lost alignment easily. Bows, low and bluff, took on water in rough weather or in bad swirls on the upper Yangtze. Cross-currents and swirls often threw them off course; gunboat commanders suggested that this could be greatly reduced by using a spoon bow with the forefoot cut away, as in upper Yangtze craft.

American business in China continued to expand. So did local instability. By November 1914, the ex-Spanish *Callao* was beyond repair, and the *Pampanga* and *Samar* soon would be. Even these ships could only travel about 200 miles above Canton, no matter what the time of year. A boat with a shallower draft could go as much as 700 miles up-river and could also travel the Southwest and North rivers. In August 1915, the Asiatic Fleet asked for three new river gunboats in the next appropriations bill to replace old Spanish boats, two on the Yangtze, the *Quiros* and *Villalobos*, and one on the West River. It argued that future gunboats would need greater speed, at least 14 knots, to pass upper Yangtze rapids as well as other internal waters with rapids, swirls, and strong currents. By 1916, the Asiatic Fleet wanted a 16-knot gunboat burning coal only.

In November 1917, the coast inspector at Shanghai, W. Ferdinand Tyler, published a table of limiting dimensions for Yangtze steamers: 210 x 35 x 4.5 feet. Minimum speed for safe navigation was 10 knots for a length of under 80 feet; 11 knots for 80 to 130 feet; 12 knots for 130 to 210 feet. These speeds would not push a boat over the main rapids without warping. Boats had to have flat bottoms and spoon bows, twin screws for safety, and if over 130 feet in length, at least two boilers. These figures defined the limits of U.S. river gunboat design.

In June 1920, the Asiatic Fleet asked for six river gunboats: two modified *Monocacy*s for the upper reaches of the Yangtze and four larger ones, 200 feet in length with a 4-foot draft. The boats would be armed with one 3-inch AA gun forward to deal with cliff-top attackers and two 3- or 6-pound guns.

Characteristics adopted in October 1922 for two projected river gunboats to be built under the FY 24 program called for two 3-inch/50 guns and six machine guns; dimensions limited to 200 x 35 x 4.5 feet; a speed of 15 knots; oil fuel; and a radius of action of at least 600 nautical miles at 10 knots. The Yangtze Patrol considered this much too large, and in July 1923, it called for a reduction to 150 x 27 x 4 feet—but an increase to 16 knots. To help reduce weight, the relatively massive 3-inch/50 could be replaced by the much smaller 3-inch/23.

Meanwhile, costs were reestimated on the basis of boat construction in China. The same amount would buy four river gunboats. The secretary of the navy decided to ask for five gunboats and then, in July 1923, for six. By late 1923, the projected group of gunboats had been split into three: the 150-footer, which would negotiate the twists of the upper river to Chungking, even at low water, to provide year-round coverage; the 200-footer, which would function as a flagship; and a 180-footer, which was considered the minimum to meet the speed and range requirements, the latter now 1,000 nautical miles at 10 knots. At this stage the 200-footer was to have been armed with two 3-inch/50 guns.

These boats would not suffice, since trade now extended beyond Chungking to Suifu and no conventional gunboat could negotiate the right-angle turns necessary to clear rocks in this stretch. The Yangtze Patrol therefore proposed an additional gunboat, a 60- to 70-foot motorboat, 18 to 20 knots. It was never built, and the big steam gunboats were not actually built until the FY 26 program.

Preliminary Design developed two series of river gunboat designs at this time. Although they were not adopted, they shaped the characteristics adopted by the General Board, and they indicate the dilemmas of these specialized craft. At first, in 1922, a 200 x 35 x 4.5–foot, 617-ton boat was designed. Even with 2,100 shp, it could not meet the required speed of 15 knots; it was limited to 14.75. Extra machinery weight could be provided by filling out the underbody and increasing displacement to 650 tons, but that would cost more speed. On 617 tons, a full 38.1 percent of normal displacement went to machinery, compared with 44.9 percent for hull, fittings, and margin. Preliminary Design argued that the required speed could be achieved only if BuEng could cut its weight. For example, if engineering weight were cut by 84 tons and shp to 1,800, then the gunboat might displace only 450 tons and make the required 15 knots at a substantially lower cost.

Pressure from the Yangtze Patrol resulted in Preliminary Design trying to cut its gunboat to 100 feet and 388 tons. In that case, only 4 tons could be devoted to machinery—BuEng wanted 200. However unpleasant, a larger design would have to be accepted. In May and June 1923, Preliminary Design tried two

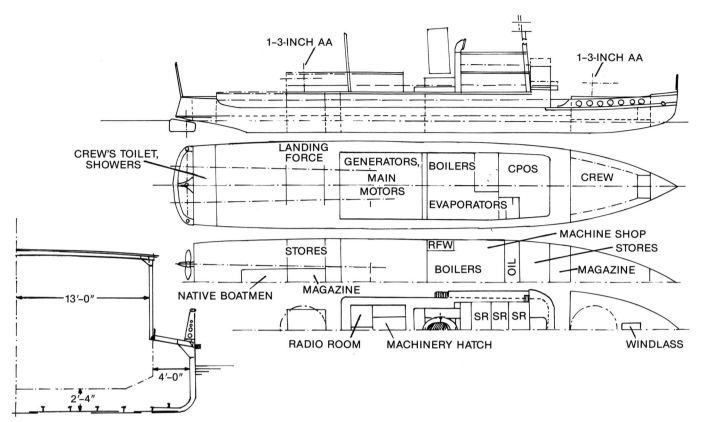

C&R's design for a FY 24 China river gunboat, 7 July 1923. It was to displace 490 tons at 180 (lwl) x 34 x 4.5 feet. Turboelectric machinery would drive surface propellers about 3,000 shp for 16 knots, and radius would be 1,535 nautical miles at 10 knots. Armament would have consisted of two 3-inch AA guns and two machine guns. Unfortunately, estimated weights added up to 528 tons for a draft of about 4 feet 9 inches. Complement would have been four officers, twelve CPOs, thirty crewmen, two native pilots, and a landing force of twelve marines; there was also accommodation for refugees. The FY 24 characteristics, to which this boat was designed, were prepared late in 1922. (Norman Friedman)

alternatives, a 200 x 34 x 4.5–foot ship capable of 16 knots in deep water (1,500 shp, using surface propellers), and a 190 x 34 x 4.5–foot ship capable of 15 knots with surface propellers.[6] The second alternative was later cut to 180 feet.

Preliminary Design preferred turboelectric drive to reduction gearing, because the propellers had to turn relatively slowly, which would entail a high gear ratio, and because repairs to gearing would be difficult in isolated areas. Turboelectric drive was also well adapted to surface (as opposed to tunnel) propellers, relatively high in the ship. Reciprocating machinery was rejected as too heavy, although some designers considered it lighter than turbines plus generators and motors. Even with the turbines, the gunboat would be unacceptably heavy, that is, would draw too much water. Turboelectric drive entailed other problems, too. For example, the diameter of the motors, if they were placed amidships, required a large beam. Further analysis showed that the proposed gunboats would devote more of their displacement to machinery than light cruisers or, in one case, destroyers. The percentage assigned to hull weight was less than that assigned any class other than destroyers.

Moreover, the projected characteristics compared poorly with those of Japanese gunboats already in service. An internal Preliminary Design memorandum of October 1923 suggested that "the adoption of any of the above suggested designs will be an admission of our inability to design and construct for our service a vessel of as moderate dimensions and power as those used by foreign governments. It will be an admission, moreover, that our principal weakness is in the weight required for reliable machinery . . . [and perhaps] an unnecessary 'holding of the candle to our own shame' as designers."

After a series of attempts to reach a satisfactory compromise between light machinery weight and reliability (with C&R and BuEng on opposite sides), characteristics for FY 26 gunboats approved in May 1924 showed diesel engines, two 1,000-bhp units de-

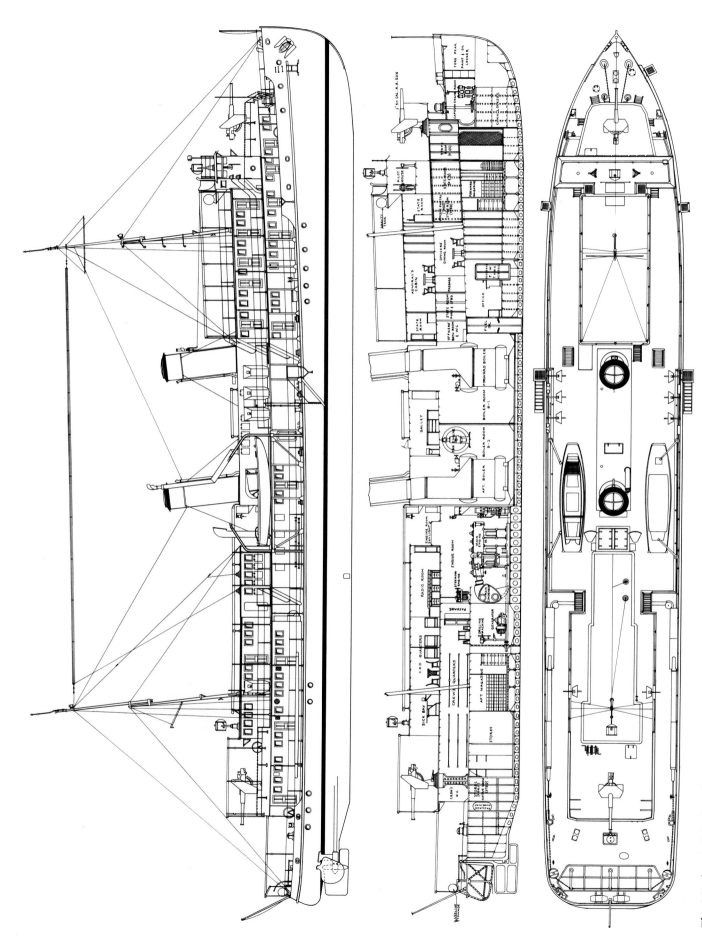

The *Mindanao* (PR 8), July 1940. (A. D. Baker III)

The *Panay* was surely the most famous of the U.S. Yangtze River gunboats. Here she is at Ichang, China, around 1932, alongside the Standard Oil pier and forward of the *Guam*. The four vertical objects fore and aft are machine-gun shields of a design apparently unique to the Yangtze Patrol.

rated to a total of 1,400 bhp to reduce wear and increase reliability. Displacement was 380 tons; dimensions in feet-inches were 180-0 x 27-0 x 4-5, with a spoon bow and a tunnel stern; and armament was two 3-inch/23 guns. The General Board decided to buy two flagships to replace the *Helena* and the converted yacht *Isabel*, two 180-footers, and two 150-footers, which, although not as capable as the larger boats, could be expected to operate above Ichang and to negotiate the river at all stages.

By this time, new surveys of the river had permitted a relaxation of the draft and speed requirements to 5 feet 3 inches and 14.5 knots. Greater draft permitted heavier machinery, and the diesels could be replaced by reciprocating steam engines. Although the diesels would have been the most reliable and modern, the lack of shore facilities was worrisome.

Only the Kiangnan Dock Company of Shanghai presented satisfactory bids. It was by far the most successful builder of commercial Yangtze steamers, and it had standard designs, including a 180-footer. All of the winning bids were class 2 (contractor's design to navy requirements): a 150-footer drawing 5 feet 3 inches and capable of 14.5 knots (the *Guam*, later renamed *Wake*, and *Tutuila*); two 180-footers drawing 5 feet 6 inches and capable of 15 knots (the *Panay* and *Oahu*); and two 198-footers drawing 6 feet and capable of 16 knots (the *Luzon* and *Mindanao*). The 150-footers were armed with two 3-inch/23s. However, the larger boats had 3-inch/50s, CinC Asiatic Fleet having argued in August 1926 that, in view of the growing sophistication of the warlords, air attacks might well be made in future.

The threat continued to grow, and in April 1933, the Yangtze Patrol asked for four more 150-footers, noting that the British had twelve river gunboats and the Japanese, in addition to other warships, eleven. They were never built, although further craft were included in some General Board long-range programs. The Yangtze was not important enough, not with the main U.S. Fleet in need of modernization. The river gunboat was carried as the lowest-priority C&R design study.

Japanese bombers sank the *Panay* on 12 December 1937, almost precipitating hostilities. The *Tutuila* moved up-river to Chungking in August 1938; she was marooned there by the advancing Japanese and ultimately, in March 1942, handed over to the Chinese as the *Mei Yuan*. Her sister *Wake* was withdrawn to Shanghai, where the Japanese captured her upon the outbreak of war. The three surviving larger gunboats were withdrawn to Manila, the two from Shanghai, the *Luzon* and *Oahu*, arriving 5 December 1941, the one from Hong Kong, the *Mindanao*, arriving on the tenth. All were scuttled to avoid capture by the Japanese.

In March 1942, with the ASW program accelerating but hardly realized, the United States obtained ten British-built corvettes, PG 62–71, under reverse lend-lease. Meanwhile, the Hyde Park Agreement of February 1942 allocated forty-eight Canadian-built escorts to the United States: fifteen Flower-class corvettes (PG 86–100); ten River-class frigates (PG 101–10); fifteen *Algerine*-class minesweepers (AM 325–39), none of which actually entered U.S. service; and eight

The USS *Surprise*, a corvette slightly larger than a PCE and rated as a gunboat, emerges after a refit at Charleston Navy Yard, 8 May 1944. Alterations have been circled: they include the two 20mm guns in the waist, the new surface-search radar (SL) on a light tripod mast, and the 3-inch/50 aft. Note the hedgehog to starboard of the bridge. The box forward of the bridge is a British-style sonar hut.

U.S. corvettes varied considerably in configuration. This is the *Prudent* (PG 96), 6 April 1945.

Fairmile-type subchasers (SC 1466–73). Of the Canadian-built ships, seven corvettes (PG 88, 90–91, 97–100) were turned over to the Royal Navy upon completion, as were eight Rivers (PG 103–10).

British-built corvettes were armed with a 4-inch/50 forward and a 3-inch/10 aft, Canadian ships with two 3-inch/50s. The latter also had a raised gun/Hedgehog platform forward.

Ten River-class frigates were ordered from Canada as PG 101–10, but only two were retained. They became the prototype U.S. Navy frigates PF 1 and 2.

As in World War I, shipbuilding resources limited escort construction. One way to increase construction was to use merchant shipyards; it was suggested that some Maritime Commission shipbuilding resources might more usefully be turned to escort than to merchant ship construction. After all, if the escorts beat the U-boats, fewer replacement merchant ships would be needed. Gibbs & Cox offered to adapt the British frigate (River-class) design to Maritime Commission methods. The commission decided to build seventy escort ships and to complete fifty during 1943; the former figure was later increased to one hundred. They became the *Tacoma* class (PG 111–210, later PF 3–102). They enjoyed a much longer range than destroyer escorts, were considerably warmer—an unfortunate attribute in the tropics—and had a much wider turning circle.

The navy protested that building any more than seventy frigates would throw its program off balance; the Maritime Commission transferred thirty-one ships (PF 72–92) to the Royal Navy as the "Colony" class. The remaining U.S. ships (PF 95–98 were canceled in December 1943) were manned by the Coast Guard. PF 17, 18, 20, 23, 24, 28–33, 40, 41, 66–69, 71, 93, 94, 99–102 operated as weather ships, a balloon hangar replacing the after 3-inch/50 gun; they had four rather than nine 20mm guns.

In the summer of 1945, with the Battle of the Atlantic over, almost all the unmodified U.S. frigates (twenty-eight ships) were transferred to the Soviet Union, which classed them as escorts (EK) rather than as "guard ships" (SKR). They were returned in 1949 and laid up in Japan; one was lost in 1948 while under Soviet control. Thirteen were reactivated for local ASW escort: PF 3–5, 7, 8, 21, 22, 27, 36, 46, 47, 51, and 70. Three were later transferred to Korea along with PF 48 and 49, which saw no postwar U.S. service, two to Thailand, two to Colombia, and the rest to Japan. Each was active for a year or more under the U.S. flag, some participating in the Wonsan landings.

B

Minor Acquired Patrol Craft: SP, YP, PY, PYc

In both world wars the U.S. Navy acquired numerous yachts and other small boats for use as patrol craft. As recounted in chapter 1, the World War I acquisition program of SPs was unusual in that a few of the motorboats had been built specifically in the hope that they would be taken over as navy patrol boats.[1]

On 17 July 1920, when the present system of ship designators was established, small patrol craft were designated YPs, or naval district yard patrol boats. Large yachts retained by the navy were designated PY (patrol yacht). At this time, ten such craft survived: five acquired in 1898 during the Spanish-American War, and six in 1917 (PY 6–11). The largest were the presidential yacht *Mayflower* (PY 1, 2,690 tons, sold in 1931);[2] the *Nokomis* (PY 6, SP 609, 910 tons, stricken in 1938); and the *Isabel* (PY 10). A destroyer-like yacht often classified as a destroyer, the *Isabel* was built by Bath Iron Works in 1917 and sold for scrap only in 1946.[3]

In theory, the commandants of the naval districts were responsible for local patrols. For example, the First Naval District, New England, included local defense forces of minesweepers, harbor patrol vessels, harbor entrance control posts, magnetic loops, and section bases; the Northern Ship Lane Patrol; the Northern Air Patrol; the Control Force for the Cape Cod Canal; and a coastal picket patrol.[4] In 1940, the Coast Guard and navy began to buy motor yachts for conversion.

The PY designation survived into World War II, and smaller yachts acquired at that time were designated PYc (patrol yacht, coastal). World War II acquisitions consisted of PY 12–32 (October 1940 through December 1942) and PYc 1–52, the latter including the former PC 454–56, 458, 460, 509, and 826 (June 1941 through March 1942). The distinction between PY and YP was not always clear, and some fairly large yachts originally designated SC were redesignated in the YP series. PY 12, the *Sylph*, had been designated YP 71; PYc 30, the *Vagrant*, had been YP 258. The designation PY 30 was not used. PYc 23, 24, and 34 became YP 261, 454, and 425; PYc 32 and 33 became the miscellaneous auxiliaries YAG 14 and 13, respectively.

The designations YP 1, 100–101, 176–77, 182, 190, 193, 201, 223, and 396–97 were not used. YP 2, 43, 68, 93, 179, 181, 185, and 214 had all been taken over in 1917 as SPs (the high numbers were translated SP numbers), and YP 3 and 4 were miscellaneous small boats.

The YP series included former Coast Guard boats taken over in 1933–36, mainly standard 75-footers;[5] navy-built 75-footers for Naval Academy training (YP 78–82, 99, 242–46, and 583–91);[6] plus a series of tuna boats built by the navy in 1944–45 for Pacific service (YP 618–46).[7] Fittingly, YP 66 was the 66-foot motor launch built in 1916–17 to illustrate proper civilian patrol boat design and construction practice. The YP designation was also applied to some former PT boats: YP 106 was the former Huckins prototype, PT 69 (the designation changed on 24 September 1941), and YP 107 was the former Higgins dream boat, PT 70 (24 September 1941). YP 110 was the Philadelphia Navy Yard's aluminum PT 8.

YP 653 was a 63-foot aircraft-rescue boat, which reverted to small-boat status in 1952.

The 20mm cannon armed almost all U.S. World War II small combatants, both those acquired commercially and those built for the navy. The 20mm was valued because it could be bolted down on any strong deck; it required external power, if at all, only to power its lead-computing sight. This is Elco's quadruple 20mm mount, the Thunderbolt. The guns are staggered so that their magazines clear each other, and a Mark 14 lead-computing sight has been installed above them. (Al Ross)

The converted yacht *Isabel* lay at the upper end of the patrol craft spectrum. She was so large (710 tons) and so fast (28.8 knots) that she served as a destroyer in World War I, although designated a yacht (PY 10). Built by Bath Ironworks, she was acquired before completion from her owner, John North Willys, and commissioned on 28 December 1917. The armament initially proposed was two 4-inch/50s, two 1-pounders, and two machine guns. However, as a destroyer she had four 3-inch/50s, two Lewis machine guns (0.30-caliber), a Y-gun, and two twin 18-inch torpedo tubes. Although this photograph is undated, it probably shows her at the end of World War I, with a destroyer-style wind deflector on her bridge. The *Isabel* was laid up in 1920, but recommissioned in 1921 for duty with the Asiatic Fleet as part of the Yangtze Patrol. She served as flagship of the Asiatic Fleet for a time in the 1930s, and shortly after the outbreak of war steamed south from Manila, having escaped damage in the bombing of the Cavite Navy Yard. She served briefly as an escort in the Dutch East Indies, and then in Australian waters for the rest of the war. By 1922 her torpedo tubes had been removed and two 3-inch/23 antiaircraft guns added; by 1930 two of her four 3-inch/50s had been landed, the remaining guns moved to her centerline.

Gaps in this series were filled by civilian yachts and motorboats. YP 105 was the ex-yacht *Sybarita*, initially designated PC 510 and redesignated on 8 October 1941. Ex-fishing boats taken over as coastal minesweepers (AMb 2–22) became YP 360–80 on 1 May 1942; similarly, AMc 143–48 became YP 381–86, and AMc 200–202 became YP 387–89 at the same time. Note that YP 150 became AMc 149, and later IX 177.

Several were accidentally lost in wartime: YP 26, 47, 72–74, 77, 88, 94–95, 128, 183, 205, 235, 270, 336, 405, 422, 426, 438, 453, 481, 492, 577, and 636.

War losses consisted of the following: YP 16–17 on 13 December 1941; 97 in March 1942; 277 in May 1942; 284 in October 1942; 345 in October 1942; 346 in September 1942; and 389 in June 1942.

Lost to the perils of the sea were YP 279–81, 289, 331, and 520.

YP 99 and 242–43 were transferred to Peru in November 1958. YP 244–45 were transferred to Chile in June 1959. All were Annapolis-type 75-footers. YP 162 and 539 were lend-leased to Panama. They were discarded in 1947.

Postwar YPs were specially built for navigation

YP 25, typical of numerous ex–Coast Guard patrol boats, presents an interesting contrast to the navy designs described in this book. It was built for the Coast Guard as CG 142 at Benton Harbor, Michigan, in 1924, and transferred to the navy in 1934. YP 25 was 74 feet 11 inches long, one of a class of 203 75-footers built for inshore patrol. The class was designed by John Trumpy, then of Mathis Yacht Building in Camden, New Jersey. The boats made 15.7 knots on trials in 1924 and carried a crew of eight. By 1941, only thirty-six remained in Coast Guard service. Coast Guard units were typically armed with one 20mm cannon or a 1-pounder and two depth-charge tracks.

and maneuvering training at the U.S. Naval Academy and later at the officer candidate school in Newport, Rhode Island. Beginning with YP 647–52, they were built by Wheeler and completed in 1945–46.[8] Five new YPs were authorized under the FY 56 program, another five under FY 57, and two under FY 59 (SCB 139: YP 654–65).[9] Further craft of this type, YP 666–67, were authorized under the FY 65 program; 668 under FY 69; 669–72 under FY 70; and 673–75 under the FY 77 program. A substantially larger replacement design was developed, and YP 676 was ordered on 15 October 1982, followed by 677–82 on 25 May 1983, 683–95 on 12 June 1984, and 696–702 on 13 September 1985.[10]

The YPs are to be available as patrol craft in wartime, and as new YPs are delivered, existing ones are being refitted as emergency minesweepers under the COOP (craft of opportunity) program or as training craft for other navy schools, with uncommissioned small craft serial numbers in place of the YP numbers. YP 659 and 660 were transferred to COOP in September and November 1985; fifteen others are to follow in FY 86–88. YP 655–56 and 667 were redesignated as uncommissioned boats in August 1985 and transferred to the surface warfare officers' school at San Diego.

Just as the U.S. Navy badly needed small craft and so had to acquire yachts, it was obliged to provide ex-yachts to its allies under lend-lease. It lent the Royal Navy the following: two large, 190-foot yachts in May 1941, a 63-foot yacht and eight motorboats in June, a 54-foot motorboat in July, a large yacht in January 1942, four stock 38-foot and twelve stock 30-foot motor cruisers in March 1942, ten used motor cruisers during the period from April to June 1942, three used motor launches in October 1942, seven 34-foot stock motor launches in November 1942, and five stock 36-foot motor cruisers from October 1943 through January 1944. None of these craft figures in U.S. PY, YP, or PYc lists. Ecuador received PY 18, a 172-foot yacht, in January 1944 and PYc 8, a 185-foot coastal yacht (USS *Opal*), in September 1943.

C

Crash Boats and Pickets

Although not strictly small combatants, American aircraft rescue boats (crash boats) are connected with this story in three ways. First, as recounted in chapter 4, they provided C&R with most of its pre–World War II fast boat experience. Similarly, much of the postwar BuShips research in fast planing craft supported future crash boats. Second, the big World War II crash boats were used as small subchasers and, after the war, were transferred abroad as fast patrol boats. Although they were not used as such, they were also proposed as the models for Vietnam-era fast patrol boats. During World War II, they also served as minor combatants for beach-jumpers. Finally, smaller wartime crash boats evolved into wartime pickets, which in turn evolved into the postwar pickets that served in Vietnamese harbor security units. Modern crash boats include a 65-foot class based on the Sewart 65-foot patrol boat Mark I.

In 1940, both the navy and the army planned enormous expansion of their air arms, which required numerous new crash boats. Like the Royal Air Force, the Army Air Force needed longer-range, rough-water boats to rescue survivors of crashed medium and heavy bombers. Naval aircraft could not fly nearly as far, so the navy tended not to develop large crash boats. Apart from the 54-foot experimental torpedo boat discussed in chapter 5 and one 72-foot yacht bought for the rough waters off San Clemente, it bought only 36- and 45-footers.

However, the Royal Air Force sought to supplement its supply of British-built crash boats, and Miami Shipbuilding developed a new 63-footer, which outwardly resembled a PT, specifically for its use. With the coming of lend-lease, the 63-footer became a standard U.S. type, and it superseded the existing 36- and 45-footers, as table C-1 shows. They in turn became available as harbor pickets, many contracts being converted directly to the new function. Conversion generally entailed fitting less powerful engines—in the 45-footer, 250-bhp Hudson Invaders for a typical maximum speed of 22.65 miles per hour at 30,669 pounds' displacement—four depth-charge slings, and a ring mount for a machine gun.

The World War II pickets (converted crash boats) were designated Marks I, II, and III. New Mark V pickets (C 7050–60, 7101–9, 7279–94, 7306–49) were ordered on 19 January 1952. Mark V was slightly longer than the earlier boats (46 feet 3.75 inches rather than 45 feet 9 inches), slightly heavier (39,000 rather than 33,850 pounds fully loaded), beamier (15 feet 3 inches rather than 13 feet 8.25 inches), considerably slower (13 rather than 20 knots), and powered by two 165-bhp diesels rather than by 250-bhp gasoline engines. It also had almost twice the payload, 6,525 as opposed to 3,850 pounds. At least some Mark III pickets survived long enough to serve in Vietnam alongside Mark Vs. Range declined from 275 nautical miles in the gasoline-powered pickets to 200 in the diesel boats, which were slower.

The pickets became obsolete when the special harbor defense units were disestablished in 1959, reflecting changing navy and national strategy. In the late 1970s, however, 32-foot commercial boats (modified racing craft, the "Sea Witch" class) were bought as pickets for Subic Bay, then for naval bases in the United States.

Both the army and the navy used 37-foot Higgins Eurekas, similar to landing craft, as pickets; navy boats were powered by a single 250-bhp Hall-Scott engine.[1] The navy also used other boats, some of them based on the cabin cruiser hull of the 36-foot crash boat; the army used 38-foot Coast Guard picket boats.

The 63-foot air-rescue boat looked like a small PT boat. It was exported to the Soviet Union as a small subchaser, the PTC, PTC 62 and PTC 64 are shown here. Twin 0.50s were mounted on the air-sea rescue version, but the 20mm cannon was limited to the PTC. These fast air-sea rescue craft were also used by beach-jumper units and special assault troops.

Picket production was resumed during the Korean War, and the U.S. Navy maintained harbor defense units throughout the 1950s. This 45-footer served a harbor defense patrol during May 1953. The guns are 0.50s. Patrols were disbanded in 1959, but the boats remained, serving a few years later in Operation Stable Door in South Vietnam.

A 37-foot picket, newly completed by Chris-Craft, displays the ring mount required for the performance of its role.

Table C-1. Crash Boats Completed, 1941–45

	1941	1942	1943	1944	1945	Total
Navy Boats						
28 Ft	1	2	—	—	—	3
36 Ft	17	5	43	2	—	57
37-Ft PK	—	207	5	—	—	212
45 Ft	13	16	32	—	—	61
45-Ft PK	—	—	188	193	88	469
63 Ft	—	67	233	191	—	491
Army Boats						
36 Ft	—	—	—	10	—	10
37-Ft PK	4	5	50	189	—	248
	—	—	6	—	—	6
38-Ft PK	—	—	—	48	—	48
	—	—	—	—	2	2
42 Ft	—	63	125	27	—	215
45 Ft	—	—	—	32	—	32
63 Ft	—	—	—	79	—	79
	—	—	—	22	5	27
72 Ft	5	—	—	—	—	5
83 Ft	—	20	12	—	—	32
85 Ft	—	—	—	130	—	—
104 Ft	—	10	90	64	—	174
	—	—	17	21	19	57

Note: In this list, PK denotes a picket boat. Among army boats, the *first* series of numbers is taken from an army small boat construction report dated 18 December 1945, the second, from the BuShips report of World War II construction. This second list includes units built late in the war when BuShips took over the entire national shipbuilding program (apart from merchant ships). The army listed dates of acquisition, not necessarily completion. Where only one line of data is listed, it is army data.

The 63-footer was designed by Dair N. Long for Miami Shipbuilding to meet Royal Air Force requirements, and the British order dates from early 1941.[2] The British initially ordered thirty-one, all for overseas use, to be delivered by 30 June 1942. The original "1799" program included ninety-one such boats. By March 1942, the U.S. Navy had decided to take over thirty-six of them, arming them with two twin 0.50-caliber machine guns and two depth-charge racks, each carrying two charges. A contract was signed only in May 1942, for forty-one boats. British boats were originally unarmed, but guns were added after the Luftwaffe began attacking them in the middle of the English Channel. The armament was strengthened after several were heavily attacked, and at least one was sunk, at Dieppe.

Through 1942, contracts totaled 142. Another 275 were ordered in 1943, along with lend-lease boats and small subchasers. Lend-lease transfers to England amounted to four in March 1942, of twenty-two ordered; six in May 1942, of twenty-four ordered; eight in December 1942, of twenty-one ordered; four in April 1943, of thirty ordered; eleven in October 1943, of eleven ordered, all for South Africa; eighteen in July and September 1943, of thirty-five ordered (the others became Russian subchasers RPC 9–14, 17–28); one in March 1944; and twenty in June and July 1944, for Australia. One was transferred in April 1944 to the Royal Air Force in Nassau, three to the Royal New Zealand Air Force in February 1945. Additionally, five were transferred to the Netherlands, in July and August 1942 and December 1943, one was transferred to Belgium in January 1943, and one to Uruguay in May 1944.

Production for the army and air force continued after the war, at least as late as 1956; about twenty boats were taken over by the navy. Some 63-footers were converted into torpedo retrievers and noise-measuring boats; total numbers are not available. Finally, many crash boats ended their careers as Mark 34 SePTars (seaborne powered targets), anywhere from 40 to 104 feet in length. They were intended to simulate Soviet-bloc missile boats and could launch BQM-34A drones as stand-ins for the Soviet STYX (SS-N-2) missile. All had been expended by mid-1970, and a new-generation 45-knot, Mark 35 boat was developed to replace them.

The army ordered its first long-range crash boats in 1940. The five design 195 72-footers were built by Greenport with 600-bhp gasoline engines. These were followed by 32 design 203 83-footers in 1942–43 (83-6 x 15-7 x 4-3 feet-inches, powered by 550- or 600-bhp gasoline engines); 130 design 379 85-footers (85 x 20-3 x 4-8, powered by two Packards); and at least 174 design 235 104-footers (104-9 x 19-2.5 x 5-3, powered by three Hall-Scott or two Packard engines). The army designated its 45-foot crash boats design 385 and also had a 42-footer variously des-

The air-sea rescue version of the 63-footer in 1943, with a well deck, extra cabin, and collapsible crane booms aft. Note the rescue net rolled up on the port side and the two tilt racks for depth charges at the stern. Some postwar versions of this craft had folding platforms at the stern and lacked weapons.

Table C-2. World War II Navy Crash Boat Characteristics

	36 Ft	45 Ft Mk I	45 Ft Mk II	63 Ft
LOA (ft-in)	36-6	45-6	45-11	63-0
Beam (ft-in)	10-7.5	13-5	13-9	15-4.25
Draft (ft-in)	3-6	3-0	3-0	3-10
Displacement (lbs)	19,400		32,350	52,000
Engines	2 Sterlings	−2 Defender−		2 Defender
BHP	300	——600——		630
Speed (kts)	30.46	—	40.07	37.3
Fuel (gals)	250	—	375	1540
Range (nm)	125	—	—	—

ignated design 200, 221, 225, and 250. Dimensions of the latter were 42-4 x 11-8.5 x 2-9 feet-inches, powered by a 500- or 600-bhp engine. One 42-footer was taken over by the navy as C-7018. Similarly, the army designated the standard 63-foot crash boat design 416. Most army boats were numbered in a P-series, although late in the war some had Q or QS designations.

The navy took over several of the army boats: fifteen 85-footers (C-3237–41, 105346–51, 105356–58, and 105365) and eighty-seven 104-footers (C 13510, 18351–53, 68734–38, 68963–69, 77420–55, 77619–21, 77623, 104566, 104743–47, and 105361–64). C-77420 and some others in its series went to the Coast Guard as CGAR-1 and so on. Some may have been bought for the army on navy account, but at least one 104-footer was used after the war for electronic tests, and one was used during the war for training in electronic countermeasures. At least four 104-footers were transferred under lend-lease, one to Cuba and three to the Dominican Republic.

Work on improved crash boats, done partly for the air force, for which BuShips acted as design and procurement agent, continued after the war. Air force

The army developed its own range of rescue boats, some of which ended in navy service. C-13590, photographed in 1945, was a 104-footer used for electronic countermeasures training. Note the TDY jammer aft and the SA and SO radars on the foremast with their "ski pole" IFF antennas. The object right aft is a small launcher for AA targets.

crash boats were transferred to the navy in 1956. There were three new designs: a 40-footer for inland and protected waters, an intermediate 50-footer to replace the 63-footer, and a 95-footer to replace the existing 85-footer for offshore work. Only the BuShips 40-footer, which came first, was built in any numbers.

Mark 1 was basically the new 40-foot personnel boat with gates in its stern through which accident victims could be pulled and with two rather than one 64HN10 diesel. The diesels ran at 225 bhp maximum, 165 bhp continuous rating, for a speed of 20 knots fully loaded at 26,330 pounds, compared with 12 knots at 23,100 pounds for the Mark 1 personnel boat. The rescue boat carried 170 gallons of fuel and had a range of 140 nautical miles at full speed, compared with 164 gallons for the personnel boat and 210 miles at full speed. The gate added 3 feet 1.5 inches to the 40-foot 2-inch overall length of the personnel boat. A later Mark 2 was slightly longer at 44 feet 2.75 inches and slightly heavier at 26,920 pounds; it had more fuel, 200 gallons for 130 miles at full power, and a different superstructure. It could carry thirteen rather than nine men.

While this boat was being designed and built, the air force bought the standard Coast Guard 40-foot utility boat in small numbers; the navy designated it the 40-foot plane personnel and rescue boat Mark 2.

After comparative tests of the two 40-footers, the 52-footer, and the standard 63-footer in October 1953, BuAer decided to adopt the air force type, but with a planked hull and with improved lines for dryness. The air force was told to hold purchases of the existing 40-foot Coast Guard boat to a minimum, limiting them to cases in which the rescue mission was almost secondary.

All of these craft had planing hulls, but speed was more important for offshore work, as the boat had to run further to get to the site of the accident. As in the 1930s, then, the postwar crash boat program became a focus of BuShips interest in advanced, high-speed, small-craft hull forms. The bureau established a project on advanced planing hulls in December 1949. Perhaps the most exotic was a planing hull in

The last of the big, fast crash boats—the air force 95-footer. Although it was turned over to the navy, and although it was considered satisfactory, this design had no impact on the Vietnam War fast combatant program begun about five years after its delivery.

which up to seven small steps cut diagonally across the hull, with only a short cut directly across the beam on the centerline. It corresponded to a 44.45 x 8.76-foot hull (about 35,000 to 45,000 pounds, 45 knots). The multistepped hull was rejected because suction behind each step caused excessive water resistance.

The air force and BuAer presented design requirements for a new generation of medium and large (50- and 95-foot) standard rescue boats in November 1950. The 95-footer would replace the existing 85-footer at a somewhat higher cost.[3] Diesel engines seemed desirable, but, as in the case of the new-generation PTs, satisfactory ones were not available. The highest-powered diesels were 600- to 750-bhp Packards, whereas the wartime gasoline PT engine was rated at 1,350 bhp, the new 3,300–cubic inch engine (postwar PT) at 1,400 to 2,000 bhp. A 50-foot crash boat could make 42 miles per hour continuous rating at 42,000 pounds' displacement using two Packard diesels (47 miles per hour maximum), or it could make 59 at 40,800 on two of the newest Packard gasoline engines (71 miles per hour maximum). Two of the standard wartime PT engines would drive the boat

Table C-3. Postwar Crash Boats

	40 Ft BuShips	40 Ft Coast Guard	52 Ft	95 Ft	65 Ft
LWL (ft-in)	40-0.75	40-0	51-9.75	94-1[1]	64-11.5[1]
Beam (ft-in)	11-3.375	11-2	16-7.5	22-4	17-2.75
Draft (ft-in)	3-8	3-2	4-7	5-8	—
Displacement (lbs)	27,500	21,460	52,405	167,473	69,400
Engines	2 Diesels 64HN10	2 GM6-71	2 Packards W-14	4 Packards W-14	2 GM 12V71
BHP	165	190	1,500	1,500	500
Speed (kts)[2]	20/18	20.3/18.7	42.8	—	24
Fuel (gals)	—	239	1,200	—	800
Range (nm/kts)	153	198/14.5	242/37.2	—	250/24

1. Overall rather than waterline length.
2. The first two speeds are presented as maximum/cruise.

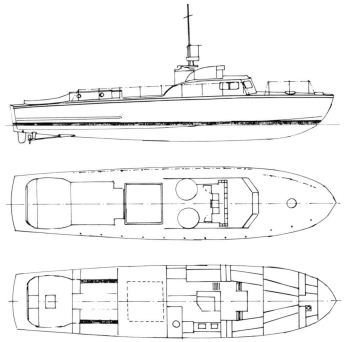

A 63-foot air-sea rescue launch (Q140) converted for special duties, at Northam, England, 10 July 1944. (Norman Friedman)

at 45 miles per hour at 38,500 pounds continuous, 55 at maximum, rating. Displacements were determined partly by fuel load, which in turn was intended for a range of 200 miles. Four Packard diesels would drive the big 95-footer at only 28 miles per hour at 110,000 pounds, 33 maximum, whereas three of the new Packard gasoline engines would drive it at 39.5 at 124,500 pounds, 51 maximum, and three wartime Packards would make 32.5 at 112,000 pounds, 42.5 maximum. In each case, range was set at 1,200 miles.

The choice of an engine was somewhat complicated by the need to open a new factory to manufacture the big 1M-3300 PT engines. However, the wartime W-14 was no longer in production. BuShips recommended a new Packard W-51 with additional power and with parts that could be exchanged for those of the W-100.

BuShips designed and Chris-Craft built the prototype of a new medium rescue boat, a 52-footer (see table C-3). Like the bureau's 40-footer, it had a stern gate, in this case a hinged transom for a rescue platform, which could be let down and submerged. Only one prototype was built.

The big 94-footer was a private design of Huron-Eddy's. Two prototypes, C-3727–28, ordered in June 1951 and completed in October 1953, were transferred to the navy under a 1956 agreement. They appear not to have had any great impact on later fast patrol boat work.

In May 1959, BuShips considered buying a folding-fin (surface-piercing) hydrofoil crash boat that would use the hull of the existing 41-foot personnel boat and be powered by a 1,000-shp Solar gas turbine. Its sea-keeping would match that of the 63-footer, but it would be 40 percent faster (40 knots cruising, 50 knots maximum).[4] Draft would be 6 feet 7 inches to the main foils (7-6 to the fixed after foil), cruising range 350 nautical miles at 40 knots. Although nothing came of this particular project, it obviously could have been applied to the fast motor gunboat designs described in chapter 9. In this case, too, hydrofoils made it possible to extend the usefulness of an existing boat hull design, the new 40-foot personnel boat, so as to lengthen its production run.

Ultimately, however, BuShips opted for a less exotic solution, a version of the 65-foot multipurpose boat designed under contract Nobs 4869. Powered by twin 12V71 diesels, it could make 24 knots. The same hull could be used as a utility boat, a personnel boat, a torpedo and weapons retriever, and a patrol boat (Mark I).[5]

D

U.S. Small Craft Weapons: Guns and Rockets

This appendix includes guns and rocket launchers but not ASW weapons or the few torpedoes and missiles tested and adopted—the SS-11/12, Penguin, Standard ARM, and Harpoon. Developments prior to 1945 are described in chapter 6, but the standardized weapons are tabulated here.

The ongoing BuOrd PT weapon-development program of 1945 was curtailed after the war to provide funds for more urgent ASW and AA programs. The only two major small-craft weapons programs of the early postwar period were the 81mm mortar Mark 2, which was not used until Vietnam, and an abortive 2.75-inch rocket gun.

China Lake developed the 2.75-inch Mighty Mouse aircraft rocket specifically for firing from streamlined pods suited to fast aircraft. An automatic rocket launcher—in effect, a rocket-firing machine gun—was developed and tested for jet fighters. Such a gun was attractive as a small-ship weapon because it could deliver substantial, high-velocity warheads without paying a heavy price in launcher weight or recoil. BuOrd began research and development about 1951, viewing the gun as a candidate weapon for future PCs and SCs. It was dropped around 1955 on the ground that the most likely air target was a fast, high-flying jet bomber dropping an antiship missile well before it got within 2.75-inch rocket range. That left only the 3-inch/50, as all alternatives were either too heavy or ineffective.

A June 1953 summary of BuShips research and development projects listed a 2.75-inch gun, the Mark 1 Mod 0, whose prototype would be ready in March 1954. It was expected to weigh about 30,000 pounds and to fire 750 rounds per minute. The gun was an adapted T110 air force weapon firing T131 ammunition.

In the long run, the mortars were far more successful. The 81mm Mark 2 mortar was first tested aboard the experimental postwar PTs. Rejected at the time, it reappeared quite successfully in Vietnam. Unlike conventional army mortars, it had a recoil-counterrecoil mechanism to reduce deck forces and a triggered firing pin that could fire the weapon when it was horizontal. It could also be fired in the conventional drop-fire mode. The Mod 1 version had a piggyback 0.50-caliber machine gun. Mod 0, the pure mortar, weighed 600 pounds (722 for Mod 1), and elevation limits were $+71.5°$ and $-30°$. Maximum range was 3,987 yards. The mortar could fire ten rounds per minute in the trigger mode or eighteen in the drop-fire mode.

The Mark 4 was a similar 60mm mortar mounted aboard some PBRs, replacing their Mark 46 mounts aft. Total weight was 177 pounds, maximum range, 1,850 to 2,000 yards. Work began in May 1966, and the first gun was test-fired in March 1967. Testing was completed in September 1967, and production began in February 1969 (preproduction examples became available in October 1968). The Mark 4 replaced the after 0.50-caliber gun in some PBRs. The 60mm mortar was well received but ineffective against short-range targets because its shell incorporated an arming delay. Field representatives from Louisville's naval ordnance station devised an adapter for a piggyback 7.62mm M60 machine gun for close-range fire in Vietnam. Over 100 adapters were shipped

The Elco "Thunderbolt" was one of several attempts to increase PT firepower. The prototype, with two 0.50-caliber machine guns as well as the four 20mm cannon, is shown aboard PT 160 for tests, late in 1942. Unlike the later production version, it did not have a Mark 14 lead-computing sight (note the simple tube sight on the left, above the operator's seat). The rectangular object on the extreme left (inside the mount) is the breech end of the left-hand 0.50-caliber machine gun. (credit: Elco photo courtesy of Preston Sutphen III)

there in September 1969; 450 Mark 4 mortars were manufactured.

A naval version of the largest army mortar, the 4.5-inch (107mm), was considered but not developed.

Work on specialized small-craft weapons resumed as the United States developed specialized craft for Vietnam under MAP. Initially, new mounts were developed for existing World War II weapons; as the notes below show, new weapons gradually evolved. Current U.S. small combatants are being armed with new-generation light weapons.

In 1960, the major small-caliber weapons were the army-type, manually-controlled 40mm Bofors, which had served through World War II; the 20mm machine cannon (the Oerlikon and the aircraft-type Hispano-Suiza, or Mark 16); the 0.50-caliber machine gun; the 0.30-caliber Browning machine gun, later often converted to 7.62mm as Mark 21; and the 81mm mortar Mark 2. The army had a new light machine gun, the 7.62mm M60, introduced in 1960, and it was developing a single-shot 40mm grenade launcher, the M79, which entered limited service in 1963.

By the late 1960s, stocks of the World War II weapons were declining, and what was then NavOrd began a program of new 20mm gun development. This problem has worsened. For example, there are barely enough single 40mm guns that can be modernized to arm the existing force of Mark 3 patrol boats.

Note that the Naval Ordnance Laboratory became lead laboratory for small craft weapons only in January 1966.

In August 1967, NavOrd proposed a five-year small-craft armament program, which would include a lightweight 3-inch/50, a new twin 40mm automatic mount, a 4.5-inch (107mm) mortar, the army Vulcan air defense system (VADS), a new 70mm grenade launcher, a 40mm multirange grenade launcher (in effect, to merge Mark 19 and Mark 20), a 30mm automatic cannon, and a new 20mm gun. Only the 20mm and 30mm guns reached the prototype stage, and not even they were adopted. The missing gun designations in the lists below probably apply to these other weapons.

The twin 40mm was to have enjoyed drastic reductions in weight through the use of a muzzle brake—standard in army 40mm weapons. The brake would absorb at least 75 percent of its recoil. Weight would also be saved through the use of an aluminum platform and a lightweight servo system, 150 rather than 1,600 pounds, for a total weight of 3,258. This is compared with the 13,000 pounds of existing weapons. The new mount was to be available in 1973.

The 70mm grenade launcher would be effective out to 3,000 yards and would double the firepower of the existing high-velocity 40mm launcher. It would be light enough to fit a 0.50-caliber socket.

The new 40mm launcher would achieve multiple ranges by venting some or none of the propelling gas, a method already in use in such weapons as the British Limbo ASW mortar.

NavOrd hoped that a new 30mm machine gun would replace the existing 20mm and 0.50-caliber weapons, achieving twice the effective range of the 0.50. The new 20mm cannon would use a 0.50 caliber–type pintle mount.

The list that follows includes all gun mounts used during the Vietnam War plus most subsequent experimental types. Note that, until about the end of World War II, Mark numbers were assigned to the mounts for each caliber, so that there were, for example, Mark 19 0.30-caliber and Mark 19 5-inch mounts. After 1945, all of these series merged, as the list shows. Similarly, machine gun Marks were amalgamated after 1945, continuing the 20mm Mark number series. They are separate from the gun mount designations.

Machine Guns

Mark 1 Original 20mm 70-caliber Oerlikon.

Mark 2 Oerlikon with two locking slots and cooling ribs.

Mark 3 Oerlikon with a single locking slot.

Mark 4 Standard World War II 20mm Oerlikon machine gun, converted from metric to English units. Effective range was 2,000 yards at 450 rounds/minute (blowback operation). Total weight was 150 pounds, length, 84 inches (muzzle velocity 2,730 feet/second).

EX-6 Experimental 20mm machine gun developed by the Naval Gun Factory during the Korean War. It was revived and revised as the Mark 22.

Mark 7 Aircraft 20mm machine gun for fixed mounting, 1,000 rounds/minute.

Mark 8 Double-barrel 20mm aircraft gun, 2,000 rounds/minute.

Mark 9 Double-barrel 20mm aircraft gun, 5,100 rounds/minute.

Mark 10 Alternative to Mark 9 with different gun mechanism.

Mark 11 Twin-barrel Hughes 20mm aircraft gun in Mark 4 pod, developed in the mid-1950s. Total weight, including the loader, was 240 pounds. Operation was recoil and gas (750 or 4,200 rounds/minute). Muzzle velocity was 3,300 feet/second and effective range, 2,000 yards.

Mark 12 Aircraft gun designed for fixed installation, 115 pounds including feeder, using blowback and gas operation (1,000 to 1,100 rounds/minute). Muzzle velocity was 3,300 feet/second.

Mark 13 Fixed, single-barrel 20mm aircraft gun.

Mark 14 Fixed, single-barrel 20mm aircraft gun.

Mark 15 Fixed, single-barrel 20mm aircraft gun.

Mark 16 20mm machine gun developed from the World War II M3 aircraft gun (Hispano-Suiza): 650

to 800 rounds/minute with combination blowback and gas operation; otherwise, performance was similar to that of the Mark 4.

Mark 17 (EX-17) A 7.62mm machine gun similar in principle to the 20mm Mark 11: it was an eight-chamber revolver with two barrels and an integral loader. The designation was assigned in April 1965.

Mark 18 A 40mm hand-cranked machine gun (grenade launcher) developed by Honeywell. It fired up to 250 rounds/minute. Muzzle velocity was 215 feet/second and effective range, 330 yards. Weight was 27 pounds. In 1952, the army began work on short-range 40mm grenade ammunition as a replacement for the standard rifle grenade, leading to production of a single-shot M 79 launcher. Mark 18 was bought as an interim weapon, to be used until Mark 20 became available. Although difficult to disassemble and inaccurate due to a jerky hand crank, the Mark 18 was often preferred over the Mark 20 because it was more reliable.

Mark 19 A 40mm blowback-operated machine gun (long-range grenade launcher) developed by the naval ordnance station in Louisville from July 1966 on. It fired 400 to 450 rounds/minute. Muzzle velocity was 800 feet/second and effective range, 1,780 yards at a 15° elevation. Total weight was 53 pounds. The longer-range grenade ammunition was initially developed by the army for its XM129 helicopter-carried launcher, used in the Cheyenne attack helicopter. The Naval Ordnance Laboratory evaluated the army XM129. It used electric drive to attain a firing rate of 350 to 450 rounds/minute, and it had a muzzle velocity of 800 feet/second (effective range, 1,760 yards at a 15° elevation). Total weight was 43 pounds. One was test-mounted aboard a PBR. In all, 750 Mark 19s (325 Mod 0s, 425 Mod 1s) were manufactured between October 1967 and October 1969. The weapon initially received mixed reviews, but it was well liked after conversion to a more reliable Mod 1 configuration. The Mark 19 was considered sturdier than the Mark 20.

Mark 20 A 40mm reciprocating-barrel machine gun (grenade launcher) developed by the naval ordnance station in Louisville from August 1966 on. It could fire up to 250 rounds/minute. Muzzle velocity was 215 feet/second and effective range, 330 yards. Weight was 28 pounds. In all, 978 were manufactured between August 1967 and the fall of 1970.

Mark 21 A 0.30-caliber Browning machine gun converted to 7.62mm caliber. Effective range was 1,200 yards, and it could fire 450 to 600 rounds/minute. By way of comparison, the M60, with similar rounds and therefore a similar performance, weighed 23 rather than 32 or 36 pounds and could fire 500 to 850 rounds per minute. M60D was specially adapted to small-craft pintle mounts.

Mark 22 The proposed replacement for Mark 16, under development in 1972 by the ordnance station at Louisville. Weight was 110 pounds. It operated by recoil/blowback and could fire 800 rounds/minute. It was rejected in favor of the Mark 29 because it could not be mounted aboard marine corps amphibious vehicles. Other candidates, which did not receive EX or Mark numbers, were the GE-120B (96 pounds, recoil-operated, 400 to 700 rounds/minute using a pulsed solenoid, or 260/400/600 rounds per minute manually selected; a candidate for the army vehicle RF weapon system); the French M621 (AME 621, 100 pounds, gas operated, 300 or 720 rounds/minute); and the Mauser Model B (107 pounds, gas operated, 900 to 1,050 rounds/minute).

Mark 23 Stoner system 5.56mm machine gun (XM207) bought for navy SEAL teams. Effective range was 3,000 yards, 750 rounds/minute. It used a link belt or 30-round box magazine.

Mark 24 A 9mm Smith & Wesson model 76 submachine gun (designation assigned March 1970).

Mark 25 A 7.62mm minigun (Vulcan GAU-2A/B), firing 2,000 or 4,000 rounds per minute, using electric drive. Total weight was 60 pounds including drive motor and delinking feeder. It was used by SEALs (see mount Mark 77). Note that a minigun on a Mark 26 mount was first tested aboard a PBR in March and April 1967, demonstrating a 500-yard effective range.

Mark 27 (EX-27) Colt Model 2 (CMG-2) 5.56mm belt-fed machine gun, 650 rounds per minute, firing from shoulder or hip or bipod and weighing 13.5 pounds empty.

Mark 28 (EX-28) A GE 30mm gun for an EX-74 mount (CPIC). It weighed 280 pounds and operated by recoil (100 to 800 rounds/minute). Muzzle velocity was 3,000 feet/second and effective range, 3,000 yards. Unsuccessful, it was replaced by the Hispano-Suiza HS 831L.

Mark 29 (EX-29) GE's 20mm small-craft gun, competing with the Mark 22 to achieve 250/550 rounds/minute. It was designed for fewer than two failures per 1,000 rounds.

Mark 30 (EX-30) A single-barrel 30mm machine gun using the same ammunition as the GAU-8 aircraft gun. It was 130 inches long and weighed 500 pounds. The natural cyclic rate was 500 rpm, but it could also fire at 250 or 125 rpm or in a single shot mode. The designation was canceled in December 1982.

Mark 31 (EX-31) An externally powered, 30mm, low-velocity gun (60 rounds per minute) produced by Hughes Helicopter. It had a 42-inch barrel (63 inches long overall) and weighed 98 pounds. It was the army XM230; its navy designation was assigned in February 1976 and canceled in December 1982.

Mark 33 (EX-33) A single-barrel, low-velocity (2,200 feet/second) 30mm gun firing 450 rounds per minute

The twin 0.50-caliber machine gun enjoyed a long life aboard U.S. small combatants. This Mark 17 is shown aboard a World War II PT. (Al Ross)

and weighing 144 pounds. The designation was assigned in 1978.

Mark 34 A 7.62mm Hughes chain gun with a 22-inch barrel (37 inches overall, including the flash hider). It fired 570 rounds per minute and weighed 30.2 pounds. The designation was assigned in 1982.

Mark 35 (EX-35) The navalized GE GAU-12 25mm Gatling gun, initially selected in 1983 to arm the SWCM but, because of rising costs, canceled in favor of the simpler (and unstabilized) Hughes chain gun. EX-35 sometimes refers to the stabilized mount as well as to the gun, a reversal of the usual heavy-gun nomenclature. However, an EX-90 designation is sometimes applied to the mount.

Mark 37 (EX-37) An externally powered, seven-barrel 30mm Gatling gun, a navalized GAU-8/A weighing 607 pounds empty or 1,800 including ammunition. It can fire 2,100 or 4,200 rounds per minute.

Mark 38 A 25mm Hughes chain gun, consisting of a 25mm gun M242 plus the Mark 88 mount. Firing rate is 1 to 550 rounds per minute, and total weight is 1,010 pounds, including ammunition. Army development and procurement were approved in December 1971.

World War II Gun Mounts

Thirty caliber

Mark 19 A 0.30-caliber tripod open mount. It was modified as the Mark 1 2.36-inch sextuple rocket launcher in some PTs and PGMs.

Mark 20 An army M41 stand-ring mount.

Mark 21 A 0.30-caliber, single-shielded scarf-ring mount. Mods 3 and 4 both weighed about 180 pounds.

Mark 22 A 0.30-caliber, single-tripod mount.

Fifty caliber

Mark 17 A 0.50 caliber, twin-scarf-ring mount used in World War II PTs. It was also used in Mark I PCFs

(Mod 5 version). Mod 5 weighed about 755 pounds, 440 pounds being concentrated in the Mark 9 shock-absorbing carriage. The mount typically carried 1,000 rounds of ammunition.

Mark 20 A twin water-cooled mount using hydraulic recoil absorbers and adjustable motorcycle-type handlebars.

Mark 21 A shielded (half-inch), 0.50-caliber, single-pedestal mount used aboard many landing craft. Mod 1 weighed 500 pounds with one M2 machine gun and a shield. This mount was adapted as the Mark 50 and Mark 51 (see below). Originally designed for the British with twin water-cooled guns, it was considerably lightened for U.S. service; air-cooled weapons were substituted.

Mark 22 A pedestal mount for twin 0.50-caliber machine guns.

Mark 23 A hydraulically-operated Frazier-Nash mount (British) used on lend-lease PTs.

Mark 24 A lightweight tripod mount, originally for the M2 water-cooled gun (Mod 1 used an air-cooled weapon) and used by PT Squadrons 31 and 32. It was also used aboard landing craft and some escort carriers.

Mark 25 An experimental PT mount using the M2 aircraft gun.

Mark 26 A 0.50-caliber (heavy barrel or aircraft-type M2) pintle mount. It could also be adapted to the 7.62mm M60D machine gun, to the 40mm Mark 18 or 20 machine gun, or to various 0.30-caliber machine guns. The pintle could be mounted in a standard tripod or directly on any horizontal surface, and there was a special stand designed to fit a torpedo launching rack on PTs. There were five different shields: ceramic, dual-hardness, or face-hardened steel. Without gun or ammunition, the Mark 26 weighed 55 to 301 pounds, and its armored shield could weigh as much as 255. The portable mount was rigid without the shock-absorbing cradle.

Mark 27 A training weapon with a 0.50-caliber sight and a 0.30-caliber gun, though the 1944 PT-boat ordnance list appears to include the Mark 27 as an alternative to the Mark 26. Total weight in either case was about 290 pounds.

Mark 32 A power-driven quadruple mount for aircraft-type 0.50-caliber guns (modified army M-45s). Sixty were obtained from the army in 1945 and fitted with their own power sources, gasoline engines, and generators with batteries. It was tested aboard several carriers, but the consensus was that it was no better than the twin free-swinging 20mm gun. Note that the M-45 was tried again during the Vietnam War.

Mark 36 A twin 0.50-caliber scarf-ring mount with ceramic armor used in PBRs. In the PBR, it carried 500 rounds of ammunition and weighed about 970 pounds, including two M2 machine guns and a Mark 9 cradle (470 pounds).

Twenty-mm Gun Mounts

Mark 2 A single mount with variable trunnion height (mechanical). It weighed 1,095 pounds with a 0.5-inch shield.

Mark 4 A single mount with variable trunnion height (mechanical). Total weight was 1,095 pounds. The manually operated trunnion tended to corrode, hence the Mark 6, with its more reliable hydraulic trunnion, and the lightweight Mark 10, with its fixed trunnion.

Mark 5 A British-designed single mount with fixed trunnion height. It weighed 1,540 pounds. All but 916 of the 6,101 produced went to Britain.

Mark 6 A single mount with variable trunnion height (hydraulic), manufactured by Pontiac. It was rejected in favor of the Mark 10. Total weight amounted to 1,691 pounds (elevated to 90°, depressed to $-15°$).

Mark 10 A single, lightweight, free-swinging mount that became standard in the middle of World War II. Total weight was 970 pounds, including a 240-pound shield. It elevated to 90°, depressed to $-15°$. Although intended for special applications, the Mark 10 supplanted all earlier Marks.

Mark 12 A scarf-ring mount for a single gun, weighing 1,234 pounds (this included twelve magazines at 32 pounds each). It was a converted 0.50-caliber Mark 17 mount.

Mark 13 A lightweight mount developed by New York Navy Yard. It weighed 884 pounds, including twelve magazines. Four were mounted in boats of MTBRon 18, but not standardized because they could not be interchanged with other 20mm mounts.

Mark 14 A lightweight mount version of Mark 10, with a low trunnion of 30 as opposed to 52 inches. Total weight, including twelve magazines, was 1,004 pounds.

Mark 15 An Elco "Thunderbolt" quadruple mount for PTs. In its original late 1942 form, it carried four 20mm cannon and two 0.50-caliber machine guns on a weight, without ammunition, of 2,550 pounds.

Mark 20 A prototype twin mount of 1,340 pounds. It led to the Mark 24.

Mark 22 A Maxson quadruple 20mm mount with four T31 aircraft guns. In all, it weighed about 2,250 pounds.

Mark 23 A triple mount developed at Pearl Harbor. Fifty were ordered but all canceled in May 1944.

Mark 24 A standard, free-swinging twin 20mm mount. Weight amounted to 1,216 pounds, including guns and shield. It elevated to 85°, depressed to $-15°$. The typical on-mount ammunition supply was two sixty-round drums.

The 40mm Bofors has been a standard U.S. small combatant weapon for forty years. This is the prototype Mark 3 Mod 7, an existing mount modified by the Naval Surface Warfare Center for one-man operation. In 1983, it was installed aboard a 65-foot patrol boat, a Sea Specter, for technical evaluation. Note that the gun is still limited to a single clip. Some foreign versions of the same basic gun incorporate an auto-loader. Mod 7 is being superseded by the Hughes Bushmaster 25mm chain gun.

Mark 25 A Mark 24 adapted by Emerson to two T31 (M3) rather than Mark 4 20mm guns.

Mark 26 A Maxson quadruple 20mm gun of 1946 with a self-contained power source. It weighed about 3,000 pounds.

Mark 27 A Mark 4 with fixed trunnion height.

Mark 28 A scarf-ring mount for 20mm aircraft cannon. It was experimental only.

37mm (army weapon with army designation)

M4 Mounted on Mark 14 (PT 565–624) or Mark 10 (PT 625–60). Weight (pounds) as follows: gun, 150; mounts, 457 and 569 (360 rounds or six magazines, respectively); ammunition weight, 312. Typically, two to three magazines were on deck or in ready boxes abaft the gun; total weights, respectively, were 919 and 1,031 pounds. Rate of fire was 150 rounds/minute, and muzzle velocity 2,000 feet/second.

M9 On a Mark 1 Mod 1 mount with magazine Mark 1. The gun weighed 398 pounds. It had a 347-pound mount; seventy-five rounds of HE and fifty rounds of AP ammunition weighed 485 pounds. Deck ready boxes abaft the gun weighed 1,230 pounds. Rate of fire decreased to 125 rounds/minute, but muzzle velocity increased to 2,900 feet/second (the gun was 104 inches long, compared with 89.5 for the M4).

Forty-mm

Mark 1 A standard World War II twin power-driven base-ring mount, weighing 13,000 to 14,000 pounds

The 81mm mortar was first tested aboard the experimental PTs in 1951, but it did not enter U.S. service until the Vietnam War. Here one is fired from a 65-foot Sea Specter patrol boat in August 1978. Note the 0.50-caliber gun in the background. Many Vietnam-era craft had combination mortar/machine-gun piggyback mounts.

without fire controls. It elevated to 90° and depressed to −15°. Because it was power-driven, it could be fitted with cutout cams to prevent it from firing into a ship's superstructure.

Mark 3 A single, unpowered, army-type (M3) 40mm/60, using an air-cooled gun. It was widely employed during World War II and afterward on PTs and PTFs. In all, it weighed 2,275 pounds, including a 275-pound gun barrel, 740-pound gun mechanism, and 1,035-pound carriage. Typically, it would be fed by two loaders with four-round clips. The recoil-operated 40mm gun itself fired 120 rounds/minute at 2,600 feet/second, with a maximum horizontal range of 9,300 yards. It elevated to 90° and depressed to −6°. Mod 4 was a power-driven conversion for small craft, such as *Asheville*-class gunboats; it weighed 4,200 pounds, elevated to 85° and depressed to −5°. During 1967–68 the Naval Ordnance Laboratory in White Oak studied possible stabilization of this mount for small combatants, and a stabilized 40mm gun mount was built and evaluated aboard the hydrofoil *High Point* (PCH-1) in 1968. Mods 5 and 6 were submarine (wet) mounts. In the early 1980s NavSea modified stock single 40mm mounts for power operation and loading, so that they could be operated by one man. They were fitted as interim weapons to 65-foot patrol boats pending the availability of either the Hughes 25mm chain gun or GE's EX-35. Contrary to published reports, they were not stabilized.

3-inch (omitting large-ship mounts)

Mark 14 A limited AA version of the 3-inch/23 "boat gun." It elevated to 65°, depressed to −15°, and weighed 1,510 pounds. Mod 1 elevated to 75°, and Mod 3 had a shield for Yangtze River gunboats.

Mark 20 An interim DP mount of World War II, using existing 3-inch/50 barrels. It weighed 7,070 or 7,430 pounds, elevated to 84° and depressed to −10°.

Mark 22 A standard World War II single, slow-firing 3-inch/50, manually trained and elevated on a pedestal mount. Total weight was about 7,500 pounds. It armed many World War II patrol craft.

Mark 26 A power-driven modification of Mark 22, weighing about 10,000 pounds.

Mark 34 The single RF 3-inch/50, which armed the *Asheville* class. It typically weighed 17,000 to 21,000

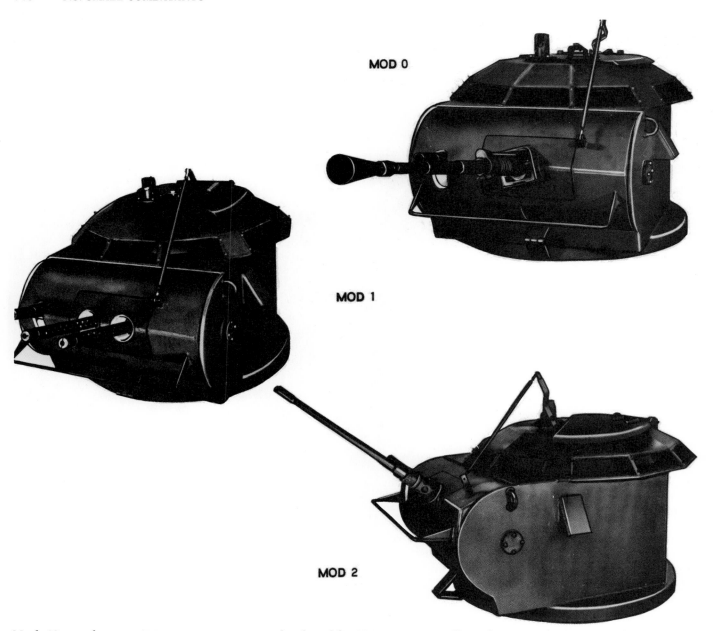

Mark 48 was the most important new weapon developed for Vietnam-era small combatants. Three principal variants are represented here: Mod 0, carrying a 20mm cannon and, to the left, a 40mm grenade launcher; Mod 1, with twin 7.62mm machine guns and a 40mm grenade launcher; and Mod 2, twin 0.50-caliber machine guns and a 40mm grenade launcher. (Aircraft Armament, Cockeysville, Maryland)

pounds and had a minimum depth within the ship of 17 feet, which explains why it could not be applied to anything smaller than an *Asheville*. Maximum rate of fire was fifty rounds per minute, and elevation limits were +85° and −15°. Maximum range was 14,000 yards. NavOrd proposed an automatic lightweight 3-inch/50 in 1967, but it was never built; the OTO-Melara 76mm/62 (Mark 75) was chosen instead.

Postwar Gun Mounts

This list includes, in parentheses, the few larger guns developed. The separate series (by caliber) appears to have continued up through Mark 37 and possibly Mark 38.

(*Mark 38* A twin 5-inch/38 for destroyers)
(*Mark 39* The single slow-firing 5-inch/54)
(*Mark 40* The wartime 5-inch/25 submarine gun)
(*Mark 41* The abortive twin slow-firing 5-inch/54)
(*Mark 42* The standard heavy RF 5-inch/54)
(*Mark 43* An abortive twin RF 5-inch/54)
(*Mark 45* The current lightweight 5-inch/54)
Mark 46 An adapted Mark 26 machine-gun mount

that could accept either the Mark 18 or the Mark 20 40mm grenade launcher piggyback on a 0.50-caliber machine gun. It was developed in 1966 for PBR use. Typical weight without weapons was 167 pounds, or 280 with a Mark 20 grenade launcher and a 0.50-caliber machine gun with flash hider. Elevation limits were +65° and −15° (−11° in some versions). Ammunition supply on the mount was typically one hundred rounds of 0.50 caliber and twenty-four rounds of 40mm.

Mark 47 A 0.30-caliber tripod modified to carry a 40mm Mark 18 grenade launcher, for SEALs and riverine assault groups.

Mark 48 A combination one-man turret mount developed by Aircraft Armaments under the direction of the naval ordnance station in Louisville, for ASPBs and later ATCs, CCBs, and monitors. There were three principal models: Mod 0, one Mark 16-4 20mm and one Mark 19-1 40mm long-range grenade launcher; Mod 1, Mark 21-1 twin 7.62mm machine guns and one Mark 19-1 40mm grenade launcher; and Mod 2, twin 0.50-caliber M2 heavy-barrel machine guns and one Mark 19-1 40mm grenade launcher. Both train and elevation were by hand crank; the Mark 48 elevated at 30° per second (five cranks/second) and trained at the rate of 8° per turn. Elevation limits were +65°, −15°. Protection was dual-hardness steel, with a full ring of nine laminated vision blocks at gunner's eye level and a fiberglass hatch cover. Total weight was about 1,850 to 1,950 pounds, including empty ammunition cans. The typical supply in a Mod 0 mount was 500 rounds of 20mm and 50 rounds of 40mm. The Mod 3 was a hydraulically operated prototype developed by the ordnance station in Louisville to allow rapid and simultaneous elevation, depression, and training of the mount as it followed targets. Mod 3 was never produced; all prototypes were converted back to Mod 0. Mod 4 carried two four-tube, 3.5-inch bazookas and thus was reminiscent of the World War II PT and PGM bazooka weapon. Mark 48 itself was developed in 1966, following tests of the army's XM-30 twin 0.30-caliber armored turret aboard a PBR and a 45-foot utility boat at the Naval Ordnance Laboratory in Maryland. For bunker-busting, some ASPBs carried Mark 47 rocket launchers (paired 3.5-inch bazookas) on their Mark 48 turrets. The Naval had chosen the 3.5-incher over the lighter M72 LAW shaped-charge rocket. Bazooka pods were also developed for the Mark 56 mount.

Mark 49 A training version of Mark 48 in mild steel.

Mark 50 A single armored 0.50-caliber machine gun mount developed by the naval ordnance station in Louisville and used in the program 4 LCM conversions. It was adapted from a 0.50-caliber Mark 21 mount and could also carry a Mark 20 grenade launcher. In that form, it weighed 2,600 pounds and carried 25 rounds of 40mm and 125 rounds of 0.50-caliber ammunition.

Mark 51 A hand-operated, armored pedestal mount for a Mark 16-4 20mm gun, used in first-generation LCM conversions (program 4). It had a circular armored shield and unarmored roof. It was a modified version of the Mark 21 0.50-caliber mount. Weight without ammunition totaled 2,600 pounds, and the mount carried 200 to 250 rounds per gun. Elevation limits were +70°, −15°.

Mark 52 A combined manual-drive M3 40mm gun and 0.50-caliber (spotter) gun developed by the naval ordnance station in Louisville for the MAP program. It required a crew of three and weighed about 8,300 pounds, including a 3,148-pound shield with access ports at its rear. Mark 52 carried 48 rounds of 40mm and 250 rounds of 0.50-caliber ammunition. It armed program 4 monitors and CCBs.

Mark 53 A twin open-pedestal mount for 0.50-caliber machine guns in shock-absorbing cradles. It was installed aboard the fast motor gunboats of the *Asheville* class. Each gun was provided with a 250-round ammunition box, and the mount weighed 200 pounds altogether.

Mark 54 A lightweight power mount, never made, which would have carried a 40mm gun and two on-mount personnel (loader and operator) for PTFs. It would have had provision for a stabilized sight. As part of the BuOrd lightweight gun-mount program described above (2,800 pounds), its designation was assigned in February 1967.

Mark 55 A pintle mount developed by the naval ordnance station in Louisville to fit the same Mark 16 stand as the Mark 26 (see above). Mod 1 carried either the Mark 18 or the Mark 20 machine gun. It weighed 35 pounds, including its stand. The pintle, without its adapter, could be mounted on standard 0.30- and 0.50-caliber tripods. Elevation limits were +70° and −40°, and the mount usually carried twenty-four rounds of ammunition.

Mark 56 A twin scarf-ring mount that could replace the earlier Mark 17 and Mark 36 scarf rings. Mod 0, which armed PBR Mark II, was a 50-inch aluminum alloy ring carrying two 0.50-caliber machine guns, trained and elevated by hand crank. Including guns and feed chutes, it weighed 585 pounds; elevation limits were +65° and −15°. It was developed by Colt Industries under contract to NAD (Naval Ammunition Depot) Crane. Mod 1 was a version armored against 0.50-caliber bullets for river minesweepers, using six periscopes for peripheral vision. Weight, including ammunition, was about 1,400 pounds. Mod 2 carried two Mark 16-4 20mm cannon, with a capacity of 225 rounds per gun (elevation limits +45°,

Bunker-busting, the major weapons problem of the Vietnam War, called forth unconventional solutions. Here a modified program 4 monitor fires one of its two flame throwers, March 1968. Note the area already burned out by the other flame thrower (lower righthand side of the picture). Unlike program 5 "Zippos," this monitor retains her forward gun mount, the flame fuel and pumps being accommodated in her decked-over mortar well.

$-15°$). It weighed about 350 pounds empty or 900 pounds including guns and ammunition. Developed by NAD Crane, it was installed aboard the USS *Tucumcari* (PGH-2). The basic mount was often adapted to alternative weapons, so that some PBRs replaced one 0.50 with a 20mm cannon. Mark 56 evaluation was completed in the fall of 1967.

A remote-control adaptation of Mark 56 was planned for the Sewart ASPB Mark II to replace manned Mark 48s; their operators were unable to pass from the armored citadel to their firing positions under armor, and therefore were considered too vulnerable. The boat would have had four remote mounts and two stabilized periscopic sights with a fire-control computer. Three of the mounts were adapted Mark 56s with combinations of the following weapons: Mark 16-4 20mm guns, XM129 40mm grenade launchers, and 0.50-caliber M2 machine guns.

In addition, the Naval Ordnance Laboratory adapted the Mark 56 to take a single GAU-2 A/B minigun in place of the usual pair of 0.50-caliber machine guns of a PBR. This version was built in February and March 1970, test-fired in April and May, and shipped to Vietnam for evaluation in May 1970. The laboratory considered it superior to the usual Mark 56, because it was lighter and the 0.50 had excessive range for use in populated areas. The minigun mount could also carry many more rounds of the lighter ammunition of this version.

Mark 57 A bulkhead mount for the Mark 21 7.62mm machine gun.

Mark 58 A lightweight bulkhead mount developed

by NAD Crane primarily for the M60 light machine gun (Mod 7 could mount the 0.50-caliber machine gun). Mode 5, 6, and 8 used an on-mount, 200-round box magazine; Mods 1, 2, 3, 4, and 9 used a larger ammunition box stored off the mount, feeding the gun through a flexible ammunition chute. Elevation and depression limits varied from mod to mod, but all except Mod 0 could elevate at least to 80°, and all could depress at least to −20°. Weight without gun and ammunition was typically 10 to 43 pounds.

Mark 62 (EX-62) Projected in 1970 as a universal remote-control gun mount carrying two weapons.

Mark 63 A hand-operated 0.50-caliber mount (Mark 21) converted to carry the Mark 19-1 40mm high-velocity grenade launcher. It was typically enclosed in part by a circular ⅞-inch armor shield with a removable plastic cover on top; LCMs had it in a well with only the armored part above deck. The mount incorporated a 300-round ammunition box and used an electric ammunition booster to help lift ammunition from the box into the gun. Weight without gun or ammunition was 2,600 pounds, elevation limits, +70° and −15°.

Mark 64 A pintle mounting for the Mark 19-1 40mm grenade-launching machine gun, similar in concept to Marks 26, 55, and 58. It normally carried a fifty-round box, but could carry a twenty-five-round box instead. Weight was 56 pounds in Mod 1, without gun or ammunition, and elevation limits were +38° and −15°.

(*Mark 65* A proposed single lightweight RF 5-inch/54)

(*Mark 66* A proposed twin lightweight rapid RF 5-inch/54)

Mark 67 A lightweight single mount developed at the ordnance station in Louisville for the Louisville-developed Mark 22 20mm gun, a candidate to replace the existing Mark 16. Mod 1 mounted the existing standard Mark 16-5. Mark 67-1 was tested aboard an ASPB in Chesapeake Bay in October 1970, and the mount was first ordered for LHA-class amphibious ships. Total weight, without gun and ammunition, was 475 pounds, compared with the standard Mark 10's 970 pounds (World War II). Elevation limits were +75° and −30°, compared with +90° and −15° for Mark 10. The mount typically carried 385 rounds of ammunition; Mark 10 carried a 60-round drum.

Mark 68 Designed specifically for small craft as a direct replacement for the existing (World War II) 20mm Mark 10 and twin Mark 24 mounts. It was a stopgap until Mark 67 became available. It carried the Mark 16-5 20mm gun, and its total weight of 775 pounds included a 237-pound shield and a 340-pound, deck-mounted stand. The deck stand was redesigned in Mod 1 to reduce overall weight by 225 pounds. Elevation limits were +70° and −45°, and the mount carried 400 rounds. Mark 68 was first tested aboard a DER in January 1970.

(*Mark 69* A proposed 8-inch/60 lightweight gun mount)

(*Mark 70* A proposed 8-inch/60 lightweight gun firing a longer round)

(*Mark 71* The lightweight 8-inch/55 gun)

(*Mark 72* Part of the Phalanx 20mm close-in weapon system)

Mark 73 (EX-73) The Emerlec 84MX mount developed for the ATC Mark 2. It could carry 20mm Mark 16, 0.50-caliber M2, or 40mm Mark 19 machine guns and was tested aboard the hydrofoil gunboat *Tucumcari* (PGH-2) in the twin 20mm configuration. EX-73 was designed for either local or remote control. In the former case, the gunner was located in a cockpit between the guns, controlling the mount electrically with a grip controller or manually with hand cranks. The mount had an unstabilized periscopic sight and vision ports. In the PGH-2 installation, ammunition was carried below the mount, within the barbette; each gun was fed from a 250-round ammunition box. Including the feed system and 500 rounds of ammunition, the system weighed about 2,200 pounds. Elevation limits were +65° and −15° (30° per second), and the mount could train at 60°/second. EX-73 was developed from an earlier Emerson Electric (Emerlec) mount, 82MX, developed and demonstrated as part of the private-venture R-ASP Mark 1 (see chapter 13). The hydraulically driven 82MX weighed 1,200 pounds and carried a 0.50-caliber M3 gun and a 40mm grenade launcher (XM129). R-ASP Mark II was to have had the Emerlec 83MX (one 20mm gun M3, one GAU-2 A/B minigun, and one XM129 grenade launcher).

Mark 74 (EX-74) The twin 30mm Emerlec gun mount developed for the CPIC. It was, in effect, an enlarged EX-73, with an on-mount capacity of 1,000 rounds per gun below deck. A planned Mod 1 version, with 400 rounds per gun, had all components entirely above deck. Total weight without guns and ammunition was about 2,800 pounds, and elevation limits were +80° and −15°; the gun trained at 80°/second. Emerlec proposed something similar, an 88MX carrying two Oerlikon 35mm guns, as an alternative to the OTO-Melara GDM-C twin 35mm mount and therefore as a weapon for both the PHM and what became the FFG 7–class frigate. Mark 74 has been widely exported, often aboard U.S.-supplied small patrol boats.

Mark 75 The OTO-Melara 76mm/62 mount installed, for example, in the *Pegasus*-class PHMs.

Mark 77 A lightweight 7.62mm Gatling gun (GAU-2 A/B minigun, designated Mark 25 in navy service) mount developed by NAD Crane for SEAL craft (MSSCs). Two experimental weapons were used suc-

Standard 21-inch PT-boat torpedo tubes aboard Elco boats of MTBRon 2, visiting the Washington Navy Yard in December 1940 or January 1941. There is a clear view of the training arcs of the forward tubes. Note that, unlike later Elcos, these boats had unmuffled exhausts. They also lacked smoke generators. The turrets are the unfortunate Dewandres.

cessfully in Vietnam from late 1967 on, and Crane was ordered to buy nine Mark 77s in February 1970. Because the requirement was urgent, the design employed stock army, air force, and GE parts. Mark 77 was tested in June 1970, and units were shipped by the end of that July. Others were later manufactured for additional small craft. The basic minigun rate of fire, 6,000 rounds per minute, was cut to 2,000 to 4,000; Mark 77 carried 3,000 rounds in four 750-round containers in an aluminum rack, feeding the gun through a flexible chute with booster assistance. All power was provided by a 28-volt aircraft battery. Weight without ammunition was 261.5 pounds, and elevation limits were +45° and −15°.

Mark 78 A twin M60 7.62mm machine gun developed by NAD Crane for the mini-ATC. There are two 600-round containers on the shield. Elevation limits are +45° and −15°. The Mark 78 designation was assigned in June 1972.

Mark 79 A twin 20mm machine gun (two Mark 16-5s in Mod 0, EX-29 in Mod 1).

Mark 80 The "Hip-Pocket II" anticruise-missile weapon, a manually operated 20mm Vulcan cannon with a stabilized computing sight. Mod 1 was controlled by a Mark 68 fire-control system, and Mod 2 was a proposed production version. The Mark 80 weighed about 5,000 pounds. This was, in effect, the army's Vulcan air defense system (VADS), a 20mm Gatling gun with a range-only radar and a gyro lead-computing sight. It was tested aboard the tanker

Canisteo (AO 99) in 1970, and the Naval Ordnance Laboratory at White Oak proposed that a VADS turret be tested aboard the hydrofoil *Flagstaff*.

Mark 81 A three-barrel XM197 Gatling gun for surface defense (Hip-Pocket II program) developed in 1972–73.

Mark 82 An M60 machine gun for 40-foot-plane personnel and rescue boats.

Mark 83 An unmanned mount for GE's GAU-8/A seven-barrel 30mm Gatling gun. Initially, the project was a private venture, given navy support in 1976. The Mark 83 is a component of the Dutch Goalkeeper CIWS system.

Mark 84 A universal gun mount for a lightweight Gatling gun (XM 197) or for several alternative weapons, including the 7.62mm machine gun. It was developed in 1977 for the Mark 3 patrol boat and weighed about 1,000 pounds. It was never adopted.

(*Mark 86* A 76mm OTO-Melara mount)

(*Mark 87* A 76mm OTO-Melara mount)

Mark 88 The Hughes helicopter (M242) 25mm chain gun, which uses M790 Oerlikon ammunition and fires 100 or 200 rounds per minute. The gun itself weighs about 230 pounds. M242 was tentatively adopted in 1985 as the new standard small boat gun.

Mark 89 An Emerlec stabilized mount for the 25mm chain gun in patrol boats. Total weight was about 1,800 pounds.

EX-90 The mount incorporating GE's EX-35 25mm Gatling gun.

(*EX-91* A universal, lightweight wheeled mount for M60, 0.50-caliber M2, and Mark 19 machine guns.)

(*EX-92* A manual mount for the 0.50-caliber XM218 aircraft machine gun in the UH-1N helicopter.)

Rocket Launchers

Mark 8 "Egg crate" launcher for 4.5-inch fin-stabilized rockets (nine per launcher). PTs also used a gravity-fed launcher.

Mark 13 Sextuple 2.36-inch (bazooka) launcher for PTs and PGMs, manufactured in 1944. Total weight was 175 pounds. It used a modified 0.30-caliber Mark 19 mount. It was initially designated the 2.36-inch Mark 1 rocket launcher, but all the rocket calibers were amalgamated in 1944–45.

Mark 50 Eight-rocket, 5-inch, spin-stabilized tube launcher for PTs, introduced in 1945. It was generally connected to the 20mm gun in the bows, which had a similar trajectory. Mark 40, which was unsuccessful, was a single-barrel, 3.5-inch launcher on a machine gun tripod.

Mark 104* A twelve-tube, 5-inch rocket launcher based on the Martin 250CH aircraft machine-gun turret, with the operator in the mount. Development appears not to have been completed. The designator was assigned in May 1945.

Mark 107 A postwar PT mount based on Maxson's quadruple 0.50-caliber machine-gun mount with nine closed-breech tubes for 5-inch SSRs. It was tested unsuccessfully aboard PT 812 in 1951. The designator was assigned in January 1946.

No higher numbers appear to have been assigned to small craft rocket weapons.

Outside these standard designations, many program 5 monitors were armed with the marine corps' standard Mark 4 105mm Howitzer gun-mount assemblies used aboard LVTs. These turrets were built of steel armor plate carrying 942-pound M49 cannon that could fire up to four rounds per minute. Total weight was 8,115 pounds, and elevation limits were +60° (59° without boost) and −4°. Maximum range was 12,300 yards for an HE shell with 43.8° elevation and a muzzle velocity of 1,550 feet/second. Note that the Sikorsky ASPB Mark II was armed, not with the existing weapon but with a new one, a Naval Ordnance Laboratory–adapted M137 howitzer (with two smaller-caliber coaxial weapons, 20mm Mark 16-4s in tests). It was operated by joystick and could be gyrostabilized. Total weight was about 8,000 pounds, and performance matched that of the Mark 4 (except that elevation limits were +55° and −7.5°). M137 was the standard army towed howitzer, and the turret was developed by Aircraft Armaments.

Similarly, other program 5 monitors carried the Mark 1 flame thrower, a modified army M10-8 (see chapter 12).

Several other weapons were tested during the Vietnam War:

- The M45C quadruple 0.50-caliber machine gun was adapted for shipborne use by White Oak under the Navy Science Assistance Program and tested in August 1971 aboard the Korean 95-foot gunboat PB-8 (ex–Coast Guard patrol boat).
- GE's three-barrel 20mm M197 Gatling gun was experimentally mounted on a Mark 10 (single 20mm) pedestal and test-fired aboard the 100-foot ex–Coast Guard tender *Brier* in June 1971. In 1972, a mount was bought by the Korean navy for testing. The Koreans placed it in production, but it was unsuccessful in their service.
- Cadillac-Gage's stabilized, armored cupola, a weapon proposed for the army M113 armored personnel carrier, was tested aboard a 45-foot utility boat in 1968. It carried a 20mm Hispano-Suiza cannon (M139) and weighed about 1,200 pounds.
- The Naval Ordnance Laboratory developed a lightweight quadruple M60 mount to replace the forward twin 0.50-caliber guns of a PBR. The

---

*In 1945, the series shifted to one beginning with 100.

It will probably never be possible to catalog all the weapons extemporized by PT crews. PT 174 and PT 168 are shown beached on a Pacific shore. PT 174 shows the shield of her unusual bow-mounted 40mm cannon, just behind the two men holding the ammunition box. Even more unusual is the 75mm recoilless rifle aboard PT 168, visible between the two boats, pointing down, mounted on a machine gun–type tripod. Recoilless rifles offered heavy firepower for their size, but their back-blast could be dangerous, and they were never official issue. The shrouded weapon forward is a 37mm cannon. (PT Museum)

The PGMs were the ultimate barge-busters. PGM 17, a converted 173-foot PC, is shown soon after completion at the Gibbs Gas Engine Company, Jacksonville, on 12 November 1944.

The water jet, or "douche," was another means of attacking bunkers. Here one mounted on an ATC, T-111-6, fires on the banks of the Vam Co Dong River, March 1969. The force of the stream causes the boat to list.

mount was to provide a higher rate of fire than a minigun with a simpler weapon. The weapon was specifically designed for simplicity of manufacture, so that it could be made in Vietnam. It was delivered to the Naval Research and Development Unit, Vietnam, in December 1970.

- The army M81E1 152mm gun/launcher (for the Shillalegh missile) was tested aboard the hydrofoil *Flagstaff* (PGH-1), as recounted in chapter 9. Total weight was about 7,700 pounds—the gun itself weighed 1,097 pounds—and elevation limits were +19° and −5°. It trained at 48°/second and achieved a maximum range of 9,800 yards with a muzzle velocity of 2,840 feet/second. The turret carried fifty rounds.
- Army 57mm and 106mm recoilless rifles were proposed and, occasionally, used in combat in Southeast Asia. They could achieve great ranges at very light weights, but their back blast was always a major problem.
- PTF-13 tested the Oerlikon twin 81mm rocket launcher in October 1967. Total weight, including eighteen rockets, was about 2,300 pounds, and maximum range was 10,900 yards. This weapon was bought and briefly used by the Ecuadorian navy.

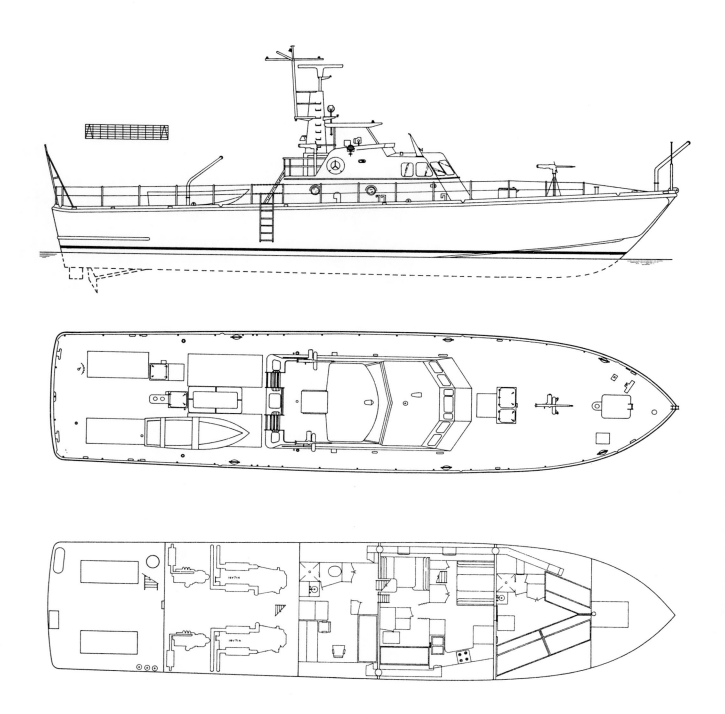

E

Small Combatants for Export

Several chapters of this book describe U.S. small combatants designed specifically for transfer abroad. Chapter 11 tells the story of Bertram boats, Sewart Swifts, and 85-footers; this appendix is limited to their successors. Unfortunately, chaotic records make it impossible to provide a complete account. It is particularly difficult to disentangle foreign naval purchases from what amount to foreign Coast Guard purchases. And, since the U.S. government acts as buyer for many foreign governments, it is difficult to identify the final owner of a particular boat. To compound the confusion, several builders may produce boats similar to a particular prototype, plans of which may become navy property. For example, Peterson built its own 50-footer, broadly similar but by no means identical to the original 50-foot Swift.

The major builders are Sewart/Swiftships, Peterson, Halter Marine, Lantana Boat, MonArk, and Robert E. Derecktor. In addition, Tacoma Boat and its Korean subsidiary built numerous craft in the Far East. In 1987, Tacoma completed two PSMM Mark 16 corvettes for Thailand. These were based on the PCG built for the Saudis.

Sewart/Swiftships

Sewart built the original Swifts. The first Sewart export, to Jamaica, was the 85-foot *Discovery Bay*; sales of the earlier 85-footers are described in chapter 12.

The Sewart company, of Berwick, Louisiana, was bought by a larger firm, Teledyne, and became its Sewart division. Then, after being bought out by several Teledyne employees, it became Swiftships, headquartered in Morgan City, Louisiana. The company claims that it has built more aluminum boats than any other in the world. It now has a Singapore division, which constructs military as well as civilian boats. It is not clear to what extent the Singapore operation builds Louisiana-designed craft; its operations are not reflected in the notes below.

- 65-foot patrol boat. In December 1966, Sewart received a contract for thirteen 65-footers (65NS671–6713): nine for Korea, one for El Salvador, one for the air force, and two for the army. Six were ordered in FY 71 (65NS711–716) for the Philippines, and then two, built to a navy design, in FY 72 (65PB721–722). Further exports of the Sewart with various engines included Antigua (one; 23 knots; 1984), Bahrein (two; 1982), Costa Rica (five; 1978), Dominica (one; 23 knots; 1984), Dubai (two; 1977), Guinea (one; 1985), Haiti (three; 1976), Honduras (five, of which two, originally ordered for Haiti, were delivered 1973–74, and three, ordered in 1979, were delivered 1980; MTU engines; 36 knots), Liberia (two; 1976), Nigeria (six ordered in 1985; MTU diesels; 1,930 bhp; four delivered in 1986), Panama (two; 1982), and St. Lucia (one; 23 knots; 1984). Except as noted, these boats had two GM 12V71 diesels for a speed of about 23 to 25 knots. The alternative MTU 331 installation drove them at about 30, although higher speeds were claimed.

A list of 65-foot aluminum boats that Swiftships has completed since July 1970 shows two survey boats for the Army Corps of Engineers, twenty aluminum utility boats, seven Mark 1 Swifts (presumably Mark 1 patrol boats), and thirty-four patrol boats (probably including the two Mark 1 Mod 1 prototypes of FY 72). They were all U.S. Navy or State Department orders. The totals exclude boats for Bahrein, Haiti, and Nigeria, which received a total of nine. The other post–FY 71 deliveries listed above total twenty-four, but part of the difference may be Mark 3 boats built for the Philippines. Total production of 65-footers may have been as high as 200 boats, since some were also used for oil field support.

The Sewart 85-foot patrol boat, the first type that company exported. (USN)

Sewart built twelve 65-footers, the *Dabur* class, for Israel. She built another thirty for her own use and exported four more to Nicaragua in 1978 and five of her original boats to Lebanese Christian forces in 1976. The Israeli *Dvora* (1978), the only ship of her class, is a *Dabur* lengthened to 71 feet; she is advertised as the smallest missile boat in the world, capable of carrying two Gabriel antiship missiles. Built as a private venture by Israel Aircraft Industries, the *Dvora* was acquired by the Israeli navy in 1979. Four ships of the same design were exported without missiles to Argentina in 1978, two to Nicaragua (embargoed in 1979 at the request of the U.S. government), and six to Sri Lanka in 1984. Sri Lanka also received eight "super *Dvora*s" in October 1986. The Taiwanese *Hai Ou* class, which carries two Hsiung Feng missiles, themselves copies of the Gabriel, is probably a *Dvora* derivative; some fifty such units were built after 1980.

All of these craft are to be distinguished from the U.S. 65-foot patrol boat; Sewart reportedly built at least twenty of the official design, with the pilothouse offset to starboard, for the Philippines, delivering them in 1972–76. Some of these may actually have been built in the Philippines. They are numbered PCF 333–52 in Philippine service. Originally, thirty-six of the boats were planned.

Typical characteristics of the Haitian boats were as follows: 65 feet-2¼ inches (overall) 58-9½ (waterline) x 18-3½ (maximum) x 14-7¾ (full load waterline) x 8-10½ (depth amidships); 55,000 pounds light load; 69,500 pounds normal load; 79,970 pounds full load. At normal load, with two General Motors 12V71T diesels and 2,300 rpm, speed was 24.0 knots (21.6 fully loaded). Total fuel supply was 1,500 gallons in four tanks (550-gallon forward tanks, 200-gallon after tanks). Swiftships sketched various alternative roles and armaments for its 65-footer, including the gunboat role (one 81mm mortar/0.50-caliber machine gun forward, two twin 0.50-caliber machine guns in scarf rings abaft and above the bridge, and a 40mm cannon aft); the patrol boat role (a single 0.50-caliber machine gun forward, an 81/0.50 aft, and twin 0.50s in scarf rings); the picket boat role (a single 0.50 forward, twin 0.50s aft and in the scarf rings atop the superstructure); the minelayer role (no gun forward, 40mm cannon aft, 0.50s in the two scarf rings atop the superstructure, and eight mines along the deck edge); and the torpedo boat role (20mm cannon forward, 40mm cannon aft, no scarf rings, and two 21-inch tubes aft). The utility boat version was entirely unarmed but could be converted to one of the armed versions. By way of comparison, the Israeli 65-footers have 20mm cannon fore and aft and two single 0.50-caliber machine guns.

Sewart also built the:
- 77-foot aluminum patrol boat, one for El Salvador and one for Guinea in 1985. Characteristics of the Salvadoran boats are as follows: 77 x 20 x 5 feet; 48 tons; 3 GM 12V71 T1 diesels, 1,200 bhp, 26 knots; one 25mm Mark 88 gun and two 0.50-caliber machine guns.
- 80-foot steel patrol boat, three for Kuwait in 1983.
- 93-foot steel patrol boat, nine for the Egyptian coast guard. Six boats were assembled in Egypt and were in service in 1985–86. Characteristics are as follows: 93 x 18.6 x 5.2 feet; 102 tons fully loaded; 2 MTU 12V331 TC92 diesels, 2,660 bhp, 27 knots; 11.7 tons of fuel; 1,000 nautical miles at 12 knots.
- 103-foot boat, one for Jamaica in 1974, the *Fort Charles*. Characteristics are as follows: 103 x 18.7 x 6.9 feet; 103 tons fully loaded; 2 MTU MB 16V538 TB90 diesels, 7,000 bhp, 32 knots; 1,200 nautical miles at 18 knots; one 20mm and two single 0.50-caliber machine guns.
- 105-foot boat, three for Burma in 1979, one for Costa Rica, two for the Colombian coast guard in 1981 and 1983, four for Ethiopia in 1976 (two more were canceled during the American arms embargo), one for Gabon in 1976, three for Honduras (one in 1977, two in 1980), and one for the Sudan in 1985. Aside from the Burmese boats, Swiftships lists a total of fourteen 105-footers procured for the U.S. Navy, presumably for export. Characteristics of the Costa Rican boat are as follows: 105 x 23.3 x 7.1 feet; 118 tons fully loaded; 3 MTU 12V331 TC 92 diesels (as refitted in 1984–85), 10,500 bhp, 36 knots. The Colombian boats have 2 MTU 12V331 TC92 diesels, 7,000 bhp, 25 knots, and displace 103 tons fully loaded; they are armed with a 40mm gun aft and two 0.50-caliber machine guns. With the same power, the Ethiopian boats are credited with a speed of 32 knots. Externally, these boats are similar to the 65-footers.
- 110-foot boat, two for the Dominican Republic in 1984, one for St. Kitts in 1985. In 1986, Swiftships listed a total of three 110-foot aluminum patrol boats, delivered to the U.S. Navy for export, plus two similar 120-footers, completed in May 1984 and October 1986 and otherwise not accounted for. St. Vincent ordered a 120-footer in August 1986. Externally, these boats are different from earlier Sewart designs. Characteristics are as follows: 110 x 24 x 6 feet; 93.5 tons fully loaded; 3 GM 12V92 TI diesels, total 2,700 bhp, 23 knots.

Swiftships also builds smaller patrol boats. They include a 34-footer (two delivered to Costa Rica in 1985), a 38-footer (thirty ordered by the Cameroons in 1986), and a 42-footer (one delivered to Costa Rica in 1985). In addition, a 42-foot aluminum patrol boat was delivered to the U.S. Coast Guard in January

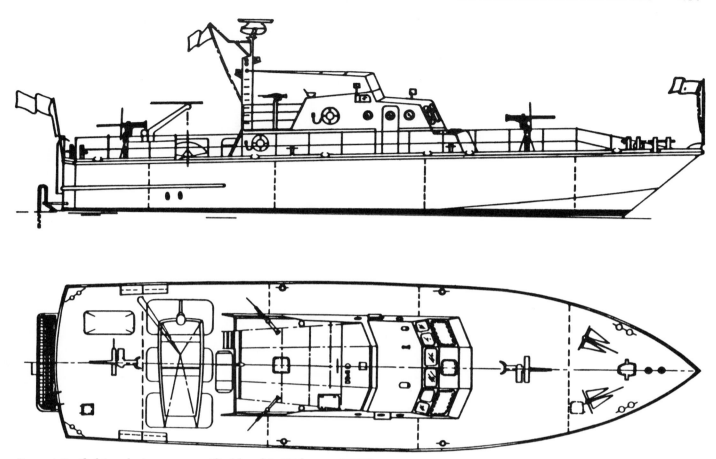

Recent Swiftships designs are typified by this 77-footer. (USN)

1971, and one was delivered to the U.S. Navy in June 1976; Swiftships claims further Coast Guard sales of this boat around the world but has supplied no details.

Peterson Builders

Peterson, of Sturgeon Bay, Wisconsin, was a major builder of U.S. motor gunboats for export and is the lead yard for the U.S. minesweeper program. The present company was formed in 1933. It supplied Iran with twenty Mark II Swifts in 1976–77. In addition, Iran ordered twenty Peterson 50-foot patrol craft in 1971, sixty-one more in 1976. All were shipped as kits, and assembly continued into the 1980s. In the early 1980s, Peterson tried, apparently unsuccessfully, to market a heavily armed version of its 50-footer in conjunction with Sperry.

Halter Marine (Halmar)

Halter Marine, in New Orleans, was founded in 1957. It claims to be the largest U.S. small craft builder, with over 1,200 craft to its name. It builds both aluminum and Corten steel hulls. As of December 1986, Halter craft for foreign navies (names in parentheses are those used by the company) included the:

- 36-foot mini-ATC ("Machete"), two for Guatemala. This is the navy design originally built by Sewart.
- 36-foot fast launch ("Barracuda"), one for the Ivory Coast in 1976. Characteristics are 36 x 12.5 x 2 feet, 6 tons, two GM 6V53PI diesels with water jets, 540 bhp, 36 knots, and a 20-man crew.
- 41-foot fast patrol boat, twenty built.
- 44-foot aluminum patrol boat, two for Ecuador plus kits for four more in late 1986.
- 47-foot fast patrol boat, two for an unspecified West Indian country.
- 50-foot aluminum fast patrol boat ("Rapier"), two for Kuwait's coast guard, twelve for Saudi Arabia's frontier forces. Characteristics are 50-0 x 15-0 x 3-4 feet-inches; 450 gallons of fuel; two GM 12-V71TI diesels, 1,350 bhp, 30 knots.
- 65-foot fast patrol boat ("Cutlass"), five for the Guatemalan navy in 1972–76; three for the Ecuadorian navy in 1976 ("Port Director" class); three for Thai customs; two for Pakistan's port authority (64-6 x 17-0 x 4-1 feet-inches, 34 tons; 1,000 gallons of fuel; two GM 12-V71T1 engines, 1,350 bhp, 26 knots). One slightly smaller boat under the same class name was built for Kuwait

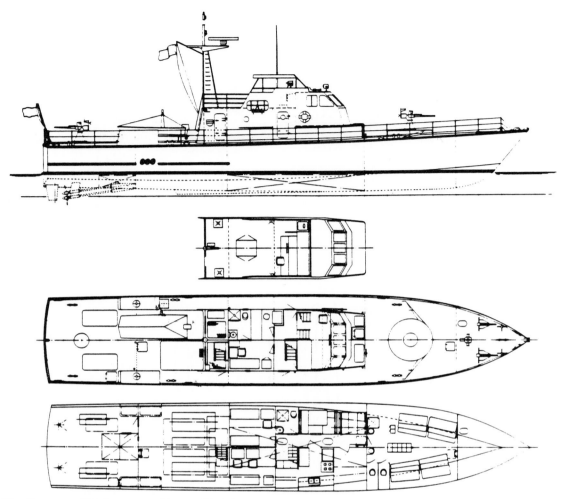

The Lantana 106-foot patrol boat. (USN)

in 1979. The hull design is based on that of a 65-foot crew boat, 112 of which have been built since 1964.
- 75-foot aluminum fast patrol boat, contract for two under negotiation in 1986 with an unnamed South Pacific state. Characteristics are 75-0 x 18-0 x 6-2 feet-inches, 44 tons; 2,750 gallons of fuel; three MTU 12V396TB93 engines, 38 knots.
- 78-foot fast patrol boat, two for Taiwan customs in 1977; two for Nigerian customs; two for an unspecified West Indian country; two pilot/search and rescue craft for Sharjah/United Arab Emirates. Characteristics are 78-0 x 18-6 x 4-8 feet-inches; 2,800 gallons of fuel; two GM 12V71T1 engines, 1,350 bhp, 24 knots. This boat is based on a crew boat hull. The first Halmar 78 was reportedly built in 1982, and twenty-three had been completed by August 1985 (presumably including nonmilitary craft).
- 85-foot fast patrol boat, one for the Saudi frontier force. Characteristics are 84-6 x 18-6 x 4-9 feet-inches; 3,000 gallons of fuel; two GM 12V71T1 engines, 1,350 bhp, 24 knots.
- 105-foot aluminum fast patrol boat (Broadsword), one for the Guatemalan navy in 1976. Characteristics are 105-0 x 20-5 x 6-3 feet-inches; 5,400 gallons of fuel; two GM16V149TIB engines, 3,200 bhp, 29 knots.

Halter and Bell jointly developed a series of surface-effect craft, one of which was built for the U.S. Coast Guard and tested by the U.S. Navy and three of which (110-foot) were sold to the Coast Guard.

Lantana Boatyard

Lantana Boatyard was established in 1965 and began building military boats in 1980. Its patrol boats are intended primarily for civil action and search and rescue and for Third World navies wishing to patrol their exclusive economic zones. Three designs have been developed:
- The 106-foot patrol boat, one to Grenada in 1984, two to Honduras in 1983 and 1986, and one to Jamaica in 1985. This is a fast, 33-knot boat meant for five-day missions. It has a radius of 1,500 nautical miles. The design was begun in February 1982, the first boat completed in October

1983. Characteristics for the Jamaican boat are 106 x 20.5 x 6.9 feet, 48.2 tons light, 75 tons fully loaded; three MTU 8V396TB93 diesels, a total of 3,600 bhp, 33 knots maximum, 30 knots cruising. The others have three GM 16V92TIs for 3,900 bhp and 35 knots.
- The 68-foot coastal patrol boat. Sales include one for equatorial Guinea in 1986 (delivery date is November 1987); the boat is armed with two 0.50-caliber and two 7.62mm machine guns.
- The 40-foot river patrol boat, six for El Salvador in 1987 and eight for Honduras in 1986. Characteristics are 36 x 10 x 1.7 feet; 8.16 tons; two Caterpillar 3208TA diesels, 630 hp, 26 knots (22 sustained); two 0.50-caliber and two 7.62mm machine guns. The boats are aluminum, with Kevlar armor. The design is comparable to that of the PBR, but it uses propellers in tunnels instead of pump-jets, on the ground that they are easier to maintain and less easily clogged.

MonArk Boats

MonArk, located in Monticello, Arkansas, is a relative newcomer to the combatant craft field. Recent products include:
- The 18-foot chase boat, fifteen for Venezuela in 1985.
- The 21-foot boat for river and lake patrol, ten for Venezuela in 1984.
- The 26-foot "Stinger," two for the Republic of Guinea, which were delivered in 1985. Characteristics are 26-0 x 10-8 x 3-0 feet-inches, 6,000 pounds; 125 gallons of fuel; two 155-bhp outboards, 35 knots maximum, 32 knots cruising speed; range 180 miles at 32 knots.
- The 40-foot patrol boat, nine for the U.S. Air Force. Characteristics are 40-0 x 12-8 x 3-6 feet-inches; two Detroit diesel 8V71T1 engines, 28 knots maximum, 24 knots in sea state 4; 350 gallons of fuel, range 340 miles at 24 knots. MonArk also built nine 40-footers with twin 6V71-N diesels, 25 miles per hour, for the Haitian navy (3812-VCF class).
- The 42-foot patrol boat, twelve for the Venezuelan national guard in 1984. Characteristics are 41-2 x 14-4½ x 3-10 feet-inches with twin Detroit Diesel 8V92TI diesels.

The company has also built a widely publicized air force parachute-training boat.

R.E. Derecktor

Robert E. Derecktor of Newport, Rhode Island, is a major builder of yachts and medium Coast Guard cutters. The boats listed below were built at its Mamaroneck, New York, yard.

Twelve patrol boats were delivered to the Venezuelan national guard in 1982 and 1984. The first six were ordered in 1980, the second six in 1982. Three of the first group had small helicopter pads and three had ramps to land small vehicles. Patrol boat characteristics are 76.9 x 16.0 feet; about 50 tons; two GM 12V92MTI diesels, 1,950 bhp, 28.5 knots.

The list above is by no means exhaustive. American Shipbuilding and Designs of Miami delivered three small patrol boats to Peru in 1982. Characteristics are 4.8 tons, 27 knots, 33 feet (length). They are cabin cruisers built of GRP with Kevlar armor, with two Perkins ST-6-354-4M diesels, two shafts, 480 bhp.

Bertram delivered four Enforcers to Jordan in 1974, four to Haiti in 1974, and twenty to Egypt in 1980. It also delivered two smaller patrol boats to Jordan in 1974, twelve 36-foot patrol boats to Venezuela in 1980, and ten 43-foot patrol boats to Venezuela in 1983.

Baycraft delivered fourteen 40-foot patrol craft to Ecuador in 1979–80.

Camcraft of New Orleans delivered three 100-ton patrol craft to El Salvador in 1974–75 and five 70-ton patrol craft to the United Arab Emirates in 1975.

Cougar USA of Miami Beach separated from its British parent firm in 1986; both build extremely fast craft as ocean-racer lines. In 1987, the company announced that an unspecified number of its 42-foot seagoing patrol boats, 30-foot catamaran river-patrol boats, and 30-foot ambulance craft had been sold to a Latin American country. The 42-footer is powered by two Ford diesels and can exceed 50 knots.

Hope/Progressive Shipbuilding of Houma, Louisiana, delivered a river patrol boat to Bolivia that was 67 feet long and powered by two GM diesels.

Magnum Marine of North Miami Beach, Florida, delivered twenty-seven 27-foot patrol boats to Kuwait in 1977–78.

Neville Boatyard in Louisiana is reported to be building two 50-foot patrol craft for Ecuador. Four more are to be built in Ecuador with U.S. assistance.

Phoenix Marine of Florida delivered four 25-foot boats to the Bahamas in 1981–82.

Many of these yards, located on the Gulf of Mexico, are in the oilfield support-craft business, which has been badly hurt by the decline in oil prices. Export military craft may be their only viable alternative products.

F

Fates and Other Notes

The following list is divided into gunboats (PG), river gunboats (PR), frigates (PF, originally PG), motor gunboats (PGM, some of which were later PG), hydrofoil gunboats (PGH), hydrofoil missile boats (PHM), subchasers (SC/PC/PCE/PCS), hydrofoil subchasers (PCH), eagle boats (PE), and torpedo boats (PT/BPT/PTC/RPC/PTF). The name of the building yard is always given before the ship's name. For compactness, dates of decommissioning and recommissioning are generally not included; in any case, very few of the ships here were ever recommissioned after decommissioning.

Note that numbers in dates stand for month, day, and year and for month, year when there are only two numbers.

Key

Acq	Acquired
BU	Broken up
Canc	Canceled
CG	Coast Guard
D	Destroyed
FLC	Foreign Liquidation Commission (disposed by)
MC	Maritime Commission (transferred to)
Res	To reserve ("mothball") fleet
Ret	Returned
RN	Royal Navy
Str	Stricken
WAA	War Assets Administration (disposed by)
WL	War loss
WL(B)	War loss to aircraft (bombing or other)
WL(G)	War loss to gunfire
WL(M)	War loss to a mine
WL(T)	War loss to torpedo

	Laid Down	Launched	Completed	Fate
GUNBOATS (PG)*				
Cramp Shipbuilding				
1 *Yorktown*	5.14.87	4.28.88	4.23.89	Sold 9.21
N. F. Palmer				
2 *Concord*	5.88	3.8.90	2.14.91	Public Health Service quarantine ship 1914–29 Sold 6.29

Some of the old gunboats served throughout World War II. *Top:* The *Sacramento* (PG 19), returning from the Asiatic Fleet, steams up the East River to the Brooklyn Navy Yard in January 1940. She flies her homeward-bound pennant and junk sail (on the mainmast). The *Sacramento* was armed with two 4-inch/50 single-purpose guns. *Below:* The similar gunboat *Tacloban* (formerly *Tulsa*, PG 22) at the end of World War II. These were the only older gunboats to survive the war. The *Tulsa* began her war duty in the Philippines, where her crew constructed a depth-charge rack so that she could serve as a convoy escort. She received British sonar, degaussing, depth-charge throwers, and 20mm guns in Australia in the spring of 1942, then operated as a submarine training target at Freemantle and as a convoy escort and PT-boat tender in the South Pacific. This unusual history was revealed by her armament, which in 1945 included not only five 3-inch/50 dual-purpose guns, six 20mm Mark 4s, and six 0.50-caliber machine guns, but also four K-guns and a special depth-charge rack for six 200-kg Dutch charges. She was renamed in 1944 to free the name *Tulsa* for a new heavy cruiser which, in the event, was never completed. (Photos by Theodore N. Silberstein and D. M. McPherson)

---

*PG numbers were reserved for all numbered gunboats in existence in 1918, but some ships were sold before the formal hull number system became effective. Also, note reclassifications as unclassified ships (IX).

462 U.S. SMALL COMBATANTS

	Laid Down	Launched	Completed	Fate
Colombian Iron Works				
3 *Petrel*	8.27.87	10.13.88	12.10.89	Sold 11.20
N. F. Palmer				
4 *Bennington*	6.88	6.3.90	6.20.91	Sold 11.10 Severely damaged by boiler explosion 7.21.05
Bath Iron Works				
5 *Machias*	2.91	12.8.91	7.20.93	Mexico 1920 as *Agua Prieta*
6 *Castine*	2.91	11.5.92	10.22.94	Merchant ship 1921
Newport News				
7 *Nashville*	8.9.94	10.10.95	6.25.97	Sold 10.20.21
8 *Wilmington*	10.8.94	10.19.95	5.13.97	IX 30 6.26.22 *Dover* 1.41 Scuttled 1947
9 *Helena*	10.11.94	1.30.96	5.24.97	Sold 7.7.32
Crescent				
10 *Annapolis*	4.96	12.23.96	7.20.97	IX 1 7.1.21 Merchant ship *Keystone State* 4.40 BU 1950
Bath Iron Works				
11 *Vicksburg*	3.96	12.5.96	7.8.97	CG 5.2.21 as *Alexander Hamilton* Sold 3.46
12 *Newport*	3.96	12.5.96	10.5.97	IX 19 7.1.21 Str 10.31
J. H. Dialogue				
13 *Princeton*	5.96	6.3.97	5.27.98	Sold 11.19
Union Iron Works				
14 *Wheeling*	4.11.96	3.18.97	8.10.97	IX 28 7.1.21 Sold 10.46
15 *Marietta*	4.13.96	3.18.97	9.1.97	Sold 3.20
Mare Island Navy Yard				
16 *Palos*	4.28.13	4.23.14	6.24.14	PR 1 6.15.28 Sold 6.3.37
Gas Engine/Power				
17 *Dubuque*	9.22.03	8.15.04	5.31.05	Sold 11.23.46
18 *Paducah*	9.22.03	10.11.04	8.31.05	MC and merchant ship 12.19.46 BU 1950
Cramp Shipbuilding				
19 *Sacramento*	4.30.13	2.21.14	4.26.14	MC 8.23.47 Merchant ship *Fermina*
Mare Island Navy Yard				
20 *Monocacy*	4.28.13	4.27.14	5.20.14	PR 2 on 6.15.28 Str 2.10.39
Charleston Navy Yard				
21 *Asheville*	6.9.17	7.4.18	7.6.20	Asiatic Fleet 1941 WL 3.3.42
22 *Tulsa*	12.9.17	8.25.22	12.3.23	Renamed *Tacloban* 11.44 Asiatic Fleet 1941 Sold 10.12.46
Harlan and Hollingsworth				
23 *Nantucket*	1973	1976	11.27.76	IX 18 7.1.21 Ex-*Rockport*, ex-*Ranger*, renamed 2.20.18 Survivor of the Old Navy Designated PG-23 in 1920
John Roach				
24 *Dolphin*	10.11.83	4.12.84	12.8.85	Sold 2.25.22 Built as a dispatch boat or small cruiser
25–26				Not used
City Point Works				
27 *Marblehead*		10.90	8.11.92	1.8.94	Sold 8.5.21 Former small cruiser
28–34				Rerated CL 16–22 8.8.21 PG 36 became CL 23
Howaldt, Kiel				
35 *Topeka*				Acq 4.2.98 Rated as gunboat IX 35 on 7.1.21 Sold 5.30 Built for Portugal 1881 Sold to Peru Laid up in the Thames until 1895 Sold to Japan but British government held up delivery
Manila Slip				
37 *Callao*				Acq 11.9.99 YFB 11 (ferry) 6.21.21 Sold 9.23 Spanish War prize 208 tons
Cadiz				
38 *Elcano*		1.28.84		Acq 11.9.99 D in test 10.9.28 Spanish War prize Originally purchased by War Department (army) 560 tons
Manila Slip				
39 *Pampanga*		1.28.88		Acq 11.9.99 D 11.21.28 Spanish War prize Originally purchased by War Department 201 tons
Hong Kong and Kowloon Dock				
40 *Quiros*		1.28.95		Acq 2.21.00 D in test 10.16.23 Spanish War prize Originally purchased by War Department About 350 tons
Manila Slip				
41 *Samar*				Acq 11.9.99 Sold 1.11.21 Spanish War prize About 210 tons
Hong Kong and Kowloon Dock				
42 *Villalobos*		1.28.96		Acq 2.21.00 D in test 10.9.28 Spanish War prize Originally purchased by War Department 347 tons
Kiangnan Dock and Engine Works				
43 *Guam*	10.17.26	5.28.27	12.28.27	PR 3 on 6.15.28
44 *Tutuila*	10.17.26	6.14.27	3.2.28	PR 4 on 6.15.28
45 *Panay*	12.18.26	11.11.27	9.10.28	PR 5 on 6.15.28
46 *Oahu*	12.18.26	11.26.27	10.22.28	PR 6 on 6.15.28
47 *Luzon*	11.20.26	9.12.27	6.1.28	PR 7 on 6.15.28
48 *Mindanao*	11.20.26	9.28.27	7.10.28	PR 8 on 6.15.28

	Laid Down	Launched	Completed	Fate
New London Shipyard				
49 *Fulton*	10.2.13	6.6.14	12.2.14	Sold 6.6.35 Former submarine tender AS 1
New York Navy Yard				
50 *Erie*	12.17.34	1.29.36	7.1.36	WL 11.12.42
Charleston Navy Shipyard				
51 *Charleston*	10.27.34	2.25.36	7.8.36	To Massachusetts Maritime Academy 3.25.48
Bath Iron Works				
52 *Niagara*				Acq 10.40 AGP 1 on 1.13.43 Ex-*Hi-Esmaro* Completed 1929
Germania Werft				
53 *Vixen*				Acq 11.13.40 MC 1.21.47 Ex-*Orion* Completed 8.29
Newport News				
54 *St. Augustine*				Acq 12.5.40 WL 1.6.44 Ex-*Noparo* Completed 4.29
Pusey and Jones				
55 *Jamestown*				Acq 12.40 AGP 3 in 1.13.43 Ex-*Alder* Completed 1927
Bath Iron Works				
56 *Williamsburg*				Acq 4.41 AGC 369 in 11.10.45 Ex-*Aras* Completed 1930 Merchant ship *Anton Bruun* 1963
Germania Werft				
57 *Plymouth*				Acq 11.4.41 WL 8.5.43 Ex-*Alva* Completed 1931
Bath Iron Works				
58 *Hilo*				Acq 11.41 AGP 2 in 1.13.43 Ex-*Moana* Completed 1931
59 *San Bernardino*				Acq 1.20.42 MC 10.15.46 Ex-*Vanda* Completed 11.28 Merchant ship *Vanda* 1946
Krupp Iron Works				
60 *Beaumont*				Acq 1.23.42 MC 2.20.47 Ex-*Carola* Completed 10.30 Merchant ship *Elpetal* 1947
Great Lakes Engineering				
61 *Dauntless*				Acq 1.21.42 MC 6.10.46 Ex-*Delphine* Completed 5.21
Smith Dock				
62 *Temptress*	7.9.40	10.17.40	2.18.41	Acq 3.21.42 Ret 8.26.45 Ex–HMS *Veronica* Merchant ship *Verlock* Lost 1.47
J. Crown				
63 *Surprise*	10.23.39	6.5.40	9.12.40	Acq 3.24.42 Ret 8.26.45 Ex–HMS *Heliotrope* Merchant ship *Heliolock* 1946 *Ziang Teh* 1947 Rearmed as PRC *Lin I*
Harland and Wolff				
64 *Spry*	11.14.39	4.6.40	5.21.40	Acq 5.2.42 Ret 8.26.45 Ex–HMS *Hibiscus* Merchant ship *Madonna* 1947
65 *Saucy*	10.30.39	2.14.40	4.5.40	Acq 4.30.42 Ret 8.26.45 Ex–HMS *Arabis* HMS *Snapdragon*, then merchant ship *Katina* 1947
66 *Restless*	10.30.39	2.24.40	4.8.40	Acq 3.15.42 Ret 8.26.45 Ex–HMS *Periwinkle* Merchant ship *Perilock* 1947
67 *Ready*	10.30.39	3.21.40	5.6.40	Acq 34.12.42 Ret 8.23.45 Ex–HMS *Calendula* Merchant ship *Villa Cisneros* 1948
Cook Welton				
68 *Impulse*	4.13.40	9.18.40	3.3.41	Acq 3.16.42 Ret 8.22.45 Ex–HMS *Begonia* Merchant ship *Begonlock* 1946
Fleming and Ferguson				
69 *Fury*	3.26.40	9.5.40	1.4.41	Acq 3.17.42 Ret 8.22.45 Ex–HMS *Larkspur* Merchant ship *Larkslock* 1946
Harland and Wolff				
70 *Courage*	11.14.39	4.20.40	6.4.40	Acq 3.18.42 Ret 8.23.45 Ex–HMS *Heartsease* Ex-*Pansy* Merchant ship *Roskva* 1951 Lost 12.58 as *Seabird*
Grangemouth Dockyard				
71 *Tenacity*	10.31.39	7.8.40	10.16.40	Acq 3.4.42 Ret 8.26.45 Ex–HMS *Candytuft* Merchant ship *Maw Hwa* 1947
Krupp Iron Works				
72 *Nourmahal*			6.17.28	CG 8.21.40 Acq 3.3.42 CG 12.29.43 Decomm 5.30.46
W. and A. Fletcher				
73–84				Not used
85 *Natchez*				AGS 3 in 6.2.42
Collingwood Shipyard				
86 *Action*	1.6.42	7.28.42	11.22.42	MC 10.18.46 Laid down as HMS *Comfrey* Merchant ship *Arne Presthus* 1952
87 *Alacrity*	1.6.42	9.4.42	12.10.42	MC 9.22.47 Laid down as HMS *Cornel* Merchant ship *Rio Marina* 1948
88 *Beacon*	4.27.42	10.31.42	5.19.42	MC 9.30.47 Laid down as HMS *Dittany* RN 3.43 Merchant ship *Olympic Cruiser* 1950
Kingston Shipbuilding				
89 *Brisk*	2.28.42	6.15.42	12.5.42	MC 10.18.46 Laid down as HMS *Flax* Merchant ship *Brisk* 1947

	Laid Down	Launched	Completed	Fate
90 *Caprice*	6.16.42	9.28.42	5.20.43	Sold 12.10.46 Laid down as HMS *Honesty* and RN 5.43
Midland Shipyard				
91 *Clash*	6.1.42	11.18.42	6.1.43	MC 1.15.48 Laid down as HMS *Linaria* and RN 6.43 Merchant ship *Porto Offuro* 1948
Morton Engine and Drydock				
92 *Haste*	12.11.41	8.22.42	3.30.43	MC 9.22.47 Laid down as HMS *Mandrake* Merchant ship *Porto Azzuro* 1949
93 *Intensity*	12.11.41	8.5.42	3.25.43	MC 10.18.46 Laid down as HMS *Milfoil* Merchant ship *Olympic Promoter* 1950
94 *Might*	11.28.41	7.15.42	12.22.42	MC 10.18.46 Laid down as HMS *Musk* Merchant ship *Olympic Explorer* 1950
95 *Pert*	7.23.42	11.27.42	7.20.43	MC 10.18.46 Laid down as HMS *Nepeta* Merchant ship *Olympic Leader* 1950
96 *Prudent*	8.14.42	12.4.42	8.16.43	MC 9.22.47 Laid down as HMS *Privet* Italian *Elbano* 1949
Kingston Shipbuilding				
97 *Splendor*	10.1.42	4.1.43	7.27.43	MC 11.19.46 Laid down as HMS *Rosebay* and RN 7.43 Merchant ship *Benmark* 1947
Collingwood Shipyard				
98 *Tact*	6.12.42	11.14.42	6.21.43	MC 10.18.46 Laid down as HMS *Smilax* and RN 6.43 Argentine *Republica* 1946
99 *Vim*	8.11.42	4.1.43	9.20.43	Sold 5.7.47 Laid down as HMS *Statice* and RN 9.43
Midland Shipyard				
100 *Vitality*	6.26.42	5.1.43	8.30.43	Sold 5.7.47 Laid down as HMS *Willowherb* and RN 8.43
101				4.15.43 PF 1 (see below)
102				4.15.43 PF 2 (see below)
Canadian Vickers				
103	4.20.42	9.26.42	4.30.43	MC 3.17.47 Completed as HMS *Barle*
104	5.4.42	10.24.42	5.10.43	MC 5.5.48 Completed as HMS *Cuckmere*
105	5.29.42	11.9.42	5.30.43	Sold 12.10.46 Completed as HMS *Evenlode*
106	8.24.42	12.5.42	6.20.43	MC 3.17.47 Completed as HMS *Findhorne*
107	9.14.42	12.5.42	7.17.43	Sold 12.15.46 Completed as HMS *Inver*
108	10.1.42	4.29.43	8.12.43	Sold 11.13.46 Completed as HMS *Lossie*
109 *Parrett*	10.31.42	4.29.43	8.31.43	MC 5.5.48 Completed as HMS *Parrett*
110	11.18.42	5.27.43	9.9.43	Sold 12.22.46 Completed as HMS *Shiel*

PG 111–210
(renumbered PF 3–102 4.15.43)

Asheville Class (PG)*

Tacoma Boat Building

	Laid Down	Launched	Completed	Fate
84 *Asheville*	4.15.64	5.1.65	8.6.66	Str 1.31.77 To Massachusetts Maritime Academy
85 *Gallup*	4.27.64	6.15.65	10.22.66	Transfer to Taiwan reported 1981, but not transferred as of 1986 Str 1.31.77 but restored to list 7.17.81
86 *Antelope*	6.1.65	6.18.66	11.4.67	Environmental Protection Agency 1977
87 *Ready*	6.7.65	5.12.67	1.6.68	Str 10.1.77 To Massachusetts Maritime Academy
88 *Crockett*	6.18.65	6.4.66	6.24.67	Environmental Protection Agency 1977
89 *Marathon*	6.21.66	4.22.67	5.11.68	Str 1.31.77 To Massachusetts Maritime Academy
90 *Canon*	6.28.66	7.22.67	7.26.68	Transfer to Taiwan reported 1981, but not transferred as of 1986 Str 1.31.77 but restored to list 7.17.81
92 *Tacoma*	7.24.67	4.13.68	7.14.69	Colombia's *Quito Sueno* 5.83
94 *Chehalis*	8.15.67	6.8.68	8.11.69	Reclassified as high-speed tug *Athena I* for NSRDC 8.75
96† *Benicia*	4.14.69	12.20.69	4.25.70	Korea's *Paek Ku 51* on 10.2.71
98† *Grand Rapids*	5.20.69	4.4.70	9.5.70	Reclassified as high-speed tug *Athena II* for NSRDC 10.77
100 *Douglas*	8.8.69	6.19.70	2.6.71	Str 10.1.77 Plans for reclassification as high-speed tug *Athena III* for NSRDC FY 83 but discarded instead 12.84.

Julius Petersen

	Laid Down	Launched	Completed	Fate
93 *Welch*	8.8.67	7.25.68	9.8.69	Colombia's *Albuquerque* 5.83
95 *Defiance*	10.3.67	8.24.68	9.24.69	Turkey's *Yilderim* 2.73 Lost 4.11.85
97 *Surprise*	5.24.68	12.7.68	10.17.69	Turkey's *Bora* 2.73
99 *Beacon*	7.15.68	5.17.69	11.21.69	Planned transfer to Greece (1977) canc and retained as of 1986, although str 4.1.77
101 *Green Bay*	11.4.68	6.14.69	12.5.69	Planned transfer to Greece (1977) canc and retained as of 1986, although str 4.1.77

RIVER GUNBOATS (PR)

1 *Palos*
   (ex–PG 16)
2 *Monocacy*
   (ex–PG 20)

*Originally PGM; reclassified 4.1.67. When reclassified, these craft repeated earlier gunboat numbers.
†Originally laid down 4.14.68 (PG 96) and 6.13.68 (PG 98) but had to be laid down again because of fire damage in the yard.

	Laid Down	Launched	Completed	Fate
Kiangnan Dock and Engine Works				
3 Wake	10.17.26	5.28.27	12.28.27	Captured by Japanese 12.7.41 and became *Tatara* China's *Tai Yuan* 1946 Former *Guam*
4 Tutuila	10.17.26	6.14.27	3.2.28	China's *Mei Yuan* 3.42
5 Panay	12.18.26	11.11.27	9.10.28	WL(B) 12.12.37
6 Oahu	12.18.26	11.26.27	10.22.28	WL 5.4.42
7 Luzon	11.20.26	9.12.27	6.1.28	WL 5.5.42 Salvaged by Japanese as *Karatsu* WL 2.3.45
8 Mindanao	11.20.26	9.28.27	7.10.28	WL 5.2.42

FRIGATES (PF)

	Laid Down	Launched	Completed	Fate
Canadian Vickers				
1 Asheville	3.10.42	8.22.42	12.1.42	Sold to Argentina as *Hercules* 6.13.46 Laid down as HMS *Adur*
2 Natchez	3.16.42	9.12.42	12.16.42	Sold to Dominican Republic as *Juan Pablo Duarte* 7.29.47 Laid down as HMS *Annan*
Kaiser Cargo				
3 Tacoma	3.10.42	7.7.43	11.6.43	Korea 10.8.51 *Taedong*
4 Sausolito	4.7.43	7.20.43	3.4.44	Korea 1953 *Imchin*
5 Hoquiam	4.10.43	7.31.43	5.8.44	Korea 10.8.51 *Naktong*
6 Pasco	7.7.43	8.17.43	4.15.44	Japan 1953 *Kashi*
7 Albuquerque	7.20.43	9.14.43	12.20.43	Japan 1953 *Tochi*
8 Everett	7.31.43	9.29.43	1.22.44	Japan 3.53 *Kiri*
9 Pocatello	8.17.43	10.17.43	2.18.44	Sold 9.22.47
10 Brownsville	9.14.43	11.14.43	5.6.44	Sold 9.30.47
11 Grand Forks	9.29.43	11.27.43	3.18.44	Sold 5.20.47
12 Casper	10.17.43	12.27.43	3.31.44	Sold 5.20.47
13 Pueblo	11.14.43	1.20.44	5.27.44	Sold to Dominican Republic as *Presidente Peynado* 9.22.47
14 Grand Island	11.27.43	2.19.44	5.27.44	FLC to Cuba as *Maximo Gomez* 6.16.47
American Shipbuilding (Lorain, Ohio)				
15 Annapolis	5.20.43	10.16.43	6.10.44	FLC to Mexico as *Usumacinta* 11.24.47
16 Bangor	5.20.43	11.6.43	7.29.44	FLC to Mexico as *Tehuantepec* 11.24.47
17 Key West	6.23.43	12.29.43	7.11.44	Sold 5.3.47
18 Alexandria	6.23.43	1.15.44	7.31.44	Sold 4.18.47
American Shipbuilding (Cleveland)				
19 Huron	3.1.43	7.3.43	11.27.43	Sold 5.15.47 Merchant ship *Jose Marcelino* 1948
20 Gulfport	5.5.43	8.21.43	12.16.43	Sold 11.13.47
21 Bayonne	5.6.43	9.11.43	8.30.43	Japanese *Buna* 1953
Butler Shipbuilders				
22 Gloucester	3.4.43	7.12.43	10.24.43	Japanese *Tsuge* 1953
23 Shreveport	3.8.43	7.15.43	11.15.43	Sold 8.15.47
24 Muskegon	5.11.43	7.25.43	11.26.43	Sold to France as *Mermoz* 3.26.47
25 Charlottesville	5.12.43	7.30.43	11.23.43	Japanese *Maki* 1953
26 Poughkeepsie	6.3.43	8.12.43	12.31.43	Japanese *Momi* 1953
27 Newport	6.8.43	8.15.43	12.31.43	Japanese *Kaede* 1953
28 Emporia	7.14.43	8.30.43	5.24.44	Sold to France as *Le Verrier* 3.26.47
29 Groton	7.15.43	9.14.43	6.7.44	FLC to Colombia as *Almirante Padilla* 3.26.47
30 Hingham	7.25.43	8.27.43	5.31.44	Sold 8.15.47
31 Grand Rapids	7.30.43	9.10.43	6.3.44	Sold 4.14.47
32 Woonsocket	6.12.43	9.27.43	6.29.44	FLC to Peru as *Galvez* 9.4.47
33 Dearborn	6.15.43	9.27.43	6.20.44	Sold 7.8.47
Consolidated Shipbuilding				
34 Long Beach	3.19.43	5.5.43	8.31.43	Japanese *Shii* 1953
35 Belfast	3.26.43	5.20.43	11.20.43	Lost 11.17.48 to Soviets
36 Glendale	4.6.43	5.28.43	9.30.43	MDAP to Thailand as *Tachin* 10.29.51
37 San Pedro	4.17.43	6.11.43	10.22.43	Japanese *Kaya* 1953
38 Coronado	5.6.43	6.17.43	11.16.43	Japanese *Sugi* 1953
39 Ogden	5.21.43	6.23.43	12.18.43	Japanese *Kusu* 1953
40 Eugene	6.12.43	7.6.43	1.14.44	FLC to Cuba as *Jose Marti* 6.16.47
41 El Paso	6.18.43	7.16.43	11.30.43	Sold 10.14.47
42 Van Buren	6.24.43	7.27.43	12.15.43	Sold 8.15.47
43 Orange	7.7.43	8.6.43	12.31.43	Sold 9.17.47
44 Corpus Christi	7.17.43	8.17.43	1.28.44	Sold 10.3.47
45 Hutchinson	7.28.43	8.27.43	1.31.44	FLC to Mexico as *California* 11.24.47
46 Bisbee	8.7.43	9.7.43	2.14.44	MDAP to Colombia as *Capitan Tono* 2.13.52
47 Gallup	8.18.43	9.17.43	2.28.44	MDAP to Thailand as *Prasae* 10.29.51
48 Rockford	8.28.43	9.27.43	3.4.44	Korean *Apnok* 10.23.50 Total loss 5.21.52
49 Muskogee	9.18.43	10.18.43	3.14.44	Korean *Duman* 10.23.50
50 Carson City	9.28.43	11.13.43	3.23.44	Japanese *Sakura* 1953
51 Burlington	10.19.43	12.7.43	3.31.44	Colombian *Almirante Brion* 1953
Froemming Brothers				
52 Allentown	3.23.43	7.3.43	12.2.43	Japanese *Ume* 4.53
53 Machias	5.8.43	8.22.43	11.23.43	Japanese *Nara* 1953
54 Sandusky	7.8.43	10.5.43	2.4.44	Japanese *Nire* 1953
55 Bath	8.23.43	11.14.43	2.29.44	Japanese *Matsu* 12.53
Globe Shipbuilding				
56 Covington	3.1.43	7.15.43	8.3.44	FLC to Ecuador as *Guayas* 8.28.47
57 Sheboygan	4.17.43	7.31.43	4.30.44	Sold to Belgium as *Victor Billet* 3.19.47
58 Abilene	5.6.43	8.21.43	5.31.44	Sold to Netherlands as weather ship *Cirrus* 5.7.47
59 Beaufort	7.21.43	10.9.43	6.17.44	Sold 4.11.47

		Laid Down	Launched	Completed	Fate
60	Charlotte	8.5.43	10.30.43	6.28.44	Sold 5.13.47 Merchant ship *Bahia* 1948
61	Manitowoc	8.26.43	11.30.43	9.27.44	Sold to France as weather ship *Le Brix* 3.26.47
62	Gladwyne	10.14.43	1.7.44	7.26.44	FLC 11.24.47 Originally *Worcester* Mexican *Papaloapan*
63	Moberly	11.3.43	1.26.44	8.5.44	Sold 12.2.47 Originally *Scranton*

Leathem D. Smith Shipbuilding

		Laid Down	Launched	Completed	Fate
64	Knoxville	4.15.43	7.10.43	12.9.43	Sold to Dominican Republic as *Presidente Troncoso* 9.22.47
65	Union Town	4.21.43	8.7.43	12.30.43	Sold 6.19.47 Originally *Chatanooga* Argentine *Sarandi*
66	Reading	5.25.43	8.28.43	2.26.44	Sold to Argentina as *Heroina* 6.23.46
67	Peoria	6.4.43	10.2.43	3.14.44	FLC to Cuba as *Antonio Maceo* 6.6.46
68	Brunswick	7.16.43	11.6.43	5.4.44	Sold 4.9.47
69	Davenport	8.7.43	12.8.43	5.18.44	Sold 6.6.46
70	Evansville	8.28.43	11.27.43	6.13.44	Japanese *Keyaki* 10.53
71	New Bedford	10.2.43	12.29.43	6.30.44	Sold 11.47

Walsh Kaiser

		Laid Down	Launched	Completed	Fate
72	Machias	4.1.43	7.14.43	10.15.43	HMS *Hallowel*, later *Anguilla* Ret 5.46
73		4.3.43	7.26.43	11.4.43	HMS *Hammond*, later *Antigua* Ret 5.46
74		4.30.43	8.6.43	11.24.43	HMS *Hargood*, later *Ascension* Ret 5.31.46
75		4.7.43	8.17.43	12.6.43	HMS *Hotham*, later *Bahamas* Ret 6.11.46
76		5.11.43	8.27.43	12.18.43	HMS *Halsted*, later *Barbados* Ret 4.13.46
77		4.23.43	9.6.43	12.31.43	HMS *Hannam*, later *Caicos* Argentine *Santisima Trinidad* 1947
78		7.15.43	8.22.43	1.20.44	HMS *Harland*, later *Cayman* Ret 4.22.46
79		7.27.43	9.14.43	1.25.44	HMS *Harnam*, later *Dominica* Ret 4.23.46
80		8.7.43	9.21.43	2.5.44	HMS *Harvey*, later *Gold Coast* *Labuan* Ret 5.2.46
81		8.17.43	9.27.43	8.12.44	HMS *Holmes*, later *Hong Kong* *Tobago* Ret 5.13.46 Egypt 1950 BU 1956
82		8.28.43	9.27.43	8.31.44	HMS *Hornby*, later *Montserrat* Ret 6.11.46
83		9.7.43	10.6.43	7.31.44	HMS *Hoste*, later *Nyasaland* Ret 4.13.46
84		9.7.43	10.10.43	7.25.44	HMS *Howett*, later *Papua* Ret 5.13.46 Egypt 1950 BU 1956
85		9.14.43	10.15.43	7.6.44	HMS *Pilford*, later *Pitcairn* Ret 6.11.46
86		9.22.43	10.20.43	2.19.44	HMS *Pasley*, later *St. Helena* Ret 4.23.46
87		9.28.43	10.25.43	7.18.44	HMS *Patton*, later *Sarawak* Ret 5.31.46
88		9.28.43	10.30.43	6.27.44	HMS *Pearl*, later *Seychelles* Ret 6.11.46
89		10.7.43	11.5.43	3.16.44	HMS *Phillimore*, later *Sierra Leone*, later *Perim* Ret 5.22.46
90		10.11.43	11.11.43	6.24.44	HMS *Popham*, later *Somaliland* Ret 5.31.46
91		10.16.43	11.16.43	5.15.44	HMS *Peyton*, later *Tortola* Ret 5.22.46
92		10.20.43	11.21.43	6.21.44	HMS *Prowse*, later *Zanzibar* Ret 5.31.46

American Shipbuilding (Lorain, Ohio)

		Laid Down	Launched	Completed	Fate
93	Lorain	10.25.43	3.18.44	9.16.44	French weather ship *Laplace* Lost 9.16.50
94	Milledgeville	11.9.43	4.5.44	10.16.44	Sold 4.47
95	Stamford	Canc 12.31.43			
96	Macon	Canc 12.31.43			
97	Lorain	Canc 12.31.43			
98	Milledgeville	Canc 12.31.43			

American Shipbuilding

		Laid Down	Launched	Completed	Fate
99	Orlando	8.2.43	12.1.43	6.19.44	Sold 11.10.47
100	Racine	9.14.43	3.15.44	10.28.44	Sold 12.2.47
101	Greensboro	9.23.43	2.9.44	8.31.44	Sold 4.28.48
102	Forsyth	12.6.43	5.20.44	10.31.44	Netherlands' weather ship *Cumulus* 7.47

Levingston Shipbuilding

	Laid Down	Launched	Completed	Fate
103	8.20.62	6.26.63	4.24.64	Iran's *Bayandor*
104	9.12.62	8.29.63	6.19.64	Iran's *Naghdi*
105	5.1.67	1.4.68	12.10.68	Iran's *Milanian* WL 1982
106	6.12.67	4.4.68	1.28.69	Iran's *Kahnamuie* WL 1982

American Shipbuilding

	Laid Down	Launched	Completed	Fate
107	9.1.70	10.17.70	11.11.71	Thailand's *Tapi*

Norfolk Shipbuilding and Dry Dock*

	Laid Down	Launched	Completed	Fate
108	3.31.72	6.2.73	7.25.74	Thailand's *Khirirat*

MOTOR GUNBOATS (PGM)

1 (See SC 644)
2 (See SC 757)
3 (See SC 1035)
4 (See SC 1053)
5 (See SC 1056)
6 (See SC 1071)
7 (See SC 1072)
8 (See SC 1366)

*The *Oliver Hazard Perry* (FFG 7) was initially designated PF 109.

	Laid Down	Launched	Completed	Fate
9 (See PC 1548)				
10 (See PC 805)				
11 (See PC 806)				
12–15 (See PC 1088–91)				
16 (See PC 1148)				
17 (See PC 1189)				
18 (See PC 1255)				
19–28 (See PC 1550–59)				
29–32 (See PC 1565–68)				
Georgia Shipbuilding				
33	1.23.53	9.25.53	9.15.54	Philippine *Camarines Sur* 1955
34	2.23.53	11.23.53	11.22.54	Philippine *Sulu* 1955
35	5.1.53	12.22.53	1.3.55	Philippine *La Union* 1955
36	7.28.53	3.6.54	5.18.55	Philippine *Antique* 1955
37	9.4.53	5.29.54	3.2.56	Philippine *Masbate* 1956
38	12.16.53	8.21.54	5.3.56	Philippine *Misamis Occidental* 1956
Tacoma Boat Building				
39	12.17.58	4.20.59	11.29.59	Philippine *Agusan* 3.60
40	1.26.59	4.20.59	1.9.60	Philippine *Catanduanes* 3.60
41	3.13.59	6.5.59	2.9.60	Philippine *Romblon* 6.60
42	4.22.59	6.26.59	3.9.60	Philippine *Palawan* 6.60
Marinette Marine				
43	6.25.58		8.29.59	Burma's PGM 401 11.59
44	9.19.58	7.29.59	8.29.59	Burma's PGM 402 11.59
45	1.13.59	8.18.59	9.19.59	Burma's PGM 403 1.60
46	1.13.59	8.18.59	9.19.59	Burma's PGM 404 1.60
Royal Dockyard*				
47	4.1.60	11.10.60	12.19.61	Denmark's *Daphne* 12.61
48	11.15.60	5.16.61	8.30.62	Denmark's *Havmanden* 8.62
49	9.20.61	6.20.62	4.26.63	Denmark's *Najaden* 4.63
50	9.1.62	5.29.63	12.18.63	Denmark's *Neptun* 12.63
Julius Petersen				
51	8.16.60	4.8.61	5.26.61	Burma's PGM 405 8.61
52	9.16.60	4.13.61	5.28.61	Burma's PGM 406 8.61
53	10.11.60	4.21.61	8.24.61	Ethiopia's PC 13 8.61
54	11.29.60	6.14.61	8.24.61	Ethiopia's PC 15 7.62
Marinette Marine				
55	1.19.61	10.5.61	11.10.61	Indonesia's *Silungkang* 1.62
56	1.19.61	10.9.61	11.10.61	Indonesia's *Waitatiri* 1.62
57	1.19.61	10.10.61	11.10.61	Indonesia's *Kalukuang* 1.62
58	6.22.61	4.5.62	5.14.62	Ethiopia's PC 14 8.61
Martinac				
59	3.15.62	12.21.62	3.20.63	Vietnam's *Kim Quy* 5.63
60	3.15.62	9.28.62	3.11.63	Vietnam's *May Rut* 5.63
61	5.18.62	11.16.62	4.9.63	Vietnam's *Nam Du* 5.63
62	8.23.62	2.15.63	5.1.63	Vietnam's *Hao Lu* 7.63
63	10.8.62	3.27.63	5.21.63	Vietnam's *To Yen* 7.63
Marinette Marine				
64	3.13.62	8.31.62	9.17.62	Vietnam's *Phu Du* 2.63
65	3.13.62	9.19.62	9.26.62	Vietnam's *Tien Moi* 2.63
66	3.12.62	9.20.62	10.15.62	Vietnam's *Minh Hoa* 2.63
67	3.7.62	8.30.62	9.12.62	Vietnam's *Kien Vang* 2.63
68	3.8.62	8.28.62	9.5.62	Vietnam's *Keo Ngua* 2.63
69	9.21.62	4.17.63	5.7.63	Vietnam's *Dienh Hai* 2.64
70	9.21.62	4.15.63	5.14.63	Vietnam's *Truong Sa* 4.64
Julius Petersen				
71	10.20.64	5.22.65	7.17.65	Thailand's T 11 2.66
72	10.27.64	6.26.65	8.4.65	Vietnam's *Thai Binh* 1.66
73	11.16.64	7.16.65	9.11.65	Vietnam's *Thi Tu* 1.66
74	12.11.64	7.29.65	9.11.65	Vietnam's *Song Tu* 1.66
75	12.23.64	8.12.65	10.16.65	Ecuador's *Lae Quito* 11.65
76	3.10.65	8.26.65	10.16.65	Ecuador's *Lae Guayaquil* 11.65
77	5.14.65	9.17.65	11.2.65	Dominican Republic's *Betelgeuse* 12.65
78	6.5.65	10.25.65	11.19.65	Peru's *Rio Sama* 9.66 To Peruvian CG 1975
79	8.24.65	12.28.65	4.22.66	Thailand's T 12 11.71
80	10.13.65	3.21.66	6.2.66	Vietnam's *Tay Sa* 10.66
81	10.18.65	5.18.66	7.9.66	Vietnam's *Phu-Quoi* 4.67
82	2.16.66	7.5.66	8.29.66	Vietnam's *Hoang Sa* 4.67
83	6.7.66	10.4.66	12.8.66	Vietnam's *Hon-Troc* 4.67 Philippine *Basilan* 12.75
(See *Asheville* class above for PGM 84–90, 92–101, later reclassified PG)				
91	6.7.66	10.4.66	12.9.66	Vietnam's *Tho Chau* 4.67
102	7.12.66	10.29.66	12.2.66	Liberia's *Alert* 3.67
103	2.9.67	8.21.67	10.17.67	Iran's *Parvin* 12.67
104	10.7.66	5.4.67	8.9.67	Turkey's AB-21 12.67
105	12.1.66	5.25.67	8.24.67	Turkey's AB-22 12.67
106	12.8.66	7.7.67	10.20.67	Turkey's AB-23 12.67
107	9.12.66	4.13.67	7.5.67	Thailand's T 13 3.68
108	2.22.67	9.14.67	12.6.67	Turkey's AB-24 4.68

*Five more of this class were built to Danish account at the Royal Dockyard. They were all of Danish design.

468   U.S. SMALL COMBATANTS

	Laid Down	Launched	Completed	Fate
Rio de Janeiro Navy Yard				
109				Brazil's *Piratini* 11.70
110				Brazil's *Piraja* 3.71
Sima (Callao, Peru)				
111				Peru's *Rio Chira* 6.72 To Peruvian CG 1975
Julius Petersen				
112	7.1.68	4.24.69	6.24.69	Iran's *Battraam* 8.69
113	8.12.68	4.4.69	5.10.69	Thailand's T 14 9.69
114	9.25.68	4.29.69	6.17.69	Thailand's T 15 9.69
115	11.8.68	4.24.69	7.22.69	Thailand's T 16 9.69
116	12.13.68	6.3.69	8.19.69	Thailand's T 17 10.69
117	1.31.69	6.24.69	9.30.69	Thailand's T 18 12.69
Rio de Janeiro Navy Yard				
118				Brazil's *Pampeio* 6.71
119				Brazil's *Paraja* 7.71
120				Brazil's *Penedo* 9.71
121				Brazil's *Poti* 10.71
Julius Petersen				
122	6.30.69	4.1.70	8.19.70	Iran's *Nahid* 8.70
123	8.21.69	5.4.70	9.6.70	Thailand 12.70
124	1.23.70	6.22.70	10.15.70	Thailand 10.70

HYDROFOIL GUNBOATS (PGH)

	Laid Down	Launched	Completed	Fate
Grumman				
1 *Flagstaff*	7.15.66	9.1.68	3.7.68	CG 1976
Boeing				
2 *Tucumcari*	9.1.66	6.30.67	9.14.68	Wrecked 11.16.72 BU 10.73

HYDROFOIL MISSILE BOATS (PHM)

	Laid Down	Launched	Completed	Fate
Boeing				
1 *Pegasus*	5.10.73	11.9.74	7.9.77	
2 *Hercules*	9.13.80	4.13.82	3.13.83	
3 *Taurus*	1.30.79	5.8.81	10.7.81	
4 *Aquila*	7.10.79	9.16.81	12.19.81	
5 *Aries*	1.7.80	11.5.81	9.18.82	
6 *Gemini*	5.13.80	2.17.82	3.13.83	

SUBCHASERS (SC/PC/PCE/PCS)*

	Laid Down	Launched	Completed	Fate
Naval Station, New Orleans				
1			10.1.17	Sold 7.20.21
2			1.8.18	Sold 10.29.30 to the City of New Orleans
3			1.23.18	Sold 10.4.20
4			2.19.18	Sold 3.19.20
New York Navy Yard				
5			8.18.17	France C-5
6			8.19.17	Sold 6.24.21
7			8.18.17	France C-2
8			8.18.17	France C-1
9			8.18.17	France C-3
10			8.18.17	France C-4
11			8.18.17	France C-6
12			9.29.17	France C-7
13			9.29.17	France C-9
14			9.29.17	France C-10
15			9.29.17	France C-11
16			9.29.17	France C-8
17			8.17.11	Sold 6.24.21
18			11.10.17	To War Department 1920
19			10.19.17	Sold 6.24.21
20			10.18.17	To War Department 1920
21			10.19.17	Sold 6.24.21
22			10.16.17	CG 11.14.19
23			10.16.17	D in accident 1920
24			10.22.17	Sold 6.24.21
25			10.16.17	Sold 6.24.21
26			10.19.17	Sold 7.20.21
27			11.8.17	CG 11.14.19
28			12.22.17	France C-24
29			12.22.17	France C-23
30			3.30.18	France C-33

*For World War I subchasers, the completion date given is generally the commissioning date, which explains why so many are identical. Actual completion dates were not available. In the case of subchasers transferred to France, the date is the date of departure for France, which might be some weeks or even months after completion. The French numbered these craft *in order of arrival* in a C (*chasseur*) series; the survivors had their designators changed to CH about 1930. France obtained one more subchaser, a ship begun on speculation as the *Charles S. Seabury* (which may have been the builder's name). Her contract was dated 25.9.18, and she was accepted on 20.2.19 as the *Victoria*. She never went to Europe but was retained at St. Pierre and Miquelon.

No laying down or launch dates are available for the World War I subchasers.

All control craft were redesignated as such (SCC/PCC/PCEC/PCSC) on 20 August 1945. Surviving PC/PCE/PCS were named in February 1956.

FATES AND OTHER NOTES 469

	Laid Down	Launched	Completed	Fate
31			5.18.18	France C-46
32			3.30.18	France C-41
33			10.5.18	France C-42
34			1.9.18	Sold 6.24.21
35			1.23.18	Sold 6.24.21
36			1.23.18	Sold 6.24.21
37			2.1.18	Sold 6.24.21
38			2.1.18	Sold 12.19
39			3.2.18	Sold 6.24.21
40			2.13.18	Sold 10.14.24
41			2.19.18	Sold 5.11.21
42			3.2.18	Sold 6.24.21
43			5.16.18	Sold 6.24.21
44			4.3.18	Sold 6.24.21
45			3.1.18	Sold 6.24.21
46			3.16.18	Sold 6.24.21
47			3.27.18	Sold 6.24.21
48			3.27.18	Sold 6.24.21
49			3.27.18	Sold 6.24.21
50			4.19.18	Sold 6.24.21
51			4.23.18	Sold 6.24.21
52			4.23.18	Sold 6.24.21
53			4.30.18	D 1920
54			5.11.18	Sold 6.24.21
55			11.3.17	Sold 6.24.21
56			5.6.18	Sold 6.24.21
57			5.6.18	Sold 12.12.35
58			5.6.18	D in accident 5.2.19
59			5.11.18	Sold 6.24.21
60			5.18	D in collision 10.1.18
61			5.16.18	Sold 6.24.21
62			5.11.18	Sold 3.6.22
63			5.16.18	Sold 7.22.31
64			5.16.18	To water barge Reclassified YW-97 11.30.42 Sold 3.11.43
Mathis Yacht Building				
65			1.11.17	France C-13
66			1.11.17	France C-14
67			12.22.17	France C-22
68			3.15.18	CG 1.15.20
69			2.16.18	Sold 12.9.22
70			2.16.18	CG 1.15.20
71			3.28.18	Sold 5.26.21
72			3.21.18	Sold 11.4.21
73			3.20.18	Sold 5.26.21
74			3.20.18	Sold 5.26.21
Hiltebrant Dry Dock				
75			1.11.17	France C-16
76			1.11.17	France C-15
77			12.5.17	Sold 6.24.21
78			11.14.17	Sold 6.19
79			12.5.17	Sold 6.24.21
80			12.18.17	Sold 6.24.21
81			12.18.17	Sank 8.6.20
82			2.21.18	Sold 6.19
83			2.21.18	Sold 6.24.21
84			2.21.18	Sank 8.6.20
85			2.21.18	Sold 6.24.21
86			2.21.18	Sold 6.24.21
87			2.21.18	Sold 6.24.21
88			3.1.18	Sold 6.24.21
89			3.1.18	Sold 6.24.21
Elco				
90			11.14.17	Sold 8.11.20
91			12.5.17	Sold 6.24.21
92			12.5.17	Sold 6.24.21
93			12.5.17	Sold 12.20.21
94			12.24.17	Sold 1919
95			12.24.17	Sold 6.24.21
96			12.18.17	Sold 3.4.24
97			1.18.18	Sold 6.24.21
98			2.19.18	Sold 10.14.24
99			3.3.18	Sold 5.11.21
100			3.2.18	Sold 6.24.21
101			3.2.18	Sold 6.24.21
102			3.12.18	To War Shipping Administration 1.3.47
103			3.7.18	Lent to Michigan State Naval Militia 7.1.26 Lent to Buffalo Council, Boy Scouts of America, 9.25.36 Damaged in storm and sank at pier 9.39
104			3.12.18	Sold 12.22.22
105			11.9.17	Sold 7.20.21

470 U.S. SMALL COMBATANTS

	Laid Down	Launched	Completed	Fate
Charleston Navy Yard				
106			12.31.17	Sold 11.10.21
107			1.19.18	Sold 6.24.21
108			2.12.18	To War Department 9.18.19
109			2.12.18	Sold 6.24.21
110			3.30.18	Sold 6.24.21
111			3.30.18	Sold 6.24.21
112			4.22.18	Sold 6.24.21
113			5.3.18	Sold 6.24.21
Naval Station, New Orleans				
114			3.28.18	Sold 6.24.21
115			5.3.18	Sold 6.24.21
Norfolk Navy Yard				
116			11.14.17	Sold 6.24.21
117			11.17	D in accident 12.22.17
118			11.26.17	Sold 6.24.21
119			11.19.17	Sold 7.20.21
120			10.4.17	Sold 7.20.21
121			10.16.17	Sold 6.24.21
122			10.21.17	Sold 6.24.21
123			11.5.17	Sold 6.24.21
124			12.11.17	Sold 5.11.21
125			12.27.17	Sold 6.24.21
126			1.14.18	Sold 6.24.21
127			1.15.18	Sold 6.24.21
128			1.18.18	Sold 6.19
129			1.17.18	Sold 6.24.21
130			12.21.17	Sold 5.11.21
131			1.19.18	Sold 6.24.21
132			2.2.18	Collision 6.5.18
133			3.13.18	Sold 7.20.21
134			3.13.18	Sold 7.20.21
135			3.13.18	Sold 6.24.21
136			3.19.18	Sold 6.24.21
Hodgdon				
137			12.14.17	Sold 6.24.21
138			1.24.18	Sold 6.24.21
139				Canc
Hartman-Greiling				
140			10.5.18	France's C-43
141			12.22.17	To have been ceded to France Lost in collision with SC-174 near Philadelphia
Rocky River Dry Dock				
142			3.30.18	France's C-34
143			11.10.17	Sold 9.8.36
Vinyard Boat Building				
144			3.30.18	Sold to State of Florida 2.3.23
145			3.13.18	Sold 7.7.21
146			3.30.18	France's C-35
L. E. Fry				
147			12.13.17	Sold 2.25.22
148			12.10.17	Sold 6.24.21
Dubuque Boat and Boiler Works				
149			1.15.18	Sold 6.24.21
150			1.15.18	Sold 6.24.21
Gibbs Gas Engine				
151			12.14.17	Sold 6.24.21
152			1.17.18	Sold 6.24.21
153			2.14.18	CG 11.22.19
154			2.15.18	Sold 11.16.26
155			4.11.18	CG 11.22.19
F. M. Blount				
156			10.30.17	Sold 11.6.24
157			3.13.18	Sold 6.24.21
158			3.13.18	Sold 2.26.21
159			9.17.17	Sold 11.16.26
Howard E. Wheeler				
160			12.22.17	France's C-25
161			3.30.18	France's C-36
162			5.10.18	France's C-44
163			5.18.18	France's C-48
164			3.23.18	Sold 6.24.21
165			3.23.18	D in accident 8.25.20
166			3.23.18	Sold 6.24.21
167			4.8.18	Sold 6.24.21
168			4.8.18	Sold 6.24.21
Matthews				
169			3.30.18	France's C-37
170			12.22.17	France's C-29
171			12.22.17	France's C-39 Delayed en route, arrived with 3.30.18 group
172			12.22.17	France's C-26

	Laid Down	Launched	Completed	Fate
173			3.30.18	France's C-31
174			12.22.17	France's C-40 Damaged in collision, started again 3.30.18
175			3.30.18	France's C-32
176			3.30.18	France's C-38
177			10.31.17	Redesignated SC-405 to replace SC-405 (see below), then to France as C-99 on 12.18.18 SC-405 redesignated SC-177 Sold out of service under that designation 1921
178				Sold 6.24.21
International Shipbuilding and Marine Engine				
179			12.24.17	Sold 1920
180			4.8.18	D in accident 7.15.20
181			4.27.18	Sold 6.24.21
182			5.6.18	Sold 6.24.21
183			4.27.18	CG 1921
184			4.27.18	Sunk 8.9.19
185			4.27.18	Lost 6.30.40
186			4.27.18	Sold 10.28.26
187			5.18	Collision 8.4.18
188			3.13.18	Scrapped 7.2.24
General Shipbuilding and Aero				
189			3.13.18	Sold 5.9.21
190			3.13.18	Sold 11.10.21
191			3.21.18	Sold 3.12.24
192			4.15.18	To Sea Scouts 5.15.37
193			4.5.18	Sold 6.24.21
194			4.5.18	Sold 6.24.21
195			4.15.18	Sold 11.12.21
196			4.15.18	Sold 6.24.21
197			4.15.18	CG 11.14.19
198			4.15.18	Sold 5.11.21
199			4.14.18	CG 11.14.19
200			4.17.18	Sold 6.24.21
201			4.23.18	Sold 6.6.22
202			4.23.18	Sold 6.24.21
203			4.25.18	CG 11.21.19
204			4.25.18	Sold 7.27.22
205			4.25.18	Sold 6.1.21
206			4.25.18	Sold 6.1.21
207			4.26.18	Sold 4.21.20
208			5.15.18	Sold 6.24.21
Mathis Yacht Building				
209			3.18	WL(G) 8.27.18
210			3.18.18	Sold 4.23.30
211			11.28.18	Sold 6.24.21
212			3.5.18	Sold 6.24.21
213			3.18.18	Sold 6.24.21
Alex McDonald				
214			12.1.17	Sold 2.21.27
215			12.24.17	Sold to Italy 6.28.19
216			2.14.18	Sold 5.11.21
217			2.19.18	Sold 6.24.21
Newcomb Lifeboat				
218			2.9.18	Sold 7.21.21
219			2.19.18	D in accident 10.9.18
220			3.13.18	Sold 6.24.21
221			3.13.18	Sold 6.24.21
222			3.13.18	Sold 5.11.21
New York Yacht, Launch and Engine				
223			12.5.17	Sold 5.11.21
224			10.27.17	Sold 9.8.36
225			12.10.17	Sold 5.11.21
226			12.24.17	Sold 7.20.21
227			12.24.17	Sold 10.14.24
228			1.23.18	Sold 7.20.21
229			1.23.18	CG 8.14.42
230			2.8.18	Sold 6.24.21
231			2.8.18	CG 8.14.42
232			2.8.18	Sold 12.12.23
233			2.21.18	War Department 9.2.19
234			2.21.18	War Department 9.2.19
235			3.2.18	Sold 6.24.21
236			3.2.18	Sold 6.24.21
237			3.7.18	Sold 5.18.23
238			3.12.18	Sold 6.24.21
239			3.19.18	Sold 6.24.21
240			3.20.18	Sold 5.11.21
241			4.8.18	Sold 5.11.21
242			4.8.18	Sold 5.11.21
Eastern Shipyard				
243			1.11.17	France's C-17
244			12.10.17	Sold 5.11.21

	Laid Down	Launched	Completed	Fate
245			3.7.18	Sold 6.24.21
246			3.7.18	Sold 6.24.21
247			3.20.18	Sold 5.11.21
248			1.17.18	Sold to Italy 6.28.19
Chance Marine Construction				
249			5.18.18	France's C-47
250			5.17.18	Sold 6.24.21
Camden Anchor-Rockland Machine				
251			12.29.17	Sold 5.19.23
252			3.7.18	Sold 9.8.36
George Lawley				
253			3.7.18	Sold 12.9.22
254			11.15.17	Sold 6.24.21
255			11.19.17	Sold 6.24.21
256			11.19.17	D in accident 11.1.19
257			11.28.17	Sold 2.25.22
258			11.28.17	Sold 6.24.21
259			12.15.17	Sold 6.24.21
260			1.12.18	Sold 10.14.24
261			2.9.18	Sold 1.25.21
262			2.9.18	Sold 6.5.20
263			2.9.18	Sold 5.18.23
264			2.9.18	Sold 12.20.21
265			2.9.18	Sold 6.24.21
266			4.1.18	Sold 1.25.21
267			4.5.18	War Department 9.18.19
268			4.1.18	CG 1.17.20
269			4.1.18	Sold 6.24.21
270			4.1.18	Sold 9.25.22
271			3.7.18	Sold 6.24.21
272			3.7.18	Sold 6.24.21
Mare Island Navy Yard				
273			3.26.18	Sold 9.25.22
274			3.30.18	Sold to Cuba 11.5.18
275			3.30.18	War Department 12.9.19
276			4.9.18	War Department 6.26.20
277			4.9.18	Sold 9.25.22
278			3.30.18	Sold 9.25.22
279			4.18.18	War Department 10.2.19
280			4.18.18	War Department 10.6.19
281			4.18.18	War Department 10.6.19
282			4.22.18	Lost 6.11.20
283			4.22.18	Sold 3.13.22
284			4.22.18	Sold 3.24.23
285			4.22.18	Sold 3.25.27
286			5.6.18	Sold 3.13.22
287			5.6.18	Sold 2.25.24
Puget Sound Navy Yard				
288			6.19.18	Sold 3.13.22
289			6.19.18	Sold 3.13.22
290			6.19.18	Sold 5.9.21
291			3.27.18	Sold 9.3.20
292			3.27.18	Sold 5.3.21
293			3.13.18	Sold 2.6.22
294			3.25.18	Sold 3.13.22
295			4.13.18	Sold 4.10.22
296			4.13.18	Sold 9.3.20
297			4.13.18	Sold 9.25.20
298			4.13.18	Sold 9.25.22
299			4.25.18	Sold 9.25.22
300			4.25.18	Sold 6.22.21
301			4.29.18	Sold 4.10.22
302			1.11.18	Sold to Cuba 11.9.18
303			5.4.18	Sold 9.25.22
304			5.4.18	Sold 4.8.22
305			5.11.18	Sold 4.10.22
306			2.27.18	Department of Justice 12.17.30
307			5.11.18	War Department 6.26.20
308			2.23.18	Sold 9.25.22
309			5.18.18	Sold 1.20.21
310			5.18.18	Sold 2.4.22
311			5.18.18	Cuba 11.5.18
312			5.18.18	Cuba 11.5.18
Robert Jacob				
313			10.5.18	France's C-45
314			12.22.17	France's C-27
315			1.11.17	France's C-19
316			1.11.17	France's C-20
317			1.11.17	France's C-21
Luders Marine Construction				
318			12.22.17	France's C-28
319			12.22.17	To have been ceded to France Disappeared at sea 1.18.18

	Laid Down	Launched	Completed	Fate
320			12.1.17	Sold 7.7.27
321			1.8.18	Sold 2.18.20
322			3.7.18	Sold 3.8.22
Kyle and Purdy				
323			12.5.17	Sold 5.11.21
324			12.13.17	Sold 6.24.21
325			12.13.17	Sold 6.24.21
326			11.14.17	Sold 11.8.35
327			12.10.17	Sold to Italy 6.19
Great Lakes Boat Building				
328			10.20.17	Sold 11.8.35
329			10.20.17	Sold 9.25.22
Burger Boat				
330			2.8.18	War Shipping Administration 10.8.46
Smith and Williams				
331			3.13.18	Sold 4.29.21
332			3.13.18	Sold 1.29.24
Barret Shipbuilding				
333			1.26.18	CG 2.4.20
334			3.4.18	CG 2.4.20
335			4.18.18	CG 11.22.19
336			5.6.18	To City of New Orleans 10.30.20
L. E. Fry				
337			12.24.17	Sold 6.24.21
338			12.24.17	Sold 9.19.22
American Car and Foundry				
339			2.16.18	Grounded 9.19
340			2.16.18	D in accident 10.6.23
341			3.22.18	Sold 4.5.27
342			3.15.18	Sold 5.26.21
343			3.15.18	D in accident 5.15.19
344			3.23.18	Sold 5.26.21
345			3.25.18	Sold 5.26.21
346			3.25.18	Sold 4.28.20
College Point Boat				
347			1.11.17	France's C-18
348			1.11.17	France's C-12
349			11.14.17	Sold 5.26.21
350			3.30.18	France's C-30
351			1.8.18	Sold 10.19.20
352			3.2.18	Sold 6.24.21
353			3.20.18	Sold 3.18.36
354			3.2.18	Sold 6.24.21
355			3.12.18	Sold 3.13.22
356			4.8.18	Sold 6.24.21
357			6.9.18	France's C-54
358			6.9.18	France's C-56
359			6.9.18	France's C-60
360			9.27.18	France's C-73
361			6.9.18	France's C-57
362			6.9.18	France's C-49
Elco				
363			6.9.18	France's C-50
364			6.9.18	France's C-51
Gibbs Gas Engine				
365			6.9.18	France's C-70
366			6.9.18	France's C-71
367			6.9.18	France's C-72
368			10.24.18	France's C-95
369			10.24.18	France's C-96
370			10.24.18	France's C-97
Hiltebrant Dry Dock				
371			6.9.18	France's C-61
372			6.9.18	France's C-58
373			6.9.18	France's C-62
374			6.9.18	France's C-52
375			6.9.18	France's C-55
Kyle and Purdy				
376			9.27.18	France's C-74
377			6.9.18	France's C-59
378			6.9.18	France's C-63
379			6.9.18	France's C-53
380			9.27.18	France's C-75
Mathis Yacht Building				
381			6.9.18	France's C-64
382			6.9.18	France's C-69
383			9.27.18	France's C-76
384			9.27.18	France's C-77
385			10.24.18	France's C-80
Matthews				
386			10.24.18	France's C-81
387			10.24.18	France's C-82
388			10.24.18	France's C-90

	Laid Down	Launched	Completed	Fate
389			10.24.18	France's C-91
390			10.24.18	France's C-92
391			10.24.18	France's C-93
392			10.24.18	France's C-89
New York Yacht, Launch and Engine				
393			6.9.18	France's C-66
394			6.9.18	France's C-65
395			6.9.18	France's C-67
396			6.9.18	France's C-68
397			9.27.18	France's C-78
398			10.24.18	France's C-79
399			10.24.18	France's C-83
400			10.24.18	France's C-86
401			10.24.18	France's C-98
402			10.24.18	France's C-85
Rocky River Dry Dock				
403			10.24.18	France's C-87
404			10.24.18	France's C-88
405			10.24.18	To have been ceded to France as C-84 but while laid over at Bermuda had an unauthorized trial 11.3.18 and damaged rudder on rock Redesignated SC-177 Commissioned in the U.S. Navy on 12.18.18 She and SC-177 only U.S. ships to swap hull numbers Sold 11.12.21
406			10.24.18	France's C-94
Camden Anchor-Rockland Machine				
407			1.16.19	Sold 4.21.20
408			2.11.19	Sold 6.6.22 Not commissioned
Chance Marine Construction				
409			2.3.19	Sold 6.24.21 Not commissioned
410				Canc when 40 percent complete
Clayton Ship and Boat Building				
411			5.1.19	Sold 1.30.20
412			5.1.19	MC 8.7.46
College Point Boat				
413			1919	War Department 4.30.20 Sold 6.24.21
414			1919	War Department 9.2.19
415			1.8.19	CG 12.16.19
416			1.13.19	Sold 6.24.21
417			1.31.19	CG 10.21.19 Not commissioned
418			1919	Sold 3.4.20
Great Lakes Boat Building				
419			11.22.18	Sold 4.27.27 Not commissioned
420			1919	Sold 2.3.20 Not commissioned
Hiltebrant Dry Dock				
421			1919	Sold 6.24.21 Not commissioned
422			1919	Sold 5.11.21 Not commissioned
423			1919	Sold 5.11.21 Not commissioned
424			12.21.18	Sold 5.19.23
425			12.23.18	Sold 5.11.21
Mathis Yacht Building				
426			1.8.19	Sold 6.6.22
427			1.8.19	Sold 8.17.21
428			1.8.19	To City of Baltimore 5.16.21
429			12.28.18	Sold 5.26.21
430			1.15.19	Sold 12.20.21
Matthews				
431			10.29.19	To CG upon completion To navy 9.1.21 commissioned MC for disposal 12.9.46
432			4.1.18	MC 7.27.45
433			11.25.18	D in accident 1.29.38
Alex McDonald				
434			1.11.19	Sold 4.22.20
435			1.27.19	CG 10.21.19 Not commissioned
436			2.4.19	Sold 6.1.21 Not commissioned
Rocky River Dry Dock				
437			3.1.19	MC 3.21.47
438			1919	CG 10.29.19 Not commissioned
Howard E. Wheeler				
439			1.3.19	Sold 2.25.22
440			1.20.19	Scrapped 8.42
441			2.28.19	Sold 6.26.22
442				Canc 1918
Naval Station, New Orleans				
443			8.26.19	Sold 1.29.24
444			8.26.19	Sold 3.24.23
445–48				Canc 11.20.18
Luders Marine Construction (SC)				
449	7.10.39	5.14.40	9.13.40	MC 7.29.49
American Car and Foundry (SC)				
450	8.9.39	3.14.40	5.2.40	MC 5.9.47

FATES AND OTHER NOTES 475

	Laid Down	Launched	Completed	Fate
Defoe Shipbuilding (PC)				
451	9.25.39	5.23.40	8.10.40	MC 12.5.46
452	3.14.40	8.23.41	5.1.44	*Castine* (IX 211) 3.10.45 MC 1.27.47
Fisher Boatworks (SC)				
453	9.24.40	5.3.41	8.12.41	CG 11.20.45 Haiti's *16 Aout 1946* 1947
454				Acq 8.12.40 *Impetuous* (PYC 46) 7.15.43 MC 6.14.45 Built by Robert Jacob 1915 Ex-*Arlis*
455				Acq 9.20.40 *Patriot* (PYC 47) 7.15.43 MC 3.14.45 Built by Herreshoff Manufacturing 1930 Ex-*Katoura*
456				Acq 8.23.40 *Persistant* (PYC 48) 7.15.43 MC 1.31.46 Built by New York Yacht, Launch, and Engine 1931 Ex-*Onwego*
457				Acq 8.16.40 Lost 8.14.41 Built by Dawn Shipbuilding 1939 Ex-*Trouper*
458				Acq 9.18.40 *Retort* (PYC 49) 7.15.43 MC 10.19.45 Built by Reed 1929 Ex-*Evelyn R II*
459				*Turquoise* (PY 18) 1.20.41 Ecuador 1.29.44 Ret/ sold 5.49 Built by Newport News 1921 Ex-*Entropy*
460				Acq 9.4.40 *Sturdy* (PYC 50) 7.15.43 MC 8.6.45 Built by Consolidated Shipbuilding 1930 Ex-*Elda*
George Lawley (PC)				
461	7.10.41	12.23.41	3.18.42	Decommissioned 11.51 *Bluffton* Str 1957
462	7.22.41	1.24.42	4.15.42	PCC MC 2.20.47
463	8.1.41	2.27.42	4.29.42	PCC Decommissioned 3.47 D in testing 7.53
464	8.8.41	2.27.42	5.15.42	Sold 1.12.47
465	8.19.41	3.28.42	5.25.42	Res 3.47 *Paragould* Venezuela's *Pulpo* 4.11.61
466	9.1.41	4.29.42	6.3.42	PCC Decommissioned 3.47 *Carmi* Str 7.1.60
467	10.22.41	4.29.42	6.20.42	Norway 9.16.42
468	1.1.42	5.30.42	7.10.42	Netherlands' *Queen Wilhelmina* 8.6.42 Nigeria's *Ogoja*
469	1.30.42	6.10.42	7.13.42	PCC D in test 6.49
470	2.27.42	6.27.42	7.31.42	Res 3.47 *Antigo* Str 7.1.60
Defoe Shipbuilding (PC)				
471	4.21.41	9.15.41	10.6.41	France's *Eveille* BU 1959
472	7.1.41	11.14.41	12.6.41	France's *Ruse* BU 1959
473	8.18.41	11.19.41	12.6.41	France's *Ardent* WL 7.7.44
474	8.20.41	12.5.41	12.19.41	France's *Indiscret* BU 1960
475	9.11.41	12.16.41	12.27.41	France's *Resolu* BU 1951
476	10.2.41	1.1.42	2.9.42	MC 7.1.48
477	11.26.41	1.29.42	2.19.42	Sold 1.17.47
478	12.17.41	2.20.42	2.25.42	Sold 11.14.46
479	1.14.42	3.10.42	4.7.42	MC 12.5.46
480	1.22.42	3.25.42	4.7.42	France's *Emporte* BU 1959
481	2.3.42	3.31.42	4.24.42	France's *Effronte* BU 1953
482	2.16.42	4.9.42	4.30.42	France's *Enjoue* 7.15.44 WL(T) 1.9.45
Consolidated Shipbuilding (PC)				
483	12.31.40	10.25.41	3.9.42	Res 6.46 *Rolla* Venezuela's *Camaron* 4.11.61
484	4.7.41	12.6.41	3.31.42	Res 3.47 *Cooperstown* Venezuela's *Togogo* 4.11.61
485	5.1.41	12.20.41	4.20.42	Korea's *Han Ra San* 1.52 Sunk in typhoon 1.64
486	10.25.41	1.25.42	5.7.42	*Jasper* Str 5.1.59
487	12.6.41	2.28.42	5.26.42	Res 12.49 *Larchmont* Venezuela's *Mejillon* 7.24.61
Sullivan Drydock and Repair (PC)				
488	3.17.41	12.20.41	8.11.42	*Eureka* (IX 221) 4.25.45 Sold 13.3.46
489	3.17.41	12.20.41	7.13.42	MC 11.5.48
Dravo (PC)				
490	5.9.41	10.18.41	2.9.42	China's *Wu Sung* 8.27.48
491	5.9.41	12.6.41	3.25.42	FLC 2.13.48
492	6.25.41	12.29.41	4.17.42	China's *Fukiang* 6.30.48
493	8.9.41	1.24.42	5.8.42	FLC 5.10.47
494	8.29.41	2.7.42	5.23.42	FLC 1.11.48
495	6.10.41	12.30.41	4.23.42	FLC 5.10.47 Thailand's *Sarasin*
Leathem D. Smith Shipbuilding (PC)				
496	4.24.41	11.22.41	2.21.42	WL 6.4.43
Westergard Boat Works (SC)				
497	3.7.41	7.4.41	4.15.42	France's CH 96 3.18.44
498	3.12.41	7.21.41	4.29.42	France's CH 142 3.18.44
Fisher Boat Works (SC)				
499	2.24.41	10.24.41	3.13.42	CG 8.20.45
500	2.27.41	10.11.41	3.31.42	USSR 6.9.45
Seabrook Yacht (SC)				
501	4.29.41	1.24.42	4.30.42	*Racer* (IX 100) 4.21.43 Used by ASW Sound School, Pearl Harbor Str 6.5.46 Sold 12.20.46
502	5.6.41	1.31.42	6.3.42	MC 3.21.48
Rice Brothers (SC)				
503	1.30.41	3.14.42	4.13.42	France's CH 112 11.4.44
504	1.30.41	3.21.42	4.23.42	MC 9.30.46
Luders Marine Construction (SC)				
505	2.24.41	2.23.42	4.10.42	MC 5.6.47
506	3.24.41	1.30.42	4.27.42	France's CH 113 9.27.44

	Laid Down	Launched	Completed	Fate
Mathis Yacht Building (SC)				
507	2.6.41	6.30.41	1.19.42	France's CH 85 3.29.44
508	2.10.41	7.25.41	3.27.42	France's CH 95 3.18.44
509				Acq. 10.7.40 Valiant (PYC 51) 7.15.43 MC 6.15.45 Built by Herreshoff Manufacturing 1929 Ex-*Vara*
510				Acq 10.18.40 YP 105 9.22.41 MC 8.30.46 Built by Consolidated Shipbuilding 1927 Ex-*Sybarita*
American Cruiser (SC)				
511	8.1.41	4.1.42	6.27.42	CG 8.18.45
512	8.12.41	3.26.42	7.10.42	CG 10.24.45
Quincy Adams Yacht Yard (SC)				
513	5.16.41	1.20.42	3.28.42	MC 10.21.46
514	6.14.41	3.7.42	4.12.42	MC 3.25.48
Elizabeth City Shipyard (SC)				
515	4.24.41	9.20.41	4.16.42	France's CH 121 10.10.44
516	5.7.41	10.11.41	5.5.42	France's CH 81 3.19.44
517	5.22.41	10.11.41	5.4.42	France's CH 82 3.16.44
518	6.2.41	11.12.41	5.11.42	D 7.13.48
Vinyard Shipbuilding (SC)				
519	6.2.41	3.14.42	4.21.42	France's CH 83 3.27.44
520	7.15.41	4.18.42	5.25.42	MC 9.25.47
Annapolis Yacht Yard (SC)				
521	5.5.41	2.1.42	4.15.42	Foundered 7.10.45
522	5.14.41	2.18.42	5.1.42	France's CH 111 9.27.44
523				Acq 11.25.40 YP 77 4.8.43 Lost in accident 5.4.42 Built by Consolidated Shipbuilding 1931
Mathis Yacht Building (SC)				
524	4.29.41	9.19.41	4.13.42	France's CH 101 10.7.44
525	7.3.41	11.21.41	4.30.42	France's CH 102 10.8.44
526	7.29.41	12.30.41	5.11.42	France's CH 114 9.25.44
527	9.4.41	1.21.42	5.28.42	FLC 5.25.47
528	10.24.41	2.17.42	6.13.42	FLC 2.26.47
529	11.24.41	3.16.42	6.26.42	France's CH 84 3.31.44
Westergard Boat Works (SC)				
530	3.27.41	8.16.41	5.12.42	France's CH 115 11.17.44
531	3.27.41	9.4.41	5.27.42	MC 2.27.48
Luders Marine Construction (SC)				
532	4.19.41	4.7.42	5.15.42	France's CH 102 11.16.44
533	5.5.41	4.13.42	6.5.42	France's CH 104 10.2.44
534	5.23.41	4.27.42	7.10.42	France's CH 122 10.28.44
535	6.18.41	7.9.42	7.21.42	France's CH 143 11.16.44
Julius Petersen (SC)				
536	4.29.41	3.5.42	4.1.42	CG 2.19.46
537	4.29.41	3.21.42	4.27.42	USSR 5.26.45
538	5.17.41	4.1.42	5.9.42	USSR 8.18.45
539	5.17.41	4.7.42	5.28.42	CG 12.4.45
Robinson Marine Construction (SC)				
540	7.28.41	4.6.42	4.22.42	CG 4.18.45
541	8.1.41	4.11.42	5.5.42	CG 10.31.45
Defoe Shipbuilding (PC)				
542	2.26.42	4.20.42	5.22.42	France's *Tirailleur* 9.30.44 BU 1958
543	3.11.42	4.24.42	5.26.42	France's *Volontaire* 6.8.44 BU 1964
544	3.24.42	4.30.42	6.5.42	Brazil's *Guapore* 9.24.42
545	3.31.42	5.8.42	6.27.42	France's *Goumier* 6.27.42 Morocco's *Agadir* 1965
546	4.9.42	5.14.42	7.8.42	France's *Franc Tireur* 10.3.44 BU 1953
547	4.16.42	5.22.42	7.24.42	Brazil's *Gurupi* 9.24.42
548	4.24.42	5.29.42	8.1.42	MC 4.19.48
549	5.1.42	6.6.42	8.1.42	PCC MC 4.23.48
Leathem D. Smith Shipbuilding (PC)				
550	6.6.41	3.8.42	4.28.42	France's *Vigilant* 6.15.44 BU 1959
551	7.22.41	4.12.42	5.15.42	France's *Mameluck* 10.10.44 BU 1958
Sullivan Drydock and Repair (PC)				
552	5.20.41	2.13.42	7.27.42	MC 12.5.46
553	12.20.41	5.30.42	10.10.42	Res 1.47 *Malone* Str 1957
554	12.20.41	5.2.42	9.1.42	Brazil's *Goiana* 19.29.43
555	2.13.42	5.30.42	2.5.43	PCC Sold 2.12.47
Luders Marine Construction (PC)				
556	10.1.41	6.23.42	8.31.42	France's *Carabinier* 10.19.44 BU 1958
557	10.31.41	8.2.42	10.15.42	France's *Dragon* 10.25.44
558	10.31.41	9.13.42	11.19.42	WL 5.9.44
Jeffersonville B/M (PC)				
559	10.14.41	2.12.42	5.12.42	France's *Voltigeur* 10.6.44 Stranded and hulked
560	11.25.41	3.17.42	6.17.42	Res 1.47 *Oberlin*
561	1.30.42	5.1.42	7.11.42	Brazil's *Grauna* 11.30.43
562	2.13.42	6.4.42	8.5.42	France's *Attentif* 6.30.44 BU 1953
Consolidated Shipbuilding (PC)				
563	12.20.41	3.17.42	6.16.42	PCC Sold 11.29.46
564	1.25.42	4.12.42	6.30.42	*Chadron* Korea's *Sol Ak* 1.22.64
Brown Shipbuilding (PC)				
565	8.14.41	2.27.42	5.25.42	*Gilmer* Venezuela's *Alcatraz* 6.62
566	8.14.41	3.21.42	6.15.42	*Honesdale* Venezuela's *Calamar* 7.61
567	9.15.41	4.11.42	6.27.42	Res 7.46 *Riverhead* Str 3.15.63 To U.S. Air Force

FATES AND OTHER NOTES 477

	Laid Down	Launched	Completed	Fate
568	9.15.41	4.25.42	7.13.42	*Altus* Str 3.15.63 To U.S. Air Force Philippine *Nueva Viscaya* 3.68
Albina Engine and Machinery (PC)				
569	9.11.41	1.22.42	5.8.42	*Potoskey* Str 7.1.60
570	9.11.41	1.5.42	4.18.42	MDAP to Thailand as *Longlom* 12.52
571	9.27.41	2.12.42	5.22.42	Decommissioned 1.47 *Anoka* Str 11.1.59
572	9.27.41	2.28.42	6.17.42	*Tooele* Venezuela 1.11.61
Dravo (PC)				
573	10.14.41	3.5.42	6.13.42	MC 6.23.48
Dravo (PC)				
574	12.30.41	3.30.42	7.2.42	Sold 11.6.46
575	2.24.42	5.5.42	8.8.42	FLC 3.47 Thailand's *Thayanchon*
576	4.3.42	6.13.42	9.10.42	MC 7.21.48
577	5.6.42	7.25.42	10.15.42	D in accident 11.12.48
Albina Engine and Machinery (PC)				
578	12.20.41	4.29.42	7.13.42	PCC Sold 4.14.48
579	1.5.42	4.29.42	8.24.42	*Wapakoneta* Str 7.1.60
580	1.22.42	4.29.42	9.26.42	*Malvern* Indonesia's *Hiu* 5.60
581	2.12.42	7.8.42	10.9.42	*Manville* Indonesia's *Torani* 5.60
582	2.21.42	7.15.42	10.22.42	PCC *Lenoir* Venezuela's *Albatros* 6.62
Defoe Shipbuilding (PC)				
583	5.8.42	6.12.42	8.18.42	MC 7.9.48
584	5.14.42	6.18.42	8.8.42	D 12.14.45
585	5.22.42	7.8.42	8.22.42	MC 7.9.48
586	5.29.42	7.15.42	9.19.42	Res 1.50 *Patchogue* Str 4.1.59
587	6.5.42	8.1.42	9.5.42	MC 4.12.48
Leathem D. Smith Shipbuilding (PC)				
588	11.22.41	5.3.42	6.22.42	Res 1.47 *Houghton* Str 4.1.59
589	3.9.42	6.7.42	7.23.42	PCC Res 1.47 *Metropolis* Str 4.1.59
590	4.14.42	7.4.42	9.20.42	D 2.23.46
591	5.8.42	8.2.42	10.10.42	France's *Spahi* 10.17.44 BU 1959
Dravo (PC)				
592	3.31.42	6.27.42	11.6.42	Res 1.50 *Towanda* Str 4.1.59
593	4.21.42	8.22.42	12.5.42	China's *Fuchiang* 8.58
594	5.1.42	9.7.42	2.22.43	WAA 12.3.47
595	5.19.42	10.9.42	4.10.43	France's CH 132 China 8.58
Commercial Iron Works (PC)				
596	4.18.42	8.8.42	1.23.43	Sold 11.14.46
597	5.9.42	9.7.42	2.15.43	Res 1.47 *Kerrville* Str 1957
598	5.23.42	9.7.42	3.5.43	PCC Sold 11.29.46
599	8.2.42	9.26.42	5.15.43	MC 4.14.48
Consolidated Shipbuilding (PC)				
600	2.28.42	5.9.42	8.17.42	Korea's *Myo Hyang San* 1.21.52
601	3.17.42	5.23.42	8.31.42	Res 1.47 *Arcata* Str 7.1.60
602	4.12.42	6.13.42	9.15.42	Res 1.47 *Alturas* Str 7.1.60
603	5.9.42	6.30.42	9.30.42	Postwar reserve training ship Res 10.49 *Solvay* Str 1.7.60
Luders Marine Construction (PC)				
604	2.16.42	10.24.42	3.9.43	Brazil's *Guaiba* 6.11.43
605	3.20.42	11.19.42	5.28.43	Brazil's *Gurupi* 6.11.43
606	4.14.42	1.8.43	8.6.43	Res 1.47 *Andrews* Str 1957
607	7.1.42	2.11.43	8.27.43	Brazil's *Guajara* 10.19.43
Brown Shipbuilding (PC)				
608	1.8.42	5.16.42	8.18.42	MC 7.1.48 Mexico's GC 31 1952 Str 3.64
609	1.8.42	5.30.42	9.7.42	FLC 5.10.47 Thailand's *Khamronsin*
610	2.24.42	6.19.42	9.28.42	D 6.7.50
611	2.24.42	6.29.42	10.26.42	MC 7.1.48
Gibbs Gas Engine (PC)				
612	6.30.42	9.7.42	4.30.43	MC 3.25.48
613	7.7.42	10.27.42	6.1.43	MC 12.10.47 Dominican Republic's *27 de Febrero*
614	7.7.42	12.23.42	7.8.43	MC 3.17.48 Mexico's GC 32 1952 Str 3.64
615	7.7.42	2.17.43	7.15.43	MC 6.21.48
George Lawley (PC)				
616	2.27.42	7.4.42	8.19.42	Thailand's *Tongpliu* 6.4.52
617	3.29.42	7.18.42	8.28.42	Res 6.49 *Beeville* Str 1957
618	4.29.42	8.1.42	9.6.42	EPC *Weatherford* Str 11.1.65 Salvage training hulk until 1.11.68 Used as target
619	4.29.42	8.15.42	9.16.42	Res 1.47 *Dalhart* Venezuela's *Gaviota* 7.5.61, later *Pulpo*
Nashville Bridge (PC)				
620	2.27.42	8.12.42	12.20.42	Res 1.47 *Bethany* Str 1957
621	3.4.42	5.22.42	11.20.42	France's *Fantassin* 10.31.44
622	5.22.42	9.7.42	2.21.43	Greece's *Vassilefs Georghios II* 6.44
623	6.23.42	9.24.42	3.29.43	Sold 11.26.46
Jeffersonville B/M (PC)				
624	3.19.42	7.4.42	8.31.42	YW 120 6.30.44 YWN 120 21.3.49 France 11.22.44 Ret 3.21.49
625	5.2.42	7.22.42	9.17.42	France's *Grenadier* 10.16.44
626	6.5.42	8.18.42	10.13.42	France's *Lansquenet* 11.11.44 BU 1958
627	6.5.42	9.7.42	10.23.42	France's *Cavalier* 10.28.44 BU 1951
Quincy Adams Yacht Yard (SC)				
628	8.25.41	4.15.42	4.29.42	SCC MC 5.6.47
629	9.3.41	4.27.42	5.15.42	SCC MC 3.8.48

478 U.S. SMALL COMBATANTS

	Laid Down	Launched	Completed	Fate
Mathis Yacht Building (SC)				
630	12.22.41	4.18.42	8.5.42	SCC MC 3.9.48
631	1.27.42	6.19.42	8.19.42	MC 3.5.48
632	2.23.42	6.25.42	9.2.42	D 3.9.48
633	3.20.42	7.3.42	9.9.42	D 4.4.46
634	4.27.42	7.10.42	9.26.42	USSR 6.10.45
635	6.6.42	10.14.42	10.23.42	CG 10.19.45
Vinyard Boat Building (SC)				
636	8.29.41	5.14.42	7.11.42	Lost 10.9.45
637	12.24.41	6.10.42	7.31.42	FLC 10.27.48
Elizabeth City Shipyard (SC)				
638	8.11.41	12.20.41	6.27.42	France's CH 116 9.27.44 WL(M) 8.21.45
639	8.16.41	12.31.41	6.21.42	France's CH 93 5.22.44
640	8.30.41	1.17.42	7.11.42	MC 12.16.46
641	11.7.41	2.12.42	7.10.42	MC 3.30.48
Julius Petersen (SC)				
642	10.28.41	5.30.42	8.4.42	CG 1.24.46
643	11.21.41	6.10.42	8.5.42	USSR 8.17.45
644	11.29.41	6.27.42	10.2.42	PGM 1 12.10.43 FLC 5.20.47
645	2.1.42	7.26.42	9.2.42	MC 3.5.48
Robinson Marine Construction (SC)				
646	9.23.41	5.21.42	7.1.42	USSR 5.26.45
647	9.23.41	6.27.42	8.3.42	USSR 5.26.45
Delaware Bay Shipbuilding (SC)				
648	10.10.41	4.18.42	7.11.42	China 6.30.48
649	10.16.41	4.18.42	7.31.42	France's CH 91 5.15.44
Westergard Boat Works (SC)				
650	10.21.41	4.30.42	7.20.42	FLC 1.23.47
651	11.4.41	5.14.42	8.8.42	France's CH 135 11.16.44
Julius Petersen (SC)				
652	12.5.41	4.18.42	6.16.42	FLC 5.20.47
653	12.15.41	4.18.42	7.6.42	CG 10.30.45
Westergard Boat Works (SC)				
654	9.18.41	5.4.42	7.18.42	D 12.18.45
655	10.11.41	5.12.42	8.8.42	France's CH 144 10.23.44
Snow Shipyards (SC)				
656	12.6.41	5.2.42	8.6.42	CG 1.17.46 Indochina's GC 7
657	12.20.41	6.22.42	8.31.42	USSR 6.5.45
American Cruiser (SC)				
658	12.8.41	6.6.42	8.12.42	MC 1.30.48 Indochina's GC 3 3.49
659	12.16.41	6.29.42	10.1.42	CG 10.19.45
Burger Boat (SC)				
660	12.24.41	6.20.42	9.18.42	USSR 6.5.45
661	12.30.41	6.26.42	10.6.42	USSR 5.26.45
Fisher Boatworks (SC)				
662	11.4.41	4.11.42	5.29.42	CG 10.11.45
663	11.4.41	4.22.42	7.1.42	USSR 6.5.45
Dachel-Carter Shipbuilding (SC)				
664	11.19.41	5.21.42	8.8.42	FLC 12.46
665	11.25.41	5.12.42	8.10.42	Sold 6.21.50
Weaver Shipyards (SC)				
666	10.16.41	3.17.42	9.2.42	France's CH 134 9.27.44
667	10.16.41	4.4.42	9.20.42	SCC MC 3.5.48
Daytona Beach Boat Works (SC)				
668	11.5.41	3.12.42	7.2.42	Sold 11.27.46
669	11.17.41	3.21.42	7.17.42	MC 3.26.47
Inland Waterways (SC)				
670	11.17.41	7.6.42	8.21.42	CG 3.19.46
671	11.17.41	8.8.42	10.7.42	Sold 12.11.46
Thomas Knutson Shipbuilding (SC)				
672	10.29.41	5.2.42	6.15.42	CG 10.11.45
673	11.7.41	5.16.42	7.6.42	USSR 6.5.45
Hiltebrant Dry Dock (SC)				
674	10.31.41	3.13.42	8.12.42	USSR 5.26.45
675	11.13.41	3.19.42	9.10.42	USSR 6.10.45
W. A. Robinson (SC)				
676	1.31.42	6.23.42	7.27.42	France's CH 105 10.9.44
677	2.27.42	7.27.42	8.19.42	Scrapped 3.1.46
Thomas Knutson Shipbuilding (SC)				
678	3.10.42	8.17.42	11.6.42	Schools and organizations 4.9.48
679	3.4.42	8.29.42	12.18.42	Indochina 3.14.51 Spain's *Candido Perez* 1957
680	3.17.42	9.7.42	1.30.43	MC 11.13.46
681	3.24.42	9.30.42	3.6.43	MC 11.13.46
American Cruiser (SC)				
682	4.8.42	9.18.42	11.1.42	CG 10.11.45
683	4.9.42	10.14.42	11.21.42	Norway's *Hessa* 10.21.45
684	7.6.42	11.21.42	12.11.42	CG 1.12.46
685	7.9.42	3.25.43	5.12.43	USSR 7.19.45
686	11.30.42	4.15.43	6.26.43	SCC D 12.20.45
687	1.21.43	6.5.43	8.18.43	USSR 5.26.45
Annapolis Yacht Yard (SC)				
688	4.25.42	8.22.42	11.28.42	MC 10.9.46
689	4.27.42	8.15.42	11.2.42	MC 9.24.46

	Laid Down	Launched	Completed	Fate
690	5.22.42	9.7.42	12.16.42	France's CH 106 10.24.44
691	5.31.42	9.19.42	12.30.42	France's CH 133 9.29.44
Calderwood Yacht Yard (SC)				
692	3.26.42	9.25.42	11.25.42	France's CH 131 11.17.44
693	5.15.42	10.24.42	1.12.43	France's CH 107 9.27.44
Daytona Beach Boat Works (SC)				
694	3.21.42	5.25.42	9.6.42	WL 8.23.43
695	3.21.42	7.4.42	10.19.42	France's CH 133 11.16.44
696	3.26.42	8.6.42	11.25.42	WL 8.23.43
697	3.31.42	8.15.42	12.24.42	France's CH 92 5.22.44
Delaware Bay Shipbuilding (SC)				
698	3.1.42	8.7.42	9.28.42	China's 6.30.49
699	3.8.42	9.7.42	10.14.42	Philippines 7.2.48
700	3.25.42	9.7.42	11.2.42	Lost in accident 3.10.44
701	4.18.42	9.17.42	12.9.42	MC 3.25.48
702	5.2.42	10.10.42	12.18.42	MC 12.16.46
703	5.19.42	11.28.42	1.5.43	FLC 10.27.48
Elizabeth City Shipyard (SC)				
704	3.7.42	4.6.42	9.24.42	China's SC 501 9.3.47
705	3.7.42	4.24.42	9.29.42	MC 9.3.46
706	3.14.42	6.26.42	10.7.42	MC 9.13.46
707	4.11.42	6.10.42	10.14.42	MC 8.13.46
708	5.3.42	6.26.42	10.23.42	China's SC 502 9.3.47
709	6.10.42	7.15.42	11.6.42	Grounded 1.21.43
Dooley's Basin and Drydock (SC)				
710	3.11.42	9.7.42	11.12.42	CG 8.23.45
711	4.8.42	10.17.42	12.26.42	CG 10.9.45
Fisher Boatworks (SC)				
712	3.16.42	7.25.42	9.10.42	SCC MC 2.27.48
713	4.18.42	7.25.42	10.22.42	USSR 5.5.45
714	4.24.42	9.17.42	11.12.42	CG 12.1.45
715	5.14.42	10.23.42	11.26.42	CG 1.9.46
716	5.18.42	11.20.42	12.11.42	Sold 5.23.50
717	7.30.42	12.14.42	5.6.43	CG 10.30.45
718	9.22.42	3.31.43	5.19.43	Norway's *Hitra* 10.12.45
719	9.22.42	4.3.43	7.10.43	USSR 8.14.43
720	11.16.42	4.10.43	7.27.43	USSR 9.24.43
721	11.16.42	4.17.43	8.7.43	USSR 9.27.43
Harbor Boat Building (SC)				
722	3.5.42	5.15.42	11.14.42	China 6.30.48
723	3.6.42	6.10.42	12.7.42	China 6.30.48
724	4.6.42	7.2.42	12.23.42	SCC MC 12.16.46
725	4.7.42	7.23.42	12.31.42	MC 12.16.46
726	4.8.42	8.12.42	1.18.43	MC 11.8.46
727	5.7.42	9.7.42	2.9.43	SCC FLC 5.47
728	5.8.42	12.16.42	3.8.43	MC 12.20.46
729	5.9.42	1.25.43	4.19.43	MC 3.8.48
C. Hiltebrant Dry Dock (SC)				
730	3.23.42	6.5.42	11.16.42	MC 12.16.46
731	3.23.42	6.17.42	11.30.42	Philippine *Cagayan* 7.2.48 Str 1956
732	4.11.42	6.27.42	12.18.42	Philippines 7.2.48
733	6.6.42	8.10.42	1.8.43	MC 4.1.48
Al Larson Boat Shop (SC)				
734	4.15.42	7.18.42	12.28.42	FLC 4.25.47
735	4.20.42	8.29.42	3.12.43	China 6.30.48
Liberty Drydock (SC)				
736	4.11.42	10.29.42	2.25.43	Philippine *Mountain Province* 7.2.48
737	4.18.42	12.15.42	5.20.43	FLC 8.13.47
Julius Petersen (SC)				
738	3.7.42	6.27.42	10.19.42	FLC 12.19.47
739	3.20.42	7.4.42	6.27.42	Philippine *Ilocus Sur* 7.2.48 Str 1956
740	3.24.42	7.14.42	11.13.42	Grounded 6.17.43
741	4.1.42	8.3.42	12.14.42	FLC 8.2.46
742	4.22.42	8.17.42	1.5.43	Philippines 7.2.48
743	4.24.42	8.26.42	2.25.43	Philippines 7.2.48
Quincy Adams Yacht Yard (SC)				
744	2.18.42	5.23.42	7.10.42	WL 11.27.44
745	3.7.42	6.15.42	8.6.42	FLC 2.9.48
746	4.1.42	7.8.42	8.16.42	FLC 4.30.48
747	4.9.42	7.28.42	9.11.42	Philippine *Surigao* 7.2.48
748	4.27.42	8.21.42	9.28.42	FLC 1.12.48
749	5.25.42	9.7.42	10.26.42	FLC 4.30.48
750	6.15.42	9.26.42	11.12.42	Philippine *Isabella* 7.2.48 Str 1950
751	7.8.42	10.15.42	11.27.42	Grounded 6.22.43
Robinson Marine Construction (SC)				
752	5.8.42	12.7.42	1.9.43	USSR 8.17.45
753	5.8.42	1.4.43	3.25.43	CG 12.4.45
754	6.6.42	2.23.43	4.25.43	USSR 8.17.45
755	6.20.42	4.3.43	5.16.43	MC 1.20.48
756	7.9.42	5.15.43	6.19.43	USSR 9.2.45
757	7.16.42	6.17.43	7.24.43	PGM 2 12.10.43 FLC 5.20.47
758	8.6.42	7.28.43	8.28.43	CG 1.25.46
759	8.20.42	8.26.43	9.25.43	MC 1.21.48

480 U.S. SMALL COMBATANTS

	Laid Down	Launched	Completed	Fate
W. A. Robinson (SC)				
760	3.14.42	8.11.42	9.10.42	SCC MC 2.4.48
761	3.14.42	8.24.42	9.24.42	MC 3.1.48
762	3.23.42	9.14.42	10.7.42	Brazil's *Jutai* 12.31.42
763	3.23.42	9.25.42	10.20.42	Brazil's *Javari* 12.7.42
764	4.3.42	10.12.42	10.30.42	Brazil's *Jurua* 12.31.44
765	4.8.42	11.12.42	12.3.42	Brazil's *Jaguaro* 2.16.43
766	4.9.42	10.27.42	11.14.42	Brazil's *Jurvena* 12.31.42
767	4.18.42	12.7.42	1.2.43	Brazil's *Jaguaribe* 2.16.43
Seabrook Yacht (SC)				
768	4.8.42	8.8.42	11.9.42	FLC 4.30.48
769	4.8.42	8.23.42	12.4.42	Philippines 7.2.48
770	7.15.42	10.18.42	12.31.42	France's CH 141 10.8.44
771	8.18.42	11.5.42	1.25.43	France's CH 124 11.17.44
Peyton (SC)				
772	5.23.42	9.7.42	4.15.43	CG 12.7.45
773	5.23.42	10.17.42	5.12.43	MC 2.11.48
774	6.11.42	10.27.42	6.28.43	USSR 8.17.45
775	6.12.42	11.28.42	7.22.43	CG 3.27.46
Commercial Iron Works (PC)				
776	8.10.42	10.27.42	3.28.43	Res 10.49 *Pikeville* Str 4.1.59
777	9.7.42	11.11.42	4.8.43	Res 1.50 *Waynesburg* Str 4.1.59
778	9.7.42	11.26.42	4.30.43	Res 3.49 *Gallipolis* Str 4.1.59
779	9.26.42	12.7.42	5.31.43	Res 3.49 *Mechanicsburg* Str 4.1.59
780	10.27.42	12.16.42	6.19.43	Res 3.49 *Maynard* Str 4.1.59
781	11.11.42	12.24.42	7.5.43	Res 3.49 *Metuchen* Str 4.1.59
782	11.26.42	12.31.42	7.19.43	Res 3.49 *Glenholden* Str 4.1.59
783	12.7.42	1.13.43	8.14.43	MC 6.16.48
784	12.16.42	1.18.43	9.17.43	MC 3.17.48
785	12.24.42	1.23.43	10.9.43	Res 3.49 *Frostburg* Str 4.1.59
786	12.31.42	2.6.43	10.30.43	Other government agencies 4.3.54 China's *Hsiang Kiang*
787	1.13.43	2.12.43	11.13.43	MC 6.7.48 Indonesia's *Alu Alu*
788	1.18.43	3.5.43	2.15.44	MC 7.8.48
789	1.23.43	3.13.43	3.6.44	MC 7.14.48
790	2.6.43	3.22.43	3.23.44	MC 6.7.48 Cuba's *Baire* 1956
791	3.13.43	4.17.43	3.28.44	MC 6.7.48
792	3.22.43	4.24.43	4.10.44	MC 7.1.48
793	4.17.43	5.22.43	5.10.44	MC 6.8.48
794	4.24.43	5.29.43	5.25.44	MC 7.1.48 Mexico's GC 34 1952 Str 3.64
795	5.22.43	7.24.43	6.10.44	MC 6.7.48
796	5.29.43	7.3.43	6.19.44	MC 6.8.48 France's *Pnom-Penh* Bought for Indochina customs 1949 but rearmed
797	7.17.43	8.21.43	7.5.44	MC 7.1.48 France's *Hue* Bought for Indochina customs 1949 but rearmed
798	7.3.43	8.14.43	7.19.44	MC 6.8.48 France's *Luang-Prabang* Bought for Indochina customs 1949 but rearmed
799	7.24.43	8.14.43	8.3.44	MC 7.7.48 Korea's *Kum Kang San* 7.48
800	7.31.43	8.28.43	9.23.44	MC 4.19.48
801	8.21.43	9.18.43	9.23.44	MC 7.1.48
802	8.14.43	9.25.43	1.6.45	PCC MC 7.14.48 Korea's *Samkaksan*
803	8.16.43	10.2.43	1.23.45	PCC MC 7.1.48
804	8.28.43	10.16.43	2.6.45	PCC MC 7.14.48
805	9.18.43	10.27.43	11.29.44	PGM 10 8.16.44 FLC 2.13.48
806	9.27.43	10.30.43	12.13.44	PGM 11 8.16.44 FLC 2.13.48
807	10.2.43	11.6.43	2.20.45	PCC MC 7.2.48
808	10.16.43	11.27.43	3.7.45	Res 3.49 *Ripley* Str 4.1.59
809	10.27.43	12.4.43	3.20.45	MC 3.15.48 Portugal's *Sal* 1948
810	10.30.43	12.11.43	4.3.45	MC 7.14.48 Mexico's G 37 7.48 Str 1965
811	11.6.43	12.18.43	4.17.45	MC 3.15.48 Portugal's *Madeira* 1948
812	11.27.43	2.11.44	5.4.45	MC 3.15.48 Portugal's *Principe* 1948
813	12.11.43	3.27.44	5.19.45	MC 6.8.48 Mexico's GC 33 1952 BU 1966
814	2.11.44	5.13.44	6.5.45	D 12.12.45
Albina Engine and Machinery (PC)				
815	10.10.42	12.5.42	1.8.43	Lost 9.28.45
816	12.5.42	1.8.43	6.9.43	MC 7.1.48
817	1.8.43	3.4.43	7.13.43	Res 3.49 *Welch* Str 4.1.59
818	3.4.43	3.30.43	8.3.43	MC 6.7.48
819	3.30.43	5.19.43	8.2.43	MC 6.8.48 Mexico's GC 37 1952 BU 1966
820	5.19.43	8.2.43	9.30.43	MC 7.1.48 Mexico's GC 30 1952 BU 1966
Leathem D. Smith Shipbuilding (PC)				
821	9.1.43	10.23.43	6.7.44	MC 6.7.48
822	10.26.43	12.27.43	5.19.44	Res 3.50 *Asheboro* Str 4.1.59
823	11.8.43	1.15.44	7.7.44	MC 2.11.46 as *Ensign Whitehead* Korea's *Paktusan*
824	3.6.44	5.10.44	8.9.44	MC 6.8.48 Mexico's GC 35 BU 1966
825	3.27.44	5.28.44	8.30.44	MC 6.8.48
826				Acq 12.27.41 *Venture* (PYC 52) 7.15.43 MC 4.12.46 Built by Consolidated Shipbuilding 1931 Ex-*Vixen*

	Laid Down	Launched	Completed	Fate
Pullman Standard Car (PCE)*				
827	10.14.42	5.2.43	7.14.43	RN's *Kilbirnie* 7.14.43 Merchant ship *Haugesund* 1947
828	11.17.42	5.15.43	7.31.43	RN's 7.31.43 *Kilbride* Ret 12.46
829	12.7.42	5.27.43	8.16.43	RN's *Kilchattan* 8.16.43 Merchant ship *Stavanger* 1947
830	12.24.42	6.13.43	8.31.43	RN's *Kilchrenan* 8.31.43 Ret 12.46
831	1.16.43	6.26.43	9.14.43	RN's *Kildary* 9.14.43 Merchant ship *Rio Vouga* 1947
832	2.5.43	7.10.43	9.27.43	RN's *Kildwick* 9.27.43 Ret 12.46
833	2.26.43	8.2.43	10.9.43	RN's *Kilham* 10.9.43
834	3.12.43	8.19.43	10.20.43	RN's *Kilkenzie* 10.20.43 Merchant ship 1947 under same name
835	3.30.43	9.3.43	10.30.43	RN's *Kilhampton* 10.30.43 Merchant ship *Georgios F* 1947
836	4.12.43	9.17.43	11.6.43	RN's *Kilmalcolm* 11.6.43 Merchant ship *Rio Agueda* 1947
837	4.23.43	10.1.43	11.13.43	RN's *Kilmarnock* 11.25.43 Merchant ship *Arion* Lost 1.5.51
838	5.4.43	10.13.43	11.20.43	RN's *Kilmartin* 12.11.43 Merchant ship *Marigoula* 1947
839	5.13.43	10.23.43	12.8.43	RN's *Kilmelford* 12.18.43 Ret 12.46
840	5.24.43	11.2.43	12.11.43	RN's *Kilmington* 12.28.43 Merchant ship *Athinai* 1947
841	6.3.43	11.9.43	12.24.43	RN's *Kilmore* 1.1.44 Merchant ship *Despina* 1947
842	6.12.43	11.14.43	1.12.44	*Marfa* Korea's *Tang Po* 12.9.61 WL(G) 1.19.67
843	6.25.43	11.24.43	1.15.44	*Skowhegan* Str 2.1.60
844	7.8.43	12.1.43	1.29.44	FLC 11.24.47 Mexico's *Sainz de Baranda* 1947 BU 1965
845	7.24.43	12.13.43	2.12.44	*Worland* To State of North Carolina 6.6.64
846	8.10.43	12.20.43	2.19.44	*Eunice* Ecuador's *Esmeraldas* 11.29.60
847	8.24.43	12.27.43	3.4.44	FLC 11.8.47 Mexico's *David Porter* 1947 BU 1965
848	9.7.43	1.21.44	3.18.44	PCER MC 1.21.47
849	9.24.43	1.31.44	3.29.44	PCER *Somersworth* Str 4.1.66 as sonar test ship EPCER Merchant ship
850	10.6.43	2.8.44	4.6.44	PCER *Fairview* Sold as merchant ship 5.1.68
851	10.18.43	2.22.44	4.23.44	PCER Naval reserve training 1946–50 Sonar test ship EPCER Decommissioned 12.68 Str 21.12.68 *Rockville* Colombia's *San Andres*
852	10.28.43	3.1.44	5.13.44	PCER *Brattleboro* Vietnam's *Ngoc Hoi* 4.29.66 Sonar test ship EPCER
853	11.16.43	3.18.44	5.31.44	PCER *Amherst* Great Lakes reserve training ship Vietnam's *Van Kiep II* 6.70
854	11.24.43	3.27.44	12.7.44	PCER MC 10.24.46
855	12.8.43	4.10.44	10.13.44	PCER Sonar test ship EPCER 1946–70 *Rexburg* Str 10.28.70
856	12.17.43	4.21.44	11.10.44	PCER *Whitehall* PCE 3.62 Great Lakes reserve training ship Str 7.1.70
857	12.21.43	5.4.44	3.30.45	PCER *Marysville* Str 7.15.70 Last PCE/PCER on U.S. Navy list
858	1.3.44	5.13.44	4.23.45	PCER CG's *C. G. Jackson* (WPG 120) 2.28.46
859	1.14.44	11.28.44	1.26.45	PCER MC 6.12.47
860	1.25.44	1.30.45	3.3.45	PCER CG's *Bedloe* (WPG 121) 4.25.46
861–66				Canc 3.21.44
Albina Engine and Machinery (PCE)				
867	7.8.42	12.3.42	6.20.43	China's *Yung Tai* 2.7.48
868	8.11.42	1.29.43	8.31.43	FLC 11.8.47 Mexico's *Virgilio Uribe* 1947 BU 1965
869	9.2.42	2.6.43	9.19.43	China's *Yung Tai* 2.7.48 *Wei Yuan*
870	11.30.42	2.27.43	10.5.43	*Dania* Korea's *Pyok Pa* 12.9.61
871	12.3.42	3.10.43	10.29.43	FLC 11.24.47 Mexico's *Blas Godinez* 1947 BU 1965
872	1.30.43	3.24.43	11.29.43	PCEC FLC 10.1.47 Cuba's *Caribe*
873	2.6.43	5.5.43	12.15.43	PCEC Korea's *Han San* 1956
874	3.1.43	5.11.43	12.31.43	*Pascagoula* Ecuador's *Manabi* 12.5.60
875	3.10.43	5.27.43	1.19.44	FLC 11.24.47 Mexico's *Tomas Marin* 1947
876	3.24.43	7.16.43	6.10.44	YDG 8 12.23.43 Decommissioned 11.47 ADG 8 11.47 *Lodestone*
877	5.6.43	8.11.43	2.14.44	PCEC *Havre* Great Lakes reserve training ship Str 7.1.70 Hulk for USN diving/salvage training school
878	5.11.43	8.26.43	3.13.44	*Buttress* (ACM 4) 6.15.44 MC 10.31.47
879	5.27.43	9.30.43	7.10.44	YDG 9 12.23.43 ADG 9.11.47 *Magnet* Decommissioned 11.47
880	8.12.43	10.27.43	4.29.44	*Ely* Great Lakes reserve training ship Str 7.1.70
881	8.11.43	11.10.43	7.31.44	Philippine *Cebu* 7.2.48
882	8.26.43	12.3.43	2.23.45	PCEC *Asheboro*
883	9.30.43	1.14.44	11.13.44	YDG 10 6.14.44 ADG 10 11.47 *Deperm*

*British PCEs were briefly designated BEC 1–15.

	Laid Down	Launched	Completed	Fate
884	10.27.43	2.24.44	3.30.45	Philippine *Negros Occidental* 7.2.48
885	2.25.44	6.20.44	4.30.45	Philippine *Leyte* 7.2.48
886	3.29.44	7.10.44	5.31.45	PCEC *Banning* Str 5.1.61 Memorial on Hood River, Oregon
887–90				Canc 11.5.43
Willamette Iron and Steel (PCE)				
891	10.28.42	4.24.43	6.15.44	PCEC designation Canc 10.15.45 Philippine *Pagasinan* 7.2.48
892	10.28.42	5.1.43	7.8.44	PCEC designation canc 10.15.45 *Somerset* Korea's *Yul Po* 12.9.61
893	10.27.42	5.8.43	7.25.44	FLC 11.20.47 Cuba's *Siboney*
894	12.7.42	5.15.43	8.10.44	*Farmington* Burma's *Yan Taing Aung* 5.31.65
895	12.2.42	5.18.43	10.30.44	*Crestview* Vietnam's *Dong Ha II* 11.29.61
896	12.2.42	5.22.43	11.27.44	PCEC Decommissioned 8.53 Korea's *Myong Ryang* 1956
897	12.16.42	8.3.43	1.6.45	Philippine *Iloilo* 7.2.48
898	12.16.42	8.3.43	1.24.45	PCEC Decommissioned 11.50 Korea's *Ok Po* 1956
899	1.11.43	8.11.43	3.17.45	*Lamar* CG 7.29.64
900	1.11.43	8.11.43	4.12.45	*Groton* Str 2.1.60
901	5.10.43	7.8.43	10.30.44	*Parris Island* (AG 72) 4.28.44 MC 1.20.48
902	1.29.43	8.28.43	4.30.45	*Portage* Great Lakes reserve training ship Str 7.1.70
903	2.18.43	9.6.43	5.16.45	*Batesburg* Str 12.9.61 Korea's *Sa Chon* 1961
904	2.18.43	9.9.43	5.31.45	*Gettysburg* Str 2.1.60
905–20				Canc 9.27.43
921–76				Canc 4.27.42
Simms (SC)				
977	4.26.42	10.10.42	11.19.42	France's CH 94 5.18.44
978	5.9.42	10.10.42	12.8.42	France's CH 145 10.26.44
979	5.23.42	12.9.42	1.9.43	France CH 146 10.27.44
980	6.18.42	12.9.42	2.5.43	MC 9.13.46
Vinyard Boat Building (SC)				
981	3.14.42	8.14.42	10.2.42	Schools and organizations 1.14.46
982	4.18.42	9.7.42	10.31.42	Philippine *Cavite* 7.2.48
983	5.14.42	11.5.42	1.16.43	MC 4.11.47
984	6.10.42	12.19.42	6.25.43	Grounded 4.9.44
John E. Matton (SC)				
985	4.3.42	9.18.42	12.17.42	CG 10.30.45
986	4.18.42	11.13.42	4.26.43	USSR 6.5.45
987	6.15.42	4.10.43	6.4.43	CG 10.11.45
988	11.7.42	6.1.43	7.12.43	CG 10.11.45
989	12.9.42	6.25.43	7.30.43	CG 10.30.45
George W. Kneass (SC)				
990	8.20.42	12.7.42	3.10.43	MC 3.9.48
991	8.20.42	1.20.43	4.30.43	MC 3.19.48
992	8.20.42	3.31.43	6.5.43	MC 10.11.46
993	11.16.42	5.1.43	6.30.43	MC 3.4.48
994	11.16.42	5.24.43	7.22.43	MC 10.2.46
995	11.16.42	6.5.43	8.14.43	MC 11.10.46
Island Dock (SC)				
996	3.30.42	9.7.42	11.5.42	CG 10.30.45
997	4.2.42	10.21.42	12.23.42	USSR 8.17.45
998	4.17.42	11.13.42	4.30.43	MC 2.27.48
999	4.29.42	12.9.42	5.17.43	SCC D 12.14.45
Dingle Boat Works (SC)				
1000	4.21.42	10.22.42	2.14.43	FLC 7.25.47 Cuba's *Oriente*
1001	5.1.42	11.5.42	3.2.43	FLC 7.25.47 Cuba's *Camaguay* Str 1960
1002	7.15.42	4.3.43	5.13.43	MC 9.6.48
Fellows and Stewart (SC)				
1003	4.8.42	7.25.42	12.24.42	CG 11.21.45
1004	4.8.42	8.1.42	1.21.43	CG 12.3.45
1005	4.13.42	8.15.42	2.24.43	MC 9.6.46
1006	4.13.42	9.7.42	5.3.43	MC 3.3.46
1007	5.4.42	11.14.42	5.3.43	USSR 8.17.45
1008	5.8.42	11.19.42	6.15.43	MC 1.12.48
1009	7.18.42	11.25.42	6.25.43	CG 4.10.46
1010	7.30.42	12.12.42	7.15.43	CG 12.6.45
1011	8.17.42	12.16.42	7.29.43	USSR 8.17.45
1012	9.7.42	12.28.42	8.18.43	D 1.1.46
Luders Marine Construction (SC)				
1013	3.23.42	7.22.42	9.19.42	CG 10.23.45
1014	2.21.42	7.26.42	8.28.42	FLC 5.20.47
1015	5.4.42	8.30.42	10.19.42	CG 10.19.45
1016	6.3.42	9.4.42	12.14.42	CG 10.11.45
1017	6.8.42	10.28.42	12.28.42	CG 10.30.45
1018	4.14.42	11.21.42	1.19.43	SCC FLC 1.28.47
1019	4.13.42	12.1.42	2.12.43	MC 5.1.46
1020	8.7.42	12.23.42	5.18.43	SCC MC 1.20.48

FATES AND OTHER NOTES 483

	Laid Down	Launched	Completed	Fate
1021	8.14.42	2.1.43	4.28.43	USSR 6.5.45
1022	9.18.42	3.23.43	6.8.43	CG 10.9.45
Mathis Yacht Building (SC)				
1023	6.15.42	11.16.42	11.18.42	CG 10.11.45
1024	6.29.42	11.28.42	12.3.42	Collision 3.2.43
1025	8.27.42	12.17.42	12.22.42	MC 3.3.48
1026	9.7.42	1.8.43	1.13.43	MC 2.11.48
1027	9.7.42	1.26.43	2.2.43	CG 12.17.45
1028	10.21.42	2.21.43	3.27.43	CG 1.8.46
Donovan Contracting (SC)				
1029	4.27.42	8.31.42	11.13.42	France's CH 123 10.30.44
1030	5.4.42	8.31.42	11.13.42	France's CH 136 10.2.44
Peterson Boat Works (SC)				
1031	4.28.42	10.3.42	1.8.43	USSR 8.17.45
1032	5.19.42	10.17.42	1.25.43	CG 11.29.45
1033	6.17.42	11.12.42	3.26.43	CG 12.4.45
1034	8.4.42	1.2.43	4.2.43	MC 3.3.48
1035	9.7.42	4.12.43	5.6.43	PGM 3 12.10.43 FLC 5.6.47
1036	9.7.42	5.10.43	5.27.43	MC 2.13.48
1037	10.31.42	5.29.43	6.24.43	CG 1.16.46
1038	11.28.42	6.12.43	7.16.43	CG 1.25.46
Rice (SC)				
1039	4.14.42	7.28.42	10.15.42	MC 2.9.48
1040	4.14.42	8.1.42	10.27.42	MC 12.31.47
1041	5.14.42	8.13.42	11.9.42	MC 2.9.48
1042	5.11.42	8.15.42	11.28.42	MC 3.5.48
1043	8.3.42	10.15.42	12.31.42	France's CH 125 10.10.44
1044	8.4.42	10.17.42	1.25.43	MC 3.25.48 France's CH 126
1045	8.19.42	11.19.42	2.18.43	MC 3.9.48
1046	8.18.42	11.21.42	3.27.43	MC 4.11.47
Ventnor Boat Works (SC)				
1047	4.24.42	11.15.42	1.30.43	MC 4.11.47
1048	4.28.42	12.12.42	4.8.43	MC 11.10.46
1049	5.2.42	5.2.43	6.9.43	SCC D 12.14.45
1050	5.4.42	5.9.43	8.4.43	MC 11.20.46
1051	5.8.42	6.19.43	7.29.43	MC 3.3.47
1052	5.14.42	6.26.43	9.6.43	MC 10.13.46
Wilmington Boat Works (SC)				
1053	4.11.42	9.7.42	3.17.43	PGM 4 12.10.43 FLC 6.9.47
1054	4.11.42	9.19.42	4.6.43	CG 2.21.46
1055	5.2.42	10.10.42	5.10.43	CG 11.27.45
1056	5.14.42	11.2.42	6.15.43	PGM 5 12.10.43 FLC 5.7.47
Gulf Marine Ways (SC)				
1057	5.7.42	11.11.42	5.18.43	Schools and organizations 7.22.48
1058	5.7.42	12.1.42	6.17.43	MC 9.19.46 Indochina's GC 8
Inland Waterways (SC)				
1059	5.20.42	11.28.42	2.25.43	MC 5.14.46
1060	7.8.42	12.1.42	4.13.43	USSR 6.5.45
Harris and Parsons (SC)				
1061	5.23.42	9.26.42	1.5.43	Norway's *Vigra* 10.21.45
1062	6.3.42	12.23.42	2.27.43	CG 10.23.45
Victory Shipbuilding (SC)				
1063	6.22.42	11.28.42	1.8.43	CG 10.23.45
1064	7.9.42	1.16.43	4.17.43	CG 10.30.45 Haiti's *Toussaint l'Ouverture* Sold 1959
Perkins and Vaughn (SC)				
1065	5.27.42	10.17.42	12.30.42	MC 2.9.48
1066	6.2.42	11.14.42	2.27.43	SCC FLC 5.47
Mathis Yacht Building (SC)				
1067	11.19.42	3.9.43	4.3.43	Foundered 11.19.43
1068	12.10.42	3.26.43	4.13.43	CG 2.28.46
1069	12.30.42	4.17.43	4.26.43	CG 12.2.45
1070	1.21.43	5.3.43	5.24.43	CG 12.2.45
1071	2.6.43	5.20.43	6.8.43	PGM 6 12.10.43 FLC 5.7.47
1072	3.10.43	6.17.43	6.28.43	PGM 7 12.10.43 Collision 7.18.44
1073	3.31.43	6.30.43	7.19.43	USSR 8.13.43
1074	4.20.43	7.12.43	7.28.43	USSR 8.13.43
1075	5.11.43	7.27.43	8.11.43	USSR 9.24.43
1076	5.29.43	8.12.43	8.23.43	
Albina Engine and Machinery (PC)				
1077	2.28.42	7.29.42	12.9.42	Decommissioned 1.47 *Edenton* Venezuela 1.11.61
1078	4.29.42	8.11.42	2.5.43	China's *Chih Kiang* 5.54
1079	4.29.42	8.25.42	3.7.43	Decommissioned 1.47 PCC *Ludington* Str 7.1.60
1080	4.29.42	8.27.42	3.29.43	PCC Sold 1.16.47
1081	7.29.42	8.29.42	4.16.43	PCC Decommissioned 1.47 *Cadiz* Str 7.1.60
1082	8.29.42	10.10.42	5.8.43	Sold 11.13.46
George Lawley (PC)				
1083	8.15.42	9.7.42	8.16.43	MC 12.19.46
1084	9.7.42	11.28.42	8.31.43	MC 8.19.46
1085	1.6.43	3.27.43	10.19.43	MC 12.10.46

	Laid Down	Launched	Completed	Fate
1086	2.13.43	4.24.43	1.19.44	Indochina's *Flamberge* 3.2.51 Cambodia's E 311 1955
1087	4.12.43	8.21.43	5.22.44	Decommissioned 1.47 *Placerville* China's *Tung Kiang* 7.15.57
1088	8.31.43	1.18.45	4.24.45	PGM 12 8.16.44 China 6.30.48 Defected to PRC
1089	1.19.45	4.12.45	6.14.45	PGM 13 8.16.44 China's *Tung Ting* 6.30.48 Later *Ling Chiang* WL(T) 1.10.55
1090	4.12.45	6.15.45	8.9.45	PGM 14 8.16.44 China 6.30.48 Defected to PRC
1091	6.15.45	7.31.45	10.11.45	PGM 15 8.16.44 China 6.30.48 Defected to PRC as *Kan Tang*
1092–105				Canc 11.5.43
1106–7				Canc 9.25.43
1108–18				Canc 9.17.43

Defoe Shipbuilding (PC)

	Laid Down	Launched	Completed	Fate
1119	6.12.42	8.11.42	11.23.42	Decommissioned 1.47 *Greencastle*
1120	6.19.42	8.24.42	12.8.42	*Carlinville* Str 4.1.59
1121	6.30.42	8.27.42	11.10.42	Philippine *Camarines Sur* 7.2.48
1122	7.11.42	9.7.42	11.30.42	Philippines 7.2.48
1123	7.22.42	9.7.42	12.24.42	FLC 8.27.47
1124	8.4.42	10.1.42	1.22.43	*Chocura* (IX 206) 2.20.45 D 10.17.45
1125	8.20.42	10.15.42	1.30.43	PCC *Cordele* Str 4.1.59
1126	8.29.42	10.31.42	2.25.43	PCC D 12.24.45
1127	9.7.42	11.2.42	2.25.43	PCC MC 7.1.48
1128	9.14.42	11.19.42	4.29.43	D 3.9.46
1129	9.23.42	12.7.42	5.8.43	WL 1.31.45
1130	10.8.42	12.10.42	5.21.43	Indochina's *Intrepide* 9.28.50 Vietnam's *Van Kiep* Str 7.1.65
1131	10.22.42	12.17.42	7.4.43	Philippine *Bohol* 7.2.48
1132	11.5.42	12.29.42	7.17.43	D 5.20.48
1133	11.25.42	1.9.43	7.31.43	Philippine *Zamboango del Sur* 7.2.48 Str 11.56
1134	12.4.42	1.18.43	8.21.43	Philippine *Batangas* 7.2.48
1135	12.11.42	2.3.43	9.11.43	Res 1.47 *Canastota* Str 4.1.59
1136	12.17.42	3.5.43	10.22.43	PCC Res 1.47 *Galena* Str 4.1.59
1137	12.29.42	3.29.43	10.2.43	PCC Res 1.47 *Worthington* Str 4.1.59
1138	1.9.43	4.19.43	8.21.43	Res 1.47 *Lapeer* Str 4.1.59
1139	1.25.43	5.10.43	10.22.43	Indochina's *Impetueux* 9.28.50 Cambodia 1954–56
1140	2.8.43	6.14.43	12.22.43	Res 1.47 *Glenwood* Str 7.1.60
1141	3.12.43	6.22.43	12.3.43	Res 3.54 *Pierre* Indonesia's *Tjakalang* 10.25.58
1142	3.31.43	8.20.43	3.1.44	Res 1.47 *Hanford* China's *Pei Kiang* 7.15.57
1143	4.17.43	9.25.43	4.26.44	Indochina's *Glaive* 3.2.51 Vietnam's *Tuy Dong* 1956
1144	5.7.43	10.4.43	5.4.44	Indochina's *Mousquet* 3.2.51 Vietnam's *Chi Lang* Str 1961
1145	6.2.43	10.27.43	5.20.44	*Winnemucca* Korea's *O Tae San* 11.21.60
1146	9.21.43	11.15.43	7.1.44	Indochina's *Trident* 3.2.51 Vietnam's *Tay Ket* Str 7.10.65
1147	10.11.43	12.2.43	7.29.44	FLC 4.30.48
1148	10.25.43	12.19.43	10.25.44	PGM 16 8.16.44 Greece's *Antiploiarkhos Laskos* 11.24.47
1149	11.6.43	1.11.44	6.10.44	*Susanville* China's *Hsi Kiang* 7.57
1150–66				Canc 9.17.43

Sullivan Drydock and Repair (PC)

	Laid Down	Launched	Completed	Fate
1167	4.3.43	7.3.43	12.2.43	Indochina 3.2.51 Vietnam's *Dong 'Da* Str 1961
1168	4.3.43	7.3.43	12.30.43	PCC China's *Ching Kiang* 5.54
1169	7.3.43	10.16.43	1.25.44	PCC Res 11.49 *Escondido* China's *Liu Kiang* 7.15.57
1170	7.3.43	10.16.43	2.19.44	*Kelso* Str 7.1.60

Leathem D. Smith Shipbuilding (PC)

	Laid Down	Launched	Completed	Fate
1171	3.12.43	5.15.43	9.3.43	Indochina's *Inconstant* 3.4.51 Cambodia's E 312 1956
1172	3.29.43	6.5.43	9.17.43	*Olney* Str 7.1.60
1173	4.21.43	6.26.43	10.8.43	Res 1.47 *Andalusia* Str 7.1.60
1174	5.18.43	7.22.43	10.18.43	Res 1.47 *Fredonia* Str 1957
1175	6.8.43	8.7.43	11.12.43	Res 1.47 *Vandalia* China's *Han Kiang* 7.15.57
1176	6.28.43	8.28.43	10.29.43	Res 1.47 *Minden* Venezuela's *Petrel* 6.62
1177	7.24.43	9.18.43	11.26.43	PCC Res 1.47 *Guymon* Str 11.1.59
1178	8.11.43	10.2.43	12.10.43	PCC Res 1.47 *Kewaunee* Str 1.1.59
1179	9.20.43	11.6.43	12.27.43	Res 1.47 *Morris* Str 7.1.60
1180	10.5.43	11.27.43	1.20.44	PCC Res 1.47 *Woodstock* Str 7.1.60

Gibbs Gas Engine (PC)

	Laid Down	Launched	Completed	Fate
1181	10.5.42	4.15.43	9.16.43	Res 2.50 *Wildwood* Str 4.1.59
1182	10.9.42	6.14.43	10.27.43	China's *Yuan Chiang* 7.54
1183	10.27.42	7.7.43	12.7.43	MC 6.7.48 Indonesia's *Tenggiri*
1184	1.8.43	8.4.43	1.24.44	MC 6.16.48
1185	3.26.43	8.27.43	4.24.44	MC 6.8.48 Thailand's *Phali*
1186	4.20.43	9.27.43	6.9.44	Res 11.49 *Ipswich* Str 4.1.59
1187	6.18.43	11.26.43	7.18.44	MC 7.1.48
1188	7.12.43	1.31.44	9.5.44	MC 3.24.48
1189	8.10.43	4.14.44	11.24.44	PGM 17 8.16.44 D 10.24.45
1190	9.1.43	6.29.44	2.6.45	MC 6.10.48

FATES AND OTHER NOTES 485

	Laid Down	Launched	Completed	Fate
Consolidated Shipbuilding (PC)				
1191	5.23.42	7.25.42	10.31.42	Res 2.50 *Bel Air* Str 4.1.59
1192	6.13.42	8.8.42	11.25.42	MC 6.18.48
1193	6.30.42	8.29.42	1.20.43	Res 7.50 *Ridgway* Str 4.1.59
1194	7.25.42	9.19.42	2.6.43	MC 6.21.48
1195	8.8.42	10.3.42	3.22.43	MC 6.7.48
1196	8.29.42	10.24.42	4.6.43	Res 12.49 *Mayfield* Str 4.1.59
1197	9.19.42	11.14.42	4.21.43	MC 6.21.48
1198	10.3.42	12.12.42	4.30.43	Res 12.49 *Westerly* Str 4.1.59
1199	10.24.42	1.2.43	5.15.43	MC 6.18.48
1200	11.16.42	1.23.43	5.28.43	D 10.21.49
1201	12.12.42	2.14.43	6.10.43	Res 1.50 *Kittery* Str 4.1.59
1202	1.2.43	2.27.43	6.22.43	MC 12.10.47 Dominican Republic's *Capitan Wenceslas Arvels* Later *Patria* Str 1962
1203	1.23.43	3.14.43	7.16.43	D 11.4.49
1204	2.27.43	4.14.43	8.4.43	MC 6.18.48
1205	3.14.43	4.29.43	8.25.43	MC 7.1.48
1206	6.9.43	7.21.43	10.25.43	MC 7.7.48
1207	6.26.43	8.18.43	11.10.43	MC 6.23.48
1208	7.21.43	9.15.43	11.23.43	China's *Li Kiang* 7.54
1209	8.18.43	10.7.43	4.28.44	*Medina* Str 4.1.59
1210	9.15.43	10.30.43	5.4.44	MC 6.8.48 Mexico's G 38 1952
Luders Marine Construction (PC)				
1211	8.11.42	3.12.43	8.14.43	MC 10.16.46 *Blue Arrow* Spain's *Javier Quiroga* 10.24.56
1212	9.26.42	4.23.43	9.16.43	Res 2.50 *Laurinburg* Str 4.1.59
1213	11.7.42	5.22.43	10.4.43	Res 10.49 *Loudon* Str 4.1.59
1214	2.22.43	6.28.43	10.26.43	D 4.12.51
1215	3.22.43	7.29.43	11.24.43	MC 6.28.48
1216	4.29.43	8.29.43	12.30.43	Res 10.49 *Elkins* Str 4.1.59
1217	5.26.43	9.26.43	4.26.44	MC 3.24.48
1218	7.2.43	10.24.43	5.26.44	MC 6.7.48 Thailand's *Sukrip*
1219	8.2.43	11.21.43	6.30.44	MC 4.5.48
1220	9.7.43	12.22.43	7.28.44	MC 6.8.48
Penn-Jersey (PC)				
1221	5.29.42	8.29.43	4.19.44	MC 7.28.48
1222	10.26.42	9.25.43	7.6.44	MC 4.22.47
1223	9.26.43	1.9.44	11.13.44	Sold 11.26.46
1224	8.31.43	7.9.44	1.8.45	MC 6.8.48 Mexico's GC 36 1952 Str 3.64
Leathem D. Smith Shipbuilding (PC)				
1225	6.10.42	9.7.42	12.12.42	Res 1.47 *Waverly*
1226	7.7.42	9.7.42	1.6.43	France's *Legionnaire* 11.6.44 BU 1958
1227	8.5.42	10.17.42	2.6.43	France's *Lancier* 11.18.44 BU 1960
1228	9.7.42	11.18.42	4.24.43	Res 1.47 *Munising* Str 1957
1229	9.7.42	12.20.42	5.20.43	Res 1.47 *Wauseon* Str 1957
1230	12.20.42	3.10.43	6.24.43	PCC Res 1.47 *Grinnell* Str 4.1.59
Sullivan Drydock and Repair (PC)				
1231	9.1.42	12.12.42	7.10.43	PCC Res 1.47 *Tipton* Str 7.1.60
1232	9.8.42	12.12.42	8.16.43	China's *Chang Kiang* 6.54 WL 8.6.65
1233	9.21.42	1.11.43	9.22.43	China's *Kung Kiang* 6.54
1234	12.12.42	4.3.43	6.19.43	Uruguay's *Maldonado* 5.2.44
1235	12.12.42	4.3.43	7.28.43	France's *Hussard* 10.26.44 Torpedo training vessel ca. 1954
1236	1.11.43	4.24.43	8.31.43	Brazil's *Graju* 11.15.43
Consolidated Shipbuilding (PC)				
1237	2.14.43	4.3.43	7.23.43	Res 12.47 *Abingdon* Str 4.1.59
1238	4.3.43	5.15.43	9.3.43	FLC 5.47
1239	4.14.43	6.9.43	9.25.43	D 2.5.46
1240	5.15.43	6.26.43	10.12.43	Res 10.49 *Culpepper* Str 4.1.59
Nashville Bridge (PC)				
1241	9.4.42	12.24.42	5.13.43	Philippine *Nueva Ecija* 7.2.48
1242	9.22.42	1.25.43	6.27.43	Res 1.47 *Port Clinton* Str 4.1.59
1243	12.9.42	4.10.43	8.29.43	FLC 8.27.47
1244	1.20.43	5.8.43	9.19.43	PCC Res 1.47 *Martinez* Str 7.1.60
1245	3.9.43	5.29.43	10.17.43	FLC 2.5.48 France 2.48(?)
1246	5.4.43	7.3.43	11.14.43	Res 1.47 *Canandaigua* Str 1957
1247	5.28.43	8.7.43	12.4.43	China's *Chialing* 6.30.48 Later *To Kiang* Str 1964
1248	6.29.43	9.18.43	1.6.44	France's *Sabre* 1.18.44 BU 1959
1249	8.4.43	11.6.43	1.30.44	France's *Pique* 2.9.44 BU 1959
1250	9.14.43	12.18.43	2.20.44	France's *Cimeterre* 3.9.44 BU 1963
Brown Shipbuilding (PC)				
1251	6.8.42	9.12.42	2.27.43	PCC Res 1.47 *Ukiah* Venezuela 1.60
1252	6.8.42	9.30.42	3.27.43	Decommissioned 1.47 *Tarrytown* Venezuela 1.61 (not placed in service)
1253	6.22.42	10.14.42	3.30.43	Thailand's *Liulom* 6.4.52
1254	6.22.42	10.31.42	4.13.43	China's *Po Kiang* 6.54
Luders Marine Construction (PC)				
1255	9.29.43	1.23.44	11.28.44	PGM 18 8.16.44 WL 4.21.45
1256	10.28.43	5.21.44	10.26.44	MC 3.15.48 Portugal's *Sao Tome*
1257	11.24.43	7.23.44	12.20.44	MC 3.15.48 Portugal's *Santiago*
1258	12.28.43	9.23.44	1.23.45	MC 6.21.48
1259	1.28.44	10.7.44	3.5.45	MC 3.15.48 Portugal's *Sao Vicente*

486 U.S. SMALL COMBATANTS

	Laid Down	Launched	Completed	Fate
Leathem D. Smith Shipbuilding (PC)				
1260	10.20.42	1.16.43	4.6.43	PCC Res 1.47 *Durango* Str 11.1.59
1261	11.20.42	2.28.43	4.30.43	WL 6.6.44
1262	1.21.43	3.27.43	6.8.43	China's *Chung Kiang* 6.54
1263	3.2.43	4.19.43	7.8.43	*Milledgeville* China's *To Kiang* 7.59
Consolidated Shipbuilding (PC)				
1264	10.7.43	11.28.43	4.24.44	MC 3.24.48
1265	10.30.43	12.19.43	5.11.44	MC 12.3.46
Quincy Adams Yacht Yard (SC)				
1266	7.28.42	11.9.42	2.12.43	MC 4.2.48
1267	8.20.42	12.12.42	2.27.43	Philippine *Alert* 7.2.48 Sunk 1956
1268	9.7.42	12.29.42	3.13.43	MC 9.23.46
1269	9.26.42	2.24.43	4.10.43	Philippines 7.2.48
1270	10.15.42	2.24.43	4.26.43	MC 11.8.46
1271	11.9.42	3.31.43	5.15.43	MC 11.27.46
1272	12.1.42	4.19.43	6.4.43	SCC MC 10.14.47
1273	12.28.42	5.8.43	6.30.43	SCC MC 3.26.47
1274	1.4.43	5.22.43	7.24.43	Philippine *Ilocus Norte* 7.2.48 Later *Malampaya Sound*
1275	1.15.43	6.5.43	8.17.43	MC 3.25.48
Elizabeth City Shipyard (SC)				
1276	9.7.42	10.27.42	3.4.43	MC 11.15.46
1277	9.7.42	10.27.42	3.11.43	MC 11.15.46
1278	9.7.42	12.7.42	3.16.43	SCC Philippines 7.2.48
1279	9.7.42	10.27.42	3.25.43	MC 9.10.46
1280	10.27.42	2.6.43	4.22.43	Scrapped 5.3.48
1281	10.27.42	2.5.43	4.30.43	SCC MC 12.6.46
1282	12.7.42	2.6.43	5.13.43	MC 9.5.46
1283	12.31.42	2.20.43	6.6.43	USSR 7.13.43
1284	2.5.43	3.26.43	6.18.43	USSR 7.23.43
1285	2.9.43	4.17.43	7.4.43	USSR 7.23.43
1286	2.22.43	5.15.43	7.20.43	USSR 8.13.43
1287	4.19.43	6.5.43	7.29.43	USSR 9.27.43
W. A. Robinson (SC)				
1288	7.27.42	12.24.42	2.12.43	Brazil's *Jacui* 5.19.43
1289	8.19.42	2.8.43	3.12.43	Brazil's *Jundiai* 4.26.43
1290	8.28.42	3.16.43	4.15.43	FLC 6.16.47 Cuba's *Las Villas*
1291	9.22.42	3.30.43	5.3.43	FLC 6.16.47 Cuba's *Habana*
1292	10.5.42	4.17.43	5.19.43	MC 3.22.46
1293	10.20.42	4.24.43	6.4.43	MC 8.12.46
1294	12.28.42	6.4.43	7.7.43	MC 9.19.47
1295	1.21.43	7.23.43	8.12.43	USSR 6.10.45
1296	2.11.43	8.20.43	9.16.43	CG 8.14.45
1297	3.19.43	9.2.43	10.8.43	CG 10.11.45
Perkins and Vaughn (SC)				
1298	9.7.42	3.19.43	4.30.43	SCC D 3.6.46
1299	9.7.42	4.10.43	6.1.43	MC 8.12.46
1300	9.7.42	5.8.43	7.3.43	MC 3.4.48
1301	9.7.42	5.15.43	8.3.43	FLC 8.16.47 Cuba's *Pinar del Rio*
Daytona Beach Boat Works (SC)				
1302	8.15.42	10.31.42	3.27.43	MC 10.9.46
1303	9.4.42	11.28.42	5.1.43	MC 9.23.46
1304	9.12.42	1.16.43	6.14.43	MC 10.2.46
1305	9.19.42	2.15.43	8.11.43	MC 1.9.46
1306	10.6.42	3.25.43	9.1.43	SCC D 1.7.46
1307	10.20.42	4.15.43	9.29.43	CG 2.12.46
1308	11.28.42	5.15.43	11.1.43	MC 4.1.48
Annapolis Yacht Yard (SC)				
1309	9.7.42	2.26.43	5.3.43	SCC Sea Scouts 9.26.46
1310	9.7.42	2.28.43	5.22.43	MC 9.24.46
1311	9.28.42	4.10.43	6.12.43	SCC MC 12.12.46
1312	10.9.42	4.26.43	7.1.43	SCC MC 3.26.47
1313	10.2.42	5.3.43	7.19.43	MC 9.18.46
1314	3.3.43	6.12.43	8.16.43	SCC MC 9.16.46
Julius Petersen (SC)				
1315	8.11.42	3.3.43	6.12.43	SCC MC 3.1.48
1316	8.18.42	3.11.43	6.18.43	SCC Army 12.11.46
1317	9.2.42	3.20.43	7.5.43	MC 11.8.46
1318	9.8.42	4.2.43	8.9.43	MC 11.10.46
1319	9.24.42	4.12.43	8.26.43	MC 3.4.48
1320	10.10.42	4.17.43	8.22.43	MC 12.31.46
Harris and Parsons (SC)				
1321	9.7.42	2.6.43	5.4.43	MC 3.22.46
1322	9.26.42	4.8.43	6.2.43	MC 1.7.48
1323	12.24.42	5.15.43	8.5.43	SCC MC 3.3.48
1324	2.4.43	7.22.43	9.24.43	USSR 6.10.45
Delaware Bay Shipbuilding (SC)				
1325	9.7.42	2.11.43	4.19.43	MC 4.2.48
1326	9.7.42	2.24.43	5.3.43	SCC MC 12.16.46
1327	10.7.42	3.24.43	6.4.43	MC 1.27.47
1328	10.9.42	4.21.43	6.30.43	MC 9.10.46

	Laid Down	Launched	Completed	Fate
Simms (SC)				
1329	10.10.42	4.19.43	5.8.43	CG 2.7.46
1330	10.30.42	4.19.43	5.29.43	MC 9.29.47
1331	12.16.42	6.2.43	7.2.43	France's CH 6 4.11.44
1332	12.16.42	7.17.43	8.6.43	MC 7.6.48
Thomas Knutson Shipbuilding (SC)				
1333	8.21.42	4.24.43	6.23.43	MC 6.23.46
1334	8.22.42	5.8.43	7.16.43	MC 8.20.48
1335	9.15.42	6.5.43	8.11.43	France's CH 52 11.12.43
1336	9.17.42	5.22.43	8.26.43	France's CH 51 11.24.43
1337	9.28.42	7.19.43	9.15.43	France's CH 71 12.29.43 Ivory Coast's *Patience* 1961
1338	10.19.42	8.7.43	9.29.43	SCC MC 1.26.48
1339	10.24.42	8.21.43	11.6.43	CG 11.15.45
1340	11.7.42	9.18.43	12.3.43	CG 10.30.45
Rice Brothers (SC)				
1341	10.21.42	1.14.43	5.25.43	SCC Scrapped 2.4.47
1342	10.21.42	1.16.43	6.14.43	Scrapped 7.3.46
1343	11.23.42	2.20.43	8.9.43	Scrapped 8.3.46
1344	11.24.42	3.27.43	8.16.43	France's CH 62 11.19.43 Senegal's *Senegal* 1962
1345	1.21.43	4.10.43	9.12.43	France's CH 61 12.23.43
1346	1.22.43	4.29.43	10.24.43	France's CH 72 1.18.44
Fisher Boatworks (SC)				
1347	2.15.43	7.15.43	8.27.43	CG 11.21.45
1348	2.17.43	8.12.43	9.25.43	CG 10.31.45 Haiti's *Admiral Killick* Str 1954
1349	4.27.43	9.11.43	10.16.43	SCC D 6.21.46
1350	4.27.43	10.6.43	11.6.43	SCC FLC 1.14.47
Vinyard Boat Building (SC)				
1351	9.7.42	3.11.43	4.30.43	Discarded 9.9.46 Dominican Republic's *30 de Marzo* Str 1965
1352	9.14.42	4.21.43	6.18.43	Scrapped 10.16.46
1353	11.5.42	6.3.43	8.4.43	Discarded 10.16.46 *Patricia* Dominican Republic's *Las Carreras*
1354	12.19.42	8.10.43	9.29.43	Scrapped 11.22.46
Luders Marine Construction (SC)				
1355	11.14.42	4.12.43	7.9.43	CG 10.11.45
1356	1.5.43	6.7.43	8.16.43	CG 10.9.45
1357	2.13.43	6.21.43	8.30.43	CG 10.24.45
Calderwood Yacht Yard (SC)				
1358	10.1.42	2.22.43	4.20.43	Department of Interior 9.30.46
1359	11.2.42	4.3.43	6.3.43	France's CH 5 8.26.44
1360	2.22.43	6.22.43	8.20.43	SCC Scrapped 10.23.47
1361	4.3.43	7.24.43	10.13.43	CG 10.3.45
Peyton (SC)				
1362	11.7.42	4.24.43	9.6.43	CG 2.8.46
1363	11.14.42	5.29.43	9.29.43	Scrapped 1.20.46
1364	11.28.42	7.10.43	11.2.43	USSR 8.17.45
1365	4.24.43	9.29.43	1.2.44	USSR 8.17.45
Wilmington Boat Works (SC)				
1366	11.2.42	5.1.43	5.1.43	PGM 8 12.10.43 FLC 5.47
1367	11.25.42	5.27.43	9.10.43	CG 1.10.46
1368	11.28.42	6.21.43	10.9.43	Scrapped 1.30.48
1369	12.9.42	7.19.43	11.12.43	CG 7.7.48
Fellows and Stewart (SC)				
1370	12.7.42	5.20.43	9.6.43	Scrapped 1.12.48
1371	12.10.42	5.22.43	9.30.43	Scrapped 1.20.48
1372	12.19.42	6.8.43	11.1.43	Scrapped 2.9.48
1373	12.23.42	6.19.43	11.1.43	CG 1.30.46
1374	12.26.42	7.5.43	11.14.43	Scrapped 3.3.48
1375	12.29.42	7.31.43	11.24.43	Scrapped 1.20.48
Wheeler Shipbuilding (PCS)				
1376	10.13.42	4.3.43	7.9.43	Res 12.49 *Winder* Str 1957
1377	2.3.43	10.14.43	1.18.44	MC 1.18.47
1378	2.13.43	12.18.43	2.8.44	Res 3.54 *Provincetown* Str 1957
1379	2.25.43	2.4.44	3.4.44	PCSC MC 5.21.48
1380	2.27.43	2.25.44	3.21.44	Res 6.54 *Rushville* Str 5.9.57 Sold 29.4.58
1381	3.8.43	3.14.44	4.8.44	MC 18.1.47
1382	3.19.43	3.16.44	3.29.44	MC 30.4.47
1383	3.27.43	6.23.44	8.4.44	Res 12.49 *Attica* Str 1957
1384	4.9.43	8.7.44	9.6.44	Res 3.54 *Eufala* Str 1957
1385	5.8.43	8.26.44	10.4.44	*Hollidaysburg* Str 7.1.70 Last PCS on USN list
1386	5.15.43	9.28.44	11.3.44	*Hampton* Str 7.1.59
1387	5.22.43	10.10.44	11.25.44	*Beaufort* Sunk as target 7.15.67
1388	12.7.42	7.17.43	12.11.44	*Littlehales* (AGS 7) 3.20.45 Sold 2.50
1389	2.27.43	9.18.43	1.8.44	PCSC MC 4.7.48
1390	3.12.43	9.13.43	2.2.44	PCSC MC 6.10.48
1391	3.31.43	10.8.43	2.29.44	PCSC MC 4.5.48
Robert Jacob (PCS)				
1392	4.5.43	10.13.43	3.25.44	Res 3.54 *Deming* Str 1957
1393–95				Canc 9.27.43
1396	3.23.43	8.7.43	3.29.44	*Dutton* (AGS 8) 3.20.45 Sold 2.50

	Laid Down	Launched	Completed	Fate
South Coast (PCS)				
1397	5.1.43	9.25.43	5.11.44	MC 9.10.47
1398				Canc 9.27.43
1399	8.7.43	4.15.44	11.25.44	Laid down as YMS 450 Philippine *Tarlac* 2.7.48
1400	9.22.43	5.20.44	12.26.44	Minesweeper (AMS 59) 9.16.47 MC 9.17.47 USN 4.14.52 PCS 4.11.53 *Coquille* 1.2.56 Decommissioned 3.54 Str 1957 Laid down as YMS 451
1401	3.3.44	7.22.44	1.13.45	*McMinnville* 1.2.56 Str 8.1.62
Colberg Boat Works (PCS)				
1402	3.24.43	8.11.43	1.13.44	PCSC MC 3.7.47
1403	4.20.43	9.28.43	2.17.44	PCSC Philippine *Laguna* 2.7.48
1404	5.15.43	11.12.43	3.30.44	*Armistead Rust* (AGS 9) 3.20.45 FLC 27.10.48
Greenport Basin (PCS)				
1405	5.1.43	8.21.43	1.31.44	Department of Interior 9.8.46
1406–12				Canc 9.27.43
Stadium Yacht Basin (PCS)				
1413	1.25.43	7.24.43	11.22.43	*Elsmere* Str 10.1.60
1414	2.11.43	9.14.43	11.24.43	MC 12.21.47
1415–16				Canc 9.27.43
Dachel-Carter Shipbuilding				
1417	3.13.43	8.7.43	12.18.43	MC 6.10.47
1418	4.14.43	9.4.43	1.26.44	PCSC D in accident 12.15.45
1419	6.18.43	10.31.43	3.4.44	MC 1.18.47
1420	6.26.43	11.20.43	4.2.44	MC 1.21.47
William F. Stone (PCS)				
1421	5.1.43	11.1.43	2.19.44	PCSC MC 4.7.48
1422	6.12.43	1.24.44	4.24.44	MC 4.25.47
Burger Boat (PCS)				
1423	12.22.42	5.22.43	11.1.43	*Prescott* Str 3.1.62
1424	2.3.43	6.19.43	11.24.43	MC 11.21.46
Hiltebrant Dry Dock (PCS)				
1425	1.22.43	7.20.43	2.3.44	MC 1.21.47
1426	3.8.43	8.31.43	8.31.44	Korea's *Su Seong* 9.6.52 Ret 4.63
1427–28				Canc 9.27.43
Gibbs Gas Engine (PCS)				
1429	2.12.43	7.16.43	2.9.44	PCSC MC 4.7.48
1430	2.27.43	8.6.43	3.3.44	MC 6.10.47
1431	5.12.43	11.2.43	3.24.44	OpDevFor 3.46 *Grafton* Str 6.1.65 Mercantile
1432–40				Canc 9.27.43
Harbor Boat Building (PCS)				
1441	2.25.43	9.15.43	4.4.44	MC 11.13.47
1442	3.10.43	11.6.43	5.27.44	MC 4.5.48
1443				Canc 9.27.43
1444				*Conneaut*
San Diego Marine Construction (PCS)				
1445	1.30.43	6.19.43	3.3.44	Korea's *Kum Seong* 26.5.52
1446	1.30.43	7.31.43	4.14.44	Korea's *Mok Seong* 26.5.52
1447				Canc 9.27.43
1448	4.30.43	6.14.44	11.21.44	Korea's *Hwa Seong* 9.6.52 Laid down as YMS 476
Burger Boat (PCS)				
1449	2.10.43	7.17.43	12.17.43	MC 1.12.48
1450	2.16.43	8.14.43	1.15.44	Department of Interior 9.8.46
Tacoma Boat Building (PCS)				
1451	3.12.43	7.3.43	1.13.44	MC 1.27.47
1452	5.4.43	8.28.43	2.21.44	PCSC MC 3.12.45
1453–54				Canc 9.27.43
Mojean/Ericson (PCS)				
1455	3.17.43	9.4.43	2.8.44	PCSC MC 4.8.48
1456				Canc 9.27.43
Ballard				
1457	3.28.43	9.4.44	2.26.44	*John Blish* (AGS 10) 3.20.45 Sold 10.2.50
1458	4.25.43	11.6.43	5.17.44	*Derickson* (AGS 6) 6.10.44 Never commissioned Coast and Geodetic Survey, Department of Commerce 12.11.48
Western Boat Building (PCS)				
1459	3.6.43	7.2.43	12.15.43	MC 2.7.47
1460	4.9.43	8.28.43	2.28.44	PCSC MC 4.7.48
Bellingham Boatyard (PCS)				
1461	3.25.43	9.15.43	2.1.44	PCSC D in accident 12.26.45
1462–63				Canc 9.27.43
1464				*Medric* (AMC 203) 1.10.45
1465				*Minah* (AMC 204) 1.10.45 AMCU 14 7.3.52 MHC 14 7.2.54
Leblanc Shipbuilding (SC)				
1466	10.27.41	7.27.42	9.26.42	Mexico 11.20.43
1467	10.31.41	8.3.42	9.26.42	Scrapped 1.31.46
1468	11.28.41	7.10.42	9.26.42	Scrapped 1.21.48
1469	11.30.41	8.12.42	9.26.42	Mexico 11.20.43
1470	10.20.41	6.17.42	10.23.42	*Panther* (IX 105) 6.26.43 Sold 2.13.47
1471	10.28.41	7.2.42	10.23.42	Mexico 11.20.43
1472	1.30.42	11.24.42	12.5.42	Scrapped 3.4.48
1473	5.20.42	11.26.42	12.5.42	Scrapped 4.21.48

	Laid Down	Launched	Completed	Fate
Quincy Adams Yacht Yard (SC)				
1474	6.3.43	1.11.44	4.8.44	SCC Sold 2.14.47
1475	6.3.43	2.8.44	4.29.44	USSR 7.1.44
1476	6.15.43	3.8.44	5.17.44	USSR 7.1.44
1477	7.7.43	3.22.44	6.9.44	USSR 7.19.44
1478	7.21.43	4.24.44	7.13.44	USSR 9.2.44
1479	8.17.43	6.5.44	8.24.44	USSR 10.3.44
Rice Brothers (SC)				
1480	4.19.43	10.1.43	4.9.44	USSR 5.17.44
1481	4.15.43	10.16.43	5.16.44	USSR 6.27.44
1482	5.18.43	12.14.43	6.27.44	USSR 8.8.44
1483	5.21.43	12.16.43	7.23.44	USSR 9.9.44
Daytona Beach Boat Works (SC)				
1484	7.10.43	10.27.43	4.10.44	USSR 5.9.44
1485	7.24.43	11.30.43	5.29.44	USSR 7.8.44
1486	8.5.43	1.15.44	7.21.44	USSR 8.22.44
1487	9.23.43	3.11.44	9.30.44	USSR 10.12.44
Elizabeth City Shipyard (SC)				
1488	7.10.43	8.24.43	4.5.44	USSR 5.19.44
1489	8.4.43	9.25.43	4.11.44	USSR 5.25.44
1490	8.14.43	10.16.43	5.6.44	USSR 6.15.44
1491	8.25.43	10.30.43	7.14.44	USSR 8.30.44
Simms (SC)				
1492	6.23.43	12.8.43	4.22.44	USSR 6.7.44
1493	6.28.43	2.22.44	6.12.44	USSR 7.28.44
1494–95				Canc 9.17.43
Vinyard Boat Building (SC)				
1496	6.5.43	12.1.43	4.7.44	USSR 5.20.44
1497	8.23.43	5.4.44	7.10.44	USSR 8.23.44
Thomas Knutson Shipbuilding (SC)				
1498	8.14.43	3.21.44	5.15.44	USSR 6.24.44
1499	9.4.43	4.27.44	8.2.44	USSR 9.22.44
1500–1501				Canc 9.17.43
Calderwood Yacht Yard (SC)				
1502	7.14.43	11.27.43	6.1.44	USSR 7.18.44
1503	8.14.43	5.22.43	8.22.44	USSR 9.30.44
Donovan Contracting (SC)				
1504	7.29.43	4.16.44	5.24.44	USSR 7.11.44
1505	7.30.43	4.30.44	6.18.44	USSR 8.20.44
1506	8.7.43	5.14.44	7.17.44	USSR 9.21.44
Harris and Parsons (SC)				
1507	7.15.43	1.26.44	5.15.44	USSR 7.4.44
1508	8.12.43	5.8.44	7.22.44	USSR 9.6.44
1509				Canc 9.17.43
Perkins and Vaughn (SC)				
1510	7.3.43	12.9.43	5.18.44	USSR 7.7.44
1511	7.12.43	4.3.44	7.3.44	USSR 8.15.44
1512	7.27.43	6.15.44	9.26.44	USSR 10.21.44
1513–16				Canc 9.17.43
Julius Petersen (SC)				
1517	8.14.43	4.12.44	5.20.44	USSR 7.6.44
1518–20				Canc 9.17.43
1521–45				Canc 9.1.43
Consolidated Shipbuilding (PC)				
1546	11.28.43	1.30.44	6.3.44	*Grosse Point* Korea's *Kum Chong San* 11.60
1547	12.4.43	2.8.44	7.7.44	Decommissioned 1.47 *Corinth*
1548	12.19.43	2.13.44	2.13.44	PGM 9 2.4.44 D 12.27.45
1549	1.30.44	3.12.44	7.22.44	China's *Chien Tang* 6.30.48
1550	2.13.44	4.11.44	10.25.44	PGM 19 8.16.44 FLC 6.16.49
1551	3.12.44	5.7.44	12.2.44	PGM 20 8.16.44 China's *Pao Ying* 6.30.48 Later *Ying Chiang* WL(T) 1.20.55
1552	4.11.44	5.25.44	11.21.44	PGM 21 8.16.44 Greece's *Antiploiarkhos Pezopoulos* 11.24.47
1553	5.7.44	6.25.44	12.22.44	PGM 22 8.16.44 Greece's *Ploiarkos Meletopoulus* 11.24.47
1554	5.25.44	7.16.44	12.4.44	PGM 23 8.16.44 WAA 12.2.47
1555	6.25.44	8.13.44	1.17.45	PGM 24 8.16.44 WAA 12.2.47
1556	7.16.44	9.6.44	2.2.45	PGM 25 8.16.44 Greece's *Plotarkis Arslandoglou* 12.11.47
1557	8.13.44	9.25.44	2.23.43	PGM 26 8.16.44 China's *Hung Tse* 6.30.48 Later *Ou Chang* Str 1964
1558	9.6.44	10.28.44	3.17.45	PGM 27 8.16.44 D 12.24.45
1559	9.25.44	11.19.44	4.7.45	PGM 28 8.16.44 Greece's *Plotarkis Blessas* 12.8.47 Sold 1963
Leathem D. Smith Shipbuilding (PC)				
1560	11.29.43	2.3.44	3.29.44	France's *Coutelas* 4.15.44 BU 1963
1561	12.28.43	2.23.44	5.4.44	France's *Dague* 5.15.44 BU 1964
1562	1.17.44	3.4.44	5.25.44	France's *Javelot* 6.5.44 BU 1951
1563	2.4.44	3.24.44	6.15.44	Philippine *Negros Oriental* 7.2.48 Lost in typhoon 11.62
1564	2.24.44	4.19.44	7.19.44	Philippine *Capiz* 7.2.48
1565	5.11.44	7.16.44	10.30.44	PGM 29 8.16.44 Greece's *Plotarkhis Chantzikonstandis* 12.11.47

490 U.S. SMALL COMBATANTS

	Laid Down	Launched	Completed	Fate
1566	5.29.44	8.12.44	11.18.44	PGM 30 8.16.44 MC 4.7.47
1567	7.18.44	9.23.44	12.21.44	PGM 31 8.16.44 Other government agencies 3.54 China's *Chu Kiang*
1568	8.14.44	10.14.44	1.18.45	PGM 32 8.16.44 FLC 10.27.47
1569	9.26.44	12.9.44	2.24.45	Decommissioned 1.47 *Anacortes* Vietnam's *Van Don* 11.60
1570–75				Canc 9.23.43
1576–81				Canc 11.5.43
1582–85				Canc 9.23.43
Commercial Iron Works (PC)*				
1586	7.31.41	2.21.42	7.28.42	MC 3.18.48
1587	8.18.41	3.12.42	8.15.42	MC 3.18.48
1588	12.30.41	4.6.42	9.2.42	MC 5.6.48
1589	1.28.42	4.18.42	9.5.42	WAA 12.3.47
1590	2.21.42	5.9.42	9.21.42	Res 3.54 Str 9.17.54
1591	3.12.42	5.23.42	10.10.42	MC 3.18.48
1592	4.6.42	6.20.42	10.27.42	WAA 12.19.47
1593	11.24.41	3.28.42	8.14.42	MC 9.16.46
1594	12.26.41	4.25.42	8.14.42	MC 7.29.46
1595	1.16.42	5.26.42	8.31.42	MC 10.21.46
1596	2.9.42	6.13.42	9.18.42	MC 7.30.46
1597	2.26.42	7.11.42	10.10.42	MC 11.8.46 Dominican Republic's *Engage Cibas Constitucion* Str 1969
1598	12.19.41	5.10.42	12.7.42	MC 6.11.47
Jakobson Shipyard (PC)†				
1599	5.11.42	9.7.42	2.4.43	PCC MC 3.18.48
Nashville Bridge (PC)‡				
1600	10.15.41	2.28.42	9.1.42	MC 6.15.48
1601	10.18.41	3.5.42	10.2.42	PCC MC 6.15.48
Penn-Jersey (PC)§				
1602	10.21.41	5.29.42	4.10.43	MC 6.15.48
1603	11.19.41	9.7.42	6.16.43	D 10.24.45
General Shipbuilding and Engineering Works (PCE)				
1604	12.18.52	7.30.53	5.4.54	Netherlands' *Fret*
1605	3.2.53	3.6.54	8.5.54	Netherlands' *Hermelijn*
1606	8.3.52	5.1.54	12.2.54	Netherlands' *Vos*
Avondale Marine Ways (PCE)				
1607	11.15.52	1.2.54	26.3.54	Netherlands' *Wolf*
1608	12.1.52	30.1.54	6.11.54	Netherlands' *Panter*
1609	12.10.52	3.20.54	6.11.54	Netherlands' *Jaguar*
Dubigeon (PC)				
1610	2.54	5.1.54	7.21.55	France's *Fougeux*
F. C. La Mediterranée (PC)				
1611	11.53	5.4.54	11.3.55	France's *Opinaitre*
A. C. Provence (PC)				
1612	1.54	6.26.54	9.10.55	France's *Agile*
Dubigeon (PC)				
1613		9.27.54		Portugal's *Maio*
Normand (PC)				
1614		2.9.55		Portugal's *Porto Santo* Str 1973
F. C. La Mediterranée (PC)				
1615	1954	12.21.54	1.56	Yugoslavia's PBR 581
Brest Navy Yard				
1616	12.17.53	9.30.54	8.23.55	Italy's *Vedetta* Intended for Germany, but transferred to Ethiopia as *Belay Deress* 1.57 Ret to USN Sold to Italy Transferred 2.3.59
Normand (PC)				
1617		6.7.55		Portugal's *Sao Nicolau*
Dubigeon (PC)				
1618			3.12.57	Germany's UW 12 Tunisia's *Sakiet Sidi Youssef* 6.70
Navalmeccanica (PCE)				
1619		7.18.54	6.1.55	Italy's *Albatros*
1620		9.19.54	10.23.55	Italy's *Alcione*
1621		11.21.54	12.29.55	Italy's *Airone*
1622		1.9.55	1.31.57	Denmark's *Bellona*
C. Del Tirreno (PCE)				
1623		12.19.54	7.30.55	Denmark's *Diana*
1624		6.25.55	8.28.56	Denmark's *Flora*
C. N. Di Taranto (PCE)				
1625		9.12.54	8.10.55	Denmark's *Triton*
Navalmeccanica (PCE)				
1626		7.31.54	10.2.56	Italy's *Aquila*
Rijkswerft Willemsoord‖				
1627	9.12.53	2.24.54	8.6.54	Netherlands' *Balder*
1628	10.10.53	4.24.54	8.9.54	Netherlands' *Bulgia*

*Ex-fleet sweepers AM 86–94.
†Ex–AM 95.
‡Ex–AM 96–97.
§Ex–AM 98–99.
‖MDAP for Dutch navy.

	Laid Down	Launched	Completed	Fate
1629	2.24.54	7.21.54	12.1.54	Netherlands' *Freyr*
1630	4.24.54	10.2.54	2.3.55	Netherlands' *Hadda*
1631	7.21.54	12.1.54	3.23.55	Netherlands' *Hefring*
South Coast*				
1632	12.31.53	7.31.54	3.1.55	Thailand's SC 31 Later SC 7
1633	2.22.54	9.30.54	4.29.55	Thailand's SC 32 Later SC 8
1634	4.14.54	12.24.54	5.27.55	Thailand's SC 33 BU 3.62
Arsenal de Alfeite (PC)				
1635		5.2.56	12.27.56	Portugal's *Brava*
1636		5.2.56	4.11.57	Portugal's *Fogo* Str 1973
Est. Navals de Viana do Castelo (PC)				
1637		7.10.56	5.17.57	Portugal's *Boavista*
Gunderson Brothers†				
1638	1.22.63	12.5.63	5.9.64	Turkey's *Sultanhisar*
1639	3.7.63	7.9.64	3.17.65	Turkey's *Demirhisar*
1640	3.11.63	5.14.64	8.27.64	Turkey's *Yarhisar*
1641	5.29.63	5.14.64	10.29.64	Turkey's *Akhisar*
1642	8.5.63	11.24.64	5.19.65	Turkey's *Sivrihisar*
Golcuk Dockyard (PC)				
1643		12.64	7.65	Turkey's *Kochisar*
Helsingor Werft‡				
1644	9.25.64	5.20.65	6.30.66	Denmark's *Peder Skram*
1645	12.18.64	9.8.65	4.16.67	Denmark's *Herluf Trolle*
Asmar§				
1646			11.27.71	Chile's *Papudo*

EAGLE BOATS (PE)‖

Ford Motor

	Laid Down	Launched	Completed	Fate
1	5.7.18	7.11.18	10.28.18	Sold 6.11.30
2	5.10.18	8.19.18	10.28.18	Sold 6.11.30
3	5.16.18	9.11.18	11.11.18	Sold 6.11.30
4	5.21.18	9.15.18	11.5.18	Sold 6.11.30
5	5.28.18	9.23.18	11.8.18	Sold 6.11.30
6	6.3.18	10.16.18	11.10.18	D as target 11.30.34
7	6.8.18	10.5.18	11.10.18	D as target 11.30.34
8	6.10.18	11.11.18	10.27.19	Sold 4.1.31 as mercantile *Fortitude*
9	6.17.18	11.8.18	8.14.19	Sold 5.26.30
10	7.6.18	11.9.18	10.27.19	D 8.19.37
11	7.13.18	11.14.18	5.29.19	Sold 1.16.35
12	7.13.18	11.12.18	11.6.19	Sold 12.30.35
13	7.15.18	1.9.19	4.12.19	Sold 5.26.30
14	7.20.18	1.23.19	6.19.19	D as target 11.22.34
15	7.21.18	1.25.19	6.11.19	Sold 6.14.34
16	7.22.18	1.11.19	6.5.19	To CG 1919 as *McGourty*
17	8.3.18	2.1.19	7.3.19	Wrecked 5.22.22 Str 10.11.23
18	8.5.18	2.10.19	8.7.19	Sold 6.11.30
19	8.6.18	1.30.19	6.25.19	D 8.6.46
20	8.26.18	2.15.19	7.28.19	To CG 1919 as *Scally*
21	8.31.18	2.15.19	7.31.19	To CG 1919 as *Bothwell*
22	9.5.18	2.10.19	7.17.19	To CG 1919 as *Earp* Ret 6.23 Str 7.2.36
23	9.11.18	2.20.19	6.19.19	Sold 6.11.30
24	9.13.18	2.24.19	7.12.19	Sold 6.11.30
25	9.17.18	2.19.19	6.30.19	Lost 6.11.20
26	9.25.18	3.1.19	9.15.19	Sold 8.29.38
27	10.22.18	3.1.19	7.14.19	Sold 6.4.46
28	10.23.18	3.1.19	7.28.19	Sold 6.11.30
29	11.18.18	3.8.19	8.15.19	Sold 6.11.30
30	11.19.18	3.8.19	8.14.19	To CG 1919 as *Carr*
31	11.19.18	3.8.19	8.14.19	Sold 5.18.23
32	11.30.18	3.15.19	8.30.19	Sold 3.3.47
33	12.4.18	3.15.19	8.30.19	Sold 6.11.30
34	1.8.19	3.15.19	8.29.19	Sold 6.9.32
35	1.13.19	3.22.19	8.21.19	Sold 6.7.38
36	1.22.19	3.22.19	8.18.19	Sold 2.27.36
37	1.27.19	3.24.19	8.29.19	Sold 6.11.30
38	1.31.19	3.29.19	7.30.19	Sold 3.3.47
39	2.3.19	3.29.19	9.10.19	Sold 6.7.38
40	2.7.19	4.5.19	9.15.19	D as target 11.19.34
41	2.10.19	4.5.19	9.15.19	Sold 6.11.30
42	2.13.19	5.17.19	10.3.19	Sold 6.11.30
43	2.17.19	5.17.19	9.18.19	Sold 5.26.30
44	2.20.19	5.24.19	9.30.19	Str 5.14.38
45	2.20.19	5.17.19	10.2.19	Sold 6.11.30
46	2.24.19	5.24.19	9.29.19	Sold 12.10.36
47	3.3.19	6.19.19	9.27.19	Sold 12.30.35

* MDAP for Thai navy.
† PC for Turkey under MAP.
‡ PC for Denmark.
§ PC for Chile.
‖ Except for Eagle 22, eagles in Coast Guard service were sold in 1922–23.

	Laid Down	Launched	Completed	Fate
48	3.3.19	5.24.19	9.23.19	Sold 10.10.46
49	3.4.19	6.14.19	9.19.19	Sold 9.20.30
50	3.10.19	7.18.19	9.25.19	Sold 6.11.30
51	3.10.19	6.14.19	9.19.19	Sold 8.29.38
52	3.10.19	7.9.19	9.24.19	Sold 8.29.38
53	3.17.19	8.13.19	10.11.19	Sold 5.26.30
54	3.17.19	7.17.19	9.29.19	Sold 5.26.30
55	3.17.19	7.22.19	9.30.19	Sold 3.3.47
56	3.25.19	8.15.19	10.15.19	WL 4.23.45
57	3.25.19	7.29.19	9.30.19	Sold 3.5.47
58	3.25.19	8.2.19	10.8.19	Str 6.30.40
59	3.31.19	4.12.19	9.9.19	Sold 8.29.38
60	3.31.19	8.13.19	10.15.19	Sold 8.29.38

HYDROFOIL SUBCHASER (PCH)

Boeing

		Laid Down	Launched	Completed	Fate
1	*High Point*	2.27.61	8.17.62	9.3.63	

TORPEDO BOATS

PT Boats*

Miami Shipbuilding

	Laid Down	Launched	Completed	Fate
1	7.12.39	8.16.39	11.20.41	Small boat C 6083 12.24.41
2	8.19.39	9.30.39	11.20.41	Small boat C 6084 12.24.41 Str 7.17.47

Fisher Boatworks†

	Laid Down	Launched	Completed	Fate
3	8.1.39	4.18.40	6.20.40	RN MTB 273 4.41
4	8.5.39	4.19.40	6.20.40	RN MTB 274 4.41

Higgins Industries

	Laid Down	Launched	Completed	Fate
5	8.1.39	11.4.40	3.1.41	RN MTB 269 4.41
6(i)			8.14.40	Finland 1940, later RN MGB 68 Original Sparkman and Stephens design
6(ii)	5.15.40	10.29.40	2.3.41	RN MTB 270 7.41 Ret 3.45 Small boat C 104340 6.1.45 Out of service 12.14.45 Used by BuOrd to test a Martin twin 50-cal aircraft turret Str 7.17.47

Philadelphia Navy Yard

	Laid Down	Launched	Completed	Fate
7	8.29.39	10.31.40	2.25.41	RN MTB 271 4.41
8	12.29.39	10.29.40	2.25.41	YP 110 10.14.41 Retained after the war Was to have become RN MTB 272

Scott Paine

	Laid Down	Launched	Completed	Fate
9			6.17.40	RN MTB 258 4.41

Elco

	Laid Down	Launched	Completed	Fate
10	2.26.40	8.20.40	11.4.40	RN MTB 259 4.41 WL 6.14.42
11	4.1.40	10.7.40	11.8.40	RN MTB 260 4.41
12	4.9.40	10.18.40	11.12.40	RN MTB 261 4.41 WL 1945
13	4.17.40	10.25.40	11.15.40	RN MTB 262 4.41 WL 2.24.43
14	4.26.40	11.7.40	11.22.40	RN MTB 263 4.41
15	5.6.40	11.15.40	12.5.40	RN MTB 264 4.41 WL 5.12.43
16	5.13.40	11.23.40	12.20.40	RN MTB 265 4.41
17	5.24.40	12.2.40	12.12.40	RN MTB 266 4.41 WL 4.17.44
18	5.31.40	12.9.40	12.17.40	RN MTB 267 4.41 WL 4.2.43
19	6.7.40	12.16.40	12.23.40	RN MTB 268 4.41
20	10.14.40	3.14.41	6.9.41	Str 12.22.44
21	12.2.40	4.11.41	6.13.41	Str 10.11.43
22	12.10.40	4.18.41	6.24.41	Beached and abandoned Adak 6.11.43
23	12.17.40	4.24.41	6.25.41	Small boat C55047 10.6.43‡
24	12.27.40	4.30.41	6.26.41	Small boat 12.9.44 for experimental firing tests Str 7.17.47
25	1.6.41	5.5.41	6.11.41	Small boat C55048 10.6.43
26	1.30.41	5.10.41	6.18.41	Small boat C55049 10.6.43
27	2.12.41	5.15.41	6.27.41	Small boat 12.9.44
28	2.20.41	5.20.41	6.30.41	Stranded Alaska 1.12.43
29	2.28.41	5.24.41	7.2.41	Str 12.22.44
30	3.7.41	5.28.41	7.3.41	Sold 1.3.47
31	3.13.41	6.2.41	7.8.41	Grounded and sunk 1.20.42
32	3.19.41	6.6.41	7.10.41	Scuttled 3.13.42
33	3.25.41	6.10.41	7.11.41	Grounded and sunk 12.15.41
34	3.29.41	6.14.41	7.12.41	WL(B) 4.9.42
35	4.3.41	6.19.41	7.16.41	Scuttled 4.12.42 Under overhaul and was destroyed when Cebu invaded
36	4.8.41	6.21.41	8.27.41	Small boat C73994 4.15.44

*German S-boats obtained by the U.S. Navy at the end of World War II were not assigned PT numbers, although they were equivalent to PTs. *S116* became C105179 and was declared surplus on 2.27.48. *S218* became C105180 and was tested at the naval engineering experimental station in Annapolis starting in April 1947; she was declared surplus on 2.18.48. *S225* became C105181 and was declared surplus on 4.21.48.
†PT 3–9 were retained in Canada as air-sea rescue craft.
‡Some or all of the PTs reclassified as small boats in 1943–44 were transferred to the army for covert operations in the New Guinea area. Boats so transferred were stripped of their torpedo tubes, armored, and fitted with additional weapons.

	Laid Down	Launched	Completed	Fate
37	4.12.41	6.25.41	7.18.41	WL(G) 2.1.43
38	4.17.41	6.28.41	7.18.41	Small boat C68730 2.6.44
39	4.22.41	7.2.41	7.21.41	Small boat 10.20.44 To BuOrd for destruction 1.45
40	4.25.41	7.7.41	7.22.41	Small boat C73995 4.15.44
41	4.30.41	7.8.41	7.23.41	Scuttled 4.15.42 Used as a gunboat on Lake Lanao, Mindanao, after all torpedoes and gasoline expended
42	5.5.41	7.12.41	7.25.41	Str 12.22.44
43	5.14.41	7.16.41	7.26.41	WL(G) 1.10.43
44	5.17.41	7.18.41	7.30.41	WL(G) 12.12.42
45	5.21.41	8.6.41	9.3.41	Small boat 4.15.44
46	6.2.41	8.9.41	9.6.41	Small boat C74095 4.29.44
47	6.4.41	8.14.41	9.9.41	Small boat 10.20.44 At Pearl Harbor for dehydration and equipment tests 12.16.44 Surplus 3.21.47
48	6.6.41	8.21.41	9.15.41	Small boat 10.20.44
49*	6.12.41	8.26.41	1.20.42	RN MTB 307 2.42
50	6.17.41	8.29.41	1.22.42	RN MTB 308 2.42 WL 9.14.42
51	6.23.41	9.3.41	2.9.42	RN MTB 309 2.42 D10.26.45
52	6.26.41	9.8.41	2.10.42	RN MTB 310 2.42 WL 9.14.42
53	7.1.41	9.12.41	2.23.42	RN MTB 311 2.42 WL 5.4.43
54	7.7.41	9.16.41	2.23.42	RN MTB 312 2.42 WL 9.14.42
55	7.10.41	9.22.41	2.28.42	RN MTB 313 2.42
56	7.15.41	9.25.41	2.28.42	RN MTB 314 2.42 WL 9.14.42
57	7.18.41	9.30.41	3.7.42	RN MTB 315 2.42
58†	7.22.41	10.3.41	3.10.42	RN MTB 316 2.42 WL(G) 7.17.43
59	7.26.41	10.8.41	3.5.42	Small boat 10.20.44 At Pearl Harbor for dehydration and equipment tests 12.16.45 Surplus 3.21.47 Sold 5.5.48
60	7.30.41	10.11.41	2.25.42	Str 4.21.44
61	8.2.41	10.15.41	2.19.42	Small boat C68371 2.16.44
62	8.6.41	10.20.41	2.10.42	Str 1.20.45
63	8.9.41	10.23.41	2.7.42	D in accident 6.18.44
64	8.13.41	10.28.41	1.28.42	Str 1.20.45
65	8.16.41	10.31.41	1.24.42	Str 1.20.45
66	8.20.41	11.5.41	1.22.42	Small boat 2.20.45
67	8.23.41	11.8.41	1.17.42	D in accident 3.17.43
68	8.27.41	11.13.41	1.13.42	Grounded 9.30.43
Huckins Yacht				
69			6.30.41	YP 106 9.24.41 Sold 1.24.47
Higgins Industries				
70			6.30.41	YP 107 9.24.41 Sold 9.19.46
71	12.2.41	5.4.42	7.20.42	D 10.24.45
72	3.2.42	7.6.42	7.23.42	D 10.24.45
73	12.25.41	7.24.42	8.12.42	Grounded 1.14.45
74	12.15.41	8.3.42	8.26.42	D 11.23.45
75	1.3.42	8.6.42	8.28.42	D 11.23.45
76	2.1.42	8.11.42	8.31.42	D 11.23.45
77	4.18.42	8.15.42	9.3.42	WL(G) 2.1.45
78	4.20.42	8.15.42	9.5.42	D 11.23.45
79	4.23.42	8.22.42	9.8.42	WL(G) 2.1.45
80	4.27.42	8.22.42	9.21.42	D 11.23.45
81	5.7.42	9.22.42	1.11.43	D 11.23.45
82	5.1.42	9.7.42	11.28.42	D 11.23.45
83	5.3.42	9.7.42	12.9.42	D 11.23.45
84‡	5.5.42	9.7.42	12.3.42	D 11.23.45
85	4.29.42	9.7.42	12.7.42	USSR 2.43 WL 6.22.45
86	5.9.42	10.17.42	12.9.42	USSR 2.43
87	5.13.42	10.22.42	12.13.42	USSR 2.43
88	5.16.42	10.23.42	12.19.42	RN MTB 419 4.43
89	5.29.42	10.28.42	12.15.42	SU 2.43
90	6.2.42	10.28.42	12.18.42	RN MTB 420 4.43
91	6.5.42	11.3.42	12.17.42	RN MTB 421 4.43 D in explosion 5.15.46.
92	6.5.42	11.21.42	12.29.42	RN MTB 422 4.43
93	6.17.42	11.19.42	12.30.42	RN 4.43 as target tower No MTB or MGB number
94	6.23.42	11.20.42	12.29.42	RN MTB 423 4.43
Huckins Yacht				
95	1.3.42	5.27.42	7.23.42	D 9.26.45
96	1.29.42	6.25.42	8.22.42	D 9.13.45
97	1.30.42	7.9.42	8.29.42	D 9.13.45
98	3.13.42	7.31.42	9.19.42	Sold 10.14.46
99	5.27.42	8.29.42	9.29.42	Sold 10.14.46
100	6.26.42	9.7.42	11.14.42	Sold 10.14.46
101	7.9.42	9.26.42	11.14.42	Sold 10.10.46
102	8.10.42	11.12.42	11.30.42	Sold 10.21.46

*PT 49–58 became BPT 1–10 7.3.41.
†PT 59–68 were ordered as BPT 11–20, became PT 59–68 on 12.12.41. See the end of this list for BPT 21–68.
‡PT 85–94 became RPT 1–10, although not all were transferred to the Soviet Union.

494 U.S. SMALL COMBATANTS

	Laid Down	Launched	Completed	Fate
Elco				
103	1.24.42	5.16.42	6.12.42	D 11.4.45
104	1.29.42	5.30.42	6.19.42	D 11.4.45
105	2.5.42	6.4.42	6.26.42	D 11.4.45
106	2.12.42	6.9.42	6.30.42	D 11.6.45
107	2.16.42	6.13.42	7.3.42	D in accident 6.18.44
108	2.27.42	6.17.42	7.7.42	D 11.11.45
109	3.4.42	6.20.42	7.10.42	WL(R) 8.2.43
110	3.11.42	6.24.42	7.14.42	Collision 1.26.44
111	3.17.42	6.27.42	7.16.42	WL(G) 2.1.43
112	3.21.42	7.3.42	7.18.42	WL(G) 1.11.43
113	3.25.42	7.3.42	7.23.42	Grounded 8.8.43
114	3.30.42	7.6.42	7.24.42	D 10.28.45
115	4.3.42	7.9.42	7.29.42	D 11.9.45
116	4.7.42	7.13.42	7.30.42	D 11.11.45
117	4.10.42	7.15.42	8.4.42	WL(B) 8.1.43
118	4.14.42	7.18.42	8.6.42	Grounded 3.17.43
119	4.22.42	7.21.42	8.8.42	D in accident 3.17.43
120	4.22.42	7.23.42	8.12.42	D 10.26.45
121	4.25.42	7.25.42	8.27.42	WL(B) 3.27.44
122	4.29.42	7.28.42	8.15.42	D 10.28.45
123	5.2.42	7.31.42	8.18.42	WL(G) 3.27.44
124	5.6.42	8.3.42	8.20.42	D 11.11.45
125	5.9.42	8.6.42	8.22.42	D 11.11.45
126	5.13.42	8.10.42	8.27.42	D 11.24.45
127	5.16.42	8.13.42	8.29.42	D 10.26.45
128	5.21.42	8.17.42	9.1.42	D 11.10.45
129	5.26.42	8.19.42	9.4.42	D 10.28.45
130	5.30.42	8.22.42	9.7.42	D 10.28.45
131	6.3.42	8.26.42	9.9.42	D 11.10.45
132	6.6.42	8.28.42	9.11.42	D 11.10.45
133	6.10.42	8.31.42	9.16.42	WL(G) 7.15.44
134	6.13.42	9.2.42	9.17.42	D 11.9.45
135	6.17.42	9.4.42	9.21.42	Grounded 4.12.44
136	6.20.42	9.7.42	9.23.42	Grounded 9.17.43
137	6.24.42	9.10.42	9.25.42	D 10.24.45
138	6.27.42	9.12.42	9.29.42	D 10.24.45
139	6.30.42	9.15.42	10.13.42	Sold 10.9.46
140	7.3.42	9.17.42	10.3.42	Sold 10.9.46
141	7.7.42	9.19.42	10.6.42	Sold 10.9.46
142	7.9.42	9.22.42	10.9.42	D 10.28.45
143	7.13.42	9.25.42	10.13.42	D 10.28.45
144	7.15.42	9.28.42	10.15.42	D 10.28.45
145	7.18.42	10.14.42	10.17.42	Grounded 1.4.44
146	7.21.42	10.3.42	10.20.42	D 10.26.45
147	7.24.42	10.6.42	10.23.42	Grounded 11.19.43
148	7.27.42	10.9.42	10.27.42	D 11.4.45
149	7.30.42	10.12.42	10.29.42	D 10.28.45
150	8.1.42	10.15.42	11.2.42	D 10.26.45
151	8.5.42	10.17.42	11.4.42	D 10.26.45
152	8.7.42	10.20.42	11.6.42	D 10.26.45
153	8.10.42	10.23.42	11.10.42	Beached 7.2.43
154	8.13.42	10.27.42	11.13.42	D 11.24.45
155	8.15.42	10.30.42	11.20.42	D 11.24.45
156	8.19.42	11.2.42	11.18.42	D 11.24.45
157	8.21.42	11.4.42	11.20.42	D 11.24.45
158	8.24.42	11.7.42	11.23.42	Beached 7.2.43
159	8.27.42	11.10.42	11.24.42	D 11.24.45
160	8.29.42	11.13.42	12.7.42	D 11.24.45
161	9.2.42	11.16.42	11.28.42	D 11.24.45
162	9.4.42	11.19.42	12.2.42	D 11.24.45
163	9.7.42	11.21.42	12.4.42	D 11.11.45
164	9.9.42	11.24.42	12.8.42	WL(B) 8.1.43
165	9.12.42	11.27.42	12.11.42	WL(G) 5.23.43
166	9.15.42	11.30.42	12.15.42	WL(B) 7.20.43
167	9.18.42	12.3.42	12.17.42	D 11.11.45
168	9.21.42	12.5.42	12.22.42	D 11.11.45
169	9.24.42	12.8.42	12.23.42	D 11.11.45
170	9.26.42	12.14.42	12.28.42	D 11.11.45
171	9.29.42	12.14.42	12.30.42	D 11.11.45
172	10.2.42	12.17.42	1.2.43	Grounded 9.7.43
173	10.5.42	12.20.42	1.28.43	WL(G) 5.23.43
174	10.7.42	12.23.42	1.6.43	D 11.11.45
175	10.9.42	12.26.42	1.8.43	D 11.11.45
176	10.13.42	12.29.42	1.12.43	D 11.11.45
177	10.16.42	12.31.42	1.14.43	D 11.11.45
178	10.19.42	1.2.43	1.16.43	D 11.11.45
179	10.21.42	1.5.43	1.19.43	D 11.11.45
180	10.24.42	1.7.43	1.22.43	Sold 5.46
181	10.27.42	1.9.43	1.26.43	D 11.11.45
182	10.30.42	1.13.43	1.30.43	D 11.11.45
183	11.2.42	1.15.43	2.2.43	D 11.11.45
184	11.5.42	1.19.43	2.4.43	D 11.11.45

	Laid Down	Launched	Completed	Fate	
185	11.7.42	1.21.43	2.6.43	D 11.11.45	
186	11.11.42	1.23.43	2.9.43	D 11.11.45	
187	11.13.42	1.26.43	2.12.43	Sold 5.46	
188	11.17.42	1.28.43	2.17.43	D 10.28.45	
189	11.19.42	1.30.43	2.19.43	D 10.28.45	
190	11.21.42	2.2.43	2.19.43	Sold 5.46	
191	11.25.42	2.5.43	2.24.43	Sold 5.46	
192	11.27.42	2.8.43	2.25.43	D 10.26.45	
193	11.30.42	2.11.43	2.27.43	Grounded 6.24.44	
194	12.3.42	2.13.43	3.3.43	D 10.26.45	
195	12.5.42	2.17.43	3.6.43	Sold 5.46	
196	12.9.42	2.19.43	5.3.43	D 10.26.45	
Higgins Industries*					
197	6.23.42	9.7.42	1.5.43	USSR WL 2.15.43	
198	6.25.42	9.16.42	1.7.43	Ret FLC 9.3.46	
199	6.27.42	9.7.42	1.23.43	D 10.24.45	
200	6.29.42	9.16.42	1.23.43	Lost (acc) 2.22.44	
201	6.30.42	10.3.42	1.20.43	RN MGB 181 10.44	To Yugoslavia 8.45
202	7.2.42	10.3.42	1.23.43	WL(M) 8.16.44	
203	7.7.42	10.3.42	1.23.43	RN MGB 189 10.44	
204	7.8.42	10.8.42	1.23.43	RN MGB 182 10.44	To Yugoslavia 8.45
205	7.9.42	10.8.42	1.25.43	RN MGB 190 10.44	
206	7.11.42	10.16.42	1.30.43	RN MGB 177 10.44	
207	7.13.42	10.16.42	2.2.43	RN MGB 183 10.44	To Yugoslavia 8.45
208	7.16.42	10.22.42	2.10.43	RN MGB 184 10.44	To Yugoslavia 8.45
209	7.20.42	10.24.42	2.9.43	RN MGB 185 10.44	To Yugoslavia 8.45
210	7.23.42	10.24.42	2.10.43	RN MGB 191 10.44	
212	8.10.42	12.7.42	2.11.43	RN MGB 192 10.44	
213	8.14.42	12.6.42	2.12.43	RN MGB 187 10.44	To Yugoslavia 8.45
214	8.19.42	12.16.42	2.15.43	RN MGB 178 10.44	
215	8.21.42	12.31.42	2.26.43	RN MGB 179 10.44	
216	9.2.42	12.30.42	2.26.43	RN MGB 180 10.44	Turkey 3.47
217	9.3.42	1.16.43	3.11.43	RN MGB 188 10.44	To Yugoslavia 8.45 Lost 11.14.46
218	9.7.42	1.12.43	3.1.43	WL(M) 8.16.44	
219	9.7.42	1.16.43	3.4.43	Foundered 9.43	
220	9.29.42	1.25.43	3.9.43	D 11.26.45	
221	9.29.42	1.16.43	3.11.43	D 11.26.45	
222	10.7.42	2.3.43	3.15.43	D 11.26.45	
223	10.8.42	3.4.43	3.13.43	D 11.26.45	
224	10.24.42	3.1.43	3.15.43	Sold 5.46	
225	10.26.42	3.13.43	3.20.43	D 11.19.45	
226	11.3.42	3.11.43	3.20.43	Sold 5.46	
227	11.16.42	3.16.43	3.25.43	D 11.19.45	
228	11.13.42	3.17.43	3.29.43	Sold 5.46	
229	9.7.42	11.2.42	3.25.43	Sold 5.46	
230	9.7.42	11.3.42	3.29.43	D 11.19.45	
231	9.22.42	11.4.42	3.31.43	Sold 5.46	
232	9.22.42	11.8.42	3.31.43	Sold 5.46	
233	9.23.42	11.8.42	4.7.43	D 11.19.45	
234	9.28.42	11.8.42	4.14.43	Sold 5.46	
235	10.1.42	11.14.42	4.17.43	D 11.26.45	
236	10.5.42	11.14.42	4.19.43	Sold 5.46	
237	10.8.42	11.14.42	4.22.43	Sold 5.46	
238	10.12.42	11.14.42	4.22.43	Sold 5.46	
239	10.15.42	11.21.42	4.30.43	D in accident 12.14.43	
240	10.17.42	11.21.42	4.30.43	Sold 5.46	
241	10.21.42	11.28.42	5.12.43	D 11.26.45	
242	10.23.42	11.28.42	5.14.43	D 11.26.45	
243	10.26.42	12.22.42	5.14.43	D 11.26.45	
244	10.28.42	12.22.42	5.24.43	D 11.26.45	
245	10.30.42	12.22.42	5.19.43	Sold 5.46	
246	11.2.42	12.23.42	5.24.43	D 11.26.45	
247	11.4.42	1.8.43	5.25.43	WL(G) 5.5.44	
248	11.6.42	1.8.43	5.26.43	D 11.24.45	
249	11.9.42	1.8.43	5.28.43	Sold 5.46	
250	11.11.42	1.15.43	5.31.43	Sold 5.46	
251	11.12.42	1.15.43	5.31.43	WL(G) 2.26.44	
252	11.16.42	1.15.43	6.14.43	D 11.24.45	
253	11.18.42	1.28.43	6.14.43	Sold 5.46	
254	11.20.42	1.28.43	6.16.43	Sold 5.46	
Huckins Yacht					
255	8.29.42	12.5.42	2.25.43	Sold 9.1.48	
256	9.7.42	1.8.43	2.26.43	Sold 9.1.48	
257	9.28.42	2.6.43	3.27.43	Sold 4.11.47	
258	11.12.42	2.27.43	3.29.43	Sold 4.11.47	
259	12.5.42	4.10.43	5.25.43	Sold 10.2.46	
260	1.8.43	4.29.43	5.26.43	Sold 10.2.46	
261	2.8.43	5.22.43	6.26.43	Sold 11.4.47	
262	2.27.43	6.5.43	7.7.43	Sold 10.2.46	

*PT 197 and 198 became RPT 11 and RPT 12; the latter, serving as a target tower, was transferred to the UK rather than to the Soviet Union.

	Laid Down	Launched	Completed	Fate
263	4.10.43	7.3.43	7.31.43	Sold 4.11.47
264	4.29.43	7.28.43	9.7.43	Sold 7.7.47
Higgins Industries				
265	11.23.42	2.6.43	8.24.43	USSR 11.43
266	11.25.42	2.6.43	8.25.43	USSR 11.43
267	12.2.42	2.16.43	8.26.43	USSR 11.43
268	12.5.42	2.16.43	8.27.43	USSR 11.43
269	12.18.42	2.16.43	8.28.43	USSR 11.43
270	12.21.42	2.20.43	8.30.43	USSR 11.43
271	12.23.42	2.20.43	8.31.43	USSR 11.43
272	12.30.42	2.25.43	9.1.43	USSR 11.43
273	1.5.43	2.25.43	9.3.43	USSR 11.43
274	1.7.43	3.4.43	9.13.43	USSR 11.43
275	1.11.43	3.4.43	9.13.43	USSR 11.43
276	1.13.43	3.4.43	9.20.43	USSR 11.43
277	1.16.43	3.10.43	7.7.43	D 11.26.45
278	1.20.43	3.10.43	7.9.43	Sold 12.46
279	11.28.42	4.11.43	6.28.43	Collision 2.11.44
280	12.5.42	4.10.43	6.28.43	D 11.26.45
281	12.23.42	4.13.43	7.8.43	D 11.26.45
282	1.26.43	3.12.43	7.10.43	D 11.26.45
283	2.1.43	3.12.43	7.12.43	WL(G) 3.17.44
284	2.6.43	3.23.43	7.21.43	D 11.26.45
285	2.8.43	3.23.43	7.16.43	D 11.26.45
286	2.11.43	3.26.43	8.4.43	D 11.26.45
287	2.13.43	3.26.43	7.22.43	D 11.26.45
288	2.17.43	3.30.43	7.23.43	D 11.26.45
289	2.19.43	3.30.43	9.20.43	USSR 12.43
290	2.22.43	3.30.43	9.22.43	USSR 12.43
291	2.26.43	4.6.43	9.25.43	USSR 12.43
292	2.27.43	4.6.43	9.24.43	USSR 12.43
293	3.1.43	4.6.43	9.25.43	USSR 12.43
294	3.3.43	4.16.43	9.28.43	USSR 12.43
295	3.5.43	4.16.43	10.15.43	Sold 8.27.48
296	3.8.43	4.27.43	10.18.43	Sold 7.8.48 Had been scheduled for sale to a foreign government*
297	3.10.43	4.27.43	10.20.43	D 11.26.45
298	3.12.43	5.10.43	10.26.43	Sold 5.46
299	3.15.43	5.10.43	10.26.43	Sold 5.46
300	3.17.43	5.17.43	10.29.43	Kamikaze 12.18.44
301	3.19.43	5.17.43	11.4.43	D 8.15.45
302	3.23.43	5.19.43	11.9.43	Sold 3.19.48
303	3.25.43	5.19.43	11.29.43	Sold 7.20.48
304	3.27.43	5.27.43	11.23.43	Sold 7.19.48
305	3.30.43	5.27.43	12.8.43	Sold 6.18.48
306	4.1.43	5.27.43	12.3.43	Sold 4.1.48
307	4.3.43	6.1.43	12.2.43	Sold 3.19.48
308	4.6.43	6.1.43	1.24.44	Sold 7.19.48
309	4.9.43	6.11.43	1.26.44	Sold 6.18.48
310	4.14.43	6.11.43	1.27.44	Sold 3.12.48
311	4.20.43	6.19.43	1.25.44	WL(M) 11.18.44
312	4.27.43	6.19.43	1.29.44	Sold 6.21.48
313	5.1.43	7.1.43	1.31.44	Sold 7.19.48
Elco				
314	12.11.42	2.20.43	3.11.43	Sold 6.23.48
315	12.15.42	2.23.43	3.15.43	Sold 6.8.48
316	12.17.42	2.25.43	3.17.43	Sold 12.29.47
317	12.19.42	2.27.43	3.19.43	Sold 6.8.48
318	12.23.42	3.3.43	3.23.43	Sold 5.46
319	12.26.42	3.5.43	3.26.43	Sold 5.46
320	12.30.42	3.9.43	3.27.43	Kamikaze 11.5.44
321	1.2.43	3.11.43	3.30.43	Stranded 11.10.44
322	1.6.43	3.13.43	4.2.43	Stranded 11.23.43
323	1.8.43	3.17.43	4.6.43	Kamikaze 12.10.44
324	1.12.43	3.19.43	4.8.43	D 11.10.45
325	1.14.43	3.23.43	4.10.43	D 11.10.45
326	1.16.43	3.25.43	4.13.43	Sold 5.46
327	1.19.43	3.27.43	4.15.43	D 11.10.45
328	1.22.43	3.31.43	4.17.43	Sold 5.46
329	1.25.43	4.2.43	4.21.43	D 11.10.45
330	1.28.43	4.7.43	4.23.43	D 11.10.45
331	1.30.43	4.8.43	4.27.43	D 11.10.45
332	2.3.43	4.10.43	4.29.43	Sold 5.46
333	2.5.43	4.13.43	5.1.43	Sold 5.46
334	2.8.43	4.15.43	5.5.43	Sold 5.46
335	2.10.43	4.19.43	5.7.43	Sold 5.46
336	2.13.43	4.22.43	5.12.43	D 11.6.45
337	2.17.43	4.24.43	5.14.43	WL(G) 3.7.44
338	2.19.43	4.28.43	5.18.43	Stranded 1.28.45
339	2.22.43	5.1.43	5.22.43	Stranded 5.27.44

*Others in this category were PT 302–10, 312–13, 450–52, 456–61.

	Laid Down	Launched	Completed	Fate
340	2.25.43	5.5.43	5.25.43	Sold 5.46
341	2.27.43	5.7.43	5.28.43	Sold 5.46
342	3.2.43	5.11.43	5.31.43	Sold 5.46
343	3.5.43	5.13.43	6.1.43	Sold 5.46
344	3.8.43	5.15.43	6.7.43	Sold 5.46
345	3.11.43	5.19.43	6.8.43	D 11.9.45
346	3.13.43	5.21.43	6.10.43	WL(B) 4.29.44
347	3.17.43	5.26.43	6.15.43	WL(B) 4.29.44
348	3.20.43	5.28.43	6.17.43	Sold 5.46
349	3.23.43	6.1.43	6.18.43	Sold 5.46
350	3.25.43	6.4.43	6.22.43	D 11.9.45
351	3.27.43	6.8.43	6.25.43	D 11.9.45
352	3.31.43	6.10.43	6.20.43	D 11.9.45
353	4.3.43	6.12.43	7.2.43	WL(B) 3.27.44
354	4.6.43	6.16.43	7.6.43	Sold 5.46
355	4.9.43	6.18.43	7.8.43	Sold 5.46
356	4.12.43	6.22.43	7.10.43	Sold 5.46
357	4.15.43	6.25.43	7.13.43	Sold 5.46
358	4.17.43	6.29.43	7.23.43	Sold 5.46
359	4.21.43	7.2.43	7.16.43	Sold 5.46
360	4.23.43	7.6.43	7.20.43	Sold 5.46
361	4.27.43	7.8.43	7.27.43	Sold 5.46
Harbor Boat Building				
362	11.16.42	5.4.43	5.18.43	D 11.4.45
363	11.30.42	5.5.43	6.5.43	WL(G) 11.25.44
364	12.19.42	5.19.43	6.8.43	D 11.4.45
365	1.4.43	5.20.43	6.12.43	D 11.4.45
366	1.18.43	6.5.43	6.8.43	Sold 5.46
367	2.1.43	6.7.43	6.23.43	D 11.4.45
Canadian Power Boat				
368				Acq 11.21.42 Stranded 10.11.44
369				Acq 11.21.42 D 11.4.45
370				Acq 11.21.42 D 11.4.45
371				Acq 11.21.42 Stranded 9.19.44
Elco*				
372	4.29.43	7.10.43	8.3.43	Sold 5.46
373	5.1.43	7.13.43	8.5.43	Sold 5.46
374	5.4.43	7.16.43	8.6.43	Sold 5.46
375	5.6.43	7.20.43	8.10.43	Sold 5.46
376	5.10.43	7.23.43	8.12.43	Sold 5.46
377	5.12.43	7.27.43	8.14.43	Sold 5.46
378	5.15.43	7.30.43	8.17.43	Sold 5.46
379	5.18.43	8.3.43	8.20.43	Sold 5.46
380	5.20.43	8.6.43	8.24.43	Sold 5.46
381	5.22.43	8.10.43	8.26.43	Sold 5.46
382	5.26.43	8.12.43	8.28.43	Sold 5.46
383	5.28.43	8.14.43	8.28.43	Sold 5.46
Robert Jackson				
384	5.15.43	10.2.43	5.6.44	RN MTB 396 5.44
385	5.15.43	10.9.43	5.13.44	RN MTB 397 5.44 With 387 in bulk sale to Egypt 1.47
386	5.25.43	10.15.43	5.25.44	RN MTB 398 5.44
387	7.2.43	11.5.43	6.1.44	RN MTB 399 6.44
388	8.17.43	12.17.43	6.13.44	RN MTB 400 6.44
389	8.17.43	1.14.44	7.8.44	RN MTB 401 7.44
390	9.14.43	1.22.44	6.29.44	RN MTB 402 6.44
391	9.14.43	4.13.44	7.11.44	RN MTB 403 7.44
392	10.18.43	3.9.44	7.22.44	RN MTB 404 7.44
393	10.18.43	6.13.44	8.11.44	RN MTB 405 8.44
394	11.15.43	5.19.44	8.2.44	RN MTB 406 8.44
395	11.15.43	6.22.44	8.30.44	RN MTB 407 8.44 To U.S. Army 4.1.46
396	12.20.43	7.28.44	9.9.44	RN MTB 408 9.44
397	12.20.43	8.7.44	10.11.44	RN MTB 409 10.44
398	1.26.44	8.22.44	10.6.44	RN MTB 410 10.44
399	1.26.44	9.1.44	10.30.44	RN MTB 411 10.44
Annapolis Yacht Yard				
400	5.26.43	9.17.43	1.17.44	USSR 1.44 Ret 6.16.54
401	5.26.43	9.23.43	1.17.44	USSE 2.44 Ret 5.27.54
402	7.7.43	9.28.43	1.20.44	USSR 2.44 Ret 5.27.54
403	7.8.43	10.2.43	1.20.44	USSR 2.44 Ret 5.27.54
404	7.15.43	10.11.43	1.24.44	USSR 2.44 Ret 5.27.54
405	7.14.43	10.13.43	1.24.44	USSR 2.44 Ret 8.11.54
406	6.23.43	10.14.43	1.31.44	USSR 2.44 Ret 7.2.55
407	6.23.43	10.18.43	1.31.44	USSR 2.44 Ret 7.14.55
408	7.10.43	10.21.43	2.4.44	USSR 2.44 Ret 5.27.54
409	7.9.43	10.22.43	2.9.44	USSR 2.44 Ret 5.27.54
410	9.22.43	11.27.43	2.23.44	USSR 3.44 Ret 7.9.55
411	9.22.43	12.1.43	2.23.44	USSR 3.44
412	10.1.43	12.18.43	2.29.44	USSR 3.44
413	10.1.43	12.27.43	2.29.44	USSR 3.44 Ret 7.2.55

*Ordered as RPT 1–12.   Redesignated on 11.21.42.

	Laid Down	Launched	Completed	Fate
414	10.27.43	2.14.44	3.9.44	USSR 6.44
415	10.27.43	2.26.44	3.9.44	USSR 6.44
416	10.18.43	2.18.44	3.14.44	USSR 6.44
417	10.18.43	2.16.44	3.18.44	USSR 8.44
418	10.26.43	2.22.44	3.23.44	USSR 8.44
419	10.27.43	2.21.44	3.23.44	USSR 8.44
420	12.4.43	3.14.44	4.5.44	USSR 7.44
421	12.4.43	3.16.44	4.13.44	USSR 8.44
422	12.23.43	3.24.44	4.13.44	USSR 7.44
423	12.23.43	3.27.44	4.21.44	USSR 8.44
424	2.28.44	4.24.44	5.5.44	USSR 7.44 Ret 7.5.55
425	2.25.44	4.28.44	5.9.44	USSR 7.44 Ret 7.5.55
426	2.24.44	5.2.44	5.16.44	USSR 7.44 Ret 7.6.55
427	2.24.44	5.6.44	5.19.44	USSR 7.44 Ret 7.6.55
428	2.22.44	5.13.44	5.25.44	USSR 10.44
429	2.21.44	5.19.44	5.31.44	USSR 10.44

Herreshoff Manufacturing

	Laid Down	Launched	Completed	Fate
430	6.4.43	1.24.44	2.12.44	USSR 2.44 Ret 7.14.55
431	6.12.43	1.25.44	2.26.44	USSR 3.44 Ret 7.14.55
432	6.15.43	1.26.44	2.29.44	USSR 3.44 Ret 7.14.55
433	6.18.43	2.26.44	3.17.44	USSR 6.44
434	6.24.43	3.3.44	3.17.44	USSR 6.44
435	6.26.43	3.4.44	3.17.44	USSR 6.44
436	7.1.43	3.4.44	3.30.44	USSR 9.44
437	7.6.43	3.10.44	3.30.44	USSR 9.44
438	7.10.43	3.27.44	4.6.44	USSR 10.44
439	7.22.43	3.29.44	4.27.44	USSR 10.44 Ret 7.6.55
440	7.23.43	4.5.44	4.13.44	USSR 8.44
441	7.26.43	4.7.44	4.20.44	USSR 10.44
442	11.30.43	4.15.44	4.27.44	USSR 10.44 Ret 7.2.55
443	12.1.43	4.15.44	5.4.44	USSR 9.44 Ret 7.6.55
444	12.14.43	4.22.44	5.4.44	USSR 9.44
445	12.14.43	4.29.44	5.19.44	USSR 7.44
446	12.20.43	5.10.44	5.25.44	USSR 10.44 Ret 7.6.55
447	12.21.43	5.9.44	5.18.44	USSR 7.44
448	12.30.43	5.24.44	6.7.44	USSR 7.44 Ret 7.14.55
449	12.30.43	5.18.44	5.25.44	USSR 10.44

Higgins Industries

	Laid Down	Launched	Completed	Fate
450	5.24.43	7.25.43	2.3.44	Sold 6.25.48
451	6.2.43	7.29.43	1.31.44	Sold 3.31.48
452	6.12.43	8.10.43	2.5.44	Sold 7.19.48
453	6.21.43	8.16.43	3.14.44	Sold 5.46
454	6.29.43	8.26.43	2.8.44	Sold 5.46
455	7.9.43	8.31.43	2.15.44	Sold 5.46
456	7.20.43	9.27.43	3.9.44	Sold 6.21.48
457	7.29.43	9.29.43	3.15.44	Sold 7.20.48
458	8.9.43	10.6.43	3.18.44	Sold 6.21.48
459	8.19.43	10.12.43	3.23.44	Sold 6.18.48
460	8.30.43	10.23.43	3.27.44	Sold 6.21.48
461	9.13.43	10.30.43	3.28.44	Sold 6.21.48
462	9.25.43	11.13.43	3.29.44	Sold 5.46
463	10.6.43	11.27.43	4.1.44	Sold 5.46
464	10.15.43	12.7.43	4.5.44	Sold 5.46
465	10.25.43	12.22.43	4.13.44	Sold 5.46
466	11.3.43	12.23.43	4.17.44	Sold 5.46
467	11.13.43	1.8.44	4.20.44	Sold 5.46
468	11.23.43	1.22.44	4.24.44	Sold 5.46
469	12.3.43	2.5.44	4.28.44	Sold 5.46
470	12.13.43	2.16.44	4.29.44	Sold 5.46
471	12.23.43	2.21.44	5.5.44	Sold 5.46
472	1.6.44	3.4.44	5.11.44	Sold 5.46
473	1.18.44	3.14.44	5.16.44	Sold 5.46
474	1.31.44	3.22.44	6.9.44	D 11.2.45
475	2.9.44	4.1.44	6.15.44	Sold 5.46
476	2.21.44	4.12.44	6.16.44	Sold 5.46
477	3.1.44	4.20.44	7.1.44	D 11.2.45
478	3.10.44	4.29.44	6.28.44	Sold 5.46
479	3.20.44	5.9.44	7.11.44	Sold 5.46
480	3.29.44	5.17.44	7.15.44	Sold 5.46
481	4.10.44	5.29.44	7.22.44	Sold 5.46
482	4.19.44	6.5.44	8.3.44	Sold 5.46
483	4.28.44	6.15.44	8.12.44	Sold 5.46
484	5.8.44	6.23.44	8.30.44	Sold 5.46
485	5.13.44	7.18.44	8.29.44	Sold 5.46

Elco

	Laid Down	Launched	Completed	Fate
486	7.27.43	10.16.43	11.25.43	Small boat C 105335 8.27.46
487	7.29.43	10.21.43	1.10.44	Small boat C 105336 8.27.46
488	8.2.43	10.23.43	11.23.43	Sold 5.46
489	8.5.43	10.27.43	11.26.43	Sold 5.46
490	8.7.43	10.29.43	11.29.43	Sold 5.46
491	8.11.43	11.2.43	11.30.43	Sold 5.46
492	8.14.43	11.4.43	12.3.43	Sold 5.46

	Laid Down	Launched	Completed	Fate
493	8.17.43	11.6.43	12.6.43	WL(G) 10.25.44
494	8.20.43	11.10.43	12.9.43	Sold 5.46
495	8.24.43	11.12.43	12.13.43	Sold 5.46
496	8.36.43	11.18.43	12.14.43	Sold 5.46
497	8.30.43	11.20.43	12.18.43	Sold 5.46
498	9.2.43	11.20.43	12.21.43	USSR 3.44
499	9.4.43	11.23.43	12.23.43	USSR 12.44
500	9.8.43	11.26.43	12.27.43	USSR 12.44
501	9.10.43	12.1.43	12.29.43	USSR 1.44
502	9.14.43	12.4.43	12.31.43	USSR 1.44
503	9.17.43	12.8.43	1.5.44	USSR 12.44
504	9.20.43	12.13.43	1.11.44	USSR 12.44
505	9.23.43	12.14.43	1.13.44	Reclassified to small boat 8.27.46 but canc Sold 9.25.47
506	9.25.43	12.17.43	1.15.44	USSR 1.45
507	9.29.43	12.21.43	1.18.44	USSR 3.44
508	10.2.43	12.24.43	1.21.44	USSR 1.45
509	10.6.43	12.29.43	1.25.44	WL(G) 8.9.44
510	10.9.43	1.3.44	2.4.44	USSR 12.44
511	10.13.43	1.5.44	2.7.44	USSR 12.44
512	10.18.43	1.7.44	2.9.44	USSR 12.44
513	10.21.43	1.12.44	2.12.44	USSR 12.44
514	10.23.43	1.14.44	2.14.44	USSR 3.44
515	10.27.43	1.19.44	2.17.44	USSR 3.44
516	10.30.43	1.21.44	2.24.44	USSR 4.45
517	11.2.43	1.25.44	2.25.44	USSR 4.45 Ret 7.2.55
518	11.5.43	1.28.44	2.29.44	USSR 4.45
519	11.9.43	2.1.44	3.1.44	USSR 4.45
520	11.11.43	2.7.44	3.7.44	USSR 4.45
521	11.15.43	2.8.44	3.11.44	USSR 4.45
522	11.18.43	2.11.44	3.17.44	Sold 5.46
523	11.22.43	2.15.44	2.8.44	Sold 5.46
524	11.25.43	2.18.44	3.24.44	Sold 5.46
525	11.29.43	2.24.44	3.30.44	Sold 5.46
526	12.2.43	2.29.44	4.5.44	Sold 5.46
527	12.6.43	3.3.44	4.11.44	Sold 5.46
528	12.8.43	3.7.44	4.17.44	Sold 5.46
529	12.11.43	3.14.44	4.22.44	Sold 5.46
530	12.16.43	3.21.44	4.27.44	Sold 5.46
531	12.30.43	3.28.44	5.4.44	Sold 5.46
532	12.23.43	4.3.44	5.11.44	Sold 5.46
533	12.29.43	4.8.44	5.17.44	Sold 5.46
534	1.4.44	4.14.44	5.24.44	Sold 5.46
535	1.7.44	4.21.44	5.29.44	Sold 5.46
536	1.12.44	4.28.44	6.3.44	Sold 5.46
537	1.17.44	5.4.44	6.9.44	Sold 5.46
538	1.21.44	5.11.44	6.15.44	Sold 5.46
539	1.25.44	5.17.44	6.21.44	Sold 5.46
540	1.29.44	5.23.44	6.28.44	Sold 5.46
541	2.3.44	5.30.44	7.5.44	Sold 5.46
542	2.7.44	6.3.44	7.10.44	Sold 5.46
543	2.10.44	6.9.44	7.15.44	Sold 5.46
544	2.15.44	6.16.44	7.21.44	Sold 5.46
545	2.19.44	4.26.44	9.8.44	Sold 9.3.46
546*	6.1.43	8.17.43	9.3.43	Sold 5.46
547	6.3.43	8.21.43	10.4.43	Sold 5.46
548	6.5.43	8.24.43	9.7.43	Sold 5.46
549	6.9.43	8.26.43	9.10.43	Sold 5.46
550	6.11.43	8.28.43	9.14.43	Sold 5.46
551	6.15.43	9.1.43	9.16.43	Sold 5.46
552	6.17.43	9.4.43	10.16.43	USSR 4.45
553	6.21.43	9.7.43	10.18.43	USSR 4.45
554	6.24.43	9.10.43	10.21.43	USSR 5.45
555	6.28.43	9.14.43	10.26.43	WL(M) 8.23.44
556	7.1.43	9.16.43	10.28.43	USSR 4.45
557	7.3.43	9.18.43	10.30.43	Small boat C 105338 8.27.46
558	7.7.43	9.21.43	11.2.43	Sold 3.12.48
559	7.10.43	9.24.43	11.4.43	Small boat C 105339 8.27.46
560	7.14.43	9.28.43	11.6.43	USSR 5.45
561	7.16.43	10.1.43	11.9.43	USSR 4.45
562	7.20.43	10.5.43	11.11.43	USSR 4.45
563	7.23.43	10.7.43	11.22.43	USSR 4.45
Higgins Industries				
564				Acq 9.2.43 Sold privately 7.2.48
Elco				
565	4.28.44	9.20.44	12.8.44	Sold 5.46
566	5.9.44	9.29.44	12.13.44	Sold 5.46
567	5.10.44	10.10.44	12.15.44	Sold 5.46
568	5.13.44	10.17.44	12.19.44	Sold 5.46

*PT 546–63 ordered as RPT 13–30. Redesignated on 5.10.43.

	Laid Down	Launched	Completed	Fate
569	5.26.44	10.25.44	12.23.44	Sold 5.46
570	6.2.44	10.31.44	12.29.44	Sold 5.46
571	6.8.44	11.9.44	1.13.45	Sold 5.46
572	6.14.44	11.15.44	1.19.45	Sold 5.46
573	6.21.44	11.22.44	1.23.45	Sold 5.46
574	6.27.44	12.5.44	2.7.45	Sold 5.46
575	7.1.44	12.11.44	2.7.45	Sold 5.46
576	7.7.44	12.16.44	2.12.45	Sold 5.46
577	7.13.44	1.30.45	2.20.45	Sold 5.46
578	7.19.44	12.29.44	2.24.45	Sold 5.46
579	7.27.44	2.7.45	3.1.45	Sold 5.46
580	8.5.44	2.12.45	3.3.45	Sold 5.46
581	8.14.44	2.16.45	3.8.45	Sold 5.46
582	8.22.44	2.21.45	3.13.45	Sold 5.46
583	8.31.44	2.26.45	3.19.45	Sold 5.46
584	9.7.44	3.2.45	3.22.45	Sold 5.46
585	9.14.44	3.7.45	3.27.45	Sold 5.46
586	9.21.44	3.12.45	3.30.45	Sold 5.46
587	9.28.44	3.16.45	4.4.45	Sold 5.46
588	10.6.44	3.21.45	4.10.45	Sold 5.46
589	10.17.44	3.26.45	4.13.45	Sold 5.46
590	10.26.44	3.30.45	4.16.45	Sold 5.46
591	11.1.44	4.4.45	4.19.45	Sold 5.46
592	11.9.44	4.7.45	4.21.45	Sold 5.46
593	11.16.44	4.11.45	4.25.45	Sold 5.46
594	11.23.44	4.14.45	4.28.45	Sold 5.46
595	11.30.44	4.18.45	5.5.45	Sold 5.46
596	12.6.44	4.21.45	5.10.45	Sold 5.46
597	12.12.44	4.25.45	5.16.45	Sold 5.46
598	12.18.44	4.30.45	5.21.45	Sold 5.46
599	12.23.44	5.3.45	5.25.45	Sold 5.46
600	12.30.44	5.7.45	5.30.45	Sold 5.46
601	1.5.45	5.11.45	6.5.45	Reclassified as BuShips equipment 12.3.48 Small boat C6083 7.24.52
602	1.10.45	5.15.45	6.8.45	Reclassified as equipment 3.12.48 with PT 603–12 Norway's *Snogg* 1951
603	1.16.45	5.21.45	6.28.45	Norway's *Sel* 1951
604	1.25.45	5.26.45	6.13.45	Norway's *Sild* 1951
605	1.31.45	5.30.45	6.18.45	Norway's *Skrei* 1951
606	2.3.45	6.5.45	6.23.45	Norway's *Snarren Lyr* 1951
607	2.7.45	6.11.45	7.4.45	Sold 5.14.47
608	2.12.45	6.16.45	7.14.45	Norway's *Springer* 1951
609	2.16.45	6.21.45	7.9.45	Norway's *Hai* 1951
610	2.21.45	6.26.45	7.19.45	Norway's *Hauk* 1951
611	2.26.45	6.30.45	7.25.45	Norway's *Hval* 1951
612	3.2.45	7.6.45	7.31.45	Norway's *Hvass* 1951
613	1.20.45	5.15.45	8.10.45	Korea's *Olpamei* 1.52 Assigned to Fifth Fleet for inactivation 3.15.51 but assignment of four PTs to Naval Forces, Far East, approved 6.51 Sent to Sasebo 12.51 Transferred 1.24.52 Lost 9.18.52 in shipyard fire Remains ret to U.S. and rebuilt for museum
614	3.8.45	7.12.45	8.14.45	Small boat C 105340 8.27.46
615	3.14.45	7.18.45	9.5.45	Small boat C 105341 8.27.46
616	3.24.45	7.24.45	9.11.45	Korea's *Kaimaeki* 1.52
617	3.29.45	7.28.45	9.21.45	Sold 10.23.47
618	4.3.45	8.3.45	9.24.45	Small boat 8.27.46
619	4.9.45	8.10.45	10.1.45	Korea's *Koroki* 1.52
620	4.13.45	8.17.45	10.5.45	Korea's *Ebi* 1.52
621	4.19.45	9.12.45	10.12.45	Sold 7.15.47
622	4.24.45	9.18.45	10.24.45	Sold 3.24.47
623–24				Canc 9.12.45 623 laid down 4.30.45
Higgins Industries				
625	5.17.44	7.12.44	12.7.44	USSR 5.45
626	5.27.44	7.18.44	12.18.44	USSR 5.45
627	6.5.44	7.26.44	1.7.45	USSR 5.45
628	6.14.44	8.3.44	1.16.45	USSR 5.45
629	6.23.44	8.9.44	1.29.45	USSR 6.45
630	7.1.44	8.16.44	2.8.45	USSR 6.45
631	7.13.44	8.23.44	2.18.45	USSR 6.45
632	7.21.44	8.31.44	3.15.45	USSR 6.45
633	7.29.44	9.9.44	3.14.45	USSR 6.45
634	8.7.44	9.18.44	3.21.45	USSR 6.45
635	8.14.44	9.26.44	3.23.45	USSR 6.45
636	8.22.44	10.4.44	3.26.45	USSR 6.45
637	8.30.44	10.12.44	4.10.45	USSR 6.45
638	9.7.44	10.19.44	4.12.45	USSR 6.45
639	9.16.44	10.27.44	4.16.45	USSR 6.45
640	9.25.44	11.4.44	4.18.45	USSR 6.45
641	10.3.44	11.14.44	4.23.45	USSR 8.45
642	10.11.44	11.24.44	4.28.45	USSR 8.45
643	10.19.44	11.30.44	5.2.45	USSR 8.45
644	10.27.44	12.11.44	5.11.45	USSR 8.45

	Laid Down	Launched	Completed	Fate
645	11.4.44	12.19.44	5.13.45	USSR 8.45
646	11.13.44	12.29.44	5.17.45	USSR 8.45
647	11.21.44	1.6.45	5.22.45	USSR 8.45
648	11.30.44	1.16.45	5.26.45	USSR 8.45
649	12.8.44	1.24.45	5.30.45	USSR 8.45
650	12.16.44	2.1.45	6.15.45	USSR 8.45
651	12.27.44	2.9.45	6.20.45	USSR 8.45
652	1.6.45	2.17.45	6.26.45	USSR 8.45
653	1.15.45	2.26.45	6.29.45	USSR 8.45
654	1.23.45	3.6.45	7.6.45	USSR 8.45
655	1.31.45	3.14.45	7.11.45	USSR 8.45
656	2.8.45	3.23.45	7.17.45	USSR 8.45
657	2.16.45	4.2.45	7.21.45	Sold 11.14.46
658	2.24.45	4.11.45	7.30.45	Small boat 8.27.46 To Point Mugu as remote-control target Remote-control target tower 12.48 Sold 6.30.58
659	3.5.45	4.19.45	8.2.45	Small boat 8.27.46 To Point Mugu as remote-control target 12.20.46 To naval inshore undersea warfare group 1 9.13.68 Target at Point Mugu 5.27.70
660	3.13.45	4.26.45	8.10.45	Small boat 8.27.46
Annapolis Yacht Yard*				
661	4.4.44	8.14.44	10.30.44	USSR 11.44 Ret 7.2.55
662	4.1.44	*	11.18.44	USSR 12.44 Ret 7.9.55
663	4.1.44	*	11.18.44	USSR 12.44 Ret 7.14.55
664	4.1.44	*	11.18.44	USSR 12.44 Ret 7.14.55
665	4.1.44	*	11.18.44	USSR 12.44 Ret 9.7.55
666	4.1.44	*	11.18.44	USSR 12.44 Ret 7.14.55
667	4.1.44	*	11.18.44	USSR 12.44 Ret 7.19.55
668	4.1.44	*	11.18.44	USSR 12.44 Ret 7.20.55
669	4.1.44	*	11.18.44	USSR 12.44 Ret 7.14.55
670	4.1.44	*	11.18.44	USSR 12.44 Ret 7.2.55
671	4.1.44	*	11.18.44	USSR 12.44 Ret 7.2.55
672	4.1.44	*	11.18.44	USSR 12.44 Ret 7.9.55
673	4.1.44	*	11.18.44	USSR 12.44 Ret 7.9.55
674	4.1.44	*	11.18.44	USSR 12.44 Ret 7.14.55
675	4.1.44	*	11.18.44	USSR 12.44 Ret 7.14.55
676	8.26.44	10.31.44	12.6.44	USSR 12.44 Ret 7.9.55
677	8.30.44	11.3.44	12.6.44	USSR 12.44 Ret 7.9.55
678	9.5.44	11.17.44	12.13.44	USSR 1.45 Ret 7.2.55
679	9.5.44	11.23.44	12.13.44	USSR 1.45 Ret 7.2.55
680	9.5.44	11.25.44	12.19.44	USSR 3.45 Ret 7.9.55
681	9.5.44	11.29.44	12.19.44	USSR 3.45 Ret 7.20.55
682	9.5.44	12.5.44	12.27.44	USSR 3.45 Ret 7.20.55
683	9.5.44	12.11.44	1.1.45	USSR 3.45
684	11.2.44	1.10.45	2.12.45	USSR 4.45
685	11.3.44	1.12.45	2.14.45	USSR 4.45
686	11.22.44	2.5.45	2.23.45	USSR 4.45
687	11.30.44	2.6.45	2.23.45	USSR 4.45
688	12.8.44	2.10.45	3.2.45	USSR 5.45
689	11.30.44	2.8.45	3.3.45	USSR 5.45
690	12.8.44	2.14.45	3.7.45	USSR 5.45
691	12.9.44	2.17.45	3.9.45	USSR 5.45
692	1.15.45	3.2.45	3.19.45	USSR 5.45
693	1.15.45	3.5.45	3.21.45	Sold 8.15.46
694	2.8.45	3.14.45	4.5.45	Sold 8.15.46
695	2.8.45	3.17.45	4.5.45	Sold 8.15.46
696	2.19.45	3.26.45	4.14.45	Sold 8.15.46
697	2.19.45	3.30.45	4.18.45	Sold 8.30.46
698	2.20.45	4.3.45	4.21.45	Sold 8.30.46
699	3.7.45	4.12.45	5.1.45	Sold 8.21.46
700	3.8.45	4.16.45	5.3.45	Sold 10.21.46
701	3.17.45	4.20.45	5.9.45	Sold 9.12.46
702	3.17.45	4.26.45	5.14.45	Sold 8.30.46
703	4.3.45	5.4.45	5.22.45	Sold 6.25.46
704	4.4.45	5.7.45	5.26.45	Sold 6.25.46
705	4.4.45	5.11.45	5.30.45	Sold 8.12.46
706	4.18.45	5.21.45	6.8.45	Sold 7.17.46
707	4.18.45	5.25.45	6.14.45	Sold 6.25.46
708	4.27.45	6.2.45	6.23.45	Sold 8.12.46
709	4.27.45	6.6.45	6.27.45	Sold 7.17.46
710	5.12.45	6.18.45	7.4.45	Sold 6.25.46
711	5.12.45	6.19.45	7.11.45	Sold 7.17.46
712	5.14.45	6.23.45	7.18.45	Sold 9.13.46
713	5.26.45	6.29.45	7.19.45	Sold 9.11.46
714	5.26.45	7.4.45	7.23.45	Sold 9.11.46
715	6.6.45	7.9.45	7.29.45	Cuba's R41 8.46
716	6.6.45	7.17.45	8.2.45	Cuba's R42 7.46
717	6.22.45	7.23.45	8.12.45	Sold 6.26.47
718	6.22.45	7.26.45	8.15.45	Sold 6.27.47

*Asterisks indicate boats delivered in knocked-down form, hence not launched.

502 U.S. SMALL COMBATANTS

	Laid Down	Launched	Completed	Fate
719	6.22.45	8.1.45	8.24.45	Sold 6.27.47
720	7.5.45	8.8.45	8.29.45	Sold 6.18.47
721	7.5.45	8.11.45	9.4.45	Sold 7.21.47
722	7.16.45	8.17.45	9.6.45	Sold 8.15.47
723	7.16.45	8.27.45	9.14.45	Sold 4.12.48
724	7.31.45	9.8.45	9.21.45	Sold 7.18.47
725	7.31.45	9.8.45	10.1.45	Sold 8.18.47
726	7.31.45	9.14.45	10.6.45	Sold 8.16.47
727	8.10.45	9.20.45	10.13.45	Sold 6.20.47
728	8.10.45	9.25.45	10.20.45	Sold 8.18.47
729	8.18.45	9.29.45	10.27.45	Small boat 11.16.45 To OpDevFor 12.14.45 Surplus 3.31.47
730	8.18.45	9.29.45	10.30.45	Small boat 11.16.45 MC 11.18.47
Elco				
731	4.4.44	8.4.44	9.19.44	USSR 10.44
732	3.30.44	*	10.3.44	USSR 12.44 Ret 7.14.55
733	4.5.44	*	10.10.44	USSR 12.44 Ret 7.14.55
734	4.11.44	*	10.10.44	USSR 12.44 Ret 7.20.55
735	4.16.44	*	10.24.44	USSR 12.44 Ret 7.20.55
736	4.23.44	*	10.24.44	USSR 12.44 Ret 7.19.55
737	4.28.44	*	11.14.44	USSR 12.44 Ret 7.19.55
738	5.2.44	*	11.14.44	USSR 12.44
739	5.8.44	*	11.21.44	USSR 12.44 Ret 7.2.55
740	5.12.44	*	11.21.44	USSR 12.44 Ret 7.2.55
741	5.18.44	*	11.25.44	USSR 12.44
742	5.23.44	*	11.25.44	USSR 12.44 Ret 7.9.55
743	5.29.44	*	11.30.44	USSR 12.44 Ret 7.14.55
744	6.3.44	*	11.30.44	USSR 12.44 Ret 7.9.55
745	6.7.44	*	12.9.44	USSR 1.45 Ret 7.9.55
746	6.12.44	*	12.9.44	USSR 1.45 Ret 7.2.55
747	6.18.44	*	12.14.44	USSR 1.45
748	6.22.44	*	12.14.44	USSR 1.45 Ret 7.2.55
749	6.27.44	*	12.23.44	USSR 1.45
750	7.2.44	*	12.23.44	USSR 1.45 Ret 7.9.55
751	7.7.44	*	12.30.44	USSR 1.45 Ret 7.9.55
752	7.12.44	*	12.30.44	USSR 1.45 Ret 7.9.55
753	7.17.44	*	1.4.45	USSR 1.45
754	7.23.44	*	1.4.45	USSR 1.45
755	7.28.44	*	1.10.45	USSR 2.45
756	8.4.44	*	1.10.45	USSR 2.45 Ret 7.14.55
757	8.7.44	*	1.20.45	USSR 2.45 Ret 7.14.55
758	8.12.44	*	1.20.45	USSR 2.45 Ret 7.14.55
759	8.17.44	*	1.26.45	USSR 2.45 Ret 7.14.55
760	8.21.44	*	1.26.45	USSR 2.45 Ret 7.9.55
761–90				Canc 8.14.45
761	3.19.45			
762	5.10.45			
763	5.15.45			
764	5.19.45			
765	5.24.45			
766	5.30.45			
767	6.5.45			
768	6.11.45			
769	6.16.45			
770	6.22.45			
771	6.28.45			
772	7.4.45			
773	7.10.45			
774	7.17.45			
775	7.23.45			
776	7.30.45			
777	8.3.45			
778	8.9.45			
Higgins Industries				
791	3.31.45	5.5.45	9.19.45	Sold 10.11.46
792	3.29.45	5.12.45	9.20.45	Sold 10.31.46
793	4.9.45	5.21.45	10.23.45	Sold 10.11.46
794	4.17.45	5.29.45	10.24.45	Sold 10.31.46
795	4.25.45	6.15.45	10.25.45	Small boat 11.16.45 Surplus 3.31.48 Sold 4.10.50
796	5.3.45	6.23.45	10.26.45	Small boat 11.16.45 To OpDevFor To Naval Mine Countermeasures Station, Panama City, Florida 7.48 as high-speed tug To PT boat museum in Memphis 1970
797–802				Canc 9.7.45
803–8				Canc 8.27.45
797		5.11.45		
798		5.19.45		
799		5.28.45		
800		6.14.45		
801		6.22.45		
802		6.30.45		

	Laid Down	Launched	Completed	Fate
803	7.12.45			
804	7.23.45			
805	8.3.45			
Electric Boat				
809	6.27.49	8.7.50	2.9.51	*Guardian*, escort for presidential yacht 11.1.59
Bath Iron Works				
810	12.1.48	6.2.50	11.24.51	Str 11.1.59 Res after 8.57 Reinstated as PTF 1 12.21.62
John Trumpy				
811	3.17.49	11.30.50	3.6.51	In reserve after 8.57 Str 11.1.59 Reinstated as PTF 2 12.21.62
Philadelphia Navy Yard				
812*	10.20.49	2.1.51	4.25.51	Service craft 11.1.59

British PT (BPT) Boats for Lend-Lease†

	Laid Down	Launched	Completed	Fate
Annapolis Yacht Yard				
21	9.5.41	5.2.42	3.2.43	RN MTB 275 3.43
22	9.5.41	5.16.42	11.10.42	RN MTB 276 11.42
23	11.5.41	8.11.42	11.10.42	RN MTB 277 11.42
24	11.17.41	8.14.42	11.10.42	RN MTB 278 11.42
25	11.19.41	8.14.42	11.10.42	RN MTB 279 11.42
26	11.19.41	8.13.42	11.26.42	RN MTB 280 11.42
27	12.9.41	8.12.42	12.2.42	RN MTB 281 12.42
28	12.11.41	8.11.42	12.15.42	RN MTB 282 12.42
Herreshoff Manufacturing‡				
29	12.23.41	2.26.43	3.18.43	RN MTB 287 3.43 WL 11.24.44
30	12.24.41	3.9.43	3.18.43	RN MTB 288 3.43 WL 7.21.43
31	1.12.42	3.22.43	4.12.43	RN MTB 289 4.43
32	1.15.42	4.6.43	4.12.43	RN MTB 290 5.43
33	3.31.42	4.19.43	5.10.43	RN MTB 291 5.43
34	4.1.42	5.3.43	5.10.43	RN MTB 292 5.43
35	4.8.42	5.17.43	6.18.43	RN MTB 293 6.43
36	4.8.42	6.7.43	7.8.43	RN MTB 294 7.43
Robert Jacob§				
37	12.26.41	7.27.42	2.17.43	RN MTB 295 2.43
38	12.31.41	8.20.42	3.3.43	RN MTB 296 3.43
39	1.8.42	7.27.42	2.6.43	RN MTB 297 2.43
40	1.13.42	11.18.42	3.30.43	RN MTB 298 3.43
41	4.2.42	8.20.42	3.17.43	RN MTB 299 3.43
42	4.2.42	11.18.42	4.28.43	RN MTB 300 4.43
Harbor Boat Building‖				
43	12.30.41	9.7.42	2.10.43	RN MTB 301 2.43
44	12.31.41	12.14.42	2.10.43	RN MTB 302 2.43
45	1.9.42	12.14.42	2.27.43	RN MTB 303 2.43
46	1.10.42	1.6.43	2.27.43	RN MTB 304 2.43
47	2.11.42	1.14.43	3.9.43	RN MTB 305 3.43
48	2.12.42	1.14.43	3.31.43	RN MTB 306 3.43
Annapolis Yacht Yard#				
49	3.16.42	9.1.42	4.3.43	RN MTB 283 4.43
50	5.13.42	9.2.42	4.3.43	RN MTB 284 4.43 Lost as cargo on passage to India 9.9.43
51	5.13.42	9.4.42	4.3.43	RN MTB 285 4.43 Lost as cargo on passage to India 9.9.48
52	5.22.42	9.5.42	4.3.43	RN MTB 286 4.43
53**	8.20.42	2.23.43	9.30.43	RN MTB 363 USSR 2.43 Ret 6.16.54
54	8.20.42	2.24.43	11.18.43	RN MTB 364 USSR 2.44
55	8.20.42	2.25.43	12.15.43	RN MTB 365 USSR 2.44 Ret 6.16.54
56	8.21.42	2.26.43	11.29.43	RN MTB 366 USSR 2.44 Ret 6.16.54
57	8.21.42	2.26.43	11.19.43	RN MTB 367 USSR 2.44 Ret 16.6.54
58	9.7.42	2.28.43	11.22.43	RN MTB 368 USSR 2.44 Ret 7.9.55
59	9.7.42	2.20.43	11.22.43	RN MTB 369 USSR 2.44 Ret 6.16.54
60	9.7.42	2.15.43	11.29.43	RN MTB 370 USSR 2.44 Ret 7.2.55
61	2.19.43	5.15.43	10.20.43	RN MTB 371 10.43 WL 11.24.44
62	2.20.43	5.17.43	9.30.43	RN MTB 372 10.43 WL 7.23.44
63	3.1.43	6.14.43	10.9.43	RN MTB 373 10.43
64	3.2.43	6.15.43	10.20.43	RN MTB 374 10.43
65	3.3.43	6.15.43	11.19.43	RN MTB 375 11.43
66	3.4.43	6.19.43	12.2.43	RN MTB 376 12.43
67	3.4.43	6.23.43	12.2.43	RN MTB 377 12.43
68	3.6.43	7.17.43	12.15.43	RN MTB 378 12.43

* PT 813–22 were built in Denmark under MDAP for the Royal Danish Navy. They were not of U.S. design.
† Other than transferred PTs.
‡ Ordered 11.24.41.
§ Ordered 11.24.41.
‖ Ordered 11.28.41.
First four ordered 11.24.41, others 7.22.42.
** BPT 53–60 were RPT 53–60.

	Laid Down	Launched	Completed	Fate
PTC Boats				
Elco				
1	6.13.40	1.3.41	2.13.41	RN MGB 82 7.41
2	6.20.40	1.11.41	2.17.41	RN MGB 83 7.41
3	6.26.40	1.20.41	2.20.41	RN MGB 84 7.41
4	7.2.40	1.25.41	3.7.41	RN MGB 85 7.41
5	7.9.40	2.1.41	2.17.41	RN MGB 86 4.41
6	7.15.40	2.8.41	2.20.41	RN MGB 87 4.41
7	7.23.40	2.15.41	3.7.41	RN MGB 88 4.41
8	7.26.40	2.21.41	3.8.41	RN MGB 89 4.41
9	8.1.40	3.3.41	3.14.41	RN MGB 90 4.41 D in accident 7.16.41
10	8.7.40	3.5.41	3.20.41	RN MGB 91 4.41
11	8.13.40	3.12.41	3.27.41	RN MGB 92 4.41 D in accident 7.16.41
12*	8.23.40	3.20.41	3.28.41	RN MGB 93 4.41
Trumpy				
37†	6.2.43	10.16.43	10.16.43	USSR 10.43
38	6.2.43	10.16.43	10.22.43	USSR 12.43
39	6.30.43	10.27.43	10.30.43	USSR 12.43
40	6.30.43	11.5.43	11.17.43	USSR 12.43
41	7.12.43	11.16.43	11.20.43	USSR 12.43
42	7.12.43	11.24.43	11.29.43	USSR 12.43
43	7.31.43	12.4.43	12.8.43	USSR 12.43
44	7.31.43	12.14.43	12.17.43	USSR 1.44
45	8.18.43	12.20.43	12.28.43	USSR 1.44
46	8.18.43	12.28.43	1.7.44	USSR 1.44
47	10.18.43	1.7.44	1.10.44	USSR 1.44
48	10.18.43	1.15.44	1.19.44	USSR 1.44
49	11.24.43	1.25.44	1.19.44	Small boat 8.1.43
50	11.25.43	2.2.44	2.5.44	Small boat 8.1.43 Torpedo retriever 6.7.51 Sold 4.4.58
51	12.13.43	2.13.44	2.19.44	Small boat 8.1.43 To army as crash boat
52	12.13.43	2.19.44	2.23.44	Small boat 8.1.43 Torpedo retriever 3.45 Sold 7.60
53	12.30.43	2.23.44	2.26.44	Small boat 8.1.43
54	12.30.43	3.10.44	3.13.44	USSR 8.44
55	1.18.44	3.18.44	3.21.44	USSR 8.44
56	1.18.44	3.27.44	3.31.44	USSR 8.44
57	2.3.44	4.7.44	4.13.44	USSR 8.44
58	2.3.44	4.17.44	4.21.44	USSR 8.44
59	2.19.44	4.26.44	5.2.44	USSR 8.44
60	2.19.44	5.5.44	5.9.44	USSR 7.44
61	3.10.44	5.21.44	5.29.44	USSR 7.44
62	3.10.44	6.2.44	6.6.44	USSR 7.44
63	3.29.44	6.15.44	6.22.44	USSR 7.44
64	3.29.44	6.29.44	7.10.44	USSR 8.44
65	4.17.44	7.18.44	7.21.44	USSR 8.44
66	4.17.44	8.1.44	8.4.44	USSR 8.44
RPC Boats				
Miami Shipbuilding				
1	1.20.43	2.16.43	4.26.43	Small boat
2	1.25.43	2.17.43	5.7.43	Small boat
3	2.20.43	3.15.43	5.7.43	Small boat
4	2.25.43	3.18.43	5.7.43	Small boat
5	2.27.43	3.21.43	5.7.43	Small boat
6	3.2.43	3.21.43	5.22.43	Small boat
7	3.4.43	3.25.43	5.22.43	Small boat
8	3.8.43	3.27.43	5.22.43	Small boat
9	3.22.43	4.9.43	6.4.43	Small boat
10	3.27.43	4.14.43	6.9.43	Small boat
11	3.30.43	4.22.43	6.10.43	Small boat
12	3.31.43	4.24.43	6.16.43	Small boat
13	4.5.43	4.27.43	6.19.43	Small boat
14	4.7.43	4.29.43	6.20.43	Small boat
15	4.9.43	5.3.43	6.29.43	Small boat
16	5.10.43	5.31.43	6.30.43	Small boat
17	5.12.43	5.31.43	6.29.43	USSR
18	5.14.43	6.3.43	6.30.43	Converted to torpedo retriever Collided with crash boat 4.17.53
19	5.18.43	6.5.43	6.30.43	Converted to torpedo retriever 1945 Sold 1962
20	5.19.43	6.8.43	7.14.43	USSR
21	5.21.43	6.10.43	7.15.43	USSR
22	5.22.43	6.14.43	7.16.43	Converted to torpedo retriever 1945 Sunk under tow 11.56
23	5.25.43	6.11.43	7.18.43	USSR
24	7.6.43	7.30.43	8.19.43	Small boat
25	7.9.43	8.2.43	8.20.43	Small boat

*PTC 13–24 became PT 33–44. PTC 25–36 became PT 57–68.
†PTC 37–66 were redesignated RPC 51–80.

	Laid Down	Launched	Completed	Fate
26	7.10.43	8.6.43	8.23.43	Small boat
27	7.13.43	8.6.43	9.5.43	Small boat
28	7.16.43	8.7.43	9.5.43	Small boat
29	7.19.43		8.22.43	Small boat
30	5.27.43	6.22.43	7.29.43	USSR 12.43
31	5.29.43	6.24.43	7.29.43	USSR 12.43
32	5.31.43	6.28.43	7.29.43	USSR 12.43
33	6.2.43	6.30.43	7.29.43	USSR 12.43
34	6.4.43	7.2.43	7.30.43	USSR 12.43
35	6.7.43	7.7.43	7.29.43	USSR 1.44
36	6.10.43	7.10.43	7.30.43	USSR 12.43
37	6.14.43	7.12.43	7.31.43	USSR 12.43
38	6.16.43	7.13.43	7.31.43	USSR 12.43
39	6.21.43	7.17.43	7.31.43	USSR 12.43
40	6.24.43	7.19.43	8.16.43	USSR 1.44
41	6.28.43	7.22.43	8.17.43	USSR 1.44
42	6.29.43	7.26.43	8.18.43	USSR 1.44
43	7.1.43	7.28.43	8.18.43	USSR 1.44
44	7.24.43	8.14.43	9.1.43	USSR 1.44
45	7.27.43	8.17.43	9.2.43	USSR 1.44
46	7.29.43	8.18.43	9.4.43	USSR 1.44
47	7.31.43	8.20.43	9.7.43	USSR 1.44
48	8.2.43	8.21.43	9.8.43	USSR 12.43
49	8.4.43	8.25.43	9.9.43	USSR 12.43
50	7.21.43	8.12.43	9.8.43	Small boat

PTF Boats

	Laid Down	Launched	Completed	Fate
1 (see PT 810)				Str 8.1.65 Sunk as target
2 (see PT 811)				Str 8.1.65 Sunk as target
Westermoens*				
3	2.62	10.62	11.62 (1.1.63)	Str 1973 Sold 12.77
4	8.62	12.62	1.63	WL 12.1.65
5			3.64 (3.1.64)	Sold 6.77
6			3.64 (3.1.64)	Str 1977 Sold 12.77
7			3.64 (3.1.64)	Str 1977 Sold 12.77
8			3.64	WL 1966
9			3.64	WL 1966
Boatservice				
10	11.64	3.65	3.65 (4.22.65)	Sold 12.79 Had been stored 1.77 for possible foreign sale
11	12.64	4.65	5.65 (7.7.65)	Sold 6.77
12	2.65	5.65	5.65 (7.7.65)	Sold 12.77
13	3.65	6.65	5.65 (8.31.65)	Str 1972 Redesignated 80PB6513 Sold 1.77
14				Leased to South Vietnam Lost in South Vietnamese service 1966
15				Leased to South Vietnam Lost in South Vietnamese service 1966
16			3.66	WL 1966
John Trumpy				
17	5.18.67	11.17.67	7.1.68	Stored after 1.77 for possible foreign sale Offered to Buffalo museum 6.78
18	7.10.67		7.1.68	Stored 1.77 for possible foreign sale Sold 1979
19	11.22.67	5.29.68	10.7.68	Stored 1.77 for possible foreign sale Sold 1979
20	1.23.68	8.7.68	10.7.68	To Point Mugu as target 2.77
21	6.3.68	11.4.68	5.14.69	To Point Mugu as target 2.77
22	8.30.68	1.23.69	5.14.69	Holed and run ashore 3.76 Too badly damaged for future use
Sewart Seacraft				
23	4.5.67	11.3.67	3.13.68	Stored 12.78
24	4.8.67	1.16.68	3.13.68	Issued to swimmer weapon-system group to be sunk in a movie 9.84 Did not sink Sold 1985 to a movie producer
25	5.10.67	2.22.68	4.8.68	Deactivated 1.23.78 for conversion to Osprey 990 program with twin 5,000-shp gas turbines and water-jet pumps Disposed of instead 3.80
26	5.12.67	3.7.68	4.8.68	For disposal at San Diego 7.84

*Date in parentheses is date "in service."

PT-7
NAVY YARD, PHILA., PA
July 11-1940
787-40

Notes

Chapter 1

1. Steam torpedo boats are described in a companion volume, *U.S. Destroyers: An Illustrated Design History* (Annapolis: Naval Institute Press, 1982). Like the small combatants of this volume, they were not an integral part of the fleet, but being the direct ancestors of modern destroyers, they are treated in the other book. Early gunboat development is discussed in a companion volume on cruisers, *U.S. Cruisers: An Illustrated Design History* (Annapolis: Naval Institute Press, 1984). That book also includes a detailed study of the *Erie* and the *Charleston* (PG 50 and 51), which were, in effect, small slow cruisers.

2. Until 1945, there was another, perhaps more traditional category of small combatant, the gunboat on a foreign station. It too lay outside the fleet as such but was clearly an adjunct essential to U.S. foreign policy.

3. Resistance is a *force*, measured in pounds. *Power* is the product of force and speed, so that two similar ships at similar speed-length ratios show the same resistance (pounds per ton) but not the same amount of hp per ton.

4. The power rating at which internal combustion engines could replace steam moved up as the former developed. The German navy powered its three "pocket battleships" with diesels, but reportedly they suffered from the resulting vibration. In 1939, it planned to power future battleships, of the H class, with large diesels. The U.S. Navy never built any large diesel warships, but in 1941, it planned to build an experimental diesel-powered destroyer as a prototype of future high-powered diesel installations. In 1944, the Germans actually launched a diesel destroyer, the *Z 51*, and planned others. Gas turbines are now the favored internal-combustion power plant for large, fast ships. At the other end of the scale, it was diesels rather than steam turbines that met with success in the World War II PC program.

5. The others were the Allison division of General Motors, the Hall-Scott Motor Car Company, and the Vimalert Company. Allison was the prospective diesel manufacturer, but its engine never materialized.

6. A later attempt to develop a larger submarine "pancake" was much less successful, and several postwar submarines had to be lengthened to accept more conventional power plants.

7. Such seals also make it possible to maintain cushion pressure using less lift power; this results in more efficient air-cushion craft.

8. The hump is the rise in resistance the boat must overcome before it moves fast enough to plane.

9. In 1915, the British firm of Thornycroft compared a number of high-speed hull forms in preparation for CMB construction. Tank tests showed that a single-step hull was best (for speed), a V-bottom (hard-chine) hull worst, and the sea sled intermediate.

10. For information on early hydrofoils, see H. F. King, *Aeromarine Origins* (London: Putman, 1966), pp. 50–64, and H. Fock, *Fast Fighting Boats, 1870–1945* (Annapolis: Naval Institute Press, 1978).

11. Ladders consisted of several small foils arranged vertically. The boat initially took off using the lift generated by all the foils at a relatively low speed. As it accelerated, foils lifted out of the water in succession, those remaining in the water sufficing because of the higher speed. This reduction in total foil area also reduced net drag. The ladder allowed for flexibility in payload, since total foil lift could vary over such a wide range. Ladder foils, owing to their complexity, are no longer used.

12. The German V-foil boats were designed by Baron Hans von Schertel, who founded Supramar AG in Switzerland after World War II. The U.S. Navy considered buying Supramar boats in the early 1960s but decided instead to continue with its own fully submerged foil type. The first Supramars were PT 10 (passenger transport, 10 metric tons) passenger ferries used on Lake Maggiore after 1953. Larger boats were built under license in Italy by Carlo Rodriquez, who was later a partner of Boeing's in Alinavi, building Boeing-designed hydrofoil gunboats under license.

13. Configurations based on other small-scale prototypes were also tried: a four-foil, surface-piercing design, the *Hypockets*, and a split, submerged-foil configuration, the *Halobates*. The former was efficient but had only a limited range of operating speeds because of its fixed-foil area. Moreover, its basic design severely limited size. The Hook-type *Halobates* was considered inefficient, and its foils were heavier and extended further out to each side than a single continuous foil.

14. As of this writing, it appears that surface-effect or even wing-in-ground machines are favored for speeds over about 50 knots, at which supercavitating foils must be used. That hydrofoils were not efficient at such speeds was not always obvious. In June 1961, BuShips ordered a 100-knot test vehicle, FRESH-1 (foil research supercavitating hydrofoil), from Boeing after a design competition held in the same year. The craft, launched in February 1963, was powered by a large turbofan engine. FRESH-1 flipped over on its initial trials. Although repaired, modified, and accepted in July 1964, it was laid up shortly thereafter when the supercavitating foil program was abandoned.

15. The Grumman hydrofoils used an aircraft planform with the main foils forward. It was well adapted to a conventional propeller, since the vertical drive shaft could run down the single after strut. The *Sea Legs*, navy-designed hydrofoils such as the *High Point*, and later the Boeing-designed boats had a canard

Hard-chine hulls and turtlebacks were characteristic of many pre–World War II motor torpedo boats. PT 7 is under construction at the Philadelphia Navy Yard, 11 July 1940, and Thornycroft's Q III, a 65-footer, runs trials before delivery to the Philippine government in 1939. Representative of British practice derived from the CMBs of World War I, Q III was one of few foreign-built fast craft to serve under U.S. control. (Thornycroft)

planform with the main foils aft, partly because that gave better support to heavy engines at the stern. There is no current consensus as to which planform is better.

16. This formulation was made explicit after World War II. See J. L. Gaddis, *Strategies of Containment* (New York: Oxford University Press, 1982). The basic national strategy was similar to the classic British balance-of-power approach.

17. Until the late 1930s, it was generally accepted that the U.S. garrison could not hold out for long and that the Japanese would inevitably be able to land. However, as chief military advisor to the Philippine government after 1936, General Douglas MacArthur claimed that a force of torpedo boats could prevent a Japanese landing. Later he lobbied successfully for a large force of U.S. heavy bombers as a local deterrent. As a result, the U.S. war plan changed to provide for an extended defense of the Philippines, and additional materiel was shipped there in 1940–41. Later events suggested that the earlier judgment was the better one.

18. Note that coastal ASW declined in importance as forward-area ASW operations showed greater promise in the 1950s. Current U.S. maritime strategy involves what amounts to forward fleet engagements against both Soviet submarines and Soviet land-based antiship (missile) bombers. This is exactly analogous to the classic fleet strategy, and it has the same impact on U.S. interest in coastal defense.

19. The strategy based on a concentrated battle fleet was adopted as an alternative to the traditional coastal defense/antitrade strategy after a bitter struggle. Thus, apart from any bearing they had on the concentrated battle fleet, there could be little interest in new coastal defense craft other than submarines in the pre-1914 U.S. Navy. The key event in the rise of fleet strategy was a review initiated in 1889 by Secretary of the Navy B. F. Tracy and influenced by many of Mahan's ideas. For more information, see two other volumes in this series: Friedman, *U.S. Cruisers: An Illustrated Design History* (Annapolis: Naval Institute Press, 1984) and *U.S. Battleships: An Illustrated Design History* (Annapolis: Naval Institute Press, 1985).

20. At the same time, about 1948, the converted *Fletcher*-class destroyer was proposed as the successor to the wartime destroyer escort.

21. In 1959, the Bureaus of Ordnance and Aeronautics merged to become the Bureau of Naval Weapons (BuWeps).

22. Reportedly, BuShips recruited several experienced yacht designers around 1964, when it became clear that many small craft designs would be required.

Chapter 2

1. Convoy ASW was very different. The chief indication of a submarine was the track of its torpedo, and the usual escort tactic was to run back down the track, peppering the presumed submarine position with depth charges. Speed and good visibility were clearly more important than silencing. However, it was understood that a combination of an effective hydrophone and good sea-keeping, such as a destroyer had, would be valuable.

2. This frequency range was chosen in hopes of excluding self-noise and sea noises. However, it corresponded to a range of long wavelengths, about 4 to 10 feet, so that any directional device at a similar frequency had to be large. Modern active sonars operate at substantially higher frequencies, "low frequency" corresponding to about 3,000 to 5,000 Hertz (3 to 5 kc).

3. Four of these boats, built by Britt Brothers of Lynn, Massachusetts, became patrol boats 1, 2, 4, and 5 (SP 45, 409, 8, and 29). There was no patrol boat 3. Swasey himself contributed no. 1. Britt also built the 45-foot *Boy Scout* (SP 53) and *Dionra* (SP 66) to Swasey's 45-foot navy design.

4. Roosevelt went so far as to warn his wife to move their family inland should the Germans begin shelling American ports.

5. The reality was that the newer U-cruisers, which displaced as much as 2,000 tons on the surface, mounted two 5.9-inch guns. Speed did not exceed 16 knots, though it was 17 in some smaller, oceangoing U-boats.

6. See Friedman, *U.S. Destroyers: An Illustrated Design History* (Annapolis: Naval Institute Press, 1981), p. 42, for details of this design.

7. At this time, the board commented that the problem had been nearly solved some months before when the U-boats operated close inshore, but that now, when they were operating 400 to 500 miles from Great Britain, the usual converted fishing craft could not cope. Hence, the series of radical measures such as the North Sea mine barrier taken at this time.

8. Although no U.S. design from World War II required such drastic measures, the Japanese adopted similar practices in their emergency-built coastal escorts (*kaibokan*), landing ships, and standardized cargo ships and tankers, as did the Soviets in their T-301 class coastal minesweepers.

9. Late in December, Chief Constructor Taylor decided to eliminate knuckles at the waterline, that is, to fair all the cross-sections, slightly complicating construction. Although the knuckles seemed to have little effect in model tests, Taylor suspected that on a full-scale ship they might add resistance by producing eddies. The model basin estimated that this fairing saved 4 percent in resistance.

10. For a more complete account of Ford's travail, see D. A. Hounshell, "Ford Eagle Boats and Mass Production," in M. R. Smith, ed., *Military Enterprise and Technological Change* (Cambridge: MIT Press, 1985).

Chapter 3

1. PC 451 had two 75-kw units. By way of contrast, a PCE had one 100-kw AC, two 60-kw AC, and three 20-kw DC diesel generators (PCEs 867–69, 871–73, 877, 891–97 had three 60-kw units). A PCS had one 30-kw and one 60-kw AC diesel generator, and an SC, two 20-kw DC diesel units.

2. Of these ships, PC 454–56 became coastal yachts PYc 46–48 on 15 July 1943; PC 457 was lost on 14 August 1941 without having been redesignated; PC 458 became PYc 49 on 15 July 1943; PC 459 became PYc 18 on 20 January 1941; PC 460 became PYc 50 on 15 July 1943; PC 509 became PYc 51 on 15 July 1943; PC 510 became YP 105 on 22 September 1941; PC 523 became YP 77 on 10 January 1941; and PC 826 became PYc 52 on 15 July 1943.

3. Wartime directives were issued by the VCNO, Admiral F. J. Horne.

4. The board was formed in 1931. In June 1941, it was given cognizance over patrol vessels, coastal minecraft, gate vessels, and net tenders, reflecting the new importance of such coastal craft. The General Board retained power over PTs.

5. The "1799" program consisted of 250 destroyer escorts, 48 fleet minesweepers (AM), 150 motor minesweepers (YMS), 150 PCEs, 300 LSTs, 10 LSDs, 300 LCTs, 500 Higgins landing boats (LCP), and 91 miscellaneous craft—salvage vessels, harbor craft, rescue tugs, motor launches, and crash boats. The program was approved on 6 February 1942 with the proviso that any of these vessels could be taken over by the U.S. Navy before completion. A total of 381 were, by August 1943. Note that as of January 1942, the estimated requirements of the British—with their designations in parentheses where applicable—amounted to 30 escort carriers (BAVG), 520 destroyer escorts (BDE), 150 PCEs, 52 PT boats (BPT), 80 fleet minesweepers (BAM), 250 motor minesweepers (BYMS), 17 artillery transports (BAPM, later LSD), 300 LSTs, 400 LCTs, 150 Higgins LCMs, 400 Higgins "Eureka" landing boats (LCP), 91 aircraft rescue boats, 74 72-foot YPs (British HDMLs), 30 boom vessels, 17 salvage vessels (BARS), 23 oceangoing tugs (BAT), and 4 depot and repair ships.

6. The 1 September directive initially canceled all 72 SCs approved on 16 April 1943. However, the VCNO ordered 35 retained on 1 October 1943.

7. The first new fleet minesweepers, the *Ravens* (AM 55 and 56), had paired Fairbanks-Morse 900-bhp geared diesels (38RD8). Succeeding ships each had four diesels in a diesel-electric arrangement. The austere 180-footers had paired engines such as Alco 539s (865 bhp), in effect half the power plant of a 220-foot *Auk* (AM 57).

8. In 1965, the Spanish navy built three diesel-engined versions of the 83-footer armed with Mousetraps and a 20mm cannon.

Chapter 4

1. The CNO, Admiral William H. Standley, was acting secretary of the navy because the incumbent, Claude Swanson, was ill.

Chapter 5

1. Andrew Jackson Higgins was a flamboyant New Orleans boat builder who specialized before the war in shallow-draft craft. He is probably best known for having designed the standard U.S. landing craft of World War II, which will be described in a subsequent volume in this series. Higgins claimed that he supplied both the Coast Guard and the rumrunners during Prohibition, and that this experience guided his PT boat designs.

2. Higgins boats built after about 1943 had flat buttocks.

3. So did five boats Higgins built for Finland, RBs 1–5. Taken over by the U.S. government, they were transferred under lend-lease as MASBs 69–73.

4. Scott-Paine's brochure offered one quadruple 0.303-caliber turret and one 0.8-inch single-gun turret. It claimed that power drive was needed in so lively a boat, although the turrets were not gyro-stabilized.

5. At the time, BuShips estimated that 500 nautical miles at 20 knots would require 1,575 gallons, but the Elco boat had a capacity of 3,000, some of which might be traded for military weights.

6. It is clear in retrospect that the key issue was the much greater displacement of the Higgins boat. Since the Higgins was generally rated as roughly equivalent to the Elco in displacement, it is difficult to see why the boat at MTBSTC weighed 10 tons more. In tests a year later, a Higgins at the same displacement as an Elco showed much the same performance.

7. MTBSTC disagreed. The one great advantage of the Higgins was its maneuverability; it could "almost reverse course in [its] own wake." The best solution would be a rudder-angle indicator, already proposed by operating personnel but rejected by BuShips. The indicator was fitted to Elco boats.

8. The Elco boat had full exhaust mufflers in the transom at the stern. The Higgins had a side exhaust and could exhaust under water to muffle its noise. However, the silencing effect was nullified when the boat rolled. The Higgins arrangement precluded the installation of Elco-type mufflers.

Chapter 6

1. In January 1944, MTBRon 25 requested 40mm guns for PTs 345 and 349–51. Two more boats, PTs 352–53, in Panama awaiting shipment, were already so fitted. The boats of this squadron, PTs 344–55, were fitted with experimental sextuple 2.36-inch (bazooka) rocket launchers on converted 0.30-caliber machine gun mounts. PTs 344 and 346–47 had four Mark 13 torpedoes and no 40mm gun aft; the others had two torpedoes and the 40mm gun aft. All had an experimental 37mm mount forward. Late in 1943, 40mm guns were also installed in PT 314 (Melville) and PTs 328 and 330 of MTBRon 21, South Pacific. Precise dates of authorization and installation are not known.

2. This list is probably not exhaustive. There is evidence, for example, that PT 109 (MTBRon 2, South Pacific) mounted a single-shot 37mm cannon before its loss on 2 August 1943. The report of three Rendova boats is dated 9 August.

3. In an early Higgins, for example, the motor-generators cut in at 1,200 rpm (main engines). On patrol at low speed, the main engines typically operated closer to 600 rpm, and the entire load was borne by the auxiliary unit. Operating personnel complained that the auxiliaries were noisy and should not be run continuously. By 1943, new low-speed cut-in motor-generators were being installed.

Chapter 7

1. Wartime PTs were transported overseas as deck cargo. They were unloaded and made operational at established bases, so they did not go directly from tanker or freighter into combat. By way of contrast, a postwar amphibious task force had to unload any accompanying small combatant craft in the combat area. That would imply transport in the well deck of an LSD (landing ship, dock), which normally carried large landing craft (LCU).

2. A new Mark 34 torpedo director was being tested in February 1945. Existing boats had a Mark 33, which worked on the assumption of a "collision course" attack. It was not considered satisfactory for postwar PTs. The available alternative was the submarine torpedo data computer (Mark 3), which was limited to a ship speed of 25 knots, a target speed of up to 40 knots, a torpedo speed of 25 to 60 knots, and a range of about 8,000 yards. Solution time could not exceed 240 seconds, which restricted torpedo range to 6,000 yards for a 45-knot torpedo, or 1,500 yards per minute times four minutes.

3. Justifying procurement of this new engine in March 1945, BuShips argued that even though barge-busting had reduced the value of speed, acceleration was still vital. The new engine could be installed in existing direct-drive boats: all Higgins boats and the Elco PTs 613–24 and PTs 761–90. The first four test engines were then scheduled to be delivered on 1 December 1945 for installation in the last Elco boat, PT 790. The boat, however, was canceled.

In the spring of 1945, the comparative performance of Packards was as follows:

	Standard Engine (W-14)		Under Test (W-50)		Design (W-100)	
	BHP	RPM	BHP	RPM	BHP	RPM
Emergency (15 min)	1,500	2,500	1,800	2,800	2,500	2,800
Maximum (1 hour)	1,350	2,400	1,500	2,500	2,000	2,500
Continuous	950	2,000	1,050	2,000	1,500	2,000

The W-50 was sometimes described as a 1,850-bhp engine.

4. Fifty-two thousand pounds was divided into 37,000 pounds of ordnance, 10,000 of armor, and 4,200 of electronics. Wartime PTs typically carried about 30,000 pounds of ordnance (see table 7-1).

5. The "CIC" was actually a combined radar, radio, and chart room. It was to be large enough to hold four men wearing life jackets as well as a chart table and desk space for radios and radars. It also had to have easy access to the bridge.

6. The new boats were designed to accommodate at least three officers and nineteen enlisted men and to provide a suitable margin for future growth in armament and electronics.

7. One officer, however, argued that the 40mm, if mounted forward of the bridge, would obstruct vision and blow smoke and gas into the eyes of the helmsman and skipper. He favored the existing arrangement of one 37mm forward and one lightweight 40mm aft, but he also wanted rockets; four torpedo racks (two torpedoes to be carried normally, four in fleet engagements); four twin 0.50-caliber machine guns (two mounts on each side); three 20mm guns (one on the centerline forward of the bridge, one on each side just abaft the forward torpedo); two 0.30-caliber machine guns (one on each bridge wing); two depth charges for attacking submarines or, at close range, surface ships; and one mortar.

8. In a planing boat, the area amidships riding the surface of the water would show particularly violent acceleration, the stern—directly under which the hull would be immersed—much less so.

9. It is not entirely clear whether this was for an aluminum boat. An 11 February 1946 design sheet, giving estimated weights for the equipment of the new ship, suggested that a full load figure of 205,000 pounds for a wooden PT would be equivalent to 179,000 for an aluminum hull: 20,000 would be deducted for structural weight and 14,000 for unnecessary fuel to drive that extra weight; 4,000 would be added back for extra structure.

10. Mark 107 was a nine-tube 5-inch rocket launcher. It was spin-stabilized, locally controlled, and power driven, and had a closed breech to avoid back blast. It used the same mounting base and weighed about the same as the 40mm single mount.

11. Later, BuOrd specified one of two existing pedestal mounts, the Mark 24 or the Mark 25. The 35mm gun was the planned replacement for the existing twin 20mm.

12. Admiral C. F. Brand, code 300, actually saved the program at this point. He avoided any further discussion with the CNO for fear that the latter would cancel the project. Funds would surely be cut, and it seemed best simply to wait two months before

awarding any contracts. However, it also seemed unlikely that Annapolis Yacht would be able to retain its technical men another two months pending contract award. Brand recommended early awards to Annapolis and Philadelphia.

13. Aircraft engines had to be used because specially designed marine gas turbines were much too heavy. For example, a plant then planned for a destroyer escort weighed 180,000 pounds for 10,000 shp, or 18 pounds/shp.

Chapter 8

1. Conventional World War II submarines were generally limited to well under 10 knots at maximum underwater power (one hour rate) and usually operated at much slower speeds. The German type VIIC, for example, was rated at 7.5 knots, so that a 17-knot subchaser enjoyed an adequate 10-knot margin. Postwar fast submarines, rated at 16 or even 18 knots under water, could evade ASW ships by heading submerged into seas. Hence, the usual postwar requirement for a sustained speed of 25 or 30 knots in a seaway, a speed performance no conventional small craft could match. The obvious next step was more threatening: a 25-knot closed-cycle or nuclear submarine.

2. Wartime Hedgehogs could be tilted from port to starboard; the on-mount operator could be operated manually to compensate for roll, like a gun. The postwar Mark 15 was power-stabilized in roll and pitch.

3. It did require substantial beam, which is why the postwar export version of the PC had a single, trainable Mark 15 Hedgehog.

4. The location was later abandoned because a bow sonar suffered from "quenching" as it rose when the ship pitched in a heavy sea.

5. The first low-frequency sonar, the SQS-4, was ready for testing only in the summer of 1950. At this stage, it operated at 10 to 14 kc (later 8) and had a 4- or 5-foot diameter transducer. It was expected to detect a submarine at 3,000 to 3,500 yards and was eventually credited with a nominal range of 5,000. The SQS-4 represented the upper limit of what could be installed in a World War II destroyer escort hull.

6. Reliability was not a major issue, because damage severe enough to stop one propeller would probably put both shafts out of action. Preliminary Design had recently decided this issue in the case of the *Dealey*-class ocean escort.

7. The single Mark 15 was more than three times as heavy as a single Mark 11, but the double Mark 11 required twice the ammunition, so the associated net weight was greater. The total topside weight of the double Mark 11 was still 5,500 pounds less than that of the single Mark 15. However, all of these projectors had such low trajectories that the centerline Mark 15 would have to be mounted on a raised platform to clear the forward 3-inch gun. It could clear a twin 40mm gun on the forecastle. Alternatively, the ship could be lengthened 7 feet and the Mark 15 installed at forecastle-deck level.

8. The SQS-17 was a scanning sonar, which sent out a pulse in all directions at once, and was therefore typical of U.S. postwar practice. The SQS-9 sent out a 40-degree searchlight beam, scanning it in 5-degree steps to give a full sweep every two minutes; it was credited with an accuracy of 3 degrees. This searchlight feature made it possible to use a much smaller transducer.

9. BuPers suggested figures as high as two officers and forty-nine enlisted men; Preliminary Design complained that a 136-foot PCS could not take more than two officers and thirty-eight enlisted men, and that any further growth would require a much larger hull. Preliminary Design considered the requested complements the single major force driving up size and therefore cost; according to an April 1957 memo, it planned to press for sufficient reductions to make the 95-foot Coast Guard boat suffice for the SC mission and the 136-foot hull suffice for the PCS mission. Otherwise the SC would grow to 105 feet at the waterline, the PCS to 150 feet.

10. The Soviets actually completed their first "November"-class nuclear attack submarine in August 1958. It encountered operational problems, and the U.S. Navy did not consider it fully effective until 1963. The 1965 date was used in several internal navy studies, written in the late 1950s, of the implications of expected Soviet nuclear submarine development.

11. According to the minutes of a BuShips hydrofoil presentation before the antisubmarine plans and policies group (Op-312) on 4 August 1958, the ASW potential of hydrofoils was first recognized around 1954. There were also indications of Soviet interest in a 50-ton ASW hydrofoil at about this time. The earliest U.S. proposals dated from January 1955, and in February 1956, ONR began operational evaluation of various hydrofoil systems for ASW. Even so, later official accounts of the U.S. hydrofoil program show that, prior to about 1957, work concentrated on landing craft, which could expect to operate in calmer water.

12. ONR tested the AQS-10 dipping sonar using the XCH-4 experimental hydrofoil. These tests in turn inspired the PC(H) sonar performance estimates.

13. To supplement but not replace the existing navy marine gas turbine program, the BuShips program included the development of a special lightweight engine: the Pratt & Whitney FT4A, designed for a maximum output of 30,000 shp (20,000 continuously). It was also available for other high-performance craft such as the Seahawk destroyer and the new air-cushion (surface-effect) vehicles.

14. Grumman built experimental supercavitating foils for FRESH-1, but that boat was retired in 1964 and they were never fitted. Fully supercavitating foils promised high maximum speeds but their lift probably would not suffice at any but high hull speeds. Grumman tried to solve the problem with a new concept, Transit, in which cavitation began at the foil tips and then moved smoothly towards the roots. Transit foils were manufactured for FRESH-1 but were never fitted.

15. For a time in 1962–63, there was also interest in a hydrofoil electronic-deception craft tentatively designated AGHR. It could use its open-ocean speed to carry electronic deception equipment far from a task force. As of 1963, several AGHRs were proposed for the FY 66 and later programs.

Chapter 9

1. The earliest MAP transfers included Elco PTs for Norway. The Mutual Defense Assistance Program of FY 52 originally included six Elco PTs to be built for Denmark. However, the Danes elected to buy boats modeled on the wartime German S-type (U.S. accounting numbers were PTs 813–18). Three more boats, PTs 819–21, were built in the 1960s.

2. The French numbered their amphibious craft in blocks within an L9000 series: 9000 for LSTs, 9010 and 9050 for LSMs, 9020 for LSSLs (and for some LSILs), 9030 for MDAP-supplied LSILs (with 9029 used later because the series became too lengthy), 9040 for purchased LSILs, 9060 and 9080 for British-type LCTs, 9090 for ex–U.S. LCTs (later redesignated LCUs), 9100 for the single LCC, and 9160 for LCM monitors. A headquarters LCI was numbered 9055, and 9059 was an ex-British LCG (landing craft, gun).

3. In January 1947, the French mounted the first riverine operation of the war, with two LCTs, one LCI, and four LCMs, a force that was to be typical. A 1948 operation employed four LCMs and two LCAs carrying two companies of troops. These forces may be compared with the later U.S. riverine assault squadrons described in chapter 13. By 1950, the French had about 165 ships and craft in Indochina, including one light cruiser (and occasionally a carrier), seven corvettes, one repair ship, and three LSTs. The remainder were small craft, forty of them organized into six naval assault divisions. The French also had 72-foot, ex-British harbor defense motor launches (HDML); in 1954, twenty-two out of a total of twenty-four served in Vietnam, and by 1954 seven had been armored.

4. They were designated Vedettes 661–87. Some surviving records indicate that the French also received their six 83-foot ex–U.S. Coast Guard cutters in 1951.

5. STCAN was a corruption of STCN (*service technique construction navale*), the French equivalent of BuShips. It has also been rendered "St. Cann."

6. As of 1955, the Vietnamese navy consisted of one PC, three YMSs, two LSSLs, four LSILs, four LCUs, twenty-eight LCMs, twenty-one LCVPs, and two YTL river minesweepers. The Vietnamese marines operated 12 LCMs, 77 LCVPs, and over 170 smaller craft in three river companies. At this time, the Cambodian navy had one PC, one small patrol boat (YP), three LCMs, and twenty-four armored LCVPs. Other craft of the former French Indo-Chinese

naval forces were out of commission and in U.S. hands in the area. They were later transferred to Vietnam.

7. The metacentric height of the Coast Guard 95-footer was 4.3 feet, compared with 1.83 for the 110-foot wooden PGM (ex-SC) and 4.5 for PGM 39. An internal BuShips memo grumbled that this absurd situation came of using a private (Tacoma Boat) rather than an in-house design.

8. They resembled midget PT boats. SV stood for *servicio de vigilancia*, or Coast Guard. Details were as follows (dimensions are given in feet-inches): 40-8 (LOA) 38-4 (LWL) x 12-0 (WL) x 3-0 (maximum); two GM 6-71 diesels, 470 bhp for 25 knots. These craft carried a single, water-cooled 0.50-caliber machine gun forward and had a prominent radar mast aft.

9. Sewart, later bought by Teledyne, operated briefly as Teledyne-Sewart or as the Sewart division of Teledyne; then it was bought out by several employees and became Swiftships. See appendix E for the latter part of its history.

10. These boats all resembled the later Swift. They displaced 18,500 pounds (40 x 12 x 3 feet) and made 30 knots on twin General Motors diesels.

Chapter 10

1. PRC torpedo boats sank several Taiwanese warships, the most notable of them the destroyer escort *Tai Ping* in December 1954, two motor gunboats in January 1955, and one LST in August 1958. The Taiwanese reaction was to obtain six torpedo boats, four commercially from the United States and two from Japan.

2. The LCSRs were built in FY 61 by reprogramming money originally earmarked to buy conventional landing craft (LCM6s). Because they were intended to support frogmen (UDTs) clearing the way for amphibious assaults, they had to fit the well deck of an LSD. However, they were also ideal for inserting small numbers of guerrillas, and as a result, they were the first boats assigned to the new SEALs. Characteristics included a range of about 200 nautical miles, a sustained speed of 30 knots, and a capacity for twenty-two men and 2,500 pounds of equipment or explosives to be carried through sea state 3 for six to eight hours at high speed. Armament would consist of two machine guns of the heaviest possible caliber. In 1964, United Boat Builders completed fourteen LCSRs (C-1310-1319 and C-5842-5845), of which four, C-1310–12 and C-1317, were transferred to Israel in 1975. C-1313, assigned to the Naval Air Development Center at Key West, burned and sank in September 1974, soon after its arrival. It was replaced by the last active unit, C-1313, which was approved for disposal in 1984.

3. PT 809 had had new engines installed, low-powered diesels for special duty on the Potomac as the presidential chase boat. One World War II PT survived at the mine defense laboratory in Panama City, but it was considered unsuitable for reactivation.

4. BuShips considered two domestic diesel engines, the turbocharged 3,000-bhp Fairbanks-Morse 38A6 3/4 (17,000 pounds) and the 900-bhp Curtis-Wright 12V142 (4,050 pounds), four of which would be needed per shaft. At this time, the Fairbanks-Morse had been run at 3,000 bhp only enough to demonstrate that it could actually produce that much power. The British Deltic was in widespread service, and it was considerably lighter.

5. The Eagle was built specifically for United Aircraft to test new marine gas turbines, gears, propellers, and controls. It used an innovative all-titanium supercavitating propeller. Powered by a 3,200-shp FT 12 gas turbine derived from the JT 12 aircraft engine, the Eagle achieved 55 mph (about 48 knots) on trials in 1965–66. It resembled a Swift but with a small stack for the gas turbine exhaust. Following the successful Eagle trials, United Aircraft ordered the 80-foot Double Eagle (two FT 12 A-3 gas turbines) from General Dynamics Quincy; it was launched in June 1966 and evaluated as a possible U.S. Navy patrol boat. Reported maximum speed was over 50 miles per hour, or around 44 knots.

6. The Garrett 990 weighed about a third as much as the Deltic—5,000 as opposed to 15,750 pounds wet, including the gear box—but was less efficient. Preliminary estimates suggested that, propelled by two Garretts, an Osprey could achieve 49.5 knots fully loaded (224,130 pounds), compared with 31.5 knots for the Deltic-powered boat (245,630 pounds). On the same fuel load of 9,450 gallons, the gas turbine boat could make only 543 nautical miles at full power, compared with 771 for the Deltic boat. Note that the 990 was substantially lighter than other gas turbines. The Avco TF-25 (2,500 shp), TF-35 (3,100 shp), and TF-40 (3,600 shp) would each have weighed about 10,000 pounds wet, with gearing. The Deltic in turn was lighter per pound than its nearest diesel rival, the German MTU M820V672 (14,561 pounds for 2,835 bhp).

7. PGM 91 was an MDAP gunboat for Vietnam.

8. M20 had air, surface, and torpedo capability; M22 eliminated torpedo control; M26 eliminated both air and torpedo control. In October 1962, BuWeps asked to procure eight M22s. The U.S. Mark 87 was a modified M22; the current Mark 92 series used in the *Perry*-class frigate and in the *Pegasus*-class missile hydrofoil has been derived from the Dutch M20 series.

9. In 1984, however, Stinger missiles, successors of the Redeye, were deployed as point defenses aboard U.S. ships in the Mediterranean.

Chapter 11

1. Even before Kennedy's call for action, OpNav's strategic plans division, prompted by events in Vietnam and Laos, made a study of the special requirements of counterinsurgency warfare. Completed in March 1961, the study proposed the SEALs, visualizing them not so much as an elite counterguerrilla force but as a specialist unit that would spread counterinsurgency ideas through the fleet. On 15 June 1961 the CNO, Admiral Arleigh Burke, issued a formal directive establishing two SEAL teams, each of ten officers and fifty enlisted men. Training began immediately.

2. This was Project Agile. Initially concentrating on Southeast Asia as a pilot project, Agile soon started seeking technological solutions to the problems of Vietnam. ARPA in turn brought in numerous civilian organizations such as RAND. The widespread use of existing civilian research companies contrasts with the ad hoc use of new civilian laboratories in World War II. ARPA's involvement as a lead agency points to a major difference between Vietnam and earlier American wars: strategic direction by the civilian Office of the Secretary of Defense rather than by the service chiefs.

3. Around 1967, the marines actively considered a division-scale landing in North Vietnam to relieve pressure on the South.

4. Operation Plan 34A, first proposed by the JCS in May 1963 as a series of hit and run, covert, commando operations, was elaborated by CinCPac in December 1963 as a means of putting pressure on North Vietnam. The twelve-month program began on 1 February 1964. It is important to this history because it motivated the initial purchase of Swifts and also the early use of the PTFs described in chapter 10.

5. In 1963, there were two river assault groups at Saigon and one each at Mytho, Vinh Long, Can Tho, and Long Xuyen. In 1965, Groups 21 and 27 were based at My Tho; 22 and 28 at Nha Be; 23 at Vinh Long; 24 at Saigon; 25 and 29 at Cantho; and 26 at Long Xuyen.

6. The 55-foot command junk was designed by a U.S. advisor, Lieutenant Commander W. E. Hanks, USNR, working with the Combat Test and Evaluation Center. Its lines were derived from those of a "deadrise" fishing and oyster boat from the Chesapeake Bay, and it was powered by one standard American 6V71 225-bhp diesel for a maximum speed of 12 knots (endurance was 1,050 nautical miles). Dimensions were as follows, given in feet-inches: 55-9 (LOA) × 15-9 × 2-8 (maximum, with center board up) and displacement, 19 tons. Sails were removed in 1966, when all remaining unpowered junks were discarded. Standard armament was initially one 0.50 and one 0.30-caliber machine gun, but South Vietnam's CNO recalled the 0.50s in May 1963, replacing many of them with 0.30-caliber machine guns. They were later restored, and a 60mm mortar (Mark 4) was added toward the end of the war. Altogether, forty-six command junks were built.

7. The proposed solution, a dunking sonar, never materialized.

8. The *Mustang* was numbered 51NS691 in the FY 69 series, but she was probably purchased locally much earlier. It was typical for such boats to be registered only when they were reported, which might be many years after their purchase. The *Mustang*'s official number at Panama City was 293-641. The three Operation Plan 34A boats do not appear in official navy records; they may

have been bought through the CIA. Many other early Vietnam-era craft such as the 222 SSBs do not appear in these records either.

9. Data are somewhat incomplete, in that lists of U.S. small craft include several unspecified experimental boats that may date from this period. The RCVP plans from which the drawings in this chapter are adapted have survived, but no photographs of the RCVP have come to light.

10. In Vietnam, a boat was classified as a junk if a water buffalo could stand athwartships, as a sampan if it could not.

Chapter 12

1. They had originally been assigned to Operation Pla 34A.

2. The 82-footer was not too different in size and sea-keeping from the export PGM; the latter would have been used had it been available in time in sufficient numbers. As in the case of the Swift, the most important advantage of the 82-footer was that it was available when urgently needed.

3. The LST-borne helicopters (UH-1B) were initially to have been army-manned, but in fact they were lent by the army and manned by navy pilots provided from the CVS air group reductions directed for the FY 66 program.

4. In February 1969, expecting to use Swifts up-river for Sealords, Admiral Zumwalt urgently sought armor and more firepower. He hoped to adapt the minigun (GUA-2B/A, a 7.62mm Gatling), training personnel in-country and installing the gun with locally manufactured bow pintles so it could be removed. Advised that tests would be required and that the gun would not be available until FY 70, beginning 1 July 1969, Zumwalt agreed to abandon it and rely on existing armament plus the automatic, long-range grenade launcher (Mark 19). At least one PCF was actually fitted with a minigun.

5. Dimensions in feet-inches were 38-9 x 23-0 x 16-6 (height) on the cushion. The air cushion itself was 28.6 x 17.25 feet (493 square feet), and the skirt could be lifted hydraulically at four points so that the craft could turn or "slide" over a surface.

Typical weights were as follows:

	Design	Combat	
Gross Weight	15,865	19,500	
Empty	9,578	10,084	
Fuel	2,120	3,074	
Payload	3,967	5,550	
Structure	8,679	8,609	(forward ballast tanks, 70 pounds, removed)
Cushion Pressure	31.8	36.5	per square foot

6. The PACVs always used the call sign Monsters afterward.

7. That was why Swifts were initially rejected for Game Warden: also too high in the water, they were good targets. Some operating inland during Sealords were severely punished. The craft were also considered unsuited to inland waters because of their relatively deep draft of 5 feet and because their propellers were not protected by skegs. On rivers, they lost their usual advantages of high speed and maneuverability. Their engines had to operate at low rpm to avoid overloading, and their generator coolers became clogged with weeds and debris. Engines overheated and were extremely noisy. A report late in the war suggested that the surprise of Sealords was that the Swifts had not done much worse, that they had been saved only through the skill and alertness of their crews.

8. Hughes proposed a towed ocean surveillance array in February 1962. Its WQR-1(XN-1) was tested off Hawaii on 30 May 1965, towed by the fleet tug *Hitchiti* (ATF 103) at up to 14 knots. It successfully detected a snorkeling submarine at one convergence zone and a surfaced submarine at an even greater range. Hughes promised ranges as great as 30 nautical miles at tow speeds of about 10 knots, and the array could be towed at up to 20 knots. This work ultimately led to some of the current tactical and surveillance arrays, such as those installed in the towed-array ocean surveillance ships (T-AGOS).

Chapter 13

1. The refuelers were conventional LCMs carrying 1,200-gallon fuel tanks and two 55-gallon drums of lubricating oil. Some ATCs were also fitted as refuelers. In addition to their own 450-gallon fuel tanks, monitors and CCBs each carried auxiliary fuel tanks of 1,000 gallons to refuel other craft.

2. The original concept called for four to six days of operations followed by two to three days of rest and refit. The riverine force proved so successful that instead it was used continuously. Typically, transits to and from the objective area took two to six hours, and boats had to patrol and deliver fire support throughout the operation itself. Upon returning to base, crews repaired their boats in preparation for another operation a day later. Under these circumstances, the 75-percent availability rate, which included boat overhauls, was a considerable achievement.

3. High-hardness steel was also much less expensive; dual-hardness steel cost about $275 per pound in 1967. According to Richard Hartley, the program manager, total program cost fell from about $400 million to $330 million.

4. A loaded 57mm recoilless rifle weighed 44 pounds, equivalent to the weight of an RPG-2 plus eleven rounds. This was also the weight of a longer-range RPG-7 with six rounds. The RPG could penetrate 3.5 times as much armor as the rifle.

5. In the same action, M-92-2 was hit by a Heat round that penetrated the opening between the 20mm and 0.50-caliber mounts, killing the captain and cox. An ATC took an RPG-7 round through its bow ramp, which severed the ramp winch cable and wounded thirty army personnel in the well deck. Another ATC was hit near the waterline by an RPG-7. It triggered on bar armor, penetrated the bulkhead, and created a shrapnel effect in the empty well deck.

6. A preliminary requirement was probably issued earlier. In June 1966, A. G. Eldredge, then equipment coordinator for unconventional warfare in NavShips, published an article on riverine and special warfare craft in the *Naval Engineers Journal*. He presented a Food Machinery Corporation (FMC) concept for an "armored river patrol craft" using interchangeable weapons pods, including one with a 25mm cannon and one with a pair of 90mm recoilless cannon. He also illustrated a United Aircraft (Sikorsky) armored patrol boat, which was *not* obviously related to that company's later ASPB Mark II. Uniflite and Emerson Electric later built their own private-venture, 36-foot R-ASP (riverine assault support patrol) armed with two remote-control gun mounts. Each mount could carry up to three weapons—a 20mm machine gun, a 40mm grenade launcher, or a 7.62mm minigun. The controls were located in a pilothouse protected against 0.50-caliber armor-piercing projectiles and 57mm shaped charges. R-ASP was demonstrated at Bellingham on 20 June 1968. Its mounts and directors were possible candidates for second-generation craft (see appendix D for details on these weapons).

7. Silencing was ordered in response to an urgent request from commander, amphibious forces, Pacific, in May 1967. A formal operational requirement followed in November and led to boat alterations for the Swift and Mark I "plastic" (PBR), which quieted it to the level of the PBR Mark II. The ASPB was also silenced. Mufflers and water-cooled exhausts had been used to silence LCM(6)s. However, ATCs could not be silenced because their helicopter platforms, which weighed 10,000 pounds, consumed too much buoyancy. Given the ATCs' operational profiles (an entire riverine force could not be very quiet), they were not worth silencing.

Chapter 14

1. For example, the first element of SEAL team 2 to operate in Vietnam was mobile training team 10-62 (1962).

2. This two-stage delivery concept was originally developed by the UDTs. The special feature of the LCSR was the new Fulton

swimmer-retrieval system; the fast-moving boat snagged a line held by swimmers, pulling them onto a towed sled that could then be hauled through the craft's transom.

3. In 1964, the standard allowance was fifteen swimmer support boats (SSB) per SEAL team. The standard existing boat was a plastic Mark IV (a plastic Mark V was being tested). After thirty had been manufactured, the Mark IV was rejected because it lacked sufficient freeboard aft to bolt down its heavy outboard motor. The standard SSB of the Vietnam War was a 16-foot Boston whaler, also used for harbor patrols and UDT and explosive disposal. Typically, the SSB was steered from amidships, had an enlisted crew of two, and was armed with a single M60 machine gun forward. Photographs of the Seafox suggest that the current U.S. swimmer delivery boat is a rubber raft. Late in the Vietnam War, the SEALs also used the Kenner ski barge, which was powered by two Evinrude outboards.

4. For example, SEAL team 2 sent Detachment Alfa, with an officer in charge and two platoons each of two officers and ten enlisted men, to Vietnam in January 1967. By early 1969, each team had six detachments in Vietnam.

5. Characteristics of the navy-designed prototype were 36-3 (feet-inches) x 11-0 x 3-0; 23,962 pounds; two ST6-J70 gas turbines of 565 shp each; 25 knots; and 200 nautical miles at 25 knots.

6. At this time, the STABs were nearly unique in Vietnam for having gasoline engines. They consumed as much as 50 gallons per hour, and a special motorgas pontoon barge had to be provided for them. The STAB engine was also unique; the squadron's approach to the spares problem was to bring two of every important item with it. Even so, after June 1970, maintenance was difficult because spares had to be taken from the support barge on the Mekong, where spares had been concentrated when the squadron operated as a single unit, to two naval support activities servicing separate divisions at Dong Tam and Nha Be. STABs were not turned over to the Vietnamese; with so many unique features, they couldn't easily be adapted to the existing logistical system.

7. The propeller versus water jet controversy raged throughout the Vietnam War. One designer stated that everyone who had propellers wanted water jets, and everyone who had water jets wanted propellers. Propellers, unless they were in tunnels, resulted in deeper draft; water jets could be clogged by vegetation and other obstructions.

8. Commander Ferguson argued, too, that sometimes existing rules of engagement virtually negated the ambush tactic, since it was necessary to obtain permission by radio telephone from the local Vietnamese authorities before opening fire. That might take as much as ten or twenty minutes, and voice radio could not always be entirely secure. On one occasion, a WBGP STAB was attacked preemptively while awaiting permission to fire; three crewmen died, a fourth was severely wounded, and the embarked patrol officer was slightly injured.

9. It might be argued that this change reflected an actual U.S. decision to risk widening the war by sealing the Cambodian border. Thus what seemed on the STABRon scale to be a loss of U.S. aggressiveness was, on a larger scale, a more belligerent and even self-confident policy.

Chapter 15

1. Although the PHM designers started with the PC(H) as a proven or state-of-the-art vehicle, a NavSec PHM design history of March 1971 reported "so many unanticipated deficiencies that the PC(H) was transferred to special status as a research platform under NSRDC." The hydrofoil special trials unit was established at Puget Sound Naval Shipyard as an NSRDC field activity to develop and debug the AGEH and the PC(H). Once the latter had been debugged, the PHM designers tried to avoid any major departure from its proven configuration.

2. Fascination with the tattletale role can be traced to a perception, at this time, that the Soviets were multiplying their surface (anticarrier) missile ships. Without specialized and inexpensive U.S. tattletales, the mere existence of these Soviet combatants would strip carrier task forces of their AA and ASW screens. The basis of the perception was the appearance in Krivak-, Kresta II-, and Kara-class ships of large numbers of relatively short missile tubes, then believed to contain a new short-range antiship missile, the SS-N-10. In fact, there were no SS-N-10; the tubes contained the SS-N-14, an ASW weapon similar in concept to the Anglo-Australian Ikara.

3. Note that, contrary to popular belief, this was not a consequence of the sinking of the Israeli destroyer *Eilat* by a Soviet-supplied, Egyptian SS-N-2 or STYX missile. The sinking occurred later, on 21 October 1967.

4. The hydrofoil-borne tactical deception mission was proposed as early as 1963. Because a hydrofoil enjoyed large-ship sea-keeping qualities on a small displacement, it could inexpensively simulate a larger ship. And because its own signature was so small, an imposed large-ship signature could be particularly effective. It is not clear whether any specialized deception hydrofoil was ever designed. One problem was endurance; another was that a flying hydrofoil generally traveled much faster than the large ships it simulated—indeed, its good sea-keeping came from that high speed.

5. The two-mode sonar could either operate at depth or trail at shallow depth. It was tested by the *High Point* around October 1970 at 42 knots. Without the sonar, PXH-A would rely on CASS (SSQ-50) and DICASS (SSQ-62) command-activated active sonobuoys (nondirectional and directional, respectively). It would carry seven CASS, put over the side by hand, and twenty DICASS, fired 10,000 yards by mortars, plus two twin Mark 32 torpedo tubes (eight torpedoes). The CASS would be used as search sensors, a single DICASS as a prosecution sensor. The concept was based in part on tests conducted in 1958–59 in which properly protected sonobuoys survived after being fired from a 5-inch/38 gun.

6. Italy bought smaller *Sparviero*-class hydrofoils modeled on the Boeing PGH *Tucumcari* instead; Japan is ordering three of them. Both countries use fast attack craft in narrow straits. The *Sparviero* carries a PHM payload—one 76mm cannon and two Otomat missiles—on a much smaller displacement of 63 tons. Her range is limited, 400 nautical miles at 45 knots, and reportedly she is underpowered, or slightly too heavy for her Proteus engine. More recent gas turbine development has filled in the gap between the Proteus and the next more powerful engine, and the *Sparviero*s may have their engines replaced with Allison 570KFs.

7. The navy did not formally submit its request for a program change in favor of the hydrofoil until December 1970. It argued that, although the November deadline for budget changes had passed, it was vital that funds be allocated for FY 72, both to demonstrate U.S. commitment to the NATO program and to begin advance procurement and detailed design toward an FY 73 purchase of three to five of what were then called PGHs.

This was clearly the PHM. According to an attached sheet of characteristics, it would displace 150 tons, be powered by two Proteus 3,300-shp gas turbines to achieve 48 knots foil-borne, and have an endurance of 600 nautical miles foil-borne, 1,500 hull-borne, in calm water. Armament was listed as four Standard or Exocet missiles with a Mark 87 or Contraves control system, one OTO-Melara 76mm, and two 0.50-caliber machine guns. It seems likely that some informal agreement was reached before Admiral Zumwalt even tentatively committed the navy to sponsor project group 6 earlier in 1970, so the December paper may not tell the entire story.

8. As of March 1971, OpNav was considering accelerating procurement by adding FY 72 money to buy eight boats in each of FY 73, 74, and 75, six in FY 76. In fact, two boats were funded in FY 73, four more in FY 75.

9. The preferred mix was two semiactive and two antiradiation standard missiles.

10. A gun duel between a PHM, armed with a 3-inch gun and totally unarmored, and a destroyer, armed with a 5-inch gun and not dependent on delicate, computer-controlled foils for survival, might seem ludicrous. However, the PHM was so maneuverable that it could be expected to escape damage while approaching a destroyer (one study showed a kill probability of only 0.10). The fast-firing 3-inch gun could do enormous damage during a short, steady firing run, thanks in part to the steadiness of the PHM firing platform.

11. To reach the U.S. endurance requirement, the PHM had to be redesigned with MTU diesels for hull-borne propulsion.

Chapter 16

1. Remarkably, none of these documents mentions the mini-ATC or the LASSC, both of which were then being designed.
2. According to a marginal note on a CPIC requirements sheet of May 1971, "Only NATO wants hydrofoils. Even Australia is not interested."

An amount of $1.069 million was provided in June 1971, largely for the 30mm gun program, another $2.75 million in navy research and development (6.3) funds in July 1971. In July, NavOrd asked for $3.4 million to cover further ordnance development and production in FY 71–73. OpNav transferred $1.99 million of 6.3 funds into the small boat program, and in September 1971, NavShips received $1.0 million for CPIC construction. At this time, the Koreans were considering building production CPICs before the prototype had been evaluated, a course OpNav deplored. However, in November 1971, PMS 300 submitted a proposal for the construction of twenty-nine CPICs in FY 74–77. The Koreans wanted to build eight CPICs in the United States and eight in Korea. Costs continued to rise; one CPIC was tentatively included in POM-74 (May 1972) at $4.66 million.

3. The CPIC itself was sometimes described as a cut-rate *Asheville* capable of doing much the same job at a lower price, thanks to a superior hull form. The CPIC designers sometimes claimed that it could do 80 percent of the PHM mission at only 20 percent of the cost.

Appendix A

1. N. Friedman, *U.S. Cruisers: An Illustrated Design History* (Annapolis: U.S. Naval Institute, 1984).
2. In particular, immediately after the United States annexed the Philippines, U.S. gunboats helped police them.
3. Almost certainly as "boiler plate"; the New Navy ships had all been built of steel as opposed to iron.
4. The C&R design called for a 1,700-ton ship, 235 x 41 x 12 feet, to make 16 knots and to carry 250 rounds per 4-inch gun. Cramp offered 1,300 tons, 215 x 37 x 11 feet, 14 knots, and 175 rounds per gun. As built, the *Sacramento* displaced 1,140 tons, had dimensions of 210 x 40.75 x 11.5 feet, and could make only 12.5 knots.
5. The *Ashuelot* was lost on rocks in 1883; the *Monocacy* survived, albeit she was decrepit until condemned in 1904. Each drew 9 feet. The only other early shallow-draft ship on the station was the tug *Palos*, built in 1866, which drew 10 feet. She was sold in 1893.
6. These propellers worked at the surface rather than well submerged. The alternative, for a very shallow-draft craft, is a propeller working in a tunnel that fills with water as the craft moves.

Appendix B

1. The most complete available list of SPs is to be found in the November 1918 edition of the official *Ships' Data: U.S. Naval Vessels*, pp. 266–415L. Space does not permit its reproduction here. Data on YPs has been compiled from *Ships' Data* 1945 and 1952, the latter including complete lists of stricken craft but omitting their characteristics.
2. The *Mayflower* was built in 1896 in Scotland for Ogden Goelet and was purchased by the navy in 1898 as a patrol vessel. She served as headquarters for U.S. forces in Puerto Rico in 1900, then as flagship off Panama during the 1903 revolution that resulted in the creation of the Canal Zone. She served as presidential yacht from 1906 through 22 March 1929, when she was decommissioned as an economy measure, and was sold after having been damaged by fire in 1931. In July 1942, she was repurchased and operated by the Coast Guard (WPG 183), becoming a Palestine refugee ship in 1948 and then the Israeli patrol and training ship *Maoz* (1950–55). The next U.S. presidential yacht was the *Potomac* (AG 25), formerly the Coast Guard 165-foot patrol boat *Electra* (1935).
3. The *Isabel* was acquired as SP 521 before completion from her owner, car manufacturer John N. Willys. She served as a destroyer (convoy escort) during World War I, and from 1921 through 1941 she was with the Asiatic Fleet, including the Yangtze Patrol. She was attacked at Cavite on 8 December 1941 but survived to reach Australia, serving through World War II as the ASW escort and submarine training ship at Freemantle. In 1945, she displaced 1,045 tons fully loaded (745 standard), had dimensions in feet-inches of 220-11 (WL) 245-3 (OA) x 27-8 x 8-6, and carried a complement of 6 officers and 121 enlisted men and, at that time, an armament of two 3-inch/50 guns. She was powered by two Parsons turbines (two Normand express boilers) of 8,400 shp and capable of 26 knots. Her World War I battery was four 3-inch/50s and two 18-inch torpedo tubes. By 1945, she was armed with two 3-inch/50s (wet, or submarine, guns), four single 20mm guns, two 0.50-caliber machine guns, two depth-charge racks (eight charges each), and two Mark 6 K-guns.
4. The Coastal Picket Patrol consisted primarily of borrowed or requisitioned rather than purchased craft. It began with Alfred Stanford, commodore of the Cruising Club of America, who offered his members' services to the navy in 1941. The offer was made formally to the eastern sea frontier in March 1942, and by 27 April, the club had found seventy seagoing yachts and one hundred smaller ones. The navy initially rejected them but soon reconsidered. On 4 May 1942, Admiral King asked the Coast Guard Auxiliary to take over this "hooligan navy." The Coast Guard Reserve took over 971 former yachts, motorboats, and fishing craft, all of which were to be capable of keeping the sea for up to forty-eight hours, and armed with one machine gun and four depth charges. By January 1943, Coast Guard 83-footers were appearing in numbers, and King ordered the patrol cut by 35 percent. It was abolished entirely on 1 October 1943.

Coastal pickets were the World War II equivalents of the World War I SPs. They are listed in *The Belgian Shiplover*, nos. 87–89 (1962). See also R. Scheina, *U.S. Coast Guard Cutters & Craft of World War II* (Annapolis: U.S. Naval Institute, 1982).

There was also a Coast Guard Auxiliary force for harbor patrol consisting of craft manned and operated by their owners. At its peak, this force consisted of 17,477 boats divided into 584 flotillas and manned by 54,000 volunteer auxiliaries.

5. The Coast Guard 75-footer was 74-11 x 13-7 x 4-0 feet-inches and 43 tons (light) with two Sterling gasoline engines and total bhp of 400. There were YP 5–17, 19–40, 45–49, 52, 54–55, 59–60, 67. There were also larger boats, YP 41–42, 50, 56, 61–64, 69, of 98 x 23 x 7 feet, about 190 tons, with two Winton diesels and a total bhp of 300, as well as smaller ones, YP 30, 44, 51, 53, 65, of 65 x 13 feet and 90 tons. The 75-foot Coast Guard boats were built in 1924–26. Contract plans were prepared by John Trumpy, who was then employed by Mathis Yacht Building of Camden, N.J. The Coast Guard built 209 75-footers but retained only 36 after Prohibition, acquiring another 6 during World War II. The 65- and 98-footers were named craft built during Prohibition and transferred in 1934.
6. Data were as follows: 75 x 16 x 5 feet, 60 tons fully loaded, with 320-bhp diesels and a speed of 12 knots.
7. These were 128 x 30-6 x 14 feet-inches with a bhp of 560. All were built on the West Coast in 1944–45.
8. Data were as follows: 83 overall, 78 waterline x 16 x 6 feet; 70 tons fully loaded; 1,200-bhp diesels, 19 knots.
9. Data were as follows: 80-5 x 18-9 x 5-4 feet-inches, 69.5 tons, with 2-shaft, 660-bhp diesels; 13.5 knots.
10. Data were as follows: 108 overall, 101 on waterline x 22.8 x 6.0 feet, 163 tons, two GM 12V71 diesels, 875 bhp, 12 knots.

Appendix C

1. Army picket boats were described as "37-foot design 243" with dimensions in feet-inches of 36-6 x 10-7 x 5-10. In effect, they were an armed version of the Higgins Eureka landing craft, powered by 165- or 250- or 300-bhp engines and numbered in a J-series.
2. The original Miami inboard profile drawing was dated 15 December 1940.
3. In January 1951, BuShips estimated that a new 85-footer, without engines, would cost $210,000, compared with $225,000 to $250,000 for a new 95-footer. A new series-produced 50-footer would cost about $75,000 and could be built in seven rather than

The *Asheville* (PF 1) lay at the upper end of the patrol craft spectrum. She is shown on 22 August 1943, about half a year after completion, with U.S. weapons but with a British Type 271 radar. In 1944 she tested the British Squid antisubmarine weapon (which was being considered for U.S. service), and from September 1945 through January 1946 she served as a radar experimental ship with the new Operational Development Force.

eight months. In each case, the first unit of a new design could be delivered within about twelve months, at a higher cost—$300,000 for the 95-footer, $100,000 for the 50-footer.

4. It was expected to operate at full throttle in a 4-foot sea and with reduced throttle in a 6-foot sea.

5. Aside from patrol boats, six, 65TR671–76, were completed as torpedo and weapons retrievers between October 1967 and July 1968, and two as air-sea rescue boats in November and December 1969 (65AR681–82). None was completed as a personnel (PE) or utility boat. The utility boat design incorporated a self-draining cargo compartment aft, which could take any of a variety of special electronic vans.

Notes on Sources

The information in this book is taken primarily from the official, declassified records of the U.S. Navy. In the special case of small, fast combatants, the records are limited for two reasons. First, many of the craft were private designs, the origins of which were never officially recorded. For example, it appears that Sewart Seacraft never submitted a standard weight statement for its Osprey PTF. Second, many official records have been lost or appear to be inaccessible. For both reasons, recourse sometimes had to be made to interviews with designers and to unofficial published material on a scale unusual for this series. Such interviews have been noted in the pages that follow.

The official records fall into three major categories: those held by the Operational Archives and the Ships Histories (SH) branch of the Washington Navy Yard; those held by the National Archives (NARS); and records retained by the navy, both at the Federal Record Center (FRC) in Suitland, Maryland, and at current naval activity headquarters. In a few cases, as noted below, private individuals and builders provided material. Small combatant craft do not receive the kind of archival attention normally accorded their larger relatives, and research was complicated by the destruction of many small-craft records, among them those lost when Preliminary Design's small-craft section moved from Washington to Norfolk.

NARS now holds General Board files previously maintained by the Operational Archives, including the hearings and files 420-12 (gunboats) and 420-14 (patrol craft, including PT boats and subchasers). It also holds the records of the secretary of the navy (SecNav), which after 1915 were combined with those of the CNO as group 80 (CNO/SecNav). The formerly classified CNO/SecNav files were particularly useful for the periods 1917–18 and 1941–45. For the period from about 1936 through 1941, they contain much material relating to early PC and PT requirements and proposals. NARS holds the main C&R/BuShips correspondence files (record group 19) through 1945, as well as the individual ship design files of the Preliminary Design branch for the period from approximately 1912 to 1922. Later preliminary design files are scattered among several files at FRC.

The Operational Archives retains files keyed to major historical projects, including Samuel Eliot Morison's *History of United States Naval Operations in World War II*, the official PT-boat history *At Close Quarters*, by Captain R. J. Buckley, and the history of the navy in Vietnam currently being written, in particular volume 2, *From Military Assistance to Combat, 1959–65*, by E. J. Marolda and O. P. Fitzgerald. The archives also holds the records of OpNav's war plans division, some of which are cited in chapters 3 and 5. Perhaps most importantly for the history of the riverine war, the Operational Archives holds the "command histories" that have to be compiled on a regular basis. Those used in this book include the SEAL histories, the histories of various elements of the riverine force, and the monthly summaries issued by Naval Forces, Vietnam.

FRC holds both the later Preliminary Design files and the files of the SCB.

SEA 941, responsible for buying and allocating small naval craft, maintains invaluable lists of earlier C-numbered craft and navy-built vessels (numbers without prefixes), as well as lists of craft bought after FY 64. In many cases, history cards remain for discarded vessels. SEA 941 also has contract cards describing the World War II small-craft program. I am grateful to both the Preliminary Design branch of NavSea and to the NavSea combat systems engi-

The big gunboats were, in effect, small cruisers and thus could serve as convoy escorts. The USS *Marietta* is shown in dazzle camouflage, probably in European waters, in 1918. Of her original armament of six 4-inch guns, two were replaced by lighter weapons, probably 3-pounders. She was sold in March 1920. (Courtesy of Paul H. Silverstone)

neering station in Norfolk (formerly NorDiv) for assistance with recent designs.

For small-craft transfers abroad, I used the international logistics program summary, declassified copies of which can be found in SH and the Operational Archives. Unfortunately, this summary has not been issued since about 1974, and even before that date it was not completely consistent or reliable. However, it does list transfers other than lend-lease vessels dating back to World War II, and it includes small craft.

For postwar small boats, I have used the data in the standard navy small-boat book. The current edition is issued by the Norfolk division of NavSea (NavSea 0900-LP-084-3010).

As in earlier volumes in this series, I have used BuOrd's wartime armament summaries and the biweekly changes published in 1942–44. Note, however, that although these documents appear to be accurate for subchasers, they are entirely inaccurate for PT boats.

Chapter 1

Notes on engine development are taken largely from a postwar manuscript history of U.S. Navy research and development, based on official sources and held by the Operational Archives. The account of U.S. tests of the Bell hydrofoil is derived from G. Adamson and D. van Patten, "Motor Torpedo Boats: A Technical Study," in the *Proceedings of the U.S. Naval Institute*, July 1940. The comment that porpoising probably killed the Bell craft is taken from a 1953 SNAME paper, "An Appraisal of Hydrofoil Supported Craft," by T. M. Buermann, P. Leehey, and J. J. Stilwell, presented in November 1953. This paper announced the discovery that submerged-foil craft enjoy a degree of inherent stability, hence that they can survive the failure of their autopilots, at least in calm seas. It includes a photograph of the Richardson submerged-foil dinghy and summarizes developments up to 1953.

Chapter 2

SecNav file 20353 describes proposals to take over civilian motorboats for wartime use. It includes files on individual motorboats offered to the wartime navy. The file also includes material on the 45- and 66-foot boats built for the navy (schedule 9069) and on the Greenport motor torpedo boat (schedule 9009). Notes on the SPs are derived primarily from the lists in *Ships Data* for 1918 and 1919.

The NARS preliminary design series (entry 449) includes the 110- and 123-foot subchasers as well as the eagle boat, the 250-footer, the 80-footer, and the 50-footer. However, the 50- and 250-foot plans are taken from a set of C&R microfilms showing plans in numerical order. For the eagles, I also used the flat contract design file (entry 448, box 34). For ASW development, including the origin of the eagle, I relied heavily on CNO/SecNav classified correspondence in NARS, particularly file C-26 (ASW). SecNav file 27219-388 describes Franklin D. Roosevelt's attempt to have the 50-footers built.

Part of this chapter is based on W. W. Nutting's *Cinderellas of the Fleet* (Standard Motor Construction Company, 1920), a volume commemorating the construction of the Elcos and the 110-footers. Another important secondary source was J. A. Furer, "The 110-foot Submarine Chasers and Eagle Boats," in *Proceedings*, May 1919. Furer was the naval constructor responsible for these craft.

Finally, much of this chapter is based on General Board files 420-15 (submarines and ASW measures) and 420-14 (subchasers). Board files on small boats also include material on the Greenport torpedo boat and on small craft built for Russia during World War I. The lists of czarist motorboats were compiled by C. C. Wright, who relied largely on the standard naval list, S. P. Moiseev, *Spisok Korablei Russkogo Parovogo i Bronensosnogo Flota, 1861–1917* (Moscow: Academy of Sciences, 1948), and to a lesser extent, on N. B. Pavlovich, *The Fleet in the First World War* (English translation published by Amerind Publishing, New Delhi, 1979).

Chapter 3

C&R file PC/S1-1 describes attempts through 1940 to design subchasers. However, considerable material on preliminary designs for subchasers came from General Board file 420-14. The notes on PC seagoing behavior were taken from BuShips file C-PC/S1-1 (1943). The same file included the gas turbine proposal. However, the story of PC 452 was taken largely from a manuscript history of navy research and development compiled from official records after World War II by the University of Pennsylvania (retained by the Operational Archives) and from the ship's history card in SH. Armament development was traced through the wartime CNO/SecNav correspondence files and through the BuOrd's armament summaries. Material on major wartime conversions came from BuShips' file PC/L9-3. District craft development board papers were found in General Board file 420-5 and in files of OpNav's war plans division.

The preliminary design file on the YMS, which became the PCS, was found in FRC record group 19 entry 63A3172.

Material on lend-lease transfers was taken largely from a BuShips summary dated 6 July 1944 and from a summary prepared by the British Admiralty delegation, Washington, 1 August 1946.

Chapter 4

The drawing of the Satterlee torpedo boat is in SecNav file 20353. The file includes an ONI study of foreign motorboats and, like General Board file 420-14, material on the Greenport torpedo boat. The file covers the decision to buy the Thornycroft boats after World War I. Material on the two torpedo-launching devices came from SecNav file 24514, in this case the file on naval districts. The Hickman material is taken from a long chronology he submitted to the navy in the 1930s in support of his proposed crash boat; one copy is in NARS (record group 19, S82-3(5)). CNO/SecNav file C-34 includes references to the fate of the 58-foot airplane lighters Hickman described. C&R file S82-3(5) in NARS describes early crash boats. This file also includes Hickman's lengthy memorandum of his early projects. However, the C numbers used in this chapter came from files in SEA 941. S82-3(15) of 1936 includes the proposed 1936 and 1937 MTB characteristics as well as correspondence with various civilian designers. The same file designation applies to the 25-foot scale-model motorboats.

SecNav file EG52-(4) includes the proposed Philippine motor torpedo boat of 1937. Design material was taken from FRC record group 19 entry 11616. Material on the Thornycroft boats actually built was provided by David Lyon of the British National Maritime Museum, Greenwich; Mr. Lyon has compiled a manuscript history of all Thornycroft-designed and -built craft (the "Thornycroft list") from that company's papers. Mr. Lyon's material was supplemented by an account of the Philippine contract in ADM 116, located in the Public Record Office in Kew, England.

The account of the World War I Shearer boat is taken from General Board files; no photograph of this craft seems to have survived, and the files do not include any data. The account of the World War II craft was taken from General Board and Preliminary Design files, together with historical material in SEA 941.

Chapter 5

For early PTs, the chief references were CNO/SecNav file PT/S1-1, the General Board files (420-14), and the official, unpublished administrative PT history by F. A. Tredinnick and H. L. Bennett, prepared shortly after World War II. The General Board file includes the report of the "Plywood Derby," the boat's evolving characteristics (including changing numbers and types of torpedoes), and sketches of the C&R designs for PT 1–6. Some material remains in Preliminary Design. Some of the comments on the boats as built are taken from a 1942 edition of PT-boat tactical doctrine (in effect, an operational manual for PT boats) in the Operational Archives. The archives also holds a copy of the 1945 draft manual among materials collected for the Buckley book.

My account of the development of the Scott-Paine crash boats is largely based on K. Phelan and M. Brice's *Fast Attack Craft: The Evolution of Design and Tactics* (London: MacDonald and Janes, 1977).

Chapter 6

The analysis of PT squadrons is taken from material in the Buckley book; the building program was extracted from S. S. Roberts, "U.S. Navy Building Programs During World War II" (*Warship International*, number 3, 1981). Information on ordnance developments came from several sources: for torpedo tube production, from the unpublished history of BuOrd in World War II held by the Navy Library in Washington; for armament changes in 1943–44, including the local gunboat conversions, from *Mosquito Bites*, a local publication in the South Pacific; from the Seventh Fleet PT information bulletins for developments there in 1944–45, including the origins of the 37mm and 75mm gun; from correspondence in wartime CNO/SecNav files S74-1/PT and S1-1/PT, which include most of the authorized changes in battery; from action reports of barge-busting and its problems in the Operational Archives; and from the Buckley book. File S74-1 includes the Elco Thunderbolt story as well as material on attempts to build lightweight 20mm mounts. The story of the origin of the torpedo-launching rack is taken from V. Chun, *American PT Boats in World War II: A Pictorial History* (Los Angeles: privately published, 1976). Chun provides no references but appears to have obtained his material by interview.

CNO/SecNav file S1-1/PT includes various comparisons of wartime trials, comments by MTBSTC and SCTC, and the official report of an extensive tour of PT bases in the south and southwest Pacific by Preston L. Sutphen, who was then vice president and general manager of the Elco naval division of the Electric Boat Company.

Also extremely valuable was a paper presented to the Naval War College on 17 March 1943, "The Motor Torpedo Boat: Past, Present, and Future History: Development, Tactics, and Accomplishments to 1943," by W. C. Specht and W. S. Humphrey. It includes the observation that the World War II Shearer boat represented a reversion to earlier concepts of a very small, nearly invisible attacker, whereas World War I PTs had already become rather large. Specht was commander of MTBSTC at the time, and his paper reflects the desire for smaller boats evident in official MTBSTC papers of the time. It entirely excludes the gunboat role. The paper by Sidney A. Peters was taken from FRC record group 19 entry 11616.

Chapter 7

The account of the early studies was taken largely from FRC record group 19 entry 61A2306, supplemented by General Board file 420-14, postwar characteristics file S1-1, and hearings. Another important source was OpDevFor's first partial report on project Op/S230/S8, "Evaluate Motor Torpedo Boats (PT 809–12)," 11 January 1954.

Chapter 8

The discussions of the SCB 90 PCE, of the postwar SC and PCS, and of the *High Point* are derived largely from FRC record group 19 entry 63A3172. Material on the design origin of the 95-foot Coast Guard boat was provided by S. G. Moritz, who was one of three coauthors, with R. Bunk and C. H. Ridnour, of a paper, "Construction, Operation, and Cost Effectiveness of Renovation: A Case Study of Coast Guard *Cape*-Class Patrol Boats," delivered at the ASNE technical symposium on patrol boats on 13–14 March 1986. Lieutenant Moritz was responsible for the summary of *Cape* (95-foot) design logic in the paper.

Chapter 9

The primary source for the French experience is an official five-volume history, *La Marine Française en Indochine de 1939 à 1955* (Paris: French Navy, Historical Section, 1972–77). I supplemented it with the accounts of U.S. advisors, located in the Operational Archives, and with a contemporary published account, Captain de Corvette de Brossard's *Dinassaut* (Paris: Editions France-Empire, 1952). Brossard explains the nature of the FOM. The ex-U.S. 83-footers do not appear in the French official history, but they are discussed and illustrated in F. Dousset, *Les Navires de Guerre Française de 1850 à Nos Jours* (Paris: Editions de la Cité, 1975). The details of the EA program were taken from contemporary accounts in the *ONI Bulletin*. The origin of the STCAN is described in the historical section of the fleet marine force's riverine warfare document (1966), mentioned below in the notes on chapter 13. For reasons unclear to this writer, the French official history explains neither the EA nor the STCAN, yet both are mentioned in passing. For the characteristics of Vietnamese riverine craft, I consulted a handbook compiled by the U.S. Navy for its Vietnamese counterpart, "Saigon Joe's Fighting Ships and Craft," dated May 1963 and held at the Operational Archives.

Chapter 10

Discussion of the *Asheville* owes much to the files in FRC record group 19, entry 68A5879. Most of the information on PTFs comes from Preliminary Design files, supplemented by an account of PTF origins in the official Vietnam history by Marolda and Fitzgerald. The account of the design of the *Osprey*s was provided by John Forster, chief designer at Sewart when they were built. Details on the *Eagle* were taken from two United Aircraft Company brochures, "This is *Eagle*" (undated, probably 1967) and "Advanced Marine Craft" (27 January 1967). Al Ross provided a copy of the official comparison sheet of possible *Osprey* performances after they received new engines; it is not dated.

Chapter 11

I used two formerly classified annotated bibliographies produced by Battelle for DARPA under Project Agile: AD 500417L and AD 507731L (1969). They include brief descriptions of efforts to counter the effects of river mines (including shock mountings and chairs) and the initial evaluation of ballistic-nylon canopies for river assault craft. The Operational Archives holds the May 1962 report of the counterinsurgency watercraft panel on requirements for riverine warfare. Preliminary Design holds the corresponding minutes of the panel's meetings. Together, the report and the minutes constitute a design history of the RPC and RCVP, and they also contain descriptions of some early aid projects. The Operational Archives copy of the watercraft panel report is attached to a catalog of available watercraft, including civilian types. The list of 1964 requirements was taken from a report by the naval ordnance test station at China Lake, "Revolutionary Warfare on Inland Waterways: An Exploratory Analysis" (Naval Weapons 8679, published as AD 357903, January 1965). The Operational Archives holds a copy of the NAGP/MACV staff study on naval craft requirements in a counterinsurgency environment (1 February 1965).

The principal published source on U.S. small craft of this period is R. T. Miller's "Fighting Boats of the United States" in *Naval Review 1968* (Annapolis: U.S. Naval Institute). R. L. Schreadly's "Naval War in Vietnam, 1950–70," in *Naval Review 1971*, provides a useful review of U.S. strategy; it is an abridged version of the official command history produced by Naval Forces, Vietnam, for the period 1963–70; it also draws on a classified history covering the period 1946–63 and dealing with U.S. efforts to build up the Vietnamese navy. My principal source for the special craft of the Vietnam period was a declassified publication of commander, Naval Forces, Vietnam, *Ships and Craft of the Vietnamese Navy* (Saigon, 1972, with later corrections); this was supplemented with NavSea's book on small boats.

Chapter 12

Most details of Market Time and Game Warden were found in contemporary reports housed in the Operational Archives. They

PC 451, the contest winner, before her armament was fitted.

include the monthly reports of commander, Naval Forces, Vietnam, from which many of the incidents described in the text have been taken. I also benefited from the archives' numerous reports, such as "Small Craft Counterinsurgency Blockade" (Naval Ordnance Laboratory, NOLTR 72-65, issued as AD 522332, February 1972), "U.S. Navy Combat Craft Operations in Vietnam" (NSRDC technical note, February 1969), and "A Review of USN Experience in Establishment and Conduct of SVN Inshore Coastal Patrol: Operation Market Time" (Westwood Research, for the Naval Research Laboratory, report WR-119-B, also issued as AD 502205L, May 1969). The story of the connection between Willis Slane and the PBR was taken from B. Cobb's "River Patrol Boat for Viet Nam," (*Yachting*, December 1966). The principal published source consulted was S. A. Schwartztrauber, "River Patrol Relearned," in *Naval Review 1970*. Schwartztrauber served as chief staff officer to TF 116 and commander, Task Force Clearwater.

Chapter 13

The summaries of Naval Forces, Vietnam, were useful, as were command histories of the riverine force. For program 5 craft, particularly LCM conversions, the program file retained by NavSea 941 proved an enormous help. Unfortunately, the program 4 file had already been destroyed. The summaries of Naval Forces, Vietnam, included trials of the helicopter pad and of the Zippo. Some of the LCM material in this chapter came from Richard R. Hartley, who was instrumental in organizing the LCM conversions. Much of the ASPB story came from Don Tempesco, who served as project engineer, and Kenneth L. Spaulding, who was directly involved in the program. Operational requirements for LCM conversions and the ASPB came from copies of message traffic in the Operational Archives (for the ASPB, called RPC in the message, the key message is MACV to CinCPac/CinCPacFlt 090600Z of February 1966); the discussion of the initial requirements is based in part on "River Warfare: A Summary of Past Experience in Southeast Asia," issued by the fleet marine force, Pacific, in March 1966. It includes the list of craft proposed by NAGP, including LCM conversion characteristics.

The Operational Archives holds the evaluation report on PACVs. The principal published source consulted was W. C. Wells, "The Riverine Force in Action, 1966–67," in *Naval Review 1969*. Wells was commander of RAF (TF 117) in 1966–67. The army's view of the logic of the riverine force is covered in W. B. Fulton, *Riverine Operations, 1966–69* (GPO: Washington, 1973, in the army series on Vietnam).

Chapter 14

The descriptions of SEAL boats are taken from SEAL command histories and appendices to the monthly reports compiled by Naval Forces, Vietnam. The SEAL command history describes the origins of the STAB and illustrates an LCM modified for SEAL insertion (the original HSSC). The account of STABRon operations and boats is taken largely from notes provided by the former STABRon commander to the Operational Archives, as noted in the text of this chapter.

Chapter 15

Material in this chapter was found in Preliminary Design files and supplemented by published papers on the PHMs.

Chapter 16

Much of the material in this chapter, including that for the design histories of the Sea Fox and Sea Specter (patrol boat Mark

Tacoma-class frigates were very nearly equivalent to destroyer escorts, but they were classed as patrol craft. The *Peoria* (PF 67) served briefly as a convoy escort and was then refitted at Charleston as an Atlantic weather ship, a balloon-inflation room replacing her after 3-inch gun. She is shown, newly converted, about June 1945; her weather service lasted about a year. PFs performed similar postwar weather duty for several other navies.

III), was supplied directly by NorDiv. I also benefited greatly from the David Taylor Ship Research and Development Center's report uu-0013, "Naval Inshore Warfare Craft Program (S 0414): Program History and Technical Accomplishments Summary Report," edited by J. L. Gore (January 1977). The account of the SWCM is based on a lecture, "*Sea Viking*, A New Kind of Patrol Boat," delivered by J. LaFemina at the ASNE technical symposium on patrol boats held on 13–14 March 1986.

Appendix A

Notes on early gunboats are taken from the annual reports of the Navy Department and from transcripts of meetings of the Board on Construction in NARS. For the China gunboats, I relied on General Board records (420-12), supplemented by the classified records of CNO/SecNav (PG/S1-1) and Preliminary Design files (FRC record group 19, entry 6019).

Appendix B

Data is derived from various volumes of *Ships Data*, particularly the 1945 and 1952 editions, the latter including valuable retrospective material.

Appendix C

For army boat data, I relied on *Army Small Boats, 1 July 1940– 31 May 1945*, issued by the Corps of Engineers at the end of 1945, a copy of which is held by NavSea 941. Total production figures on navy boats were taken from tables in BuShips' World War II administrative history. Data on postwar crash boats comes from SEA 941, and some material on the 63-footer is available in British air ministry reports now held by the public records office in Kew.

Appendix D

The principal source for Vietnam-era weapons was the ordnance catalog of advanced ship types and combatant craft, prepared by L. Lange and M. Maylack of the Naval Ordnance Laboratory at White Oak and issued in June 1972 as NOLTR 72-140; it has now been declassified. I also benefited from the ordnance nomenclature records held by SEA 62Y14.

Appendix E

Much of this appendix is based on material supplied by Halter, Lantana, MonArk, and Swiftships, supplemented by an appendix to H. Fock, *Schnellboote: Die Entwicklung von 1974 bis Heute* (Herford: Koehler Verlag, 1986) and by standard reference works, primarily *Combat Fleets* (Annapolis: Naval Institute Press). Unfortunately, many transfers of small combatants, particularly those built in the United States, go unrecorded in standard reference works because they do not appear in the major journals. Fock's list, which covers the period 1974–85, is therefore extremely valuable. These data were supplemented by material from SEA 941, and for the period prior to 1974, from international logistics program books.

Appendix F

For the period up through 1952, building dates come from declassified information in *Ships Data*; postwar issues include retrospective information on numbered ships dating back to World War I. Subchaser fates were taken from an appendix in volume 6 of the official *Dictionary of American Naval Fighting Ships*. PT fates were taken from history cards maintained by SH. Building dates for MAP and Mutual Defense Assistance Program gunboats were taken from BuShips' official monthly progress reports in SH and from material supplied by Peterson Builders.

Index

Abra (Philippine MTB), 113
Abray (Philippine MTB), 113
Active class, 212
Advanced tactical support base (ATSB), 298
AG(EH) (*Plainview*), 217
AGHS concept, 379
AGIS-162 combat-control system, 384
Agusan (Philippine MTB), 113
Aircraft carriers (fast boats), 103
Air-cushion craft, 4. *See also* Hovercraft
Air propeller (for boat), 12
Air roll (LVA-1), 293
Air-sea rescue boats, 97, 99, 105. *See also* Crash boats
Albay (ex-Spanish gunboat), 418
Alco, 110
Algerine-class minesweepers, 423
Alinavi, 262
Allison engine, 119
All-weather subchaser, 35
Aluminum hull, 119
Alvarado (ex-Spanish gunboat), 419
American Shipbuilding, 36
American Shipbuilding and Designs, 459
Amherst, 88
Ammi pontoon, 298
Anderson, Admiral G. W., 268
Annapolis (PG 10), 413, 417
Annapolis Yacht, 188
Antelope battle damage, 271
Anti-infiltration role, 392. *See also* Operation Market Time
ANVCE (advanced naval vehicles concept evaluation), 392
Aqua skimmer, 293
Aquila (PHM 4), 379, 389
Arayal (ex-Spanish gunboat), 418
Armor, for riverine craft, 228; for Vietnam service, 293
Army crash boats, 433
Army Transportation Corps, 108
ARPA (Advanced Research Projects Agency), 15, 279, 286, 392
Asheville (PG 21), 413, 418
Asheville (PG 84), 243, 255, 277, 299, 379; design, 265; hull form, 267; sea-keeping performance, 269; armament, 268–69; speed performance, 269; transfers, 272. *See also* PGM
Ashuelot (river gunboat), 418
Asiatic Fleet, request for PTs, 115
Askari (ARL 30), 329
ASPB (assault support patrol boat), 325, 391; requirements, 333, 348; original proposal, 348; protection, 349, 351; sponsons and revised lines, 349, 352; alternatives considered, 351, 353; armament, 351; initial service experience, 352; characteristics, 355; revised (program 5) design, 355
ASPB Mark II, 5; characteristics, 355; Sewart prototype, 360, 362, 363; Sikorsky prototype, 357, 361
Astor, J., 99
ASW: by fast small boats, 101; conference (OpNav), 199; destroyer, proposed, 34
ATC (armored troop carrier), converted LCM, 325, 331, 332, 333; characteristics, 330; versatility, 331. *See also* Mini-ATC and CCB
ATC/CCB, 391
ATC/CCB Mark II, 5, 347; requirements, 358; Sewart prototype, 358, 359; rejection, 359
Atlantic Hydrofoils, 396
Auk-class minesweeper (220 foot), 81
AVCO Lycoming gas turbine, TF12, 3, 5; TF14, 3, 5, 396; TF25, 3, 396; TF35, 381, 384; TF40, 386
Ayer, N., 21
Aylwin (destroyer), 20

Baldwin, C., 7
Ballistic nylon, 293, 341
Bangor-class minesweeper, 81
Bar armor, 338, 341, 342,
Barge-busting, 112, 156, 160, 164
Barrier concept, 298
Baycraft, 45
Bazooka, 75, 112, 252
Bell, Alexander Graham, 7
Bell Aerospace-Textron, 402
Bell-Halter surface-effect craft, 458
Bell SKMR-1 Hovercraft, 293
Benewah (APB 35), 327
Benson, Admiral W. S., 26
Bertram fast boat, 276
Bertram Yacht, 239, 459
Bima Samudera I, 389
Binh Xuyen forces, 280
Blackwood (British frigate), 198
Bloch, Admiral C. C., 54
Blockade operations, 298
Board of Inspection and Survey, 122, 136
Boat Support Unit 1, 276, 365
Boeing, 214, 262, 380, 386, 389; hydrofoil, 384, 386; jetfoil, 389
Bofors (40mm) gun, 160
Boston whaler, 295, 311, 365, 371, 375
Bowling, Captain S. S., 164, 185
Boyle, J., 273
Brattleboro, 88
Brave (fast patrol boat), 250
British Power Boats, 108
Broncos (OV-10 aircraft), 297
Bronstein (ocean escort), 207
Brooke (ocean escort), 207
Bucklew, Captain P. H., 282
Bulkley, Admiral J., 171
Bunker-busting, 112, 336
Bureau of Aeronautics (BuAer), 105
Bureau of Construction and Repair (C&R), 14
Bureau of Engineering (BuEng), 14; diesel engine, 119
Bureau of Ordnance (BuOrd), 14
Bureau of Ships (BuShips), 14; hydrofoil program, 215
Bureau system, 14
Burgess, J. S., 110, 112
Burgess, W. S., 48, 116
Burke, Admiral A. A., 243, 246
Bush, V., 8
Buttress (ACM 4), 85
Buy-American Act, 246

Calamianes (ex-Spanish gunboat), 419
Callao (ex-Spanish gunboat), 418

The navy took over numerous large civilian yachts to fight the coastal ASW battle in both world wars. PC 460, built by the Consolidated Shipbuilding Company in 1930 as the yacht *Elda*, was acquired in September 1940. She is shown newly refitted at Pensacola in May 1943, before radar had been installed. Note her paired Mousetraps forward and her depth-charge throwers and tracks aft. In July 1943 she was redesignated PYc 50 and renamed *Sturdy*. Ordered disposed of in October 1944, she was stricken on 27 November 1944 and returned to her former owners. The *Sturdy* displaced 330 tons and was 154 feet long (beam 24 feet, draft 8 feet 6 inches); her rated speed was 17 knots.

Camcraft, 459
Camper & Nicholson MTB, 108
Canard configuration, 213
Captured air bubble, 10
Cardinal (mine hunter), 11
Caribbean Sea, 417
Carney, Admiral R. B., 123
Castine (PG 6), 414; (IX 211, ex–PC 452), 59
Castro, F., 14
Cavitation of hydrofoil, 213
CCB (command and control boat, converted LCM), 325, 330, 334, 339, 347; arrangement, 337. *See also* LCM conversions
Chain-drag sweep, 335
Chain gun, 442
Chantry, Captain A. J., 116
Chariots, 101
Charleston (PG 51), 413
Chase I., 24, 122
China Lake, 439. *See also* Naval Weapons Center
Chinese Civil War, 221
Chokepoints, 221
Chunking, 418
CIA, 251, 279, 286
Cincinatti (cruiser), 413
Claud Jones (ocean escort), 207
CMB (Thornycroft), 6, 11, 99, 100, 103, 104, 108; trials, 103; tactics, 104; as crash boats, 105
Coastal Force (Vietnamese), 283, 284
Coastal minesweepers (AMB and AMC), 428
Coastal patrol, 22, 23
Coast Guard, coordination with, 211; boats acquired, 427; cutters, 299; 40-foot utility boat, 226, 232, 237, 239, 435; 83-footer, 93, 95, 199, 222; 95-foot cutter, 199, 200, 208, 209, 234
Cochrane, Admiral E. C., 64, 81, 179
Cockerell, C., 10
CODAG (combined diesel and gas turbine propulsion), 178, 191, 192, 248, 266, 395
CODOG (combined diesel or gas turbine) propulsion, 286, 398
COGAG (combined gas turbine and gas turbine) propulsion, 192
COMAX, 7
Combustion Engineering, 59
Commandement, converted LCM, 227; converted LCT, 229, 230, 231
Composite construction, 417
COOP (craft of opportunity) program, 429
Corvette (PG), 13, 423, 425; armament, 425
Costigan, Lieutenant (j.g.) J., 157
Counterinsurgency, 279; research and development group (DoD/CIA), 15; watercraft panel, 284
Cove (inshore minesweeper), 209
CPC (coastal patrol craft), 391, 392, 399, 401
CPIC (coastal patrol and interdiction craft), 16, 192, 308, 391, 392, 397, 398, 399; design requirements, 392; mission profile, 396; prospective operators, 396; design alternatives, 396; special hull design, 396; armament, 399; sea-keeping tests, 399; structural tests, 399
Cramp Yard, 417

Crash boats, 94, 431; wartime construction, 433; army, 433; characteristics, 434; postwar design, 435, 437; hydrofoil alternative, 437. *See also* Air-sea rescue boats
Crouch, G., 116, 119
C-tube (passive sonar), 35
Cua Viet River, 283, 297
Cuba, 250
Cummins diesel, 245, 264, 269; NVMS-1200 diesel, 234
Curtis V-1400 engine, 105
Curtis-Wright, 3-D diesel, 266; CW12V142A diesel, 268
Cyprus crisis (1964), 243

Daggett, Commander R. B., 107
Daniels, J., 21, 26, 33, 34, 103
Datu Marikudo (Philippine), 88
Dauntless (PG 61), 413
David Taylor Model Basin, 294
Davis, S., 21
DBH (large hydrofoil), 219
DD(H) concept, 217
Dealey (ocean escort), 14, 202
Defender engine, 105, 110, 112, 119
Defense Systems Acquisition Council (DSAC), 388
Defiance, (PG 95), 269
DE(H), 10
De Laval Steam Turbine Company, 48, 59
Deltic Engine, 2
Denison, 10, 217
Denver-class cruisers, 417
Depth charge, 20; trials, 35; lightweight, 209
Designations, 16
Designators (small boat), 17
Dewandre-Elco turret, 125
DFH (fast hydrofoil), 219
Diesel competition (1932), 58
Diesel power, 35
Dinassaut (*division naval d'assaut*), 223, 233, 297
Direktor, R. E., 459
District Craft Development Board, 66
Dodge Water Car, 106
Dolphin, 10
Douche, 340
Douglas (PG), 272
Dual-hardness steel, 293, 340
Dubuque (PG 17), 413 417
Dynamic Development, 10

EA (*engin d'assaut*, modified LCVP), 226, 231
Eagle (fast gas-turbine boat), 253
Eagle 32, 19; *58*, 39; *60*, 40
Eagle boat, 13, 15, 34, 47, 49; design, 37; production, 37; named, 38; features, 38; power plant, 38; instability, 39; armament, 41; wooden mockup, 41; production, 43; performance, 43; as minesweeper, 40. *See also* Patrol boat
E-boat (S-boat), 185, 189
Edison, Secretary of the Navy C., 116, 121, 122
Edo Aircraft, 188
Effekt rudder, 189
Eisenhower, D. D., 14
Elcano (ex-Spanish gunboat), 418, 419
Elco, 15, 23, 58; PT, 115, 122; "slipper," 173
Elcoplanes, 151

Emerson Electric, 399, 449
Engine noise, in riverine operations, 223, 233
Engines, marine, 2
EPCER 857
Erie (PG 50), 413
Exocet missile, 385
Experimental sonar ship, 69
Explosive motor boats, 101

Fairbanks-Morse, 48, 58
Fairchild Aircraft, 192
Fairmile "D", 181
Fairmile subchasers, 66, 81
Fast patrol boats (1959), 248
Ferguson, Lieutenant Commander J., 369
Ferocity (fast patrol boat), 248
Ferro-cement construction, 300
Fire-control systems: Mark 63, 269; Mark 87, 269, 271, 272; Mark 92, 268, 271; Mark 94, 388
First Patrol Squadron, 21
Fisher Boat Works, 119
Flagstaff (PGH 2), 10, 257, 259, 261, 262, 263, 277, 397. *See also* PGH
Flagstaff Mark II hydrofoil, 263. *See also* Shimrit
Flame throwers, 232, 337
Fly-catching, 191
FMC (river patrol boat), 292
FOM (French riverine boat), 231
Ford, Henry, 37
Forlanini, E., 7
FRESH-1 (experimental hydrofoil), 217
Frigate (PF), 13, 425; transfers, 425; Korean War service, 425; Canadian, 423
Fulton (PG 49), 413

Garcia (ocean escort), 207
Gardner, W., 33
Garrett 990 gas turbine, 253
Gasoline hazard, 192
Gas turbines, 3, 178, 192; in PC, 71
GE LM100 gas turbine, 3, 309; LM1500 gas turbine, 3, 217, 269, 386; LM2500 gas turbine, 3, 4, 386, 395; MS420 gas turbine, 267; T56 gas turbine, 192; T58 gas turbine, 192
General Board, 11, 14, 98
German hydrofoil project, 384
Gibbs & Cox, 9
Gielow, H. J., 48
GM 16-184A ("pancake") engine, 2, 73; 6V71 engine, 237; "quad" engine, 237
Graham County (LST 1176), 379
Grant, Admiral A. W., 34
Grasshopper (sprint-and-drift) tactics, 212
Gray Marine engine, 233
Greek Civil War, 221
Green Bay (PG 101), 265
Greenport Basin, 22, 100
Grenade-launcher, 336, 440
Ground-effect machine, 10
Grumman Aircraft, 10, 217, 262
Guam (Yangtze gunboat, PR 3, renamed *Wake*), 423
Guardian (ex-PT), 195
Gunboat, 413; converted from LCM(3) (World War II), 164. *See also* Monitor, converted from LCM
Gunboat, river. *See* River gunboat
Gunboat armor, 416

Gunboats, proposed for MAP (1964), 295
Gun mounts, list, 442
Guns: for subchasers, 52; M60 machine gun, 439; 20mm lightweight, 67; 37mm, 165; 40mm, 257, 335, 440, 445; 75mm, 170; 105mm, 338; 152mm, 261; rocket gun, 439; 4-inch, 418

Halobates (experimental hydrofoil), 8
Halter 36-footer, 457; 41-footer, 457; 44-footer, 457; 50-footer, 457; 65-footer, 457; 75-footer, 458; 78-footer, 458; 85-footer, 458; 105-footer, 458
Halter Marine (Halmar), 457
Hankow, 418
Hannibal (survey ship), 47
Harbor pickets, 431
Hard-chine hull form, 7, 396
Harlee, Lieutenant Commander J., 161, 223
Harpoon missile, 272, 381, 383, 385, 401, 418
Hartley, R. R., 333
Hatteras Yacht, 312
HD-4, 7
HDML (British), 233
Heavy-weather patrol craft (HWX), 308
Hedgehog (ASW weapon), 68, 197, 198, 200, 202
Helena (PG 9), 415, 419, 423
Hepburn, Admiral A. J., 47, 48
Hercules (PHM 2), 383
Herreschoff, 44, 116
Hickman, W. A., 7, 99, 103, 107, 108
Higgins, A. J., 15, 121
Higgins, dream boat (PT 70), 136; Eureka boat, 431; Hellcat (PT 564), 149, 150; PT hull form, 121; inboard profile (showing structure), 141
High Point (PCH), 8, 10, 14, 198, 209, 210, 213, 214, 218, 380; as test platform, 215. See also PC(H)
Hip Pocket II program, 450
Homer, A. P., 26
Homing torpedo, 197, 198
Hook, C., 8
HOR (Hoover-Owens Rentschler) Engine, 48
Horne, Admiral F. J., 15, 67
Houp-La (motor boat), 21
Hovercraft, 10, 284, 293, 300, 308. See also Skirt, Bell SKMR-1, PACV, and SRN-5
Hope/Progressive Shipbuilding, 459
HSSC (heavy SEAL support craft), 366, 371, 374
Huckins PT boat, 136
Huff, Lieutenant S. L., 110
Hughes towed-array sonar, 319
Hull forms. See Hybrid, Hydrofoil, Hydroplane
Hump (hydrodynamic), 6, 176
Huron-Eddy, 437
Hvass, 251
Hybrid craft (surface effect/planing), 399. See also Ski cat
Hyde Park Agreement, 423
Hydrodrome, 7
Hydrofoil, 4, 7, 8, 192, 244, 435; control, 8, 9; fully submerged, 8; V-foil, 8; program (prospective, 1959), 214; propulsion, 215; autopilot, 213, 214; for ASW, 212; crash boat, 437; for Germany, 384; PT, 194; patrol craft, 257; river patrol boat, 246

Hydrofoil Corporation, 9
Hydrogen bomb, 198
Hydronautics CPC proposal, 399, 400
Hydrophones, 19
Hydroplane, 4, 6, 396. See also planing hulls

Ichang, 418
Indiana Gear Works, 312
Indochina war, 221, 223; French amphibious forces (1946–49), 225; (1950–54), 228
Indonesian navy, 389
Invader engine, 110, 112
Isabel (former yacht), 423
ISSM missile, 272

Jacuzzi pump-jets, 315; spares problem, 315
Jiger amphibian, 293
JL (passive sonar), 50
Joint Chiefs of Staff (U.S.), 225
Junk Force (Vietnamese), 283
Junks, 284

Kamikaze attacks, 179
Kelly, Commander R. B., 164
Kennedy administration, 279
Krushchev, N., 243, 279, 280
Kiangnan Dock, 423
King, Admiral E. J., 15, 66, 107, 177
Knox (ocean escort), 207
"Komar"-class missile boat, 273, 380
Korea, 392
Korean War, 221, 365; war experience, 192
Korody Marine, 239
Kronshtadt attack, 103
K-tube (passive sonar), 20

Land, Admiral E. S., 107
Landing craft inventory (1964), 335
Landing parties, 416
Lantana 40-footer, 459; 68-footer, 459; 106-footer, 458
Lantana Boatyard, 458
Lantern, experimental hydrofoil, 9
LASSC (River Fox), 372, 376, 401
Latimer-Needham, C., 10
Lawley, G., 21, 22
LCA landing craft, in riverine war, 223
LCI 221, armament modification, 227
LCM conversions (U.S.): requirements, 333; design, 334; armament, 334, 336; internal arrangement (CCB/monitor), 337; conversion yards, 340; armor, 340; protection extemporized, 341; sponson, 341; problem of commander's vision, 341; helicopter platform, 344, 345; as modified under program 5, 346; conversion deficiencies, 346; conversion characteristics, 347; totals converted, 330; replacement program, 331; ammunition, 340. See also ATC, CCB, LCM monitor, LCM refueller, and LCMM
LCM in riverine war, 223, 224; conversions, 232; as monitor, 283
LCM monitor (U.S.), 325, 328, 329, 335, 337
LCM refueller, 325
LCM troop carrier, 235. See also ATC
LCMC, 283. See also Commandement

LCM-C (LCM-combat), 283
LCMM (converted LCM), 33, 343, 344
LCPL, 239, 279, 289, 311, 365
LCS landing craft, in riverine war, 224
LCSR, 244, 246, 277, 285, 313, 365; deficiencies, 286
LCSR(L), 366, 391
LCT landing craft, 223; in riverine war, 224
LCU landing craft, 223
LCVP, 283; in riverine war, 223; as converted by French, 231. See also EA and STCAN
LDNN (Vietnamese SEALs), 365
Leahy, Admiral W. D., 48, 52, 116
Lend-lease Act, 81
Lend-lease program, 433
Leyte (ex-Spanish gunboat), 419
Libbey, Lieutenant Commander M. S., 34
Liberty engine, 2, 52, 105
Long, D. N., 433
Long-Range Objectives (LRO) Group, 15, 192, 250, 264, 319
Low, Admiral F. S., 198
LSSC (light SEAL support craft), 366, 369, 371, 391
LSSL support landing craft, 221; transfers, 222
LSTs for riverine bases (Operation Game Warden), 317
Luders Marine, 23, 24, 58, 112; crash boat, 106
Lung Chiang (fast patrol boat), 272
Luzon (Yangtze gunboat, PR 7), 423
Lycoming gas turbine, 259. See also AVCO Lycoming engines

MAAG (Military Assistance Advisory Group), 282, 286
Machine guns, types, 440
McNamara, R. S., 251, 257, 264, 279, 299, 326
MACV (Military Assistance Command, Vietnam), 282
Magnetic anomaly detector (MAD), 213
Magnum Marine, 459
Mahan (destroyer) hull form, 53
Manned torpedo, 100
MAP. See Military Assistance Program
Marine Brigade of the Far East (French), 223
Market Time. See Operation Market Time
Marshall boats, 105
Marsh screw amphibian, 284, 293, 363
Martinac Yard, 214
Marysville (EPCER 857), 89
Maximum-effort program, 66
MB-tube (passive sonar), 20
MDAP. See Mutual Defense Assistance Program
Mekong Delta, 282; Mobile Assault Force, 325
Mercedes diesel, 237, 265, 285; 820Bb diesel, 234
Metrovick Gatric G.2 gas turbine, 192
Miami Shipbuilding, 119, 431
Mighty Mouse (2.75-inch) rocket, 439
Military Assistance Program (MAP), 221; requirement, 250
Mindanao (Yangtze gunboat, PR 8), 422, 423
Mindoro (ex-Spanish gunboat), 419
Mine Defense Laboratory, 286

Minesweepers: coastal, 428; program, 65; conversion for ASW, 207; 136-foot (YMS), 65, 66, 69, 85, 92, 93, 208; 173-foot (ex-PC class), 65; 180-foot (AM 136 or *Admirable* class), 66, 81, 207; 220-foot (*Auk* class), conversion for ASW, 209
Minesweeping motor launches (MLMS), 283
Mini-ATC, 373, 374, 376, 377, 457
MIT Flight-Control Laboratory, 9
Mobile Riverine Assault Force, 284
Mobile Riverine Force, 297; tactics, 356; compared with RAG, 357
Mobilization plan for coastal escorts, 211
Modular systems, 381, 384, 391, 401
MonArk boats, 459
Monitor (French), 223; converted from LCM, 221, 223, 228, 229; converted from LCT, 228
Monocacy (old river gunboat), 418; (PG 20), 420
Morgan, Admiral A., 8
Mortar: 60mm, 439; 81mm, 177, 253, 439; 107mm, 440
Motorboat inspection boards, 22
Motorboat reserve: U.S., 21; British, 20
Motor launches (for Britain), 23–25; drawing, 25
Motor torpedo boats: U.S., 99, 100, 101; British, 108; foreign, information concerning, 98, 110. *See also* PT boats
Mousetrap (ASW weapon), 68, 197
MSD (minesweeping drone), 333
MSR (modified ASPB minesweeper), 333, 344, 345, 356
MSSC (medium SEAL support craft), 366, 371, 449
MTB. *See* Motor torpedo boats
MTBSTC (MTB Squadron Training Center), 140, 153, 181
Mumma, Admiral A. C., 188
Mustang, 286
Mutual-Defense Assistance Program (MDAP), 221, 225; table, 226
MV-tube (hydrophone array), 20, 50
Mystery (motor yacht), 24

NACA (National Advisory Committee on Aeronautics), 7
NAGP (Naval Advisory Group), 295, 334
Napier Deltic diesel, 243, 245
Nashville (PG 7), 413, 415
Nasty-class fast patrol boat, 245, 248, 251, 254, 255, 277, 299, 316. *See also* PTF
Naval Ammunition Depot Crane (Indiana), 448, 449
Naval districts, 427
Naval ordnance station, Louisville, 439
Naval weapons center, China Lake, 273. *See also* China Lake
NavSea (Naval Sea Systems Command), 14
Neville Boatyard, 459
New York Navy Yard, 157
Ngoc Hoi (Vietnamese), 88
Nichols, Captain S. G., 188
Niedermair, J. C., 184
Nimitz, Admiral C., 185
Nitze, P., 257, 299, 326
Nixon, L., 25, 97
North Carolina (BB 55), 109
North Sea blockade (proposed), 35

North Vietnam, 243
Nourmahal (PG 72), 413
NSRDC, 399
Nuclear submarine threat, 215

Oahu (Yangtze gunboat, PR 6), 423
Ocean reconnaissance ship (PR), 319; alternative designs, 321
Oelrichs, H., 21
Oerlikon: mortar, 257; rocket launcher, 257
One-man torpedo boat, 17. *See* Shearer
Opal (PYc 8), 429
Operational Development Force (OpDevFor), 189, 191
Operation Barrier Reef, 298, 369
Operation Foul Deck, 297
Operation Game Warden, 297, 311
Operation Giant Slingshot, 298
Operation Jackstay, 326
Operation Market Time, 282, 297, 298, 380, 391, 392; craft for, 303
Operation Plan 34A, 282, 286, 301, 302, 365
Operation Search Turn, 297
Operation Silver Mace, 298
Operation Stable Door, 279, 298, 311
OpNav (Office of the Chief of Naval Operations), 15
Orange war plan, 11, 47, 55, 110, 115
"Osa"-class missile attack boat, 380
"Osprey"-type PTF, 243, 253, 256, 273, 277, 396, 397
OTO-Melara gun, 383, 385
Owasco class (Coast Guard cutters), 413

Packard engine, 2, 112, 119, 149, 173, 177, 192, 244; 1500 engine, 105; 4M2500 engine, 105; MGT-10 gas turbine, 192; diesel, 436
PACV (Hovercraft), 309; performance, 309; armament, 310; operational use, 310; deficiencies, 310
Paducah (PG 18), 417
Palos (PG 16), 420
Pampanga (ex-Spanish gunboat), 418
Panay (ex-Spanish gunboat), 419
Panay (Yangtze gunboat, PR 5), 423
Pancake engine, 2. *See also* General Motors 16-184A engine
Paragua (ex-Spanish gunboat), 418
Parris Island (AG 72), 85
Parrot's Beak, 297
Patrol boat (later eagle boat), 36–37; designs (1959), 251; 65-footer (Sea Specter), 391, 399, 405, 409, 410, 411; alternative armaments, 407; 68-footer, 408, 411
Patrol motorboats (1916), 22
PBM (patrol boat, medium), 402. *See also* SWCM
PBR (river patrol boat, or plastic), 284, 297, 305, 391; designation chosen, 313; armament, 313; performance, 313, 315; armor, 314; tactics, 315; performance, 316; repairs, 316; program, 316; production, 318; transfers, 318
PC (173-foot steel subchaser): series, 47; design, 53, 59; comments on design, 54; in Orange war plan, 55; detail design, 55; steam plant, 58; competition, 58; power plants, 58, 66, 67; steam turbine, 58; design

requirements, 59; evaluation, 64; electrical generating capacity, 64; program, 64; cancellations, 66; modifications, 67; weapons, 67; evaluation, 71; postwar versions, 71; for Indochina War, 225
PC replacement (SCB 51), 198, 199; engines, 201; alternative designs, 202; 1953 version, 207
PC 449, 47, 56
PC 450, 57, 58
PC 451, 55, 60
PC 452, 58, 61. *See also Castine*
PC 466, 62
PC 472, 65
PC 546, 64
PC 600, 68
PC 618, 69
PC 1638, 74
PC 1638 class, 75
PCC program, 69
PCE (180-foot steel subchaser): origin, 65; postwar versions, 73; design, 81; original design illustrated, 82; as delivered to Royal Navy, 83; weapons, 85; conversions, 85; weather ships, 85; experimental conversion, 89
PCE replacement (SCB 90), 198, 201; comparison with wartime PCE, 202; sketch designs, 204–5; characteristics, 206; power plant, 206; preliminary designs, 207–8; hull form (stern), 207
PCE 827 (original design), 82
PCE 893, 86
PCE 899 (weather ship), 87
PCEC program, 85
PCEC 882, 87
PCEC 886, 90
PCER 853, 88
PCF. *See* Swift
PC(H), 244, 379; concept, 212; sonar, 212; preliminary design, 215. *See also High Point*
PCH-2, 380
PCMM, 398
PCS (136-foot wooden subchaser), 66; design, 85; armament, 92; original design, 94; sonar school ships, 92; special minesweepers, 93; survey ships, 93; replacement, 209
PCS 1386, 91
PCS 1461, 92
PCS 1445 (sound school conversion), 95
PCSC program, 93
Pearl Harbor, 115
Pegasus (PHM 1), 213
Pegasus (PHM 1) class, 9; main propulsion, 4. *See also* PHM
Penguin missile, 405, 410
Perch (submarine), 58
Perfume River, 283, 297
Peterson Builders, 457
Petrel (PG 2), 413
PGH, 257, 380, 382; program, 214, 215, 265; designs, 259; speed, 259
PGM, 67, 221, 285; program, 68, 75, 237, 263, 265; postwar, 234; for Philippines, 222; evaluation, 237 (slow versions); designs (for *Asheville* class), 267; 130-foot design, 276; hydrofoil design, 276
PGM 2, 82
PGM 4, 84
PGM 7, 80, 88
PGM 9, 70

PGM 17, 72
PGM 33, 77
PGM 39, 236, 267
PGM class, 234
PGM 43 class, 234
PGM 53 class, 234
PGM 59 class, 237
PGM 71, 277
PGM 71 class, 237, 238
PGM 116, 235
Philippine Huk Rebellion, 221
Philippine islands, 11, 417
Philippine patrol boat, 107, 110, 111, 113, 116
PHM, 16; NavSec design, 382; mission, 384. See also PMH, PXH, and *Pegasus* class
Phoenix Marine, 459
Picket boats, 311
Piggyback mounts, 439
Plainview (AGEH), 10, 216, 217, 218, 219
Planing hulls, 395, 435. See also Hydroplanes
Plastic. See PBR
Plywood Derbies, 136
PM(H), 380, 381, 384; design study, 379
Port security mission, 209
Portsmouth Navy Yard, 48
Postwar patrol craft plans, 71
Power Boats crash boat, 123
Pratt, Admiral W. V., 36, 47
Pratt & Whitney: FT4 gas turbine, 3, 267; FT9 gas turbine, 3; FT-12 gas turbine, 259; ST-6A gas turbine, 383
Preparedness program, 21
President's Science Advisory Committee, 215
Project 60, 271, 382
Project Zulu, 366
Proxy warfare, 198, 200
PSMM-5 fast patrol boat, 272, 274, 275
PT 1, 118
PT 5, 119
PT 3, 120
PT 8, 120
PT 18, 127
PT 19, 135
PT 30, 135
PT 34, 129
PT 59 (gunboat conversion), 133
PT 61, 165
PT 65, 131
PT 77, 158
PT 95, 148
PT 109, 145
PT 117, 138–39
PT 168, 165
PT 174, 160
PT 200, 140
PT 209, 143
PT 211, 168
PT 268, 176
PT 330, 168
PT 462, 153
PT 515, 159
PT 556, 171
PT 588, 169
PT 596, 147
PT 631, 115, 169
PT 809, 1, 180, 181
PT 810, 177, 187, 189
PT 811, 1, 191
PT 812, 192; conversion, 192
PT boats: role, 109, 156 (1945); concept of operations (late World War II), 156; requirements, 109, 116, 134; engines, 116, 182, 191; 1945 alternatives, 184; design, 116, 178 (1945); with sketch, 179, 188 (1947); hydrodynamic scale models, 116; hydrodynamics, 116, 182, 188; design competition, 118; prototypes, 121; program, 123, 139, 154; experimental, 136, 177; proposed post-1950, 190; construction in Yugoslavia, 154; armament, 115, 121, 125, 157, 167 (1944), 180, 182, 185; tests, 171; torpedoes, 121, 157; torpedo launching, 121; torpedo racks, 157; rockets, 148, 167, 168; mortar, 166; gunboat version, 165; costs, 137; evaluation (Higgins vs. Elco), 140; trials, 140; speed, 137, 149, 173 (1945), 179; maneuverability, 137; structure, 190; strength, 136; weather damage, 134; overall evaluation, 122, 149, 151, 189; disposal, 151; distribution (1945), 157; round-bottom hull, 185; deployment, 192; disposal, 195; age distribution (1945), 173; electrical load, 173; postwar plans, 177; transfers (postwar), 177, (to Korea) 192; radar, 171; cockpit arrangement, 171–72; protection, 173; weight growth, 173. See also PTF
PTC: ASW version of PT, 93, 95, 125; sonar, 125; tactics, 125
PTC 1, 124
PTC 62, 197
PT conference (OpNav, 1941), 136
PT designation, 116
PTF, 246; program, 195, 265; characteristics, 247; armament, 249, 250; sketch designs, 249; hydrofoil version, 250. See also PT
PTF 2 (ex–PT 811), 194
PTF 6, 256
PTF 13, 257
PTH, 257
PT lend-lease program, 153
PT modernization (as PTF), 246, 247, 249, 250
PT squadron organization, 153
PT tenders (AGP), 115
Pulitzer, R., 24
Pump-jet propulsion, 293
PXH, 379, 381, 384
Pye, Admiral W. S., 109, 115

Quad engine, 2
QFB (quiet fast boat), 293, 295, 368, 391
Quiros (ex-Spanish gunboat), 418

Radar: SA, 87; SA-1, 64; SC 517A, 161, 166; SF-1, 64; SO-3, 169; SW-1C, 65, 78; SPS-5, 189
Raleigh (cruiser), 413
R-ASP (private venture river craft), 449
RCVP (river boat), 286, 287
Ready (PG 87), 271
Recoilless rifle, 214, 337
Redeye missile, 253, 266, 269, 403
Reservists, 21
Resistance (hydrodynamic), sources of, 1
Rhea (AMc 58), 415
Richardson, Admiral J. O., 116
Richardson, H. C., 7
River assault interdiction divisions (RAIDs), 298
River Assault Flotilla (RAF), proposed, 327; formed, 328
River Assault Squadron (RAS), 325
River gunboats, 413, 414, 415; proposed by General Board (1903), 419; design requirements, 419; plans purchased, 420; evaluation of first two, 420; new requirements, 420; preliminary designs, 420; fiscal pressures, 420; turboelectric drive proposed, 421; comparison with Japanese, 421; adoption of Chinese plans, 423; fates, 423
Riverine armor, 233, 333
Riverine craft: French, 223; U.S., second-generation program, 357; follow-on planning, 363; losses to North Vietnam, 363
Riverine forces, French evaluation of, 232; U.S. concept of, 325
Riverine minesweepers, 333, 343; Indochina war, 226. See also MSR, LCMM
Riverine mine threat, 227, 331, 343
Riverine operations, U.S., 329
Riverine war, French requirement for new craft in (1946), 224; (1954), 233
Riverine weapons, 233, 358
River patrol boats (PBR), 244, 247, 313. See also Plastic
Robinson Marine, 106
Rocket gun, 439
Rocket launchers, 451
Rodman, Admiral H., 26
Rodriguez, C., 262
Rohr CPC proposal, 399, 400
Rohr Marine, 402, 410
Rolls-Royce Proteus gas turbine, 244, 247, 259, 266, 268, 383
Roosevelt, F. D., 14, 16, 21, 26, 31, 33, 48, 73, 122
Royal Air Force, 431
RPC (Russian PC), 95
RPC (river patrol craft), 15, 280, 281, 286, 288, 291, 305
RSSZ. See Rung Sat Special Zone
Rule propeller, 292
Rumrunners, 15
Rung Sat Special Zone (RSSZ), 283, 297
Russia, motor torpedo boats, 26; motorboats for, 22, 25–26; subchasers (RPC), 433

Sacramento (PG 19), 413, 417
Saltonstall, J. L., 21
Samar (ex-Spanish gunboat), 419
Sandoval (ex-Spanish gunboat), 419
Sarancha missile boat, 7
Satterlee, H. L., 98
Satterlee torpedo boat, 98, 99
S-boat, 7
SC. See subchaser
SC 159, 32
SC 223, 47
SC 235, 30
SC 383, 47
SC 431, 12
SC 453 (prototype of World War II SC), 74
SC 661, 78
SC 677, 76
SC 712, 78
SC 717, 79
SC 759 (as minesweeper), 50
SC 1078, 79
SC 1049, 80
SCB (Ship Characteristics Board), 15

SCC program, 75
Schedule 9009, for motor torpedo boat, 100
Scott-Paine, H., 15, 97, 108, 122
Scruggs Boat, 106
SCTC (Subchaser Training Center, Miami), 149, 153
SC-tube (passive sonar), 20
Sea Float, 298
Sea Force (Vietnamese), 283
Sea Fox, 365, 376, 391, 402, 403, 404, 405
Sea Legs (experimental hydrofoil), 9, 21
Sealords, 282, 297, 319
SEALs, 14, 251, 279, 292, 295, 298, 365
Sea Mauler (missile), 255
Sea sled, 7, 99, 100, 107. See also Hickman
Sea Sparrow (missile), 385
Seats, shock-mounted, 294, 344
Sea Viking, 401. See also SWCM
SEPTARS (seaborne powered targets), 433
SES ("3K"), 10
Sewart 40-footer, 239
Sewart 65-footer, 431, 455; characteristics, 456; versions, 456; production abroad, 456
Sewart 85-footer, 241, 455
Sewart Seacraft, 239, 253, 399, 455. See also Swiftships
Seawolves (helicopter gunships), 297
Shanghai, 418
Shearer, W. B., 100, 101; one-man torpedo boat, 17, 97, 101
Sheridan light tank, 263
Shimrit (Israeli hydrofoil), 263
Shipping Board, 36
Shock-hardening, 343
Sidewinder (missile), 269
Sihanoukville, 282
Sims, Admiral W. S., 36
Ski cat hybrid-hull concept, 400
Skirt (for Hovercraft), 10
Skrei, 251
Slane, W., 312
Small-boat alternatives, 288
Small craft: on hand in 1966, 302; program (1938), 52; weapon development program, 440
Smith, Admiral S., 7
Solar gas turbine, 286
Sonar, 50; AQS-10, 212; QHB, 200, 202; SQG-1, 200; SQS-4, 197, 203, 209; SQS-9, 211; SQS-10 (VDS), 203; SQS-17, 209, 211, 285; SQS-17A, 237; SQS-20, 212; SQS-23, 212; SQS-33, 213; UQS-1, 207; towed array, 319; variable depth sonar (VDS), 197
Sonar dome sizes, 211; drag, 211
Sonar noise limitations, 212
SOSUS, 198, 199
Sound equipment, 49
Sparkman and Stevens, 119, 188
Sparviero (Italian hydrofoil), 262
Special Service Squadron, 413
Speed-length ratio, 1
Speedy, 380, 389
Sphinx (ARL 25), 329
Spirit of '76, 392, 395
Sprint-and-drift tactics, 210, 212
Sprugel, Lieutenant G., 157
SP series (requisitioned motor craft), 23, 427
Square-cube law, 4, 213

SRN-5 Hovercraft, 308, 309
SS-11 missile, 257, 264
SS-12 missile, 257
SS-N-2 missile, 264, 433
STAB, 365, 367, 369, 391
STABRon 20, 369
Standard ARM missile, 269, 272, 381, 385
Standard Motor Construction, 23, 31
Stark, Admiral H. J., 65, 101
STCAN (modified LCVP), 231, 232, 283, 325
Sterling engine, 106
Stinger, 377
Stinger missile, 406
Stirling Motors, 26
Stocker, Captain R., 34, 44
Storm-class fast gunboat, 264
Strategy, antisubmarine, 19
Subchaser (50-foot), 27
Subchaser (110-foot, World War I) design, 27; drawing, 28–29; engine problems, 31; performance, 30; production, 30; employment, 31; armament, 32; disposal, 32; enlarged version (proposed), 33
Subchaser (110-foot, World War II), 208; design, 55, 73; deficiencies, 55; design competition, 58; speed issue, 59; cancellations, 66; program, 73; evaluation, 73; engines, 73; armament, 73; prototype (SC 453), 73; minesweepers, 81; ferry proposal, 81; for Indochina war, 225
Subchaser (250-foot, sketch 42), proposals, 43; design, 44; power plant, 45
Subchaser, all weather, 35
Subchaser (steel), 33
Subchaser designs (1917), 36–37; (1934, 1937), 49, 51, 53
Subchaser leader, 33
Subchaser prototypes (1940–41), 54
Submarine detection board, 34
Submarines, in antisubmarine warfare, 19
Suifu, 418
Suikiang fast patrol boat, 272
Supercavitating foils, 215
Supramar PT-50 hydrofoil, 259
Surface-effect craft, 399. See also Hovercraft
Sutphen, H. R., 24, 122
SWAB (shallow-water attack boat), 17, 273
SWAL (shallow-water attack craft, light), 391, 392, 401
SWAM (shallow-water attack craft, medium), 391, 392, 401, 402
Swamp buggy, 284
Swasey, A. L., 15, 21, 66, 73, 107, 122
SWCL (special-warfare craft, light), 401. See also Sea Fox
SWCM (special-warfare craft, medium), 10, 399, 401, 406, 410; design, 402. See also Sea Viking
Swift patrol boat, 239, 284, 286, 293, 299, 301, 305, 455; program, 300, 302, 303; tactics, 302, 304; support, 303; as modified (Mark II), 305; deficiencies, 302, 305, 306; refits, 308. See also PCF
Swiftships, 402, 455; 34-footer, 456; 38-footer, 456; 42-footer, 456; 80-footer, 456; 93-footer, 456; 103-footer, 456; 105-foot *c*, 456; 110-footer, 456

Swimmer support boat, 284, 288, 289, 292
Systems Commands, 14

Tacoma Boatbuilding, 272, 274, 275, 396, 398, 399
Target illumination, 253
Tarrant Admiral W. T., 54
Tartar-Bullpup (missile), 255, 380
Task Force Clearwater, 297
Tattletale role, 271, 379
Taylor, Admiral D. W., 26, 44
Taylor, General M., 281
Tet Offensive, 297
Thai navy, 286
Thornycroft, 103; torpedo boats for Philippines, 113
Thunderbolt (quad 20mm), 160, 164, 170, 171
Tonkin Gulf, 281; incident, 251
Topeka (gunboat, later IX 35), 413
Torpedo: Mark 35, 197; Mark 43, 198, 208, 209; Mark 44, 210; fired from small boats, 99; Mark 27X (launcher), 209; Mark 32 (launcher), 209. See also PT boats
Tracy, B., 414
Training craft (YP), 429
Trim tab (variable planing surface), 189
Tucumcari (PGH), 213, 257, 258, 259, 260, 261, 277, 380, 382, 386
Tulsa (PG 22), 418
Tuna boats (YP), 427
Turning fin, 189
Tutuila (Yangtze gunboat, PR 4), 423
Type 21 submarine, 197

UDT (underwater demolition team), 285, 365
Uniflite, 312, 402
United Aircraft, river patrol boat, 292; ST6 gas turbine, 3
United Boatbuilders, 312

Van Kiep II (Vietnamese), 88
Viet Cong, 282; ambush tactics, 315, 331; weapons, 341
Vietnam, 264; geography, 282; Patrol Force, 298; war, 14; navy, 234
Vietnamization, 298
Villalobos (ex-Spanish gunboat), 418
Vimalert engine, 105, 106, 110, 112, 116, 119
Vincent, S. A., 58
Viper (sea sled), 99, 102
Von Mueller-Thomamuhl, Lieutenant D., 10
Vosper, 108, 122
Vulcan cannon, 214, 401, 440

Wake (Yangtze gunboat, PR 3), 423
War plans division (OpNav), 54, 104
Waterborne guard post (WBGP), 298, 319, 369
Water jet, 5, 312
Weatherford (EPC 618), 69, 212
Wells, J. H., 116
Wheeler Shipyard, 93
Wheeling (PG 14), 413, 417
White, J. C., 7
White, J. S., 8
Williamsburg (PG 56), 413
Wilmington (PG 8), 413, 415, 418
Winton (division of GM), 48; diesel, 58

Some motor yachts were initially included in the PC series. PC 457, ex-*Trouper*, completed in 1939 and acquired from Mr. C. A. Tilt in 1940, is shown in 1941, still wearing her civilian dark hull paint. She was 107 feet long and could make 15.5 knots on her two 500-bhp diesels; assigned armament was two 0.30-caliber machine guns and two Mark 1-1 depth charge racks. After conversion by the New York Navy Yard, she was assigned to Charleston, South Carolina (10th Naval District). On 14 August 1941, before being redesignated as a PYc or PY, she was lost near San Juan. She and the other early yachts (PC 454–460 series) were all placed in service before the threat of magnetic mines led to a general program of degaussing all district defense craft; thus, they were not degaussed.

Wizard boats, 227, 284
Wood County (LST 1178), 379
Wooden hull, 249
Wooden shipbuilding program, 66
Wright T-2 engine, 105

XCH-4 hydrofoil, 10
XCH-6 hydrofoil, 215

Yachts: acquired, 427; transferred to Royal Navy, 429
Yangtze River, 413, 415, 417, 418; limits on navigation, 420
Yarrow, 420
YDG, 85
Y-gun, 20
York (British cruiser), 100

YP 110 (ex–PT 8), 396. See also PT 8
YP designation, 427

Z-drive, 5, 10
Zippo, 338; rechargers, 340
Zodiac boat, 289, 292
Zumwalt, Admiral E., 15, 16, 219, 271, 297, 300, 309, 368, 382, 383